Electronic Communications

John J. Dulin
Victor F. Veley
John I. Gilbert

TAB BOOKS
Blue Ridge Summit, PA

FIRST EDITION
FIRST PRINTING

© 1991 by **TAB BOOKS**
TAB BOOKS is a division of McGraw-Hill, Inc.

Library of Congress Cataloging-in-Publication Data

Dulin, John J., 1921 –
    Electronic communications / by John J. Dulin, Victor F. Veley, and
John I. Gilbert.
      p.   cm.
    Includes index.
    ISBN 0-8306-7365-2 (hardbound)   ISBN 0-8306-3365-0 (pbk.)
    1. Telecommunication.   I. Veley, Victor F. C.   II. Gilbert, John
(John I.)   III. Title.
TK5101.D824   1990
621.382—dc20
                               90-43890
                                  CIP

TAB BOOKS offers software for sale. For information and a catalog, please contact TAB Software Department, Blue Ridge Summit, PA 17294-0850.

Questions regarding the content of this book should be addressed to:

Reader Inquiry Branch
TAB BOOKS
Blue Ridge Summit, PA 17294-0214

Acquisitons Editor: Roland S. Phelps
Technical Editor: David M. Gauthier
Production: Katherine G. Brown
Book Design: Jaclyn J. Boone

Contents

APPENDICES

Preface

THE PURPOSE OF THIS BOOK IS TO INVITE THE READER TO THINK AND LEARN ABOUT electronic communications; only then can a technician hope to progress in this rapidly expanding field. To achieve these goals the following format is used:

1. Each topic in a particular chapter contains a limited number of concepts, which are explored in depth.
2. At the end of each chapter there is a summary of the main topics and their related equations.
3. SI units are primarily used throughout the text.
4. The order of the chapters is a logical sequence so that there is a minimum amount of cross-referencing.
5. The number of multiple-choice questions, basic and advanced problems are the same for each chapter. These questions and problems are carefully presented in ascending order of difficulty.
6. The writing style is ''active'' and is aimed at involving you in the discussion of basic principles.

Although some of the FCC Commercial Radiotelephone examinations have been discontinued, the General License is still in force and most of its requirements are covered by the contents of this book. It is hereby emphasized that this book primarily stands on its merits as an introductory text on radio communications and as a reference manual for technicians. To this end it is divided into two parts. The first part is devoted to *Devices and Circuits*, while the second part reveals the application of these circuits in *Techniques, Equipment and Systems*. Only a knowledge of basic ac and dc principles is assumed; if you need strengthening in this area, it is recommended that you read *Ac, Dc Electricity and Electronics Made Easy, Second Edition*, and *The Benchtop Electronics Reference Manual*, both published by TAB BOOKS.

We hope that you will enjoy reading *Electronic Communications* and will find the effort to have been worthwhile.

We would like to acknowledge a considerable debt to our wives, Margy and Joyce, for typing the manuscript. Without their help and encouragement, this book would not have been possible.

Introduction

MAN'S DEVELOPMENT AND THE GROWTH OF CIVILIZATION HAVE DEPENDED UPON progress in just a few activities. Although one of the more important advances is commonly accepted as being the invention of the wheel, a truly more important advance has been the development of a system of communication, and the recording of information. Fundamentally, any form of communication involves a language which can be the spoken word, inscriptions on stone, handwriting, codes, or any other form of systematic process. This language must obey an accepted and acceptable set of rules if there is to be any information exchange.

The growth of the spoken word is shrouded in the mists of time and you can only trace the development of communication systems by the study of recorded information. A vast amount of literature has been written on this subject and it makes for very interesting reading.

The first attempts at recording information were made by the early Mediterranean civilizations in the form of a picture script: Egyptian hieroglyphics are perhaps the best known example of this form of writing. One of the main disadvantages of hieroglyphics was that one idea could be represented by several pictures, the choice being left to the author. This implies a large redundancy, which is undesirable in an efficient system. Chinese is still basically of this form, although attempts are being made to rationalize the language and eventually to Romanize the script.

Difficulties in writing complicated picture symbols as well as the lack of a common standard gave rise to phonetic writing in which certain sounds were given symbols. Therefore alphabets appeared in various forms; one of the most efficient in the communication theory sense is Hebrew, which has no vowels. In 63 B.C., the freed slave, Tyro, invented shorthand in which groups of sounds were given a symbol. It is interesting to note that modern shorthand differs little from the original invention, except that grammalogues and contractions have been introduced to increase the efficiency.

Progress in most fields came to a virtual stop during the Middle Ages and had it not been for the monasteries, much of the earlier work would have been lost. When interest revived, it did so very slowly. Roger Bacon in the 13th century was one of the first to suggest a telegraph system using lodestone, but his suggestions were so vague and impractical that it took some 500 years until Watson in 1746 sent telegraph signals over two miles of wire. Various minor experiments were made thereafter and several telegraphic systems were used. The next major step was the invention of the Morse code in 1832. Samuel Morse had long been considering the problem of increasing the efficiency of telegraphy from the points of view of minimum time for transmission and errorless reception at the receiver. He finally developed his now famous code by assigning to the letters of the alphabet combinations of dots and dashes, but not in a random manner. He counted the numbers of type in a printer's type box to discover the relative frequencies of the letters and then gave the most frequently occurring letters the smallest code combination. For example, E is • , T is – , while the least frequent letter, Z, is – – • • .

Another such system was devised by Louis Braille to enable blind persons to read. This system uses combinations of six dots arranged in two columns.

In 1833, Gauss and Weber invented the five-unit code, which is a binary code taking into account the relative frequency of the letters. In Morse, Braille, and the five-unit code, statistical data were used to formulate the code.

Telephony first appeared in a practicable form when Alexander Graham Bell invented the telephone in 1876. The effect that this invention has had upon society cannot be discussed here but it is considerable.

During the First World War, there arose the need to send messages over greater distances than telegraph lines could provide (not to mention the practical difficulties of laying them). Furthermore, there arose the need for fleets and ships to communicate with one another with some degree of secrecy. Campbell's invention of the communication systems using frequency division multiplexing was quickly developed.

Commercial and military radio stations sprang up everywhere and it soon became obvious that communication systems had to be analyzed more fully, in particular from the point of view of bandwidth occupancy and transmitted power. Even in the early 1920's, considerable problems were being encountered because of poor signal-to-noise ratios and interference between stations. Efforts were made to increase the signal-to-noise ratio while at the same time reducing the bandwidth. In 1922, J.R. Carson proved that these two requirements were incompatible. He directed his attention to frequency modulation which, by considering the spectrum of such transmissions, he proved to be an essentially wideband system. Carson's paper finally put a stop to the considerable amount of nonsense that was being written about sidebands and paved the way for practical FM using the Armstrong system, pulse modulation, and development of time division multiplex.

In parallel with this work, the problem of the speed of transmission of the signal was being investigated and in 1928 R.V. Hartley published a paper on *Factors Affecting the Speed of Transmission of Telegraph Signals*. In this paper, he proposed a method of measuring the amount of information contained in a message by the number of states the message could assume. Information in this sense has no relation to the meaning of the message. Shortly before this paper, H. Nyquist had published a paper entitled *Transmission of Information* in which he concluded that there was a relationship between the time of transmission of the signal and the bandwidth it occupied. In neither case had any consideration of the effects of noise in the system been made such as to produce a bandwidth/time/signal-to-noise ratio relationship.

In 1946, D. Gabor extended some of the ideas put forward by Nyquist and Hartley but it was left to C.E. Shannon in 1949 to relate in mathematical form the three important parameters of the communication system. He produced a formula for the maximum rate at which information could be transmitted in terms of the bandwidth and the signal-to-noise ratio. Shannon's paper, *A Mathematical Theory of Communication*, laid the foundations of modern communications theory, one of the immediate outcomes being the development of Pulse Code Modulation (PCM) which is the nearest approach to the ideal that can be made in a practical form.

The problem of noise (any unwanted disturbance or signal present in the received signal), has been receiving its fair share of attention from the beginnings of radio communication. Einstein (1905), Schottky (1918), Johnson (1928), and Nyquist (1928) were all instrumental in analyzing noise and characterizing its properties. Schottky investigated noise due to the random emission of electrons from a cathode, while Johnson and Nyquist investigated thermal noise. Nyquist's formula for the noise power is now well known.

In 1944, S.O. Rice published a paper entitled *The Mathematical Analysis of Random Noise* in which he treated noise from the statistical point of view. The paper is very thorough and is still regarded as having laid the foundations of the analysis of noise as a random error. In 1949, T. Wiener published a book - *The Interpolation, Extrapolation and Smoothing of Stationary Time Series*. Wiener was primarily concerned with the design of optimum filtering systems to produce a high signal-to-noise ratio. In the book, he presented the idea of treating noise from the Fourier approach and to do this he had to include correlation theory. It was this approach that rounded off the ways of considering noise, as (a) a frequency spectrum, (b) a randomly varying quantity such as a phasor and finally (c) a power density spectrum and a time varying function.

The practical results of all this work, which might at first sight appear to be a collection of turgid mathematical theses, are evident in the majority of modern systems, and in those that are still under development. Such innovations as matched filters, correlation detection, error correcting codes, chirp radars, FM sonar and so on are becoming commonplace and are the result of the use of communications theory and its application to the analysis of systems. At the same time modern developments such as transistors, integrated circuits, lasers, fiberoptics, and satellites have further revolutionized our methods of communication.

It follows that the modern communications engineer, whether he be designer or maintainer, must appreciate the fundamental methods of system analysis. It is of little value to apply rule-of-thumb methods to systems which are complex, although such methods must not be entirely discarded, for a considerable amount of experience is often contained in such methods.

You will find that this introductory book fully covers both the theoretical and practical aspects of modern communications without neglecting the evolutionary features of this exciting field.

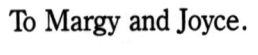

To Margy and Joyce.

1
PART

Devices and circuits

1
Resonance and filters

THIS CHAPTER BEGINS WITH EXTENSIONS OF SERIES AND PARALLEL CIRCUITS TO THE unique conditions of resonance. The vibration of mechanical devices, such as the tuning fork, at natural frequencies exemplifies resonance. In fact their vibrations can be conveniently studied through the observation of the voltages and currents in their electrical analogs—series and parallel resonant circuits.

The phenomenon of resonance allows the designer to distinguish between different frequencies and ranges of frequencies. Passive circuit configurations, each containing one or more reactive elements (L or C), can be used to filter out the unwanted frequencies and to pass signals having other ranges of frequencies without demonstrating resonance in the sense intended in the preceding paragraph.

Different types of filters constructed of passive (nonamplifying) electrical elements will be covered first, followed by other filter classes involving nonelectrical phenomena: the mechanical filter, the crystal filter, and the ceramic filter. You will then look at the basic nature of the surface acoustical wave (SAW) filter and finally, a glimpse at a greatly different method of filtering, called *digital filtering*.

These topics are organized into sections having the following titles:

- Series resonance
- Parallel resonance
- Decibels
- Basic filter principles
- Constant-k filters
- M-derived filters
- Active filters
- Mechanical vibration filters

- Surface acoustic wave filters
- Digital filters
- Impedance matching with L-sections

This chapter is concluded with some fundamental methods of impedance-matching with L-sections. You'll quickly recognize that the basic purpose of filters is not involved, but that there is a strong tie between passive filter circuits and these L-sections.

Series resonance

At one particular frequency, the reactances of a series RLC circuit are equal ($X_L = X_C$) so that the impedance diagram appears as in Fig. 1-1. This circuit is said to be *resonant* or *at resonance*, and has a number of unusual and useful properties:

1. The impedance of the circuit is equal to its resistance ($Z = R$).
2. The power factor is unity ($= 1$).
3. The current is in phase with the applied voltage.
4. Voltages appearing across the two reactances are equal in magnitude but 180° out of phase.
5. For any total applied voltage, the circuit current and the reactance voltages are at their maximum values when the frequency alone is varied to achieve resonance.
6. The resonant frequency is given by the formula:

$$f_r = 1/(2\pi \sqrt{LC})$$

In addition, the more general characteristics of the series RLC circuit are still present. Particularly, either reactance voltage might be greater than the total applied voltage. Whether or not this is true depends upon the ratio of reactance to impedance (X_L/Z or X_C/Z) which is equal to the ratio of the reactance voltage to the total applied voltage (V_L/V_T or V_C/V_T). At resonance, for which the ratios are greatest, visualize a "magnification" effect expressed as a ratio Q.

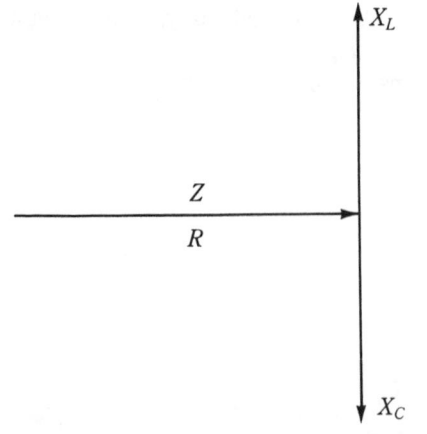

1-1 Impedance diagram for a series resonant circuit.

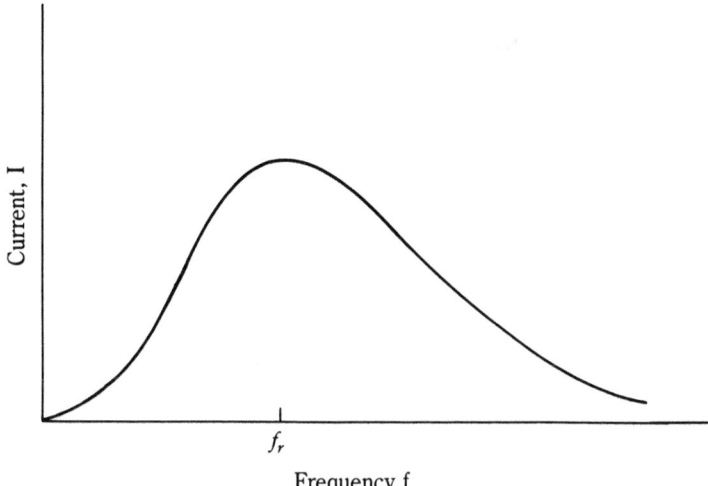

1-2 Current in a series RLC circuit as frequency is varied.

Current in a series RLC circuit depends upon the frequency shown in Fig. 1-2. As the frequency increases from a low value, the current also increases, reaching a peak at the resonant frequency, f_r, and then gradually decreasing. At f_r the current is limited only by resistance (R). At any frequency substantially less than or greater than f_r, the value of R has less relative significance in limiting the current flow, and the net reactance dominates. Therefore, at f_r a smaller value of resistance R will yield a greater peak current without greatly changing the current for frequencies appreciably below or above f_r. Figure 1-3 illustrates the circuit response for a lower value of R with the understanding that a different current scale is used than in Fig. 1-2.

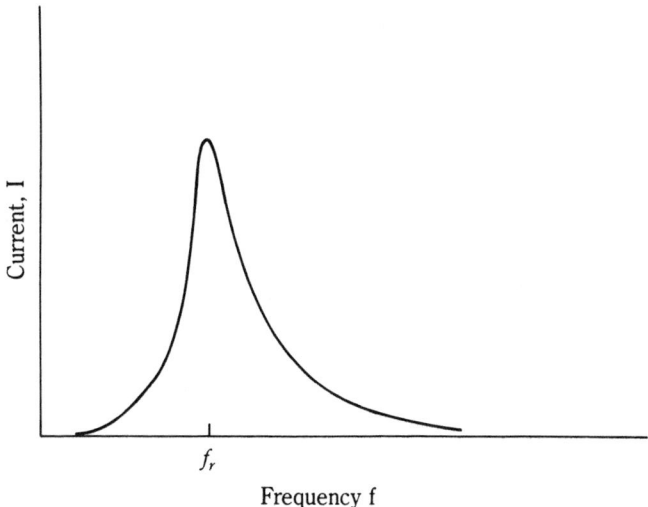

1-3 A series RLC circuit with a smaller value of R as compared with Fig. 1-2.

It should be clear that the circuit with the smaller R has a sharper response than the other. More accurately, the bandwidth is less for the second circuit, where bandwidth has a particular technical meaning. *Bandwidth* is the frequency span between the half-power points; these half-power points are shown in the two previous figures where the current is 70.7 percent of its maximum value. Figure 1-4 shows the bandwidth concept for a given circuit.

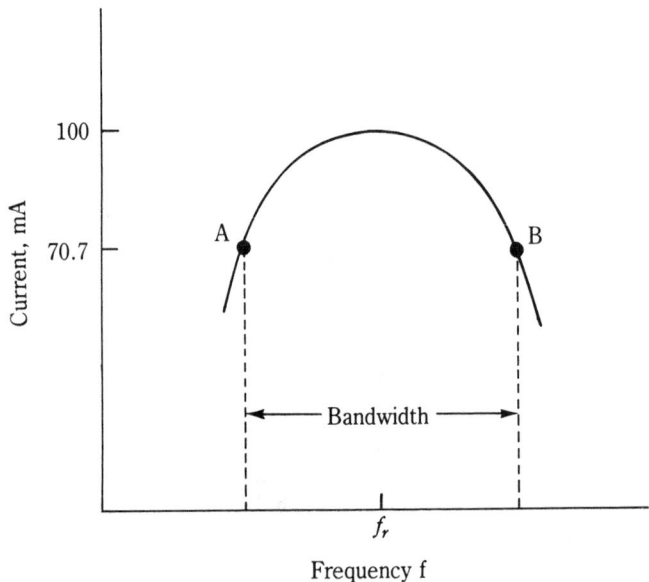

1-4   Bandwidth concept for a circuit having 100mA maximum current.

In a practical circuit, such as would be found in communications equipment, a narrow bandwidth is often desirable. The inductor, which cannot be constructed without having power losses (hence resistance), establishes a lower limit on the bandwidth. The *quality factor*, Q, for the inductor is defined to be X_L/R. For Q of the order of 10 or more the bandwidth is given by:

$$BW = f_r/Q$$

Adding a resistor to an inductor and capacitor in series has the effect of reducing Q (for the circuit) and correspondingly increasing the bandwidth.

Parallel resonance

A simple RLC parallel circuit is illustrated in Fig. 1-5. It has the same properties at resonance as the series circuit except for the following:

1. $Z = R_P$.
2. The total current is in phase with the applied voltage.
3. Voltages appearing across the two reactances are equal to the applied voltage.

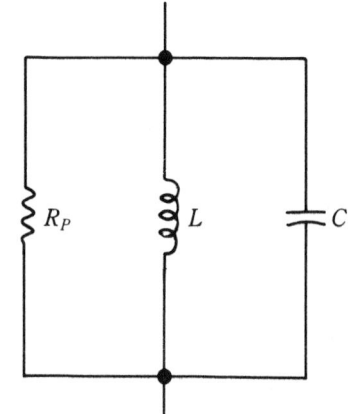

1-5　A simple RLC parallel circuit.

However, the reactance branch currents are equal in magnitude but 180° out of phase.

4. For any applied voltage, the total circuit current is at its minimum value when the frequency is varied to obtain resonance.

Note that points 2 and 6 stated previously are the same for parallel and series resonant circuits. In addition, the relationship

$$BW = f_r/Q$$

is valid for parallel resonance.

Figure 1-5 is not an exact representation of a practical inductor in parallel with a nearly lossless, practical capacitor. Figure 1-6 provides a better equivalent circuit, in which the losses in the inductor are attributed to a series resistor, R_S. Fortunately, Figs. 1-5 and 1-6 are easily reconciled when Q is of the order of 10 or more. At resonance,

$$Q = X_L/R_S = R_P/X_L$$

1-6　Practical RLC circuit representation with a "lossy" inductor.

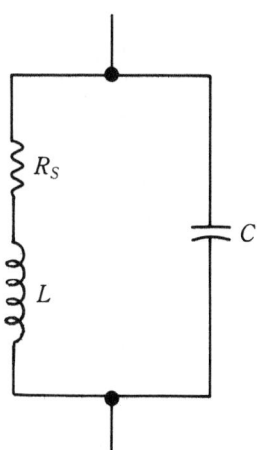

Moreover,

$$R_P = Q^2 R_S$$

The impedance of the parallel circuit is greatest (i.e., equal to R_P at resonance. Compare this with the minimum impedance, R_S, of the series circuit at resonance). Because the total current for the parallel circuit is minimum at resonance, there appears to be current amplification, because the currents in the reactive branches are each Q times that of the total current.

Decibels

In dealing with resonant circuits, filters, and the enormous range of signal levels found in electronics communications, it is convenient to use powers of ten. By this means extremely small and extremely large numbers are easily written and manipulated. Thus, a power level of 1000 W is 10^3W or 30 dBW, which means 30 decibels with respect to 1 watt. More formally, it is $10 \log_{10}(1000/1) = 30$ dBW. We recognize $\log_{10}1000 = 3$ as the *common*, or *base 10 logarithm*, of 1000, which is easily justified since $10^3 = 1000$. Usually, however, when a decimal power is involved, tables or calculators must be used to obtain the logarithm. For example, $\log_{10}500 = 2.7$ so that $10 \log_{10}500 = 27$ dB or 27 dBW if the reference level is understood to be 1 watt.

You might be interested to know that the *bel* is named for Alexander Graham Bell and was first employed as a measure of sound levels because hearing sensitivity is logarithmic in nature. The insertion of the 10 multiplier converts bels to decibels.

In addition to specified reference levels such as 1 W in the previous example and 1 mW (which yields dBm), the decibel is often applied to relate the output level to the input level for an amplifier, attenuator, or filter. An amplifier with 20 dB gain has an output power that is 100 times that of the input. Likewise, if an attenuator reduces the output power level to 0.001 times that of the input, the gain is -30dB ($10 \log_{10}0.001 = -30$). Alternatively, it is said that an attenuation of 30 dB exists.

Decibels entail comparisons of power levels for which the 10 multiplier is necessarily required. Under the assumption that impedances remain the same at the input and the output, power comparisons are still implied by decibels, but the calculations are possible using voltage or current ratios. In such situations a 20 multiplier is necessary. For example, if a 30μV input to an amplifier yields a 600mV output, the gain is $20 \log_{10}(600 \times 10^{-3}/30 \times 10^{-6}) = 86$ dB.

Basic filter principles

The circuits introduced in the previous sections of this chapter demonstrate properties that depend upon frequency. They belong to a class of electrical networks called *filters*, which are deliberately constructed to favor signals at some frequencies, while discriminating against signals at other frequencies. The favored frequencies lie in the passband, which corresponds to the frequencies between the cutoff (half-power) frequencies for the resonant circuits already described, by way of illustration. Those frequencies outside the passband can be referred to as the stopband or attenuation band.

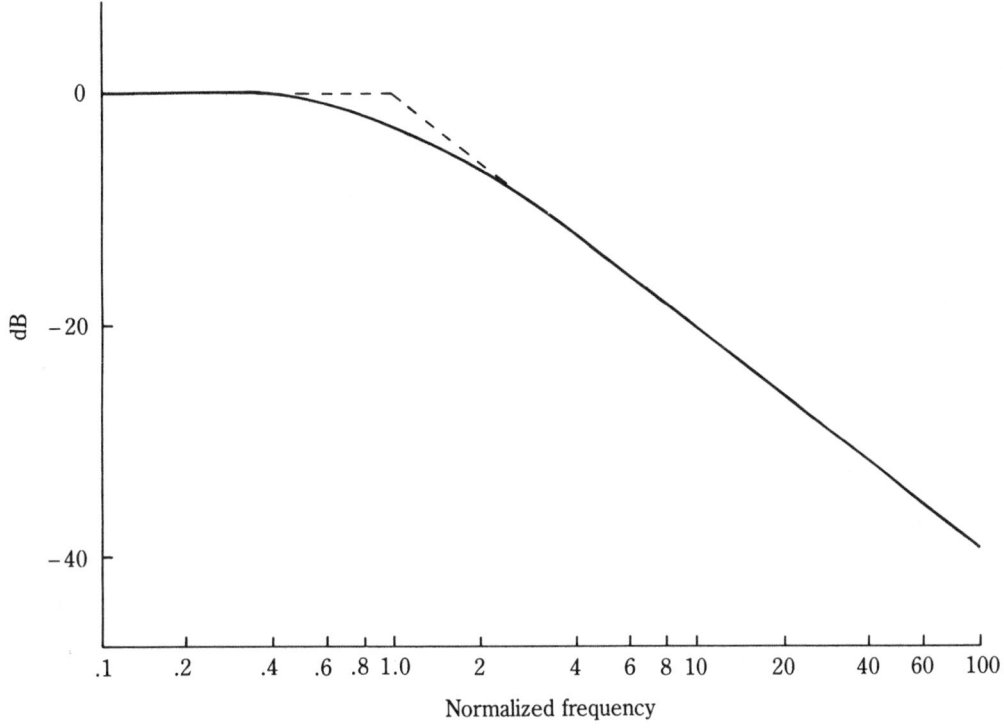

1-7. Bode plot of a low-pass filter.

A low-pass filter has a single cutoff frequency and freely passes alternating-current (ac) signals whose frequencies are less while attenuating ac signals with greater frequencies (see Fig. 1-7). A simple unregulated power supply filter fits into this category, although the emphasis is placed on the attenuation of ripple frequencies and the maintenance of satisfactory regulation.

A high-pass filter also has a single cutoff frequency, but is intended to attenuate lower frequencies while passing higher frequencies. See Fig. 1-8.

In Figs. 1-7 and 1-8, the filter responses are expressed in dB, with 0 dB representing the response in the passband. This does not mean that the output level is equal to the input level in the passband. Indeed this is rarely the case. It means that the response at any frequency is compared with that in the passband and the ratio is converted to dB.

Close examination reveals that the frequency scales are not only nonlinear but are also normalized. A normalized frequency scale signifies comparative values—more specifically, ratios—in which the cutoff frequency is the reference. A logar   scale is used, and the response is a straight line for all frequencies except those frequency. With this limitation in mind the straight line can be extended illustrations to form the Bode plot, which is also used for bandpass filt etc. The Bode plot is especially useful for constructing straight-line approximations for more complex systems having two or more points changes.

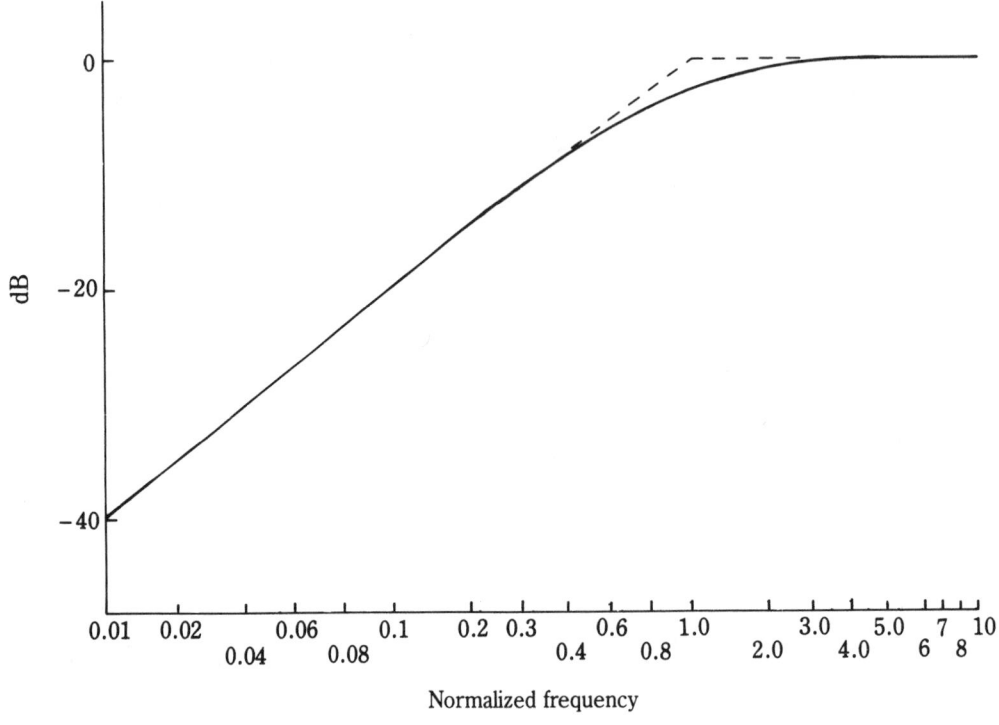

1-8 Bode plot of a high-pass filter.

Resonant circuits exhibit bandpass characteristics, as is seen in the Bode plots of Fig. 1-9, which represents four filters with different values of Q. When the circuit is rearranged, a band of frequencies can be attenuated, as is shown for the notch filter of Fig. 1-10. Like the bandpass filter, the frequency range singled out for attention can be narrowed by using a larger value of Q for the notch filter.

Practical filter applications require that frequency considerations other than the cut-off frequency(ies) be specified. Included are the phase characteristics, impedance matching, skirt steepness (Fig. 1-11), and flatness in the passband.

Constant-k filters

One form of electrical filter consists of cascaded identical sections. A half-section of such a filter is shown in Fig. 1-12A with $Z_1/2$ being the impedance of a series reactive element and $2Z_2$ being the impedance of a parallel reactive element. When properly joined together, two half L-section elements become a T-section, which is shown in Fig. 1-12B or a π-section, which is shown in Fig. 1-12C. When terminated with an *image* impedance, Z_0, the input impedance of either the T or π filter sections is also Z_0, in which $Z_0^2/Z_1Z_2 = 1 + Z_1/4Z_2$. With these terminations any number of sections can be cascaded without changing the filter input impedances.

Low-pass filters can be constructed with Z_1 being inductive and Z_2 being capacitive. ~~i~~gh-pass filters, the reactive elements of the preceding arrangement are inter-

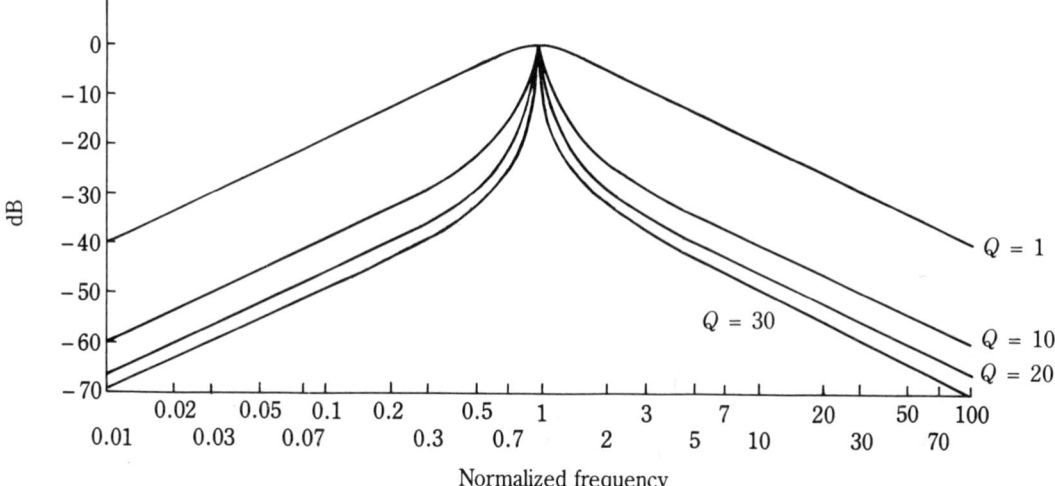

1-9   Bandpass filter response.

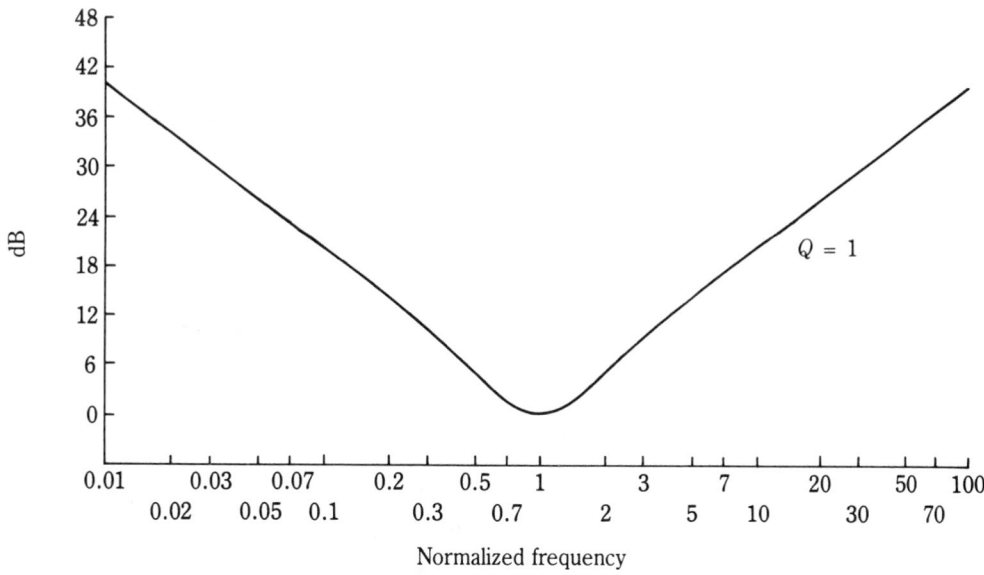

1-10   Response curve for a notch filter.

changed. Bandpass filters can have series resonant elements for Z_1 and parallel resonant elements for Z_2. If the roles are reversed, a band-stop filter is the result. Meeting the termination requirements for Z_0 described in the previous section causes the product of the two image impedances to be a constant (i.e., $k = \sqrt{Z_1 Z_2}$), giving rise to the term *constant-k filter*. When Z_1 and Z_2 are purely reactive, the product $Z_1 Z_2$ is independent of frequency. For example, if Z_1 is inductive and Z_2 is capacitive, $k = \sqrt{L/C}$.

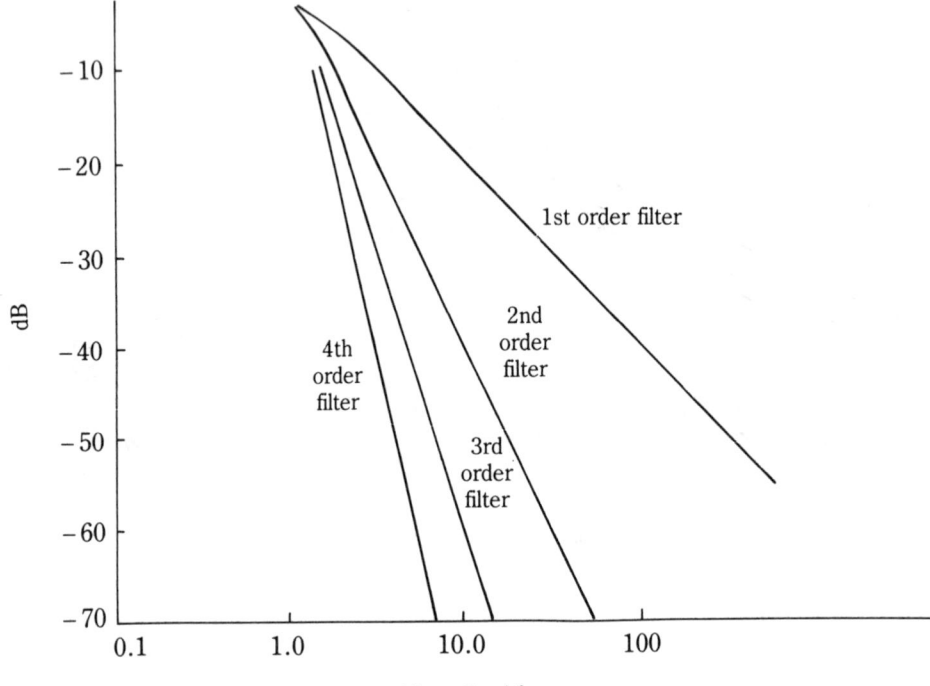

1-11 Different low-pass filter designs illustrating skirt steepness with the order of filter indicated.

The attenuation-frequency curve for this T-section, low-pass filter appears in Fig. 1-13. Here the cutoff frequency is $f_C = 1/\pi \sqrt{LC}$. Similarly, the T-section, high-pass filter attenuation-frequency curve is shown in Fig. 1-14.

One T-section of a bandpass filter, together with the attenuation-frequency curve, appears in Fig. 1-15. For this filter, the series and shunt arms are resonant at:

$$f_0 = \frac{1}{2\pi} \sqrt{LC}$$

A bandstop filter and its curves are presented in Fig. 1-16. The resonant frequencies of the arms, f_0, is given in the equation for the bandpass filter.

One basic limitation of the constant-k filter is the limited slope of the skirts near the cutoff frequencies. For this reason these filters are referred to as prototypes of a new line of filters having more desirable attenuation characteristics.

M-derived filters

An m-derived filter has the same basic configuration as its constant-k prototype. Confining your attention to the T-section, the constant-k filter of Fig. 1-12B is transformed into the m-derived filter of Fig. 1-17 through the application of a numerical parameter, *m*,

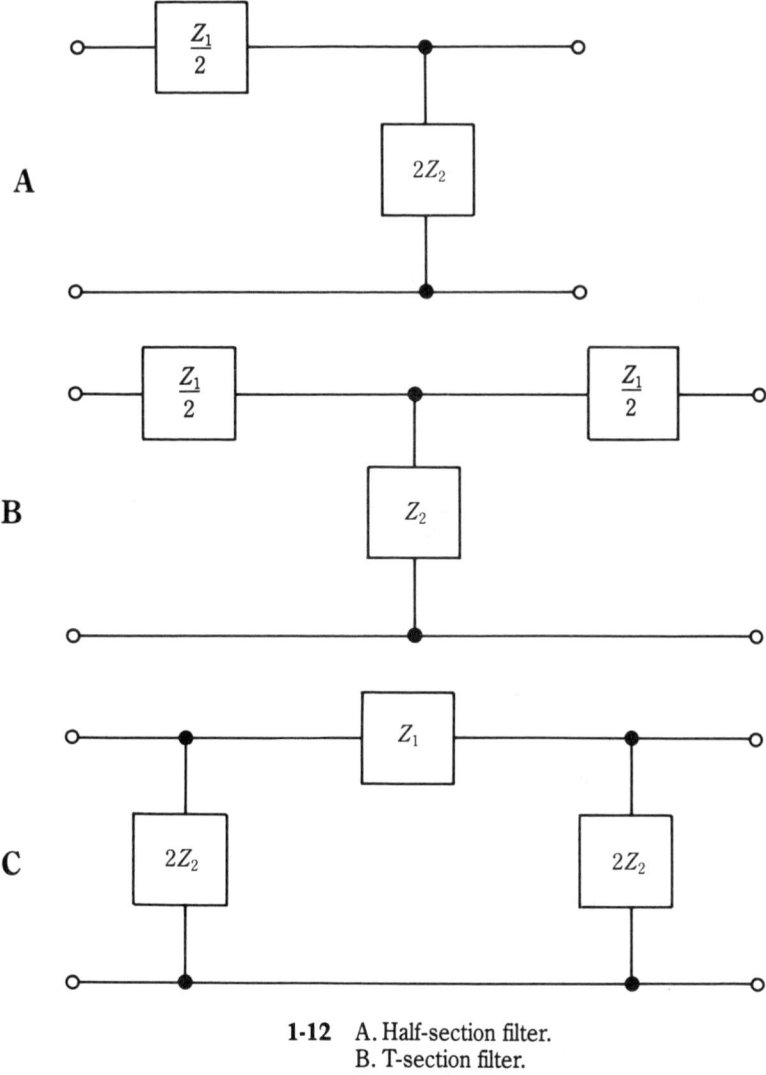

1-12　A. Half-section filter.
B. T-section filter.
C. π-section filter.

whose value is selected between 0 and 1. The substitution of the m-derived section for the constant-k section causes a beneficial change in the shape of the attenuation-frequency characteristic. In the vicinity of the cutoff frequency(ies) the slope of the curve is increased so that the attenuation is increased for frequencies immediately outside the passband. This is illustrated for a low-pass filter prototype and its m-derived section in Fig. 1-18.

Shaping of the m-derived filter attenuation-filter curve is achieved through the selection of:

$$m = \sqrt{1 - (f_0/f_\infty)^2}$$

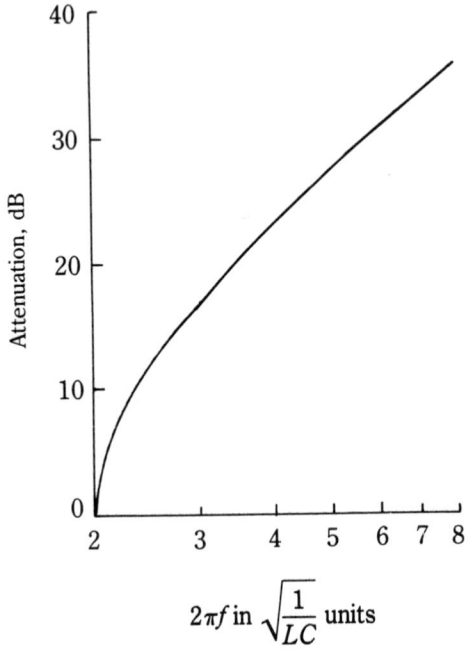

1-13   Attenuation curve for a low-pass constant-k filter.

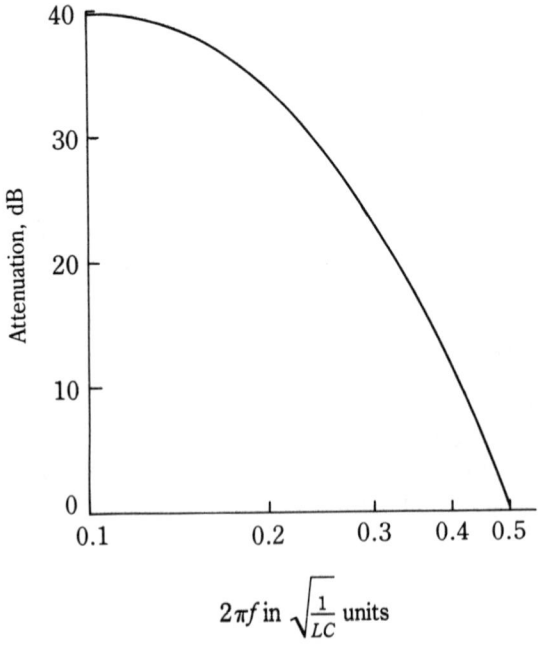

1-14   Attenuation curve for a high-pass constant-k filter.

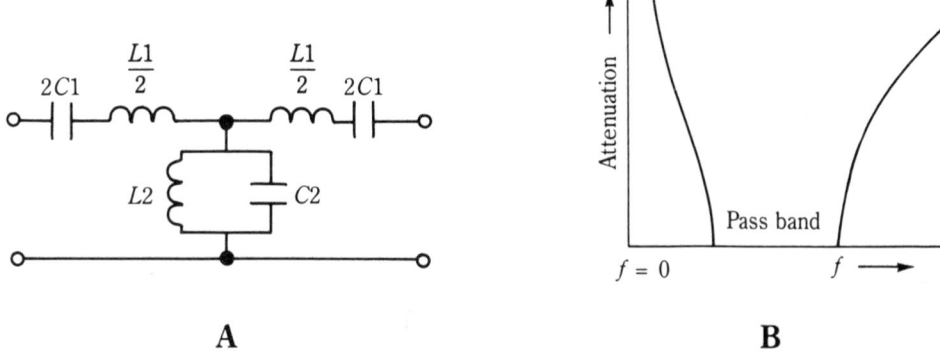

1-15 A. T-section band-pass filter.
B. Attenuation of a band-pass filter.

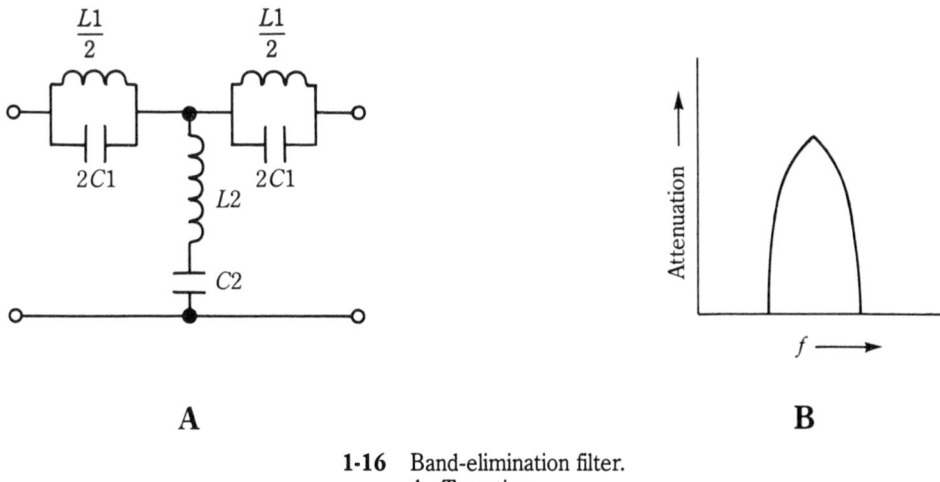

1-16 Band-elimination filter.
A. T-section
B. Attenuation

Moving the frequency, f_∞, for theoretically infinite attenuation close to f_0 causes m to be smaller and the slope near f_0 to be greater. At frequencies beyond f_∞, the attenuation for the m-derived filter is less than that for the constant-k prototype, which is the price paid for the benefit of the increased slope near f_0.

M-derived filters can be devised for high-pass, band-pass, and band-stop filters as well. The different configurations and their attenuation-frequency characteristics are summarized in Fig. 1-19. In each case L1, C1, L2, and C2 are the circuit elements for the constant-k prototypes. A designer must establish the value of m that best fits the overall performance to be achieved.

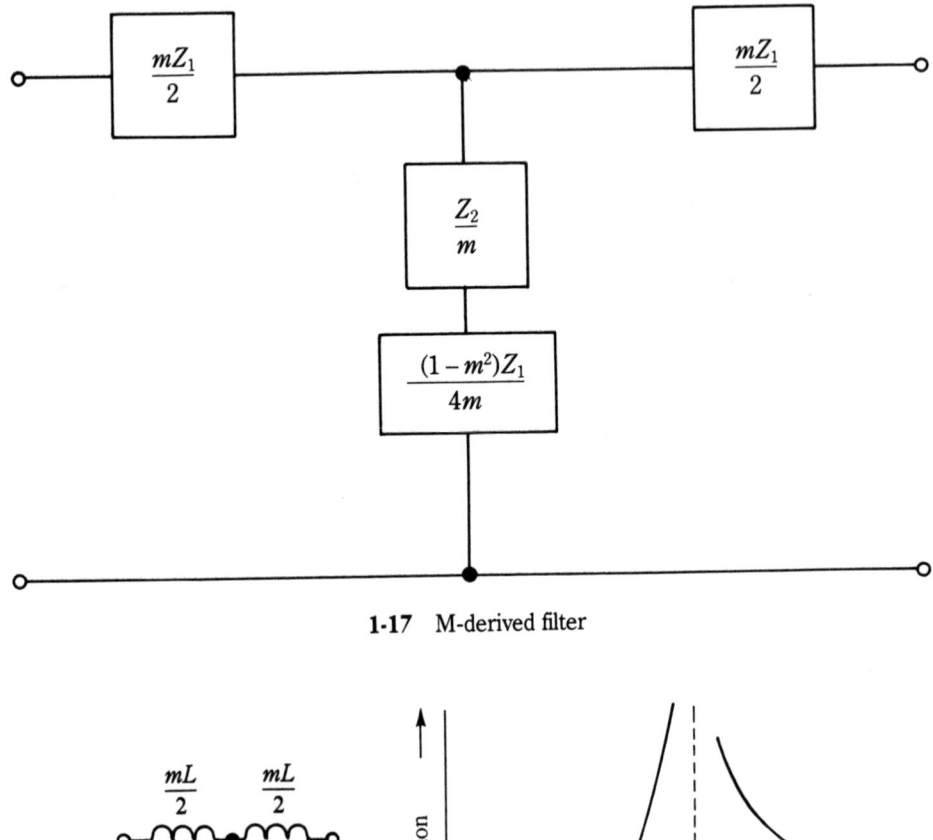

1-17   M-derived filter

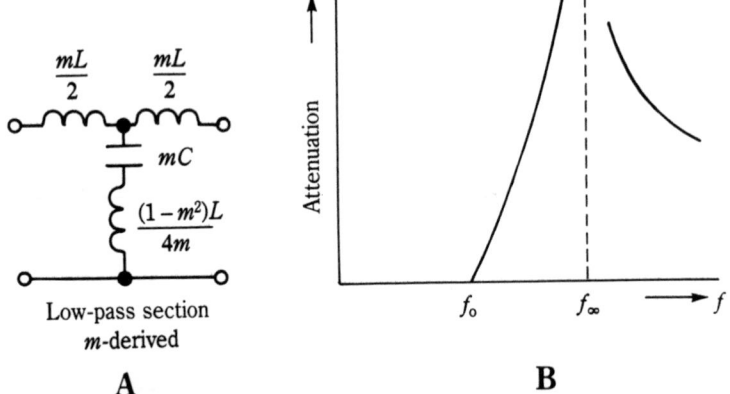

A

B

1-18   A. Low-pass m-derived section.
B. Attenuation of m-derived filter.

Active filters

Filters constructed of passive circuit elements have fundamental limitations. One of the most serious limitations stems from the inability to obtain reactive elements, especially inductors, having adequate Q to achieve the bandwidths and skirt steepness required in certain communications applications. Therefore, all of the previously introduced filters have characteristics that are modified by the presence of power losses. These losses

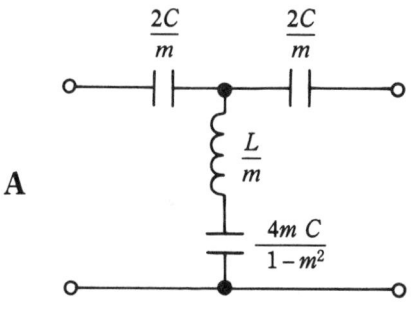

A

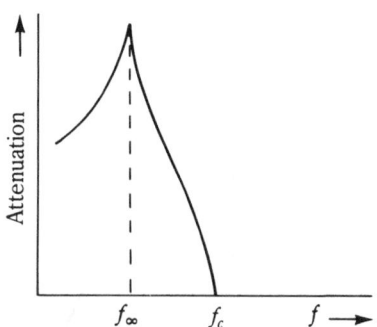

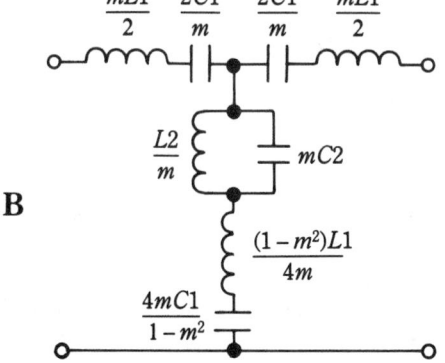

B

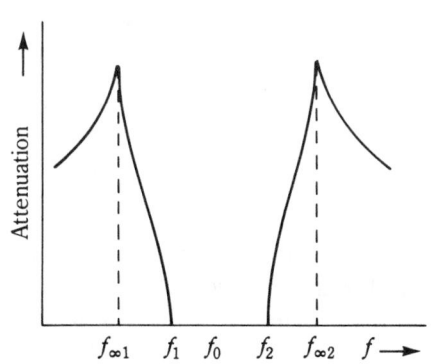

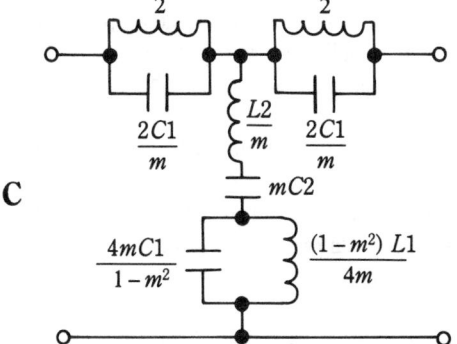

C

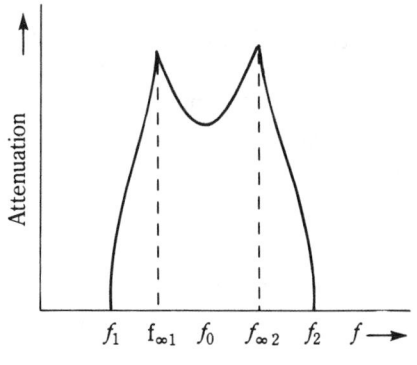

1-19 M-derived filter and attenuation.
  A. High-pass
  B. Band-pass
  C. Band-stop (notch)

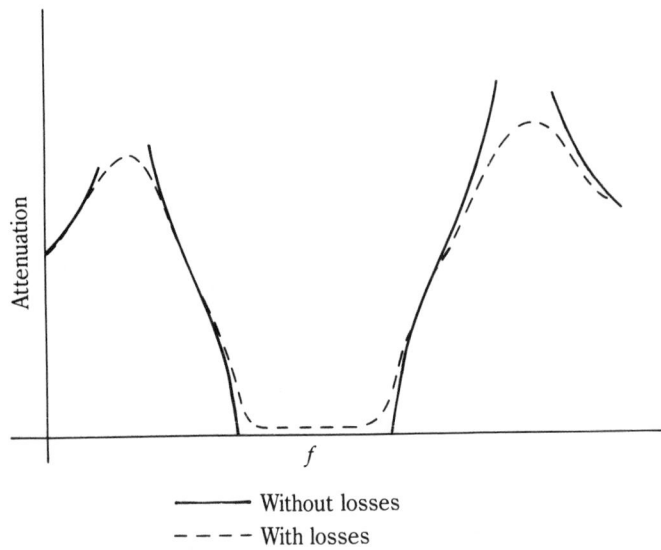

——— Without losses

– – – – With losses

1-20   Modification of attenuation by losses.

generally reduce the selective features of filters in the vicinity of the cutoff frequencies, causing the performance to be inferior to that required. Figure 1-20 demonstrates the modification of the attenuation-frequency curve of an m-derived, bandpass filter (solid lines) by the presence of inductor losses (dashed lines).

Active filters, incorporating amplifiers with feedback, can be designed to perform all filter functions using only resistors and capacitors as circuit elements. With such filters, the attenuation-frequency characteristics can be predicted more exactly. Since the losses of capacitors are comparatively low, the element values can be adjusted to any desired accuracy, and the temperature variations can be largely controlled.

Amplifiers   The subject of amplifiers will be introduced briefly at this point. An *amplifier* is a device that can cause an input voltage, current, or power level to be increased at the output terminals. Here the emphasis is placed on voltage amplifiers, but the basic principles apply to other amplifier types. Amplifiers require direct-current (dc) power supplies and are considered to be active devices as contrasted with passive circuit elements.

Amplifiers such as the differential amplifier, are represented as triangles on schematics (see Fig. 1-21) with input terminals (+ and −) on the left side and a single output terminal on the right side. The differential amplifier is most often used in the design of active filters. The (−) terminal causes a phase reversal of the input signal, whereas no phase reversal is experienced for inputs applied to the (+) terminal. The amplifier output voltage polarity depends on the net input voltage applied across the terminals, polarity being positive if the (+) terminal is positive relative to the (−) terminal, and vice versa.

A particular kind of differential amplifier, called the *operational amplifier* (op amp), is well suited to active filter applications. In all but the most demanding applications, semiconductor op amps are inexpensive, reliable, and compact, and have a number of other

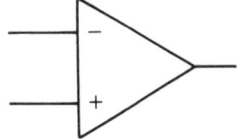

1-21 Differential amplifier symbol.

superior features. The most important of these features allows the op amp to be used so that variations of its own characteristics due to temperature change, ageing, or manufacturing tolerances have a minimal effect on the active filter performance.

Active filters also enjoy an additional cost advantage since inductors are eliminated. Because of their low output impedance and high input impedance, they can be *cascaded* (operated in tandem) freely without regard for loading effects. Active filters do require dc power supplies, which can be a serious disadvantage or of little consequence, depending on the application.

Low-pass and high-pass filters Figure 1-22 illustrates a *first-order* (single reactive element) low-pass filter in which the cutoff frequency is $1/2\pi RC$. The voltage gain, A_V, applies to the frequencies below cutoff. By trading the positions of R and C, a first-order high-pass filter having the same cutoff frequency is obtained.

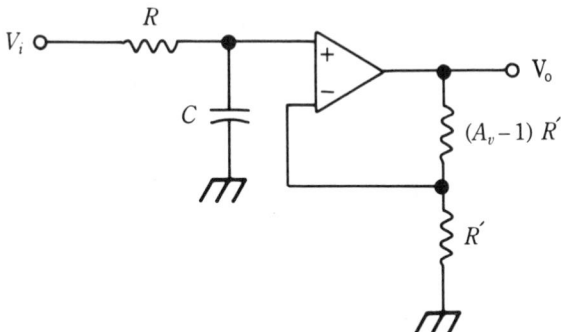

1-22 First-order active low-pass filter.

One form of second-order, low-pass filter appears in Fig. 1-23. It has a cutoff frequency of $f_C = 1/2\pi \sqrt{R1C1R2C2}$. The same cutoff frequency can be obtained for a high-pass filter by interchanging the resistors and capacitors in the circuit. Higher-order active filters are usually achieved by cascading first- or second-order filters. The objective is the attenuation of the frequency components just beyond the cutoff frequency.

The response of first- and higher-order low-pass filters is summarized in Fig. 1-11. Note that a normalized frequency scale is used in this figure so that the cutoff frequency, f_C, is 1.0 on the scale, which extends from $0.1f_C$ to $10f_C$. Each curve passes through the -3 dB (half-power) point at the cutoff frequency. Moreover, to the right the curves approach straight lines whose slopes are $-20n$ dB per decade, n being the order of the filter. The low-frequency attenuation is taken to be 0 dB, but this is an arbitrarily selected reference.

Bandpass filters Active bandpass filters require at least two reactive elements with an amplifier in a variety of configurations. Figure 1-24 illustrates one multiple-feed-

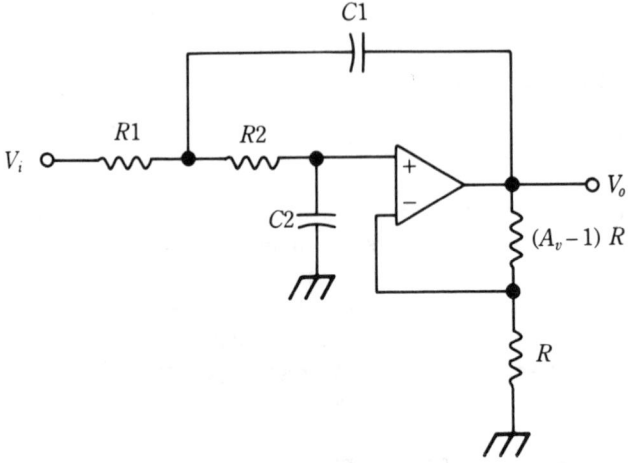

1-23 Second-order active low-pass filter.

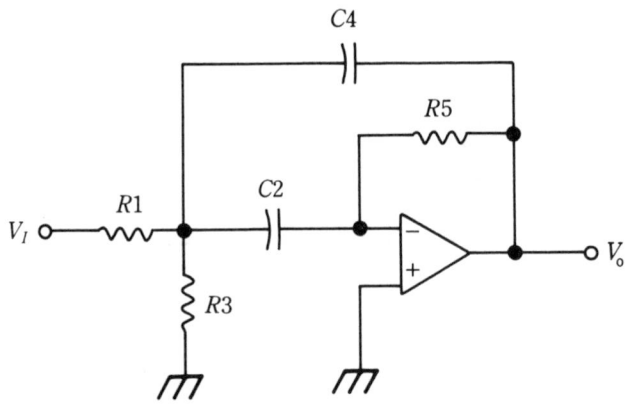

1-24 Multiple feedback active band-pass filter.

back, bandpass active filter schematic. All of the filter parameters are established by the resistor and capacitor values for an ideal operational amplifier. The same is true for practical purposes when production op amps are used. For Fig. 1-24, the center frequency $f_0 = 1/2\pi R_1 C_4$ and the voltage gain in the passband, $A_V = -Q$, in which $Q = R_5 C_2/R_1(C_2 + C_4)$. For example, with $R1 = 1$ kΩ, $C2 = 1$ μF, $R3 = 10$ kΩ, $C4 = 0.1$ μF, and $R5 = 10$ kΩ, $A_V = -10$ and $2\pi f_0 = 10,000$ rad/s.

A notch filter is intended to attenuate over a restricted range. Its function is exactly the opposite of a bandpass filter. One simple approach to the notch filter is shown in Fig. 1-25, where two amplifier stages are cascaded: the first is a bandpass filter; the other is a summing amplifier. The function of the notch filter is to combine the input voltage, V_I, and the output of the first op amp associated with the bandpass section. When the two are properly balanced through the selection of R6 and R7, the two input voltages will be amplified with equal gains for both. At the center frequency of the bandpass section, the

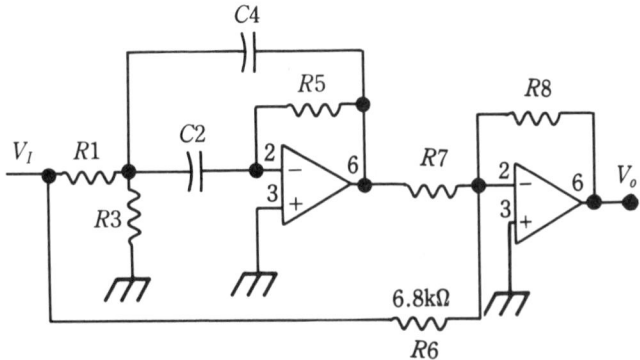

1-25 Cascaded active notch filter.

output will have undergone a phase reversal (180° phase shift) and, under ideal conditions, the two voltages will cancel to give 0 output voltage. Although perfect cancellation is impossible, significant attenuation can be expected with carefully selected resistors and capacitors.

At other frequencies the output of the bandpass filter section is greatly attenuated. Therefore, the inputs to the summing amplifier differ greatly in amplitude so that cancellation cannot take place. Note that a phase reversal occurs in the summing amplifier.

Mechanical vibration filters

Three other filter types having important characteristics are presented in this section. They are the piezoelectric crystal filter, the ceramic filter, and the mechanical filter. These three filters have one feature in common: mechanical vibration is responsible for the resonant frequency(ies) of each device and for the Q when it is incorporated into an electronic circuit.

When the body of a piezoelectric crystal is mechanically deformed through the application of forces, a potential appears across its faces. If the forces are removed abruptly, the crystal will vibrate at its own natural frequency, although the vibration decays with time because of internal energy losses, damping by the surrounding air, etc. Correspondingly, a damped sine wave voltage is produced by the crystal.

It is also true that the application of a voltage across the crystal faces will cause a mechanical deformation. When an alternating voltage of the proper frequency is applied, the periodic mechanical vibrations are sustained, and the device is the basis of a crystal oscillator. A crystal oscillator, having an electronic amplifier to convert power supply energy into alternating form to replace the energy losses, has an extremely stable frequency, especially when its temperature is accurately controlled. See chapter 6 for more information on this subject.

The piezoelectric crystal, when mounted in a holder that allows electrical contact with the crystal faces, behaves as a series-parallel circuit from the standpoint of any electronics circuitry. Figure 1-26 depicts an equivalent electrical circuit for the crystal. $L1$ and $C1$ in the one branch result in series resonance at f_S for which the two reactances are equal, so that the net reactance is 0 (see Fig. 1-27). Above f_S the net reactance is

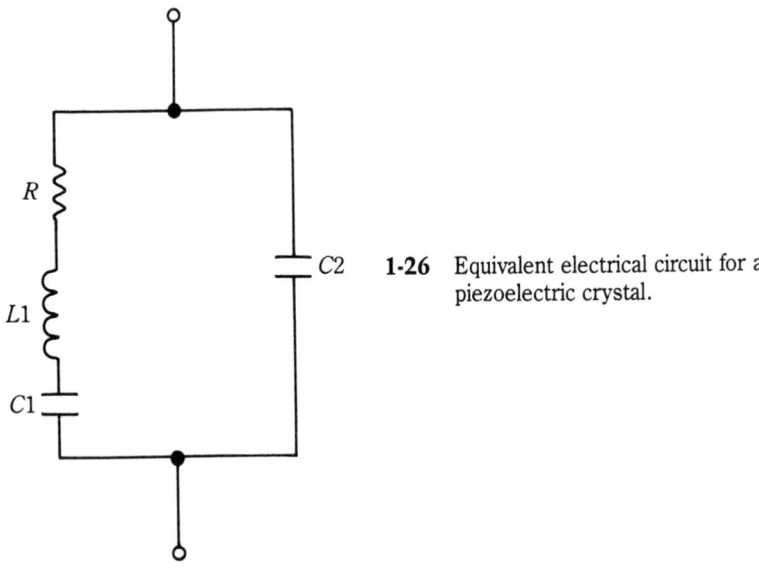

1-26 Equivalent electrical circuit for a piezoelectric crystal.

inductive so that parallel resonance occurs at f_P, when this reactance is equal to that resulting from C2. Note that C2 also includes the distributed capacitance of the wiring and the crystal holder.

In summary, the equivalent circuit arrangement of Fig. 1-26 has both a series and a parallel resonance frequency (see Fig. 1-27). Recall that the impedance at f_S is very small and that the impedance at f_P is very large. One or more crystals can be incorporated into circuits in which the resonant frequencies are staggered to achieve filtering action not possible with a single resonant circuit. Moreover, the Q of the resonant crystal at either resonant frequency is greater than that of practical electrically resonant LC circuits, thus achieving a sharper response.

Figure 1-28 is a crystal lattice filter. The crystals at the top and the bottom are selected to be as nearly identical as possible, each having reactance X_1. Also, the crystals in the diagonal arms are alike, each having reactance X_2. As seen in Fig. 1-29, series resonance occurs for one pair of crystals at the same frequency that parallel resonance occurs for the other. The combination of the resonant points in the passband of the filter achieves reasonably flat and wide response with sharp skirts to attenuate frequencies outside the passband.

Mechanical filters were developed to meet the demanding requirements of single sideband (SSB) transmitters and receivers. A basic mechanical filter is illustrated in Fig. 1-30. It consists of an input transducer, resonant metal disks, coupling (connecting) rods, and an output transducer.

The input transducer converts electrical energy to mechanical energy by using the magnetostrictive effect. This phenomenon causes a ferromagnetic rod to contract in the presence of a magnetic field. The mechanical motion of the rod end is transmitted to the metal disks and coupling rods. Each disk behaves as a series-resonant circuit, as modeled by the $L1C1$ combination of the equivalent circuit shown in Fig. 1-31. The mechanical coupling rods are represented by $C2$ in the figure. A number of disks are usually

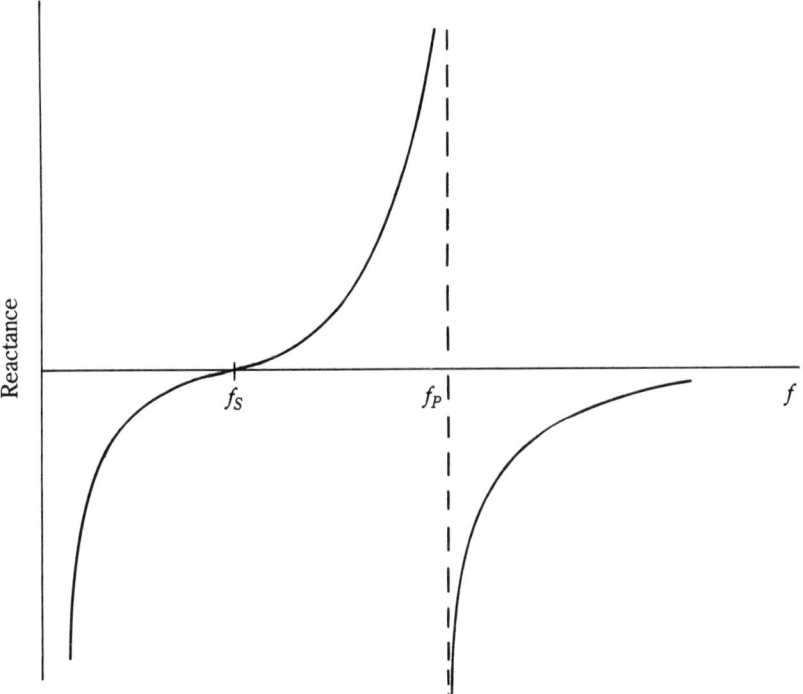

1-27 Equivalent reactance of a piezoelectric crystal equivalent circuit.

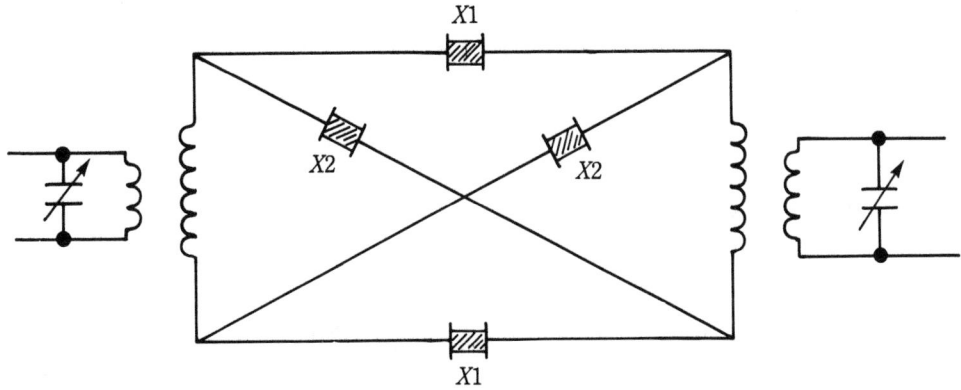

1-28 Crystal lattice filter.

constructed in the filter so that the series-parallel arrangement of Fig. 1-31 is the result. The last disk in the filter vibrates the output transducer rod, which induces a voltage in the output coil.

The excellent selectivity features of the mechanical filter are largely caused by the equivalent Q of about 10,000 for each resonant metal disk. Moreover, the number of disks determines the skirt selectivity of the overall filter. The *shape factor*, which is the

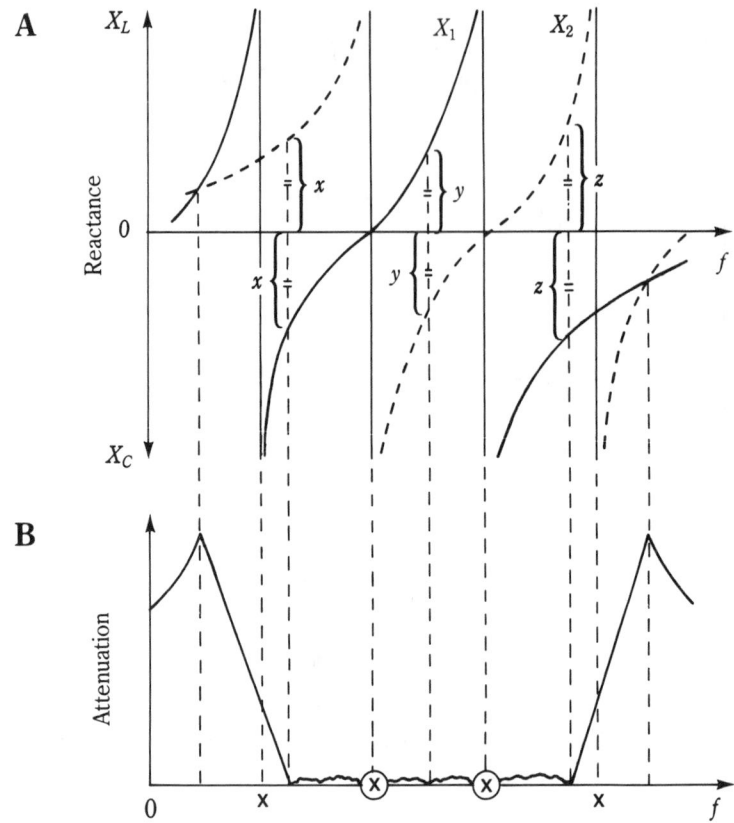

1-29 Attenuation of a crystal lattice filter.

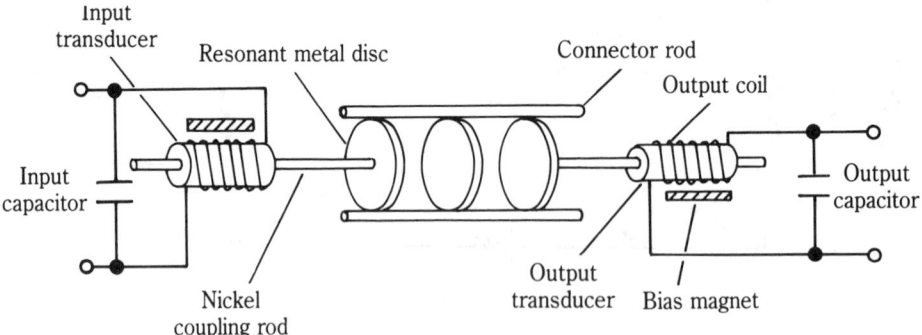

1-30 Mechanical filter.

ratio of the frequency span between the 60 dB points to that for the 6 dB points, is determined by the number of disks. Typically, six-, seven-, and eight-disk filters will have shape factors of approximately 2.2, 1.85, and 1.5, respectively.

Mechanical filters are extremely stable, so the bandwidth can be as small as about 0.1 percent of the center frequency while retaining the desired flat response, with

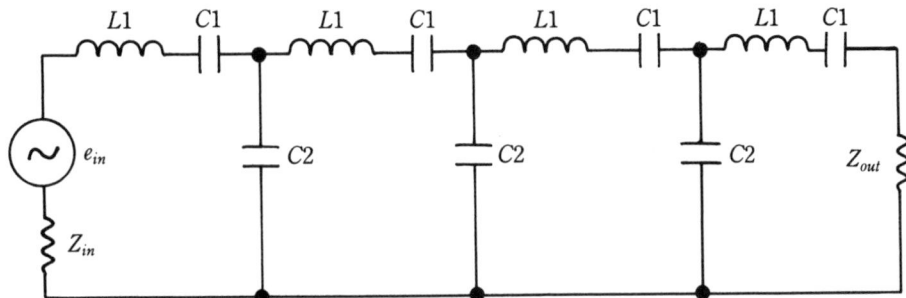

1-31 Equivalent circuit of the mechanical filter shown in Fig. 1-30.

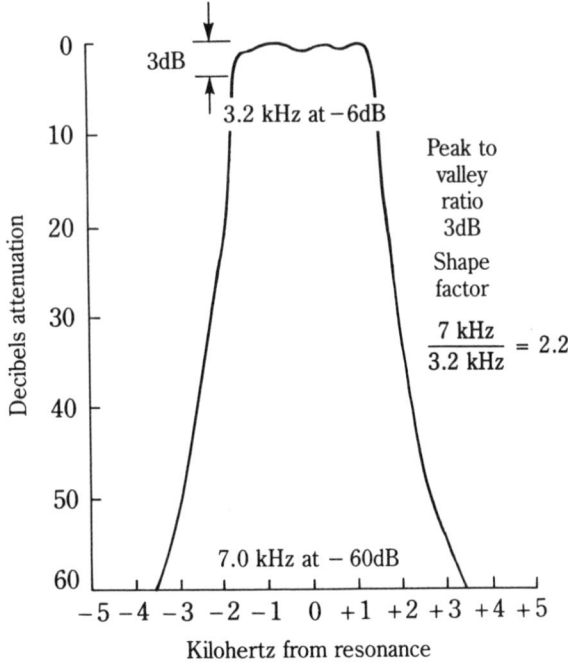

1-32 Mechanical filter response curve.

acceptable ripple amplitude within the passband (see Fig. 1-32). Mechanical filters are available with center frequencies in the range of 60 to 600 kHz.

A coaxial-cavity resonator can be the basic component of bandpass and bandstop filters having the steep skirts of the mechanical-vibration filters previously described. To reduce the size and facilitate adjustment, a cavity is filled with a ceramic dielectric so that the cavity length is inversely proportional to the square root of the dielectric constant. Low losses in the dielectric aid in the attainment of a high Q.

One design of the ceramic-filled resonator (Fig. 1-33) includes the outer conducting layer of the cavity by firing a thick film of silver onto the ceramic cylinder. Integral to the cavity formed in this manner is a trimmer capacitor (not shown) allowing fine adjust-

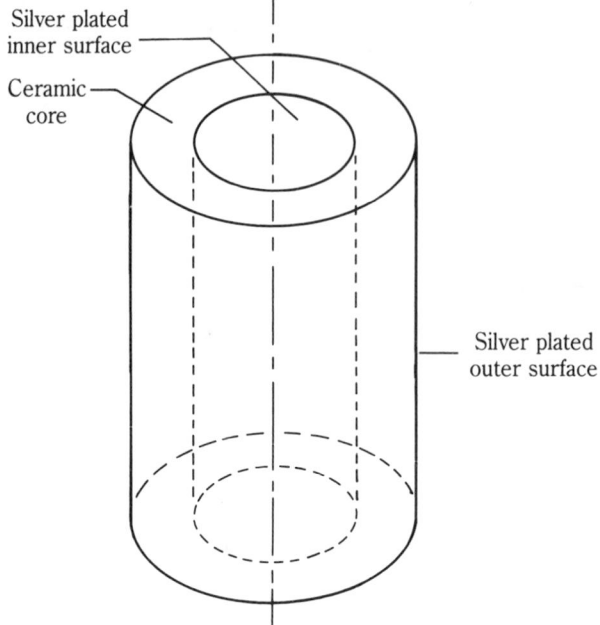

1-33 Ceramic-filled resonator.

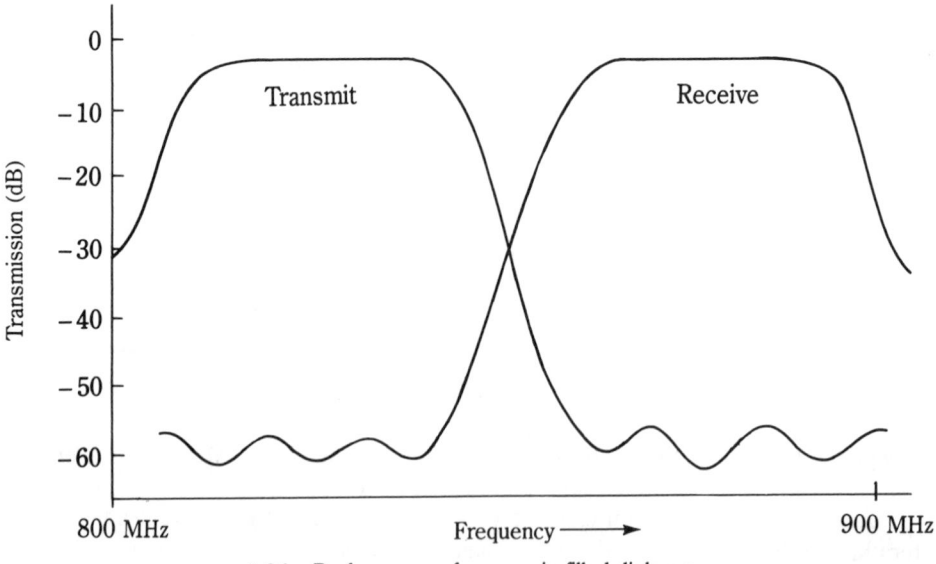

1-34 Performance of a ceramic-filled diplexer.

ments of the frequency. The term *varisonator* has been coined for the name of this design.

In a typical application to the 800 – 900 MHz mobile radio frequency range, a pair of filters constructed from the resonators forming a diplexer will have the characteristics shown in Fig. 1-34. Note that a diplexer allows transmission and reception with the same antenna without switching.

Surface acoustic wave filters

In the simplest form, a transversal standing acoustic wave (SAW) filter has two transducers attached to opposite edges of a polished piezoelectric substrate. The application of an electrical signal to one transducer causes the piezoelectric substrate surface to be distorted, which creates a series of traveling waves along the length of the substrate. The arrival of these mechanical waves at the transducer at the other end of the substrate causes the mechanical energy to be transformed back into electrical energy.

Figure 1-35 shows the interdigital conducting electrodes of the transducers that are deposited on the substrate, on the input, and on the output circuits. Because this surface wave travels from one electrode pair to another, the signal is repetitively delayed and added to itself, constructively or destructively, to determine the amplitude characteristics of the filter. The finger spacing determines the wavelength of the acoustic wave, and the number of sections governs the bandwidth.

SAW filter designs allow the amplitude and phase characteristics to be specified separately, which is particularly important for some signal processing applications. Quartz is a material suitable for the substrate that allows a wider bandwidth than the other common material, lithium niobate.

The SAW filter offers advantages for the center frequencies in the range of less than 30 to more than 500 MHz, and bandwidths in the range of 0.1 to 30 percent. These advantages include reproducibility, reliability, low cost, selectivity, and small size. Low cost is associated with volume production only because of the cost of designing the photolithography masks for their production.

Digital filters

A *digital filter* (as used here) is an implemented digital process that converts an input periodic time sequence of numerical data point samples into an output signal, also repre-

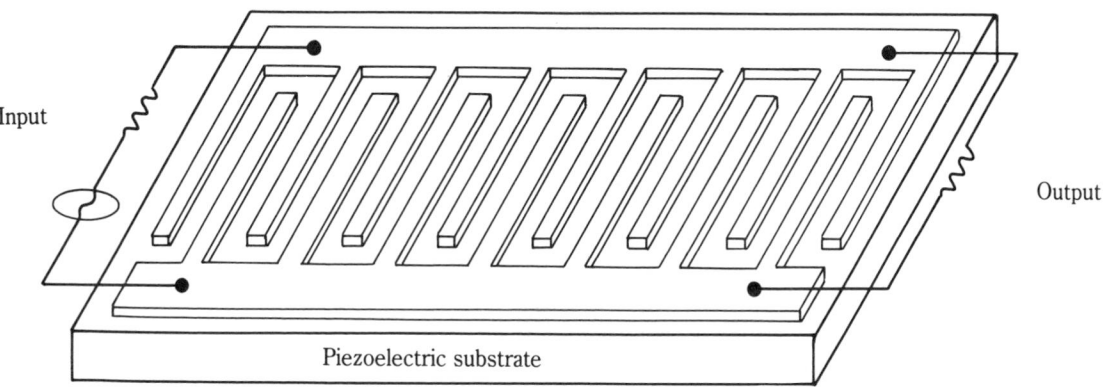

1-35 Surface-acoustic-wave (SAW) device.

sented by a number sequence. Because of its enormous versatility and power, together with present practical processing methods, digital filtering has in recent years been widely used in seismic data processing, image processing, speech processing, and many other applications too numerous to mention.

Digital filters can be realized as software on general-purpose digital computers. Alternately, they can be constructed of dedicated electronic hardware.

If the time required to process each point (number in the sequence) exceeds the time interval between adjacent input data samples, the filter cannot function as a real-time filter, and its purpose more closely resembles that of an after-the-fact data processor. However, if the digital processing can be achieved in real time, the results are comparable to those obtained using other filters with the following exceptions:

1. The input signal is known only by its sampled values. Improper sampling techniques will invalidate the number sequence as representing the input signal.

2. The processed output is also discrete, unless converted to analog form by a digital-to-analog converter (DAC).

In the manipulation of a sequence of numbers, only a limited number of elementary arithmetic or other operations are possible. For example, a finite-impulse response filter multiplies several numbers in the sequence by selected constants, and then adds the results, as indicated in the multiplier-accumulator of Fig. 1-36. In this figure, the last sample is designated as "n," the previous sample as "$n-1$," the one before that as "$n-2$," etc., with the earliest sample entering into the process designated as "0." The multiplication coefficients appear as constants labeled "k" with the sample number in the sequence as the subscript.

The next process execution is identical, with the understanding that one new sample (number in the sequence) assumes the n-subscript designation and the earlier samples have slipped exactly one place in the process. Eventually, any one sample will disappear from further processing after having played each of n parts; i.e., having in turn been multiplied by n different coefficients in the basic multiply and add process.

There is some resemblance here to the numerical determination of Fourier series components using the weighted summation of points spaced along the waveform, each

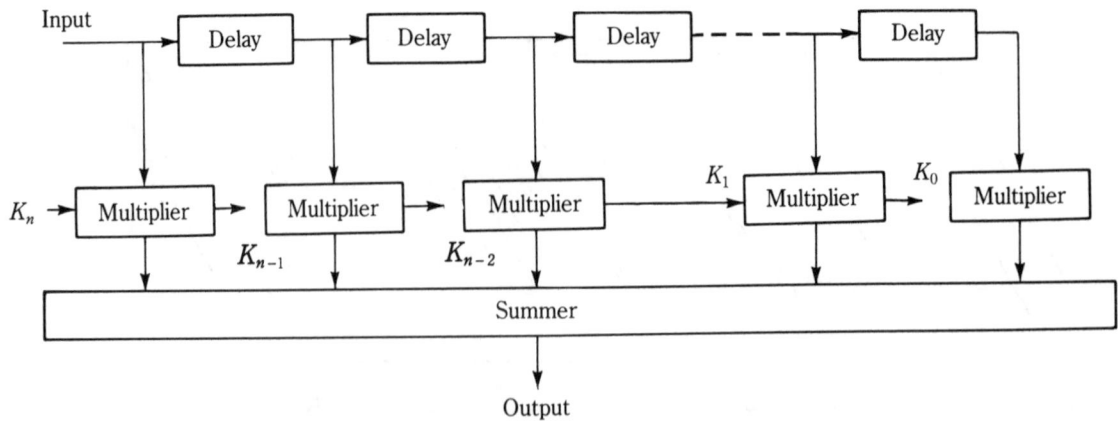

1-36 Finite impulse response filter.

multiplied by the sine or cosine of an angle corresponding to the location of the sample point. Through an extension of this idea, the numerical coefficients (multipliers) are chosen to yield summation results indicating the existence of frequency components in any desired range. Low-pass, bandpass, or high-pass digital filters are thus made possible.

Perfect response from such a filter is only possible with an infinite number of terms, that is, sample points. Perfection imposes the dual burden of an infinite processing load and an infinite total time delay. Practical filters can tolerate neither, so the number of samples processed at any time ($n + 1$ in Fig. 1-36) must be limited, and total delay also limited.

Software implementation in digital computers is possible. However, the effectiveness of computers is limited by their architecture when processing speed and total time delay are important. Through the exploitation of the symmetry of the weighting/summing steps, hardware digital processors are gaining in popularity. Single-chip digital processors include the Intel 2920, AMI2711, NEC 7722, and TM320. TRW's TDC1028 provides a high level of on-chip integration with multiplication, addition, and registers on a single chip.

In the TDC1028, circuit complexity has been dramatically reduced through use of the knowledge that the multiplication and addition are linear operations and can be interchanged (see Fig. 1-37). The TDC1028 is capable of operating at video speeds and can be used to construct either fixed or adaptive filters. Adaptation requires that the coefficients be capable of modification.

Finite impulse response filters suffer no drift of performance characteristics with time, temperature, or supply voltage. Unlike analog filters, they provide exact *repeatability*—exactly the same characteristics and exactly the same linear phase over and over.

Impedance-matching with L-sections

The main purpose of the circuits connecting a transistor's power amplifier stages and loads or sources is impedance matching. It is possible to approach impedance matching

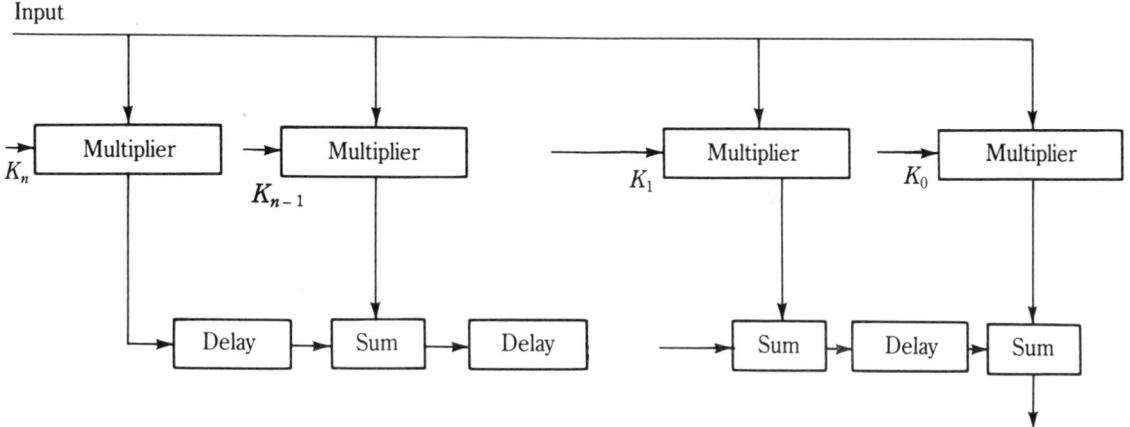

1-37   Architecture of a TDC1028 filter.

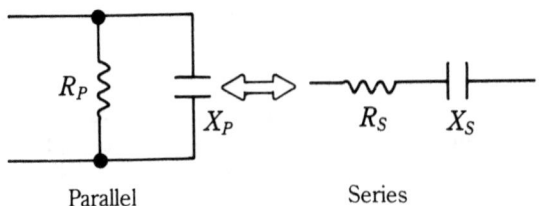

R_P
X_P
R_S
X_S
Parallel
Series

1-38 Parallel/series transformation.

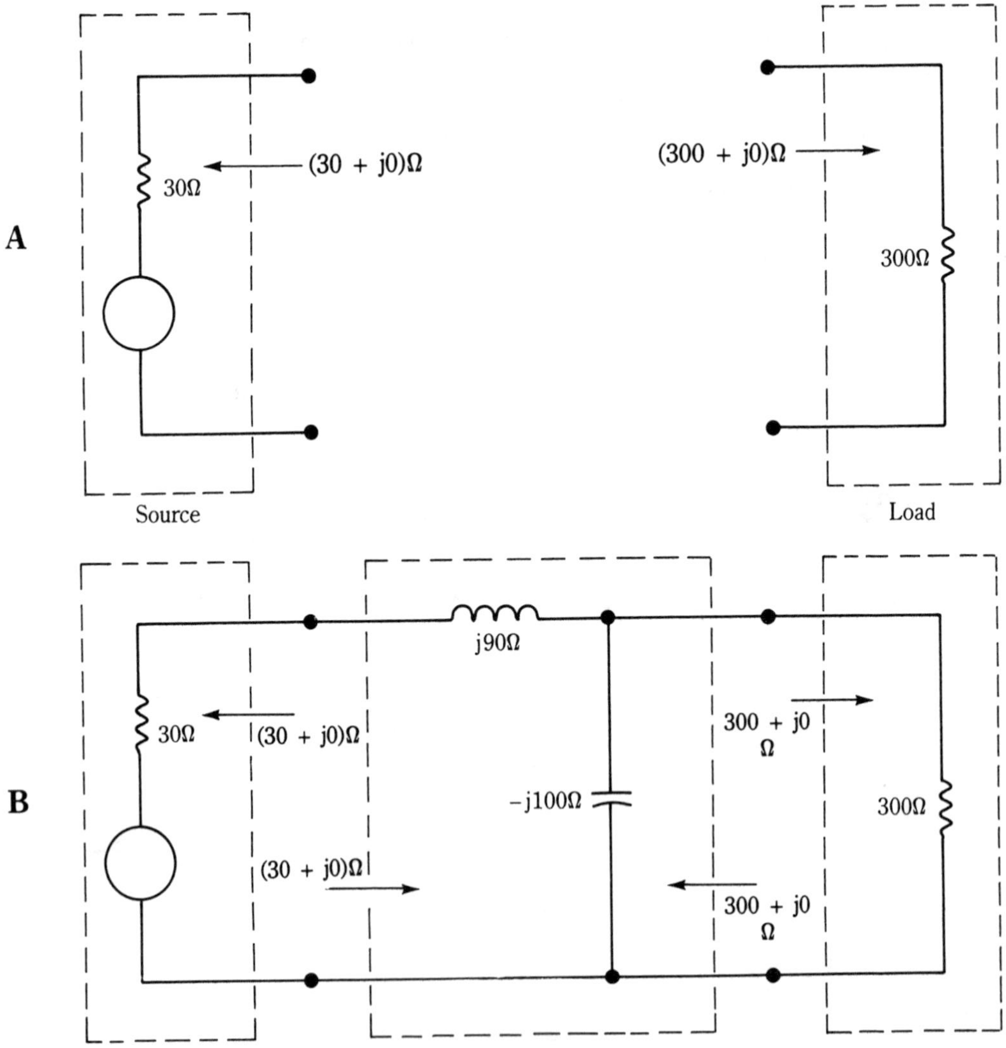

A
30Ω
(30 + j0)Ω
(300 + j0)Ω
300Ω
Source
Load
B
30Ω
(30 + j0)Ω
j90Ω
300 + j0
Ω
−j100Ω
300Ω
(30 + j0)Ω
300 + j0
Ω

1-39 Low source to high load:
A. Source and load before matching.
B. With L-section.

by means of L-sections composed of practical circuit elements. The basic transformations are from RX combinations to their series or parallel equivalents for which the following equations apply (see Fig. 1-38):

$$R_P/X_P = X_S/R_S = Q$$
$$R_P = (Q^2 + 1)R_S$$
$$X_P = (Q^2 + 1)X_S/Q^2$$

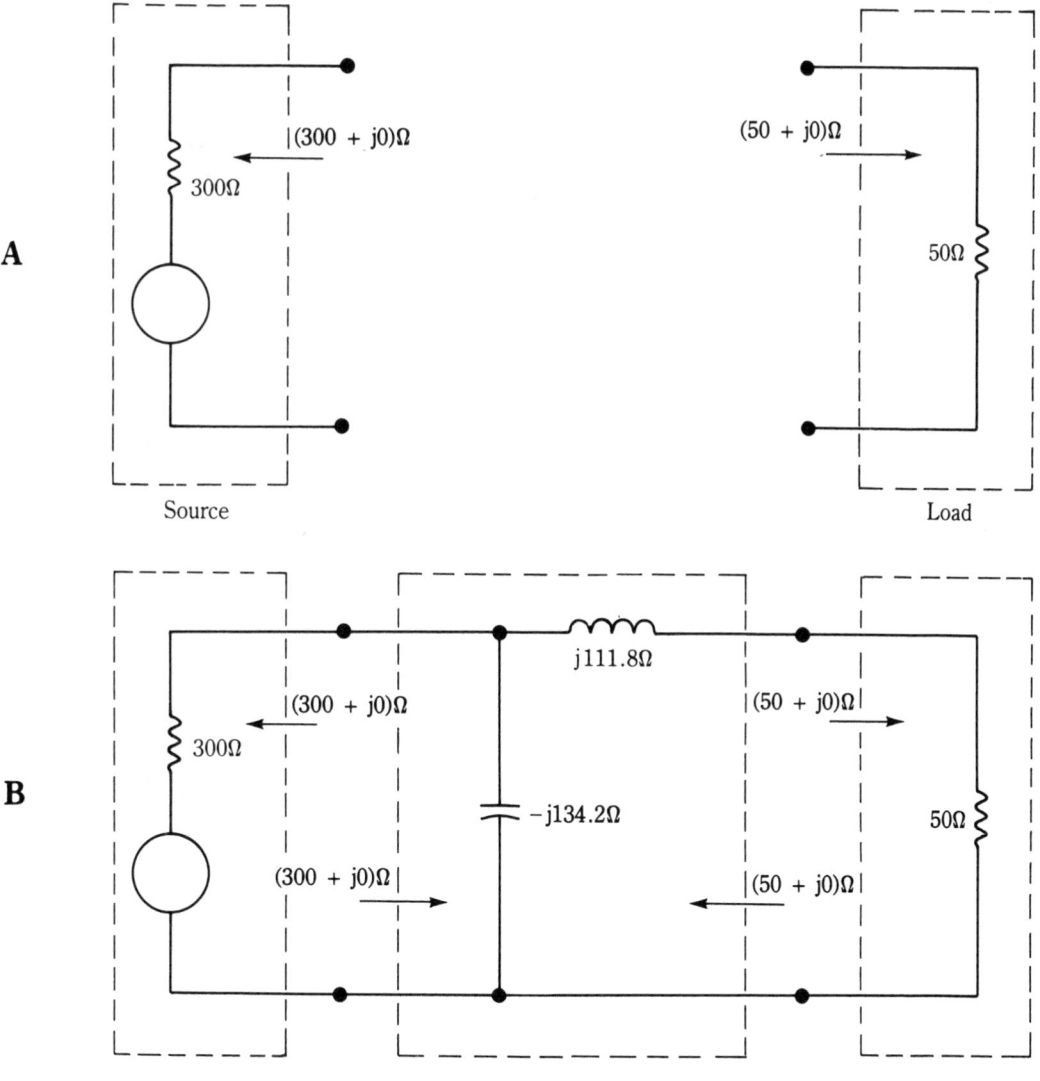

1-40   High source to low load:
A. Source and load before matching.
B. With L-section.

Low source to high load

In Fig. 1-39A an L-section is selected to match a 30 Ω source to a 300 Ω load. Here, $Q^2 + 1 = 300/30 = 10$, for which $Q = 3$. The parallel element of the L-section has a reactance of $X_P = R_P/Q = 300/3 = 100 \ \Omega$, which is chosen to be capacitive. Combining 300 Ω (resistance) and 100 Ω (capacitive reactance) in parallel gives the series equivalent $R_S = 300/10 = 30 \ \Omega$ and $X_S = 100 \times 9/10 = 90 \ \Omega$; so the series arm of the L-section is j90 Ω. The input impedance of the L-section is $(30 + j0) \ \Omega$ (see Fig. 1-39B), and the matching is complete.

High source to low load

The problem presented in Fig. 1-40A is the matching of a 300 Ω source to a 50 Ω load. Note that $Q^2 + 1 = 300/50 = 6$ is determined by these two values. Therefore, $Q^2 = 5$ and $Q = \sqrt{5}$. In configuring the L-section of Fig. 1-40B, place $X_P = 300/\sqrt{5} = 134.2 \ \Omega$ in parallel with the 300 Ω source resistance, and balance out this parallel capacitive reactance with an equivalent series reactance of $134.2 \times 5/6 = 111.8 \ \Omega$. The impedance looking into the L-section from the load is $(50 + j0) \ \Omega$ so that matching is again complete.

A

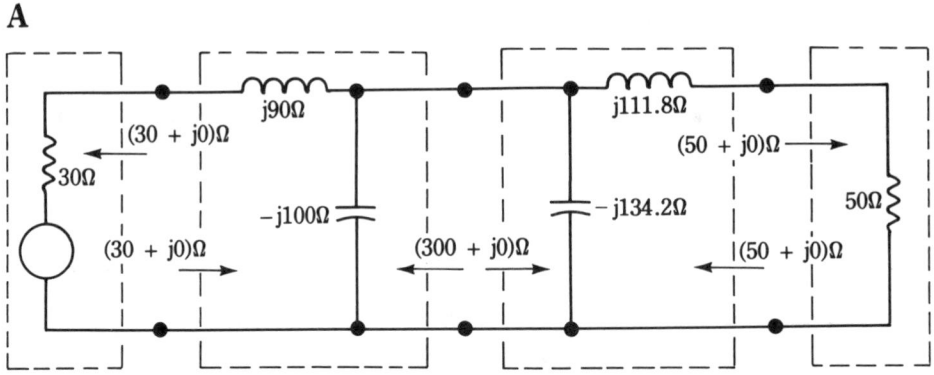

B

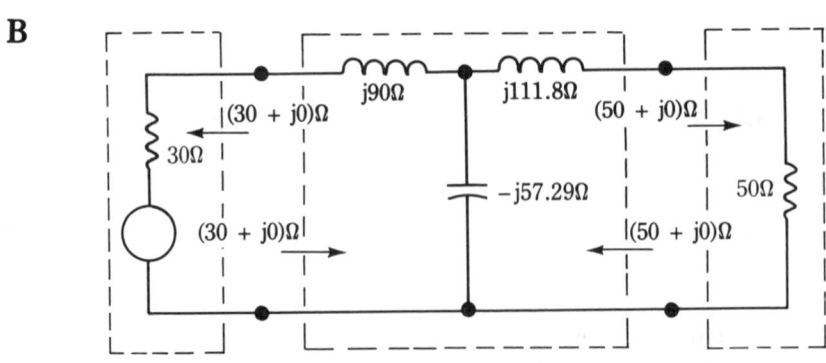

1-41   Low source to low load.

Low source to low load

The familiar T-section can be visualized as the combination of two L-sections. Suppose that you wish to match a 50 Ω load to a 30 Ω source. Note that these values incorporate the source of Fig. 1-39A to the load of Fig. 1-40A. It is then only necessary to place the respective L-sections of these two together, as shown in Fig. 1-41A, to obtain the T-section of Fig. 1-41B. Note that a designer has resources such as tables, computer programs, etc., to aid in the selection of the T-section elements. This explanation is not intended to replace such powerful tools, but rather to provide a key to understanding how the matching circuits work.

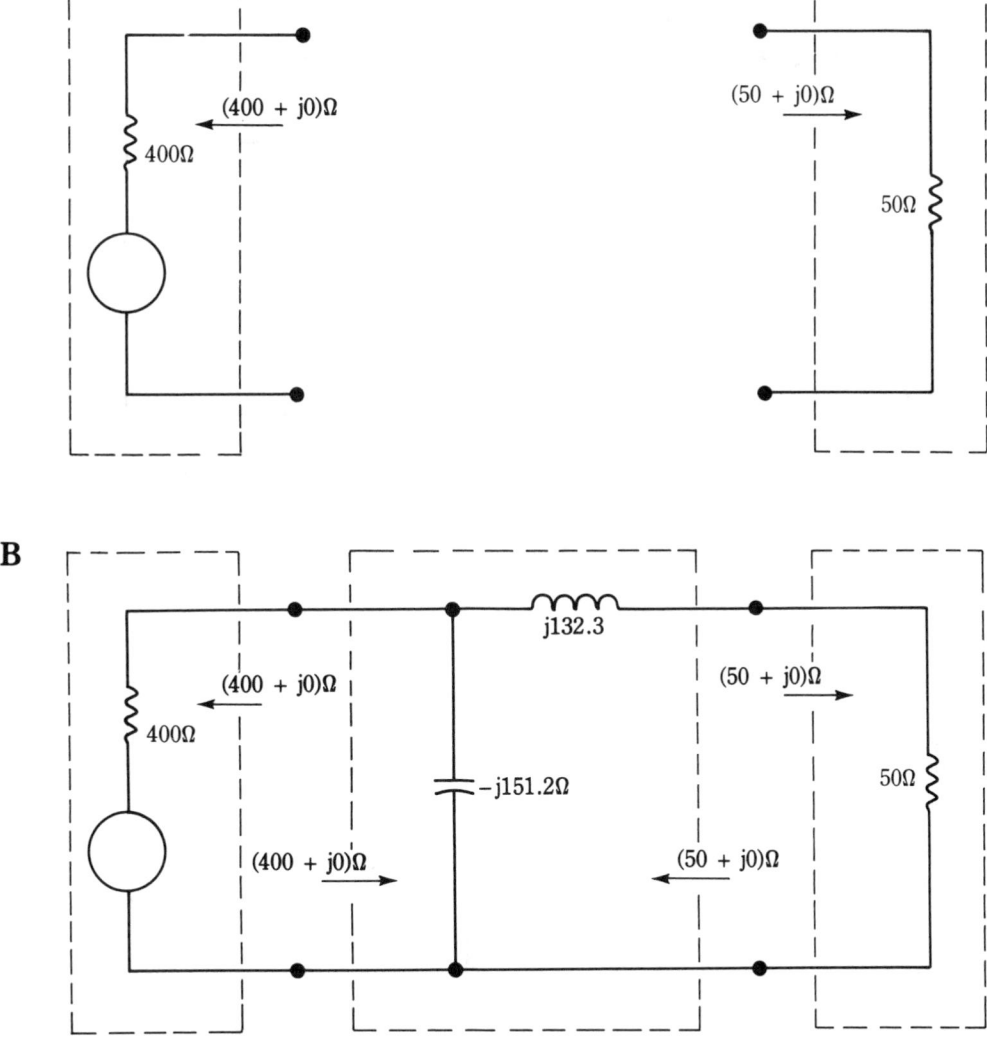

1-42   High source to high load—first L-section.

High source to high load

You might wish to follow the reasoning behind L-section selections for a different source and load resistance combination. Starting with Fig. 1-42A, see that $Q^2 + 1 = 400/50 = 8$, for which $Q = \sqrt{7}$, $X_P = 400/\sqrt{7} = 151.2\ \Omega$, and $X_S = 151.2 \times 7/8 = 132.3\ \Omega$. The matching L-section is shown in Fig. 1-42B. The L-section for Fig. 1-43A is shown in Fig. 1-43B. Finally, the results of Figs. 1-42 and 1-43 are combined in Fig. 1-44, in which the two L-sections form a single π-section.

Additional practical solutions to the matching problem using L-sections are demonstrated in the chapter on rf amplifiers. In some of these solutions the L-sections are actually cascaded to meet particular design restrictions and requirements.

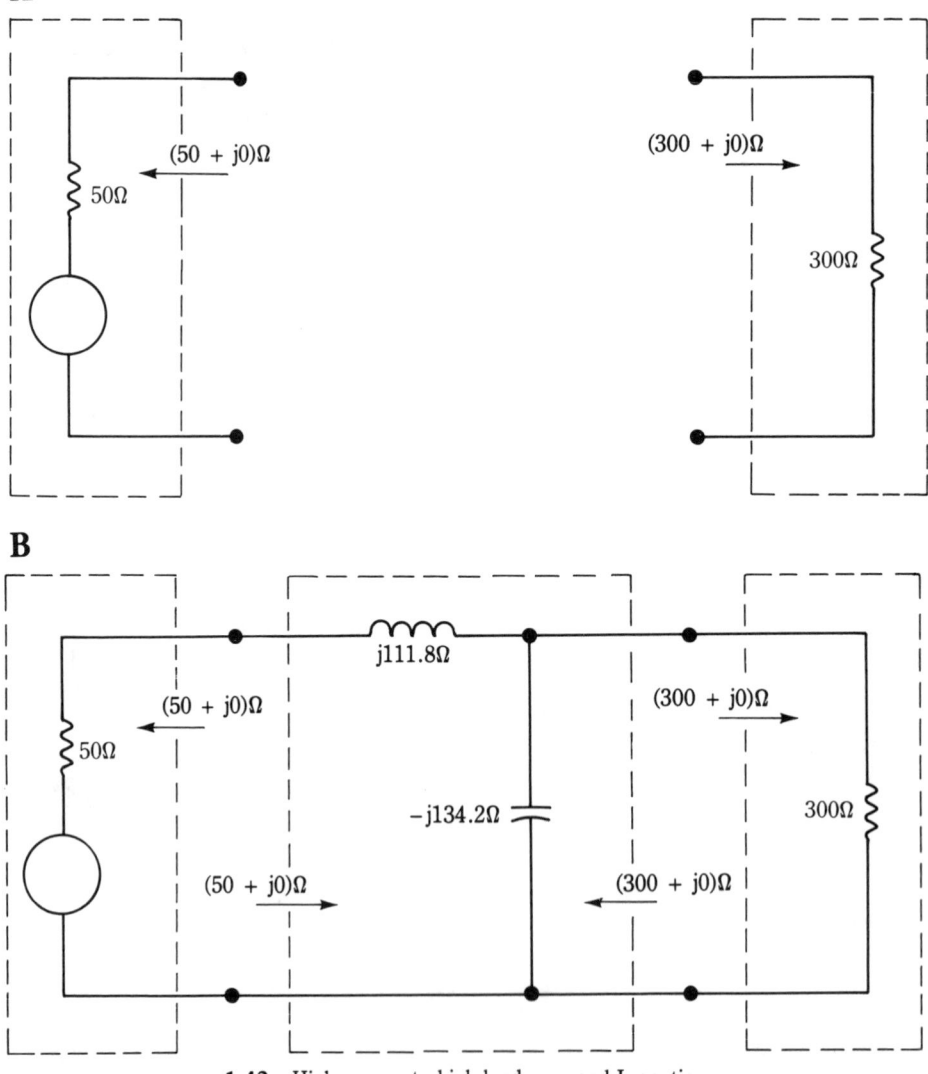

1-43   High source to high load—second L-section.

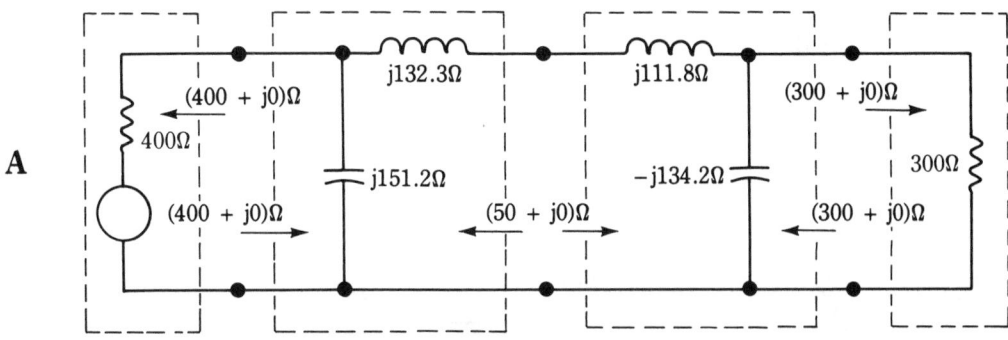

A

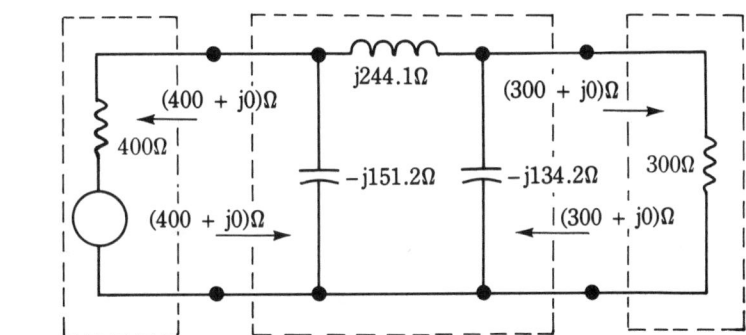

B

1-44 High source to high load—combining L-sections into a π-section.

Multiple-choice questions

1. The frequency at which series resonance occurs is:

A. $X_L = X_C$.
B. $X_L = R$.
C. $Z = R$.
D. $X_C = R$.
E. Choices A and C are correct.

2. In a series RLC circuit, $V_L = V_C$ when:

A. The total applied voltage is zero.
B. The voltage across the resistance is minimum.
C. The current leads the total voltage by 90°.
D. The power factor is zero.
E. The value of the impedance is minimum.

3. In a series RLC circuit:

A. The impedance is always greater than X_L.
B. The current lags V_L by 90°.

C. $Z = jX_L$ at the resonant frequency.
D. The current leads V_L by 90°.
E. X_L leads X_C by 90°.

4. In a series circuit, the power factor will be zero if:

   A. The applied voltage is dropped to zero.
   B. The frequency is adjusted to 1 kHz.
   C. The value of the capacitance is zero farads.
   D. $X_L = X_C$ with R not specified.
   E. This question cannot be answered with the given information.

5. At resonance, one of the following conditions cannot be true for a series RLC circuit:

   A. $Z = R$.
   B. $X_L = R$.
   C. $Z = jX_L$.
   D. The phase angle is zero.
   E. The magnitude of Z is $\sqrt{R^2 + (X_L - X_C)^2}$.

6. In a series RLC circuit:

   A. Increasing the frequency decreases the resistance.
   B. Decreasing the frequency increases the resistance.
   C. X_L and X_C both change as the frequency increases.
   D. The amount of measured current will dip at the resonant frequency.
   E. As the frequency increases, Z will always decrease in magnitude.

7. Bandwidth, as applied to a series RLC circuit, refers to:

   A. The separation of the half-power points.
   B. The resonant frequency multiplied by Q.
   C. The frequency for which $X_L = X_C$.
   D. The difference between the highest and lowest frequencies of the oscillator.
   E. The frequency range for maximum power transfer.

8. The symbol Q refers to:

   A. Resonance quotient.
   B. Qualification tests.
   C. Quick tuning.
   D. Quality factor.
   E. X_L/X_C.

9. If $f_R = 10$ kHz and $Q = 50$,

   A. Bandwidth = 200 Hz.
   B. $X_L = 50,000\ \Omega$.

C. X_C = 50,000 Ω.
D. R = 50 Ω.
E. No correct answer is given.

10. A voltage ratio of 100 is:

    A. 100 dB.
    B. 20 dB.
    C. 40 dB.
    D. 0.01 dB.
    E. Insufficient information is given.

11. If 2V is applied to a 20dB attenuator, the output level will be:

    A. 10 V.
    B. 40 V.
    C. 2000 V.
    D. 200 mV.
    E. 0.02 V.

12. At the half-power points, the gain is down:

    A. 6 dB.
    B. 2 dB.
    C. 3 dB.
    D. 0.5 dB.
    E. 0.707 dB.

13. A T-section has inductance only in the series arms and capacitance only in the parallel branch. It is a:

    A. Matched load.
    B. Bandpass filter.
    C. Low-pass filter.
    D. Bandstop filter.
    E. No correct choice is given.

14. A T-section has capacitance only in the series arms and inductance only in the parallel branch. It is a:

    A. Matched load.
    B. Bandpass filter.
    C. Low-pass filter.
    D. Bandstop filter.
    E. No correct choice is given.

15. In a Bode plot for a filter, the response is plotted as:

    A. Ohms.

B. Radians.
C. Siemens.
D. Watts.
E. Decibels.

16. In a Bode plot for a filter, the frequency scale is:

   A. Logarithmic.
   B. Inverse decibels.
   C. Decibels.
   D. Bels.
   E. Linear.

17. A π-section of a filter has at least:

   A. Three resistors.
   B. Three inductors.
   C. Three capacitors.
   D. Three elements.
   E. Two series and one parallel element.

18. A constant-k filter has series resonant elements for Z_1 and parallel resonant elements for Z_2. It is a:

   A. High-pass filter.
   B. Low-pass filter.
   C. Bandpass filter.
   D. Bandstop filter.
   E. M-derived filter.

19. What feature distinguishes a constant-k filter?

   A. $LC = k$.
   B. $\sqrt{Z_1 Z_2} = k$.
   C. $Z = k$.
   D. Attenuation $= k$.
   E. All reactances are constant.

20. At cutoff the attenuation curve is sharper for the:

   A. m-derived filter.
   B. Constant-k filter.
   C. Constant-Q filter.
   D. Variable-k filter.
   E. Lattice-ladder filter.

21. An active filter usually contains at least one:

   A. R.

B. Capacitor.

C. Amplifier.

D. PIN diode.

E. A, B, and C are true.

22. One prominent feature of an active filter is the presence of:

   A. Feedthrough.

   B. Feedforward.

   C. Feeddown.

   D. Feedback.

   E. Feedup.

23. Capacitors are preferred over inductors in active filter circuits because of:

   A. Low losses.

   B. Low cost.

   C. Low Q.

   D. A, B, and C.

   E. Both A and B.

24. An ideal op amp will have:

   A. Low gain.

   B. High power losses.

   C. Low-input impedance.

   D. High gain.

   E. High output impedance.

25. A notch filter attenuates over:

   A. All low frequencies.

   B. All high frequencies.

   C. All harmonic frequencies.

   D. A restricted range.

   E. Microwave frequencies.

26. When a piezoelectric crystal is mechanically deformed,

   A. Voltages are produced.

   B. The resistance across the faces is zero.

   C. Rectification takes place.

   D. It is ruined.

   E. The crystal warranty is invalidated.

27. The equivalent circuit of a piezoelectric crystal will include:

   A. A series resonant circuit.

   B. A parallel resonant circuit.

C. Two diodes back-to-back.
D. Both A and B.
E. A, B, and C.

28. The processing of a number sequence is the method used in:

   A. Logic gates.
   B. Digital filters.
   C. Modems.
   D. Random memory.
   E. Digital tape recorders.

29. Implementation of a digital filter is possible using:

   A. Dedicated hardware.
   B. Piezoelectric crystals.
   C. Computer software.
   D. L and C in parallel.
   E. Both A and C.

30. A SAW device has electrodes that are described as:

   A. Silicon-on-substrate.
   B. Retarded traveling wave.
   C. Polyoxidized epitaxial wafer.
   D. Micrometer siliconite.
   E. Interdigital.

Basic problems

1. At what frequency will the current in a series RLC circuit reach its maximum value for an applied voltage of 15V with $R = 500\ \Omega$, $L = 100\ \mu H$ and $C = 0.001\ \mu F$?

2. For the conditions of question 1, what are the respective element voltages?

3. Using the inductor and capacitor of question 1, determine a three-branch parallel circuit having one element in each branch and the same resonant frequency.

4. The voltage gain of an amplifier is 1200. Convert this gain into decibels, assuming equal input and output impedance levels.

5. A low-pass, constant-k filter T-section, as shown in Fig. 1-13, has 10 mH in each series arm and 0.1 μF in the shunt arm. Determine the cutoff frequency.

6. A high-pass, constant-k filter T-section, as shown in Fig. 1-14, has capacitors of 0.002 μF in each series arm and an inductor of 25 mH in the shunt arm. Calculate the cutoff frequency.

7. Refer to Fig. 1-22. Determine the cutoff frequency when $R = 330 \; \Omega$ and $C = 0.001 \; \mu F$.

8. Refer to Fig. 1-24. Determine the center frequency of this bandpass filter when $R1 = 180 \; \Omega$ and $C4 = 0.05 \; \mu F$.

9. An important, relatively new filter consists of two transducers attached to opposite edges of a polished piezoelectric substrate and is referred to as a "SAW" filter. What does SAW mean?

10. Convert a series combination of $5 \; \Omega$ resistance and $25 \; \Omega$ capacitive reactance into the parallel equivalent.

Advanced problems

1. A series RLC circuit consists of a $10 \; \Omega$ resistor in series with $L = 10 \; \mu H$, and $C = 100 \; pF$. Determine a new value of L for which the resonant frequency is one-half the original value.

2. Calculate the original bandwidth and the new bandwidth for the larger value of L in question 1.

3. Corresponding to the original circuit of question 1, let R and L in series be placed in parallel with C. Calculate the value of Q and the bandwidth for this circuit.

4. Design a circuit for question 3 having the same Q with the same L and C but with three parallel branches, each containing a single element.

5. A source whose output is 12 dBm is followed by a 31 dB gain amplifier and a 15 dB attenuator (gain = -15 dB). How much additional gain is required for a 25 W output level?

6. Refer to Basic Problem 5, which is concerned with a constant-k, low-pass filter having $L = 20 \; mH$ and $C = 0.1 \; \mu F$. Modify the design to provide an m-derived filter having a f_∞ that is 20 percent greater than f_C.

7. A bandpass filter is constructed as shown in Fig. 1-15, with $L1 = 0.191$ H, $L2 = 0.02385$ H, $C1 = 0.0663 \; \mu F$, and $C2 = 0.530 \; \mu F$. Determine the upper and lower cutoff frequencies.

8. The equivalent circuit of a piezoelectric crystal in Fig. 1-26 has $L1 = 96$ mH, $C1 = 0.01$ pF, and $C2 = 8$ pF. Estimate the resonant frequencies.

9. Briefly describe the two basic methods of implementing digital filtering.

10. Using L-sections, match a $12.5 \; \Omega$ source to a $50 \; \Omega$ load.

2
Solid-state devices

ANY PERSON FAMILIAR WITH THE DEVELOPMENT OF RADIO COMMUNICATIONS WILL recognize that many changes have occurred in the past few decades. However, a little reflection should disclose that most of these changes are directly connected with the widespread use of solid-state devices in communication equipment proper, and in all other electronic equipment connected in any way with the design, production, testing, and maintenance of communications equipment.

It is necessary, therefore, that any technically responsible person in the communications field be solidly grounded in basic semiconductor principles. Moreover, the accelerating trends of advanced device applications require that this technical knowledge be current. To demonstrate this assertion, list the semiconductor devices presently found in communications systems that were not so used only a few years ago.

In order to introduce semiconductors for communications, this chapter has been divided into a number of related sections. The first seven sections relate to semiconductor materials, their properties, and the discrete devices fabricated from them. The last four sections are concerned with one or more aspects of integrated circuits, which are generating the most significant changes. Integrated circuits are responsible for changes because they are capable of performing entire system functions with compact, reliable packages, each costing no more than the many discrete components of yesterday.

The following sections are included:

- Matter and energy
- Atomic structure
- PN junctions
- Semiconductor diodes
- Bipolar transistors
- Field-effect transistors

- Other discrete semiconductor devices
- The monolithic integrated circuit
- Digital integrated circuits
- Linear integrated circuits
- Integrated circuit arrays

Matter and energy

Matter is anything in the universe having mass and occupying space. It can be a solid, liquid, or gas. *Energy* is a physical concept associated with the ability to do work. A body suspended above the earth's surface possesses energy. An automobile speeding along a highway contains enough energy to destroy itself and its contents when halted abruptly. Deeper insight into these two ideas was provided by Albert Einstein and equates mass and energy through a mathematical equation whose proof has been dramatically demonstrated more times than you'd wish to think. Both matter and energy are closely related to the semiconductor phenomena, upon which the characteristics of semiconductor materials and devices depend.

You will presently be concerned with the structure of matter. It is now sufficient to recognize that matter is made of tiny particles, and even the largest are separated by distances vastly greater than their dimensions. Gravitational, electrical, and other forces determine how the particles interact. Energy affects the ability of these and the other physical particles in nature to join chemically and to function as electrical conductors, semiconductors, or insulators.

Atomic structure

All matter consists of chemical elements composed of structural entities called *atoms*. The substances most familiar to us are chemical compounds, which are made of linked atoms of the same or different elements. Two atoms of hydrogen and one atom of oxygen make one molecule of the compound water, for example.

An atom is often pictured as a dense nucleus surrounded by orbiting electrons, similar in principle to the sun in the solar system with its planets. An element is characterized by the amount of positive charge borne by the nucleus and, in the electrically neutral state, the equal number of negative orbiting electrons. Recognize that this representation of the atom is greatly simplified. For instance, the nucleus also contains uncharged particles called *neutrons*, which do not influence the chemical properties of the element (see Fig. 2-1).

The hydrogen atom has one orbiting electron and a nucleus with the same charge, but the opposite polarity, as shown in Fig. 2-2. More complicated atoms have additional orbiting electrons arranged in layers or shells that have letter designations (see Figs. 2-3 and 2-4). It is the number of electrons in the outer shell that determines the ability of the element to enter into chemical reactions with other elements. Figure 2-5 illustrates how the atoms of hydrogen and oxygen share outer-shell (valence) electrons so that the K shell of hydrogen has the maximum, stable number of two electrons, while the L shell of oxygen has eight electrons.

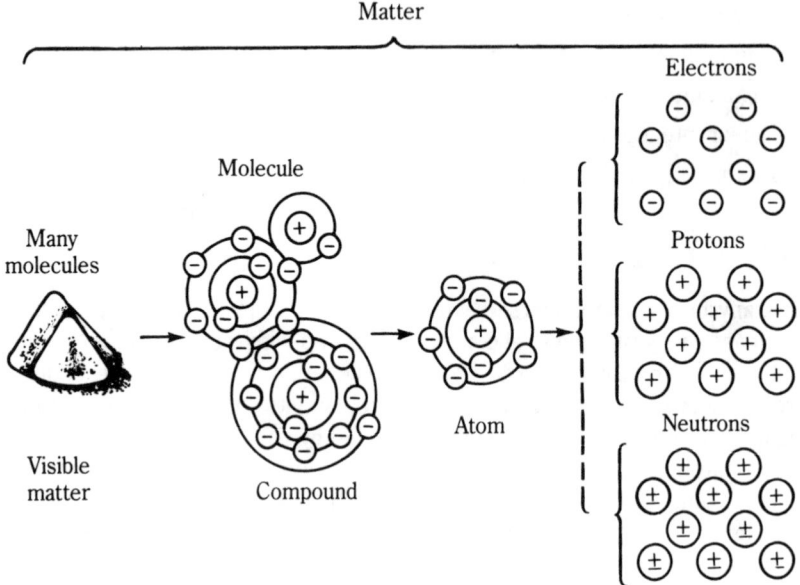

2-1   Breakdown of matter into subatomic particles.

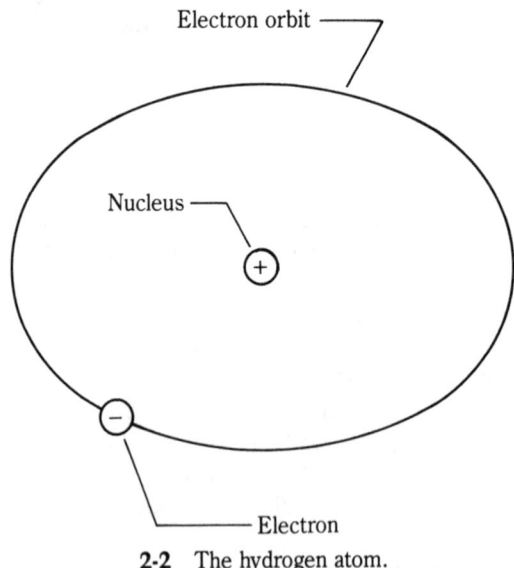

2-2   The hydrogen atom.

Electron sharing between identical atoms in a crystalline element, such as pure semiconductors, fixes the nucleus positions in a regular structure, as exemplified by the frame of a modern, high-rise building. The positions of the atoms relative to one another provide great stability because the outer shell of all atoms is completed by the electron sharing. In the absence of electrons free of the stable sharing arrangement, electrical current cannot flow and the bulk crystalline material is an insulator.

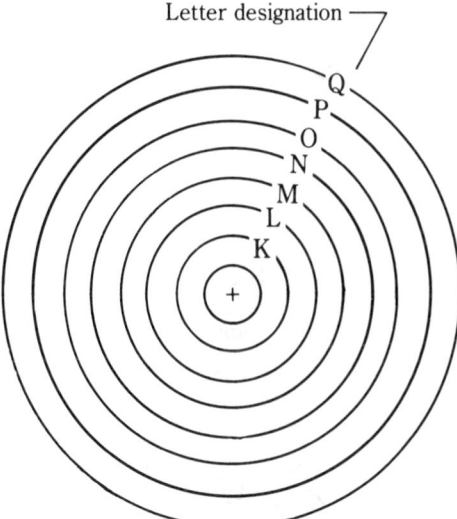

2-3 Arrangement of the shells in an atom.

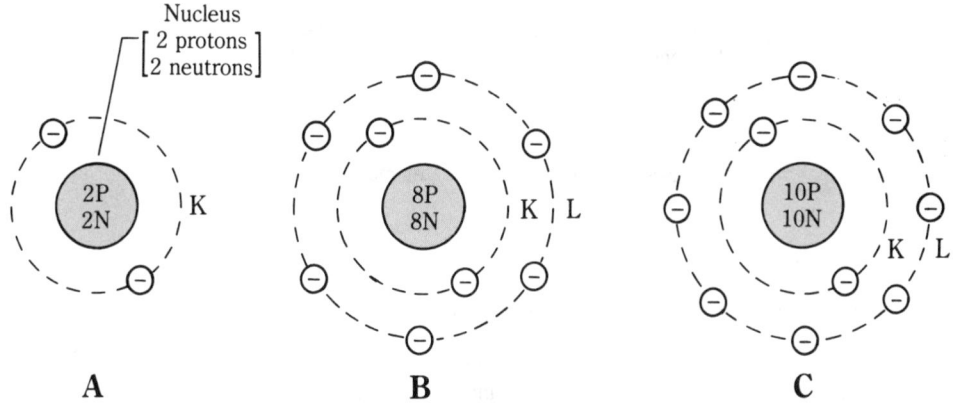

2-4 Atomic structure of various elements: A.helium, B.oxygen, C.neon.

However, when sufficient energy is supplied to the crystal, as by an increase in temperature, some electrons break the bond of sharing and are freed to move under the influence of any applied electrical field. For a semiconductor material, such as germanium or silicon, the density of free electrons is very small at room temperature, and the flow of current through the material is consequently quite small. The semiconductor material is neither a good insulator nor a good conductor, which suggests the name *semiconductor*. More specifically, the material is referred to as an *intrinsic semiconductor*.

From an energy point of view, the distinctions among an insulator, a conductor, and a semiconductor are shown in Fig. 2-6. There is a band of energy levels for each substance for which the sharing of *valence*, or outer-shell, electrons takes place. The upper, or *con-*

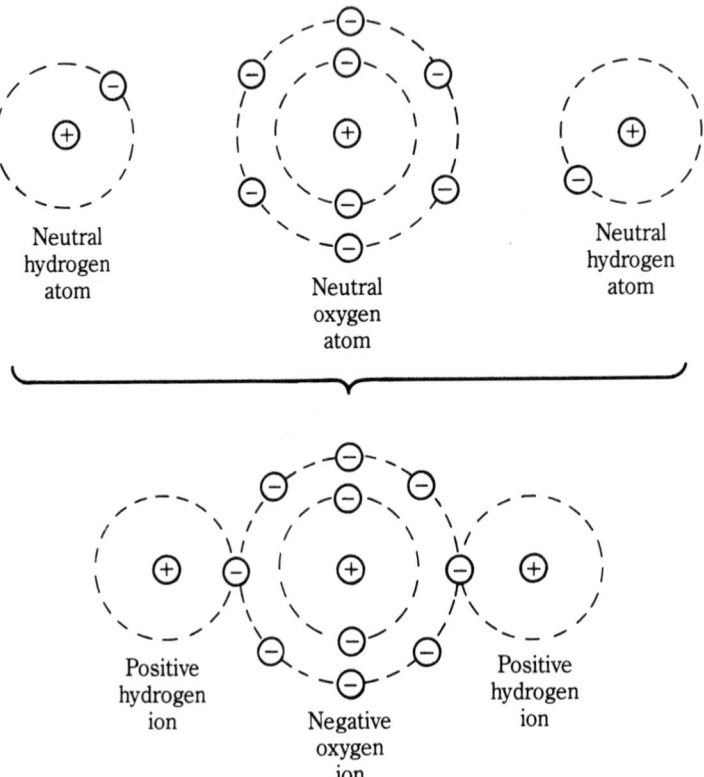

2-5 Formation of the water molecule.

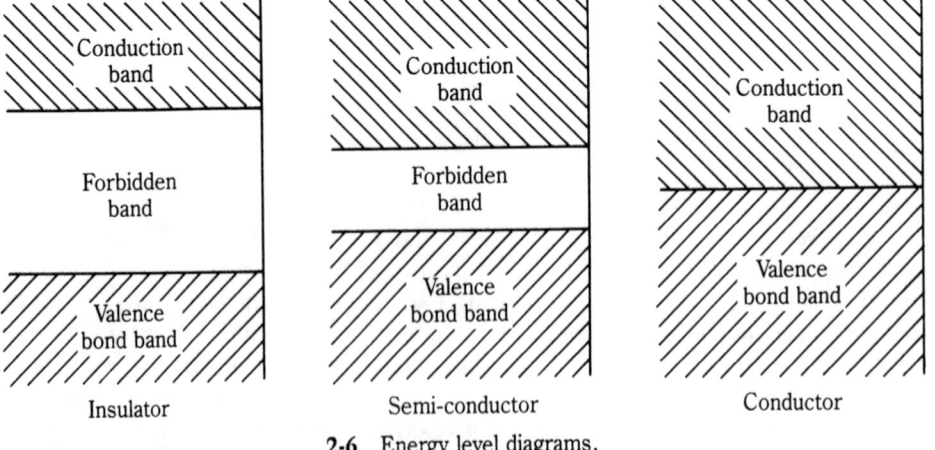

2-6 Energy level diagrams.

duction, band indicates the energy levels that allow electrons to move and cause electrical current. The gap between the two bands is labeled the *forbidden band*. In a good conductor, the valence band and the conduction band overlap, so the forbidden band does not exist, and electrons have the energy to wander freely or migrate under the influence of an applied electric field.

The most widely used semiconductor is silicon, which contains four electrons in the valence band. In the silicon crystal, an atom shares its electrons with four neighboring atoms, which in turn share theirs, so that the outer shell is complete for all. This covalent bonding between atoms is represented in Fig. 2-7. Also shown is one impurity atom of arsenic, called a *donor impurity* because it has five valence electrons, four of which enter into covalent bonding, leaving the fifth electron free.

Two points should be emphasized: One, the donor atoms are introduced deliberately, in closely controlled concentrations, as a part of the semiconductor device manufacturing process. Second, it would appear that the body of material would have an overall negative charge, but this is not so, since an arsenic atom with five valence electrons is electrically neutral. This semiconductor is called *n-type silicon* when it is doped with donor atoms.

Alternatively, it is possible to introduce some number of indium atoms, each having three electrons in the valence shell. In the crystal this provides, after covalent bonding with the neighboring silicon atoms, an unoccupied point, called a *hole*, for which an electron is always welcome. The hole need not remain fixed in one position but can move aimlessly or migrate to create a current flow under the influence of an electric field. It is much like an empty parking space in a busy parking lot; at one moment it is at one location, and later it is at another.

Although it is known that electrons actually move to fill one hole and, consequently, to create another in a different location, the action is simply explained as the movement of positively charged holes. This indium-doped silicon is p-type and the indium atoms are referred to as *acceptor atoms*.

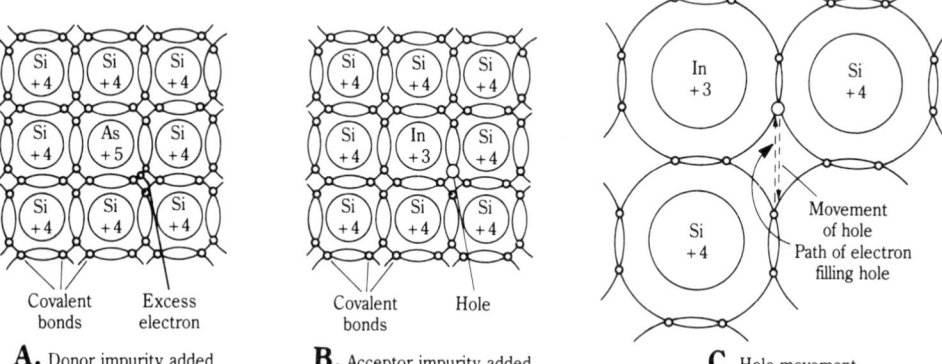

A. Donor impurity added   **B.** Acceptor impurity added   **C.** Hole movement

2-7 Silicon lattice with impurities added.

pn junctions

Many semiconductor devices are fabricated so that p-type and n-type materials in the same crystal are formed in adjacent regions. The boundary between regions is named the *junction*. In the absence of any externally applied voltage across the semiconductor junction, a number of important events occur. First, the electrons in the n-type material tend to diffuse across the junction into the p-type material. The principle is exactly the same as the diffusion of a puff of smoke; the natural tendency is to make the concentration equal in all parts of the constrained volume.

In the same manner, the holes in the p-type material diffuse across the junction boundary into the n-type material. A concentration of holes and electrons then exists adjacent to the boundary, as shown in Fig. 2-8. Their presence causes a local imbalance of charge and results in an electric field, oriented horizontally in the figure, which drives electrons to the left and holes to the right. This potential barrier, or junction barrier, causes the net current to be zero by balancing the diffusion and electric field migration effects.

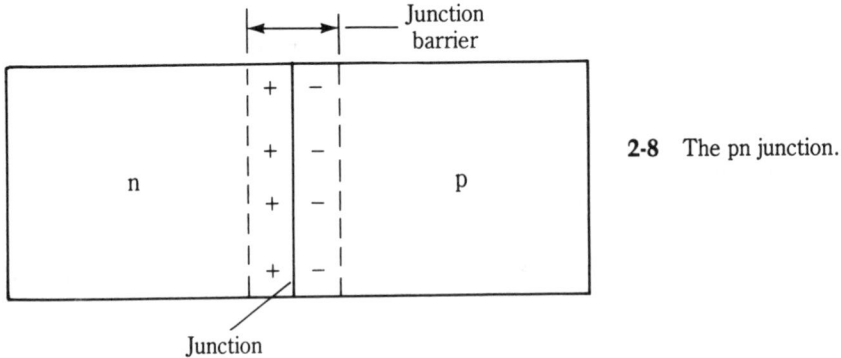

2-8   The pn junction.

The diffusion of *carriers* (holes from p-type material and electrons from n-type material) to the other side of the junction also results in a net decrease in carrier concentrations in the regions of their origin. Thus, a *depletion layer* is considered to exist on either side of the junction.

Semiconductor diodes

The application of a bias potential to the crystal with the pn junction can cause current to flow in the external circuit. See Fig. 2-9 for the condition of reverse bias applied to the junction. This reverse bias reinforces the potential barrier at the junction so that even smaller concentrations of holes and electrons due to diffusion are necessary for zero current. Alternatively, in Fig. 2-10, forward bias is applied, which reduces the potential barrier so that the diffusion of holes and electrons causes a greater current component than that caused by the electric field potential barrier. Therefore, a net current flows across the junction and passes through the external circuit.

In practice, the current with reverse bias is small but is never zero. To account for this difference, remember that a relatively small number of electrons and holes exist in

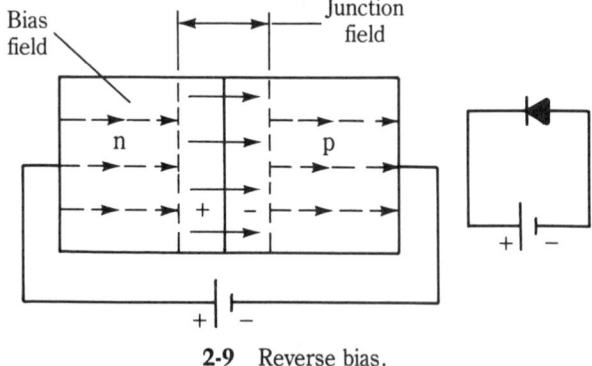

2-9 Reverse bias.

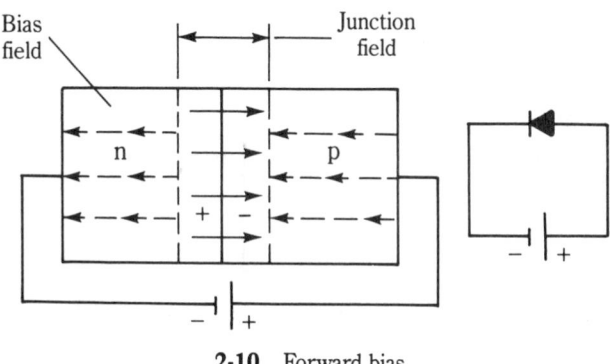

2-10 Forward bias.

the intrinsic semiconductor as a result of thermal energy. These electrons will move across the junction in the direction indicated by the bias potential to form the leakage current.

A single pn junction has the rectifying characteristics of a diode, as seen in Fig. 2-11, and is referred to as a *semiconductor diode*. Semiconductor diodes are commonly used in power supplies and in a wide variety of other applications. The diode symbol appears in Fig. 2-12; the arrow, positioned on the anode side, points in the direction of conventional current flow.

Increasing the amount of reverse bias applied to the semiconductor diode will have little effect on the flow of leakage current. However, an excessively large reverse bias will cause breakdown of the diode so that a large current will flow and the diode will possibly be destroyed.

A zener diode is specifically designed to break down at a particular reverse bias voltage to provide an accurate constant-voltage reference. Zener diodes are available in voltage ratings from a few volts to several hundred volts. They are commonly used for voltage regulation in power supplies.

A specialized diode called the *pin diode* has a layer of *intrinsic*, or undoped, semiconductor material between the p and n regions. This layer increases the time required for carriers to pass through the diode. At high enough radio frequencies the polarity of the voltage across the diode reverses before the carrier can pass through the intrinsic layer,

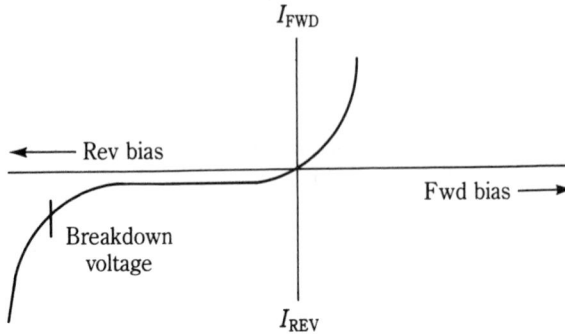

2-11 Semiconductor diode characteristic curve.

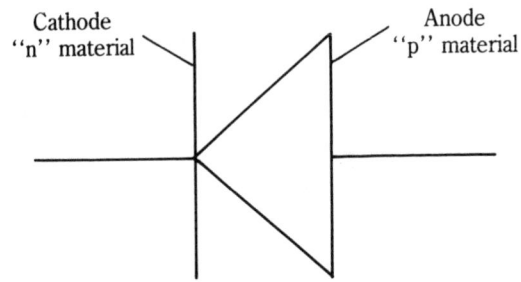

2-12 Schematic symbol for a semiconductor diode.

and attenuation takes place. For this reason pin diodes are used as components of rf attenuators.

Pin diodes are also used for photodetectors. Radiant flux of wavelengths up to about 1100 nm are absorbed in the intrinsic silicon layer and create electron-hole pairs. With reversed-bias diodes, the electric field sweeps the electrons and holes out of the depletion region, causing the flow of photocurrent. This current can develop a voltage across an external load placed in series with the diode. The voltage is proportional to the intensity of the incident light.

A point-contact diode is formed by placing a sharp metal lead, called a *whisker*, in contact with a semiconductor body. When properly biased, electrons are injected into the semiconductor so that a rectifying junction is realized. The associated capacitance is low because of the small contact area, making it useful for rf applications.

Schottky, or hot-carrier, diodes have properties similar to those of point-contact diodes. Schottky diodes are fabricated by depositing metal on silicon to form the junction.

Bipolar transistors

Although the exact geometry depends on the fabrication process employed, the junction transistor can be pictured as having a slice of one type of doped semiconductor sandwiched between two slabs of the other semiconductor. Thus, two possibilities exist as

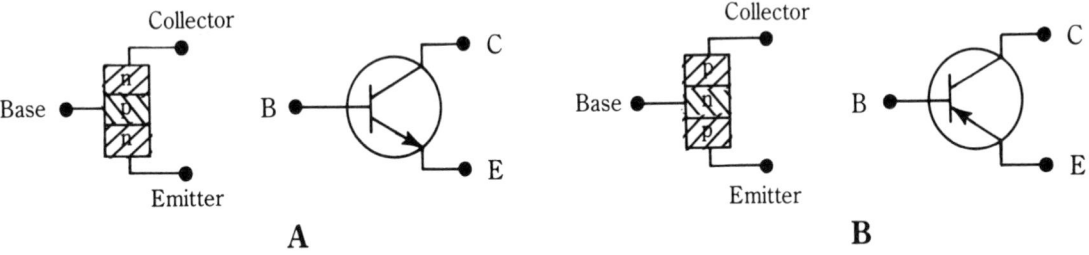

2-13 Bipolar transistor elements and symbols: A npn, B pnp

shown in Fig. 2-13. In Fig. 2-13A the middle slice is p-type, while in Fig. 2-13B it is n-type. For each type, the three different regions and associated terminals are named the *emitter* (E), the *base* (B), and the *collector* (C). The schematic symbols for the npn transistor (Fig. 2-13A) and the pnp transistor (Fig. 2-13B) are also given.

This chapter will not discuss the methods through which donor and acceptor impurities are introduced into localized regions of a single semiconductor crystal. The basic physical mechanism of operation and the properties affecting the behavior of circuits containing junction transistors will be briefly discussed.

Confine your attention now to the npn transistor; in the absence of any external connections, the activity in the vicinity of a pn junction is the same as that of the diode. The majority carriers (electrons) in the n-type collector and emitter regions diffuse across the junctions into the base. Likewise, holes in the p-type base region diffuse across the junction into the collector and emitter regions. (See Fig. 2-14.)

2-14 Unbiased two junction transistor.

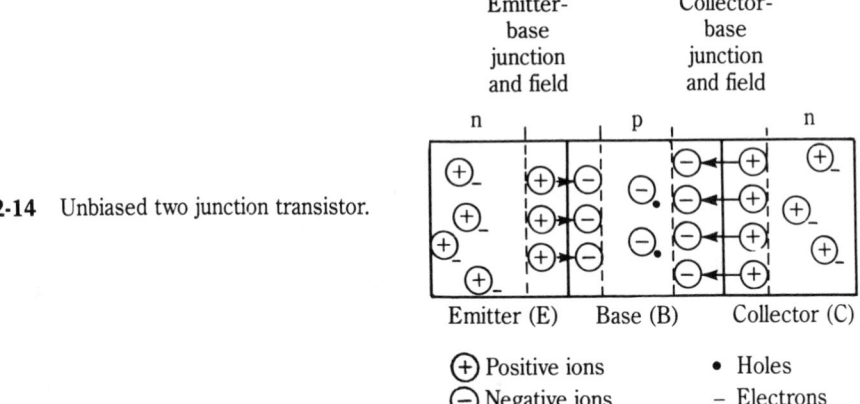

There are two consequences of the diffusion across the pn junction:

1. A depletion layer appears in the vicinity of each junction.

2. An electric field is set up at each junction opposing the diffusion and causing the net current to be zero.

In normal operation, as in an amplifier circuit, the emitter-base junction is forward biased, and the collector-base junction is reverse biased. In Fig. 2-15 the biasing voltages are labeled V_{EE} and V_{CC}. Note that this circuit is only intended to demonstrate bias polarities; it is not otherwise practical.

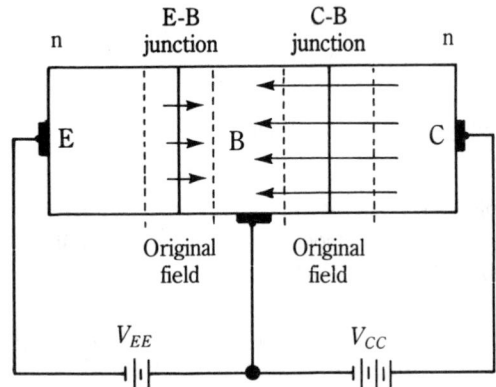

2-15 Biased two junction transistor.

In the figure, if V_{EE} is greater than about 0.6 V, the potential barrier of the emitter-base junction is overcome, and electrons move from the emitter to the base in a continuous stream. A small fraction of the electrons pouring into the base combine with the holes there. The other electrons are swept across the collector-base junction by the field in the vicinity of that junction. Note that the reverse bias imposed by V_{CC} adds to the potential barrier hill, which retards the movement of electrons (the majority carrier in the collector) from collector to base. However, the electrons in the base that originated in the emitter are urged across the junction into the collector, and from there through the external circuit and back into the emitter.

Reconsider the combination of electrons (from the emitter) with the acceptor atoms in the base, even though the number suffering this fate might be only 1 percent of the total. The capture of the electrons in the base by the neutral atoms tends to create a net negative electrical charge in the base, which opposes the movement of additional electrons from the emitter. However, under equilibrium conditions, electrons will flow out of the base and pass through the external circuit, propelled by V_{EE}, back into the emitter at the same rate as the electrons enter into combination within the base region.

It follows that any impediment to the flow of current at the base terminal, as might be established by placing a resistor in the base lead, restricts the base current, establishes a new balance in the negative charge on the base, and sets the rate at which current flows in the emitter and collector circuits. In other words, the base current controls the emitter and collector currents.

With the transistor in operation,

$$I_E = I_C + I_B$$

and the current gain,

$$\beta = I_C / I_B$$

is usually 100 (more or less) for general-purpose small transistors. Another current gain parameter is,

$$\alpha = I_C / I_E$$

which is nearly equal to, but always less than, one. Alpha and beta are related as follows:

$$\alpha = \frac{\beta}{(1 + \beta)}$$

$$\beta = \frac{\alpha}{(1 - \alpha)}$$

Field-effect transistors

The current passing through the body of a junction field-effect transistor (JFET) is controlled by the voltage applied to the gate terminal. A JFET is a tiny bar of silicon having two junctions (Fig. 2-16). The JFET in this figure has an n-type channel so that the current consists of electrons in motion between the source terminal on the left and the drain terminal on the right. The channel is flanked by p-type regions electrically connected to the gate terminal (Fig. 2-16B).

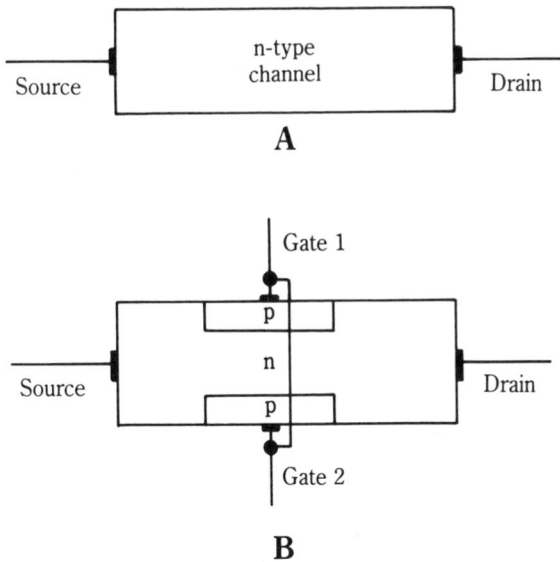

2-16 A n-type JFET, B with gates.

The biasing arrangement for the n-type JFET appears in Fig. 2-17A. The pn junctions are reverse-biased, so a depletion layer exists on either side of the channel. The thickness of the depletion layer is not uniform, being wider near the drain and narrower near the source because of the voltage drop along the length of the channel. If V_{DS} is increased sufficiently, the channel experiences *pinch-off* (see Fig. 2-17B), beyond which

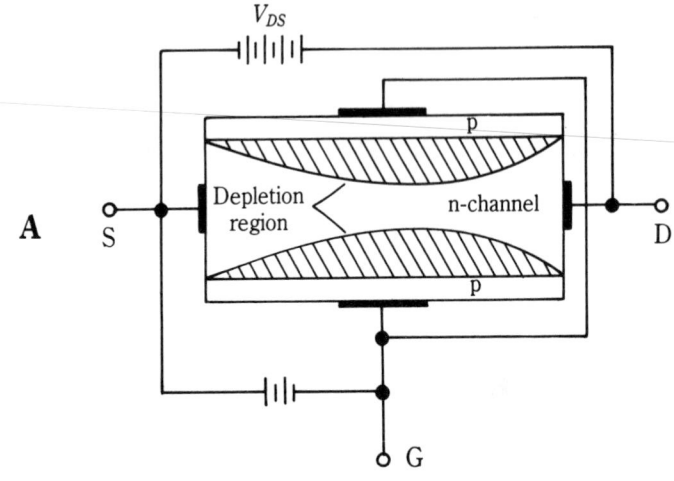

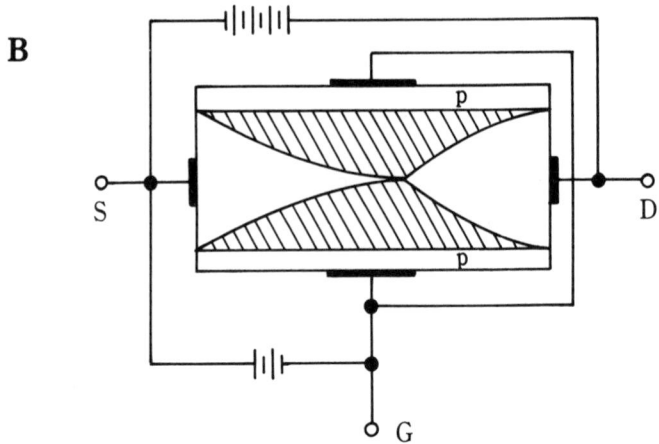

2-17 An n-channel JFET. A. ohmic region. B. saturation region.

the current rises relatively slowly with further increases of V_{DS}. Characteristics illustrating this feature are shown in Fig. 2-18.

The insulated-gate field-effect transistor (IGFET) can also function in the same way as the JFET (see Fig. 2-19). In the figure, the gate terminal is coupled capacitively to the n-channel, and the resistance at the gate is extremely high as compared with that of the reverse biased pn junction of the JFET. If V_{GS} is negative, as pictured in Fig. 2-17 for the JFET, the IGFET is operating in the *depletion mode*.

An *enhancement mode* IGFET has a positive V_{GS}, which induces an n-region between the source and the drain. In the absence of this gate-source bias, the channel does not exist.

The p-channel JFET and the IGFET operate in the same way described for JFETs, except that the carrier is the hole and the polarities of the voltages and current directions

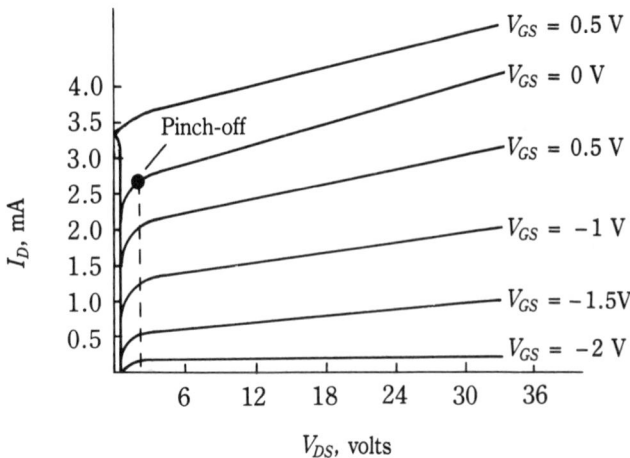

2-18 Drain family of curves for an n-channel JFET.

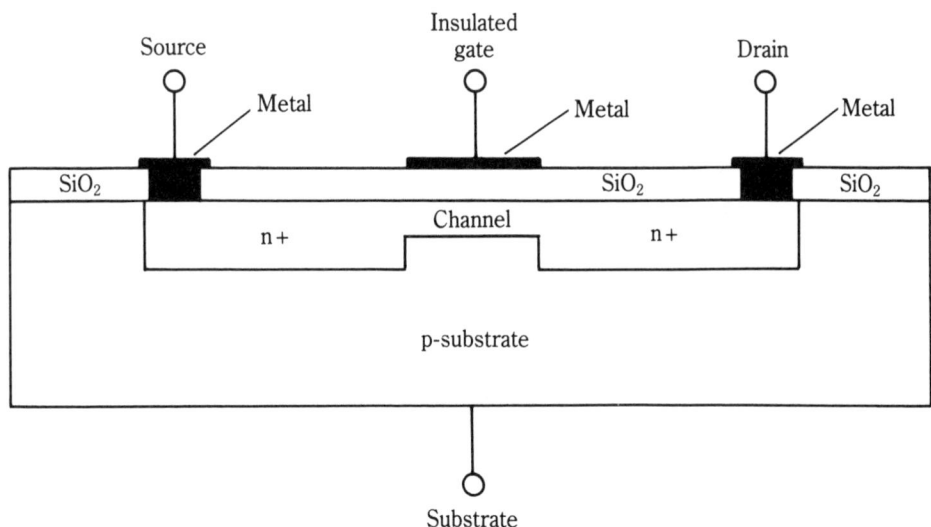

2-19 Insulated-gate field effect transistor.

are reversed. The symbols and bias polarities for the various types of field-effect transistors are summarized in Fig. 2-20.

Presently, the only commercially available IGFET is the metal-oxide-semiconductor FET (MOSFET, often designated as the MOS transistor). The MOSFET is very popular for a number of reasons, including the low price resulting from the relatively simple fabrication process and the high gate resistance. MOSFETs are vulnerable to the breakdown of the silicon-oxide insulating layer when exposed to a static electrical charge. For this reason, special precautions are necessary in the storage, handling, and installation of MOSFETs.

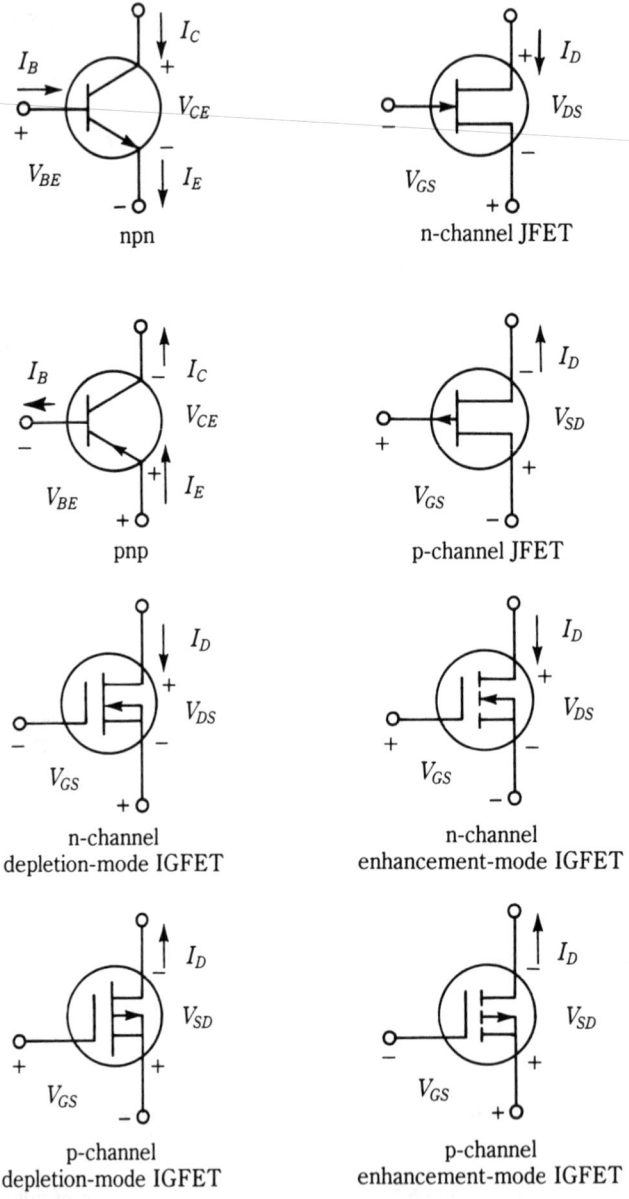

2-20   Symbols and polarities for transistors.

If the IGFET has two gate sections, similar to those of Fig. 2-17, that are not connected, but are rather brought out as separate terminals, a *dual-gate IGFET* is the result. The dual-gate IGFET allows control of the drain current to be exercised by two separate gate voltages. Available designs of the dual-gate IGFET are suitable for radio frequencies through the vhf range. Dual-gate IGFETs are used for mixers, product detectors, and agc amplifier stages.

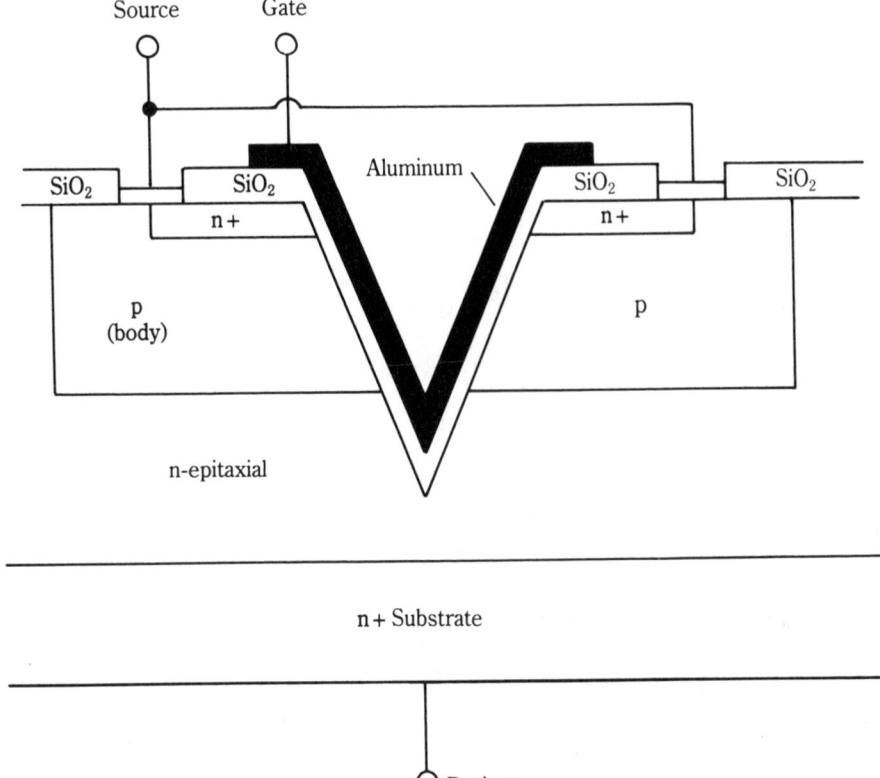

2-21 Cross-section of a VMOS transistor.

A specialized form of the MOSFET is shown in Fig. 2-21. Because of the shape of the channel cross section, it is popularly known as a *VMOS transistor*, the ''V'' designating the vertical dimension of the channel path. The VMOS provides high voltage, high current, and high-frequency performance as compared with earlier MOS designs.

Both n-channel (NMOS) and p-channel (PMOS) MOSFETs are available. The combination of the two forms in particular circuits, such as the inverter of Fig. 2-22, provides the advantage of greatly lowered power consumption, compared with inverters constructed with only one MOSFET type.

Some transistor types are commonly available in integrated circuits, even though they have been described as would-be discrete transistors. This need not present a conceptual problem since the basis for the operation of these chips is the same.

Other discrete semiconductor devices

The *silicon-controlled rectifier* (SCR) is constructed of alternate p-type and n-type silicon semiconductor layers with metallic connections to the anode, cathode, and gate (see Fig. 2-23A). The SCR symbol appears in Fig. 2-23B. Visualize the layers of the SCR as forming two transistors (see Fig. 2-24). When the base voltage of Q2, with reference to the

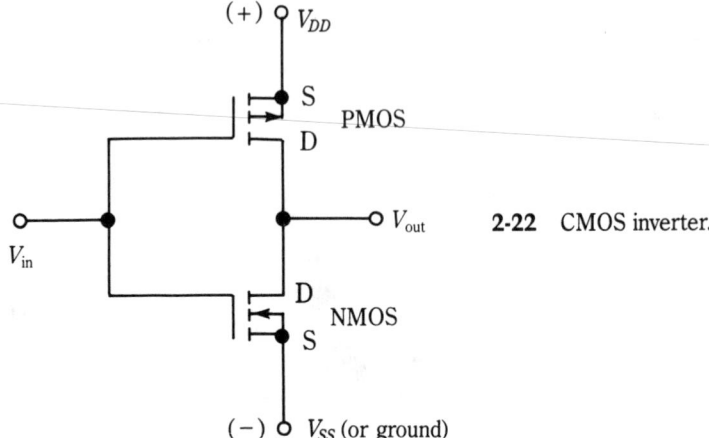

2-22 CMOS inverter.

emitter, is low, the collector current of Q2, and hence the base current of Q1, are also low, so the SCR appears as an open circuit between the anode and the cathode.

Raising the gate voltage causes gate current (base current of Q2) to flow, resulting in Q2 collector current, Q1 base current, and Q1 conduction. If the SCR voltage applied between the anode and cathode is sufficiently large with a given amount of gate current, the overall action is regenerative, the SCR voltage will drop, and the current will increase. Moreover, the gate current ceases to have any control over the SCR operation.

Figure 2-25 summarizes the SCR characteristics and emphasizes the results of having different gate currents I_{G0}, I_{G1}, and I_{G2}. V_F is the "forward" (forward bias direction) SCR anode-to-cathode voltage. Note that, for I_{G2}, the characteristic resembles that of a forward-biased diode.

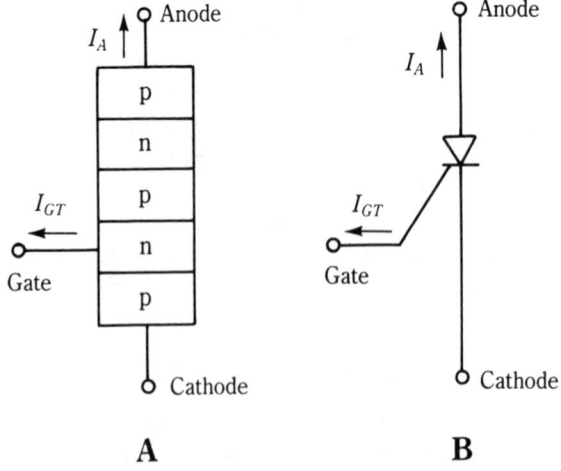

2-23 A silicon controlled rectifier (SCR).

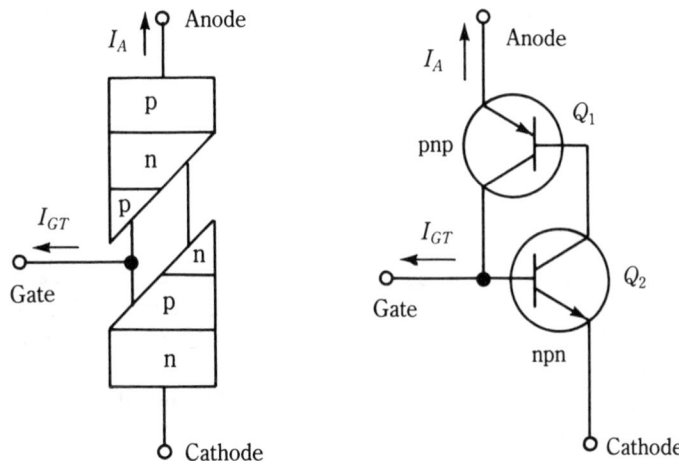

2-24 Two-transistor equivalent circuit for a SCR.

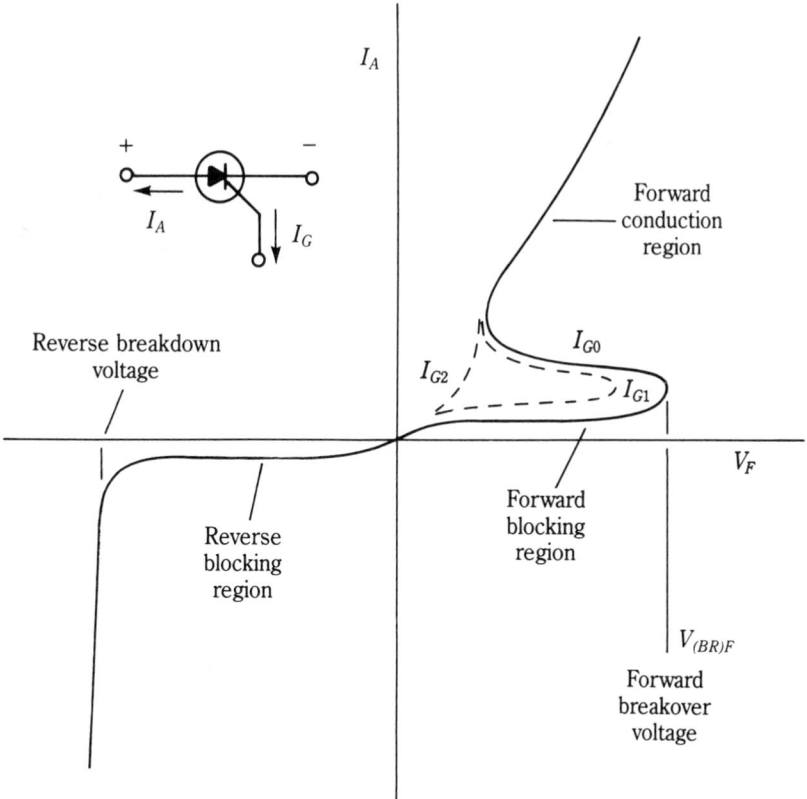

2-25 The SCR symbol and its characteristics.

When the SCR is ON, it cannot be turned OFF by removing the gate signal. Two possibilities exist for turning the SCR OFF. The first, *anode current interruption*, can be accomplished by opening the anode circuit or by placing a direct short circuit between the anode and cathode. The second method, *forced commutation*, requires that current be forced from the anode to the cathode in the direction opposite to that used to turn the SCR ON.

Silicon-controlled switch

The silicon-controlled switch (SCS) semiconductor device resembles the SCR, except that an anode gate is added (Fig. 2-26). The SCS can turn the device ON with a negative voltage, or OFF with a positive voltage. Usually the anode gate current for turn-on is much larger than the required cathode gate current. In addition to reduced turn-off time, the SCS has the advantage of increased control. The SCS is limited to low-power ratings, as compared with the SCR.

Light-activated SCR

The light-activated SCR (Fig. 2-27) has a gate allowing triggering like a typical SCR. However, with the gate open or in another condition for which conduction does not take place, the increase of light power density on a light-sensitive area of the semiconductor can cause the device to be triggered.

Diac

The bidirectional diode thyristor permits triggering with either polarity of voltage on the two electrodes, called *anodes*. Figure 2-28 shows that the breakdown voltage for the characteristic is V_{BR}. The double diode in the circuit symbol is misleading because conduction is implied, as by a forward-biased diode, without the presence of breakdown. A bidirectional characteristic is particularly useful in ac circuits. However, no direct control of breakdown (conduction) is possible.

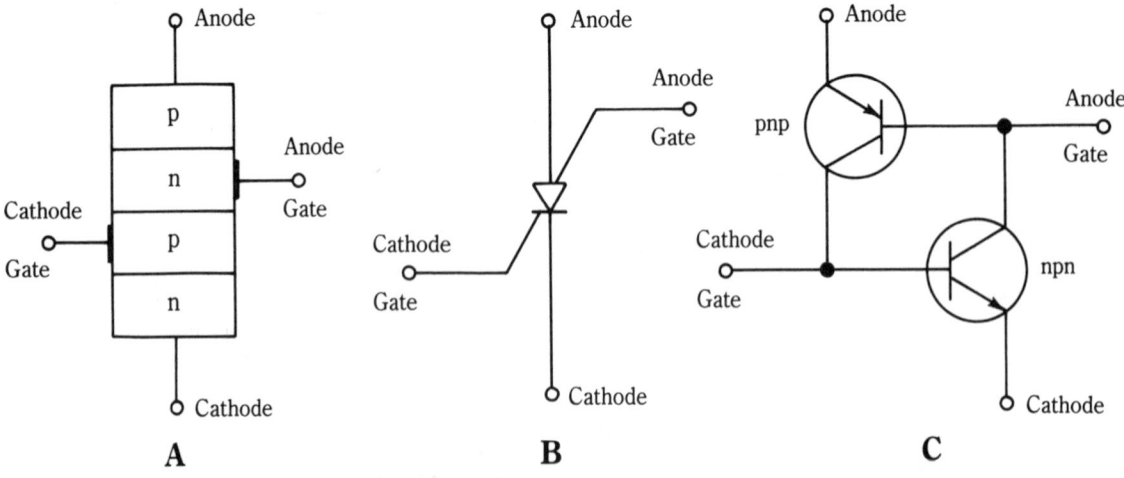

2-26   Silicon controlled switch A Construction. B Symbol. C Transistor circuit equivalent.

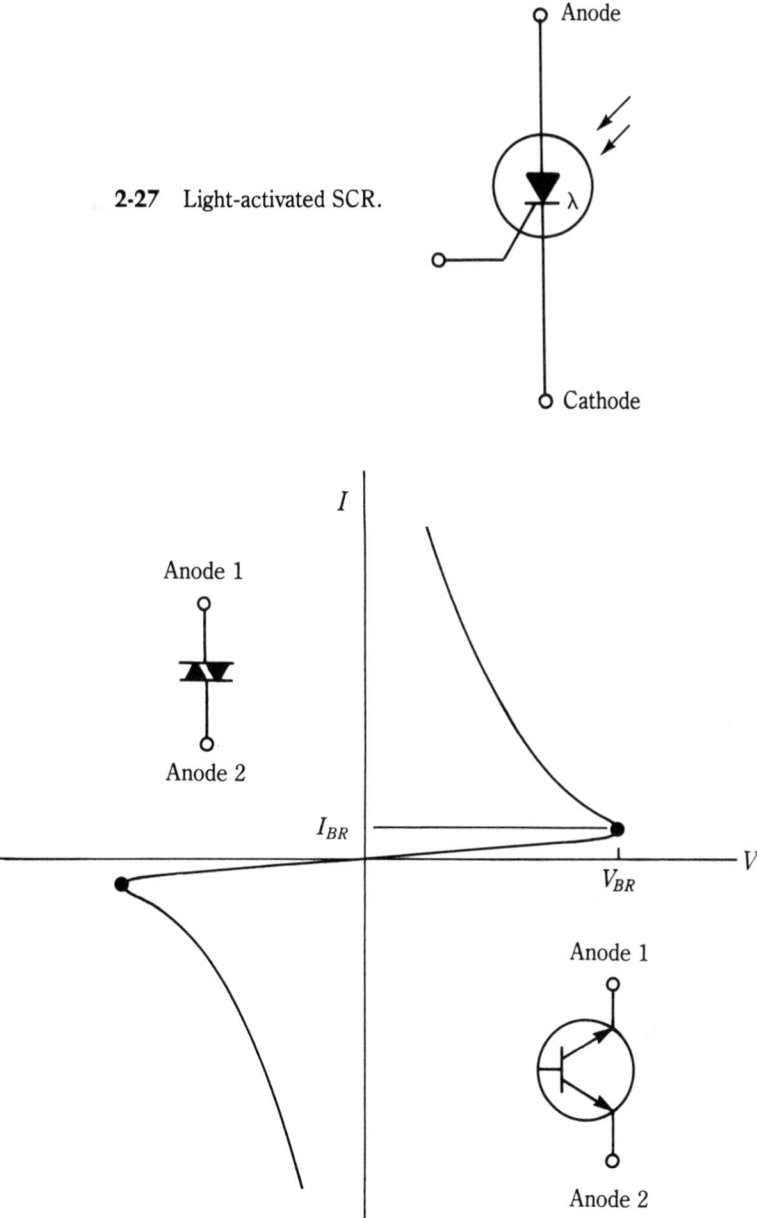

2-27 Light-activated SCR.

2-28 Diac symbol and its characteristics.

Triac

Adding a gate to the basic diac structure forms the bidirectional triode thyristor (triac). Triacs behave much like diacs, except that gate current (positive or negative) can exercise control over the breakdown point. Like the diac, the triac continues to conduct until

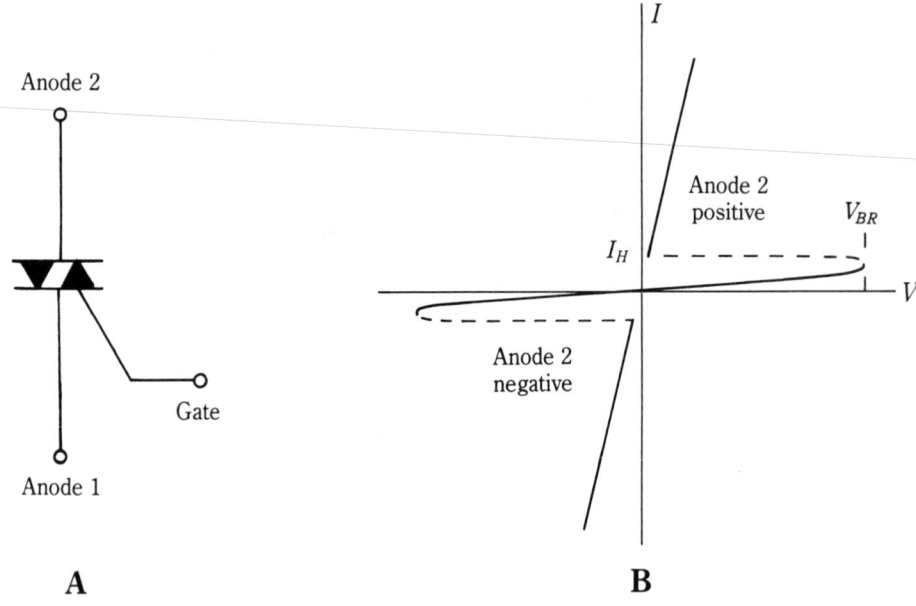

A	B

Anode 2

Gate

Anode 1

I

Anode 2 positive V_{BR}

I_H

V

Anode 2 negative

2-29 Triac: A symbol, B characteristics.

the applied voltage drops to a level of about 1 volt. For ac, conduction occurs for a partially controllable fraction of each alternation. The symbol and characteristics for triacs are shown in Fig. 2-29.

Unijunction transistor

As implied by the symbol in Fig. 2-30, the unijunction transistor (UJT) consists of a monolithic n-type semiconductor with two metallic contacts, referred to as *bases* (B1 and B2), through which the controlled current passes. A pn junction with another metallic contact, the *emitter*, completes the UJT.

Now consider the application of a particular voltage, E, relative to B1, such that the emitter current is zero (Fig. 2-31). If a source in the emitter circuit is adjusted to

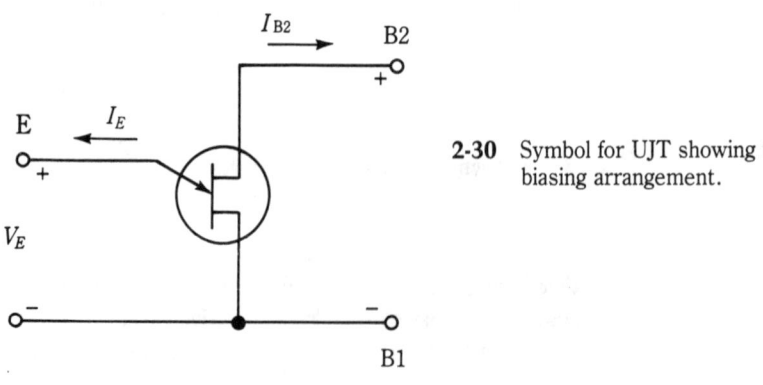

I_{B2} B2

E I_E

V_E

B1

2-30 Symbol for UJT showing the biasing arrangement.

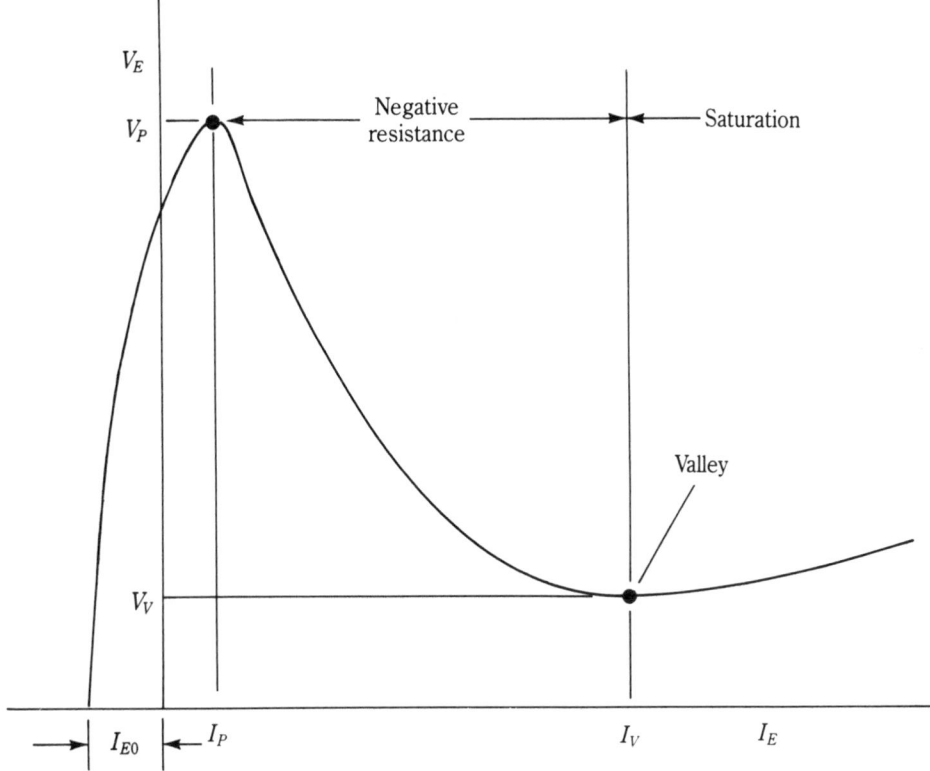

2-31 UJT static emitter characteristic.

increase I_E, V_E will at first increase and then peak at V_P. A further increase of I_E will cause V_E to decrease, until a minimum is reached at V_V. These points are indicated in Fig. 2-31. Further increases in I_E are associated with increases of V_E.

The injection of holes at the emitter greatly increases the conductivity of the slab so that the resistance between B1 and B2 becomes relatively small and I_{B2}, the controlled current, can increase greatly.

This process can be reversed. Reduction of I_E, by whatever means, can bring V_E backwards beyond the V_P point, changing the slab to a poorly conducting condition. You can see that the function of the UJT in any application is that of a triggering device whose behavior is heavily influenced by the circuit it is contained in.

The monolithic integrated circuit

A monolithic integrated circuit consists of a single crystal substrate upon which a large number of resistors, capacitors, diodes, and transistors are formed, together with the necessary insulating layers and metallic interconnections. This IC achieves the same electronic functions that once required a correspondingly large number of discrete elements. Monolithic ICs provide significant operational characteristic advantages, in addition to compactness and low cost.

The monolithic integrated circuit results from a process sequence that starts with a polished semiconductor (usually silicon) wafer that has been sliced from a single p-type grown crystal ingot. Next an epitaxial region is diffused into the substrate (Fig. 2-32). An oxidation process causes an insulating layer of silicon dioxide (SiO_2) to be deposited on the n-type layer (Fig. 2-33).

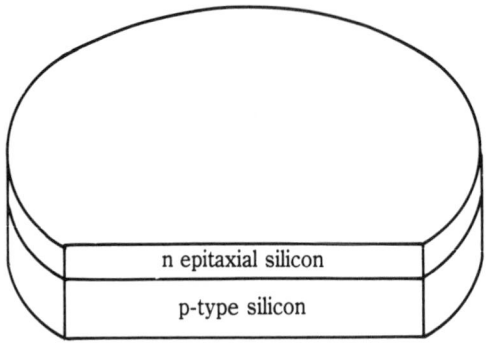

2-32   P-type wafer with n-type layer.

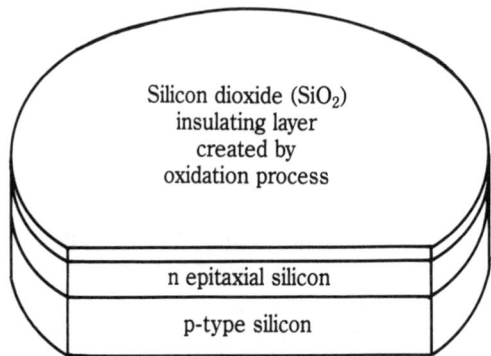

2-33   Wafer with silicon-dioxide insulating layer.

The deposition process is followed by the selective etching of the SiO_2 by means of a photographic process. Using a mask that covers the major part of the wafer area, a photosensitive layer, called *photoresist*, is exposed to ultraviolet light. The mask contains a large number of identical chip images. Every chip will ultimately be separated from the others to become an IC.

After exposure (Fig. 2-34), the unexposed photoresist is removed chemically (Fig. 2-35). Another solution will then etch the SiO_2 layer in any area not covered by photoresist (Fig. 2-36). In the next step, all the remaining photoresist material is removed by solution (Fig. 2-37). Still another diffusion process inserts heavily doped p-type regions into those areas not covered by the SiO_2 (see Fig. 2-38). The basic purpose of this step is to achieve electrical isolation of different regions within the chip.

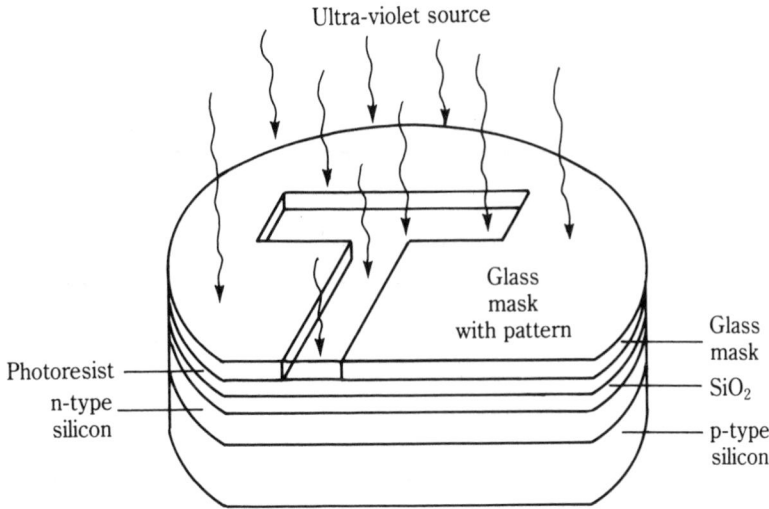

2-34 Photolithic process-mask features (not to scale).

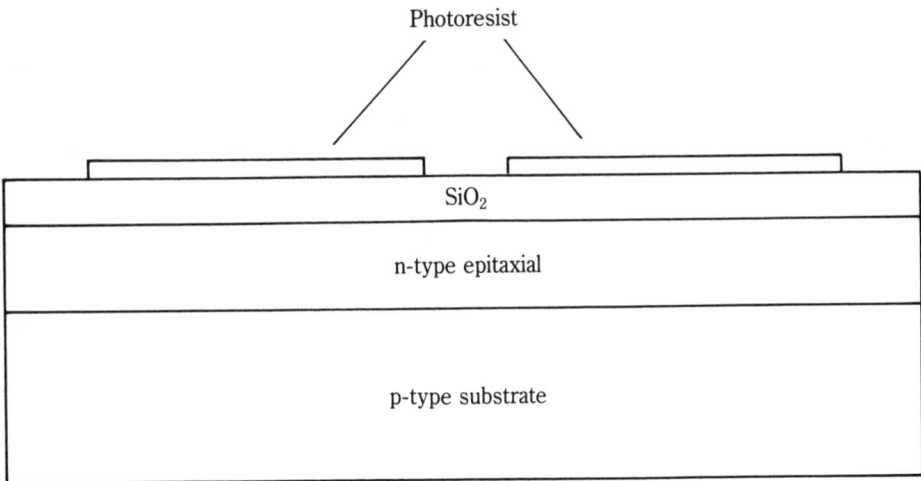

2-35 Cross-section of a chip following removal of unexposed photoresist.

Next, a series of masking, etching, and diffusion processes convert the n-type epitaxial region of Fig. 2-33 into a transistor, an example of which is shown in Fig. 2-39. The first of these masking and etching operations opens a surface area above the n-type epitaxial region, which then allows p-type diffusion to form the base. Another mask, another etch, and another diffusion of n-type impurity completes the formation of the bipolar transistor (Fig. 2-39). Note that this is one of many transistors of identical form crowded on the same substrate. Adjacent transistors are electrically isolated by reverse bias voltages between the p-type substrate and the n-type epitaxial regions making up the transistor collectors.

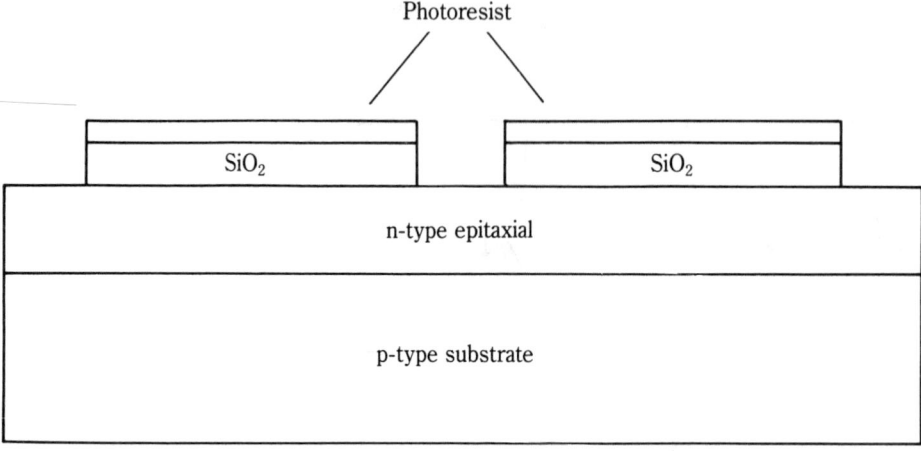

2-36   Cross-section of a chip after removal of uncovered SiO$_2$ from Fig. 2-35.

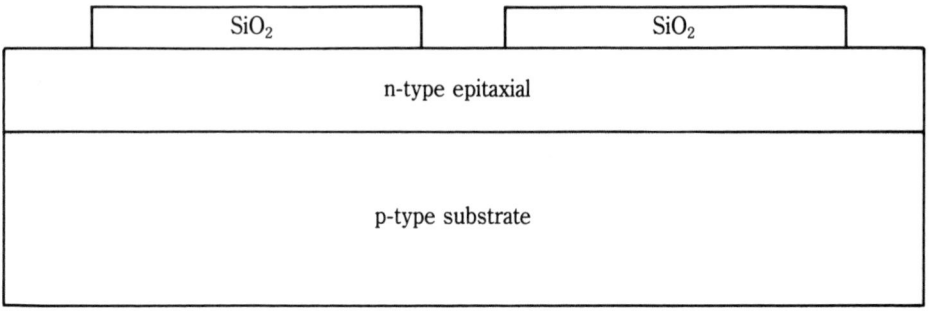

2-37   Cross-section of a chip after removal of all remaining photoresist.

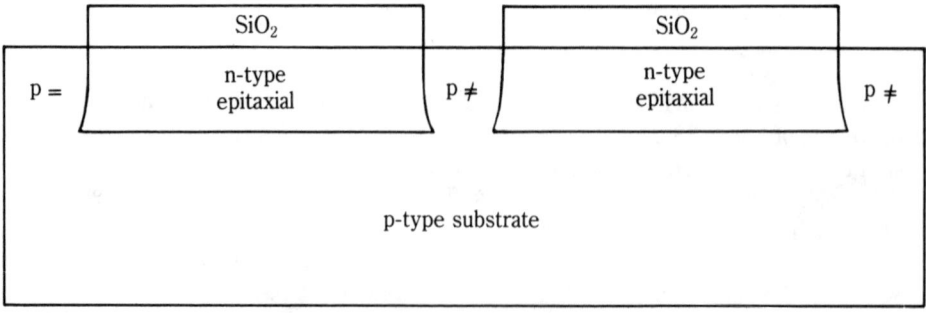

2-38   Cross-section of a chip after the isolation diffusion of Fig. 2-37.

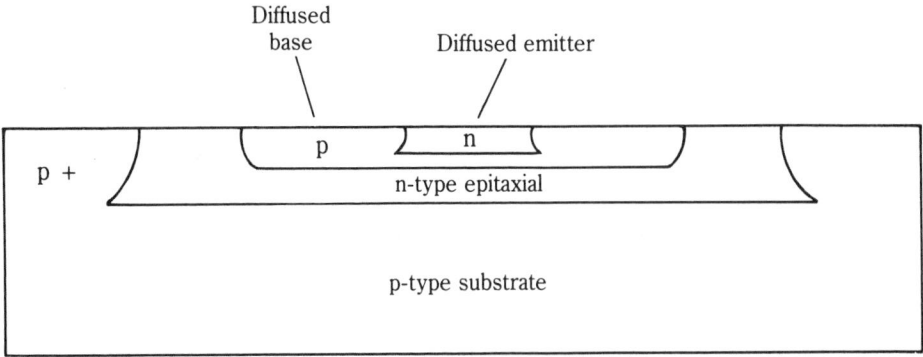

2-39 Cross-section of a transistor after the base and emitter diffusion process.

Finally, another sequence of masking and etching, combined with metal deposition, is necessary to make electrical connections to the emitter, base, and collector of each transistor. This sequence is not illustrated here because different IC packages are involved.

If each transistor, diode, or other element is to be identified separately, a correspondingly large number of connections must be made to the chip package housing these elements. At the other extreme, a large number of transistors, diodes, etc. formed on the substrate can be directly connected by metallization on the chip proper to implement more complex functions, e.g., a microprocessor.

The number of external connections to the chip, while considerable, is much smaller than the total number of anodes, cathodes, emitters, collectors, sources, gates, drains, etc., on the substrate. Packaging, a relatively simple design consideration in the pre-IC era, is absolutely vital to present and future electronic systems.

Digital integrated circuits

The key feature of digital circuits is the dependence upon two distinguishable voltage (or current) levels at the input and output terminals. One level is said to be *low* if it falls within a prescribed range. Likewise, a level is referred to as *high* when it is within a second, nonoverlapping range. For example, a bipolar transistor circuit, which by reason of design has the transistor either cut off or saturated, behaves as a digital circuit.

In general, a digital circuit can have more than a single input signal and more than a single output. However, each input and output must be either high or low. Moreover, a number of different semiconductor devices and circuits can be configured for the purpose of implementing the several logic functions.

We will start with the older logic families still used in early manufactured equipment, but no longer chosen in new designs. It is then possible to move into the vital families of today with some mention of current trends of development.

Resistor-transistor logic

Figure 2-40 illustrates one member of the resistor-transistor logic (RTL) family. If *A* is high, Q1 is saturated, and *X* is low. The same is true of *X* if *B* is high and Q2 is satu-

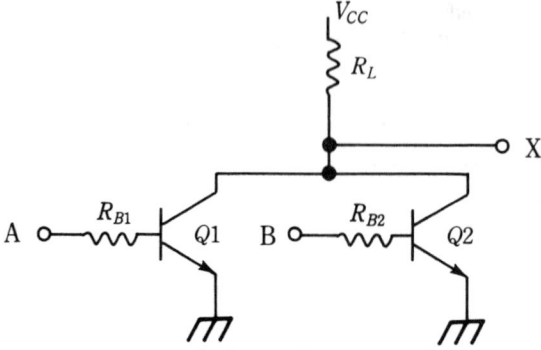

2-40 Resistor-transistor logic (RTL).

rated. The output X is also low when both A and B are high. Otherwise, with both A and B low, X is high.

Diode-transistor logic

Figure 2-41 provides an example of the diode-transistor logic (DTL) family. In this circuit, if either A or B is low, Q1 is cut off, which in turn causes Q2 to be cut off, and D will be high. If, however, both A and B are high, Q1 will conduct, providing base current for Q2, which also conducts and brings D low. The present circuit implements the NAND function, and it is known as a NAND gate.

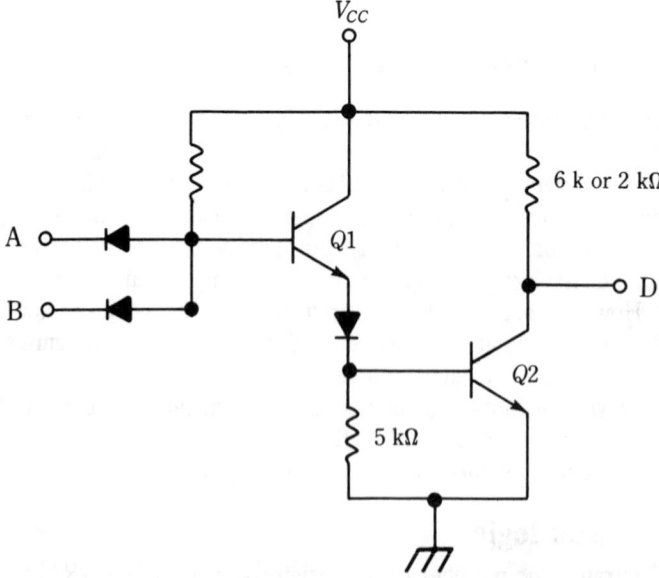

2-41 Diode-transistor logic (DTL).

Transistor-transistor logic

The transistor-transistor logic (TTL) family has perhaps been the most widely used in the past, and still appears in newly marketed equipment. It is a mature technology, exhibiting high reliability at low cost, with an extensive line of parts. Moreover, like many other mature product lines, improvements in speed and power consumption continue to appear, extending the life of TTL in the face of increasing competition from newer semiconductor technologies.

Refer to Fig. 2-42 for the circuitry of a basic TTL device. Note that Q1 is a bipolar transistor having two emitters. If either *A* or *B* (or both) is low, current will flow in the base of Q1 through the base-emitter junction(s) and the transistor will be saturated, so the collector will be at approximately 0.1 V. This causes Q2 and Q4 to be cut off and Q3 to be saturated. Therefore, the output will be high, but less than V_{CC} because of the drop across *R*3, Q3, and *D*3.

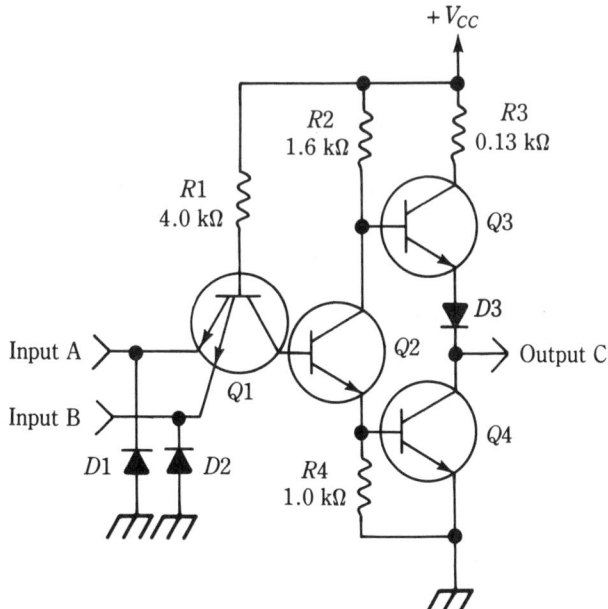

2-42 Schematic diagram of a TTL NAND gate.

When *A* and *B* are both high, the emitter voltages are greater than the collector voltage of Q1. The base-collector diode of Q1 conducts, providing base current for Q2 which conducts, causing Q4 to saturate. Meanwhile, the base of Q3 is at approximately $0.1 + 0.7 = 0.8$ V well below the 1.4 V for the base-emitter junction of Q3 and the drop across *D*3, so Q3 is off. Output *C* will then be approximately 0.1 V. The function of *D*3 is revealed in this analysis. Note that the vertical arrangement of Q3, *D*3, and Q4 is called a *totem pole*.

A variation of the totem pole replaces Q4 and *D*3 with an external pull-up resistor (Fig. 2-43). This resistor allows the outputs of several TTL circuits to be connected in order to obtain an additional logic function without supplying an additional TTL device.

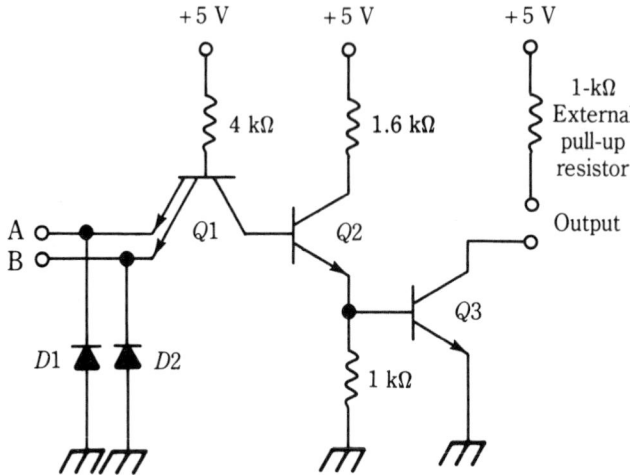

2-43 A TTL NAND gate with open-collector output.

Voltage levels for TTL are summarized in Fig. 2-44. Four limiting conditions are specified: two for output levels, and two for input levels. $V_{IH,MIN} = 2.0$ V is the specified minimum level for which the input is recognized as high. Compare this with $V_{OH,MIN} = 2.4$ V, which is the specified minimum output for high output. The difference between these levels (400 mV) is the amount of noise voltage that can be tolerated before a high output level is unacceptable. A similar noise margin exists for low output and input levels.

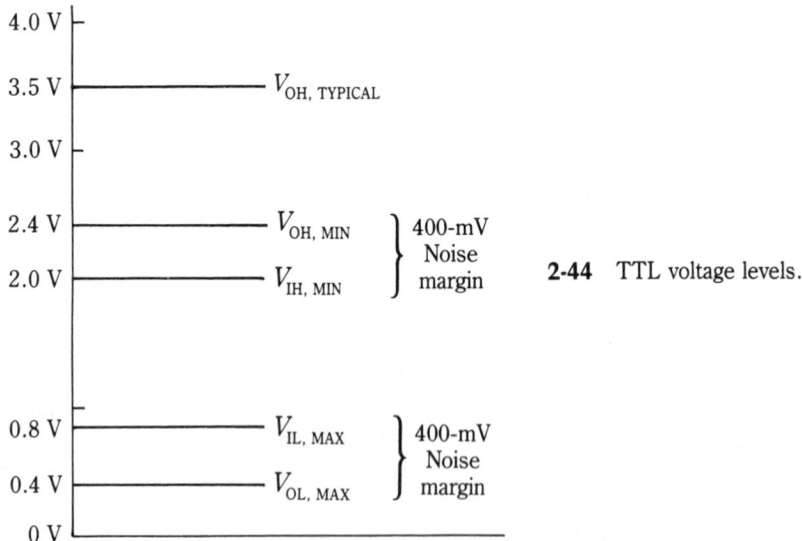

2-44 TTL voltage levels.

In addition to the standard 74XX TTL family, there are several subfamilies designed for improved performance in one respect or another. The key features of the TTL family variations are given in Table 2-1.

Table 2-1. TTL logic family characteristics.

Designation	Word description	Typical propagation delay ns (per gate)	Typical power dissipation mW (per gate)
74XX	Standard	10	10
74HXX	High-power	6	22
74LXX	Low-power	30	1
74SXX	Schottky	3	19
74LSXX	Low-power Schottky	10	2

Emitter-coupled logic

Emitter-coupled logic (ECL) is a nonsaturating form of digital logic that eliminates speed limitations resulting from transistor storage time, thereby permitting very high speed switching. The ECL circuit is a modification of the basic differential amplifier, in which the bipolar transistor emitters are connected. High input and low output impedances are characteristics of ECL.

Figure 2-45 shows part of an ECL device. In this figure, Q6 supplies a stable reference of -1.3 V to the base of Q5. If the inputs A, B, and C are all low, Q5 conducts, so its collector is low. If any one, or more, of the inputs is high, the appropriate transistor Q1 through Q3 conducts and Q5 is cut off, with the result that its collector is now high. The circuit performs the logic function of the OR gate.

Integrated injection logic

Integrated Injection Logic (I^2L) is an extension of the bipolar integrated circuit, with several distinct advantages:

1. High packing density
2. Low supply voltage and low power
3. Relatively high speed

The explanation of I^2L operation requires that the interaction of several transistors be considered simultaneously. If in Fig. 2-46 input A is high, Q2 will be saturated, lowering its collector voltage so that the collector currents of Q3 and Q5 will be diverted to become I_4 and I_6, respectively. Both Q4 and Q6 are then cut off. Note that the collectors of Q4 and Q6 correspond exactly to the same condition assumed for input A initially: they are high.

If input A is low, the collector current of Q1 is diverted from the base of Q2 so that Q2 is cut off, making I_4 and I_6 zero. Q4 and Q6 will conduct to divert current from the bases of other transistors (not shown) that are connected to the collectors of these two transistors.

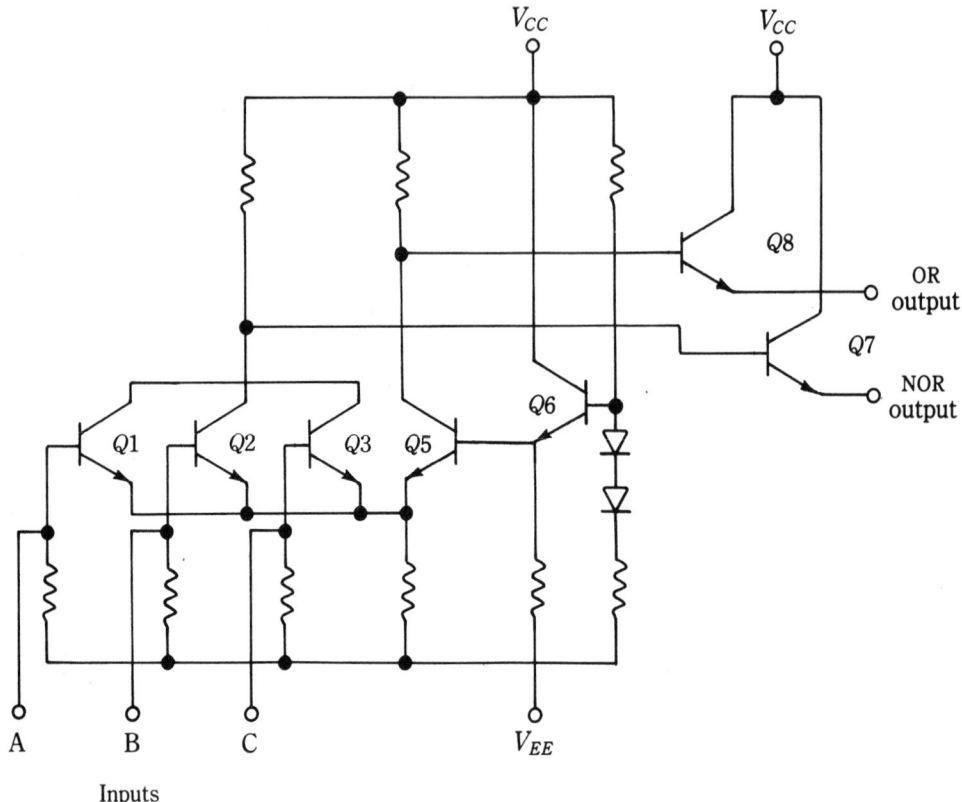

2-45   Emitter-coupled logic OR gate.

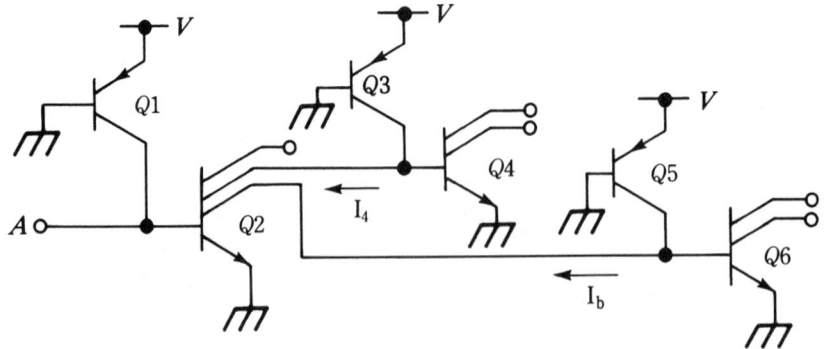

(pnp emitters connect to
common *V* supply)

2-46   Operation of basic I²L gate.

Metal-oxide-semiconductor logic

Metal-oxide semiconductor (MOS) logic circuits are based on MOS field-effect transistors (MOSFETs) and their interconnections on the IC. Three basic forms can be recognized: PMOS for p-channel, NMOS for n-channel, and CMOS for complementary, having both p-channel and n-channel field-effect transistors. Currently, NMOS is predominant, with CMOS exhibiting the most dynamic growth.

Although PMOS was the first MOS design to be produced, because of processing advantages at that time, it was soon overwhelmed by NMOS because of the higher speed and reduced chip area, referred to as the *real estate*, required for each NMOS transistor. Figure 2-47 shows two NMOS logic devices. A quick glance at the circuits reveals a strange situation. The gate of Q1 for each circuit is permanently connected to the drain. This means that Q1 behaves as a load resistor whose current is controlled by the source voltage, since the voltage between the gate and the drain is fixed.

In Fig. 2-47A, if both inputs *A* and *B* are low, Q2 and Q3 are cut off, and output *Y* will be high, the exact level being dependent on the current drawn by the load connected to the output terminal. On the other hand, if *A* is high, *B* is high, or both *A* and *B* are high, substantial current must pass through Q1, which is designed to present a relatively high resistance, and the output *Y* is low. One disadvantage of the NMOS demonstrated in this NOR gate circuit is the power dissipation on the chip when the output is low.

In Fig. 2-47B both inputs *A* and *B* must be high so that current flows through Q1 for low output, which is the requirement for the NAND gate. The same power dissipation disadvantage also exists for this NMOS circuit.

The CMOS structure (Fig. 2-48) greatly reduces the power dissipation compared to the NMOS gates just described. Assume that both *A* and *B* are low in Fig. 2-48A. Both

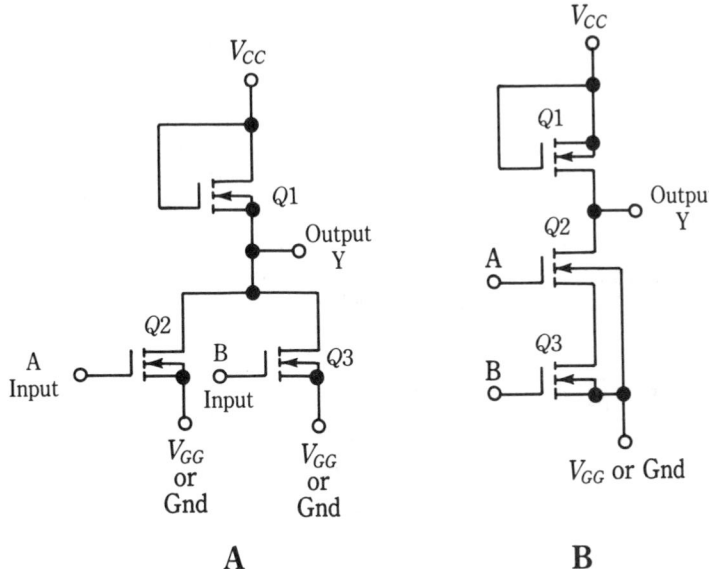

2-47 N-MOS gates A. NOR, B. NAND.

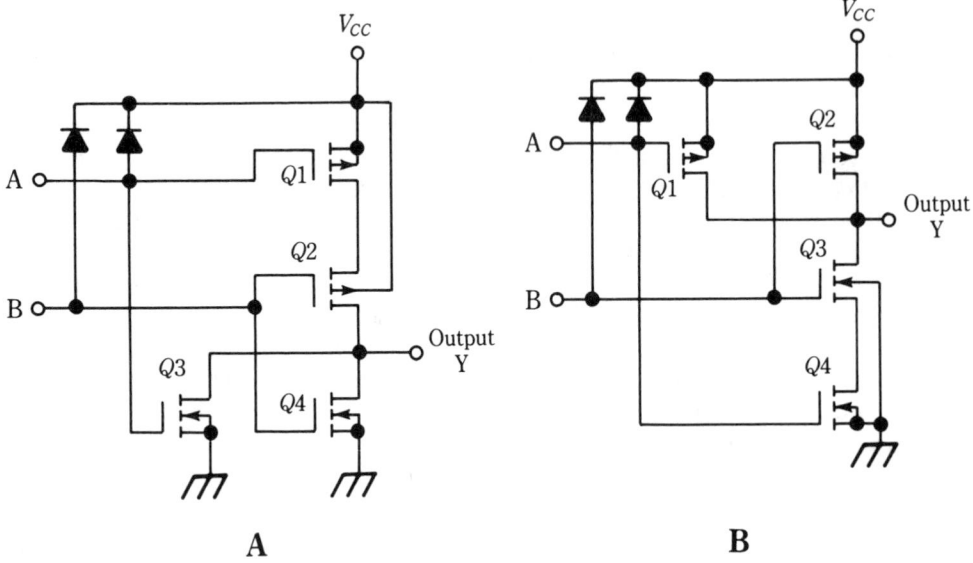

A   **B**

2-48   CMOS gates A. NOR, B. NAND.

Q3 and Q4 are cut off, since the channel is not established when the gate and source voltages are equal. However, with the same low inputs for *A* and *B*, Q1 will conduct, making its drain high so that Q2 will also conduct. Note that the amount of conduction of Q1 in series with Q2 depends on the external load connected to *Y*, but the output voltage level will be high. No significant amount of current flows through any transistor that also has a significant voltage drop between the source and the drain. This means that the power dissipation is small in all transistors for the CMOS NOR gate of the figure. Similarly, when *A* or *B*, or *A* and *B*, are high, one or both of Q1 and Q2 are cut off, so the current from the V_{CC} source is negligible with the same low power dissipation.

In Fig. 2-48B with *A* and *B* high, a conducting path will exist through Q3 and Q4, but both Q1 and Q2 will be cut off, so Output *Y* will be low. With *A*, *B*, or both *A* and *B* low, no conducting path will exist through the series combination of Q3 and Q4. There will, however, be at least one path through Q1 or Q2, so Output *Y* will be high. Like its counterpart in the CMOS NOR gate, the CMOS NAND gate does not conduct freely through any transistor element that has a significant voltage, making the power dissipation small.

Linear integrated circuits

The operational amplifier (op amp) is considered to be the most representative linear integrated circuit for two reasons: it is probably the most widely used, and it is the most flexible as a basic functional part of other linear circuits having specific purposes. An equivalent schematic diagram of the very popular and enduring 741 op amp appears in Fig. 2-49. Because of the practical need to preserve its performance integrity over a range of ambient temperatures and supply voltage variations, the circuit is necessarily more complicated than its individual, basic functions would indicate. A simplified break-

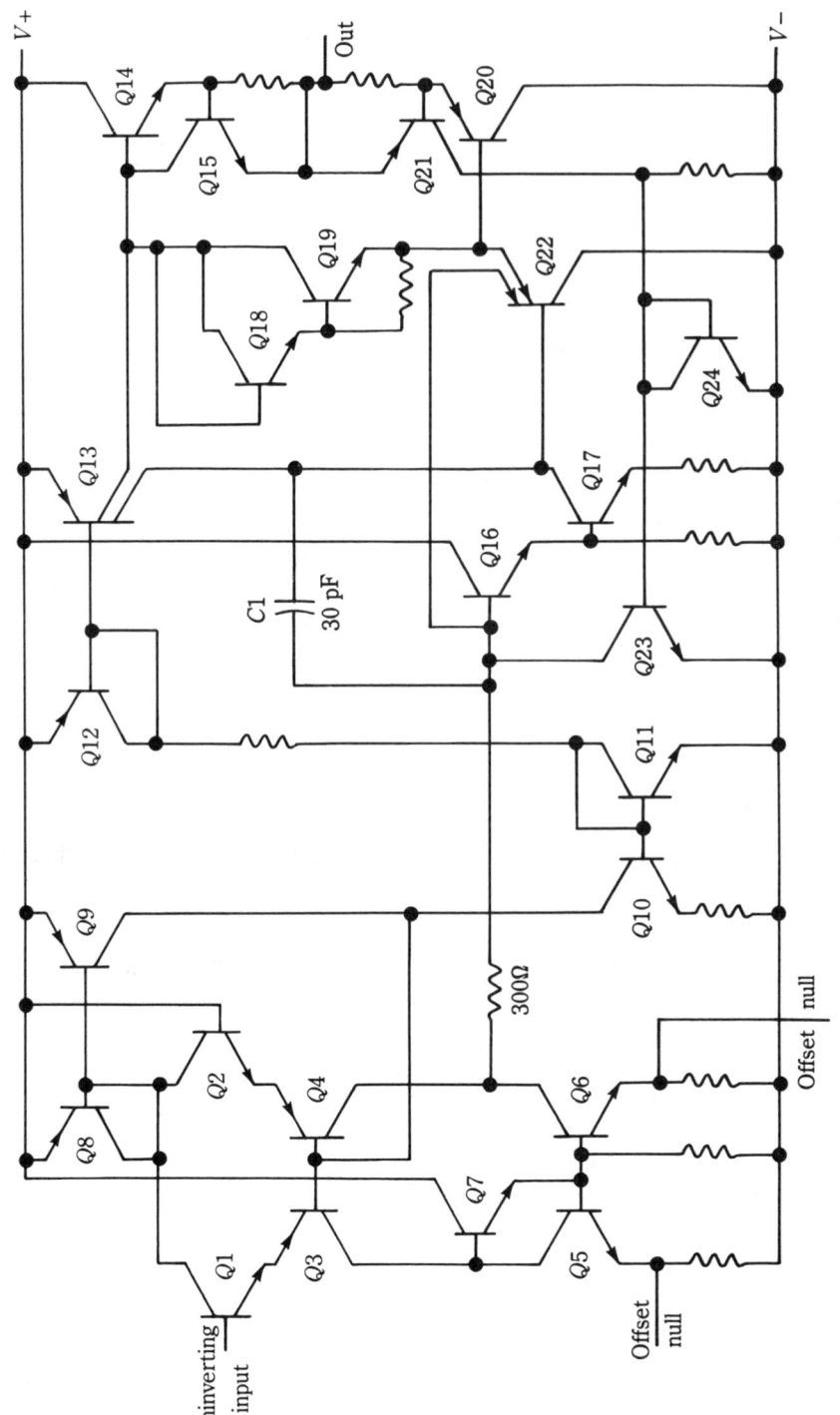

2-49 741 operational amplifier equivalent circuit.

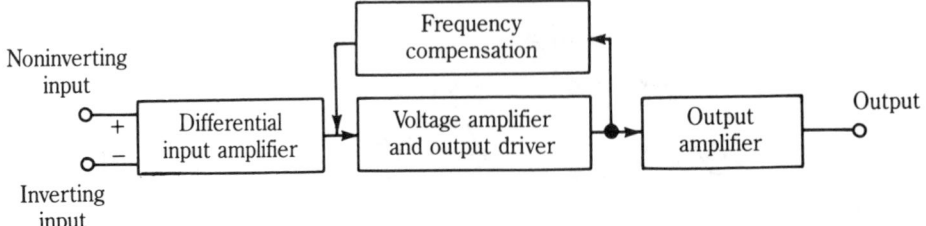

2-50 741 operational amplifier functional block diagram.

down of these separate functions appears in the diagram of Fig. 2-50. The functions are listed here:

1. The differential input amplifier includes transistors Q1 through Q4, and current sources. These current sources are involved with transistors Q5 through Q12. "Current mirrors" are implemented by some of these transistors in creating the current sources.

2. The voltage amplifier and output driver input is coupled to the collector of Q4. It includes Q13 through Q22, excluding Q14 and Q20. In addition to amplification, this section changes voltage levels to match the transistors in the output amplifier section.

3. The output amplifier section consists primarily of transistors Q14 and Q20, and associated circuitry, with a single output terminal providing a range of positive and negative output voltages with respect to the common or ground. This total range is approximately 3 volts less than the total supply voltage span.

4. The frequency compensation is accomplished by an internal feedback capacitor, $C1$. Its purpose is to impose negative feedback to reduce overall gain at higher frequencies in order to reduce the possibility of self-oscillation.

This amplifier and many other op amps require dual power supplies with equal positive (V_{CC}) and negative (V_{EE}) voltages. The input terminals are indicated as *inverting* or *noninverting*. The differential voltage between the two terminals determines the op amp output voltage level, which will be positive if the noninverting terminal is at the higher voltage level relative to ground. When the input differential voltage polarity is reversed, the output polarity is also reversed. Ideally, a zero differential voltage input will yield zero voltage output.

An ideal op amp has infinite gain, infinite input impedance, and zero output impedance. Although the 741 op amp does not meet these demanding requirements, it does have remarkable performance characteristics. For many purposes, the op amp might be regarded as sufficiently close to being ideal to justify that assumption.

A number of the actual performance data of the 741 op amp are summarized in Table 2-2. The terminology of these characteristics is briefly defined as follows:

Adjustment for input offset voltage The range of input offset voltage that can be compensated for by potentiometer adjustment. This feature is not available on all op amps.

**Table 2-2. Abbreviated electrical
characteristics of a 741 op amp**

Characteristic	Typical value
Input offset voltage	0.8 mV
Input offset current	3.0 nA
Input bias current	30 nA
Power supply rejection ratio	15 μV/V
Output short circuit current	25 mA
Power consumption	80 mW
Input impedance	6 MΩ
Large signal voltage gain	200,000
Transient response rise time	0.25 μs
Bandwidth	1.5 MHz
Slew rate	0.7 V/μs
Common mode rejection ratio	95 dB

Common mode rejection ratio (CMRR)   The ratio of the voltage gain output for differential input to the voltage gain with a common input (input terminals connected). Ideally it is infinite.

Input bias current   The input current, averaged for the two input terminals. In conformance with the concept of the ideal input impedance, the ideal input bias current is zero.

Input impedance   The ac impedance between the two input terminals, disregarding the usual inconsequential capacitance. Ideally it is infinite.

Input offset current   The difference between the input bias currents. Ideally it is zero.

Input offset voltage   The dc value of applied differential input voltage (V_d), which reduces the output voltage to zero. This input voltage is ideally zero, but cannot in practical circuits be zero because of normal circuit imbalances.

Large signal voltage gain   The open loop differential gain is a ratio of the output voltage to the differential input voltage. The value is slightly larger for small signal inputs.

Output resistance   The ratio of the change in output voltage to the change in output current for a load change with the input voltage held constant.

Output voltage swing   The maximum output voltage with a particular, specified load resistance and supply voltage.

Power supply rejection ratio   The ratio of the change in output voltage to the change in power supply voltage with the differential input voltage held constant. Ideally it is zero.

Slew rate   The maximum rate of change of the output voltage.

Op amps with special characteristics

Practical limitations of op amps have been emphasized in the preceding descriptions. Through design, one or more of the performance limitations can be minimized. Additionally, it is possible to tailor the op amp for certain general classes of applications. This aspect of op amps as linear integrated circuits is emphasized here. Other linear ICs having highly specialized applications will be introduced in the appropriate chapters of this book.

Comparator An op amp designed for open-loop applications for which saturation of either polarity is intended.

High-current operational amplifier An op amp that will deliver an output currents in amperes.

High-current operational amplifier An op amp that will deliver an output current in amperes.

Instrumentation amplifier An op amp design that enhances the performance characteristics of particular benefit in instrumentation applications, for example, CMRR.

Norton (current-differencing) amplifier A unique op amp design that incorporates a current source called a *current mirror* in the input circuit and is capable of functioning with a single positive voltage supply source.

Programmable operational amplifier An op amp having external connections (pins), permitting the control of various parameters.

Single supply operational amplifier An op amp that requires a single voltage power supply as contrasted with the usual dual supply.

Video amplifier A special op amp design permitting bandwidths up to 120 MHz with gains up to 40 dB.

Voltage follower An op amp with an internal connection that causes the output voltage to closely follow or duplicate the input voltage applied to the noninverting input terminal. The inverting input terminal is not available for use. The voltage follower serves as a buffer amplifier with high input impedance and low output impedance.

Integrated circuit arrays

Integrated circuit arrays are of two greatly different forms and purposes. Older arrays consist of a relatively small number of elements, such as transistors, on a single chip. In addition to the advantages of compactness and economy, the existence of multiple transistors on a single substrate permits precision matching. In contrast, the presently burgeoning field of large arrays is directly related to the customizing (usually by computer-aided design (CAD) methods) when making the final interconnection on the LSI chip.

Basic IC arrays

The Signetics CA3081/CA3082 is a typical array composed of seven separate npn transistors on a common substrate. These arrays can be connected in either a common-emitter (CA3081) or common-collector (CA3082) configuration with either one capable of 100 mA collector drive.

A comparison can be made with the Signetics ULN2001/03/04, which is comprised of seven silicon npn /Darlington pairs on a common monolithic substrate. Each pair will supply up to a maximum of 500 mA current with a maximum V_{CE} of 50 V.

The Fairchild μA3086 general-purpose transistor array is illustrated in Fig. 2-51. It

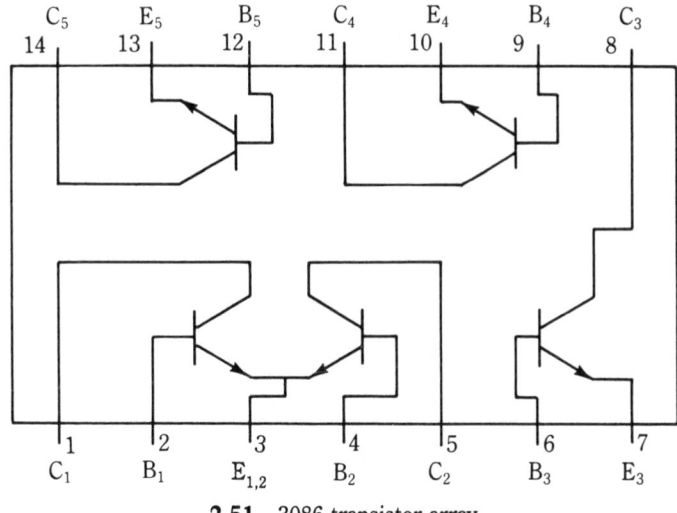

2-51 3086 transistor array.

contains a matched, differentially connected pair and three individually isolated transistors. This device is intended for low-power consumer and industrial applications.

Another array is the National LM194/LM394, which are junction-isolated, matched, monolithic npn transistor pairs. To guarantee long-term stability of matching parameters, internal clamp diodes have been added across the emitter-base junctions.

Another National product is the LM3045, LM3046, and LM3047 transistor arrays. The five individual npn transistors form one differentially connected pair, and the other three transistors are available separately.

It is possible to think of IC op amps as transistor arrays, except that resistors are added on the chip and the interconnection configuration is fixed. The inability of the user to choose the manner in which the op amp component transistors, diodes, and resistors are connected together is sufficient reason to remove them from the array category.

Gate arrays

A *gate array* is a predesigned, preprocessed matrix of transistors, which are subsequently interconnected uniquely for each application. At least one interconnection (metallization) level is required in the semicustomizing process. Among the advantages of gate arrays are lowered development time and costs, as compared to custom gate arrays, with some sacrifice in gate area and a lower transistor count.

To emphasize the preceding points, standard ICs have fixed functional properties. The overall system depends on the manner in which the standard parts are interconnected on the PC board. With very large scale integration (VLSI), the system design flexibility is limited because the multitude of individual functions are permanently committed and crowded onto fewer chips. Thus, whole families of standard parts are produced to the support requirements of single microprocessors.

Custom IC parts have the objective of reducing the parts count to a minimum, so the functional performance of one can be comparable with that of a PC board on which is mounted a number of standard, lesser-capability parts. Once designed and produced, a custom IC ordinarily cannot be used in a different application than the one for which it was intended.

Gate arrays combine the attractive features of standard parts with the ability to interconnect the constituent transistors, etc., on the chip to fulfill a system design requirement. The same gate array part can be customized for a particular application, with other users also customizing the part for their own individual applications.

Standard cells

Standard cells include basic logic elements, memory, and microprocessor CPUs. When designing with standard cells, a library of cells with automatic placement rules and software is used. With some restrictions on element placement imposed by the design automation system employed, the custom design is realized for each IC. Because of such constraints, and the continuing investment in and improvement of CAD systems, it is expected that design costs will decrease through the years with less sacrifice, compared to fully customized design.

A comparison of the three competing nonstandard-device approaches is given in Fig. 2-52. The superiority of standard-cell design for complex applications in which the production volume does not justify fully customized design is shown.

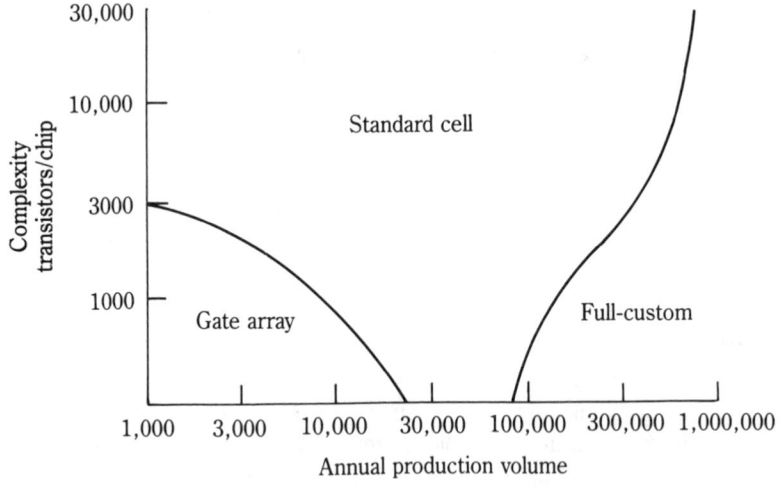

2-52   Comparison of competing design approaches.

Multiple-choice questions

1. When compared with a vacuum tube diode, a germanium diode:

   A. Can withstand higher working current.
   B. Can be used with higher working voltages.
   C. Can be used with higher rms voltage.
   D. Cannot be used with as high PIV.
   E. Both A and B.

2. An impurity having five electrons in the outer shell is added to the intrinsic semiconductor. It is called a(n):

    A. Polarizer.
    B. Donor.
    C. Negatizer.
    D. Proton.
    E. Acceptor.

3. The conductivity of an intrinsic semiconductor is greater than that of:

    A. Vacuum.
    B. Copper.
    C. Glass.
    D. Silver.
    E. Both A and C.

4. The emitter of a pnp transistor in a common-emitter configuration is analogous to what element in a vacuum tube?

    A. Plate.
    B. Screen grid.
    C. Filament.
    D. Control grid.
    E. Cathode.

5. A p-type semiconductor is formed by adding what to the intrinsic semiconductor?

    A. Free electrons.
    B. Protons and neutrons.
    C. Donor atoms.
    D. Acceptor atoms.
    E. Silicon crystals.

6. The number of electrons in the valence shell of indium is:

    A. 1.
    B. 2.
    C. 3.
    D. 4.
    E. 5.

7. A Zener diode is connected so that:

    A. The anode is positive with respect to the cathode.
    B. The cathode is positive with respect to the anode.
    C. The base is always grounded with the emitter floating.
    D. It is reverse-biased.
    E. Both B and D.

8. A p-n junction requires that:

    A. A p-type crystal and an n-type crystal be joined.
    B. P-type and n-type regions be present in a crystal.
    C. Donors and acceptors be mixed throughout the crystal.
    D. Both silicon and germanium form the crystal.
    E. None of the preceding choices is true.

9. With no applied bias voltage, a potential barrier and depletion layer are established near the pn junction of a diode because of:

    A. Infiltration.
    B. Ohm's law.
    C. Inertia.
    D. Diffusion.
    E. Adhesion.

10. In p-type silicon there is:

    A. A net positive charge.
    B. A net negative charge.
    C. An excess of free protons.
    D. An excess of free electrons.
    E. None of the preceding choices is true.

11. A depletion layer will be formed in a BJT:

    A. In the vicinity of each junction.
    B. At the base terminal connection.
    C. At the emitter terminal when biased to cutoff.
    D. At the collector terminal connection.
    E. Both B and D.

12. In the symbol for the npn transistor,

    A. The arrow points to the cathode.
    B. The arrow points toward the base.
    C. The arrow points from the cathode.
    D. The arrow points from the base.
    E. The arrow is attached to the symbol for the collector.

13. As compared with the BJT, a FET:

    A. Has more terminals.
    B. Also has an active region.
    C. Is a voltage-controlled device.
    D. Has a single charge carrier.
    E. Both C and D.

14. When the channel is formed by the application of a gate voltage, the mode is referred to as:

    A. Displacement.
    B. Depletion.
    C. Enhancement.
    D. Donor.
    E. Acceptor.

15. The "M" in MOSFET stands for:

    A. Mica.
    B. Micro.
    C. Magnetic.
    D. Miniature.
    E. Metal.

16. Identify the enhancement-mode IGFETs:

    A. Npn.
    B. NMOS.
    C. Pnp.
    D. PMOS.
    E. Both B and D.

17. An SCR can be turned off by:

    A. Reducing the gate current to zero.
    B. Reducing the gate voltage to zero.
    C. Opening the anode circuit.
    D. Placing a short-circuit between the anode and the cathode.
    E. Both C and D.

18. In fabricating a monolithic integrated circuit,

    A. A single crystal of silicon can be used.
    B. Silicon dioxide is the insulating substance.
    C. A photographic masking process is employed.
    D. Many chips are formed on the same wafer.
    E. All of the preceding choices are correct.

19. The principal families of digital integrated circuits include:

    A. PDQ.
    B. TRL.
    C. SSL.
    D. TTL.
    E. Both C and D.

20. Select the single most outstanding advantage of CMOS:

    A. None of the fabrication materials are toxic.
    B. Low impedance input and output levels.
    C. Frequency stability over a wide temperature range.
    D. Simplicity in fabrication process and low power.
    E. Frequency compensation by internal feedback.

21. Characterize the channel in CMOS devices:

    A. N-channel.
    B. P-channel.
    C. Q-channel.
    D. Ionic-channel.
    E. Both A and B.

22. The unique feature of ECL contributing to its operating speed is:

    A. The cancellation of holes and electrons at the gate.
    B. Operation of transistors without saturation.
    C. The integration of base and gate junctions.
    D. Conservative operation below the maximum temperature ratings.
    E. Both A and B.

23. The IC operational amplifier is considered to be:

    A. An n-channel PMOS differential amplifier.
    B. A linear integrated circuit.
    C. A differential amplifier with integrated injection logic.
    D. Choices A and C are correct.
    E. None of the preceding choices is correct.

24. An ideal operational amplifier will have:

    A. Infinite input impedance.
    B. Zero output impedance.
    C. Infinite gain.
    D. All of the preceding choices are true.
    E. None of the preceding choices is true.

25. Which one of the following statements is not true?

    A. The gain of the op amp decreases with frequency.
    B. The voltage comparator is similar to the op amp.
    C. Some op amps can be used as video amplifiers.
    D. An op amp has zero dc voltage gain.
    E. Many op amps require dual power supplies.

26. A practical op amp, unlike an ideal op amp,

    A. Always requires a dual power supply.

    B. In any amplifier configuration must have a gain of one or more.

    C. Cannot have a voltage gain greater than the numerical value of the supply voltage.

    D. Usually has an open-loop bandwidth of 1 kHz or more.

    E. None of the given answers is correct.

27. A monolithic integrated circuit can contain:

    A. Standard cells including logic elements and memory.

    B. A custom IC with all interconnections complete.

    C. A relatively small number of transistors.

    D. A specialized IC to perform communications and other dedicated functions.

    E. All of the preceding are true.

28. The digital logic family most widely used is:

    A. RTL.

    B. DTL.

    C. TTL.

    D. ECL.

    E. I^2L.

29. At an unbiased pn junction, electrons and holes pass across the junction because of:

    A. Depletion.

    B. Enhancement.

    C. Diffusion.

    D. The potential barrier.

    E. Both C and D.

30. CMOS logic is characterized by having:

    A. Bipolar FETs.

    B. P-channel FETs.

    C. N-channel FETs.

    D. No correct choice is given.

    E. Both B and C.

Basic problems

1. Explain the basic features of a semiconductor crystal.

2. A donor atom, such as arsenic, has five valence electrons. Describe the effect of doping intrinsic silicon, having four electrons in the valence shell, with arsenic.

3. Using the equations relating the emitter, collector, and base currents, together with the definitions of alpha and beta, show how alpha can be expressed in terms of beta.

4. What is the fundamental difference between a depletion-mode FET and an enhancement-mode FET?

5. Describe the key feature of digital circuits as contrasted with linear and analog circuits.

6. Name a number of op amp types having special characteristics.

7. Why might a Zener diode be an important element in a regulated power supply?

8. Briefly explain the particular measure, or measures, required to turn an SCR off.

9. What are the advantages of having a number of unconnected transistors in an integrated circuit array?

10. Name the essential features of a monolithic integrated circuit.

Advanced problems

1. In an intrinsic semiconductor (one free of impurities) what effect does temperature have on the movement of electrons?

2. Briefly discuss the distinction between an insulator and a good conductor, in terms of different energy levels.

3. An *acceptor atom* has three electrons in the valence shell. Explain how doping an intrinsic semiconductor with such atoms influences the flow of current.

4. Elaborate on the role of diffusion at the pn junction.

5. Explain why a semiconductor diode conducts better in one direction than it does in the other.

6. In a bipolar transistor, why does the relatively small base current effectively control the much larger collector current?

7. Explain the particular advantage of CMOS as compared with NMOS and PMOS.

8. In what ways does a practical op amp differ from the ideal op amp?

9. Define *gate array*.

10. Explain the basic role of gate arrays in the design and implementation of integrated circuits for special applications.

3

Thermionic tubes

THE PHENOMENON OF CURRENT FLOW IN AN ELECTRICAL CONDUCTOR HAS BEEN explained by the hypothesis that, in a conductor, certain electrons in the outer orbits of the component atoms are loosely held to their parent nuclei. When a voltage is applied across the conductor, a drift of the so-called *conduction electrons* results. This drift is from the negative side to the positive side of the applied voltage.

When a metal is heated, the normal random motion of the conduction electrons is intensified, and at very high temperatures, electrons will tend to leave the surface of the conductor altogether. This phenomenon is called *thermionic emission*—the name meaning simply the emission of electrons due to the application of heat. The operation of thermionic tubes depends on this phenomenon. A tube consists of a suitably heated *cathode*, or emitter of electrons, together with one or more electrodes for collecting the electrons emitted and for controlling and utilizing the resulting flow of electrons.

In order to use a heated metal as a source of electrons, the cathode must be enclosed in an evacuated envelope. This envelope is necessary for two reasons: first, most metals oxidize rapidly when heated to a high temperature in air. Secondly, if air (or any other gas) were present in the envelope, the emitted electrons, having attained sufficient velocity to leave the cathode, would collide with the molecules of the gas, causing ionization and producing undesired results.

The cathode can either be directly heated, by constructing it in the form of a wire and passing an electric current through it, in which case it is called a *filament*, or it can be indirectly heated by making it in the form of a nickel cylinder around a heater wire, through which the heating current is passed.

The heater of an indirectly heated tube is electrically insulated from the cathode, but is attached to it mechanically so that almost all the heat generated by the heater passes to the cathode. To permit this passage of heat, the insulation between the heater and the cathode must be chosen and designed with its thermal behavior as the primary consideration, and so it is usually weak electrically. For this reason, care must be taken that too high a potential difference does not develop between the heater and the cathode, or the insulation will break down. The maximum permissible heater cathode voltage for most small tubes is usually approximately 100 volts.

Indirectly heated tubes have three main advantages over the directly heated type. First, because of the high thermal capacity of the insulation between the heater and the cathode, the cathode remains at a constant temperature even when ac is used for the heater. However, the temperature of the filament of a directly heated tube when heated by ac varies between wide limits at twice the ac supply frequency. Owing to its thermal capacity, the cathode might take several minutes to reach its final working temperature, although it usually approaches this temperature after about 30 seconds.

Second, because the cathode is electrically isolated from the heater, greater flexibility in circuit design is possible. In particular, several tubes can be operated from the same heater supply yet have their cathodes connected to any desired points in their respective circuits without mutual interaction.

Third, since no heating current is passing through the cathode itself, the whole of the cathode is at the same potential.

The metal originally used for directly heated cathodes was tungsten, either pure or with about 1 percent thorium oxide, but the temperatures required were high so that the necessary heating current was large. Most cathodes in use today, except in high-power transmitting tubes, are oxide-coated. They give emission of electrons at dull red heat and are economical in supply power.

Directly heated filaments of the oxide-coated type consist simply of a wire of nickel, tungsten, or nickel alloy, coated with a preparation of barium, strontium, or calcium oxides. Indirectly heated cathodes consist of a nickel tube coated with oxide forming the cathode. Inside this tube, and insulated from it, is a stout tungsten wire that forms the heater.

This chapter covers the theory and construction of the following thermionic tubes:

- The diode
- The triode
- The instability of the triode
- The screen grid tube
- The pentode tube
- The beam power tube
- The variable-mu pentode
- The multigrid tube
- The gas tube
- Ultrahigh frequency circuit limitations

The diode

The subsequent motion of the free electrons that surround the cathode as the result of thermionic emission can be influenced by electric fields applied by means of further electrodes. Thermionic tubes are classified according to the number of electrodes they possess, the simplest being the two-electrode tube, or *diode*.

The diode consists of a cathode, either directly or indirectly heated, surrounded by a second electrode, called the *plate*. Figure 3-1 shows the assembly of the two types of diode and the conventional symbol associated with each. Figure 3-1A shows a directly heated tube with a filament, f_1f_2, and a plate, p, while Fig. 3-1B shows an indirectly heated tube having a heater, h_1h_2, a cathode, k, and a plate, p.

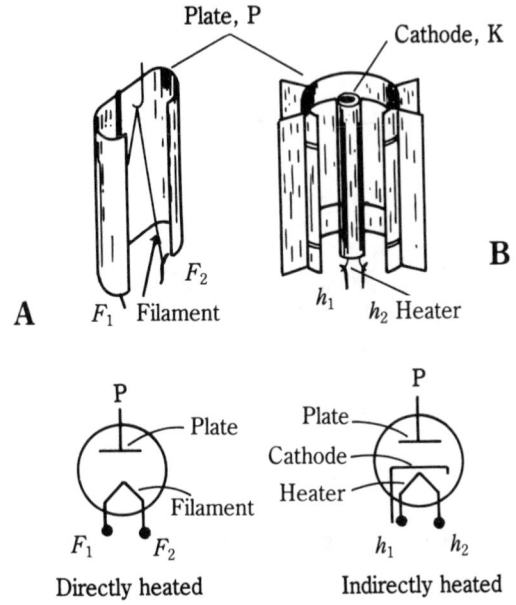

3-1 Directly heated and indirectly heated diodes.

The rate of emission of electrons into the free space surrounding the cathode can be considered to be constant, since it depends principally on the temperature of the cathode. What happens to these electrons afterwards depends on the potential of the plate relative to the cathode. If the plate is made negative with respect to the cathode, the emitted electrons will be repelled back into the cathode, and no current will flow between the two electrodes. If the plate is maintained at a positive potential relative to the cathode (Fig. 3-2), emitted electrons will be attracted toward the plate.

Notice that the flow of the electrons in the diode occurs when the plate is positive with respect to the cathode; if the polarity is reversed, the electrons are repelled back to the cathode and the current is zero. The diode therefore behaves as a one-way device or valve, since electrons can move from the cathode to the plate but not in the reverse direction. It is this one-way action that allows the diode to behave as a *rectifier* (a device for converting ac into dc).

The rate of arrival of electrons at the plate, which is the electron plate current, I_p, will be limited by the negative space charge produced by the electrons in transit between the cathode and the plate. The number of electrons in transit at any instant is sufficient to produce a negative space charge that neutralizes the attraction of the plate on the

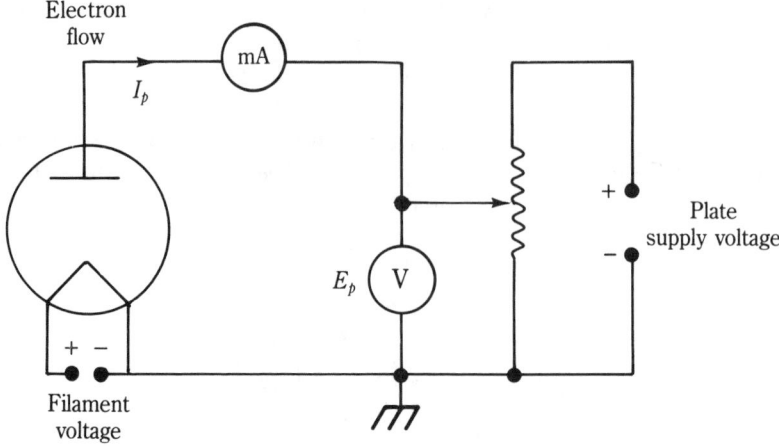

3-2 The circuit used to obtain the diode's plate characteristic curve.

electrons just about to leave the cathode. All electrons emitted in excess to this number are at once repelled back to the cathode. Where the plate current is limited by the space charge it will be dependent on the plate potential, and is substantially independent of the rate of emission of electrons by the cathode.

Figure 3-3 shows the relation between the plate potential, E_p, and the plate current, I_p, for a diode. This relation indicates that the diode behaves as a nonlinear resistance that does not obey Ohm's law. Consequently, the value of the diode's resistance is not constant but is determined by the point chosen on the characteristic curve.

Refer to Fig. 3-3, the value on the dc plate resistance, R_p, at the point Q is given by:

$$R_p = \frac{E_p}{I_p} \qquad (3\text{-}1)$$

where

R_p = dc plate resistance in kilohms (kΩ)
E_p = Plate to cathode voltage in volts (V)
I_p = Plate current in milliamperes (mA)

Because the diode is used for converting ac to dc, a varying voltage rather than a constant value will be applied across the tube. You should, therefore, be concerned with the changes in the plate voltage and the corresponding changes in the plate current. At the point Q, the ac plate resistance, r_p,—sometimes called the plate impedance—is determined by:

$$r_p = \frac{\Delta E_p}{\Delta I_p} \qquad (3\text{-}2)$$

where

r_p = ac plate resistance in kilohms (kΩ)
ΔE_p = the small change in plate voltage (V)
ΔI_p = the small change in plate current (mA)

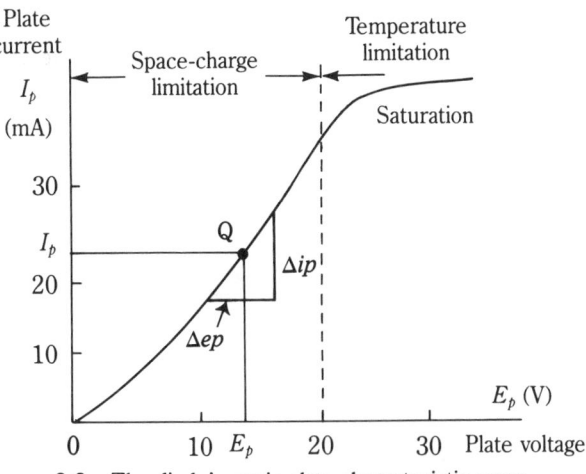

3-3 The diode's static plate characteristic curve.

Typical values of R_p and r_p are in the hundreds of ohms.

The lower part of the curve in Fig. 3-3, corresponding to the space-charge limitation of the plate current, can be represented by a three-halves power law:

$$I_p = k\,E_p^{3/2}$$
$$= k\,E_p^{1.5} \tag{3-3}$$

where k is a constant depending on the construction of the tube.

As the plate potential is raised, a point is eventually reached where the space-charge effect produced by all the electrons emitted is not sufficient to balance the attraction due to the plate, and the plate current is largely independent of the plate voltage, but will be determined by the rate of emission of electrons from the cathode, and therefore by the temperature of the cathode. The upper portion of the curve of Fig. 3-3 shows this condition.

The limiting value of the plate current—that is, the value above which it is impossible to increase the plate current by increasing the plate voltage—is called the *saturation current*. The tube, in this condition, is said to be *saturated*. Figure 3-4 shows how the saturation current is increased by raising the temperature of the cathode (for example, by increasing the heater current).

The electrons emitted from a heated cathode form a space charge around it and are attracted to the plate only when a positive plate voltage is applied. This is not strictly accurate, for some of the emitted electrons might have a sufficiently high initial velocity to enable them to reach the plate. This causes a small plate current (mA) to flow when the plate voltage is zero, or even slightly negative as illustrated in Fig. 3-5. This plate current ceases to flow when the plate potential is made slightly more negative. This negative voltage on the plate, necessary to reduce the current to zero, is normally less than 1 volt.

The diode rectifier circuit

During the first half-cycle (A) of the ac input voltage, V_i (Fig. 3-6), assume that the plate is positive with respect to the cathode. The diode therefore conducts and its electron

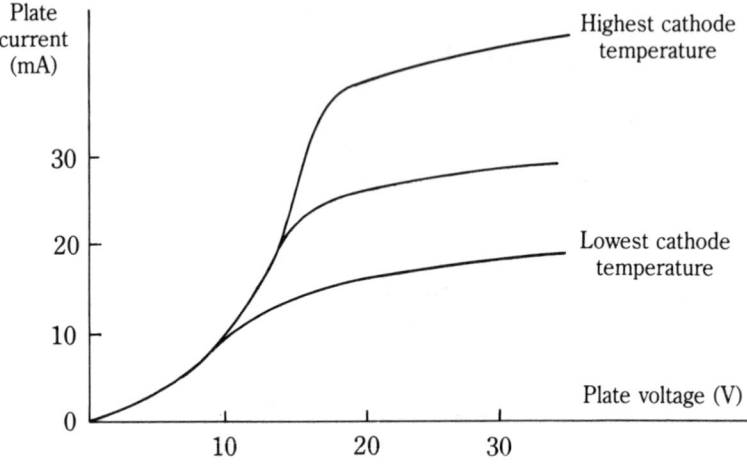

3-4   The effect of the cathode temperature on a diode's characteristic curve.

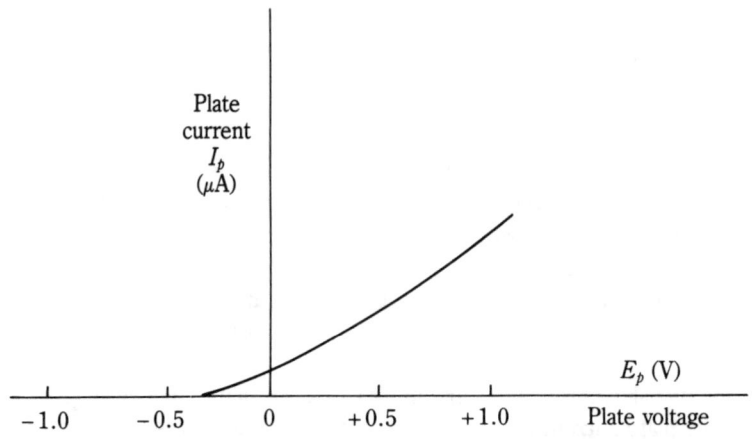

3-5   The diode characteristic curve for small plate voltages.

flow develops an output voltage, V_o, across the load, R_L. With only a small voltage drop across the diode, the V_o waveform will be approximately the same as the "A" half-cycle.

For the "B" half-cycle the plate is negative with respect to the cathode; the diode ceases to conduct and therefore the value of V_o falls to zero. The load voltage is a fluctuating dc output that never reverses polarity. Before such an output can normally be used, it must be filtered to create a steady dc voltage with a small ripple, or fluctuation, superimposed.

The triode

The diode, which is the simplest form of thermionic tube, is limited to a single function: rectification. In the triode, the flow of electrons from cathode to plate is controlled by

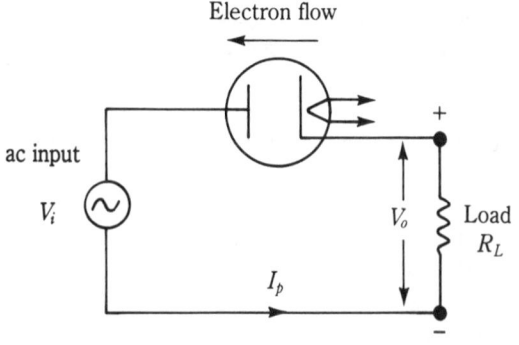

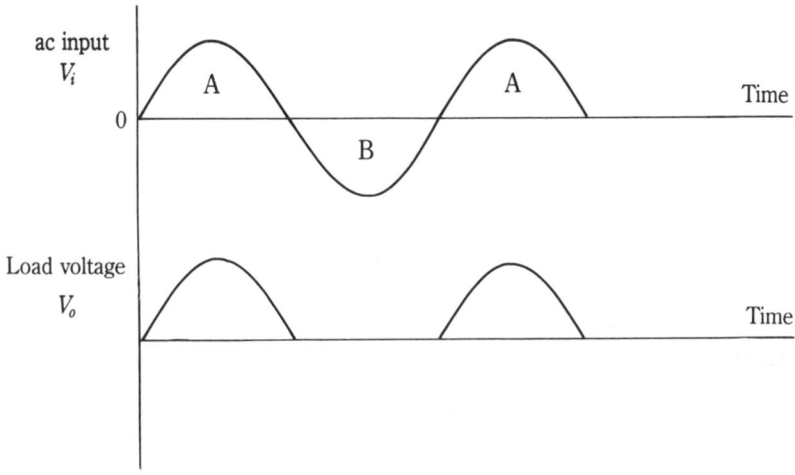

3-6 The basic principle of the diode rectifier circuit.

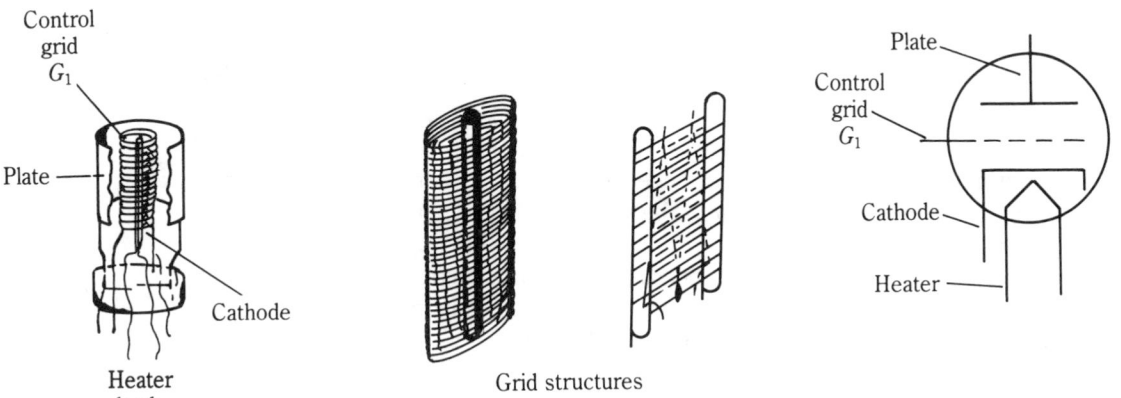

3-7 The indirectly heated triode, construction and symbol.

means of an additional electrode interposed between the cathode and the plate (Fig. 3-7). This electrode is called the *control grid* (G_1) because of the form taken in early examples of such tubes. Its modern form is in the shape of a spiral or open mesh. The grid is normally operated at a negative potential relative to the cathode, so it attracts no electrons to itself and no grid current flows; however, it tends to repel those electrons that are being attracted to the plate.

The number of electrons reaching the plate is determined mainly by the electric field near the cathode. The field in the rest of the space between the cathode and the plate hardly affects this number. Near the cathode the electrons are traveling slowly, compared with those that have already moved some distance toward the plate. The electron density in the interelectrode space will be high near the cathode, but will decrease toward the plate. The total space charge will be concentrated near the cathode, since once an electron has left this region, it contributes to the space charge for only a very brief interval of time. Therefore, the space current in the triode is determined by the electric field near the cathode produced by the combined effect of plate and grid potentials.

For a symmetrical grid structure, it can be shown that the electric field near the cathode is proportional to:

$$\left(E_g + \frac{E_p}{\mu}\right)$$

where E_g and E_p are the grid and plate potentials respectively, and μ is a constant determined by the geometry of the tube.

The total space current I_p varies with $E_g + (E_p/\mu)$ in exactly the same way that the plate current varies with the plate voltage for the space-charge limited diode. Therefore:

$$I_p = K\left(E_g + \frac{E_p}{\mu}\right)^{3/2} \tag{3-4}$$

where K is a constant, which is determined by the dimensions of the tube.

This yields:

$$I_p = 0$$

if,

$$E_g = -\frac{E_p}{\mu} \tag{3-5}$$

When the grid is made sufficiently negative with respect to the cathode, all the emitted electrons will be repelled back to the cathode and no plate current will flow. Therefore, $I_p = 0$, and the tube is said to be *cut off*. The value of this cutoff bias is then given by $E_g = -(E_p/\mu)$.

From Equation 3-4, it is clear that the value of the plate current is governed by both the grid voltage and the plate voltage. It can therefore be said that the triode is a voltage-operated device. For a particular tube you need to know:

- The degree of control that the plate voltage exercises over the plate current while maintaining the plate voltage constant

- The degree of control that the grid voltage exercises over the plate current while maintaining the plate voltage constant
- The relative controls of the grid and the plate in maintaining a constant level of plate current

This information is displayed on static characteristic curves, which show the interaction between the variables I_p, E_g, and E_p. The word *static* means that the curves are obtained under controlled laboratory conditions in which two of the three quantities are varied during the experiment, and the third quantity is kept constant. This is in contrast with *dynamic conditions*, such as in an amplifier circuit, in which all three variables are changing simultaneously.

Plate characteristics

Each characteristic curve is obtained by plotting plate current vs. plate voltage while keeping the value of grid voltage at a constant level. The circuit for conducting this experiment is shown in Fig. 3-8. Notice that the batteries that supply the necessary voltages to the triodes are labeled with the letters of the alphabet. The "A" battery is the filament supply, while the "B" battery provides the voltage to the plate/cathode circuit. The "C" or bias battery applies a voltage between the control grid and the cathode; the circuitry is such that the grid can be made either positive or negative with respect to the cathode.

Because of the lettering it is common practice (for example, in the tube manual) to use I_b and E_b, rather than I_p and E_p, to represent the plate current and the plate voltage. Similarly, I_c and E_c might be used instead of I_g and E_g as the symbols for grid current and grid voltage.

To obtain the plate characteristic for $E_c = 0$ move the slider X to the center point of its potentiometer. The slider Y is set at its lowest position so that $E_b = 0$ V. Move Y upwards and increase E_b in 10 V steps. Record the corresponding values of I_b. Then you can obtain the characteristic by plotting I_b vs. E_b. Move the slider X down so that E_c is successively -1.5 V, -3 V, -4.5 V, ... -15 V. For each value of E_c, repeat the proce-

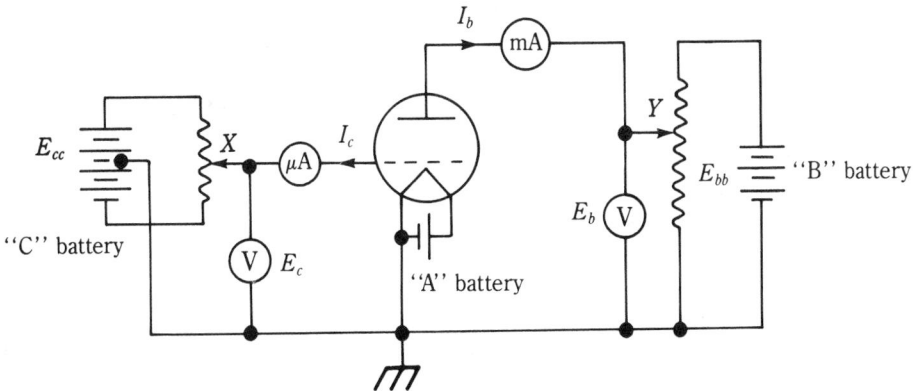

3-8   The circuit for obtaining the triode static characteristics.

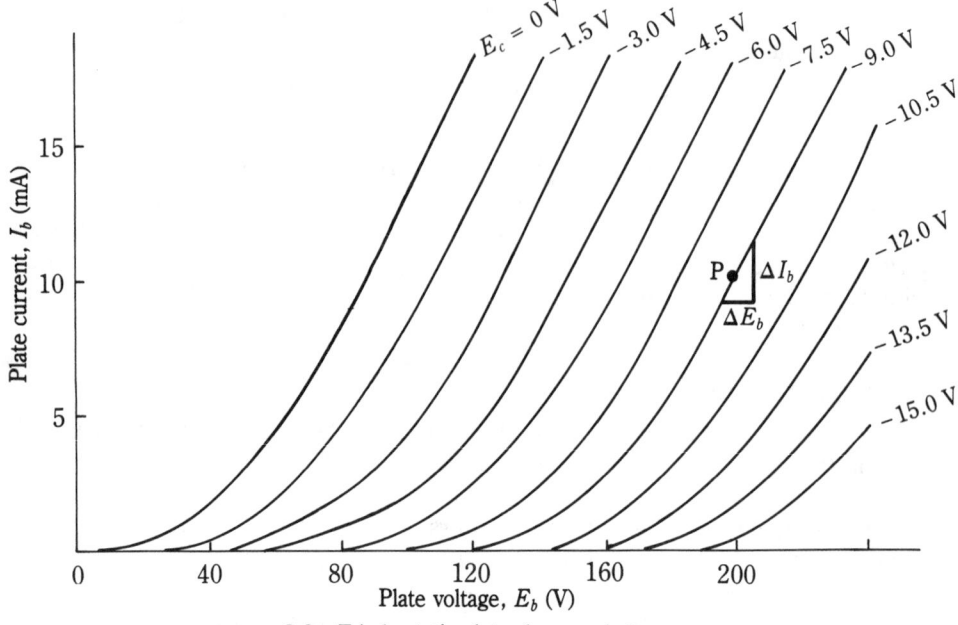

3-9   Triode static plate characteristics.

dure for the corresponding values of E_b and I_b. The result is the family of curves displayed in Fig. 3-9.

To determine the control that the plate voltage has over the plate current, you must select some operating point such as P, which is specified by $E_c = -9$ V, $I_b = 10$ mA, and $E_b = 200$ V. Consider the triangle containing the point P, a small change in plate voltage, $\Delta E_b = 9$ V approximately corresponds to a small change in plate current, $\Delta I_b = 2$ mA. Therefore,

$$\text{ac plate resistance, } r_p = \frac{\Delta E_b}{\Delta I_b} = \frac{9\ V}{2\ mA} = 4.5\ k\Omega$$

This means that a change in plate voltage of 4.5 V produces a change in plate current of 1 mA.

Since the plate characteristics are curves of differing slopes, the value of the parameter r_p is not a constant but is dependent on the chosen operating point.

Transfer or mutual characteristics

Each transfer characteristic curve is derived by plotting I_b vs. E_c, keeping the value of E_b at a constant level. Such a curve will demonstrate the control the grid voltage has over the plate current. In Fig. 3-10, the value of E_b is set to 40 V and maintained at that level. Initially X is at the center point of the potentiometer, so $E_c = 0$, which corresponds to a certain level of plate current, I_b. This means that the triode tube is a normally ON device, since it requires a negative grid/cathode voltage to cut off the plate current. The slider X is now moved downward so that E_c is successfully -1.5 V, -3.0 V, -4.5 V, ... and the corresponding values of I_b are recorded. Plotting I_b vs. E_c reveals the transfer characteristic curve for $E_b = 40$ V.

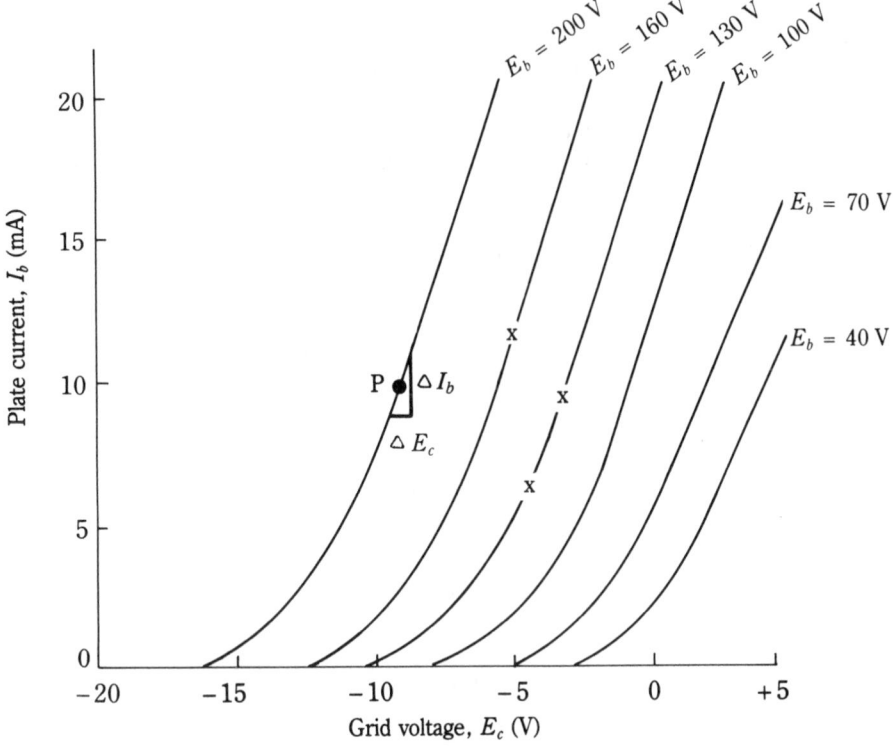

3-10 Triode static transfer characteristics.

Repeating the procedure for E_b = 80 V, 120 V, 160 V, and 200 V, reveals the family of curves shown in Fig. 3-10. Note that the transfer characteristics and plate characteristics are merely two different ways of presenting the same information. For example, the operating point, P, which you selected on the plate characteristics, is readily located on the transfer characteristics.

Consider the triangle containing the point P on the transfer characteristic. The small change of grid voltage, ΔE_c, (0.8 V approximately) corresponds to a small change in plate current, ΔI_b, (2 mA). The control that the grid voltage has over the plate current is measured by the transconductance, g_m, which is given by:

$$\text{Transconductance, } g_m = \frac{\Delta I_b}{\Delta E_c}$$
$$= \frac{2 \text{ mA}}{0.8 \text{ V}}$$
$$= 2.5 \text{ mA/V}$$
$$= 2.5 \text{ ms}$$
$$= 2500 \ \mu\text{s} \qquad (3\text{-}6)$$

This means that a 1 V change on the grid will produce 2.5 mA change in the plate current.

Now combine the results from the plate and the transfer characteristics. Starting at the operating point, P, make the grid more negative by 1 V. The plate current will fall from 10 mA down to $10 - 2.5 = 7.5$ mA. However, the plate can be brought back to 10 mA by raising the positive voltage on the plate to $200 + (2.5 \times 4.5) = 211.25$ V. In other words a 1 V negative change on the grid has been compensated by an 11.25 V positive change on the plate. This shows that the grid is 11.25 times more effective than the plate in controlling the plate current.

A comparison between the controls that the plate voltage and the grid voltage have over the plate current is measured by the amplification factor, μ. This is given by:

$$\text{amplification factor, } \mu = \frac{\Delta E_b}{\Delta E_c}$$

$$= \frac{\Delta E_b}{\Delta I_c} \times \frac{\Delta I_c}{\Delta E_c}$$

$$= r_p \times g_m \qquad\qquad (3\text{-}7)$$

For our example,

$$r_p = 4.5 \text{ k}\Omega$$

and

$$g_m = 2.5 \text{ ms}$$

Therefore:

$$\mu = 4.5 \text{ k}\Omega \times 2.5 \text{ ms}$$

$$= 11.25$$

In a strict sense, μ is a negative number since the changes in ΔE_b and ΔE_c are in opposite directions. As the grid is made more negative, the plate must be made more positive in order to keep the plate current at the same value. Consequently, μ has no units but has a value that is determined by the geometry of the tube. It is important to realize that the relationship $\mu = r_p \times g_m$ has no meaning unless all three parameters are measured at the same operating point.

Grid current

When the slider X is moved upward, the grid is made positive with respect to the cathode and can collect electrons to produce a flow of grid current. In the discussion, the flow of grid current has so far been ignored for two reasons:

1. The grid has in general been kept so negative with respect to the cathode that the grid will not collect electrons.
2. Even if the grid is slightly positive with respect to the cathode, the grid current will generally be very small (in microamps) since the grid's surface is itself small.

Grid current curves using a magnified scale can be conveniently shown on the display of the transfer characteristics (Fig. 3-11). The value of the grid current depends on the plate voltage and is small when the plate voltage is high.

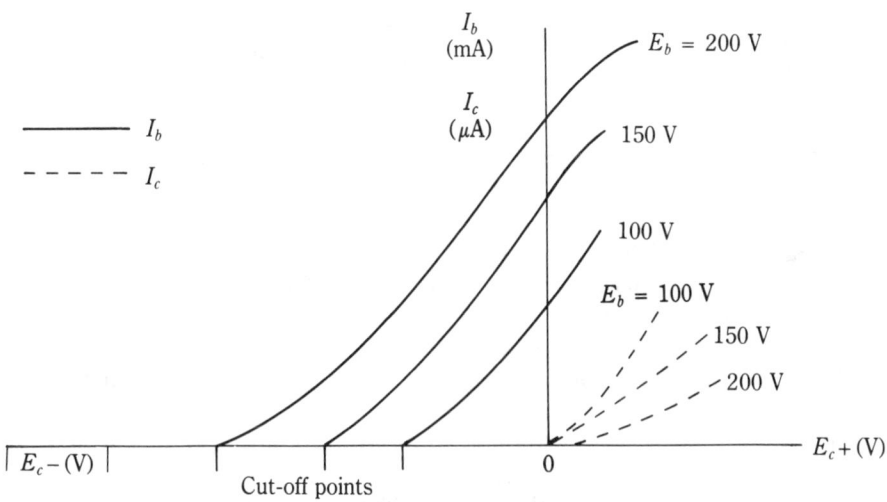

3-11 Triode I_b/E_c and I_c/E_c characteristics.

Although there are circuits that make use of grid current flow, it can be undesirable in some amplifiers for two reasons:

1. Any flow of grid current will be at the expense of the plate current. In consequence, the plate current waveform will be distorted, and the output signal will no longer be mainly controlled by the input signal.

2. Any flow of grid current implies a dissipation of power in the grid circuit, so the tube is no longer strictly a voltage-operated device.

The triode tube as an amplifier

Figure 3-12A shows a triode that is arranged as an amplifier tube in the simplest possible way. The signal to be amplified is some type of alternating voltage, e_c—for example, sine wave, pulse, sawtooth, square wave—that is applied to the control grid.

E_{cc} is a steady dc voltage supplied from a C battery, and is referred to as the *bias*. The size of the bias is such that, throughout the cycle of the signal, the grid is always negative with respect to the cathode. The plate is maintained at a high positive potential by the "B" battery, which provides a voltage, E_{bb}, in series with the tube and the load resistor, R_L. Plate current will flow and there will be a voltage drop across the amplifier load, R_L. The voltage at the plate, E_b, will therefore be less than the value of E_{bb}.

During quiescent conditions, when no signal is being applied to the grid ($e_c = 0$), the dc bias voltage, E_{cc}, will determine the value of the steady plate current, I_b, and the steady plate voltage, E_b. The relationship between E_b and I_b is:

$$E_b = E_{bb} - I_b R_L \qquad (3\text{-}8)$$

Under signal conditions, an alternating voltage is applied to the control grid, creating a fluctuating plate current that will now contain an alternating component, i_b. This component will develop an alternating voltage drop across R_L. Since at all times the sum of the voltages across the load and the tube must equal the fixed value of E_{bb}, an alternating component, e_b, will appear in the waveform of the plate voltage. This fluctuation in the voltage at the plate is the amplified output signal. To determine the actual voltages and currents in the circuit, combine the dc levels with the signal values by using the principle of superposition.

As an example, assume that $E_{cc} = -5$ V, $I_b = 8$ mA, $R_L = 10$ kΩ, and $E_{bb} = 250$ V. Notice that capital letters represent the steady quiescent levels, while lowercase letters are used for the signal values. The voltage drop across the load is 8 mA × 10 kΩ = 80 V, and therefore $E_b = +250 - 80 = +170$ V.

When a sinewave voltage, e_c, of 1.5 V peak is applied to the control grid, the triode's assumed parameters are such that the fluctuation in the plate current waveform has a peak value of 3 mA. The plate current therefore varies between 8 + 3 = 11 mA and 8 − 3 = 5 mA. These extremes correspond to plate voltages of +250 V − (11 mA × 10 kΩ) = +140 V, and +250 V − (5 mA × 10 kΩ) = +200 V. The plate voltage waveform therefore consists of a (200 + 140)/2 = +170 V dc level together with an alternating component (output signal) of (200 − 140)/2 = 30 V peak.

The input signal has a peak value of 1.5 V. The voltage gain, G_v, of the amplifier is therefore 30 V/1.5 V = 20. Since G_v is a voltage ratio, it has no units.

The waveforms of e_c, i_b, and e_b are shown in Fig. 3-12B. Notice that under the dynamic conditions in an amplifier circuit these three quantities are varying simultaneously. This is in contrast with the static or laboratory conditions that were used when deriving the tube's static parameters. In the procedure for calculating static parameters, two of the quantities were varied, while the third quantity was kept constant.

When the grid voltage, e_c, is becoming less negative in Fig. 3-12A, the plate current, i_b, is increasing, and there is a greater voltage drop across the load, R_L. Therefore, the plate voltage, e_b, across the tube increases, so e_c and e_b are 180° out of phase. This is often described as the *phase inversion*, which occurs as the signal is transferred

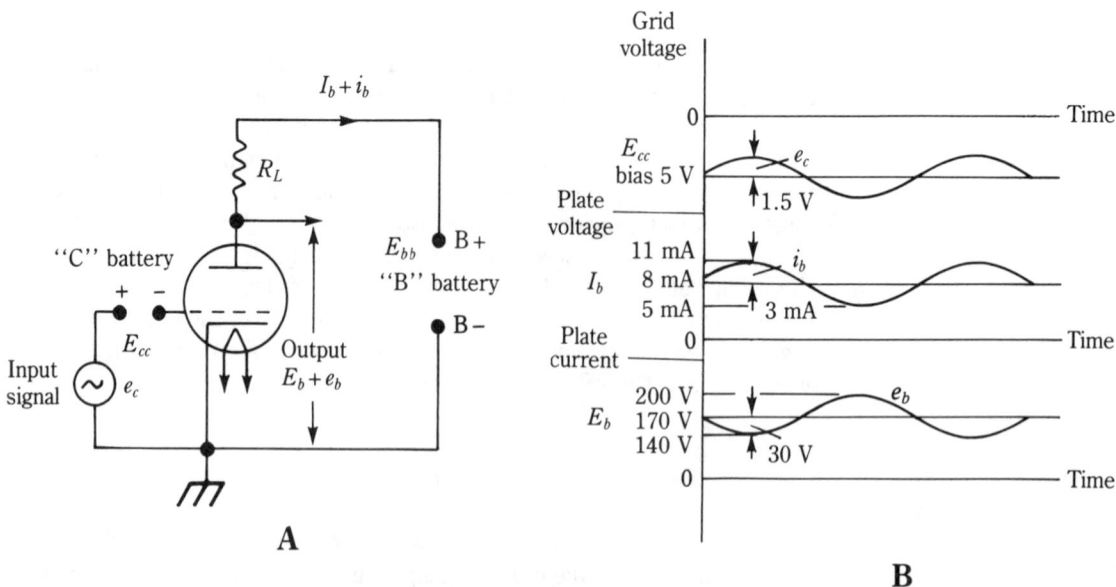

3-12   The circuit and waveforms of a triode amplifier.

through the tube from the input control grid to the output plate. The output signal is:

$$e_b = - i_b R_L \qquad (3\text{-}9)$$

where the negative sign indicates the 180° phase change between the input and the output signals.

Voltage gain formula

In an amplifier circuit, e_c, i_b, and e_b are all varying simultaneously. The signal component of plate current is therefore the result of combining the current changes produced by the variations in the grid voltage and the plate voltage.
Then,

$$i_b = g_m \, e_c + \frac{e_b}{r_p}$$

$$= g_m \, e_c - \frac{i_b \, R_L}{r_p}$$

This yields,

$$i_b = \frac{r_p g_m e_c}{r_p + R_L} \qquad (3\text{-}10)$$

$$= \frac{-\mu \, e_c}{r_p + R_L}$$

Equation 3-10 indicates that the amplifier behaves as a constant voltage source whose open circuit output is $-\mu \, e_c$ and whose internal resistance is r_p (Fig. 3-13). The output signal is:

$$e_b = i_b \, R_L \qquad (3\text{-}11)$$

$$= \frac{-\mu \, R_L \, e_c}{r_p + R_L}$$

$$\text{Voltage gain, } G_v = \frac{\text{Output signal}}{\text{Input signal}} = \frac{e_b}{e_c} \qquad (3\text{-}12)$$

$$= \frac{-\mu \, R_L}{r_p + R_L}$$

This formula only applies to small input signals when the values of μ and r_p can be regarded as approximately constant.

Amplifiers are subdivided into two classes according to the use made of the output signal. In the simplest case it is essential to develop the maximum output signal, which in turn is applied to the grid of a second triode for further amplification. Since the following grid is voltage-controlled without the consumption of power, the first stage would be called a *voltage amplifier*.

In certain cases, the main consideration is the power developed in the plate load. For example, in the final audio stage of a receiver, the loudness (volume) of the music or speech will depend on the power developed in the loudspeaker, which will act as the

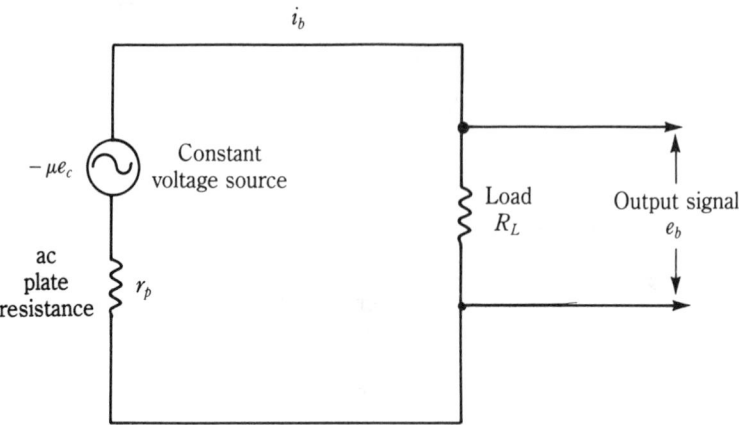

3-13   The equivalent circuit of a triode amplifier as a constant voltage source.

stage's plate load and will convert electrical power into acoustic power. This last stage, in which power is the main concern, is called a *power amplifier*.

In terms of the triode's parameters, the tubes for voltage amplifiers tend to have relatively high r_p, low g_m, and high μ, while power amplifiers require tubes with low r_p, high g_m, and low μ. As shown in Equation 3-5 a low μ means a high value of cutoff grid voltage; this will allow the amplifier to accommodate a large input signal, which is necessary to develop an appreciable power output.

The instability of the triode

Within the triode are the cathode, the control grid, and the plate. These electrodes represent conducting surfaces that are separated by the vacuum dielectric. Consequently there are three interelectrode capacitances, whose values are normally a few picofarads. Refer to Fig. 3-14. These capacitances are:

1. The grid-cathode capacitance, C_{gk}, which is effectively in parallel with the input signal circuit and is called the tube's input capacitance.
2. The plate-cathode capacitance, C_{pk}, which is effectively in parallel with R_L as far as the signal is concerned, and is called the tube's output capacitance.
3. The plate-grid capacitance, C_{pg}, which allows the output signal to drive a current back into the input circuit.

Of the three capacitances, C_{pg} has the most important effect. Clearly C_{pg} provides, particularly at high frequencies, a path between the plate and grid circuits so that the output signal at the plate can drive a current back into the grid circuit and develop a feedback voltage across the impedance of that circuit; this is commonly referred to as the *Miller effect*.

If this feedback voltage is of sufficient magnitude and is in the correct phase (positive feedback), the circuit might cease to function as an amplifier and become an oscillator. This will occur if the plate load (the tank circuit L1C1) behaves inductively. However, if

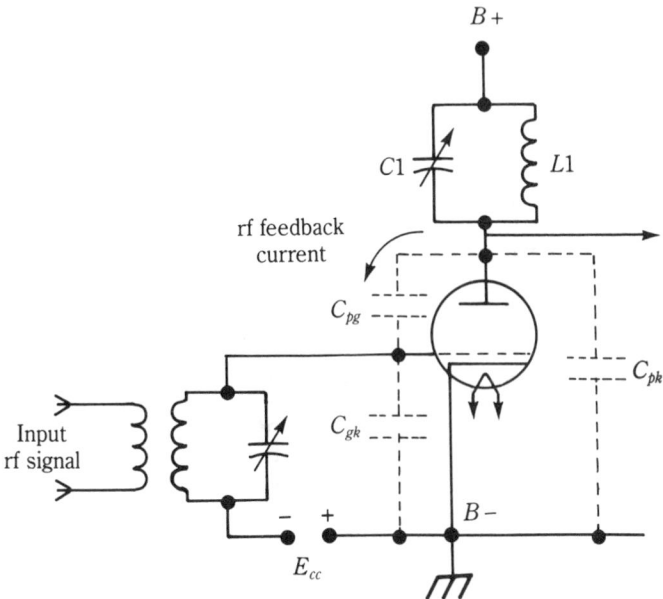

3-14   The triode rf amplifier showing the interelectrode capacitances.

the plate tank circuit is capacitive, the phase of the feedback voltage is reversed (negative feedback), and the amplifier's gain is reduced. Neither of these effects will occur when the plate tank circuit is at resonance and behaves resistively, but the circuit might drift to become either inductive or capacitive. The triode is therefore unstable as an rf amplifier unless the circuit is neutralized.

Typical neutralization circuits are shown in Fig. 3-15A and B. Here the purpose is not to eliminate the feedback through the C_{gp} but to cancel its effect by an opposite feedback through the neutralizing capacitor, C_N. In the plate neutralization circuit of Fig. 3-15A, a center tap on the plate tank coil, L1, is connected to B+ and is therefore effectively at rf ground. The points X and Y are at opposite ends of coil L1, and therefore the rf potentials at these points are equal in magnitude but 180° out of phase. The feedback from the voltage at Y through the interelectrode capacitance C_{gp} is canceled by the opposite feedback from the voltage at X through the neutralizing capacitor, C_N. During the neutralizing procedure, the value of C_N is varied until complete cancellation is achieved.

In the grid-neutralizing circuit of Fig. 3-15B, the grid coil, L2, is center-tapped and is at rf ground. The feedback from point P through the C_{gp} is taken to the top of the grid coil, while the feedback from the same point through C_N is applied to the bottom of L2. The neutralizing capacitor can then be varied until the overall feedback to the grid circuit is zero.

The evolution of the various screen grid tubes was a direct consequence of attempts to reduce the plate-control grid capacitance to such an extent that the tube would be stable when used in an r-f amplifier circuit. However, some screen-grid tubes have other advantages over triodes and are commonly employed where a voltage amplifier at audio or radio frequencies is required.

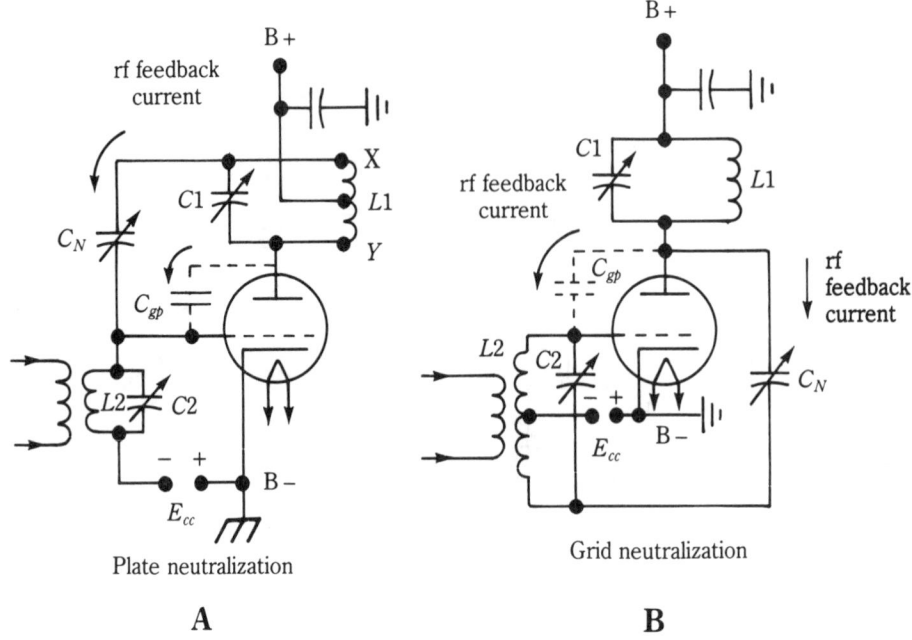

3-15 Triode neutralizing circuits.

The screen-grid tube

The first screen-grid tube, or *tetrode*, was originally introduced to overcome the ill effects of the grid-plate capacitance. These effects become apparent when a triode is used as an rf amplifier. The screen-grid tube has two grids between the cathode and plate. The grid nearer the cathode performs the same function as the grid in the triode and is referred to as the *control grid* (G_1), while the additional grid acts as an electrostatic screen between the control grid and the plate, and is therefore called the *screen grid*, or *screen*, (G_2). The screen is maintained at a high positive potential approaching that of the plate, and has a considerable effect on the electron stream between the control grid and the plate.

Consider the electric field between the electrodes in terms of its flux lines. If the screen were a solid metal plate maintained at a potential equal to that of the plate, the flux lines leaving the cathode and the grid would terminate on the screen, and there would be no electric field in the space between the screen grid and the plate. Consequently, there would be no capacitance between the plate and the screen grid, nor between the control grid and the plate.

Now consider the screen grid to be made in the form of a close mesh and to be maintained at a potential not necessarily equal to, but approaching, that of the plate. This time the screening effect will be considerable, but not perfect—although with a fine mesh structure it will be practically so. The result is that there will be capacitance between the pairs of electrodes: control grid and screen grid, screen grid and plate, and plate and control grid, though the grid-plate capacitance will be very much reduced from that in the triode. In commercial types of screen grid tube the residual control grid plate capacitance varies from 0.001 pF to 0.02 pF, as compared with 2 pF to 8 pF for a triode.

Figure 3-16 shows a typical plate characteristic for a triode, drawn under conditions of constant control grid voltage (E_{c1}) and constant screen voltage (E_{c2}). When the plate potential is zero, all the emitted electrons are attracted to the screen, giving a fairly high screen current (I_{c2}) and the plate current (I_b) will be zero. If the plate potential is increased, some of the electrons passing through the mesh of the screen are carried on by their momentum and come under the influence of the plate, to which they are attracted, giving current to the plate. This current will increase with increased plate potential. Because of the shielding effect of the screen, however, the potential of the plate will have very little effect on the electric field in the vicinity of the cathode, and an increase in plate potential will not appreciably increase the total space or cathode current, $I_k = I_b + I_{c2}$. Any increase in plate current will therefore be at the expense of the screen current.

As the plate potential increases, so will the velocity of electrons on arrival at the plate. One effect of bombarding the plate with fast-moving electrons is that other electrons are ejected by the force of impact. The quantity of these ejected electrons—or *secondary electrons*, as they are usually called—will vary with the material of the plate and the velocity of the electrons reaching the plate from the cathode, or primary electrons. In certain circumstances, as many as ten secondary electrons could be liberated by one fast-moving primary electron.

This phenomenon of the secondary emission also occurred in the diode and triode, but in those cases the secondary electrons were attracted back into the plate surface and had no effect on the tube. With the tetrode, however, the velocity of the primary electrons is sufficiently high to cause secondary emission while the plate is at a lower potential than the screen grid; there is, therefore, a tendency for the screen to collect these secondary electrons emitted from the plate. The result is an increase in the screen current at the expense of the plate current.

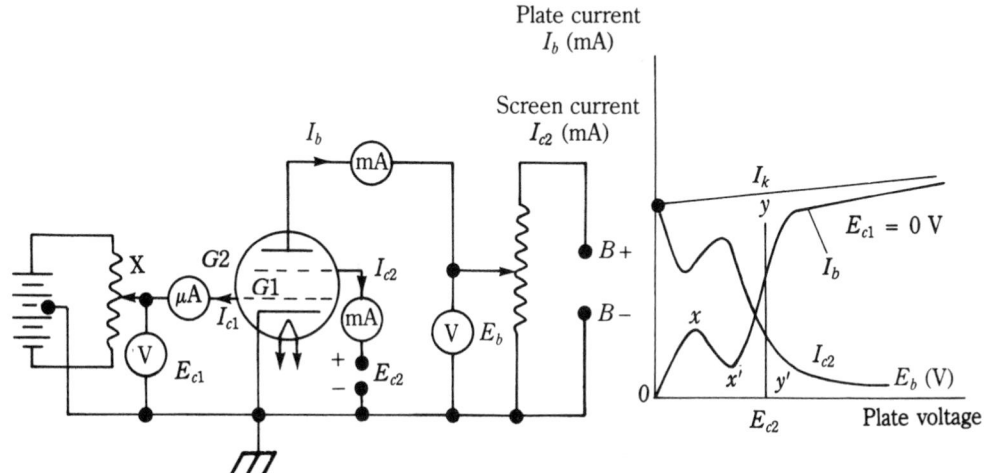

3-16 Tetrode static plate and screen characteristics.

A further increase in the plate potential will increase the velocity of the primary electrons and therefore increase the emission of the secondary electrons. If the screen is still at a higher potential than the plate, the screen will collect practically any of these slow-moving secondary electrons, with the result that the plate current will actually decrease with increased plate potential.

The increase in plate potential and decrease in plate current is represented by the portion of the XX' of the plate characteristic (Fig. 3-16). Under these operating conditions that control the XX' portion of the characteristic, the tube behaves as a negative-resistance device, since a decrease in plate voltage causes an increase in the plate current.

If the plate potential is further increased, the majority of the secondary electrons will no longer be attracted to the screen, but more and more will be drawn back into the plate. The plate current will once more increase with increased plate potential, at the expense of a decreasing screen current.

The portion of the tetrode characteristic that is useful for most purposes is the portion well to the right of the vertical line YY' in Fig. 3-16. In this region the curve becomes practically straight, and the plate current is nearly independent of the plate voltage, indicating a very high value for the ac plate resistance, r_p. The effect of the grid, however, is practically the same as if the screen and the plate together formed the collecting electrode—that is, the transconductance is of the same order as for a triode.

The required value of the screen voltage commonly obtained by connecting the screen to B+ by means of a dropping resistor of suitable value or a potentiometer is shown in Fig. 3-17A and B. The value R_{sg} can be calculated from the fact that the screen current is normally about one-fifth to one-tenth the value of the plate current. The equation is:

$$E_{c2} = E_{bb} - I_{c2} \times R_{sg} \tag{3-13}$$

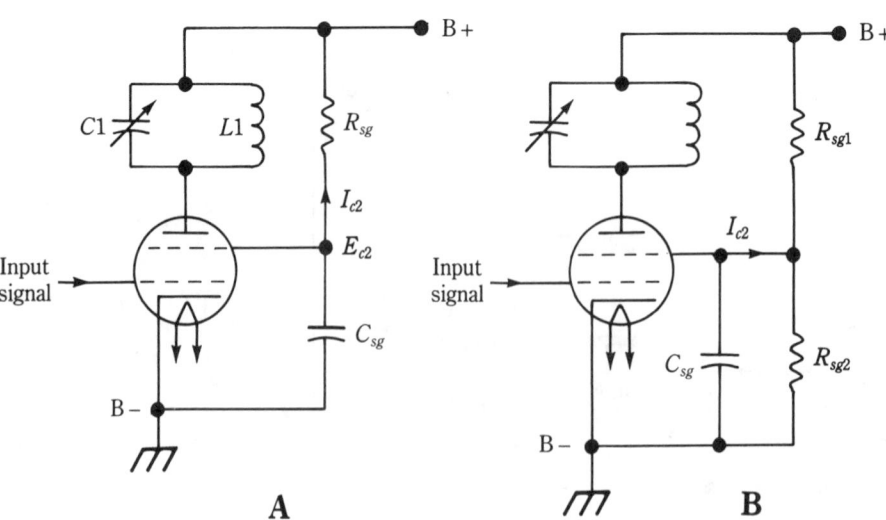

A   **B**

3-17 Methods of providing the correct screen voltage for a tetrode tube.

With an alternating signal applied between the cathode and the grid, the screen current will fluctuate just as the plate current does. The effect of a fluctuating screen voltage can be overcome by connecting the screen grid to the cathode through a capacitor, C_{sg}. This capacitor represents a negligible reactance at high frequencies, so that the screen grid and cathode will be virtually at the same alternating potential. There will be no coupling between the plate and the grid circuits, apart from the very small residual grid-plate capacitance.

Because of the restriction on the working part of the characteristic imposed by secondary emission, the screen grid tetrode is of little or no use as a power amplifier. Its use as a voltage amplifier is limited, since it can handle only a very small grid signal. These tubes are virtually obsolete.

The pentode tube

One method of reducing the effect of secondary emission is to introduce an additional electrode, in the form of a third grid, between the screen and the plate. This third grid (G_3) is called the *suppressor*, and the resulting five-electrode tube is referred to as a *pentode*. The suppressor is given a negative potential relative to the plate and the screen grid, which prevents the low-velocity secondary electrons from reaching the screen. At the same time the suppressor is usually built of open-mesh wire, so it does not interfere appreciably with the passage of the high-velocity primary electrons toward the plate. The suppressor grid is usually connected to the cathode, but since other connections might be required, the lead to the suppressor grid is usually brought out of the tube to a separate pin, and the connection is made externally. In certain cases where a pentode is suitable only as a power amplifier, the connection is made internally.

Figure 3-18A illustrates the pentode's plate characteristics that do not contain the negative resistance section associated with the tetrode. The transfer characteristics are displayed in Fig. 3-18B, and the approximate values of the tube's parameters are $r_p = 1.5$ MΩ, $g_m = 2400$ μs, and $\mu = 3600$. With such high values of r_p and μ, it is preferable to regard the tube in the equivalent amplifier circuit as a constant current generator (Fig. 3-19). Then the tube's output signal is

$$e_b = g_m e_c \times \left(\frac{1}{r_p} + \frac{1}{R_L} \right)$$

$$= \frac{(g_m e_c)}{(1/r_p + 1/R_L)}$$

$$= g_m e_c R_L \quad \text{if } r_p >> R_L \tag{3-14}$$

The voltage gain is:

$$G_v = \frac{e_g}{e_c} = g_m R_L \tag{3-15}$$

For example, if $g_m = 2400$ μS and $R_L = 33$ kΩ, the voltage gain is (2400×10^{-6}) $(33 \times 10^3) \approx 80$.

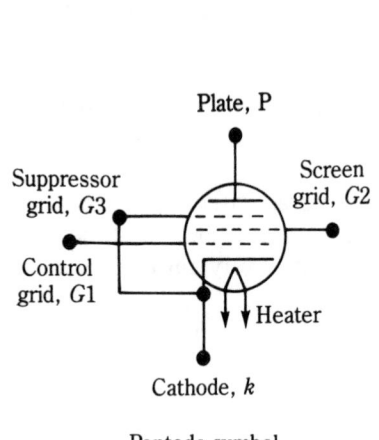

Pentode symbol

A

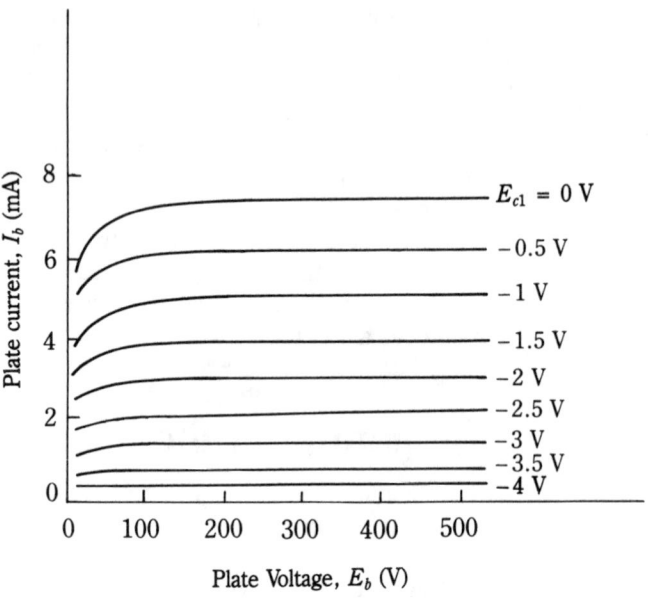

Pentode plate characteristics
curves taken at E_{c2} = +100 V

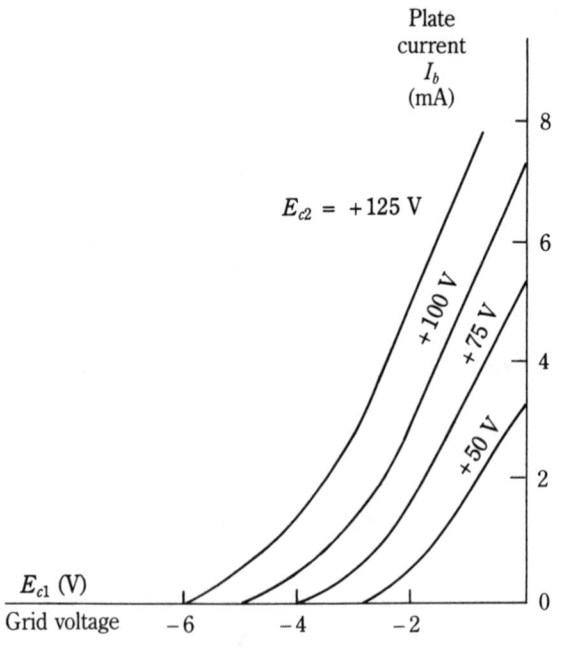

B

Pentode transfer characteristics

3-18   Pentode plate and transfer characteristics.

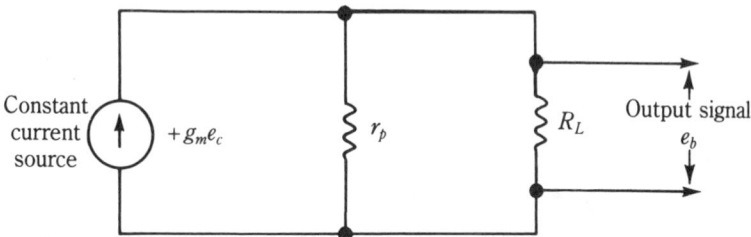

3-19 The equivalent circuit of a pentode amplifier as a constant current source.

The beam power tube

Another method of reducing the effects of secondary emission is to include additional electrodes between the screen grid and the plate. These electrodes are connected to the cathode inside the tube and will repel the electron stream. The electrodes are arranged so they concentrate the electron stream into a comparatively narrow beam, and for this reason they are usually referred to as *beam-forming plates* (Fig. 3-20A).

The concentration of the electrons into this beam, combined with a large distance between the screen grid and the plate, gives an intensified space-charge effect in the screen grid-plate space, which will repel the secondary electrons back into the plate's surface. The screen current is made small by having an open-meshed screen and an optical alignment of the control grid and the screen grid. Such a tube is referred to as a *beam power tube*; its plate characteristics are shown in Fig. 3-20B.

The variable-mu pentode

The variable-mu pentode has a control grid with an asymmetrical structure. This is normally done by varying the pitch of the control grid along its length, with the mesh of the grid being closer at the ends rather than at the center. The result is that various parts of the tube cutoff with different grid bias voltages, so that the overall cutoff comes gradually, rather than abruptly.

Figure 3-21 shows a set of transfer characteristics for a typical variable-mu pentode, plotted for a constant plate voltage but a variable screen voltage. With curved characteristics, the value of the g_m will depend on the chosen bias point. Since the voltage gain, G_V, $= g_m R_L$, this will enable the voltage gain of an amplifier to be varied over a wide range by changing the bias voltage on the control grid.

The multigrid tube

If the potential on the suppressor grid of a pentode is varied, there will be a corresponding variation in the plate current. This fact can be used to allow two independent signal voltages simultaneously to control a single plate current by applying one voltage to the control grid and the other to the suppressor. Owing to the nonlinearity of the tube, mixing between the frequencies of the two inputs will take place. Tubes used in this way are called *frequency changers*. However, under these conditions, the plate-suppressor capac-

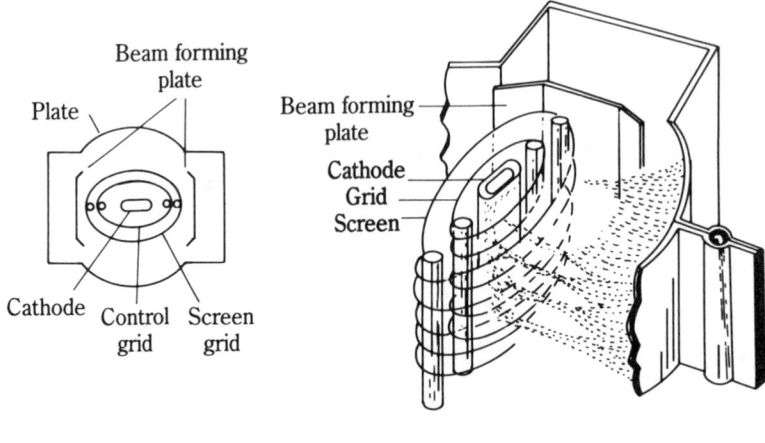

A

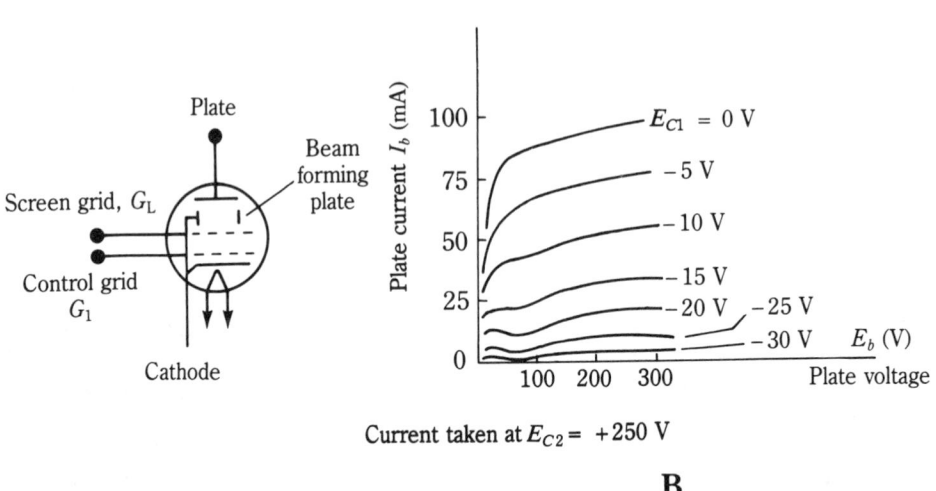

B

Current taken at $E_{C2} = +250$ V

3-20   Construction and plate characteristics of the beam power tube.

itance of a pentode has the same bad effects as the plate-control grid capacitance of a triode. Special frequency changer tubes have, therefore, been developed with electrostatic screening between the plate and the second signal grid.

The hexode

The *hexode* (Fig. 3-22A) is a tube having an extra electrostatic screen enclosed between the second signal, or *injector grid*, G_3, and the plate. This additional grid (G_4) is usually connected internally to the existing screen grid, (G_2), and is maintained at a steady positive potential so that it functions exactly the same as the screen of a tetrode. The hexode then forms a much more stable frequency changer than the pentode.

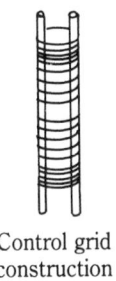

Control grid
construction

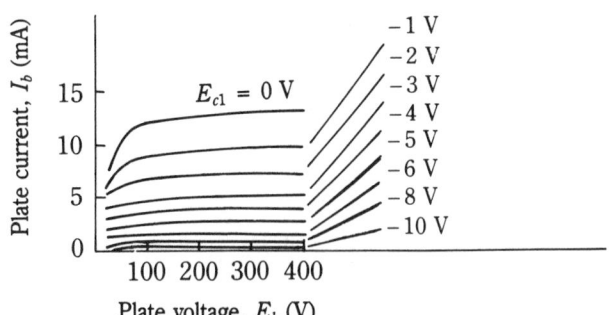

Variable-mu pentode plate characteristics
curves taken at $E_{c2} = +100$ V

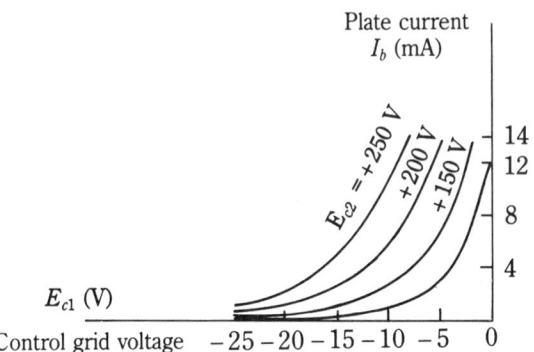

Variable-mu pentode transfer characteristics
curves taken at $E_b = +250$ V

3-21 Characteristics of the variable-mu pentode.

The heptode

Because in the hexode the screen grid, G_4, carries a positive potential and is adjacent to the plate, trouble might be experienced from secondary emission effects, as in the case of the tetrode. These effects can be overcome by the insertion of a suppressor grid between the screen grid, G_4, and the plate. The resulting tube is called a *heptode* (Fig. 3-22B).

Triode-hexodes and triode-heptodes

Frequency-changer tubes are normally used to mix two signals; one signal is generated by an oscillator coupled directly to one of its two signal grids. To reduce the number of tubes needed in a piece of equipment, hexodes and heptodes are sometimes built into an envelope that also contains a triode. This triode can be used as an oscillator and has its grid internally connected to one of the two signal grids of the hexode or heptode.

Figure 3-22C shows a representation of a triode-hexode, in which the triode grid is internally connected to the hexode grid nearest the cathode. Figure 3-22D shows a triode-heptode, in which the triode grid is connected to the third grid (G_3).

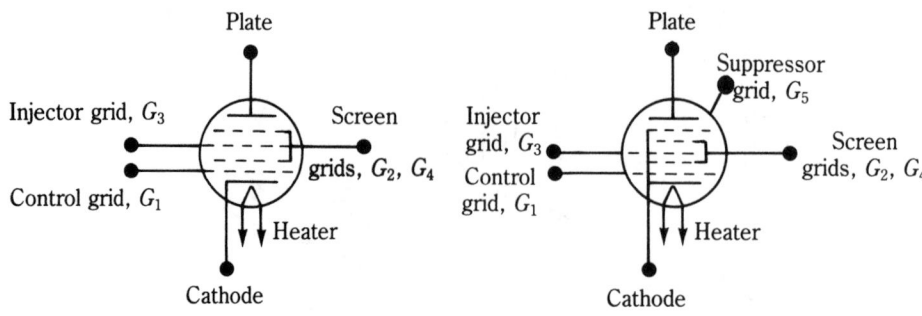

A. The hexode tube

B. The heptode tube

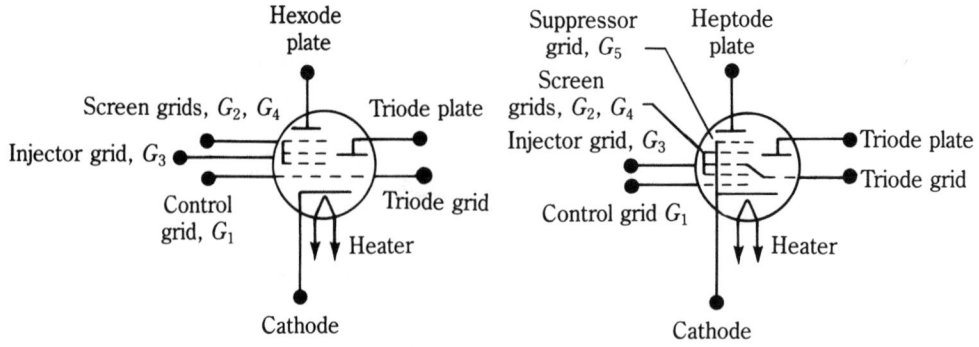

C. The triode-hexode tube

D. The triode-heptode tube

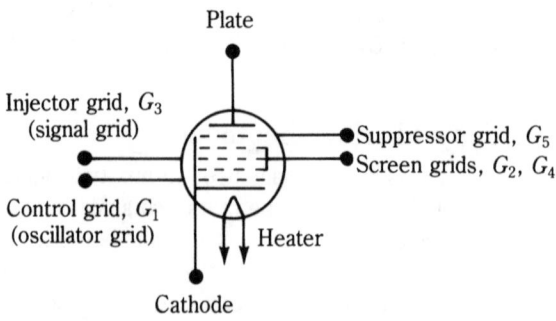

E. The pentagrid converter tube

3-22   The multigrid frequency conversion tubes.

The pentagrid converter tube

The pentagrid converter is a special single tube that combines the functions of the oscillator and the mixer section (Fig. 3-22E). The cathode, the control grid (G_1), and the screen grid (G_2) form a triode, which is used in the oscillator circuit. The frequency conversion section is made up of the screen grids G_2 and G_4; the injector grid, G_3; the suppressor grid, G_5; and the plate.

The gas tube

Mercury vapor diode

The hot-cathode gas (soft) rectifier tube contains neon, argon, or mercury vapor at low pressure.

For small voltages the E_b/I_b characteristic is little different from that of the high vacuum type (hard) tube. However, when the plate voltage reaches approximately 10 V to 15 V for mercury vapor, the electrons leaving the heated cathode have sufficient velocity to ionize the vapor particles by collision. Because of the production of the positive ions, the tube is no longer space-charge limited. A further small increase of voltage up to 22 V will cause the current to rise very quickly to a value that represents the total electron emission of the cathode (Fig. 3-23A and B).

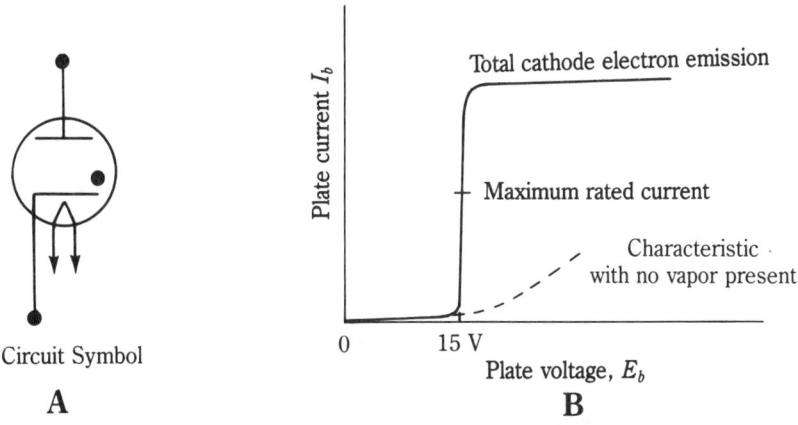

Circuit Symbol

A

B

3-23 Mercury vapor diode characteristic.

Under no circumstances should you allow the voltage drop across the mercury diode to exceed 22 V, or excessive tube dissipation will occur. In addition, the bombardment of the cathode by positive ions will disintegrate the emitting surface. For this reason the cathode must be operating at its normal temperature before you can apply the ac voltage to be rectified. Consequently the mercury vapor diode requires an initial warm-up period, which can range from 1/2 minute up to several minutes. The cathode itself can either be directly or indirectly heated.

Compared with the hard diode, the mercury vapor type has the advantage of a lower voltage drop, which does not change appreciably when plate current is flowing. The lower voltage improves the degree of voltage regulation in rectifier circuits where large changes in the load current are involved. Apart from the warm-up period, the disadvantages of the mercury vapor diode are a lower peak inverse voltage rating at which arc-back will occur, and the generation of rf interference as a result of the ionization. To reduce this interference will require shielding of the tube and filtering of the rectifier's output.

The silicon diode has tended to supersede the mercury vapor diode as a high-power rectifying device. The silicon diode is capable of passing high currents with a voltage drop of less than 1 V, so its power dissipation is low.

The voltage-regulator tube

The voltage-regulator (VR) tube is basically a cold cathode diode that contains an inert gas, such as neon, at low pressure. The plate/cathode voltage at which the gas will ionize is called the *strike voltage*, and its value depends on the gas pressure and the cathode material.

A typical characteristic for a VR tube is shown in Fig. 3-24A. Notice that the tube's strike voltage is greater than the normal operating voltage; this is because the ionization process is self-sustaining at a voltage that is lower than the value required to ignite the tube. As an example, a VR tube strike voltage can be 180 V, while its operating range might be centered on 150 V; the 30 V difference occurs when the tube ignites and an increase in the voltage drop occurs across R_s (Fig. 3-24B). If the voltage across the tube is reduced to 90 V (the extinction voltage), all ionization ceases.

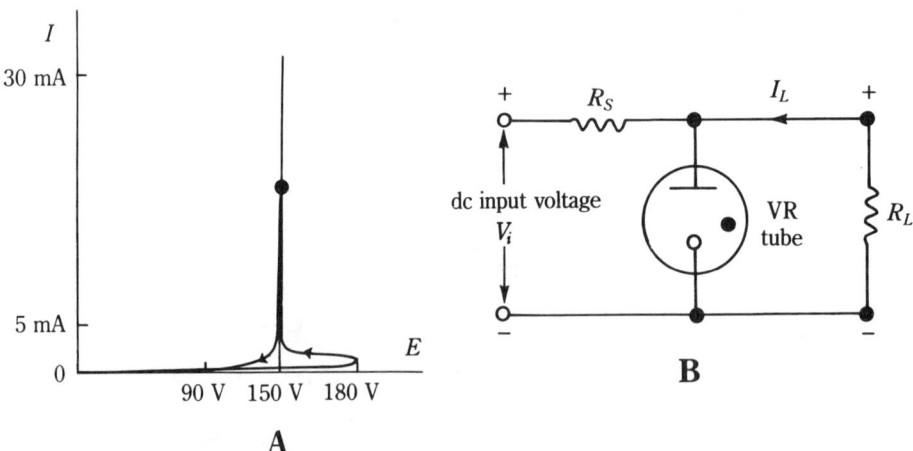

3-24 The VR tube.

Over the operating range, the voltage across the VR tube is essentially constant and is independent of the tube current. The tube can, therefore, be used to provide a voltage output that is regulated against changes in load current, I_L, and variations in the dc input voltage, V_i. A similar action can be provided by the zener diode, which is the solid-state equivalent of the VR tube.

The thyratron

The *thyratron* is equivalent to a gas-filled (for example, argon, mercury vapor) triode that can be used as a relay in control circuits. It is operated at low frequencies (less than 50 kHz), and its solid-state equivalent is the SCR.

The action of the thyratron's control electrode is similar in part to that of the control grid in the "hard" triode. By means of a negative bias, the control electrode can be held beyond cutoff so there will be no plate current and no ionization will occur. For a given positive voltage on the plate, the negative grid voltage can be reduced to the point at which ionization occurs (Fig. 3-25). As an example, the thyratron might fire at E_b = +200 V, E_c = -14V. If E_b were lowered to +150 V, the corresponding value of E_c could be reduced to -12 V. The ratio $\Delta E_b / \Delta E_c$ is called the *thyratron's control factor*, which in our example is (200 - 150)/(14 - 12) = 25. The purpose of the control grid is to determine the value of the plate voltage at which the gas will ionize.

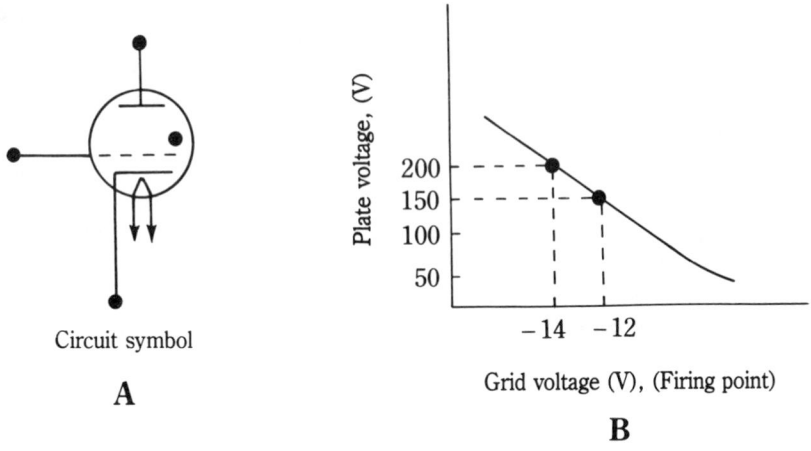

Circuit symbol

A

Grid voltage (V), (Firing point)

B

3-25   The thyratron tube.

Once the thyratron has fired, the positive ions neutralize the space-charge effect of the electrons. The result is an increase in plate current and a reduction in voltage across the tube. The grid has now lost all control over the plate current, which can only be extinguished by one of three methods:

1. Reducing the plate voltage below the extinction level

2. Reducing the plate voltage to zero

3. Reversing the polarity of the plate voltage

Ultrahigh frequency circuit limitations

As the frequency to be amplified increases, the voltage gain that can be achieved through the use of conventional electron tubes decreases until a frequency is reached where such gain is unity. The reduction of gain is caused by dielectric losses, finite values of lead inductance, interelectrode capacitances, and the effect of *transit time* (the time taken for electrons to cross from the cathode to the plate).

Dielectric losses have been decreased by tube designs that confine the dielectric material (bases, insulators, and envelopes) to positions of the tube where the dielectric stresses are minimized, and by the use of dielectric materials with the lowest possible hysteresis loss.

As the operating frequency is increased, the inductances and capacitances inherent in the tube structure become an increasing portion of the tuned circuit of an amplifier stage. This effect continues until, for all practical purposes, the external tuned circuit disappears or is obliterated by the tube and no further tuning is possible at this limiting frequency. In addition, as the operating frequency is increased, even the reactances of relatively short leads within the tube become great enough to decrease the size of the actual input signal applied to the tube's electrode.

The transit time effect is reduced by decreasing the interelectrode spacing in the tube and increasing the plate voltage. However, decreasing the interelectrode spacing increases the interelectrode capacitances; therefore, electrode dimensions must be decreased in order to maintain a low capacitance. Reducing the size of the electrodes reduces the heat-dissipation capabilities of the tube. However, the use of heatsinks will allow the operation of relatively small electrode triode tubes at dissipation levels that permit an appreciable output signal.

The effects of lead inductance, and also skin effect, must be minimized in tubes designed for uhf operation. This minimization is accomplished by the use of large-diameter leads, multiple strand leads, and *planar element construction* (the arrangement of cathode, grid, and plate in parallel planes). This type of construction allows connections to external circuitry to be made around the periphery of contact disks or over the entire surface of cylinders. Figure 3-26A illustrates a lighthouse triode using planar construction. Figure 3-26B shows the same tube with construction details. The name *lighthouse* is derived from the tube's physical appearance.

In a typical lighthouse triode, the separation between the cathode and the plate is approximately 1 mm. Its configuration allows the tube to be inserted into tunable coaxial lines that form a part of the external circuitry. All rf connections ar made to the sleeve, disk, and cap, which connect to or support the cathode, grid, and plate, respectively. This connection allows low-inductance leads, which when combined with the close spacing and low interelectrode capacitances make the tubes useful at uhf up to approximately 2 GHz. The plate is relatively small, and connection to a heatsink is required to take advantage of the rated plate dissipation.

Multiple-choice questions

1. For a conventional vacuum tube used in the uhf band,

    A. The electron transit time becomes critical.
    B. The distance between the control grid and the plate must be increased.
    C. The physical size of the tube must be increased.
    D. Only a pentode can be used because of noise effects.
    E. The lead inductance must be increased.

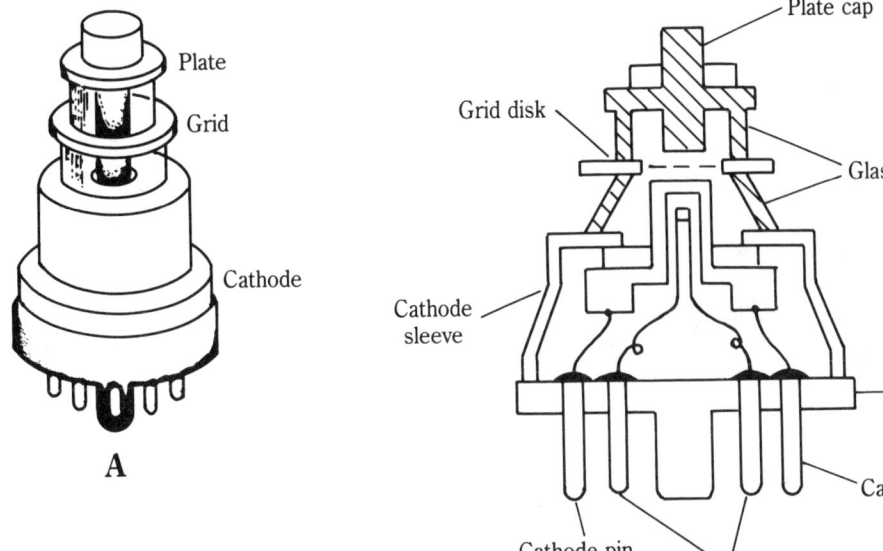

A

3-26 The lighthouse triode.

B

2. When a thyratron tube has fired, one thing that will cause it to stop conducting is:

   A. A more positive voltage on the plate.
   B. A more negative voltage on the control electrode.
   C. A more positive voltage on the control electrode.
   D. A negative voltage on the plate.
   E. The thyratron cannot be stopped unless the grid circuit is opened.

3. A tube tester is used to check a triode's transconductance, which is the ratio of:

   A. A small change in cathode current to the corresponding small change in grid current.
   B. A small change in plate current to the corresponding small change in grid voltage.
   C. A small change in plate current to the corresponding small change in grid current.
   D. A small change in plate voltage to the corresponding small change in plate current.
   E. A small change in plate voltage to the corresponding small change in grid voltage.

4. What would cause the plate current to increase in a pentode tube?

   A. A short circuit between the plate and the screen grid.
   B. An open circuit in the lead that is connected to the control grid.
   C. A short circuit between the suppressor grid and the screen grid.
   D. A short circuit between the suppressor grid and the cathode.
   E. A short circuit between the control grid and the cathode.

5. The symbol shown in Fig. 3-27 is that of a:

   A. Cold cathode gas triode.
   B. Thyratron.
   C. Lighthouse triode.
   D. Mercury vapor diode.
   E. VR tube.

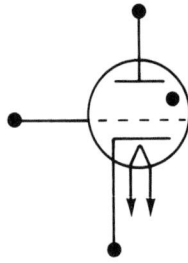

3-27   Symbol for multiple-choice question 5.

6. If a triode's amplification factor, μ, is equal to 20,

   A. The output signal voltage is twenty times the input signal voltage.
   B. The voltage gain is greater than 20.
   C. The small change in the plate current is twenty times the corresponding small change in the grid current.
   D. The small change in the plate voltage is twenty times the corresponding change in the grid voltage (to maintain the same level of the plate current).
   E. r_p/g_m is also equal to 20.

7. Why should the polarity of the filament voltage of a high power vacuum be reversed periodically when a dc filament supply is used?

   A. To distribute filament use uniformly over the active area of the filament.
   B. To lengthen the life of the filament.
   C. To reduce the plate dissipation to a minimum.
   D. To rejuvenate the emitting surface.
   E. Both A and B.

8. When an amplifier with a resistive load has a voltage gain of 50,

   A. The tube's amplification factor is equal to 50.
   B. The output signal voltage from the plate is 50 times the input signal voltage to the control grid.
   C. The tube's μ is less than 50.
   D. $\mu\, r_p/(r_p + R_L)$ is equal to 50.
   E. Both B and D.

9. The screen grid is used to:

   A. Increase the capacitance between the control grid and the plate.

B. Decrease the capacitance between the control grid and the plate.

C. Reduce the secondary emission effect.

D. Lower the tube's plate resistance.

E. Decrease the capacitance between the control grid and the cathode.

10. For maximum power transfer, the load of an amplifier should be:

    A. Ten or more times the r_p.

    B. Equal to the r_p.

    C. Five times the r_p.

    D. One-half of the r_p.

    E. Less than one-fifth the r_p.

11. The indirectly heated cathode of a diode is coated with:

    A. Thoriated tungsten.

    B. Nickel.

    C. Carbon.

    D. Strontium or barium oxide.

    E. Silicon or germanium.

12. What is a pentagrid converter?

    A. A tube with a total of five electrodes.

    B. A tube with a total of five grids.

    C. A tube that can be used for frequency conversion.

    D. A tube that requires twice as much plate voltage as a single triode.

    E. Both B and C.

13. In uhf equipment, lighthouse triodes are used because:

    A. Their interelectrode capacitance is maximized.

    B. Their electrodes are closer together than in conventional tubes.

    C. They have higher lead inductances.

    D. They dissipate more power than conventional tubes.

    E. All the above are true.

14. What is the purpose of an ''A'' battery in a vacuum tube circuit?

    A. The bias battery for the control grid.

    B. The voltage supply for the suppressor grid.

    C. The power supply for the plate circuit.

    D. The filament battery.

    E. The emergency battery.

15. Which of the following would have the most effect on decreasing the life of a vacuum tube?

    A. Too much of a grid excitation.

   B. An excessive filament voltage.
   C. A grid current that is too low.
   D. A plate resistance value that is too high.
   E. A B+ voltage that is too low.

16. Uhf tubes, such as lighthouse triodes,

   A. Are used in power supplies to improve the voltage regulation.
   B. Have special filaments to minimize their inductive effect.
   C. Take advantage of the electron transit time to achieve amplification.
   D. Always have a metal envelope to provide shielding.
   E. Have their lead inductance minimized.

17. What is one advantage of a pentode tube over a triode?

   A. Lower input impedance.
   B. Lower output impedance.
   C. Less noise internally generated.
   D. Lower plate resistance.
   E. Less control grid to plate capacitance.

18. A tube manual shows a tube to have a maximum plate dissipation of 12 watts. This means that:

   A. The input power should not be greater than 12 watts.
   B. The output power is 12 watts.
   C. The maximum power that the tube can safely dissipate is 12 watts.
   D. The output power is greater than 12 watts.
   E. The minimum power input must exceed 12 watts.

19. The triode's ac plate resistance, r_p, is the ratio of:

   A. A small change in plate current, ΔI_p, to the corresponding small change in plate voltage, ΔE_p.
   B. A small change in plate voltage, ΔE_p, to the corresponding small change in grid voltage ΔE_g.
   C. A small change in plate current, ΔI_p, to the corresponding small change in grid voltage, ΔE_g.
   D. The transconductance, g_m, to the amplification factor, μ.
   E. A small change in plate voltage, ΔE_p, to the corresponding small change in the plate current, ΔI_p.

20. The "B" battery in a vacuum tube circuit is:

   A. The filament battery.
   B. The bias for the control grid circuit.
   C. The supply voltage for the suppressor grid circuit.
   D. The power supply for the plate circuit.
   E. Used to apply a positive voltage to the control grid.

21. In a beam power tube, the deflection plates are used to:

    A. Neutralize the amplifier that employs the tube.
    B. Reduce the interelectrode capacitance between the control grid and the plate.
    C. Increase the screen grid current.
    D. Concentrate the primary stream of electrons.
    E. Deflect the electron beam away from the surface of the plate.

22. One advantage of the mercury vapor diode over the high vacuum rectifier is:

    A. Its higher peak inverse voltage rating.
    B. Its reduced rf interference effect.
    C. Its lower voltage drop when the plate current is flowing.
    D. The elimination of the need for a warmup period.
    E. The reduction in the ionic bombardment of the cathode.

23. The cold cathode VR tube:

    A. Uses a low-voltage heater circuit.
    B. Is used to counteract the effects of large relative changes in the load current.
    C. Is used to counteract the effect of large relative changes in the dc input voltage.
    D. Contains gas or vapor at high pressure.
    E. Both B and C.

24. The triode's amplification factor, μ, is:

    A. The product of ac plate resistance, r_p and the transconductance, g_m.
    B. The ratio of the small change in plate current to the corresponding small change in grid current.
    C. The ratio of the small change in plate voltage to the corresponding small change in plate current.
    D. The ratio of the small change in plate voltage to the corresponding small change in grid voltage.
    E. Both A and D.

25. The "C" battery is used:

    A. To supply dc power to the plate circuit.
    B. To apply a negative dc bias to the control grid relative to the cathode.
    C. To apply a negative dc bias to the cathode relative to the control grid.
    D. To supply a heater voltage to the cathode.
    E. To apply a positive potential to the control grid.

26. In an amplifier employing a beam power tube, the control grid current is 5 mA, the plate current is 60 mA, and the screen grid current is 10 mA. The cathode current is:

    A. 60 mA.
    B. 70 mA.

C. 15 mA.
D. 75 mA.
E. 65 mA.

27. In Fig. 3-28, the voltage at the plate is:

A. +200 V.
B. + 2000 V.
C. +2800 V.
D. +1800 V.
E. +600 V.

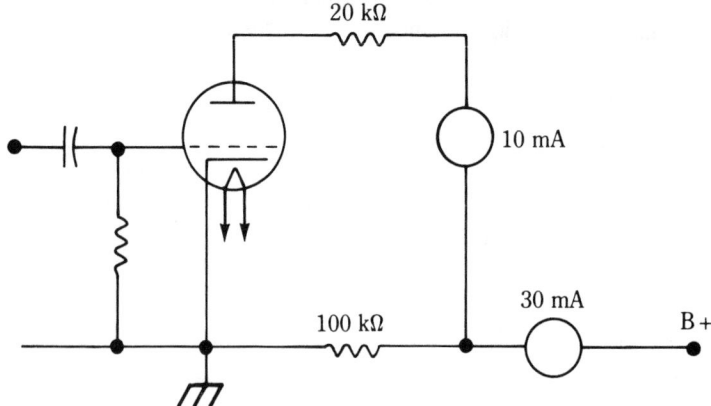

3-28   Circuit for multiple-choice question 27.

28. In a triode voltage amplifier, the ratio of R_L to r_p is:

A. More than 5:1.
B. Less than 1.2:1.
C. One to one.
D. As low as possible.
E. Extremely critical.

29. A pentode amplifier circuit uses a dropping resistor to supply the correct voltage to the screen grid. The plate supply voltage is 250 V and the required voltage at the screen grid is 200 V. If the plate current is 45 mA and the screen-grid current is 5 mA, the required value for the dropping resistor is:

A. 10 kΩ.
B. 50 kΩ.
C. 40 kΩ.
D. 4 kΩ.
E. 5 kΩ.

30. The life span of a vacuum tube will not be reduced by excessive:

    A. Plate current.
    B. Screen current.
    C. Plate voltage.
    D. Grid bias.
    E. Filament voltage.

Basic problems

1. a. For a particular diode the plate current is 15 mA when the plate/cathode voltage is 10 V. What is the value of the diode's dc resistance?

    b. If the voltage is reduced to 9.5 V, the diode current falls to 14 mA. What is the diode's ac plate resistance?

2. While maintaining the control grid voltage constant, the plate voltage of a triode tube is shifted by 20 V. If the corresponding change in plate current is 2.5 mA, what is the value of the triode's ac plate resistance?

3. The plate/cathode voltage of a triode is maintained at 220 V. When the control grid/cathode voltage is changed from -2 V to -3.5 V, the plate current falls by 4.5 mA. What is the value of the transconductance in microsiemens?

4. The plate current of a triode is 10 mA when the plate/cathode voltage is 200 V and the control grid bias is -3 V. The plate current remains at 10 mA when the plate/cathode voltage is lowered to 175 V and the bias is changed to -1.5 V. What is the value of the tube's amplification factor?

5. At a particular operating point a triode has an amplification factor of 25 and a transconductance of 5000 μS. What is the value of the ac plate resistance? If the plate supply voltage is 200 V, what is the triode's cutoff bias?

6. A triode amplifier operates with the following conditions: load resistance 20 kΩ, ac plate resistance 5 kΩ, grid bias -4 V, plate current 6 mA, plate supply voltage $+250$ V, and amplification factor 20. What is the voltage gain of the amplifier?

7. In Problem 5 the input signal is a sinewave of 2 V rms. Draw the waveform of the output voltage at the plate of the triode, indicating the voltage levels.

8. A pentode voltage amplifier operates with the following conditions: cathode current 8 mA, plate current 6.5 mA, plate supply voltage $+200$ V, and screen grid voltage $+80$ V. Calculate the required value for the dropping resistor connected to the screen grid.

9. A pentode amplifier has a 27 kΩ load resistor, a control grid bias of -3 V, and a transconductance of 3500 μS. What is the voltage gain of the amplifier?

10. In Fig. 3-24 the voltage across the VR tube is 150 V and its current is 18 mA. If R_L = 75 kΩ and R_s = 10 kΩ, calculate the value of V_i.

Advanced problems

1. A triode's operating point is defined by E_b = 250 V, I_b = 8 mA, and E_c = −5 V. When the plate voltage is changed from 240 V to 260 V, while maintaining the grid at −5 V, the plate current varies between 7.4 and 8.6 mA. What are the values of the triode's dc resistance and ac plate resistance at the operating point?

2. In Problem 1, the plate voltage is held constant at +250 V. When the control grid voltage is shifted from −4.8 V to 5.2 V, the change in the plate current is 1.0 mA. What are the values of the triode's transconductance and the amplification factor at the operating point?

3. The operating point of a triode is defined by E_b = +300 V, I_b = 12 mA, and E_c = −8 V. The value of the amplification factor is 20. If the grid voltage is changed from −8 V to −8.5 V, to what value must the plate voltage be shifted in order to restore the plate current to 12 mA? What is the triode's cutoff bias under these conditions?

4. Figure 3-29 illustrates a triode amplifier with a resistive load. If the dc plate current is 10 mA, what is the value of the plate-to-cathode voltage? Calculate the dc resistance of the tube and its power dissipation.

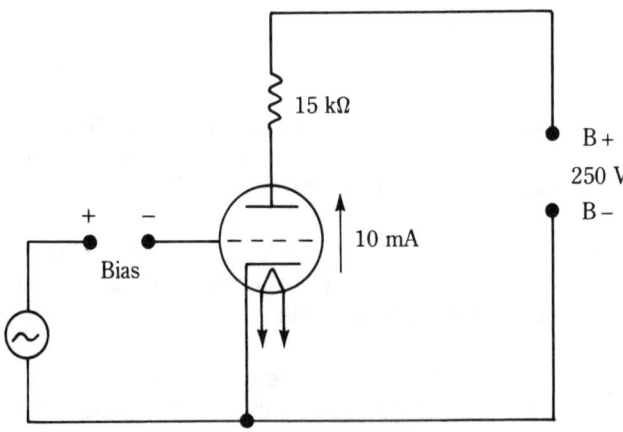

3-29   Circuit for advanced problem 4.

5. In Fig. 3-30 the triode's amplification factor is 22 and its ac plate resistance is 10 kΩ. What is the voltage gain of the amplifier? If the input signal is a sine wave whose rms value is 1.5 V, what is the peak-to-peak value of the voltage fluctuation at the plate?

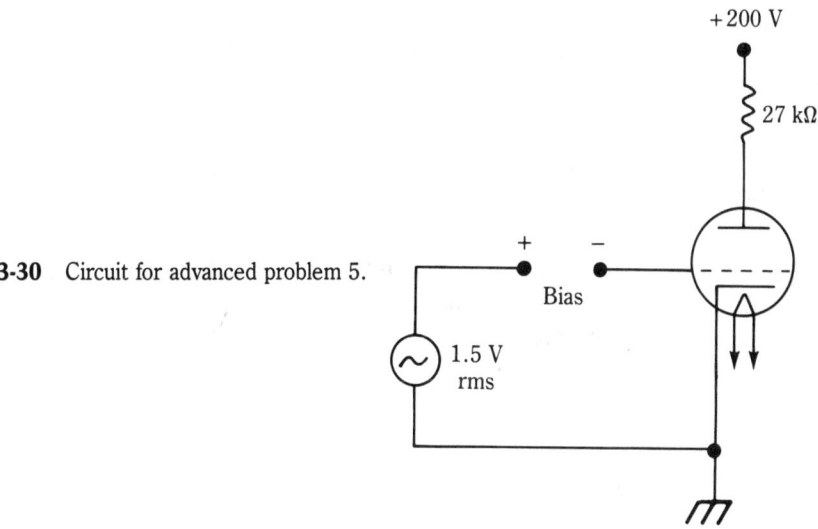

3-30 Circuit for advanced problem 5.

6. Figure 3-31 illustrates a pentode voltage amplifier. If the plate current is 3 mA and the screen current is 0.5 mA, calculate the values of the plate potential and the screen potential. If the pentode's transconductance is 2500 μS, what is the voltage gain of the amplifier?

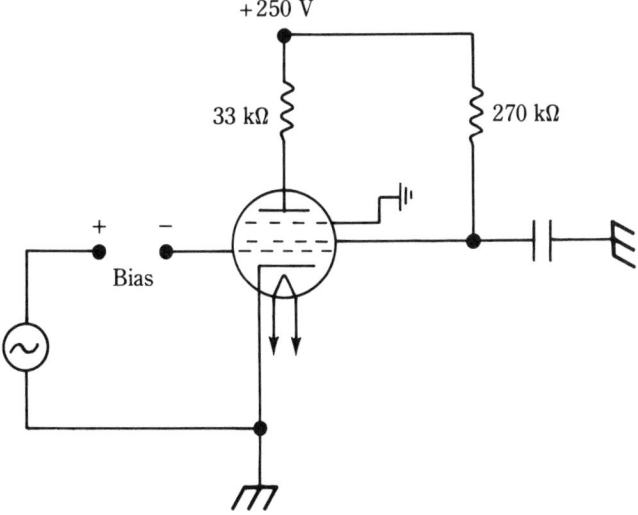

3-31 Circuit for advanced problem 6.

7. In Problem 6, the input signal is an audio tone. If the peak-to-peak value of the sine-wave voltage swing at the plate under signal conditions is 125 V, what is the rms value of the input signal on the control grid?

8. In Fig. 3-10, determine the value of the transconductance at the operating point $E_c = -3$ V and $I_b = 4.25$ mA. What is the value of the amplification factor at this same point?

9. In Fig. 3-21 determine the values of the transconductance at the points $E_b = +250$ V, $E_{c2} = +150$ V, and $E_{c1} = -5$ V, and $E_b = +250$ V, $E_{c2} = +150$ V, and $E_{c1} = -15$ V.

10. A particular thyratron fires at $E_b = +220$ V and $E_c = -12$ V. If E_c is increased to -14 V, and the thyratron's control factor is 20, what is the new value of E_b for which the thyratron will fire?

4

Basic amplifiers

AMPLIFIERS PLAY AN IMPORTANT ROLE IN MANY PIECES OF ELECTRONICS communications equipment. *Amplifiers*, as the term is used in this book, are electronic assemblies, containing one or more active devices. The purpose of amplifiers is to provide amplification of voltage, current, or power. Although the distinctions among these three classifications might be blurred on occasion, most amplifiers are easily identified by class.

This chapter introduces basic amplifiers. Subsequent chapters will not only extend the amplifier concepts, but also introduce specialized amplifier applications in communications equipment. The predominant interest is in solid-state amplifiers, which continue to increase in their range of applications. However, because the vacuum tube continues to serve in the communications field, a section has been included on vacuum tube amplifiers in the interest of completeness.

In this chapter the following additional sections will be found:

- Single-stage amplifiers
- Bias considerations
- Class-A operation
- Push-pull amplifiers
- Negative feedback
- Operational amplifiers
- Basic vacuum-tube amplifiers

Single-stage amplifiers

In Fig. 4-1 is a simplified schematic diagram of a common emitter (CE), junction transistor amplifier. It has voltage gain given by:

$$A_V = \beta R_L / (R_S + r_\pi)$$

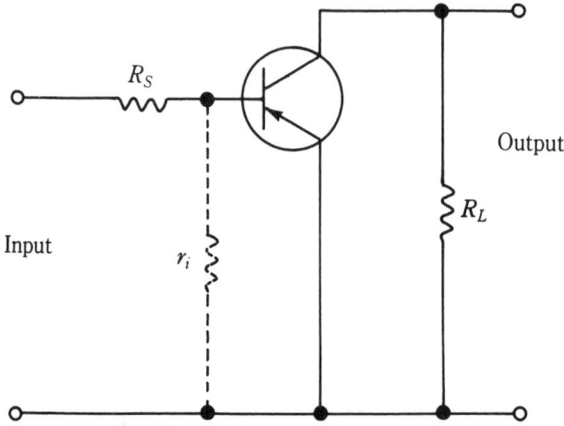

4-1 A simplified schematic of a CE amplifier with source resistance.

where

β = the ratio of collector current change to base current change
r_π = the resistance looking into the base terminal
R_S = the internal source resistance

Another amplifier configuration, the common-base (CB), is represented by the schematic diagram in Fig. 4-2. Its voltage gain is:

$$A_V = \beta R_L / [(1 + \beta)R_S + r_\pi]$$

Still another amplifier arrangement is called the common-collector (CC) or emitter follower. Its schematic is shown in Fig. 4-3 for which the voltage gain is:

$$A_V = (1 + \beta)R_L / [R_S + r_\pi + (1 + \beta)R_L]$$

Each of these three amplifier configurations has particular properties that determine its choice for a given application. Voltage gain, current gain, power gain, input impedance, and output impedance can be significantly different for each application, which permits the designer to select the configuration that most effectively meets the performance requirements for an amplifier stage. Table 4-1 summarizes some of these properties.

Table 4-1. A comparison of transistor amplifiers.

Characteristic	CB	CE	CC
Current gain (A_i)	Less than one	High	High
Voltage gain (A_v)	High	High	Less than one
Power gain (A_p)	Medium: 500	High	Low
Input resistance	Low: 30-150 Ω	500-1500 Ω	High: 20-500 kΩ
Output resistance	High: 300-500 kΩ	30-50 kΩ	Low: 50-1000 Ω
Phase shift	None	180°	None
Signal input	Emitter	Base	Base
Signal output	Collector	Collector	Emitter

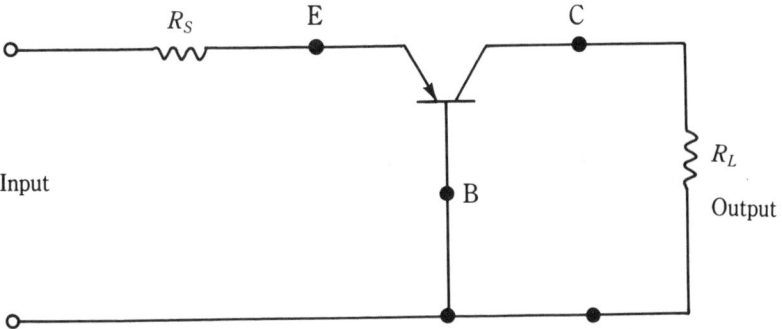

4-2 A simplified schematic of a CB amplifier with source resistance.

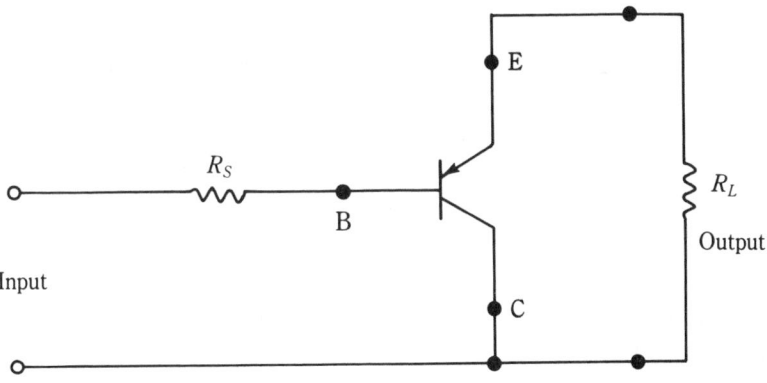

4-3 A simplified schematic of a CC amplifier with source resistance.

The simplified circuits presented thus far neglect a number of important practicalities. For example, it is necessary to establish dc bias currents within the transistor and to ensure that the bias currents retain satisfactory values under conditions of temperature change, transistor aging, substitution of production transistors whose parameters only meet rather loose specifications, etc. Additionally, as the frequency of the signal being amplified is increased, capacitances (not shown in the figures) must be added and accounted for. Practical circuits will also have altered impedances at low frequencies.

Before continuing with the bias circuits for the bipolar junction transistor, briefly consider the single-stage amplifier configuration for field-effect transistors (FETs). Three different circuit arrangements are possible for the FET. These arrangements are easily drawn by substituting the symbols of the gate for the base, the source for the emitter, and the drain for the collector, using the BJT circuits introduced previously. You must realize that the many similarities resulting from this approach mask a larger number of differences that become more apparent upon closer examination.

Bias considerations

A transistor amplifier must contain additional resistors in the circuit to establish a proper quiescent (Q) operating point. This point is superimposed on the static characteristics of

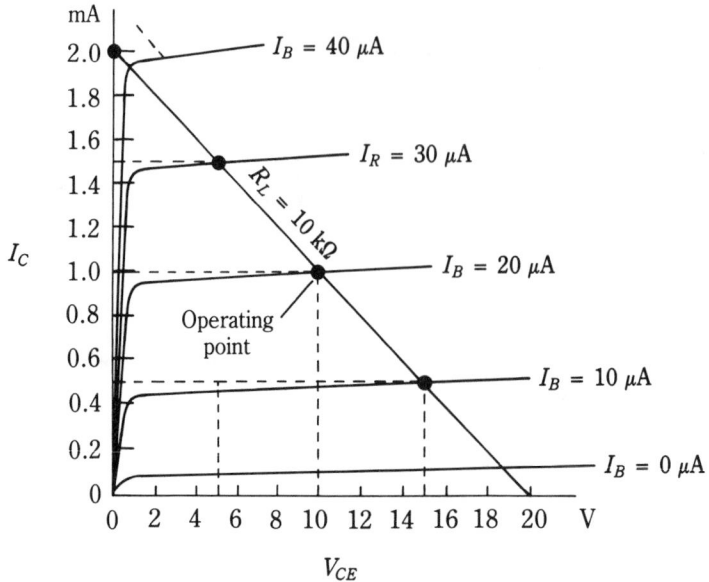

4-4   The operating point for a CE amplifier.

Fig. 4-4. The Q point lies at the intersection of the sloping, straight (load) line and $I_B = 20 \mu A$ characteristic. The position of the load line is established by the values of $V_{CC} = 20$ V and $R_L = 10$ kΩ.

The base current can be determined by one of several circuits. Figure 4-5 demonstrates fixed bias with $R_B = (20 - 0.7)/20 = 0.965$ MΩ. A fixed biasing arrangement, while easily understood and implemented, can result in unsatisfactory operation because of temperature change, individual transistor variations, etc. Another bias form seen in Fig. 4-6 establishes an operating point and partially compensates for changes in I_C resulting from the same causes. This bias circuit is said to have a good *stability factor*.

The addition of the resistors to provide a stable dc operating point alters the over-simplified amplifier circuits shown earlier, so that the parameters—that is, resistance values—appearing in the gain equations must also be altered. This alteration involves the combining of the additional resistance values in parallel with those of the earlier circuits, with some degradation of gain performance. The refinement of gain calculations by this means is not warranted for the present discussion.

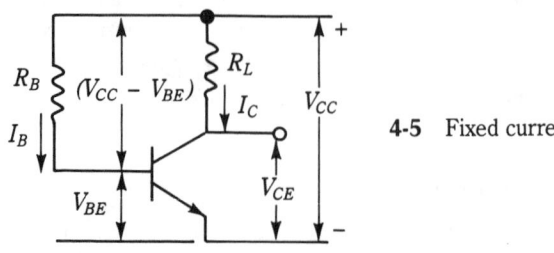

4-5   Fixed current bias.

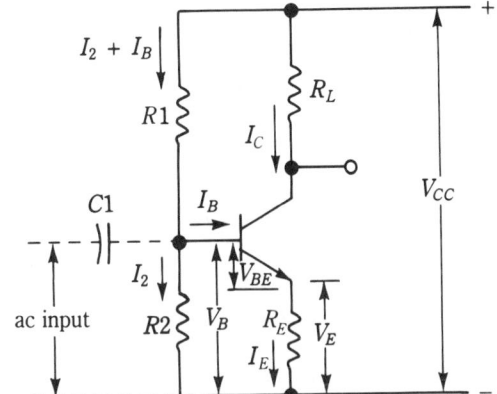

4-6 Emitter current bias.

Class-A operation

The objective of most amplifiers is to reproduce as the output the same waveform introduced at the input, except that the voltage, current, power, or impedance levels might be deliberately changed. The manner in which this takes place can be seen in Fig. 4-4.

Suppose that the base for the circuit of Fig. 4-4 is driven, such that the base current varies from 10 μA to 30 μA with a sine waveform. This variation of 20 μA, peak to peak, is centered on the dc current I_B = 20 μA, for which I_C = 1 mA and I_E = 1 mA (approximately). The ac signal voltage applied to the base can be found using r_π. The value of r_π can be calculated using r_π = 0.03 β/I_E = (0.03 × 50)/0.001 = 1500 Ω. The input wave is also sinusoidal of 20 × 1500 = 30,000 μV (peak-to-peak).

Examine the output as scaled from Fig. 4-4. The collector current varies from about 0.5 mA to 1.5 mA, a peak-to-peak variation of about 1 mA. Likewise, the output voltage from collector to emitter varies from 5 to 15 V, a 10 V, peak-to-peak sinewave variation. From these numbers, the voltage gain, A_V, is 10/0.03 = 333. Likewise, the current gain, A_I = 10^{-3}/(20 × 10^{-6}) = 50. Therefore, the power gain A_P = A_V × A_I = 333 × 50 = 16,700.

Note that the gains just calculated are for the specific purpose of demonstrating the basic principle of amplification for the common-emitter circuit. When the essential bias resistor circuits are added, the gain figures will be reduced from those previously derived.

The amplifier being discussed is a *class-A* amplifier, for which the collector current flows throughout the entire cycle. Class-A amplifiers, when properly designed, usually provide faithful, amplified reproductions of the input signal, and can be classified as high fidelity amplifiers when used as audio-frequency amplifiers. More generally, class-A amplifiers are considered to be *linear amplifiers*, which implies exact reproduction of the waveform with no distortion introduced by the amplifier.

In Fig. 4-4, the operating point was selected to allow the maximum swing of base current without distortion of the output. If the base current exceeds 40 μA, the output will be clipped, causing harmonic frequencies to be introduced (harmonic distortion). Also, simply overdriving the base with the Q point properly located or accomplishing normal operation with the Q point relocated, as might occur with a replacement transistor not having characteristics that are identical to those of the original transistor, can cause harmonic distortion.

A class-A amplifier has a relatively low theoretical maximum efficiency of 50 percent, where *efficiency* is the ratio of output signal power to input dc power. The actual efficiency might be closer to 25 percent or less for many practical class-A amplifiers.

Push-pull amplifiers

A class-B amplifier transistor is biased so that the collector current flows during one-half the signal cycle. Such an amplifier, when provided with a resonant collector circuit, can serve as a linear amplifier for an input of one frequency, the resonant frequency, only. In the absence of a resonant collector circuit, the output waveform becomes a grossly distorted version of the input. However, by placing two class-B amplifier transistors in a push-pull arrangement, each conducts nearly one-half of the cycle, and linear amplification is achieved. See Fig. 4-7 for the circuit. Note from Fig. 4-7 that a short interval of time, called the dead zone, exists when neither transistor is conducting. This result is *crossover distortion*, which can be greatly reduced by slightly biasing both transistors to conduct during the dead zone period. The circuit of Fig. 4-8 shows one method of supplying the needed additional bias.

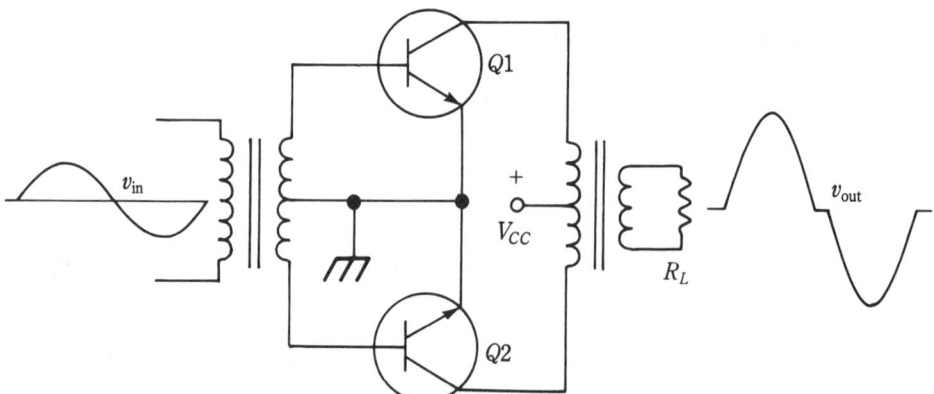

4-7   A class B push-pull amplifier and voltage waveforms.

The theoretical efficiency of the class-B, push-pull amplifier is 78.5 percent, but because of a number of practical considerations, including the need for the additional bias, an efficiency of about 50 percent is more likely.

Efficiency for the sake of saving energy is not a key feature of amplifier design. Higher efficiency is important from the standpoint of the heat-dissipation capability of the transistors, together with heatsinks and cooling equipment.

To illustrate this point, consider a superior amplifier whose efficiency is such that the heat dissipation is one-half that of a less-efficient type. An efficiency of 75 percent for the first amplifier, as compared with 50 percent for the second amplifier, exactly meets this condition. Then, if both amplifiers operate in the same temperature environment, with the same heatsinking and the same cooling arrangement, the first amplifier can

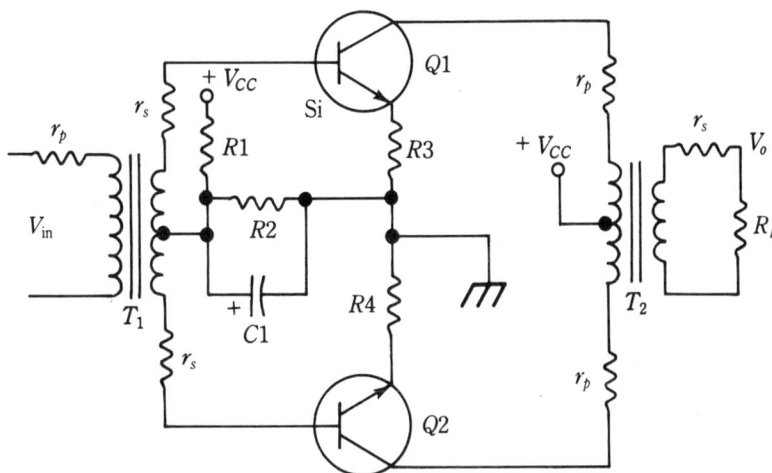

4-8 A push-pull amplifier biased to reduce crossover distortion.

deliver three times the power of the second. Expressed in another way, the power-dissipation capability of the first amplifier's transistors need only be one-third as much as that of the second amplifier's transistors for the same amount of power to be delivered. Moreover, the power supply for the first amplifier need neither be as large nor as costly. The importance of efficiency goes far beyond the cost of the energy consumed.

A phase splitter can be used to eliminate the input transformer, as shown in Fig. 4-9, in which Q2 and Q3 are VMOS field-effect transistors.

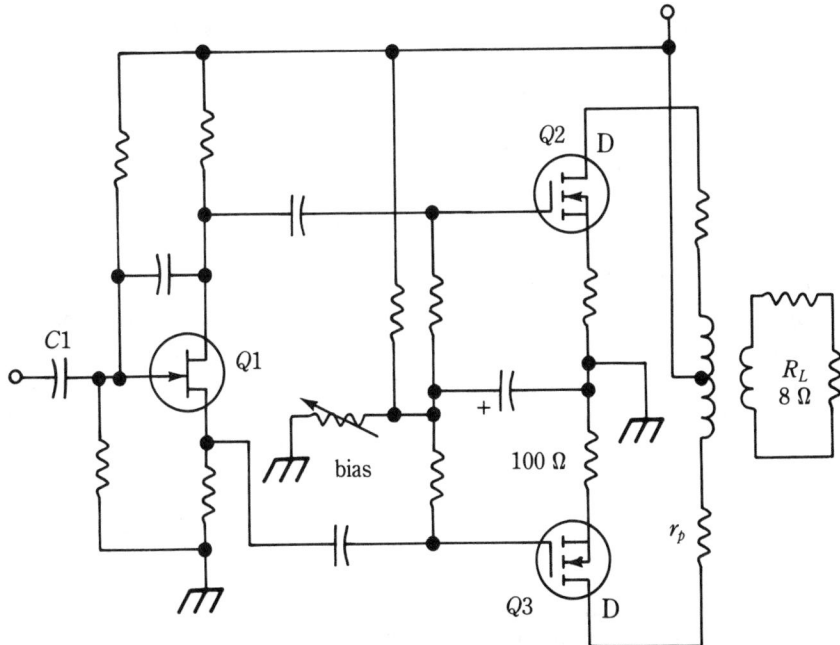

4-9 Elimination of the input transformer by means of a phase splitter.

Both transformers are eliminated by the adoption of complementary (pnp and npn) transistors, as shown in Fig. 4-10. In this circuit the base-emitter junctions of Q2 and Q3 are biased to cutoff, and the dc drop across R4 is $2V_{BE}$. With Q1 driven by an ac signal, the inverted waveform appears at the bases of the two transistors connected to each other by C2, which has low reactance at the signal frequency. Q2 and Q3 alternately conduct so that the full cycle of current passes through R_L.

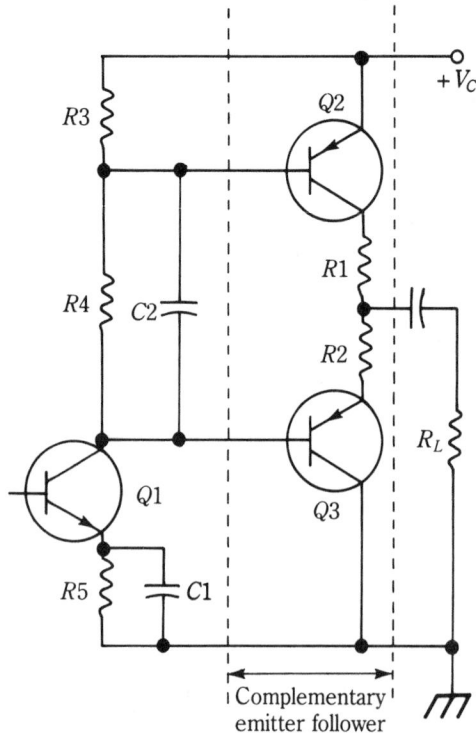

4-10   Complementary emitter follower.

Complementary emitter follower

Negative feedback

Feedback is the introduction of some fraction of the amplifier output into the input circuit. Feedback is commonly used to stabilize the dc operating point of a transistor in an amplifier circuit. The present concern is with the effect of negative feedback on certain characteristics of the amplifier: gain, input resistance, output resistance, and bandwidth.

Negative feedback can be introduced as a fraction of the output voltage placed in series with the input (voltage-series), as in Fig. 4-11A. Alternatively, the feedback voltage is proportional to the output current (current-series), as in Fig. 4-11B.

Negative feedback can also take the form of injecting current in parallel with the input current, such that the injected current is proportional to the output voltage (voltage-shunt), as in Fig. 4-12A, or to the output current (current-shunt), as in Fig. 4-12B.

A little thought reveals that negative voltage feedback (voltage-series) tends to oppose changes in the output voltage, thereby reducing the apparent output impedance

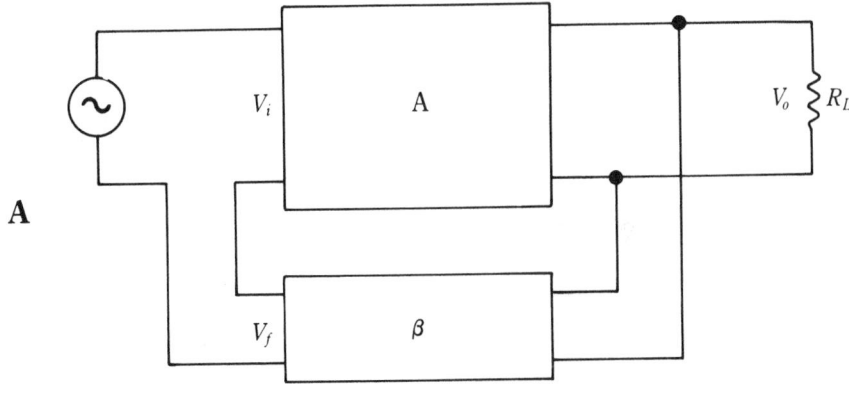

A

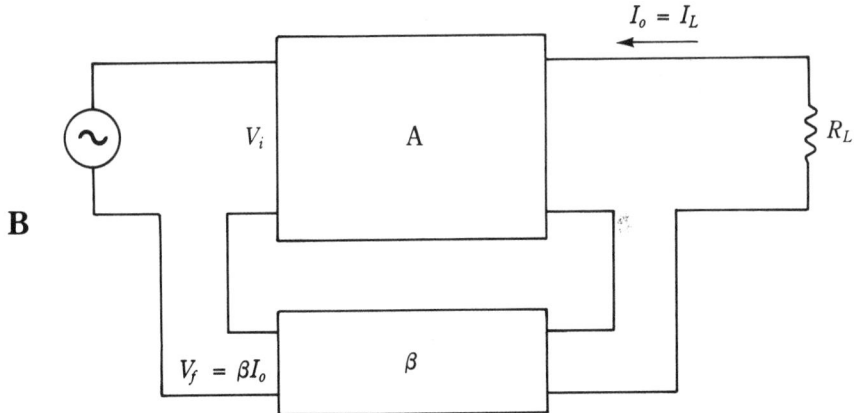

B

4-11   Negative feedback A. Voltage series. B. Current series.

of the amplifier. Moreover, this same form of feedback tends to oppose any change of amplifier current caused by a change of applied input voltage, which is the condition for which the apparent input impedance is increased. Since high input impedance and low output impedance are usually preferred by designers for the amplifier as one device in a system, the voltage-series arrangement is most often used. It will be the only configuration whose properties will be summarized here.

Concentrate on Fig. 4-11A. A is the voltage gain of the amplifier, without feedback, and β is that fraction of the output voltage, V_O, to be placed in series with the input voltage, V_I. The gain of the overall combination with feedback is then:

$$A_F = \frac{A}{1 + \beta A}$$

Moreover, the new input resistance with feedback is:

$$R_{IF} = R_I (1 + \beta A)$$

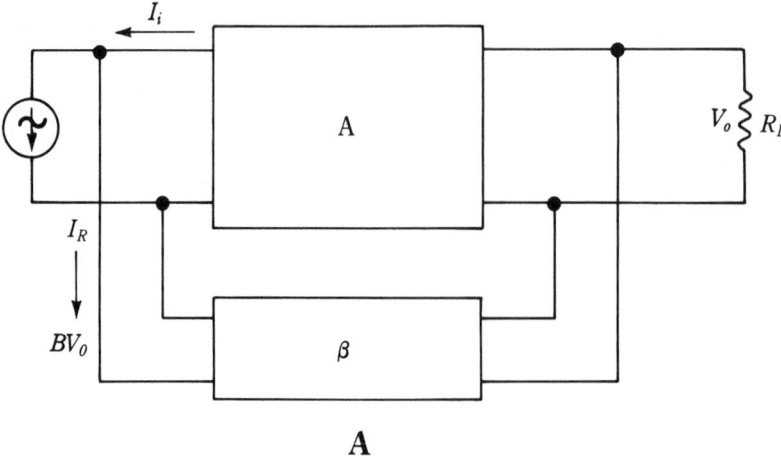

A

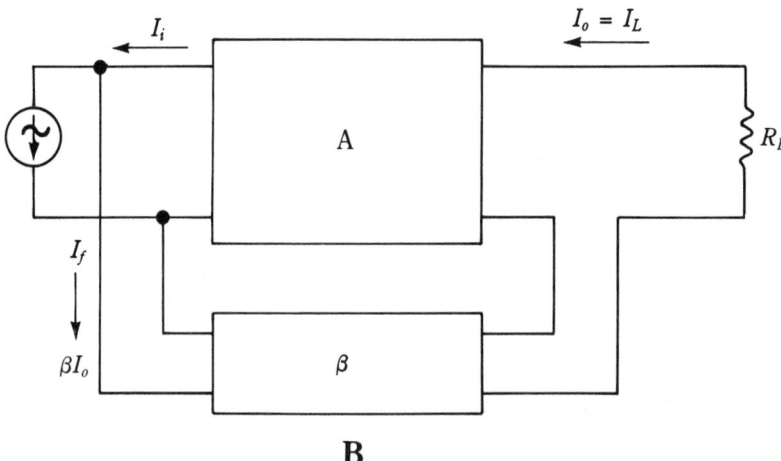

B

4-12 Negative feedback A. Voltage-shunt. B. Current-shunt.

where R_I is the input resistance of the amplifier without feedback.

Similarly, there is a different apparent output resistance, R_{OF}, with feedback, as compared with R_O, the output resistance without feedback.

$$R_{OF} = \frac{R_O}{1 + \beta A}$$

In addition, there is an increase in bandwidth such that the gain-bandwidth product remains constant. You might speak of the tradeoff of gain and bandwidth, as illustrated in Fig. 4-13. This tradeoff has even broader implications. It was already noted that A_F is inversely proportional to $1 + \beta A$. It is also true that the bandwidth is directly proportional to $1 + \beta A$. Therefore, assume A to be a fixed parameter of the amplifier; as β is deliberately changed, the gain-bandwidth product will remain constant.

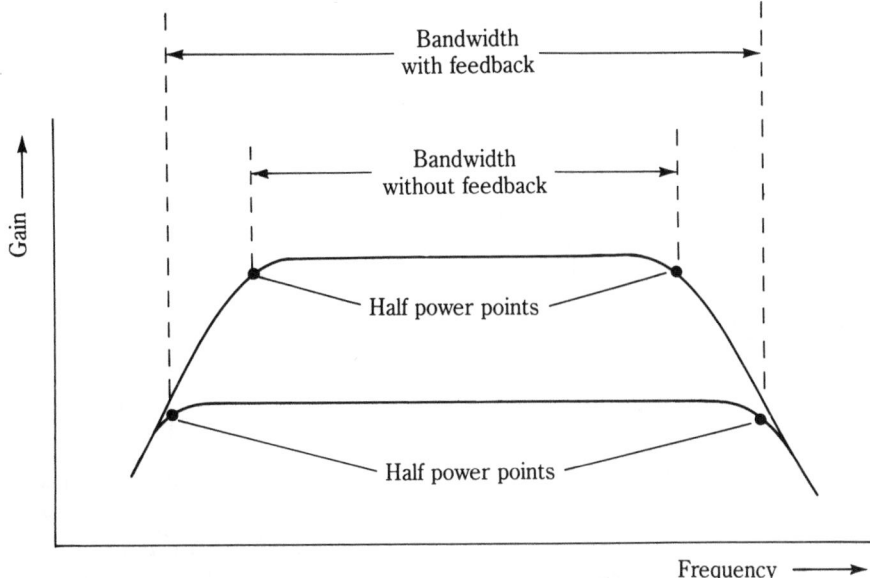

4-13   Reduction of gain and increase of bandwidth resulting from negative feedback.

Remaining constant is characteristic of operational amplifiers in their feedback amplifier (closed-loop) configuration. The operational amplifier is normally employed in this manner.

Operational amplifiers

A typical operational amplifier, or op amp, schematic diagram looks like Fig. 4-14. A double-ended power supply has $+V$ and $-V$ with respect to ground. Inputs are applied to a differential amplifier $Q1 - Q2$. The low-impedance level output at terminal 6 can be either positive or negative, depending on the relative voltage levels at the input terminals. The output voltage can swing within about 90 percent of the power supply voltages before saturation occurs. The other transistors in the circuit provide gain, establish dc voltage levels, etc.

The full schematic diagram of Fig. 4-14 is most often represented by the symbol in Fig. 4-15, which designates an inverting input terminal ($-$) and a noninverting terminal ($+$). The output is positive if the noninverting terminal ($+$) has a voltage polarity that is positive relative to the inverting terminal ($-$) and vice versa.

The op amp is seldom operated in the open-loop mode implied by Fig. 4-15. A more common, and practical, arrangement makes use of controlled feedback to establish a voltage gain given by:

$$A_V = \frac{-R_F}{R_I}$$

and is demonstrated in Fig. 4-16. The negative sign in the gain formula signifies phase inversion of the output relative to the input.

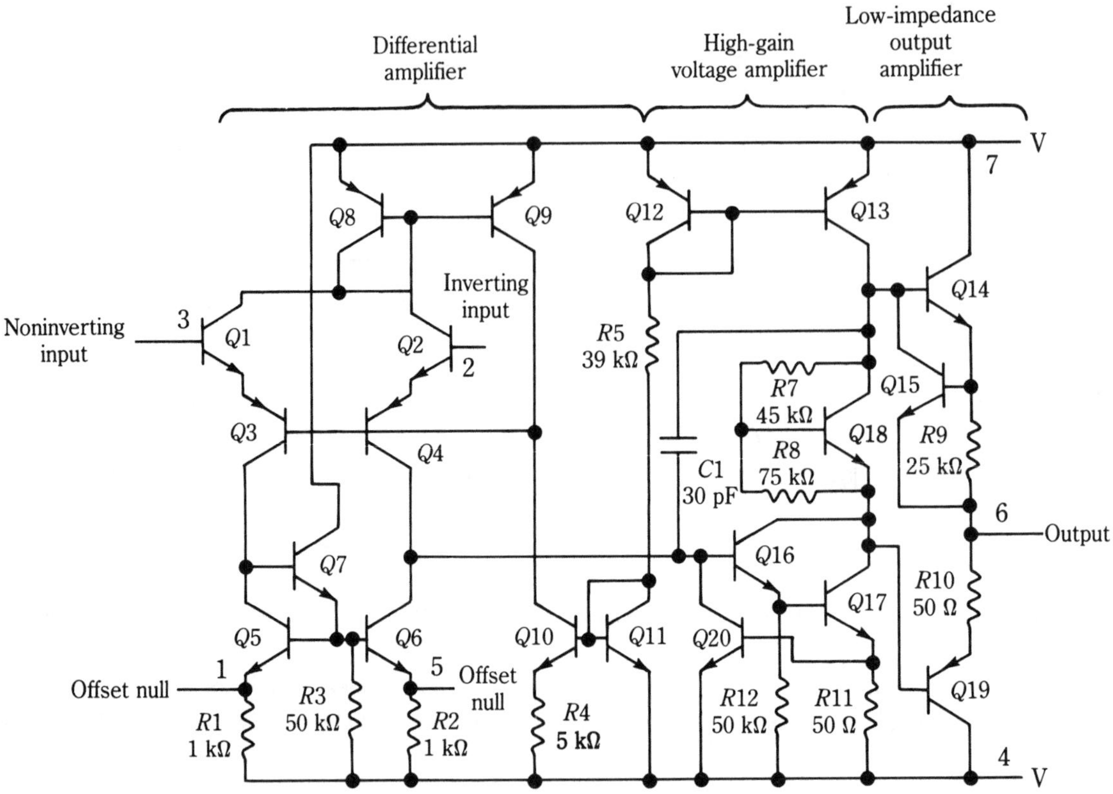

Differential amplifier

High-gain voltage amplifier

Low-impedance output amplifier

Noninverting input

Inverting input

Q8   Q9   Q12   Q13

Q1   Q2   Q14

Q3   Q4   Q15

R5 39 kΩ

R7 45 kΩ

R8 75 kΩ

C1 30 pF

R9 25 kΩ

Q18

Q16

Q17

R10 50 Ω

Q7

Q5   Q6   Q10   Q11   Q20

Output

Offset null

R3 50 kΩ

R1 1 kΩ

R2 1 kΩ

Offset null

R4 5 kΩ

R12 50 kΩ

R11 50 Ω

Q19

4-14   A typical op amp schematic diagram.

A more precise expression for gain is more complicated and includes the value of the open-loop gain. For large values of the open-loop gain, the simple formula gives a very satisfactory approximation that depends only on the values of the passive elements R_F and R_I. The input impedance of Fig. 4-16 is approximately R_I, and the output impedance is very low, its value depending on the resistances in the driver circuits of Fig. 4-14.

A noninverting alternative amplifier configuration is illustrated in Fig. 4-17. The approximate voltage gain is:

$$A_v = 1 + \frac{R_2}{R_1}$$

and no phase inversion is experienced.

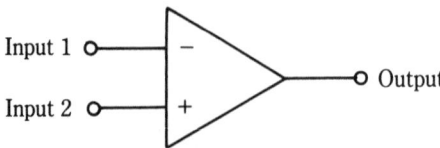

Input 1   —
Input 2   +   Output

4-15   The commonly-used symbol for op amps.

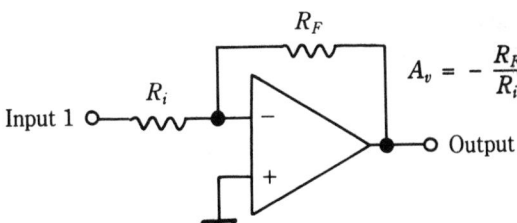

4-16 Inverting, negative feedback configuration for op amps.

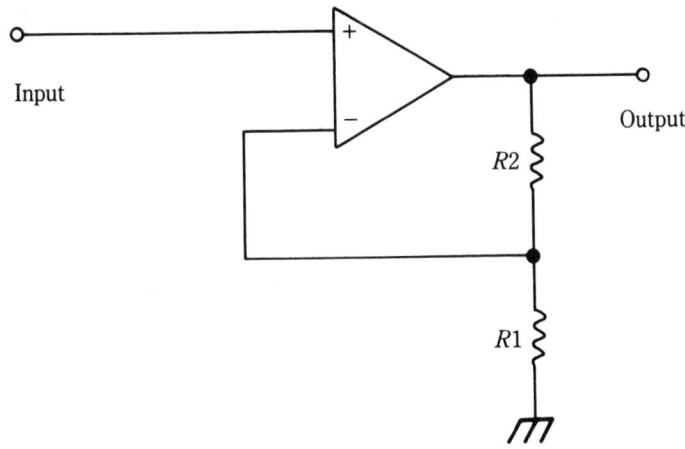

4-17 Noninverting op amp configuration.

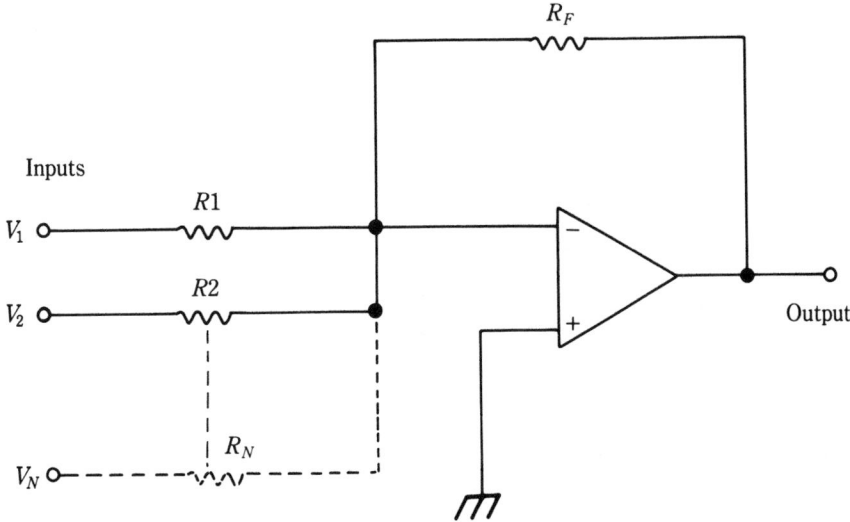

4-18 Summing amplifier.

A number, N, of input voltages can be summed linearly with the circuit of Fig. 4-18, shown for only three inputs. The output is:

$$V_{OUT} = -R_F (V_1/R_1 + V_2/R_2 + ... + V_N/R_N)$$

in which the input voltages are V_1, V_2,...V_N. These voltages can be dc of either polarity or any combination of dc, ac, and any other waveform, periodic or not. The only limitation imposed is that the op amp must not be forced into saturation at any time.

Mathematical operations are possible using the op amp in conjunction with the proper passive elements. Figure 4-19 is the basic circuit for a differentiator, shown here with a triangular wave input. By way of comparison, Fig. 4-20 is a practical circuit for an integrator, which is showing a square-wave input. The inverse natures of differentiation and integration are emphasized by the selection of input waveforms in these two figures.

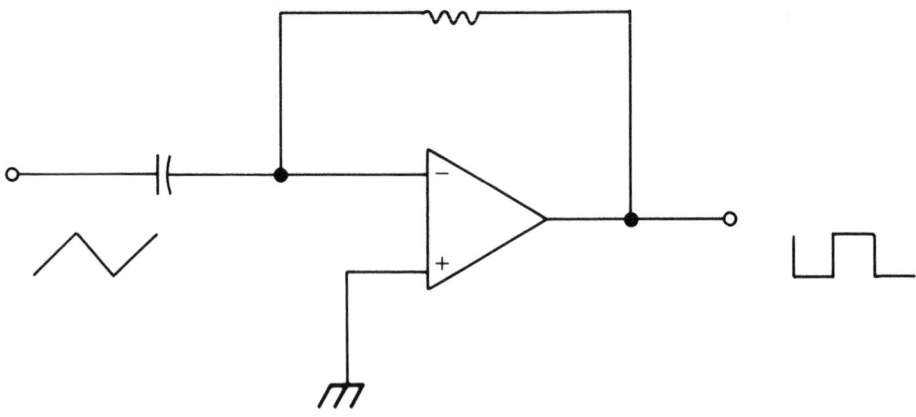

4-19 Differentiator.

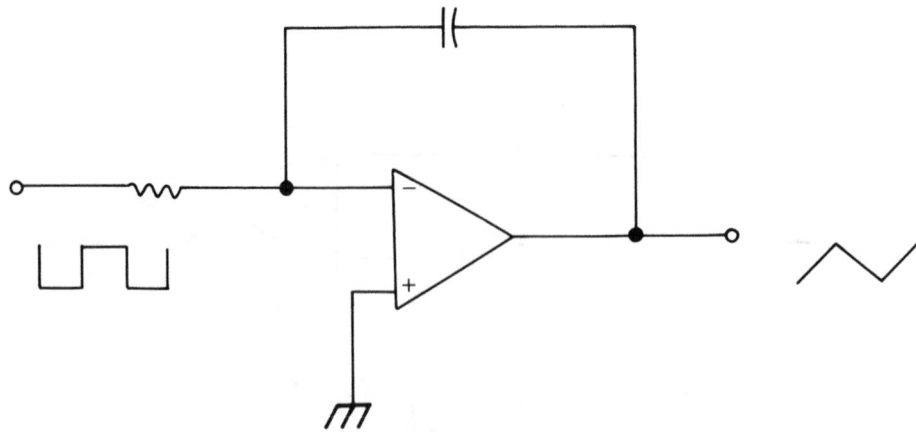

4-20 Integrator.

Basic vacuum-tube amplifiers

The effect of a load applied to a triode (and other vacuum-tubes) can be predicted by adding a graphical representation of the load to the static plate characteristics shown in Fig. 4-21. When the graph is completed, the distribution of the plate voltage supply, E_{bb}, between the load and the internal resistance of the tube under different conditions of plate current will be presented.

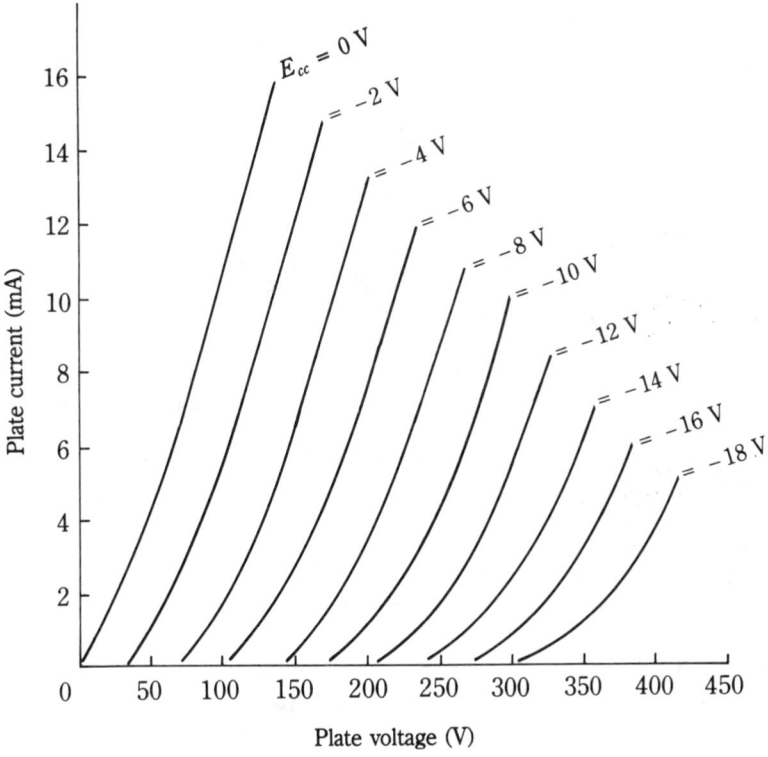

4-21 Plate family of characteristic curves for 6J5 triode.

Figure 4-22 is the same as Fig. 4-21, except for the addition of the load line, which corresponds to a selection of 25,000 Ω for R_L. Note that this selection is flexible, except that there are practical lower and upper limits for R_L beyond which the amplifying action of the circuit is considered to be defective because of low gain, distortion in the output, etc.

The extremities of the load line correspond to two particular points. With no current flow, the drop across the load resistance is zero, so the entire amount of E_{bb} (350 V) appears across the tube at point Y in the figure. Point Y is one extremity of the load line; the other point occurs for an amount of current that drops the entire amount of E_{bb}, leaving no plate voltage at point X. Naturally this current, by Ohm's law, is $E_{bb}/R_L = 350/25 = 14$ mA.

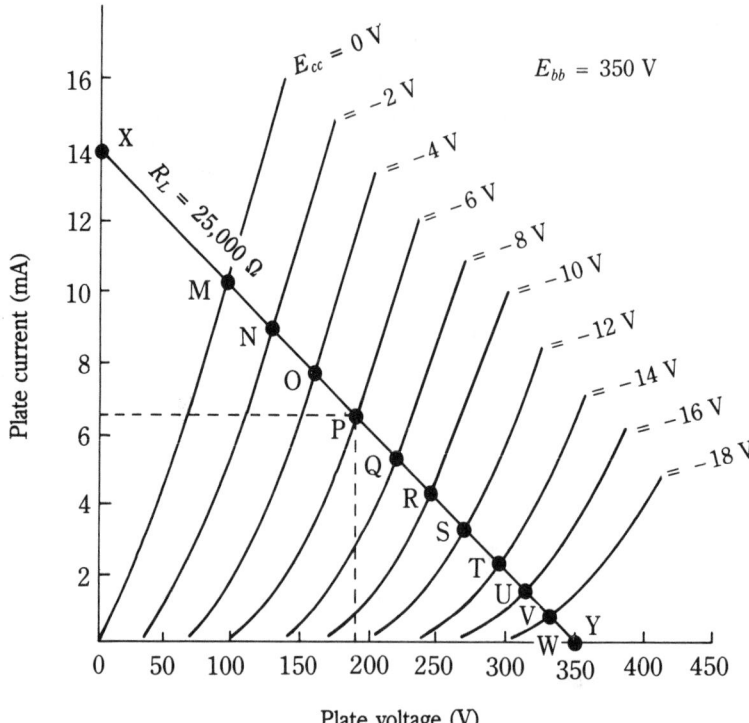

4-22 Plate family of characteristic curves for 6J5 triode including load line.

The straight line drawn between X and Y is the load line. Any condition of operation—for example, any specified control grid voltage—can only result in a point that falls exactly on this line. In order to illustrate this feature, suppose that the grid voltage is set at -6 V. (Don't be concerned about the manner in which this grid voltage is established.) The point P on the load line results from this particular selection of -6 V. Moreover, the amount of plate voltage, as read from the scale at the point directly below P, is approximately 190 V, and the plate current, read from the point on the scale to the left of point P, is approximately 6.5 mA.

Consider two additional points along the load line. For a grid voltage of -12 V, the plate voltage will be about 270 V. Also, for a grid voltage of 0 V, the plate voltage will be about 95 V. (It is easier to argue the exact results of these visual interpolations than it is to justify the time in doing so.) The plate voltage swings from a low of 95 V to a high of 190 V, a span of 95 V, as the grid swings from 0 V to -12 V, a span of 12 V. In this example, a voltage gain can be calculated as the total swing in plate voltage divided by the total swing in grid voltage, or $95/12 = 8$ (in round numbers).

The load line superimposed on the static characteristics provides a great deal of information about the operation of the vacuum tube with a load.

The preceding discussion leads to the explanation of the triode amplifier of Fig. 4-23. In this illustration can be seen a variation of the breakdown of the grid voltage into two parts. A fixed $E_{CC} = -6$ V is supplied by a dc voltage source, which is placed in series with an ac sine-wave source of 6 V peak value. Also, the plate supply $E_{bb} = 350$ V is positioned in series with R_L.

The illustration also includes equations for e_c, the instantaneous voltage between the grid and the cathode, in which e_g is the instantaneous value of the signal voltage. In the plate circuit, e_b is the instantaneous voltage between the plate and the cathode, while i_b is the instantaneous plate current.

Although Fig. 4-23 emphasizes a sine wave of signal voltage, the equations are valid for any waveshape, so long as the tube is neither cut off nor saturated—in other words, so long as the operation is confined to the load line of Fig. 4-22.

One final point: $E_{CC} = -6$ V establishes one quiescent or operating point. A different value of E_{CC} will determine still another operating point. Each is equally valid as an operating point, although the best selection for an amplifier is a point near the center of the load line, such as -6 V, which was the original selection.

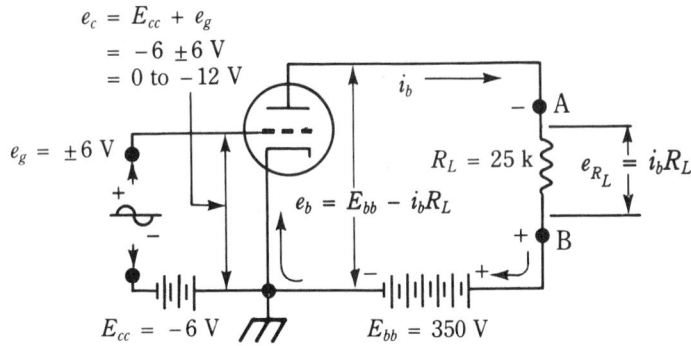

4-23   The basic circuit of a triode, including the input signal and load.

Amplifier classes

In addition to being classified by application, such as frequency range, amplifiers are usually classified in accordance with the mode of operation.

Class-A   In class-A operation, the plate current flows continuously, as illustrated in Fig. 4-24. In this illustration, a sine wave input results in a sine wave of current in the plate circuit, as shown on the dynamic characteristic. Reliance on the location of the operating point for class-A operation is easily seen. If the operating point is moved far enough toward point C, the tube will be saturated for the positive swing of the grid voltage, and the current waveform will be distorted excessively. A similar problem exists for movement of the operating point toward point A. In addition, class-A operation can be achieved only with limited grid voltage swing.

Class-B   With class-B operation, the operating point is set at, or near, cutoff (Fig. 4-25). One-half of the signal—that part shaded in the figure—is clipped and plate current flows for 50 percent of the cycle. An amplifier consisting of one tube and a resistive load will have intolerable distortion. Two tubes functioning in push-pull remedy the distortion problem for the most part. The class-B amplifier is more efficient than the class-A amplifier.

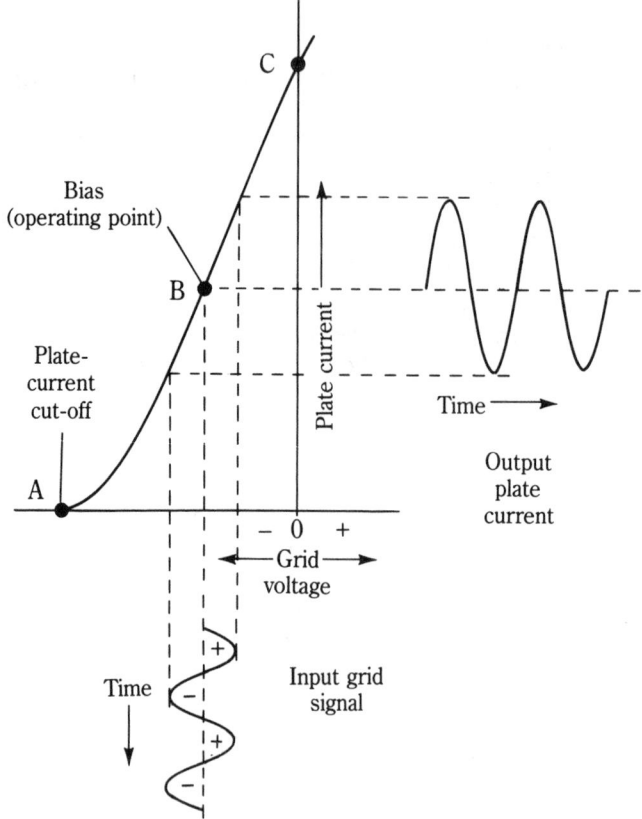

4-24 Class A operation on dynamic characteristic.

Class-AB An amplifier that operates in the region between class-A and class-B is called a *class-AB* amplifier. In a class-AB amplifier, current flows for more than a half-cycle, but for less than the full cycle, as shown in Fig. 4-26.

This illustration also reveals the distinction between class-AB$_1$ and class-AB$_2$. In class-AB$_1$ operation, the operating point is located between plate-current cutoff and the linear portion of the dynamic characteristic. To prevent the flow of grid current, the positive peak of the grid cannot exceed the fixed bias, E_{CC}. The output plate-current waveform is distorted because of the clipping action on the negative peaks. With class-AB$_2$ operation, the grid swing drives the grid positive, so grid current flows. Clipping occurs on both the positive and negative swings of the grid voltage, so the distortion is significant.

Class-C A class-C amplifier's operating point is located well beyond the plate-current cutoff point, so plate current flows for appreciably less than a half-cycle. Class-C operation is shown in Fig. 4-27; the operating point is shown at *J*. The positive peak of the input grid signal extends to point *K* on the dynamic characteristic, and the negative peak extends to point *I*. The shaded areas indicate the portions of the input grid signal that are cut off.

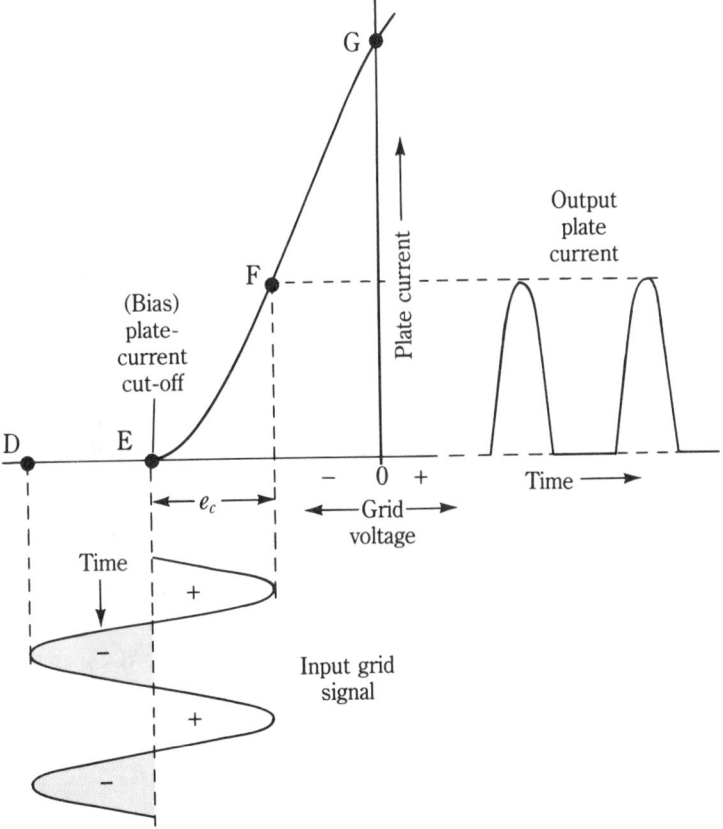

4-25   Class B operation on dynamic characteristic.

The output current waveform represents only a small part of the positive peaks of the input grid signal. If the positive peaks extend beyond point *L* on the dynamic characteristic and into the positive grid-voltage region, grid current flows. The result is clipping on the positive peaks.

Push-pull operations

The basic circuit of a push-pull triode amplifier is shown in Fig. 4-28. The bias, E_{CC}, determines the class of operation. For example, if E_{CC} corresponds to Fig. 4-25, each triode will operate in class B. Visualize this situation for push-pull operation through the construction of Fig. 4-29, in which the dynamic characteristics of identical vacuum tubes are placed together, touching at the cutoff grid voltage, but with one characteristic inverted. This arrangement causes current to flow in each triode for one-half the signal period. This means that the current through the transformer primary windings will alternate and flow continuously.

Unfortunately, there is significant distortion in the plate current waveform, as shown in the figure, because of the curvature of the dynamic characteristic near the cutoff

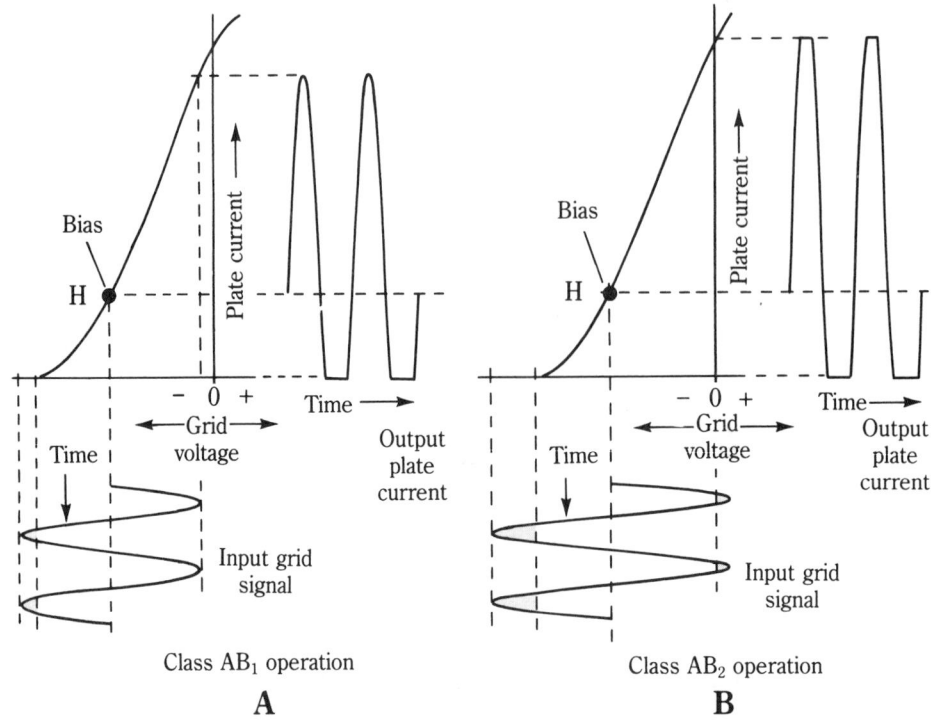

4-26 Class AB operation on dynamic characteristic.

point. If the bias is modified slightly, the effective characteristics become more nearly linear, and the distortion effects are reduced. Figure 4-30 illustrates how Fig. 4-28 is changed for the new bias, and demonstrates how the distortion is greatly reduced.

Methods of biasing

Fixed bias has appeared in all of the vacuum-tube circuits explained thus far. The method of bias most commonly used is called *self-bias*. A convenient method of obtaining self-bias for all operating modes is shown in Fig. 4-31. The dc component of plate current passing through the cathode resistor, R_K, places the cathode at a positive voltage (5 V in the figure) with respect to ground, which is equivalent to placing a negative voltage of the same amount on the grid with respect to the cathode. *Cathode bias*, as this is known, requires that R_K be bypassed by a capacitor, C_K, whose reactance at the signal frequency is small—10 percent or less of the value of R_K.

Whenever the grid drive is intended to make the grid positive for any part of the cycle, as for class-C, two possible forms of grid-leak bias are possible. In Fig. 4-32, the flow of grid current through R_g would cause a fluctuating voltage drop, except that the paralleled C_g smooths out the fluctuations, leaving only the average value (dc level) to give the bias.

A variation of this grid-leak method is found in Fig. 4-33. Capacitor C_g charges quickly when the grid goes positive during a small fraction of the grid voltage cycle. Dur-

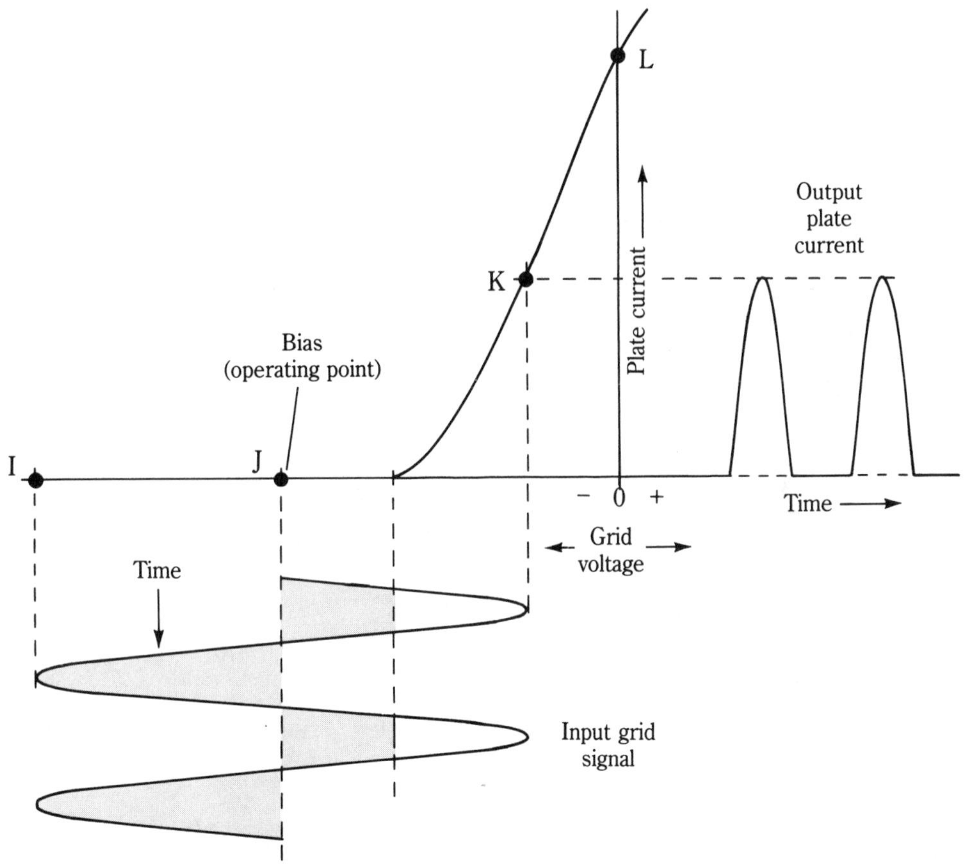

L
Output
plate
current
K
Plate current
Bias
(operating point)
I
J
− 0 +
Time
Grid
voltage
Time
Input grid
signal

4-27 Class C operation on dynamic characteristic.

V_1
i_{b1}
e_{g1}
E_{C1}
T
B
e_{b1}
e_p
i_o
Output
voltage
E_{cc}
E_{C2}
E_{bb}
A
e_{b2}
e_{g2}
C
V_2
i_{b2}

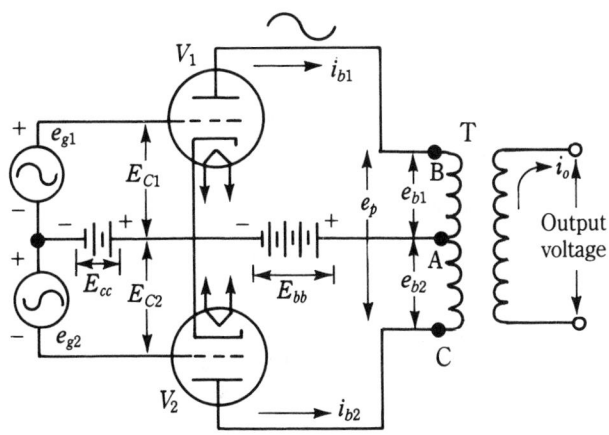

4-28 Push-pull triode amplifier circuit.

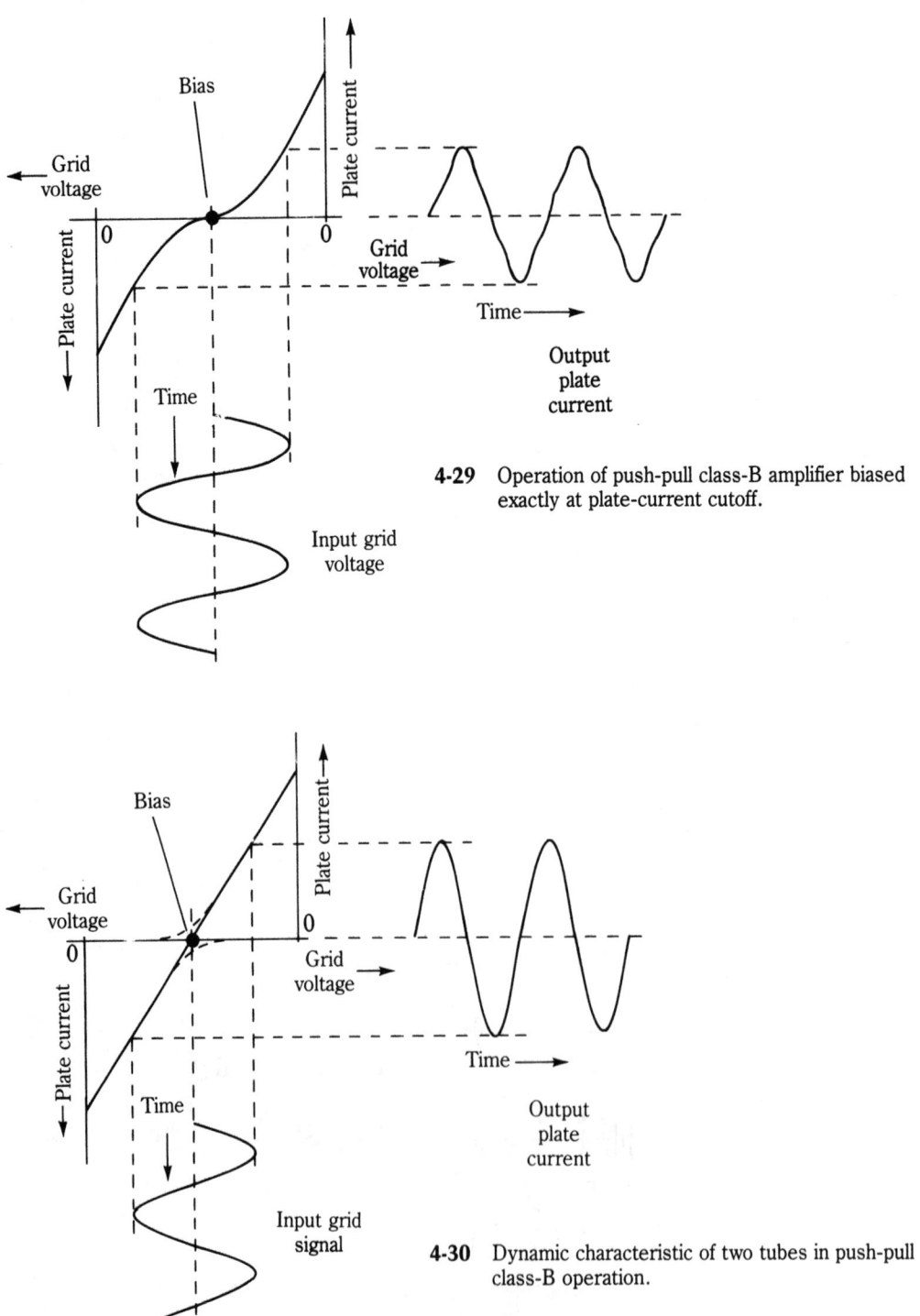

4-29 Operation of push-pull class-B amplifier biased exactly at plate-current cutoff.

4-30 Dynamic characteristic of two tubes in push-pull class-B operation.

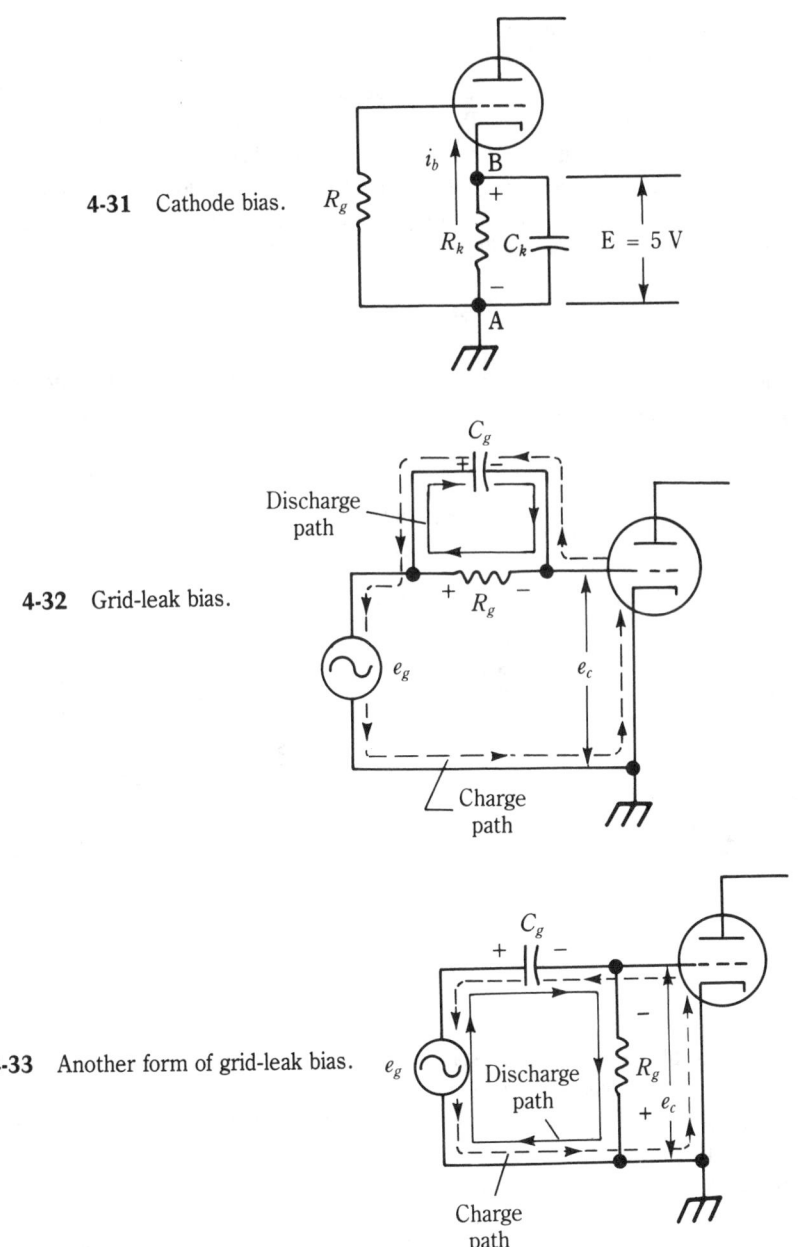

4-31 Cathode bias.

4-32 Grid-leak bias.

4-33 Another form of grid-leak bias.

ing the remainder of the cycle, the charge leaks off the capacitor through the large-value resistor, R_g, and creates a negative bias voltage at the upper end of this resistor.

Amplifier coupling circuits

When the output of one stage is coupled to the input of another, a coupling network must be used. In addition to the tube parameters, load resistance, etc., the characteristics of the coupling network determine the amplification. The coupling devices include capacitors, resistors, inductors, and transformers.

One of the most common vacuum-tube amplifier coupling networks is the R-C (resistance-capacitance) coupler. Two triode amplifier stages with R-C coupling are shown in Fig. 4-34. Concentrate for the moment on the first amplifier stage. A large voltage gain can be achieved with large R_L. Note that the selection of R_L must be coordinated with the placement of the operating point on the static characteristic, so that the swing of the grid signal voltage will not result in distortion. The voltage gain is given by:

$$A_V = \frac{\mu R_L}{r_p + R_L}$$

where μ is the amplification factor parameter of the tube and r_p is the ac plate resistance. This gain formula holds only for a total load of R_L and must be modified when the load is altered.

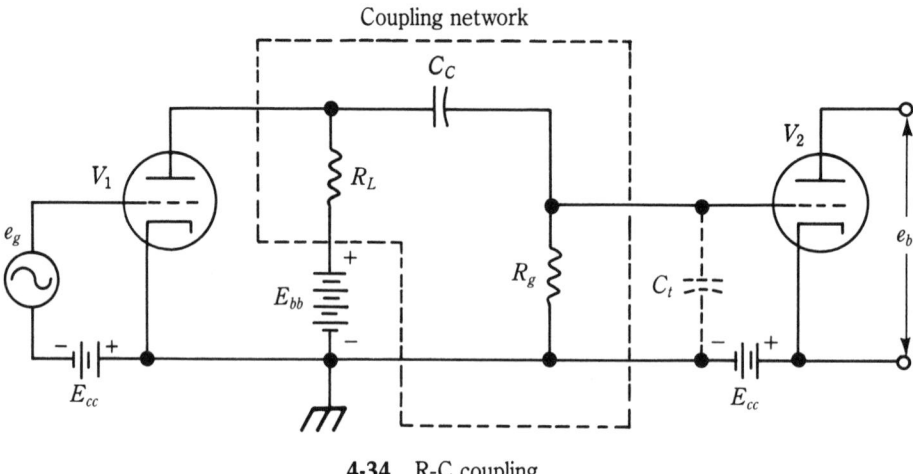

4-34 R-C coupling.

The output of the V_1 amplifier stage is coupled by means of C_C to the grid circuit of V_2. Capacitor C_C is a *blocking capacitor*, whose purpose is to isolate the relatively high positive dc voltage level in the plate circuit of V_1 from the relatively low negative dc voltage level on the grid of V_2. At signal frequencies, the reactance of C_C is small and the ac grid voltage at V_2 is equal to the ac voltage across R_L. The reactance of C_C becomes appreciable at low ac frequencies and can no longer considered negligible. Indeed, the low-frequency gain limit of the overall amplifier is dependent on the reactance of C_C, and the gain drops off at low frequencies, as seen in Fig. 4-35. The drop of gain at high frequencies shown in the illustration will be explained shortly.

In addition to the coupling capacitor, C_C, which influences the low-frequency response of the amplifier, there exists shunt capacitance in the grid circuit of V_2. In Fig. 4-36C, C_T represents the total shunt capacitance arising from the grid and the cathode of V_2 combined with the distributed or parasitic capacitance of the wiring and associated hardware.

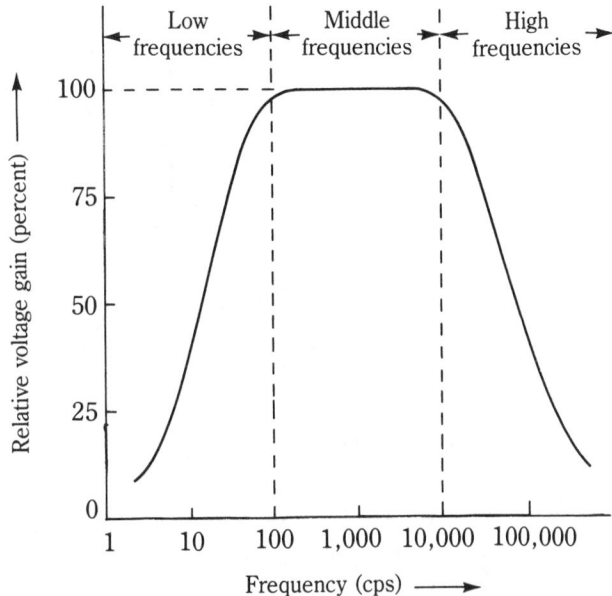

4-35 Frequency response of a typical R-C coupled audio frequency amplifier.

C_T is in parallel with R_g, whose value is approximately 1 megohm. At low or intermediate signal frequencies, the reactance presented by C_t is large compared with R_g, and the gain is not affected by its presence. However, at some frequency, the reactance will be equal to the combination of R_g in parallel with R_L, and the upper frequency limit, indicated to be 10,000 Hz in Fig. 4-35, is reached. The equivalent circuit for the frequency range appears in Fig. 4-36C, which should be compared to that of Fig. 4-36B, in which the reactance of C_T is too large to significantly alter the impedance values for the circuit.

Similarly, in Fig. 4-36A, the capacitance, C_C, is shown because it presents a significant reactance in series with R_g at low frequencies. When the reactance is equal to R_g, the lower limit of the pass band is defined. The corresponding frequency is the *half-power point*, for which the response is down 3 dB.

Return again to Fig. 4-36C, the upper half-power point occurs at the frequency for which the reactance of C_t is equal to the resistance of the two, R_L and R_g, in parallel.

At the middle, or intermediate frequencies, neither C_C nor C_t have any significant influence on the loading presented to the first amplifier. In this frequency range, the load resistance to be substituted into the formula for gain must be that of R_L and R_g in parallel. However, since R_g is normally at least one order of magnitude greater than R_L, little error results from the neglect of R_g in the application of the gain formula.

Impedance coupling

With impedance coupling the load resistance of Fig. 4-34 is replaced with an inductor (Fig. 4-37) so that R_L is replaced by Z_L, which can be represented by a resistor of relatively low value in series with an inductive reactance. The resistance, together with the plate supply voltage, establishes the operating point on the static characteristics.

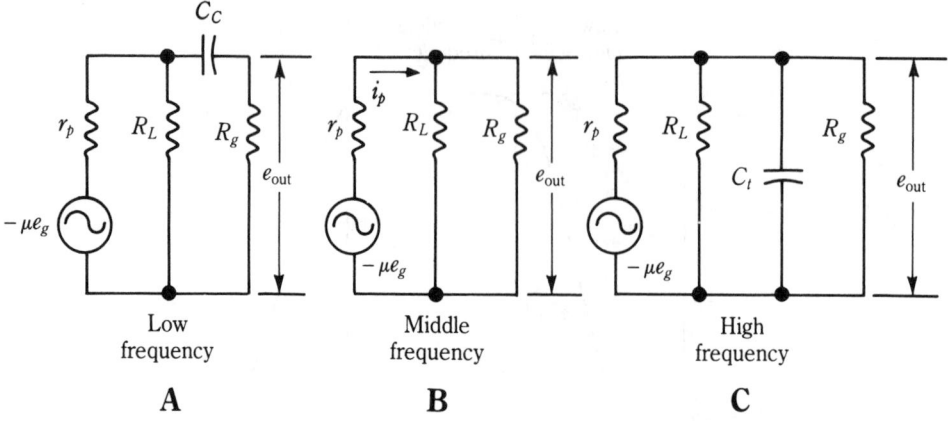

4-36 Equivalent circuits of a R-C coupled amplifier.

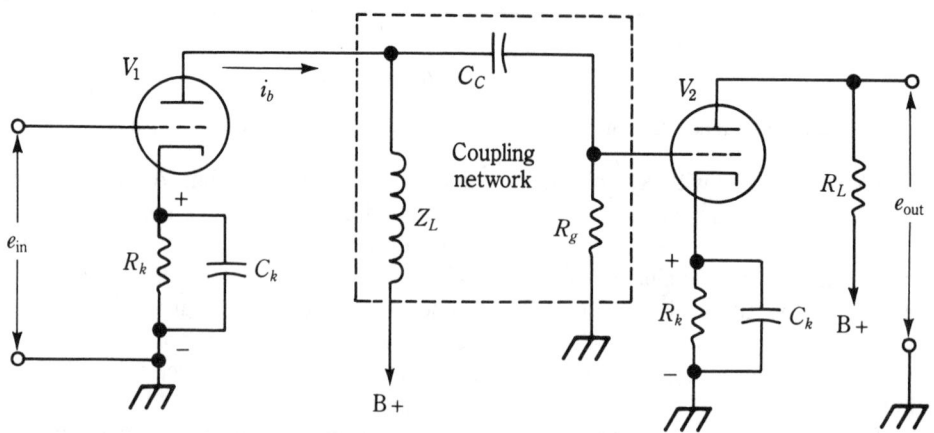

4-37 RLC coupled amplifier using triodes.

At signal frequencies, the load line has a different slope because of the inductive reactance, and the gain of the amplifier will actually increase with frequency over some range of frequencies. Eventually, as the signal frequency is increased, the existence of shunt capacitance, expressed as C_t previously, will cause the gain to fall off. But the decrease in gain occurs at higher frequencies with impedance coupling. This gain would not be the case with R_L. Proper selection of Z_L, intended to extend the upper cutoff frequency, will cause resonance of the inductance in the load and the distributed shunt capacitance at some frequency beyond the upper cutoff frequency of resistance alone.

Transformer coupling

In Fig. 4-38, the two amplifier stages are coupled by means of a step-up transformer. The amplifier has the advantage of providing increased voltage gain in the mid-frequency range. At low frequencies, however, the load presented by the transformer is dominated by the open-circuit inductance of the primary, causing the gain to be frequency-depen-

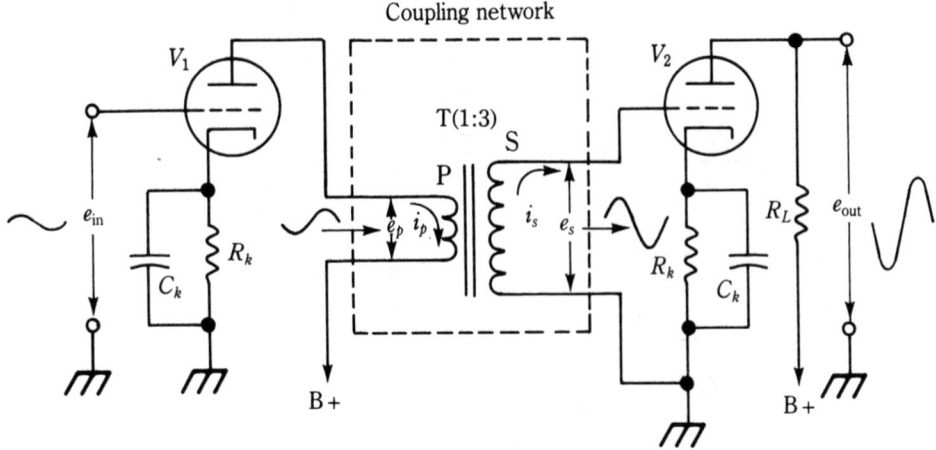

4-38   Transformer-coupled amplifier using triodes.

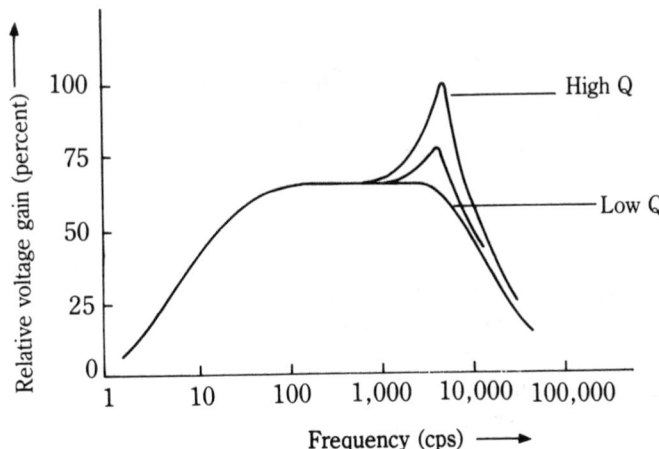

4-39   The voltage gain characteristic of a typical transformer-coupled amplifier for different values of Q.

dent. At high frequencies the inductance of the leakage flux and the distributed capacitance determine the gain, so a resonant peak is to be expected (Fig. 4-39). Transformer coupling between amplifier stages is not often used.

Tuning in the primary circuit, secondary circuit, or both the primary and secondary circuits (double-tuned as in Fig. 4-40) is more widely used for narrow-band amplifiers. As will be seen in the consideration of intermediate-frequency (i-f) amplifiers (commonly used at radio frequencies in superheterodyne receivers), a number of design parameters affect the frequency response of the amplifier as a result of the coupling method. These parameters include the individual frequencies to which the primary and secondary are tuned, the quality factors Q_p and Q_s, and the coefficient of coupling between the primary and secondary circuits.

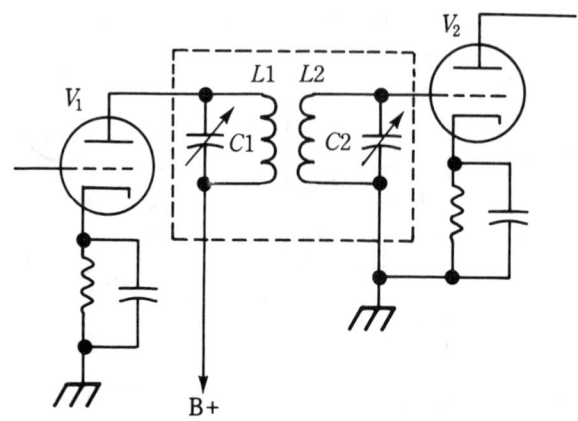

4-40 Double-tuned transformer coupling.

Multiple-choice questions

1. In the common-emitter configuration,

    A. The base and the emitter are connected.
    B. The input signal is applied between the base and the emitter terminals.
    C. The load is connected between the emitter and the base.
    D. The bias resistor is connected between the base and the collector.
    E. This arrangement can only be practical with a FET.

2. For the CE amplifier arrangement,

    A. Voltage gain is always less than one.
    B. Current gain is always less than one.
    C. The power gain is always less than one.
    D. The input impedance is low.
    E. None of the preceding choices is true.

3. In a common-emitter amplifier the voltage gain, A_V, can be increased by:

    A. Making internal source resistance, R_S, larger.
    B. Selecting a transistor with a larger beta.
    C. Increasing the load resistance.
    D. Both B and C.
    E. None of the preceding choices is true.

4. In a common-emitter amplifier,

    A. Base current is greater than the collector current.
    B. Load current is greater than the emitter current.
    C. The base voltage is always negative with respect to the emitter.
    D. Load current is equal to the emitter current.
    E. None of the preceding choices is true.

5. The voltage gain, A_V, of a common-base (CB) amplifier can be made larger by:

    A. Increasing the source resistance, R_S.
    B. Decreasing the source resistance, R_S.
    C. Increasing the load resistance, R_L.
    D. Decreasing the load resistance, R_L.
    E. Both B and C.

6. With the CC (emitter follower) amplifier,

    A. Voltage gain is always less than 1.
    B. Current gain is always less than 1.
    C. The power gain is always less than 1.
    D. The input impedance is low.
    E. None of the preceding choices is true.

7. The voltage gain, A_V, of a common-collector (CC) amplifier can be made larger by:

    A. Increasing β.
    B. Increasing R_L.
    C. Increasing the product $(\beta)R_L$.
    D. Decreasing R_S.
    E. All of the preceding choices are true.

8. A transistor biasing arrangement for an amplifier might yield an unsatisfactory operating point because of:

    A. Inductance changes at higher frequencies.
    B. Temperature changes.
    C. Power supply "hum."
    D. Variations of transistor characteristics.
    E. Both B and D.

9. The operating point for a common-emitter (CE) amplifier lies at the intersection of:

    A. The load line and the base current characteristic established by the bias circuit.
    B. The load line and the vertical line passing through the supply voltage.
    C. The load line and the horizontal line passing through the short-circuit collector current value.
    D. Both A and B.
    E. Both A and C.

10. In an amplifier it is possible to reproduce at the output the input waveform exactly, while causing the _____ to be changed.

    A. Voltage level.
    B. Current level.
    C. Power level.

D. Impedance level.

E. All of the preceding choices are true.

11. A class-A amplifier can be identified because:

A. It is 100 percent efficient.

B. Of its high standards of reliability.

C. Current flows throughout the cycle.

D. It combines the best features of the bipolar transistor and the FET.

E. Its value of Q is greater than 10.

12. Select one choice which most accurately reflects the efficiency of a class-A amplifier:

A. 10 percent.

B. 20 percent.

C. 35 percent.

D. 60 percent.

E. 100 percent.

13. Select a single word describing an amplifier for which the output waveform is a faithful reproduction of the input waveform:

A. Lumped.

B. Synthetic.

C. Linear.

D. Distributed.

E. Stereophonic.

14. A single transistor in the common-emitter (CE) configuration biased for class-B operation can be made to serve as a linear amplifier by:

A. Providing a resonant collector circuit.

B. Single frequency operation.

C. Reducing the base current to zero throughout the cycle.

D. Shorting the collector to ground.

E. Both A and B.

15. Identify an item, or items, associated with a push-pull amplifier.

A. Complementary transistors.

B. Center-tapped transformer.

C. Phase splitter.

D. Both B and C.

E. A, B, and C.

16. With push-pull, class-B operation, a dead zone exists which will result in:

A. Transistor overheating.

    B. Class-C syndrome.
    C. Crossover distortion.
    D. Power supply overload.
    E. Relay contact arcing.

17. The push-pull, class-B amplifier efficiency is:

    A. Theoretically 70.7%.
    B. Theoretically 78.5%.
    C. Practically 50%.
    D. Both B and C.
    E. About 75% for each transistor to yield a total efficiency of about 150%.

18. The principal advantage of an efficient, as compared with an inefficient, amplifier is:

    A. Less heat dissipated.
    B. Less air pollution.
    C. More output power.
    D. Less cost.
    E. Less overall consumption of semiconductor materials.

19. Which of the following forms of feedback will cause the amplifier to have reduced voltage gain, increased input resistance, and decreased output resistance?

    A. Inverted.
    B. Negative.
    C. Voltage.
    D. Both A and B.
    E. Both B and C.

20. Identify a rule relating the gain and bandwidth of a feedback amplifier with different feedback conditions.

    A. The gain is constant regardless of feedback.
    B. The gain-bandwidth product is constant.
    C. Gain is inversely proportional to feedback.
    D. The bandwidth is directly proportional to the gain.
    E. Both A and D.

21. The negative sign in the formula for voltage gain signifies:

    A. Current drawn by the negative power supply only.
    B. Power flow from the amplifier into the dc power sources.
    C. Phase inversion of the output relative to the input.
    D. An arbitrary choice of polarities that can be reversed with a different assumption.
    E. It has no significance at all.

22. A typical op amp has dual (split) power supplies of -15 V and $+15$ V respectively.

Select the maximum positive voltage that might be measured with the amplifier saturated.

A. 0 V.
B. −15 V.
C. +13.5 V.
D. +15 V.
E. −30 V.

23. The major factor influencing the output impedance of an op amp with feedback is:

A. The resistance of the op amp driver circuits.
B. R_S.
C. R_F.
D. R_I.
E. The power supply regulation.

24. A differentiator constructed using an op amp has a triangular waveform input. The shape of the output is:

A. Square.
B. Pulse train.
C. Triangular.
D. Sine.
E. Exponential.

25. An integrator is constructed using an op amp. With a square wave input, the output wave shape will be:

A. Square.
B. Pulse train.
C. Triangular.
D. Sine.
E. Exponential.

26. A particular construction is superimposed on the vacuum-tube characteristics to represent the behavior of the plate voltage and the plate current with a load when the grid voltage is varied. This construction is called:

A. Dynamic locus.
B. Load line.
C. Resistance curve.
D. Characteristic curve.
E. Kirchhoff line.

27. For a vacuum-tube, a graph of plate current plotted as a function of grid voltage is called the:

A. Dynamic characteristic.

    B. Transformation curve.
    C. Output characteristic.
    D. Transconductor line.
    E. Load curve.

28. The curvature of the dynamic characteristic as seen in Fig. 4-29 results in:

    A. Grid current flow.
    B. Plate overheating.
    C. Spurious oscillations.
    D. Waveform distortion.
    E. Premature cutoff.

29. In Fig. 4-36 C_C determines the:

    A. Low-frequency cutoff frequency.
    B. High-frequency cutoff frequency.
    C. Mid-frequency cutoff frequency.
    D. Mid-frequency gain.
    E. Self bias.

30. In Fig. 4-36 C_T determines the:

    A. Low-frequency cutoff frequency.
    B. High-frequency cutoff frequency.
    C. Mid-frequency.
    D. Mid-frequency gain.
    E. Self bias.

Basic problems

1. Calculate the voltage gain of the simplified version of a CE amplifier for which $R_L = $ 1 kΩ, R_S = 500 Ω, r_π = 100 Ω, and β = 150.

2. The transistor and load of basic problem 1 are arranged into a common-base (CB) configuration. Determine the voltage gain.

3. The transistor and load of basic problem 1 are arranged into a common-collector (CC) configuration. Determine the voltage gain.

4. A particular amplifier is biased such that the load current is barely cut off in the absence of an ac signal. Briefly describe the nature of such an amplifier.

5. A particular amplifier has a bandwidth of 50 kHz and a gain of 40. If the feedback is now altered to produce a gain of 20, what will be the new bandwidth?

6. A conventional op amp has split power supplies and differential connections with inverting and noninverting input terminals. If the open-loop gain of the op amp is

10,000, what will be the output if +1 mV is connected to the inverting terminal, and +0.7 mV is connected to the noninverting terminal?

7. A summing amplifier is arranged to add two independent voltages that are applied to the input terminals. When only one of the two is applied, with the other input terminal grounded, the amplifier output is −7.5 V. With the first terminal grounded, and the other voltage applied to the second terminal, the output is +12.5 V. Determine the output when both input voltages are applied simultaneously.

8. A vacuum-tube amplifier is to operate at frequencies above 30 Hz with cathode bias (self-bias) having a 680 Ω resistor placed in parallel with a capacitor. Calculate the least amount of capacitance that is satisfactory for this purpose.

9. Describe a basic operating requirement for grid-leak bias.

10. A 6J5 triode vacuum-tube has a plate resistance of 7700 Ω and a μ of 20. If an amplifier is formed using this tube in which R_L is 25 kΩ, what is the voltage gain of the amplifier?

Advanced problems

1. Referring to basic problem 1, find the current gain of the given CE amplifier. Note: The current gain is the ratio of the load current to the source current.

2. Explain briefly, why the unplanned change in the operating point of a transistor amplifier is undesirable.

3. Using the characteristic and load line of Fig. 4-4, determine the value of β for the given operating point. Hint: Use a range of I_B from 10 μA to 30 μA for this purpose.

4. A voltage amplifier has a voltage gain of 1000 without feedback. If ¹/₁₀ (10 percent) of the output voltage is introduced in series with, and in opposition to, the input, what will be the gain of the overall feedback amplifier in this negative feedback configuration?

5. The amplifier of advanced problem 4 has R_I = 500 kΩ and R_O = 10 kΩ before being connected as a feedback amplifier. What will be the input and output resistances after the amplifier is connected as a feedback amplifier with β equal to 0.1?

6. Refer to Fig. 4-16 for which R_F = 5 kΩ, and R_I = 1 kΩ. Let a dc source of +1.5 V be connected in series with a 0.3 V (peak) sine wave as the input. Determine the maximum and minimum instantaneous voltages expected at the output.

7. The feedback path resistor and input resistor of advanced problem 6 are reconnected into the noninverting configuration of Fig. 4-17. Solve for the answers called for in advanced problem 6 for the conditions given here.

8. Design the circuit necessary for an op amp in the inverting configuration to have a voltage gain of five with a 10 kΩ input resistance.

9. Both differentiators and integrators can be constructed using op amps. Consider an integrator placed in tandem following a differentiator. A triangular wave is introduced as the input to the first amplifier. What will be the waveshape of the second amplifier output? Comment as to what will happen if the order of the two amplifiers is reversed with the same triangular waveform introduced into the first amplifier.

10. In Fig. 4-36 a capacitor, C_C, appears. Why is the capacitor deliberately inserted into the circuit?

5
rf amplifiers

THIS CHAPTER IS CONCERNED PRIMARILY WITH RADIO-FREQUENCY POWER AMPLIFIERS, which are the key elements of transmitters. Small signal-amplifiers are covered first, as an introduction to the topics that differentiate the design and operation of basic amplifiers from amplifiers in the rf range.

Only a few isolated examples of vacuum-tube amplifier applications other than rf power amplifiers are now found, except in older equipment. Nevertheless, the vacuum tube continues to appear in new rf power amplifiers, especially in the most powerful broadcast transmitters. Transistor rf power amplifiers have increasingly high power and frequency ratings, but there is no reason to believe that the vacuum tube will soon be completely displaced. The vacuum tube, as applied to rf power amplifiers, is covered in this chapter as follows:

- rf small-signal amplifiers
- Class-A and class-B rf power amplifiers
- Class-C rf power amplifiers
- Other high-efficiency classes
- Vacuum-tube rf power amplifiers
- Vacuum-tube class-C rf power amplifiers

rf small-signal amplifiers

rf amplifiers are specialized, compared to the basic amplifiers of chapter 4 in two ways:

1. Design practices and component values differ because of the higher radio frequencies. Distributed parameters, which have been disregarded in considering basic amplifiers assume important roles in rf amplifier design and operation.

2. rf amplifiers are intended to amplify a relatively narrow band of frequencies. For this reason they employ tuned LC circuits or other devices capable of exhibiting resonance and filtering action.

In this section, you will be concerned with small-signal rf amplifiers such as are found in signal generators and receivers, and in the low-power-level stages of rf transmitters, as a preliminary introduction to the primary subject of power amplifiers. Such a small-signal voltage amplifier appears in Fig. 5-1. This amplifier has a parallel LC combination in the base circuit and another parallel combination in the collector circuit. Each amplifier is adjusted to the same particular frequency by modification of the inductances, commonly accomplished by the mechanical movement of a powdered-iron core, or *slug*, within the windings proper. This adjustment feature is typical of intermediate frequency (i-f) amplifiers found in superheterodyne receivers. The center rf frequency is fixed, and tuning is carried out during receiver alignment. The bias arrangement is exactly the same as that previously introduced for class-A amplifiers.

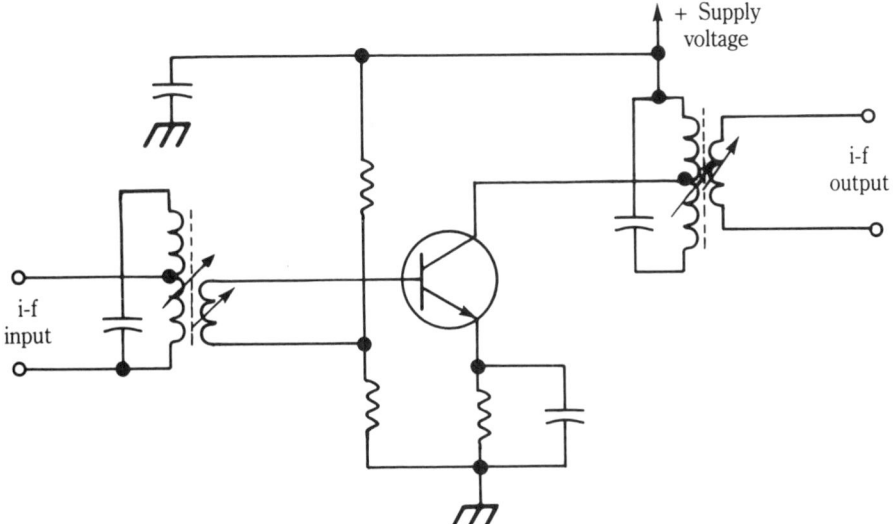

5-1 Inductance-tuned i-f amplifier.

Figure 5-2 is a variable-frequency rf amplifier with two parallel-tuned circuits as before. Tuning is accomplished by the adjustment of C1 in the secondary of the input-circuit transformer and by C2 in the secondary of the output transformer.

Distributed capacitances and inductances arising from the interconnection of elements on the printed-circuit board, together with the internal capacitances in the transistor, can cause regeneration (positive feedback) within the amplifier and undesirable oscillation. Capacitance between the higher-signal-level collector and the lower-signal-level base of Fig. 5-1, for example, is usually involved in amplifier instability and sustained oscillation. The effects of such capacitance can be canceled by the selective feedback of output signals of the proper phase to *neutralize* the regenerative feedback. Capacitor C_N in Fig. 5-3 is intended for this purpose.

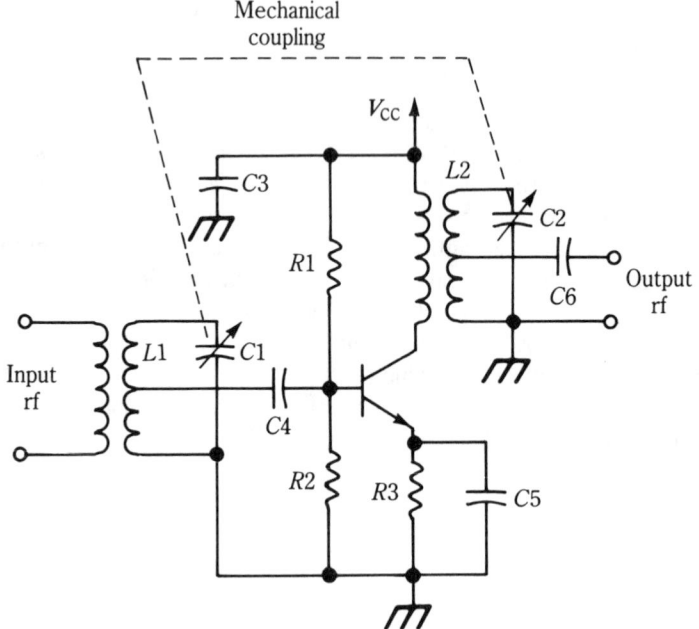

5-2 rf amplifier with capacitor tuning.

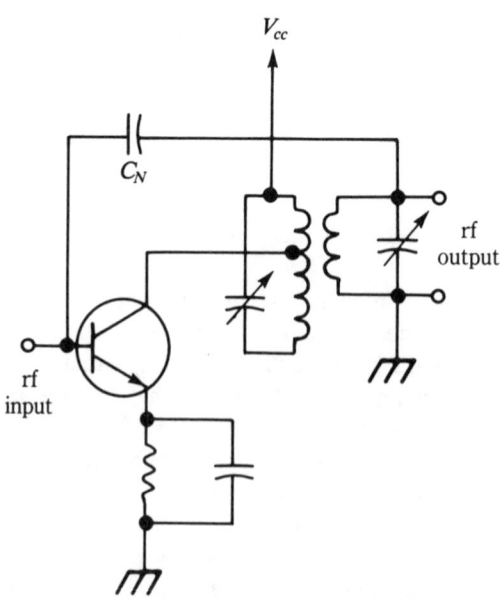

5-3 rf amplifier with neutralizing capacitor.

Another function of rf amplifiers will be introduced at this point. Assume that a multiple of a given radio frequency is needed. This situation is frequently encountered in rf systems, and the solution is frequency multiplication. Typically, a stable frequency originates in a crystal-controlled oscillator. Multiplication is achieved by operating the amplifier in class C so that the transistor current is rich in harmonics, and then tuning the output to that harmonic frequency. Figure 5-4 is an example of such a *frequency multiplier.*

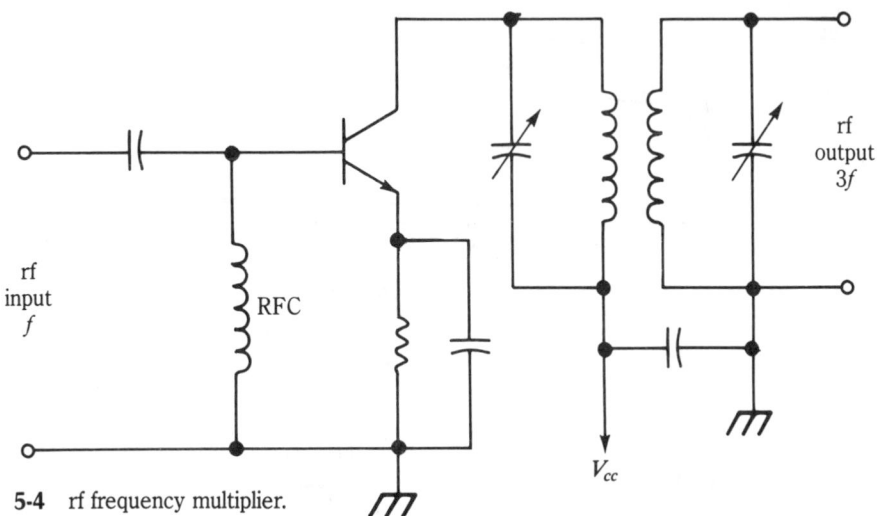

5-4   rf frequency multiplier.

Class-A and class-B rf power amplifiers

A class-A linear, rf power amplifier can be used where power gain is required, and accurate reproduction of the amplitude and phase of the modulated rf signal is necessary. The power amplifier's operation is not greatly different from that of the small-signal class-A voltage amplifier, but the circuit configuration is different, both to accommodate the voltage and current levels required and because of the relatively narrow bandwidth containing the carrier and sideband components. This narrow bandwidth permits unwanted signals, which arise from amplifier nonlinearities, to be reduced by filters. The undesirable products arising from amplifier nonlinearities include intermodulation-distortion products, as well as parasitic and subharmonic spurious products.

Bipolar junction transistors (BJTs) have been extensively used in the past for class-A rf power amplifiers. The vertical MOSFET, popularly known as the VMOS, is being widely used in new equipment design. Regardless of type, all rf power transistors must be operated within the maximum voltage and current ratings, which are interpreted with considerably more difficulty in rf applications.

In the class-A amplifier, the operating point is established in the active region for the BJT (Fig. 5-5A) or the field-effect transistor (Fig. 5-5B). Moreover, the load presented to the transistor, which usually has a low impedance value, and the base (or gate) drive must be able to keep operation within the active region at all times.

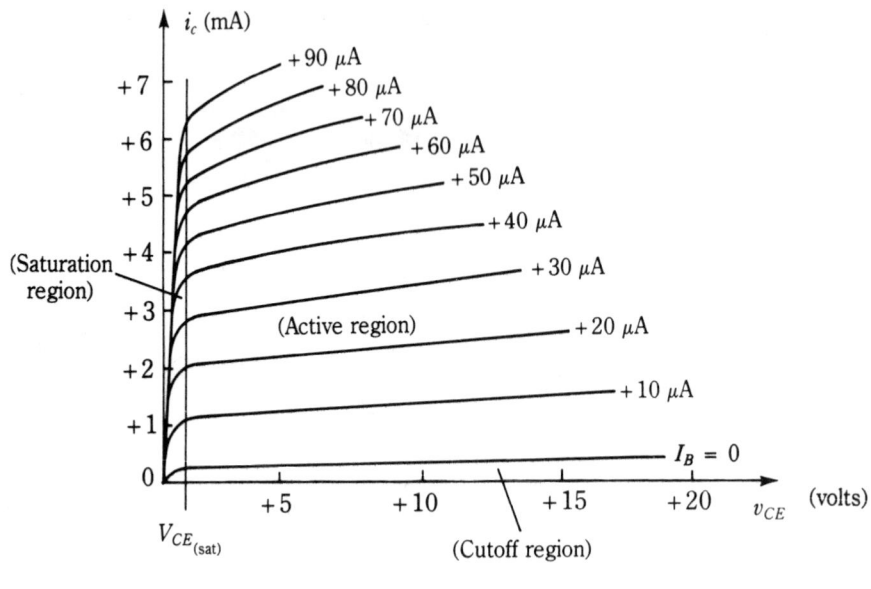

A

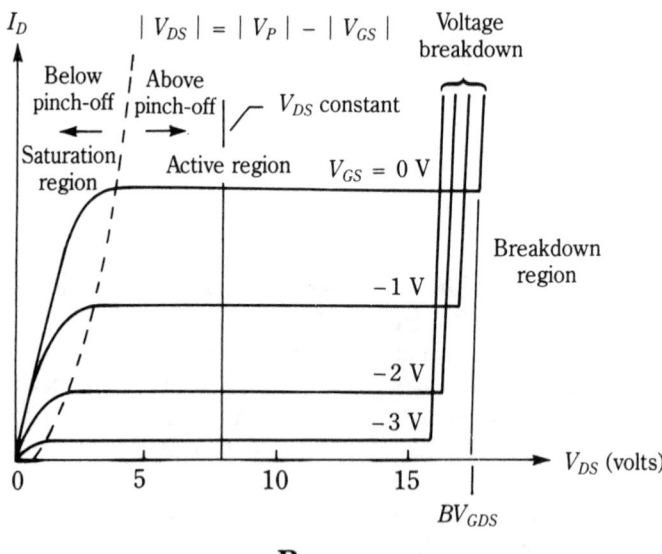

B

5-5 A. pnp collector characteristics.
B. JFET drain-source characteristics.

An npn transistor amplifier appears in Fig. 5-6. Capacitor C_B blocks the dc supply voltage from the load and tuned circuit. L and C are resonant at the carrier frequency, thereby presenting a high value of resistance in parallel with R_L at the resonant frequency and a low impedance path to ground as harmonic and subharmonic frequencies.

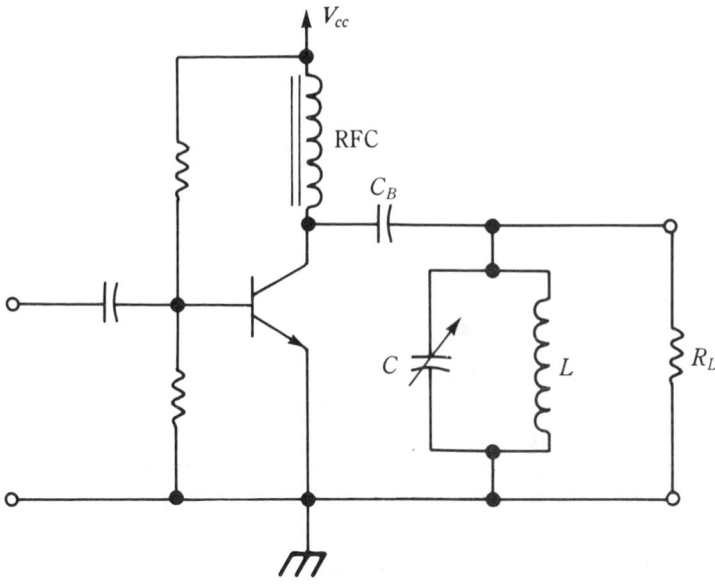

5-6 Class A rf amplifier.

Collector (drain) efficiency (η) for the amplifier is the ratio of the output signal power to the power delivered by the dc source to the collector (drain) circuit. Its theoretical limit is 50 percent for the class-A amplifier.

Because of its higher efficiency, the class-B amplifier is more frequently used for medium- and high-power applications than is the class-A amplifier. Like the audio frequency class-B amplifier, the rf class-B power amplifier appears in two basic versions: the push-pull amplifier and the complementary amplifier. With either configuration, a tuned parallel circuit in the output circuit, similar to that of Fig. 5-6, will reduce the existence of unwanted frequencies. Recall that the theoretical maximum efficiency for the class-B amplifier is 78.5%.

Power limitations imposed by maximum practical current and voltage variations are also affected by the transistor parameter frequency dependency. The high-frequency current gain figure of merit (f_T) defines the frequency for which beta (h_{fe}) is unity for the BJT. Clearly, an amplifier must be operated below f_T if it is to have any practical value of current gain at all. More insidiously, at frequencies well below f_T charge storage effects can cause collector waveform distortion with the consequent reduction in efficiency.

At higher frequencies, loads are more likely to have significant reactive components. Impedance matching is also more difficult to achieve, and must ultimately require fine tuning and adjustment. Also, the person knowledgeable in vacuum-tube rf amplifiers will discover that current levels are much higher for the semiconductor amplifier because of the lower impedance levels, which further contribute to losses and lowered efficiency.

Intermodulation distortion is the result of amplifier nonlinearities. For modulated rf signals, those products near the carrier frequency cannot easily be filtered and should be minimized through good design practices. Gain reduction at high frequencies (previously mentioned) and crossover effects (see chapter 4) are two causes of intermodulation distortion. Operation beyond the active region into saturation or cutoff are also serious causes. Recall that crossover effects are usually avoided by adjusting each transistor bias for the push-pull amplifier so that it conducts for more than one-half of each cycle, which actually corresponds to class-AB, as the term is usually applied to amplifiers.

Current drive for the BJT amplifier is selected to reduce the nonlinearities inherent in voltage drive. The FET has a square-law characteristic, which introduces distortion that can be reduced by proper bias selection. Introduction of the negative feedback arrangements tends to stabilize the amplifier against parameter variations, and can reduce the distortion effects.

Class-C rf power amplifiers

Class-C amplifiers provide efficient operation with simple design and are very effective for continuous wave (CW) transmitters, amplitude modulation (AM), and frequency modulation (FM). Class-C amplifiers cannot be used where linear amplifiers are required to preserve the amplitude and phase relations of an established, modulated carrier. Thus, a class-C amplifier, together with a modulator, will produce an amplified rf wave with amplitude variations corresponding to the modulating waveform, i.e., double-side-band, full-carrier AM. However, this amplified, modulated AM waveform cannot be followed by another class-C amplifier. If it is, the amplitude variations in the output will not faithfully follow those variations in the input.

Figure 5-7 illustrates a class-C amplifier based on a FET. The amplifier is similar to the classic class-C vacuum-tube amplifier because the gate drive is sufficient to saturate

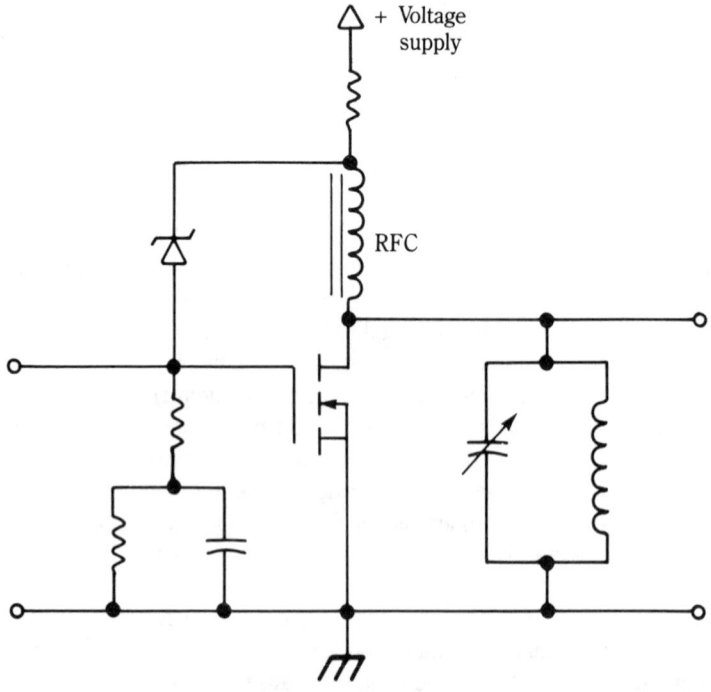

5-7 Class C rf amplifier.

the transistor for a time interval somewhat less than one-half cycle, and is cut off for more than one-half cycle. Operation in the active region usually occupies only a small fraction of a cycle, conventionally expressed as a *conduction angle*. As the conduction angle decreases toward zero (as graphed in Fig. 5-8), signifying conduction for a smaller fraction of the cycle, the efficiency increases toward 100%, which is a desirable but impossible design goal.

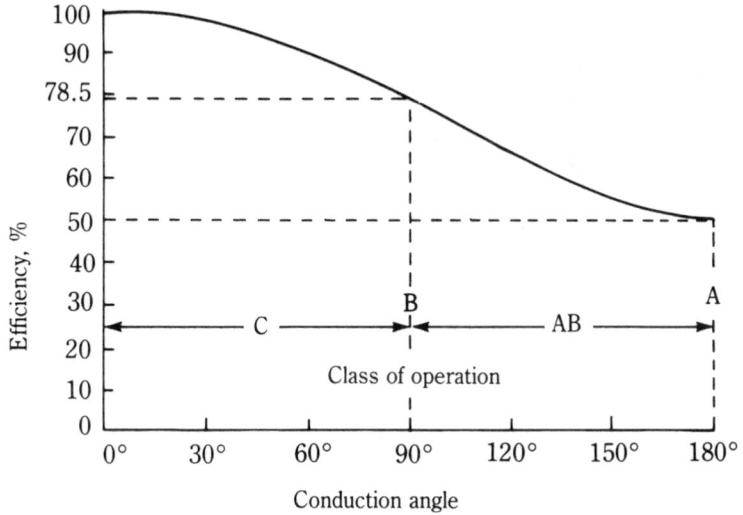

5-8 Efficiency of class-C operation.

To understand how the class-C amplifier achieves a high level of efficiency, consider the nature of its operation. During the major part of the cycle, the transistor is either saturated or cut off. In this states either the drain voltage or the drain current is zero, or nearly so, and therefore the power dissipation during both of these times is essentially zero. It is only when the transistor is in transition between these two extreme states that both the current and voltage are both significantly different from zero, so their product, the power, is no longer negligible.

The circuit of Fig. 5-7 cannot be conveniently adapted to the BJT. One problem of several is caused by the need for impractical circuit elements in the parallel tuned circuits.

Figure 5-9 shows the modified class-C amplifier for the BJT. You must recognize the need for selecting impedance-matching networks for a given load. Figure 5-10 contains a single L-network for impedance matching. C and $L2$ are selected to transform load resistance, R_L, into the resistive component, R_C, of the collector load impedance. In practice $L1$ and $L2$ are provided by a single inductor.

In the event that the impedance transformation cannot be realized with a single L-section of practical elements, it might be necessary to employ two or more L-sections in tandem (see Fig. 5-11). The basic concepts of impedance matching with L-sections is covered in chapter 1.

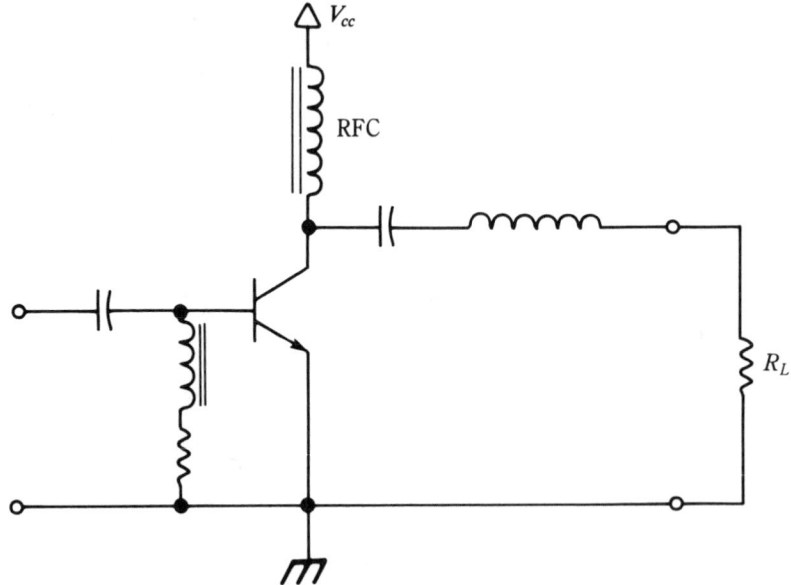

5-9   Modified class-C amplifier.

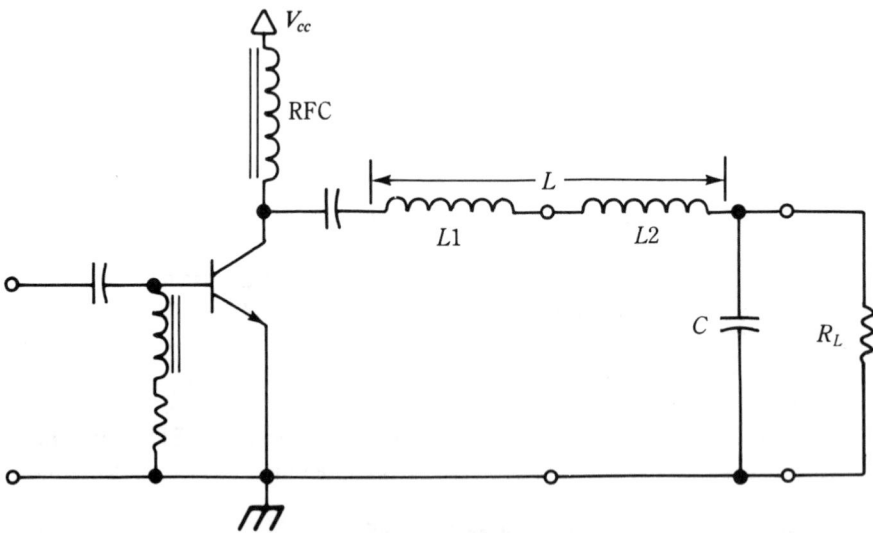

5-10   Single L-section output matching.

Other high-efficiency classes

There is a continuing need for rf power amplifiers whose efficiencies exceed that of the class-C amplifier. The principal advantage of higher efficiency in fixed installations is reduced cost because of hardware savings for the same output power. All the high-effi-

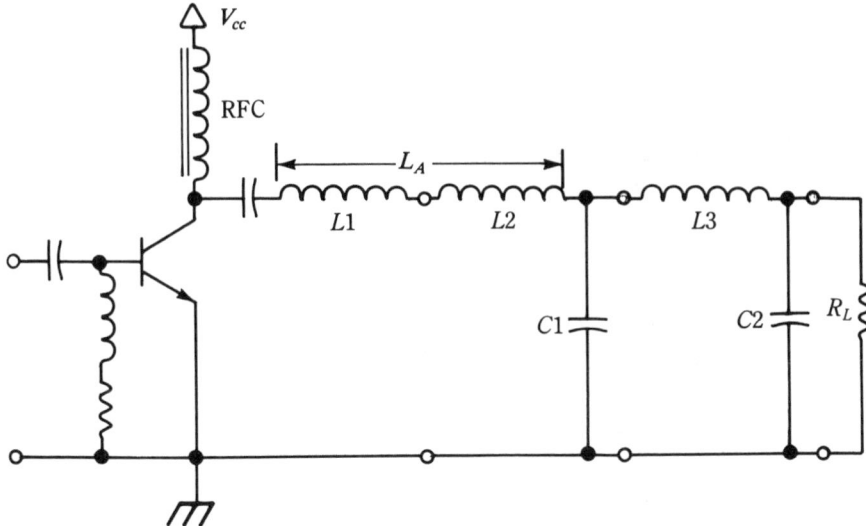

5-11 Double L-section output matching.

ciency rf amplifier classes strive to reduce the proportion of time operation takes place in the transistor-active region. Most of the amplifier types introduced in this section depart from the more familiar amplifier categories—e.g., class A, class AB, and class B—and these categories move a giant step beyond class C by deliberately introducing switching action. After all, a switching transistor is never intended to operate in the active region. Unfortunately, the amplifier must operate in the active region in making the switching transition. How fast it is able to make that transition determines, to a large extent, the power dissipated.

Class-D amplifier

Figure 5-12 illustrates complementary voltage switching for a class-D power amplifier. A square-wave drive is applied to the transformer primary causing $Q1$ and $Q2$ to be in complementary states, each alternating between saturation and cutoff. This causes an essentially square wave (0 to V_{CC}) to be applied to the LC series combination whose resonant frequency is chosen to be that of the square wave. A sine wave current of that chosen frequency then passes through the load, R_L, together with greatly attenuated harmonics.

An alternate voltage-switching configuration with similar operating characteristics appears in Fig. 5-13. In Fig. 5-14 is a transformer-coupled, current-switching class-D power amplifier.

Class-E amplifier

The class-E amplifier is closely related to the class-D amplifier, but uses only one driving transistor, which is switched ON and OFF at the rf (Fig. 5-15).

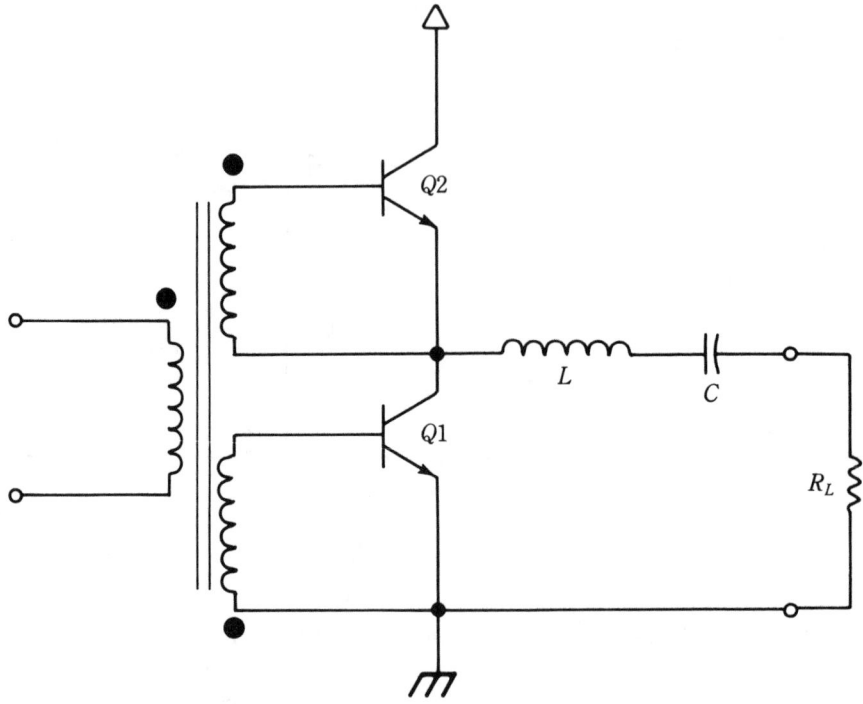

5-12 Voltage-switching class-D power amplifier.

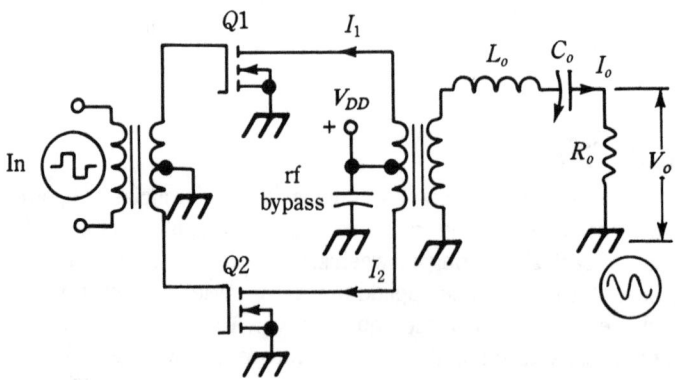

5-13 Voltage-switching class-D power amplifier with power FETs.

Class-F amplifier

Of all the high efficiency amplifiers, the class-F amplifier has been in use the longest time, and might be recognized as bearing a number of names, including: biharmonic, polyharmonic, and multiresonator, among others.

By placing one, or more, additional resonant circuits in the amplifier, as in Fig. 5-16, the collector voltage can be modified to more nearly resemble that of the class-D or

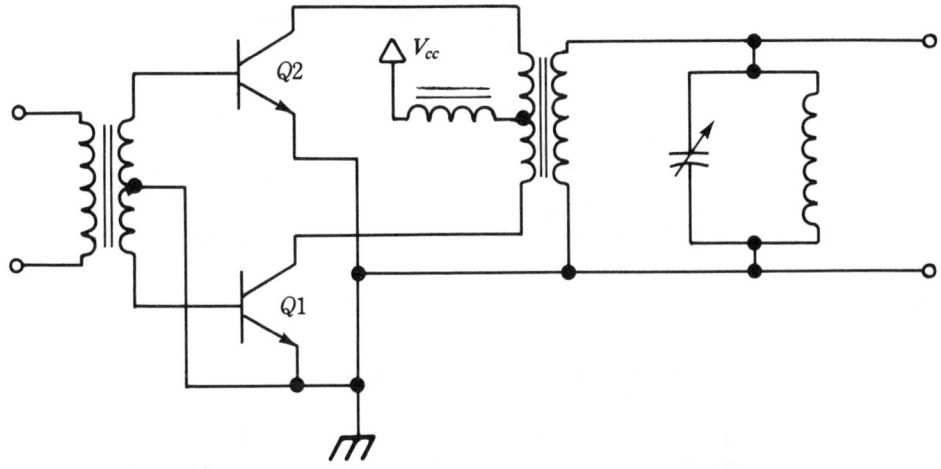

5-14   Transformer-coupled, current-switching class-D rf power amplifier.

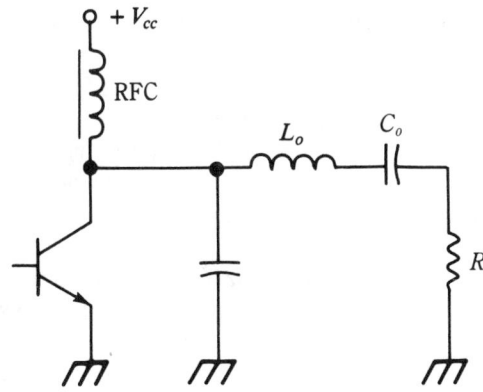

5-15   Class E power amplifier.

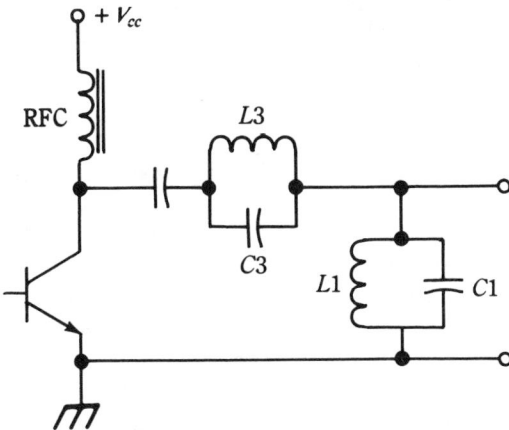

5-16   Class F power amplifier with third harmonic filtering.

class-E amplifier, thereby contributing to the efficiency. Concentrating on Fig. 5-16, assume that the $L1C1$ combination is tuned to the carrier frequency. Clearly, the load voltage will be sinusoidal at that frequency.

If the $L3C3$ parallel combination is tuned to the third harmonic, an additional component is then added to the fundamental harmonic to make up the collector voltage. This additional component flattens the top of the collector voltage waveform and steepens the sides, as compared to the sine wave.

However appealing this relatively simple innovation might appear, you should be aware that resonant circuits are lossy, and that the theoretical efficiency cannot be realized in this, or any of the other circuits presented in this chapter.

Vacuum-tube rf power amplifiers

Vacuum-tube amplifiers continue to appear in modern high-power radio transmitters and there is no reason to believe that the vacuum-tube will disappear from new transmitters in the near future. Moreover, the high quality of modern transmitting equipment based on mature vacuum-tube technology guarantees their useful existence for many years to come. It seems necessary to include a reasonably comprehensive account of vacuum-tube equipment and its operation here.

Some of the classes of amplifiers previously explained for solid-state active devices were originated for vacuum-tubes, and are continued in current applications. Most commonly classes-A, AB, B, and C, the latter for high-power, high-efficiency transmitters, are in widespread use. With minor changes in wording, the definitions of these classes already given for transistor amplifiers, are still applicable. Figure 5-17 compares these classes using the dynamic transfer characteristic. Note that the vertical axis indicates the amount of plate current, I_b, while the horizontal axis shows the control grid voltage, E_g. The format in this figure might be unfamiliar to you. Look, for example, at the class-A operation, you see that an operating point is established at point *"a."* The sinusoidal grid voltage drive is about this point and is reflected in sinusoidal variation in the plate current.

Continuing, the operating point for class-B is at the origin of the graph, so that the considerably larger grid voltage variation results in plate current for only one-half a cycle. Class-AB has an operating point between the origin and point *"a."* You can see that plate current does not flow for the entire cycle. The bias established for class-C results in plate current for less than one-half cycle.

Vacuum-tube class-C rf amplifiers

Figure 5-18 is the schematic diagram of a class-C vacuum-tube amplifier with a tank-circuit (parallel resonant circuit) load. This particular arrangement is referred to as *series fed,* since the plate supply source, pictured as a battery, is placed in series with the inductor, $L1$, of the tank. $C1$ couples the input signal from the preceding stage, thereby isolating the dc level existing there from that required in the grid circuit of V_1. The input signal appears across the grid resistor, $R1$. The positive bias on the cathode is adjusted for class-C operation by means of potentiometer $R2$. Capacitor $C2$ bypasses the rf signal around the bias supply. The tank circuit consists of $C3$ and $L1$ in parallel. Capacitor $C5$

Collector or plate current waveforms

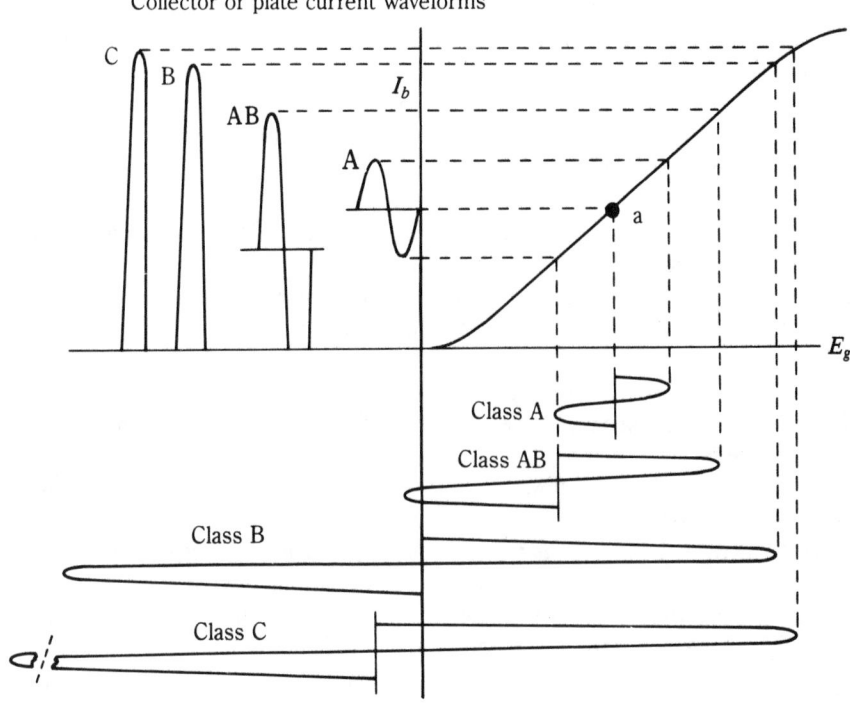

5-17 Classes of operation.

5-18 Class C vacuum-tube amplifier.

bypasses the rf tank current around the plate supply whose voltage is E_{bb}. $C4$ couples the output rf signal to the next stage, if one exists.

To better visualize the operation of the amplifier, assume that the frequency of the signal applied to the grid circuit is equal to the resonant frequency of the tank circuit. When the grid voltage reaches a level that causes plate current to flow, the current will pass through the inductor. Cutoff of the plate current by the reduced level of grid voltage cannot immediately stop the inductor current that continues to flow and charge the capacitor in parallel with it. Eventually the capacitor voltage will reach its maximum level as the inductor current slows to a halt and slowly reverses as the capacitor begins to discharge through it. This discharge action continues until the reversed flow of current reaches its maximum, as the capacitor voltage becomes zero and reverses polarity, because the inductor current slows but continues to flow in this same reverse direction. Eventually, the capacitor will become completely charged with the opposite polarity as the inductor current again slows to zero and reverses, flowing in the original direction. At some time with this direction of inductor current, the plate will begin to conduct, causing additional inductor current to flow, and bringing the cycle to the same point in which these events started.

In summary, the *flywheel action* of the tank circuit causes sinusoidal voltages of alternate polarities to appear across the capacitor. Likewise, and with a 90° phase difference in comparison with the voltage, the inductor current will also be alternating, except that, while flowing in one direction, the inductor current receives a boost in the same direction because of the pulse of plate current. Visualize that the total energy stored in the tank electric and magnetic fields remains constant. Whatever losses are suffered as this energy alternates between the two fields, losses are replenished by the short duration of plate current flow at exactly the right time in the cycle.

The efficiency, as has been stated in slightly different form for the semiconductor amplifier, is the ratio of the ac power to the dc power supplied to the plate circuit. Thus, if 451.5 W of useful power is delivered to the load, and the dc input power to the plate circuit is 620 W, then the efficiency is:

$$\text{Efficiency} = 451.5 \times 100\ /\ 620 = 72.8\%$$

This means that the power dissipated by the vacuum-tube is the difference: 620 − 451.5 = 168.5 W. Note that the filament power heating the cathode, which is considerable, is not accounted for in the preceding calculations.

Frequency multiplication

In the class-C amplifier, the pulse of plate current has an appreciable harmonic content, in addition to its fundamental component, which is determined by the frequency of the grid drive voltage. Because of this appreciable harmonic content, the class-C amplifier can be caused to generate output power at a frequency that is a harmonic of the input signal. A class-C amplifier operated in this manner is called a *frequency multiplier*.

Three conditions must be met in order that the amplifier be an effective voltage multiplier:

1. The tank circuit must be tuned to a harmonic of the fundamental drive voltage

2. The grid must have a strong negative bias

3. A strong grid drive voltage is necessary

Condition 1 requires no further explanation, except that a tuned circuit is necessary to select the desired harmonic of the fundamental input signal frequency. Conditions 2 and 3, in combination, ensure that a short pulse of plate current flows. Such a pulse is rich in harmonics. The tank circuit will respond to the harmonic for which it is tuned, and will supply a sine wave at that frequency to the load. The fundamental, and the undesired harmonics, will see low impedances at their respective frequencies, and will produce very small load voltage components at these frequencies.

Another way of visualizing the operation is based on the flywheel effect already mentioned. If, for example, the tank is tuned to the third harmonic, plate current pulses will flow every third cycle of the tank circuit voltage. Because of the energy stored in the inductor and capacitor fields, energy that is partially renewed every third cycle, will cause the circuit to continue oscillating at its own, natural resonant frequency. Actually, following the plate current pulse boost, the tank will experience several cycles of damped oscillations, until the next plate current pulse. This means that each cycle amplitude will be less than that of the preceding cycle as energy is transferred, a little at a time, into the load from the oscillating magnetic and electric fields. This suggests that a practical limit exists for the frequency multiplication factor. When properly designed, the frequency multiplier produces an acceptable sine wave with a relatively small decrease in wave amplitude between plate pulses.

Linear rf power amplifiers

A linear amplifier can be operated in class-A or -AB single-ended, or class-B, single-ended or push-pull configuration. Class-A rf power amplifiers, due to their low efficiency, are primarily used in low-power applications. Class-B power amplifiers provide greater power output with increased efficiency, but require well regulated power supplies. Class-AB amplifiers represent a compromise in power and efficiency between class A and class B. Linear amplification cannot be achieved with class-C operation.

A push-pull amplifier consists of two electron (vacuum) tubes whose grid and plate signals are 180° out of phase. The two tubes are operated class A, class AB, or class B. Less distortion with a greater power output and plate efficiency is obtained in push-pull operation.

A push-pull triode amplifier is shown in Fig. 5-19. The upper and lower sections of the circuit are similar. Triodes $V1$ and $V2$ are the same tube types, and are reasonably well matched with regard to characteristics. (Assume perfect matching for the purposes of this discussion.) The two grid signals, e_{g1} and e_{g2}, have the same amplitudes but are 180° out of phase with each other. The bias and plate supplies are indicated by batteries of the proper polarities. Transformer T couples the plate circuits of the tubes to the load (not shown) which is connected across the leads of the secondary. The primary of T is center-tapped at point A so that its signal-frequency voltages e_{b1} and e_{b2} are equal, but opposite in phase. Signal currents i_{b1} and i_{b2} are shown with waveforms to indicate their 180° phase difference.

With no input signal voltage ($e_{g1} = e_{g2} = 0$), the secondary signal current i_0 is zero. This does not imply that no current flows as the dc grid bias is in place and the quiescent plate current, whose value depends upon the class of operation, exists. Note that the primary windings of T provide low resistance paths for the dc components of the plate currents. Moreover, since these currents flow in opposite directions in the two halves of the primary, there is no tendency to saturate the magnetic core of the transformer.

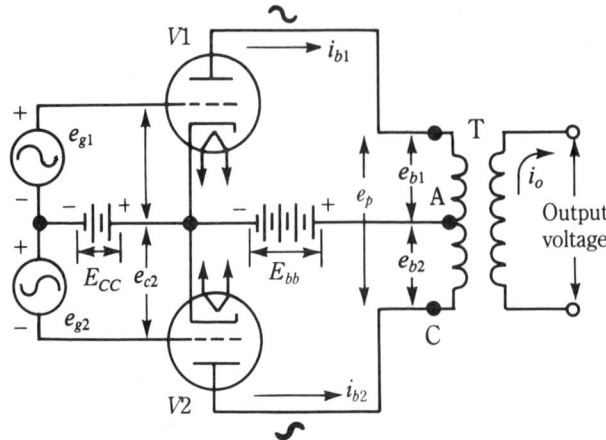

5-19 Push-pull triode amplifier circuit.

There are two common methods of obtaining a 180° phase reversal of the input grid signals. The first is the transformer method. This method (not shown in Fig. 5-19) has the opposite secondary leads of a center-tapped transformer connected to the two control grids, while the center-tap of the transformer is connected to the negative terminal of the grid bias supply in Fig. 5-19. The other common method uses a phase splitter amplifier. The phase splitter consists of a tube having equal plate and cathode load resistors. When operated, equally amplified signals will appear across these two resistors, except that the two output signals will be 180° out of phase with each other.

Dynamic characteristics for push-pull operation

The dynamic characteristics for two tubes operating in push-pull is constructed from the individual dynamic characteristics of the two tubes. Figure 5-20 shows a dynamic characteristic of two tubes operating in class-A push-pull. It is obtained in the following manner. The dotted curve labeled V_1 is the dynamic characteristic of one tube and V_2 is the dynamic characteristic of the other. These dynamic characteristics are identical and, because the plate currents of the tubes are 180° out of phase with each other, one characteristic is inverted—right to left and top to bottom—and superimposed on the other with their horizontal axes common.

The characteristics are then lined up so that the bias voltage of one tube meets the bias voltage of the other. For example, if $E_{CC} = -5$ V, this value of voltage appears at the same place on both grid-voltage axes. The resultant dynamic characteristic is obtained by algebraically adding the corresponding values of plate current for different values of grid signal voltage. This yields the composite dynamic characteristic for push-pull operation shown as the solid line passing through the bias point in Fig. 5-20.

A little thought reveals that there is a cancellation effect of the curvature inherent in the individual dynamic characteristics of the tubes. The result is a nearly straight-line composite dynamic characteristic so that little distortion is experienced with wide swings of the grid voltages. If similar swings of grid voltage are attempted with single-tube operation, the curvature of the individual characteristic will cause substantial distortion.

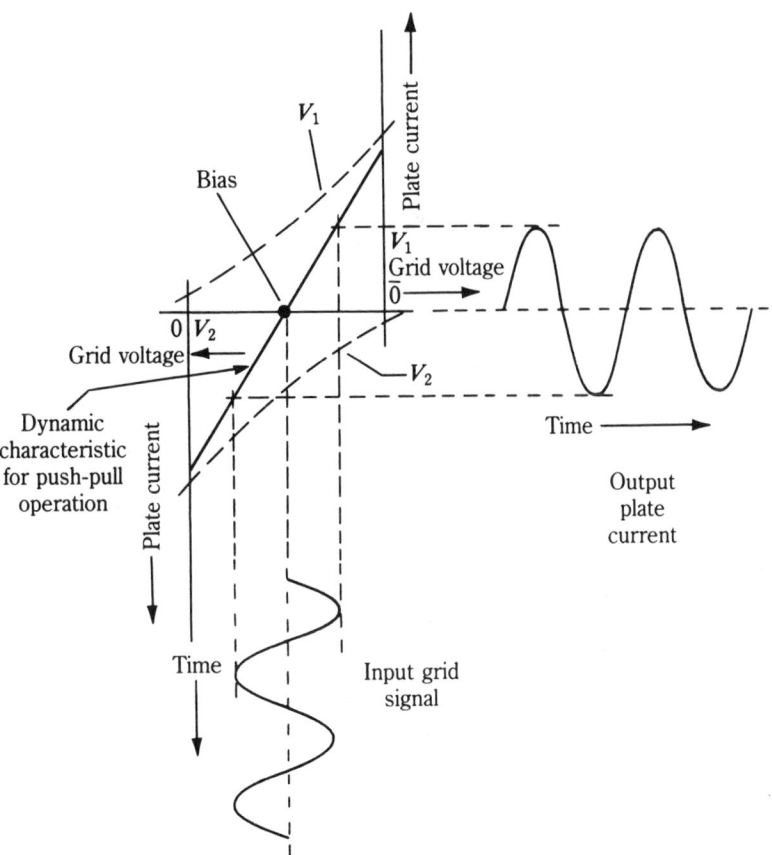

5-20 Dynamic characteristics of two tubes in push-pull class-A operation.

Since the efficiency is dependent on the amount of grid drive, and the distortion is worsened by larger drive voltages, the efficiency for push-pull class-A operation will be greater than that for single-tube operation with acceptable amounts of distortion. An efficiency of 30% for push-pull as compared with 20% for single-tube operation is typical.

Class-AB push-pull operation is also used and the analysis of operation is very similar to that explained for class A. The principal difference is the extended length of the resultant characteristic that permits a greater grid swing and yields efficiencies up to about 55%.

The resultant, composite characteristic of class-B push-pull operation, is shown in Fig. 5-21. The bias voltage is nearer to plate-current cutoff than in class-AB push-pull operation. Again, in this case, the characteristic approaches a straight line in shape so the plate-current waveform is fairly undistorted. Note that the bias is not adjusted exactly to the plate-current cutoff value. If exact cutoff bias were used, distortion would occur as seen in Fig. 5-22. Here, the resultant dynamic characteristic has an S shape which causes severe distortion of the output waveform. The plate efficiency of a class-B push-pull amplifier is about 60 to 65%. Table 5-1 summarizes the principal features of the different classes of amplifiers.

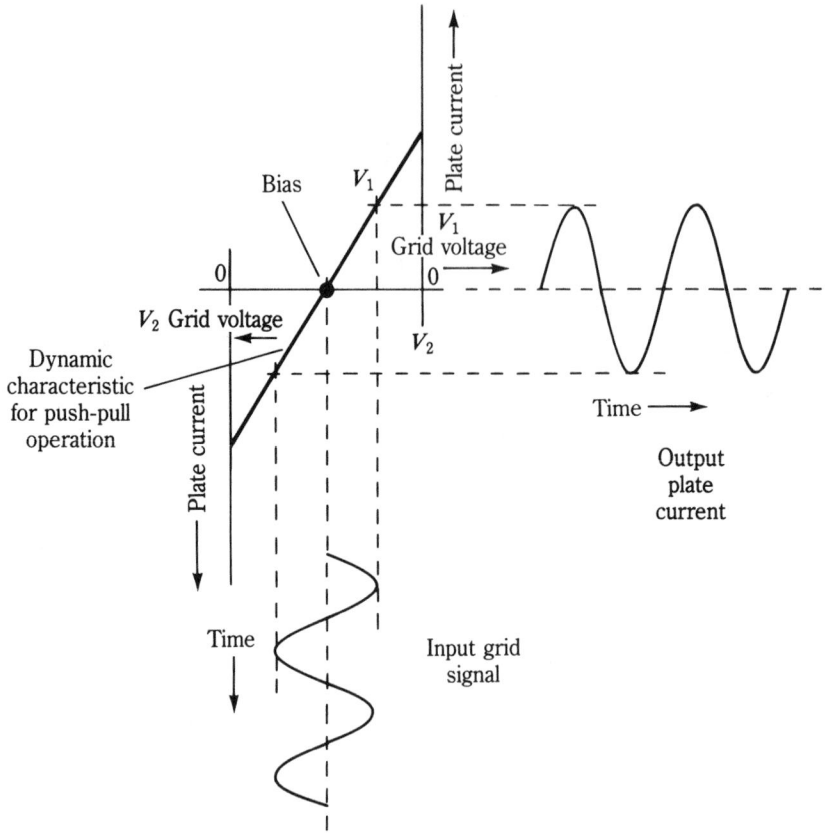

5-21 Dynamic characteristic of two tubes in push-pull class-B operation.

Methods of biasing

In the vacuum-tube amplifier circuits described so far, the grid bias has been supplied by batteries representing fixed dc supplies, in the interest of simplifying the explanations. This type of bias, when used in a practical circuit is referred to as *fixed bias*. However, the type of bias most often found in use is *self-bias* in which the bias is developed across a resistor by the action of the tube. The level of this bias voltage depends on the plate current or the grid voltage drive, depending on the type of self-bias employed. A *combination bias* is a mixture of fixed bias and self-bias.

The most easily understood self-bias, called *cathode bias*, depends upon the flow of plate current through a resistor placed between the cathode and common (ground) as seen in Fig. 5-23. Assume, in this figure, that the voltage drop across R_K is five volts. This makes the cathode five volts positive with respect to the grid so that the grid is negative five volts with respect to the cathode. Capacitor C_K has the function of bypassing the signal frequency components of plate current with the objective of having no significant ac drop in the impedance of the $R_K C_K$ parallel combination. In the absence of C_K, signal frequency current passing through R_K will cause degeneration of the amplifier

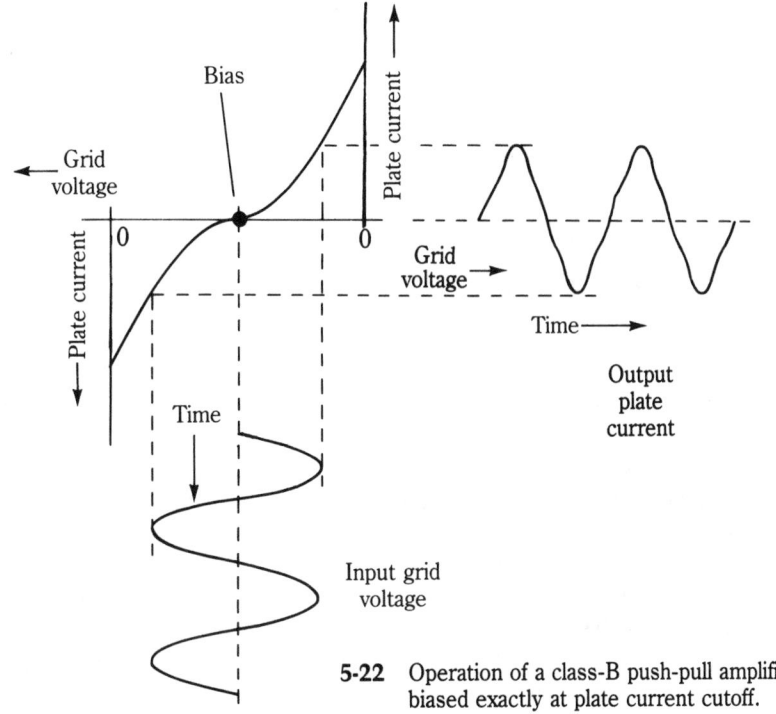

5-22 Operation of a class-B push-pull amplifier biased exactly at plate current cutoff.

Table 5-1. Amplifier characteristics.

Class	Location of operating point on dynamic characteristic	Relative distortion	Relative power output	Approximate percentage of plate efficiency
A single-tube		Low	Low	Under 20%
A push-pull	On linear portion	Very low	Moderate	20 to 30%
AB single-tube		Moderate	Moderate	40%
AB push-pull	Between linear portion and plate-current cut-off	Low	High	50 to 55%
B single-tube		High	High	40 to 60%
B push-pull	At vicinity of cut-off	Low	Very high	60 to 65%
C single-tube	About 1½ to 4 times plate-current cut-off	Very high	Very high	60 to 80%

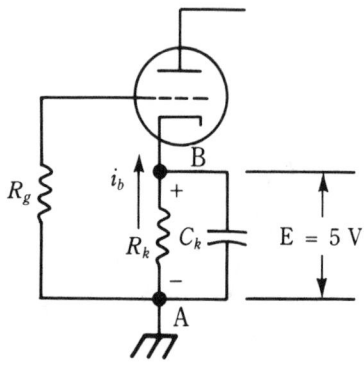

5-23 Cathode bias for a triode vacuum-tube.

gain, because the cathode voltage will have a signal frequency component in phase with the applied grid voltage. The net grid-to-cathode voltage will thereby be reduced to the detriment of the effective gain of the amplifier. It is common design practice to select C_K so that its reactance is about 10% of R_K at the lowest signal frequency to be amplified.

Grid-leak bias, or signal bias, is illustrated in Fig. 5-24. Look at Fig. 5-24A, the grid drive provided by e_g must be sufficiently great to make the grid positive at its peak. When the grid voltage is positive, grid current will flow and cause C_g to charge with the polarity shown. The charging will normally occur during a small fraction of the signal cycle. During the remaining part of the signal cycle, C_g will discharge in the direction shown and the drop across R_g will place the negative bias voltage on the control grid. This arrangement is particularly beneficial whenever the level of grid voltage drive is not known in advance because the bias level is self-adjusting. That is, in the event a greater signal voltage is applied to the grid, the grid will be driven more positive, a greater grid current will flow and the charge on C_g will be greater, so the bias level will be correspondingly greater.

The other version of grid-leak bias works in a similar way. Refer to Fig. 5-24B. The positive excursion of the grid voltage causes a charge to be placed on C_g in the same manner described in the preceding paragraph. The difference results from the discharge of the capacitor through R_g, which is located directly in parallel with C_g in the circuit.

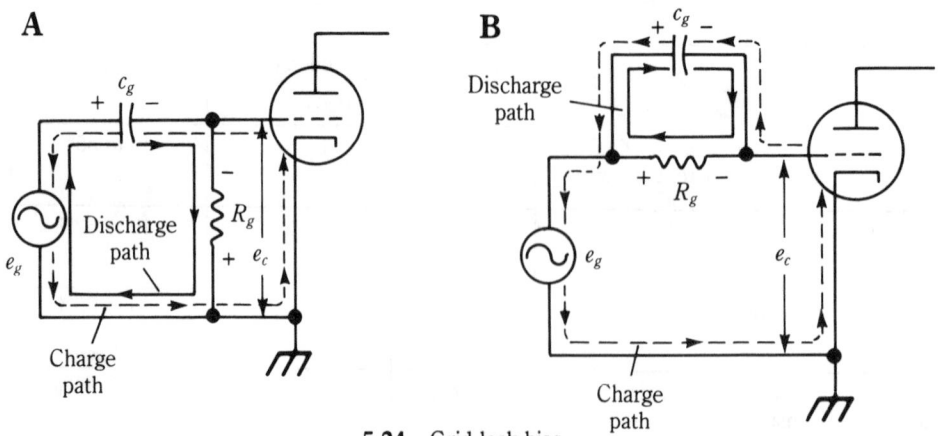

5-24 Grid-leak bias.

Nevertheless, the polarity of the grid bias is the same for either of the two circuits. Also this circuit causes the grid bias voltage level to automatically adjust to fit the grid drive should it increase to decrease for any reason.

Amplifier coupling methods

A single stage of power amplification normally is not sufficient for radio transmitters. To obtain the necessary gain, several stages must be connected together. The output of one stage then becomes the input of the next in cascading the amplifier stages. The basic methods of coupling are:

- Resistance-capacitance
- Impedance coupling
- Transformer coupling
- Link coupling (which is a special form of transformer coupling)

Each of these coupling types has been briefly described in chapter 4 except for link coupling, which is illustrated in Fig. 5-25. In this illustration two tuned circuits are shown, one in the output circuit of the driver and another in the input circuit of the power amplifier. A low impedance rf transmission line having a coil of one or two turns at each end is used to couple the plate and grid tank circuits. The coupling links or loops are attached to each circuit at its cold end (point of minimum rf potential). Circuits that are cold near one end are called *unbalanced circuits*. Link coupling systems normally are used where the two stages to be coupled are separated by a considerable distance. One side of the link is grounded in cases where capacitance coupling between stages must be eliminated or where harmonic suppression is important.

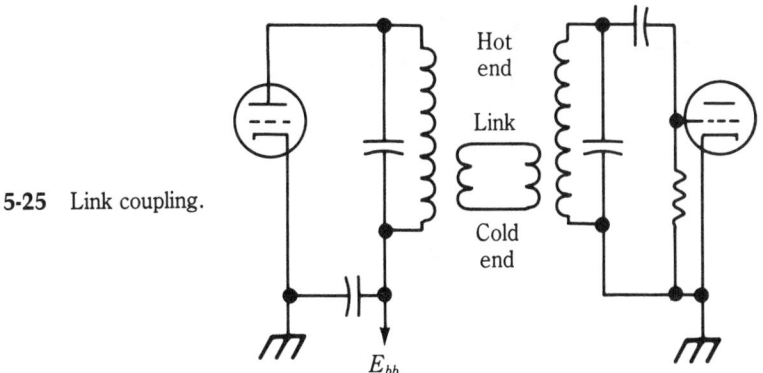

5-25 Link coupling.

Neutralization

Any amplifier will oscillate if sufficient energy having the same frequency and the same phase as the input signal is fed back from the output circuit to the input circuit. Feedback of the proper phase for oscillation (regenerative feedback) can take place through the grid-plate capacitance of a tube, or through external capacitance or inductive coupling between the output and input circuits. Although regenerative feedback is deliberately employed in oscillator circuits, it is undesirable in amplifier applications because of the resulting distortion and spurious oscillation. It is possible to eliminate these oscillations by a process called *neutralization*. In neutralization, a network is included in the amplifier that feeds back to the input a voltage of proper amplitude and polarity to cancel the regenerative feedback. Two of the principal neutralization methods are the plate neutralization system and the grid neutralization system.

Plate neutralization is shown in Fig. 5-26. This is a typical transformer-coupled rf amplifier in which a tapped primary for transformer T_2 is used. The voltage induced in L_2 of the primary is 180° out of phase with the induced voltage in L_1. The tapped point of this transformer is placed at rf ground by the low reactance of the rf bypass capacitor $C4$. Therefore, the rf voltages measured at points A and B are of opposite phase and of the same magnitude if the transformer is center-tapped. C_F represents the internal feedback capacitance. C_N is the capacitor through which the neutralizing signal is coupled to cancel the effects of C_F.

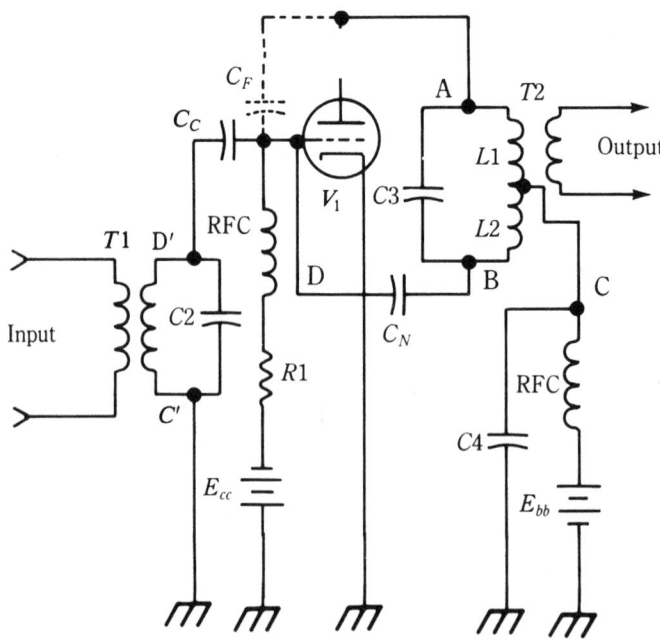

5-26   A plate neutralization circuit.

Another circuit that provides a means of neutralizing the effects of feedback capacitance is the grid neturalization circuit shown in Fig. 5-27. It differs from plate neutralization in that the split tank circuit is located in the input circuit. Note that the neutralizing signal coupled through C_N is connected to the opposite end of the transformer secondary as compared to the signal that is coupled through C_F. Since the two signals are in phase, but coupled into opposite ends of the secondary winding, the two are capable, with proper adjustment of C_N, of causing cancellation or neutralization.

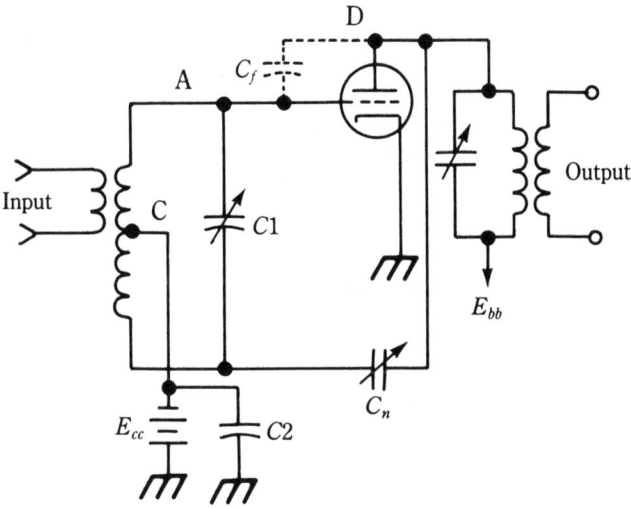

5-27 A grid neutralization circuit.

Multiple-choice questions

1. In rf amplifiers consideration must be given to parameters that can usually be neglected at lower frequencies. These are known as:

A. Alien parameters.
B. Distributed parameters.
C. Unknown parameters.
D. Hidden parameters.
E. Lost parameters.

2. The bandwidth of rf amplifiers is usually:

A. Relatively narrow.
B. Relatively wide.
C. Below 1 kHz.
D. From dc to the carrier frequency.
E. No wider than the carrier frequency.

3. The tuning of rf amplifier stages can be accomplished by varying the:

A. Transconductance.
B. Capacitance.
C. Inductance.
D. Resistance.
E. Both B and C.

4. When inductance is manually adjusted in an rf circuit to change the resonant frequency, how is the change usually accomplished?

    A. The number of turns in the winding is changed.
    B. The diameter of the conductor is altered.
    C. The winding direction is changed from clockwise to counterclockwise.
    D. The position of a powdered iron slug is changed.
    E. The winding direction is altered from counterclockwise to clockwise.

5. Regeneration (positive feedback) in an amplifier can be caused by:

    A. Distributed capacitance.
    B. Interconnection inductance.
    C. Internal transistor capacitance.
    D. All of the preceding are true.
    E. None of the preceding is true.

6. The deliberate construction of a feedback path containing a capacitor and its adjustment to prevent self-oscillation is known as:

    A. Deemphasis.
    B. Neutralization.
    C. Defuzing.
    D. Naturalization.
    E. Gain compensation.

7. One objective in building an amplifier with significant distortion is to accomplish:

    A. Self-oscillation.
    B. Neutralization.
    C. Frequency multiplication.
    D. Damping.
    E. Both A and B.

8. The purpose of a frequency multiplier is to:

    A. Provide a lower multiple frequency.
    B. Provide a difference frequency.
    C. Provide a modulated frequency with one set of sidebands.
    D. Provide the multiple of an original reference frequency.
    E. Establish an accurate frequency standard.

9. A bipolar transistor can be operated in the:

    A. Active region.
    B. Saturation region.

C. Cutoff region.

D. Above-pinchoff region.

E. D is the only incorrect answer.

10. Name a particular kind of FET that is currently being widely used in new power amplifier designs.

A. Junction FET.

B. Darlington pairs.

C. Cathode follower.

D. Gas diodes.

E. VMOS.

11. A certain frequency is specified for which the beta of a transistor is one. This frequency is designated as:

A. f_r.

B. f_1.

C. f_2.

D. f_T.

E. f_C.

12. At higher frequencies the problem of impedance matching is made more difficult by the presence of:

A. Reactive elements.

B. Resonance.

C. Antiresonance.

D. Electrons.

E. Holes.

13. One source of waveform distortion at higher frequencies is caused by:

A. Carbon granularity.

B. Charge storage.

C. Sticky electrons.

D. Crystal imperfections.

E. Radioactive ionization.

14. An increased collector (drain) efficiency can signify:

A. Increased supply power.

B. Decreased supply power.

C. Increased output signal power.

D. Decreased output signal power.

E. Either B or C, or both, will cause increased efficiency.

15. In attempting to realize more efficient amplification through switching, the ultimate improvement appears to be limited by:

  A. The slope of the load line.
  B. Power supply regulation.
  C. Transformer saturation.
  D. Switching transition speed.
  E. Capacity of the heatsink.

16. The operating point of a class-A power amplifier must be established:

  A. Beyond pinch off.
  B. In the active region.
  C. Below saturation.
  D. At cutoff.
  E. Above the noise region.

17. For the class-A power amplifier the collector (drain) efficiency is a maximum of:

  A. 25%.
  B. 50%.
  C. 75%.
  D. 78.7%.
  E. 90%.

18. How can Class-B efficiency be achieved in power amplifiers without excessive distortion products?

  A. Center-tapped transformers.
  B. Complementary transistors.
  C. Bias to cutoff.
  D. Field-effect transistors.
  E. Tuned circuits.

19. Impedance matching of transistor rf amplifiers to their loads can often be conveniently done with:

  A. L-sections.
  B. Parallel LC circuits.
  C. AF transformers.
  D. Series LC circuits.
  E. Darlington pair transistors.

20. One high-efficiency amplifier class has a part of the output circuit tuned to the rf third harmonic. Name it.

  A. Class-A.
  B. Class-B.
  C. Class-C.

D. Class-D.
E. Class-F.

21. Vacuum-tube amplifiers are commonly used in what amplifier class or classes?

   A. Class-A.
   B. Class-B.
   C. Class-C.
   D. Class-F.
   E. All of the answers are true.

22. The resonant plate circuit of the class-C vacuum-tube amplifier is called the:

   A. Resonator.
   B. Amplexor.
   C. Wave smoother.
   D. Tank.
   E. Holding circuit.

23. In order to have a class-C amplifier operate efficiently as a frequency multiplier:

   A. The tank circuit should be tuned to the harmonic frequency.
   B. The grid circuit should have a strong negative bias.
   C. A strong grid voltage drive is necessary.
   D. Both A and C.
   E. A, B, and C.

24. Linear vacuum-tube amplifiers can be operated:

   A. Class-A.
   B. Class-B.
   C. Class-C.
   D. Class-AB.
   E. A, B, and D.

25. Phase reversal for the input to a push-pull amplifier can be obtained by a:

   A. Phase splitter.
   B. Center-tapped transformer.
   C. VMOS FET.
   D. Complementary transistor.
   E. Both A and B.

26. In a vacuum-tube amplifier the use of separate power supplies (batteries), to establish the grid bias is known as:

   A. Cathode bias.
   B. Fixed bias.
   C. Override bias.

D. Self-bias.

E. Negated bias.

27. Cathode bias of a vacuum-tube amplifier requires adding something in the cathode circuit. It is:

A. Resistance.

B. Inductance.

C. Capacitance.

D. Both A and B.

E. Both A and C.

28. Cathode bias cannot be used with what class of operation?

A. Class-A.

B. Class-B.

C. Class-C.

D. Class-AB.

E. None of the choices is true.

29. Grid-leak bias and cathode bias are both forms of:

A. Variable bias.

B. Fixed bias.

C. Adjustable bias.

D. Self-bias.

E. Class-C bias.

30. Neutralization in a vacuum-tube amplifier stage prevents oscillation due to positive feedback through the:

A. Power supply.

B. Cathode resistor.

C. Grid-plate capacitance.

D. Screen grid.

E. Suppressor grid.

Basic problems

1. What are the bandwidth requirements of an rf amplifier?

2. Specify how the inductance or capacitance can be varied to change the resonant frequency.

3. Explain the basic principle of neutralization.

4. What is the fundamental requirement of an amplifier circuit in order that it be the basis of frequency multiplication?

5. A class-B push-pull amplifier draws 8 A, exclusive of bias resistor arrangements in the base circuit, from a 24 V supply. If the rf power generated is 100 W, what is the collector efficiency of the amplifier?

6. In what way are transistor power amplifiers less efficient than the vacuum-tube amplifiers?

7. Explain why the class-F amplifier is more efficient than the class-C amplifier.

8. The power delivered by the power supply of a class-C vacuum-tube amplifier is 50 kW with an input of 61.35 kW. Calculate the efficiency.

9. Briefly explain why a linear amplifier cannot be operated class C.

10. Explain the effect of placing too little capacitance in parallel with the resistor for cathode bias.

Advanced problems

1. In the output circuit of a frequency multiplier, a parallel resonant circuit is tuned to the third harmonic of the 10 MHz input frequency. Assuming a Q of 100 and an impedance at the resonant frequency of 100 kΩ, find the impedance at the fundamental frequency.

2. Why are distributed parameters (the inductance of leads and the capacitance between leads and other conducting surfaces), more important as operating frequencies are increased?

3. Discuss the limitations of class-C amplifiers with regard to their use as the final power amplifier in practical transmitters.

4. Explain the role of the conduction angle in establishing the efficiency of a class-C amplifier.

5. Explain the essential feature of a class-D power amplifier.

6. Describe the flywheel action of the class-C rf tank circuit.

7. A 50-kW, class-C rf amplifier requires 65.5 kVA at a 95 percent power factor. Calculate the efficiency.

8. Explain how the dynamic characteristic for two vacuum-tubes operating in push-pull is constructed.

9. The plate current of a vacuum-tube is 8 mA, and the grid bias for proper operation is −6 V. Design the circuit for cathode bias in the frequency range 50 Hz to 16 kHz.

10. A class-C amplifier is to use grid-leak bias. If the grid current during positive grid peaks averages 5 mA during 1 percent of the cycle, how much resistance is required for a -5 V grid bias?

6

Oscillators

AN AMPLIFIER IN A RADIO FREQUENCY (rf) TRANSMITTER IS RESPONSIBLE FOR RAISING the power level of its input signal. Trace backwards from the transmitter's final amplifier, you will ultimately arrive at a stage capable of producing a continuous rf power output without any input from a proceeding stage. Such a stage is called an *oscillator*, its primary function is to generate a given waveform (sine, square, pulse, sawtooth, triangular, etc.) at a constant peak-to-peak value and at a specific frequency. These desirable traits are achieved by the partial conversion of a dc input power into an ac output power.

1. The voltage feedback from the output circuit is the signal to the input circuit. It must therefore provide the correct action at the input so as to aid the operation of the circuit. This is known as positive or regenerative feedback.

2. The amount of the positive feedback must be sufficient to compensate for the losses naturally occurring in the circuit.

An oscillator circuit must also contain a dc supply for its operation, an active device (op amp, transistor or tube), a frequency determining network, and a positive feedback path. The various oscillators can then be classified according to the method of providing the positive feedback.

The topics of positive feedback and the individual oscillators are covered in the following sections:

- Positive or regenerative feedback
- Reaction oscillators
- Direct feedback oscillators

- Indirect feedback oscillators
- Frequency stability
- The crystal and the piezoelectric effect
- Crystal oscillators
- RC sine wave oscillators
- Negative resistance oscillators
- Relaxation oscillators
- Parasitic oscillations

Positive or regenerative feedback

Positive feedback can be used to increase the gain of an amplifier circuit. This is illustrated in Fig. 6-1 where V_i is the input signal from the preceding stage. Such a signal is applied between the base of a transistor (or the control grid of a tube) and ground. The output signal, V_o, appears between the collector (or plate) and ground and a fraction, beta (β), of this output is fed back to the input circuit so this feedback voltage, $+\beta V_o$, is in phase with V_i. β is called the feedback factor, which can either be expressed as a decimal fraction or a percentage.

In order for the feedback to be positive, there must be a total of 360° phase shift (equivalent to zero phase shift) around the feedback loop, base→collector→base (or grid→plate→grid). The total signal voltage when applied between the base and emitter

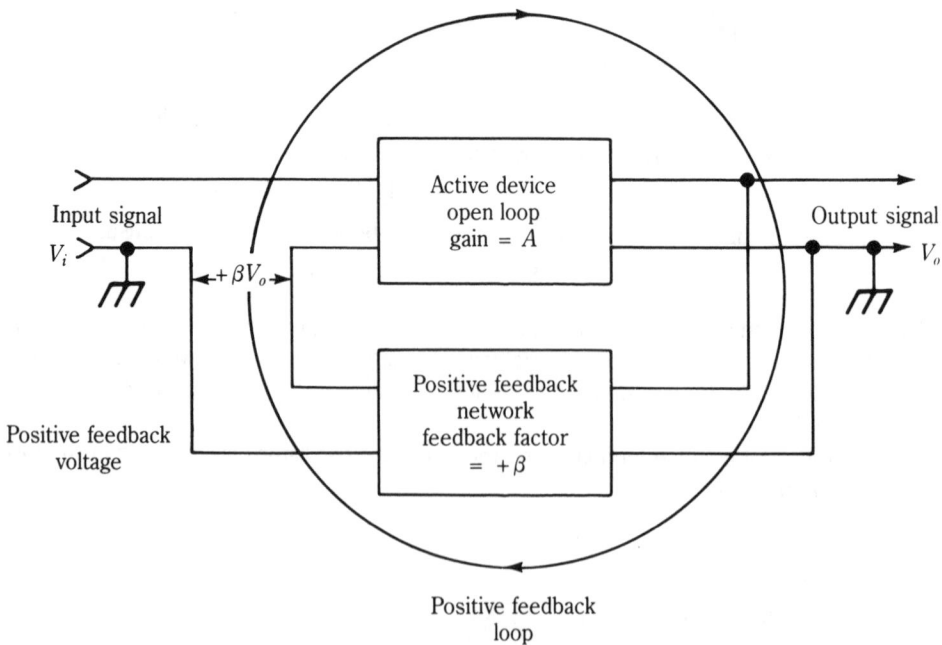

6-1   Amplifier employing positive feedback.

(or control grid and cathode) is the sum of the input signal, V_i, and the positive feedback voltage, $+\beta V_o$. Since the voltage gain (open loop gain) of the active device is A,

$$\text{Output signal, } V_o = A \times (V_i + \beta V_o) \tag{6-1}$$

This yields

$$A' = \frac{V_o}{V_i} = \frac{A}{1 - A\beta} \tag{6-2}$$

where A' is the overall voltage gain with the positive feedback present (closed loop gain). There are three possible conditions that are associated with Equation 6-2.

1. If A and β are chosen so that the value of $A\beta$ is less than 1, then A' is greater than A and the amplifier's gain has been increased as the result of the positive feedback. Such is the case with the so-called regenerative amplifier.

2. If $A\beta = 1$ (for example $A = 10$ and $\beta = 0.1$ or 10%), A' is theoretically infinite. This means that the circuit can provide a continuous output without any input signal from the previous stage. This is the condition for a stable oscillator.

3. If $A\beta < 1$, the oscillator is unstable. The output, V_o, increases which tends to reduce the value of A until the equilibrium condition of $A\beta = 1$ is reached.

The condition for oscillation is therefore $A\beta = 1\underline{/0°}$; sometimes referred to as the *Barkhausen* or *Nyquist* criterion. The inclusion of " $\underline{/0°}$ " in the polar value of $A\beta$ means that the resultant phase shift around the loop is zero degrees and consequently the feedback is positive. By contrast an angle of 180° would indicate that the feedback is negative.

In a sinewave oscillator the frequency determining network is (in most cases) some form of tuned LC circuit (Fig. 6-2A). If a charged capacitor is connected across an inductor, the capacitor will discharge and establish a magnetic field around the inductor. When the discharge is completed, the magnetic field collapses and a counter emf is induced in the coil. This emf will recharge the capacitor whose voltage has its polarity reversed when compared with the original condition. The sequence of events is then repeated so that an alternating sinewave current exists between the inductor and the capacitor. This is sometimes referred to as the *flywheel effect* which occurs at the frequency, f_o, of free oscillation. The formula for f_o is:

$$\text{Frequency, } f_o = \frac{1}{2\pi}\sqrt{\frac{1}{LC} - \frac{R^2}{4L^2}}$$

$$\approx \frac{1}{2\pi\sqrt{LC}} \text{ Hz} \tag{6-3}$$

where R is the resistance of the coil.

Owing to the power dissipated as heat in the coil's resistance, the flywheel current is in the form of a damped oscillation (Fig. 6-2B) that gradually decays to zero. In order to sustain a continuous oscillation, you must supply positive feedback to the tuned circuit and thereby compensate for the losses that naturally occur.

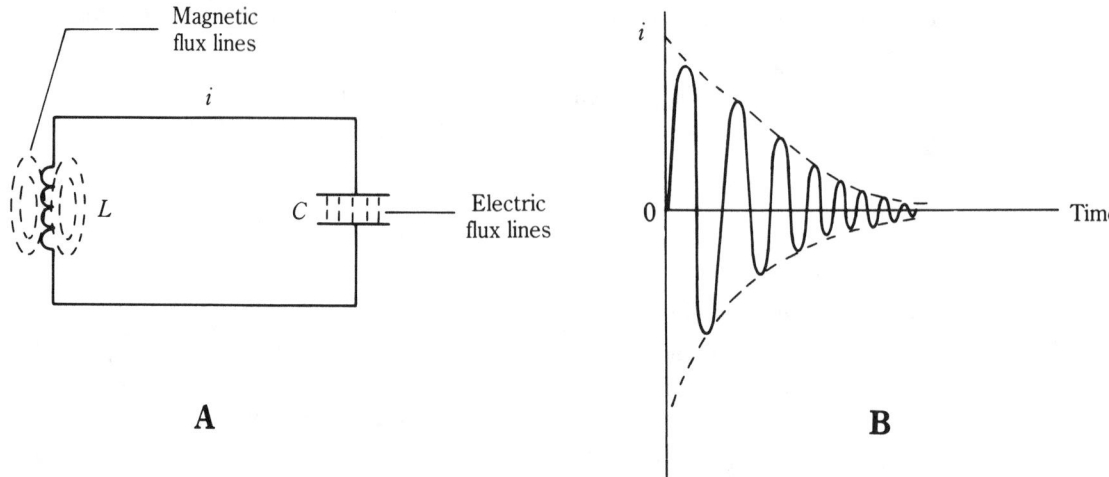

6-2   Damped oscillation in an LC circuit.

Figure 6-3 shows the principle of positive feedback in an oscillator circuit. Assume that the input signal is 1 V rms and the voltage gain of the active device is 10, the output signal is 10 V rms: If the feedback factor is 1/10 or 10%, the 10 V output signal will be responsible for creating the 1 V input signal (this does not mean that there is only 9 V left at the output signal!). This argument sounds rather like the chicken and the egg so the

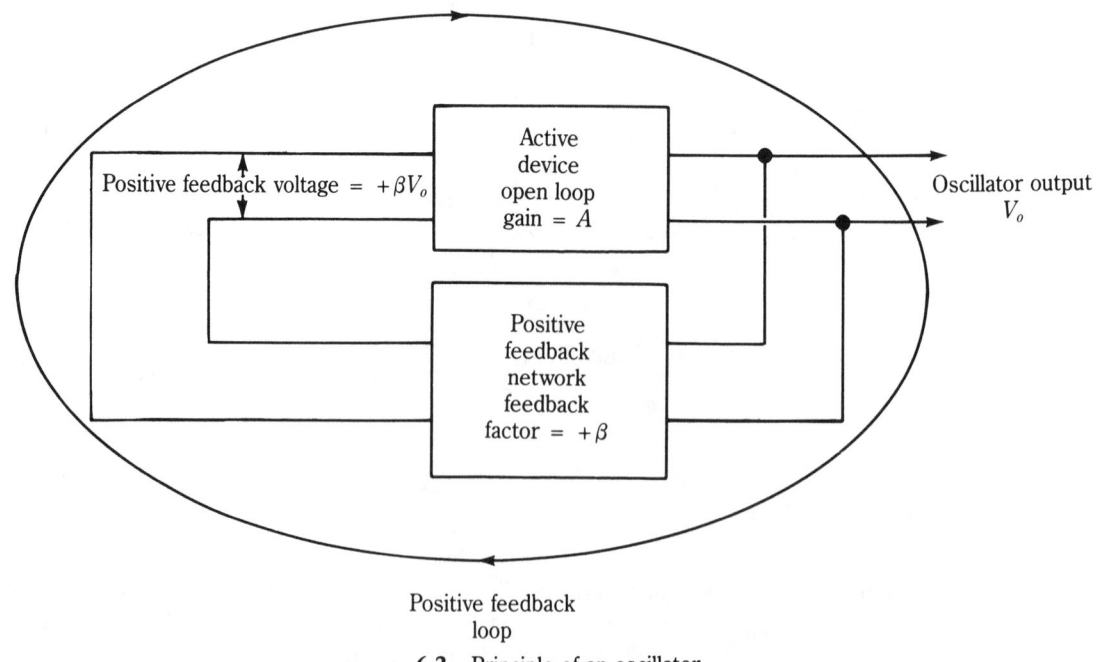

6-3   Principle of an oscillator.

question arises "How does the circuit get started in the first place?". There are two possible answers:

1. When the dc power supply is applied, the resulting transient conditions will cause the tuned circuit to be shocked into oscillation. The positive feedback takes control and builds up the oscillation.
2. All active devices are inherently noisy. Since the noise is spread through the frequency spectrum, it will contain a component at the frequency of oscillation. This component will trigger the positive feedback network increasing the oscillation until the equilibrium condition of $A\beta = 1 \underline{/0°}$ is reached.

You have studied the requirements for an oscillator circuit to operate successfully. Now look at the various oscillator circuits and see how the requirements are fulfilled.

Reaction oscillators

In reaction oscillators the positive feedback is provided by *mutual* (inductive) *coupling*, otherwise referred to as transformer action. One oscillator circuit which falls into this category is the *tuned collector* (tuned plate) *type* whose circuit diagrams are shown in Fig. 6-4A, and B.

When the switch, S, is closed in Fig. 6-4A and the V_{CC} voltage is applied to the circuit, the transistor is forward biased by the inclusion of R, and consequently there is a surge of collector current that shocks the tank circuit into oscillation at its resonant frequency. The magnetic flux surrounding $L1$ cuts $L2$ and induces a signal voltage which is fed to the base. Since we are using a common-emitter arrangement, there is 180° phase change between the signal voltage at the base and the output signal voltage at the collector. The coils $L1$ and $L2$ are normally wound close together on a common former; and provided the windings are in the same direction, there will be a further 180° phase change between the output collector signal voltage applied to $L1$ and the induced feedback signal from $L2$ to the base. Consequently there is a total of 360° or zero phase shift around the loop of base→collector→base and the feedback is positive. The degree of feedback will be primarily determined by the amount of spacing between the coils and, provided the feedback is sufficient, the oscillation will build up to the equilibrium condition. The frequency of the oscillator's output between the collector and ground is:

$$\text{Output frequency, } f_o = \frac{1}{2\pi\sqrt{L_1 C}} \text{ Hz} \qquad (6\text{-}4)$$

The capacitor, C, is a signal bypass capacitor that prevents the forward bias on the base from being shorted to ground through the lower portion of $L2$. The tapping point on $L2$ provides the feedback voltage with an impedance match between the inductor and the low input impedance of the transistor. Because the transistor is heat-sensitive the parallel combination of R_E and C_E stabilizes the circuit against temperature variations.

The tube version shown in Fig. 6-4B is a tuned plate oscillator whose action is similar to that of the solid-state version. There is a 180° phase change from the control grid to the plate owing to the amplifying action of the pentode. The mutual coupling between

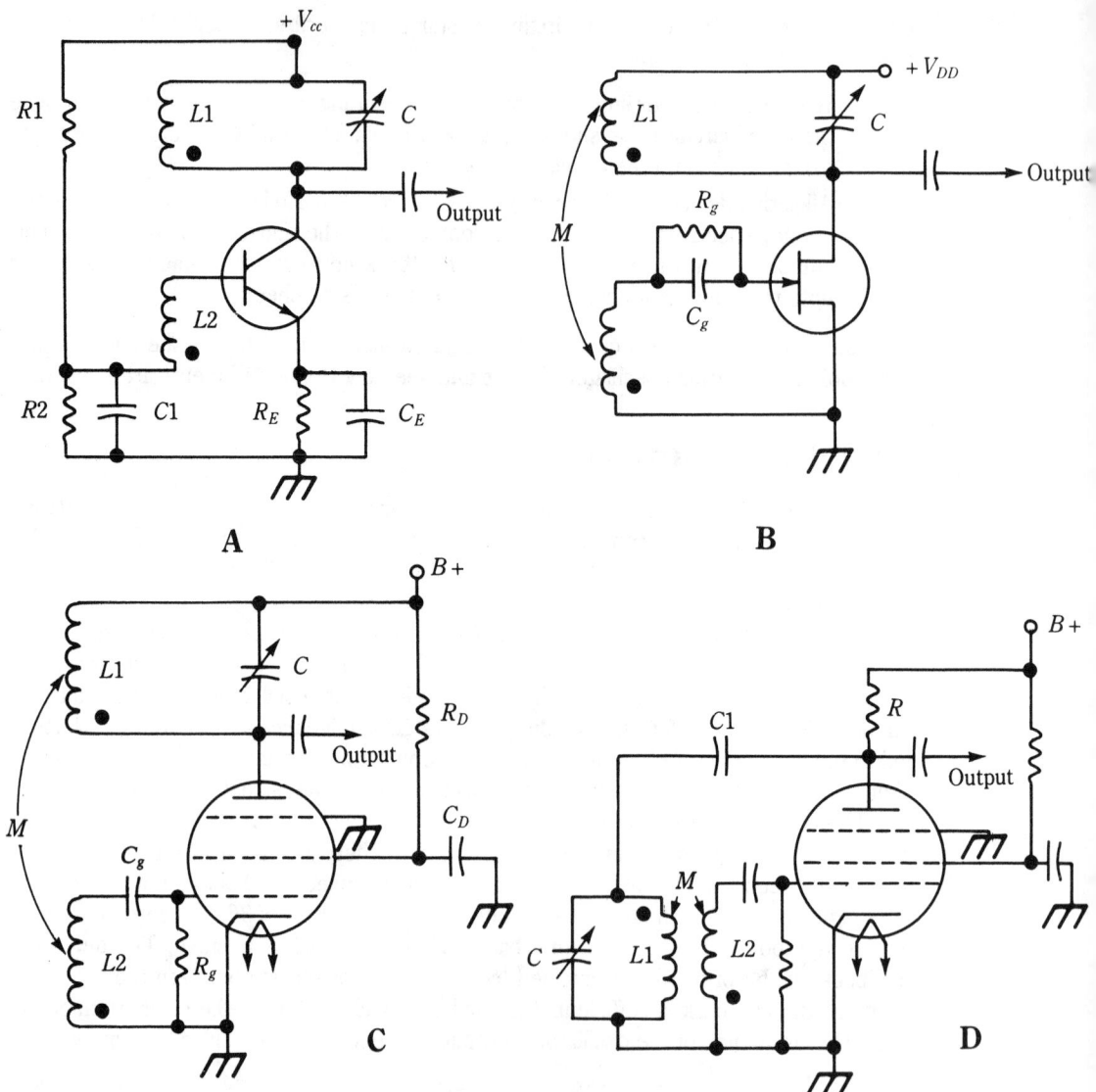

6-4   Examples of the tuned output oscillator.

$L1$ and $L2$ completes the loop and introduces another $180°$ phase change so that the feedback is positive.

Mathematical analysis of an oscillator circuit reveals two results—the generated frequency, and the critical amount of feedback required to sustain a continuous oscillation. For the tuned plate oscillator the equations are:

$$\text{Generated frequency, } f_o = \frac{1}{2\pi\sqrt{L_1 C}} \times \sqrt{1 + \left(\frac{R}{r_p}\right)} \text{ Hz} \qquad (6\text{-}5)$$

$$\text{Critical mutual inductance, } M = \frac{L_1 + CRr_p}{\mu} \tag{6-6}$$

where R = rf resistance of the coil, $L1$.

These equations involve the ac plate resistance, r_p, and the amplification factor, μ. Since both these parameters depend on the value of the plate supply voltage, any fluctuation of this voltage will cause the generated frequency to change. A poorly regulated and filtered power supply will therefore produce a form of frequency drift known as *dynamic instability*. For this reason it is common practice to use a separate well regulated power supply to operate the oscillator circuit.

The tuned plate circuit is an example of a variable frequency oscillator (VFO), which can operate over a complete frequency range by adjusting the value of $C1$. For example, if $C1$ is reduced, Equation 6-5 shows that the output frequency is increased. However, a smaller value of $C1$ lowers the required critical value of M (Equation 6-6). It follows that, since the mutual inductance is constant, the oscillation will increase until a new equilibrium condition is obtained. This tendency for the output to increase as the frequency is raised is common to most oscillator circuits.

Tube oscillators commonly use grid leak or signal bias which is provided by the combination of C_g and R_g (Fig. 6-5). During part of the feedback voltage cycle the control grid is driven positive with respect to the cathode causing grid current to flow and charge the capacitor toward the peak value of the grid signal. When the grid current ceases, C_g discharges through R_g but their time constant is designed to be high compared with the period of the oscillation. Consequently the capacitor discharges slowly and the resulting bias is approximately equal to the peak value of the grid signal. However, should the oscillation cease, the bias will vanish and a heavy plate current will flow. The presence of oscillation can therefore be detected by measuring:

1. a dc bias across either C_g or R_g
2. a low dc level of plate current
3. a dc current through R_g with the direction of the electron flow from the grid to ground

Grid leak or signal bias is ideal for oscillators because:

1. There is initially no bias, so a large transient plate current flows to start off the oscillation.
2. As the oscillation builds up, the bias also increases and the circuit is naturally brought up to its equilibrium condition of $A\beta = 1 \underline{/0°}$. If the oscillation is sufficiently strong, the resulting bias could provide class-C conditions with its high plate efficiency (up to 85%). However class-C operation can also generate appreciable harmonics.
3. Any changes that occur in the feedback loop will alter the amount of the feedback voltage to the control grid. This will change the amount of the signal bias and the compensating action will tend to stabilize the amplitude of the rf output between the plate and ground. For the same reason grid-leak bias commonly appears in rf power amplifier stages.

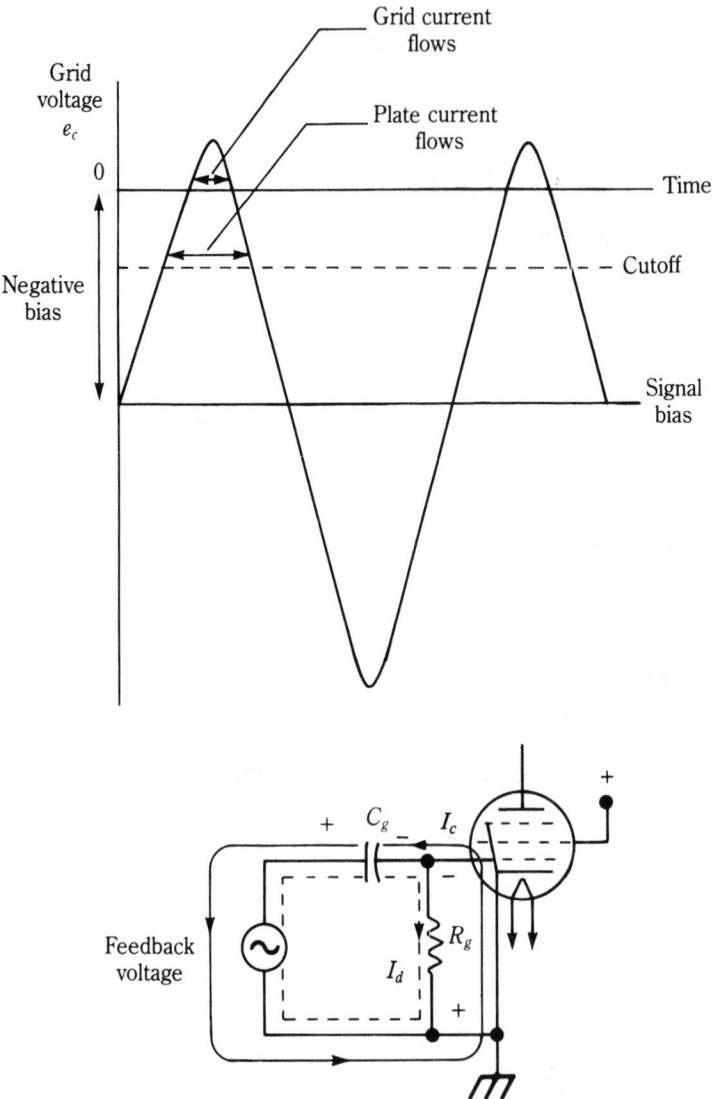

6-5   Principle of signal bias.

The value of the tuned circuit's Q is one of the principal factors in determining the oscillator's degree of frequency stability. If, for any reason, a phase shift occurs in the feedback loop, the oscillator changes its frequency in order to restore the condition $A\beta$ = 1 $\underline{/0°}$. A high Q circuit produces a large change of phase angle for a small shift in frequency and therefore the oscillator that contains such a circuit, will possess a high degree of frequency stability.

Shunt or parallel feeding of an oscillator circuit is one method of improving the frequency stability. The circuits of Fig. 6-4A, and B are series-fed in the sense that the tuned circuit and the active device are connected in series across the dc supply. In the shunt-fed arrangement (Fig. 6-4C) the tuned circuit is in parallel across the active device whose load also contains the resistor, *R* (an rf choke is sometimes used instead of a resistor). As a result the inductor of the tuned circuit is isolated from the dc level of the collector current and the tuned circuit, as a whole, is not affected by the collector's dc voltage; the capacitor, *C*, is included to prevent the dc level of plate voltage from being shorted to ground through *L2*. In a series-fed circuit any fluctuation in the dc power supply due to poor filtering and/or regulation is bypassed by the tuned circuit and is applied directly to the active device; the result is dynamic instability. However, in the shunt-fed circuit, the fluctuation in the dc power supply is divided between *R* and the active device. Therefore, less of the fluctuation reaches the active device and the dynamic instability is reduced. However, as far as the oscillation is concerned, the resistor R is across the tuned circuit and lowers its Q; as a result there is a greater amount of frequency drift.

The *Armstrong oscillator* is another type of oscillator that relies on mutual coupling between *L1* (the so-called tickler coil) and *L2* for its feedback; its series npn transistor version is shown in Fig. 6-6A, while Fig. 6-6B illustrates an n-channel JFET Armstrong circuit which includes signal bias. The tuning capacitor no longer forms part of the collector load but is instead connected across the coil in the base circuit. The method of providing the positive feedback is the same; however, if the output is still taken from the collector, its waveform will contain appreciable harmonics due to the nonlinear operation of the active device. To avoid the harmonics the output can be taken from the base circuit, but the size of this output will be less because it does not receive the transistor's amplification between the base and the collector.

In each of the reaction oscillator circuits only one of the coils is primarily involved in determining the generated frequency. This can be a disadvantage in generating a low frequency where large values of inductance are required.

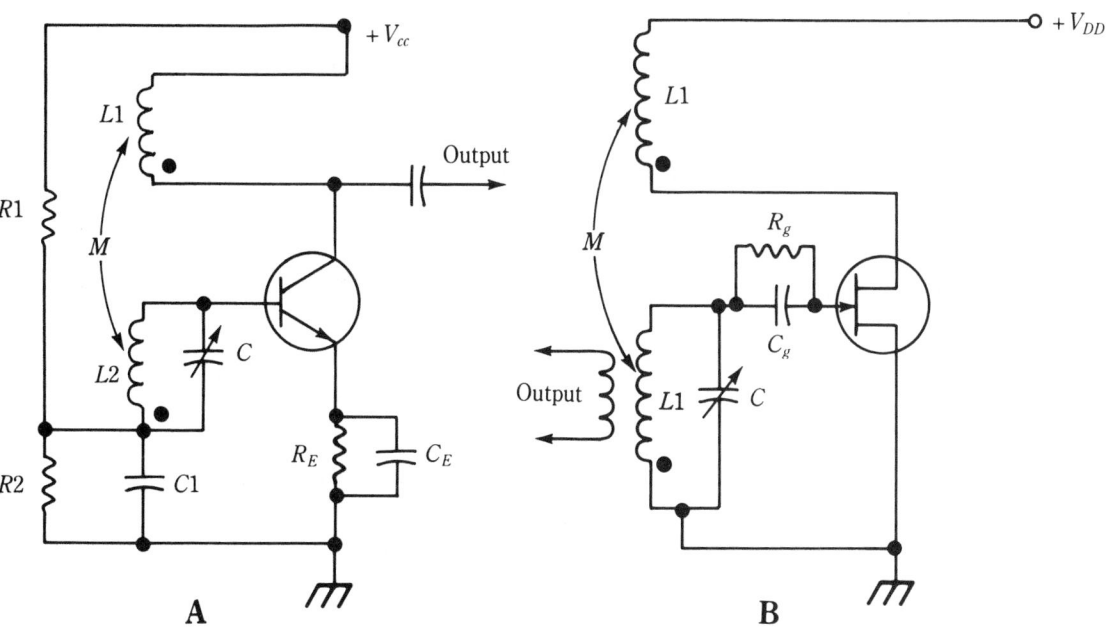

6-6   Examples of the Armstrong oscillator.

Direct feedback oscillators

In Hartley oscillators the feedback voltage is developed across a component that is common to both the output and input circuits. Figure 6-7A, B, and C shows three types of Hartley oscillator. The transistor and tube versions are series-fed but the FET arrangement is shunt-fed. However, all three circuits contain a tapped coil which is the distinguishing feature of this type of oscillator. This coil is common to both the input and output circuits, since the portion L1 belongs to the output circuit, while L2 is part of the input circuit.

Look at the BJT transistor arrangement in detail. The capacitor, C1, is a bypass capacitor to prevent the rf current from entering the V_{cc} supply. The tapping point, E, is effectively at rf ground and is therefore at emitter potential as far as the oscillation is concerned. The top of L1 (point C) is directly joined to the collector while the bottom of L2 (point B) is connected to the base through the bypass capacitor, C2, which prevents the V_{cc} voltage from being applied to the base through L2. Because the npn transistor is arranged in the common-emitter (CE) configuration, there is 180° phase change from the base to the collector as the result of the amplifying action. Since points C and B are at opposite ends of the tapped coil, these points carry rf voltages that are 180° out of phase with respect to the tapping point, E (rf ground). Consequently there is a further 180° phase change across the coil, and a total shift of 360° around the loop of base→collector→base. The feedback is therefore positive with the feedback voltage developed across L2 and applied between the base and emitter.

The amount of feedback, and hence the oscillator's output from the collector, can be easily determined by the position of the tapping point on the coil. The higher the position of the tapping point, the greater the amount of feedback and the higher the oscillator's output. Because the feedback is easily controlled, the Hartley circuit oscillates readily over a wide range of frequencies.

The generated frequency is approximately:

$$\text{Frequency of oscillation, } f_o = \frac{1}{2\pi\sqrt{(L_1 + L_2 + 2M)C_1}} \text{ Hz} \qquad (6\text{-}7)$$

where M is the mutual inductance between L1 and L2.

Notice that although there is normally mutual coupling between L1 and L2, the circuit will still oscillate through direct feedback even if L1 and L2 are shielded from each other. The frequency is determined by both L1 and L2 and therefore for a given total inductance, the Hartley oscillator generates a lower frequency than either of the reaction oscillators previously discussed. For this reason the Hartley circuit is commonly used in the lower rf ranges where the frequency can extend as high as tens of MHz.

Owing to nonlinear operation of the transistor the oscillator will generate harmonic voltages that appear between the collector (point C) and rf ground (point E). In Fig. 6-8 these harmonic voltages are divided between C and L2. For the high harmonic frequencies the reactance of C1 is low while the reactance of L2 is high. Consequently there is appreciable feedback across L2 and the Hartley oscillator's output is rich in harmonics. In some applications the presence of harmonics is a definite disadvantage, while in other circuits they serve a useful purpose.

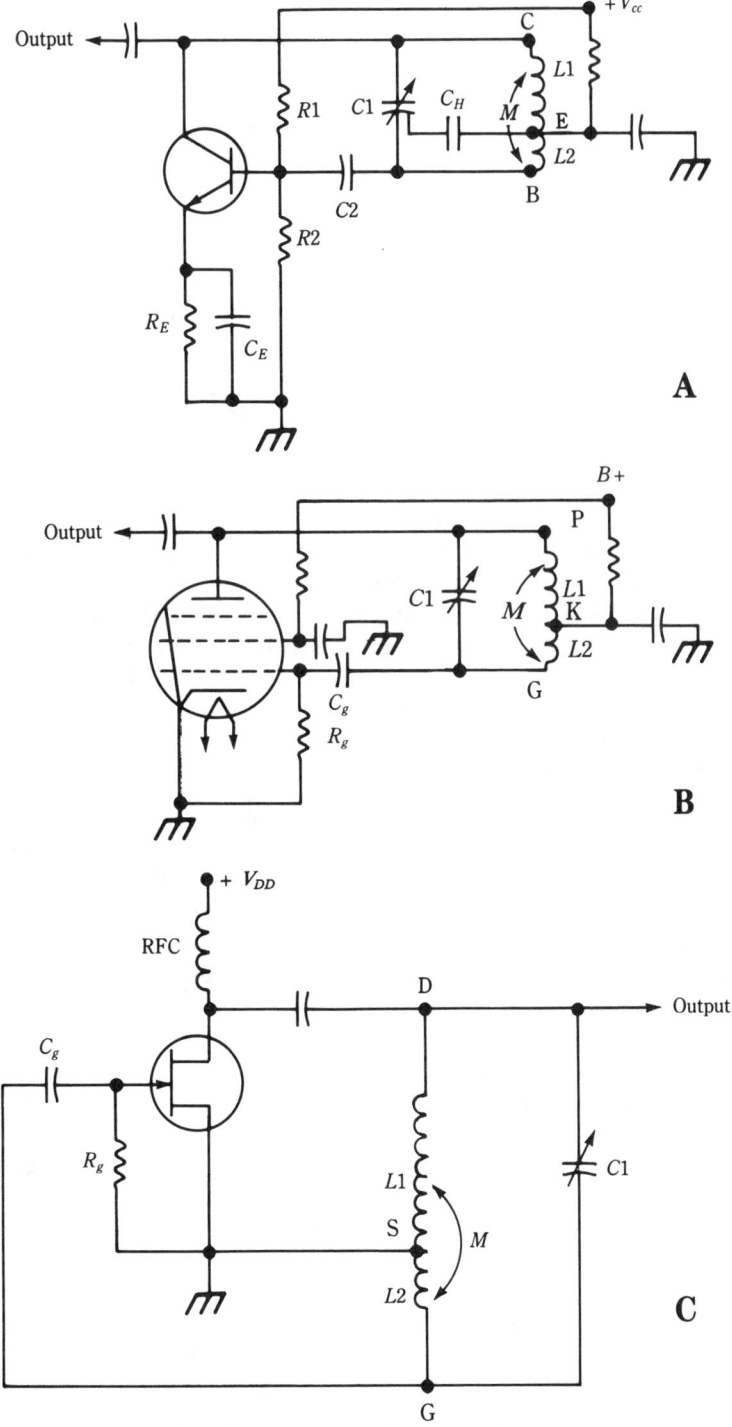

6-7 Examples of the Hartley oscillator.

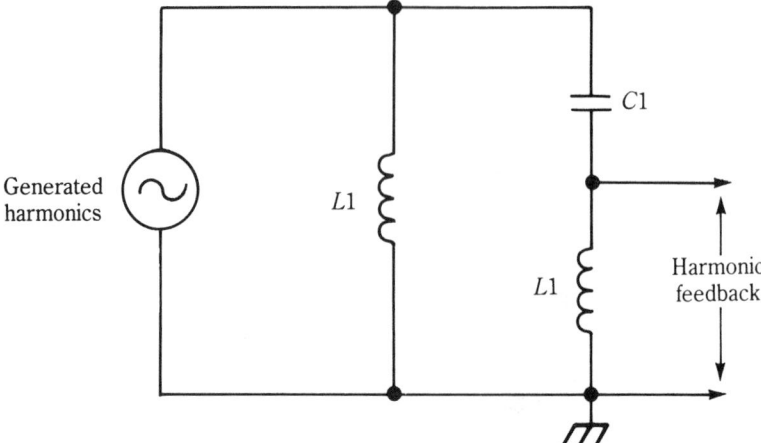

6-8 Harmonic feedback in the Hartley oscillator.

Notice that when the variable capacitor, $C1$, is manually tuned, the presence of the hand will introduce a so-called *hand capacitance*, C_H, between the capacitor's rotor and ground. As shown in Fig. 6-7A, C_H is in parallel with $C2$ and will therefore affect the generated frequency; when the hand is removed there will be a shift in the output frequency. To avoid this problem the tuning capacitor can be remotely controlled; as an example, the tuning control and the rotor can be connected by a belt.

The Colpitts oscillator

The Colpitts oscillator is very commonly used, especially in the high-frequency range of hundreds of MHz. Four versions (transistor, tube, FET, and op amp) are shown in Fig. 6-9A, B, C, and D, and in each there is a capacitive voltage divider $C1$, $C2$, which is the distinguishing feature of the Colpitts oscillator. In the Hartley circuit the direct feedback was taken across the inductor, $L2$, while in the Colpitts oscillator the positive feedback is developed across the capacitor, $C2$. It is common practice to vary the values $C1$ and $C2$ so their ratio remains constant and the feedback factor does not change with frequency.

All the arrangements with the single active device are shunt-fed, since the active device could not conduct if the positive dc supply were series connected to the lead joining the capacitors $C1$ and $C2$. $C3$ is included to prevent the dc collector voltage from being applied to the base of the BJT.

Examine the n-channel JFET version in detail, there is 180° phase change from the gate to the drain due to the amplifying action of the device. Because the points D (drain) and G (gate) are connected to opposite ends of the tuned circuit, these points will carry rf voltages that are 180° out of phase with respect to the grounded point, S (source). Consequently there is a further 180° phase change across the tuned circuit and a total of 360° (zero) phase shift around the positive feedback loop gate→drain→gate. The components C_g, and R_g provide signal bias by the use of gate current.

In Fig. 6-9A the capacitors $C1$, and $C2$ are in series across the coil L, while the tuning capacitor, $C3$, is in parallel. Therefore, ignoring the parameters of the active device, the output frequency from the drain is given by:

$$\text{Frequency of oscillation, } f_o = \frac{1}{2\pi\sqrt{LC_T}} \text{ Hz}$$

where

$$C_T = C_3 + \frac{C_1C_2}{C_1 + C_2} \tag{6-8}$$

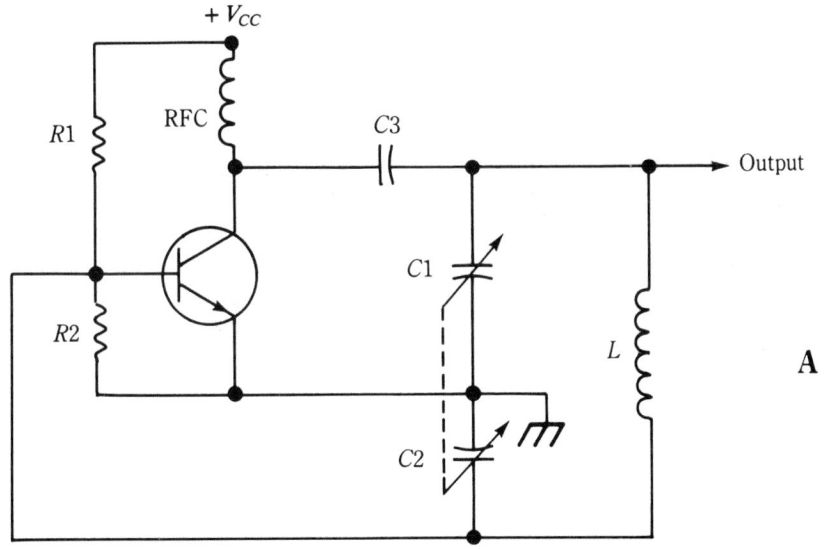

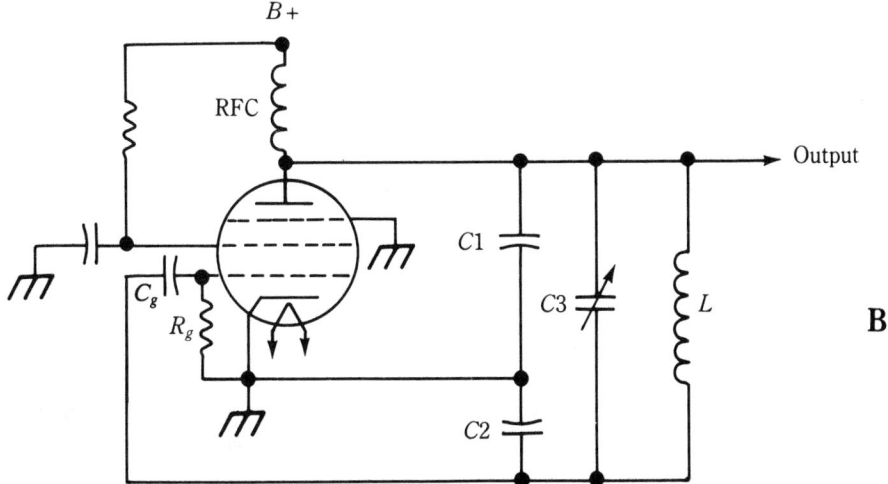

6-9 Examples of the Colpitts oscillator.

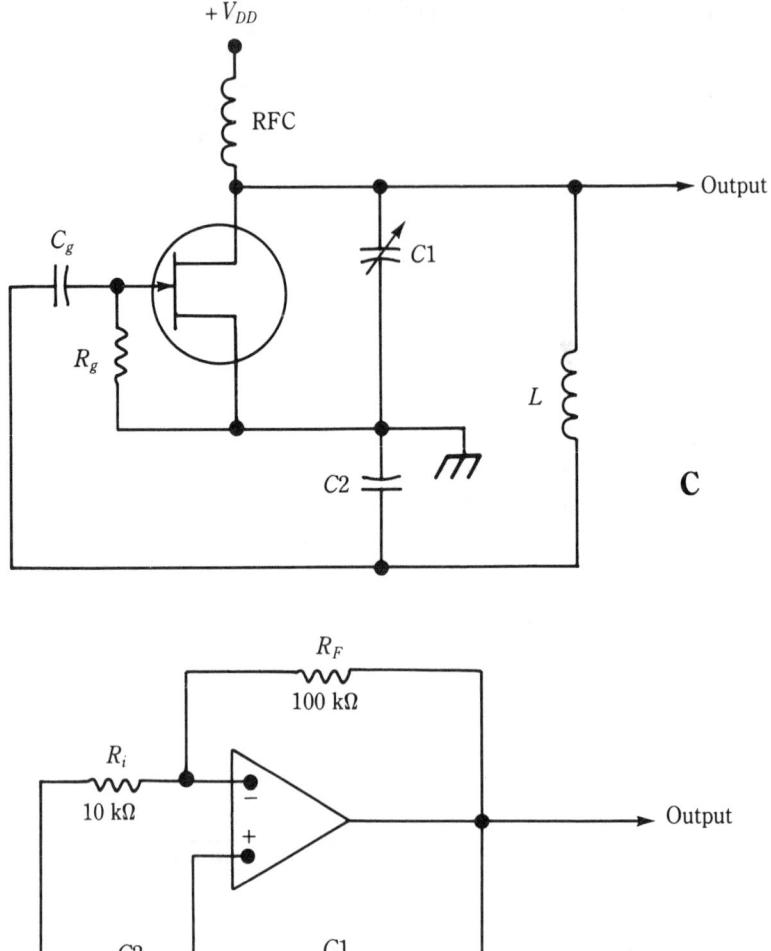

6-9 Continued

The capacitor $C3$ can be omitted, and the circuit can then be tuned by varying either $C1$ or $C2$ (or both). However, any changes in the capacitive divider might alter the amount of feedback and the amplitude of the oscillator's output.

The harmonics generated by the nonlinear operation of the JFET appear between points D and S (Fig. 6-10). These voltages are divided between L and $C2$, and since the

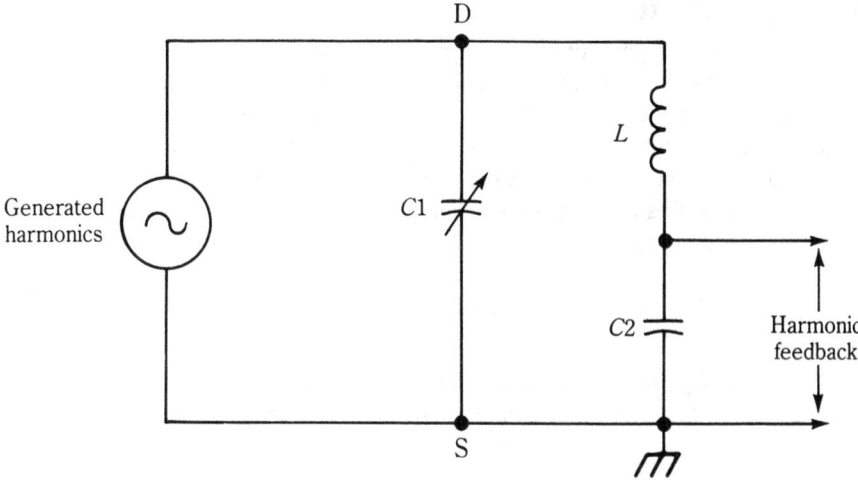

6-10 Harmonic feedback in the Colpitts oscillator.

reactance of L is high for the harmonic frequencies while that of $C2$ is low, there will be little harmonic feedback developed across $C2$. Consequently the output waveform is virtually devoid of harmonics; this is a major advantage in some applications.

The capacitors $C1$, $C2$, and $C3$ occupy the same positions as the interelectrode or the junction capacitances of the active devices. It is therefore possible to leave out $C1$, $C2$, and $C3$, and the circuit will continue to oscillate through the internal capacitances. Since these capacitance values are typically a few picofarads, the generated frequency is extremely high (hundreds of MHz) and can be varied by slug-tuning the inductor; such an arrangement is called an *ultraudion oscillator* (Fig. 6-11).

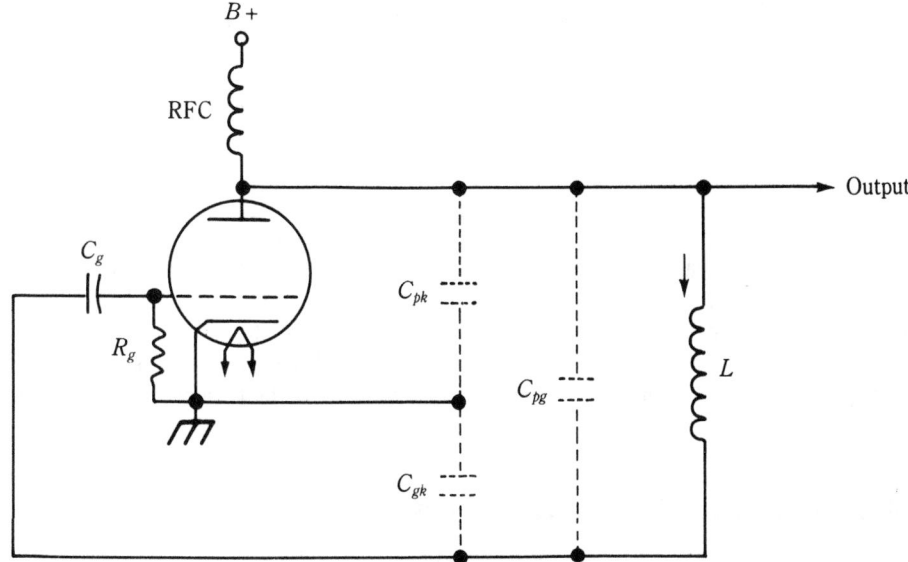

6-11 The ultraudion oscillator.

The Clapp oscillator

The common base arrangement of the Clapp oscillator is shown in Fig. 6-12. It is a modification of the Colpitts circuit and is capable of a greater degree of frequency stability. This stability is achieved by using high values for $C1$ and $C2$ (thousands of picofarads), which form the capacitive divider and provide positive feedback. By contrast $C3$ is a low-value capacitor (tens of picofarads) in series with a high-value inductor, L. Because $C1$, $C2$, and $C3$ are in series with L in determining the generator frequency, the values of $C1$ and $C2$ can be ignored and the frequency of the collector output is:

$$\text{Frequency of oscillation, } f_o = \frac{1}{2\,\pi\,\sqrt{LC_3}}\ \text{Hz} \qquad (6\text{-}9)$$

The ratio of L to C_3 can be designed to have a large value; this will result in a high Q which is the major factor in determining the circuit's frequency stability.

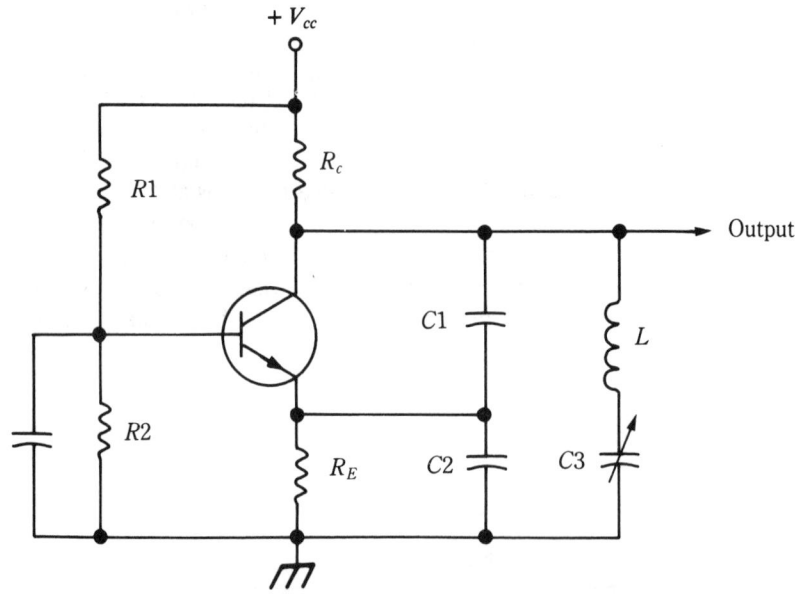

6-12 The Clapp oscillator.

In the common base arrangement there is zero phase change between the emitter and the collector voltage due to the transistor's amplifying action. The oscillation at the collector is divided between $C1$ and $C2$, and with the lower plate of $C2$ grounded, there is no phase change between this voltage and the feedback voltage developed across $C2$ and applied back to the emitter. With zero phase shift around the loop of emitter→collector→emitter, the feedback is positive.

Indirect feedback oscillators

The indirect feedback oscillator is an oscillator whose feedback is through a component that couples together the input and output circuits, but which does not form part of

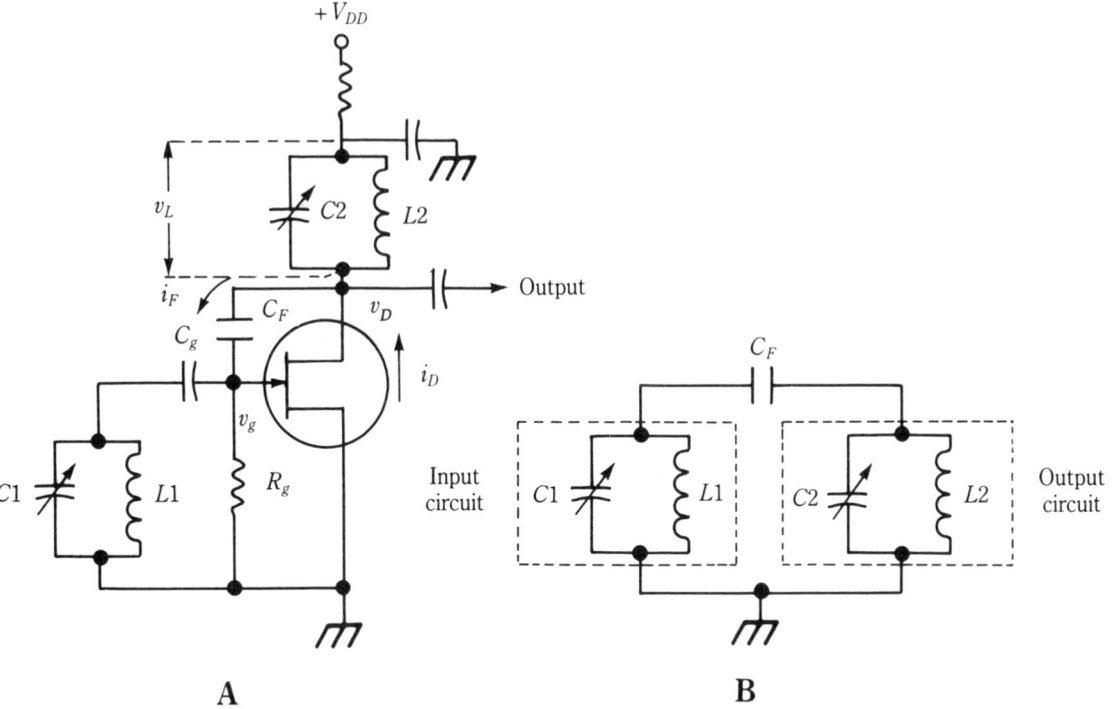

6-13 The tuned output - tuned input oscillator.

either circuit. In general this type is referred to as the *tuned-output/tuned-input oscillator* but its specific versions are the tuned-plate/tuned-grid, tuned-collector/tuned-base and the tuned-drain/tuned-gate whose series arrangement is shown in Fig. 6-13A. The feedback from the drain to the gate is through the internal capacitance C_{DG} but if this feedback is insufficient, a low value additional capacitor, C_F, can be added externally. As shown in Fig. 6-13B, C_F indirectly couples the input and output circuits but is not included in either.

In order to provide positive feedback around the loop of gate→drain→gate, it is necessary that both tank circuits should behave inductively with respect to the generated frequency. This is illustrated in the phasor diagram of Fig. 6-14. The gate oscillation, v_g, is the horizontal reference phasor and since the drain tank circuit is tuned slightly above the generated frequency, the drain current, i_d, will lag v_g. The voltage, v_L, across the drain tank circuit leads i_d by nearly 90° and is 180° out of phase with the output drain oscillation, v_d. Because the feedback capacitance is small, the feedback current, i_f, leads v_d by approximately 90°. To complete the loop the feedback current develops the original voltage, v_g, across the gate tank circuit which is also tuned above the output frequency, f_o, and behaves inductively. The equivalent inductances of the tank circuits and the feedback capacitance then determine the value of f_o.

Since none of the tank circuits is actually resonant at the frequency, f_o, this circuit is difficult to adjust and is therefore comparatively rare. Its main application is in a push-pull version (Fig. 6-15) to be used at frequencies in the low end of the ultra-high frequency

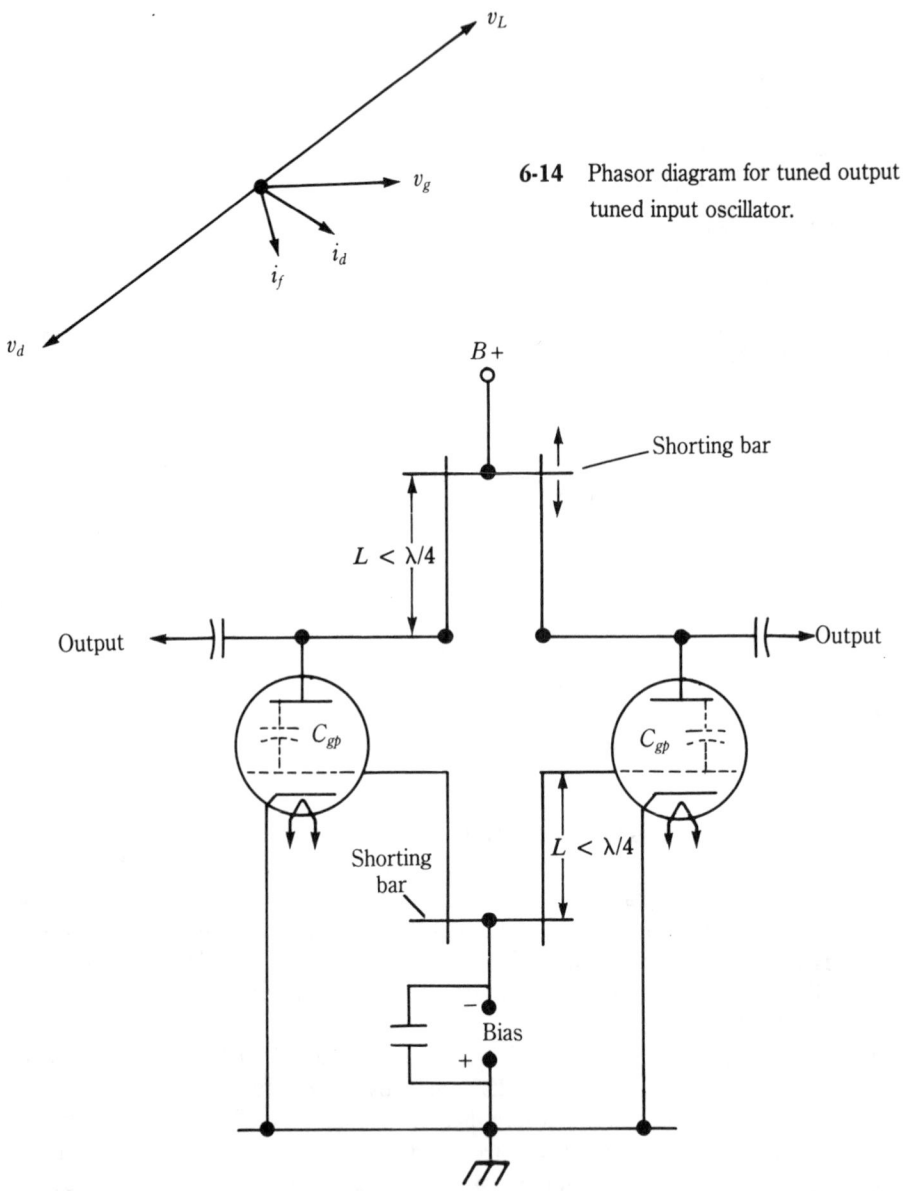

6-14   Phasor diagram for tuned output - tuned input oscillator.

6-15   The push-pull line oscillator.

(uhf) band. The tank circuits are replaced by shorted sections of twin line whose lengths are slightly less than a quarter-wavelength so that they behave inductively.

Frequency stability

All of the circuits discussed so far in this chapter have been variable frequency oscillators (VFO) in which the output is capable of being tuned over a complete range. If a variable capacitor is used for this purpose, the ratio of its maximum (limited by physical size) to minimum capacitance is approximately 10:1. Since the resonant frequency, f_r, is given by:

$$\frac{1}{2\pi\sqrt{LC}}$$

the ratio of the maximum to the minimum frequency is $\sqrt{10}$:1 or approximately 3:1. As an example the AM broadcast band ranges from 535 kHz to 1605 kHz; its frequency ratio is exactly 1605:535 = 3:1. The tuning capacitor of an AM receiver typically has a range of 30 pF (including stray capacitance) to 300 pF and can therefore cover the complete band.

The Federal Communications Commission (FCC) requires stringent tolerances for transmitters in the AM broadcast band. For example, the Los Angeles station KFWB has an assigned frequency of 980 kHz while its permitted tolerance is ± 20 Hz, so the transmitted frequency must not exceed 980 kHz $+20$ Hz = 980.02 kHz or fall below 980 kHz -20 Hz = 979.98 kHz. A VFO suffers from various forms of frequency drift and is not capable of providing the necessary frequency stability. The solution is to use some form of crystal oscillator that is much more stable.

The frequency instability of a VFO is due to a variety of reasons:

1. Temperature and humidity variations will alter the values of L and C in the frequency determining network; low values are most affected by this form of frequency drift. The most common remedies are to enclose the oscillator in a thermostatically-controlled oven or to provide the oscillator with a fan and/or adequate ventilation. Some oscillators require a long warm-up period (up to several hours) before they are sufficiently stable. It is also possible to use temperature compensating components; for example, a single capacitor might be replaced by a number of capacitors, some with positive temperature coefficients, others with negative coefficients so that the overall temperature coefficient is virtually zero.

2. As previously discussed, poorly regulated and filtered power supplies will cause dynamic instability, which is a form of undesirable modulation. The solution is to use high-quality power supplies; in particular the oscillator can be provided with a separate dc voltage which is not used for any other stage.

3. Some values associated with the active device and the L and C components might be subject to vibration. The effect can be reduced by antivibration mountings.

4. Any fluctuation in the heater voltage of tube oscillators can change the tube parameters and cause the output frequency to drift.

5. In oscillators the input impedance of the following stage is shunted across the LC circuit. Any changes in this impedance will therefore affect the generated frequency. For this reason an oscillator is often loosely coupled to a buffer stage that has a high and constant input impedance and is a class-A or class-AB amplifier with low gain.

The electron-coupled oscillator

The requirement for a buffer stage can be eliminated by using an electron-coupled oscillator (Fig. 6-16). This is a tube circuit that has no equivalent transistor, FET, or op amp version. The screen grid behaves as the oscillator's plate in an inverted arrangement where both the grid and the cathode carry rf voltages but the screen grid is at rf ground. In our example the cathode, control grid, and the screen grid form the active device of a Colpitts circuit; the only coupling between the oscillator section and the output plate circuit is the electron stream through the tube. Consequently the input impedance of the following stage can have little effect on the oscillator circuit and its frequency.

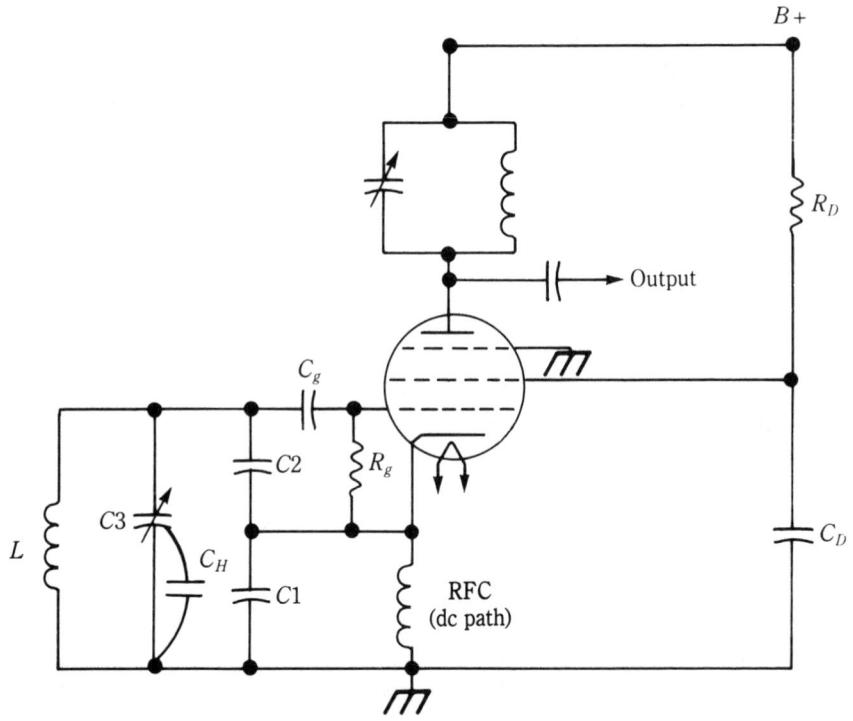

6-16   The electron coupled Colpitts oscillator.

If the grid-leak bias provides class-C operation, the waveform of the plate current will contain strong harmonics. The plate tank circuit can then be tuned to a particular harmonic so that the process of frequency multiplication can be achieved at the oscillator stage. For example the oscillator circuit might generate 1 MHz, but the plate tank circuit can be tuned to the third harmonic so that the output frequency to the following stage is 3 MHz. This enables the oscillator's tuned circuit to be operated at a comparatively low frequency, and allows the use of higher-value L, and C components, which are less subject to frequency drift. Notice also that since the rotor of the capacitor, $C3$, is connected directly to ground, this oscillator's tuned circuit will not be affected by hand capacitance.

The crystal and the piezoelectric effect

In order to obtain a higher degree of frequency stability required by broadcast transmitters a crystal oscillator is used. A properly-cut crystal possesses the characteristics of a resonant circuit and can therefore be used in place of a tuned circuit as the frequency controlling element.

Before discussing the use and operation of a crystal in a circuit it would prove advantageous to discuss some of the properties and the manufacture of the various types of crystal. Controlling the frequency by means of crystals is based on the piezoelectric effect. When certain crystals are compressed or stretched in specific directions electric charges appear on the surface of the crystal. Conversely, when such crystals are placed between two metallic surfaces across which a difference of potential exists, the crystals expand or contract. This interrelation between the electrical and the mechanical properties of a crystal is termed the *piezoelectric effect*. If a slice of crystal is stretched along its length, so that its width contracts, opposing electrical changes appear across its traces and a difference of potential is generated. If a slice of crystal is compressed along its length, so that its width expands, the charges across its faces will reverse polarity. Consequently if alternately stretched and compressed, the slice of crystal becomes a source of alternating voltage.

The effect is reversible so that if an alternating voltage is applied across the faces of a crystal, it will vibrate mechanically. The amplitude of these vibrations will be vigorous when the frequency of the ac voltage is equal to the natural mechanical frequency of the crystal. Such fundamental frequencies are governed by such factors as the dimensions of the crystal and in particular its thickness. If all mechanical losses are overcome, the vibrations at this natural frequency will sustain themselves and generate an electrical oscillation at a constant frequency. Accordingly, a crystal can be substituted for the tuned tank circuit in certain solid-state and tube circuits (Fig. 6-17).

Practically all crystals exhibit the piezoelectric effect, but only a few are suitable as the equivalent of tuned circuits for frequency-control purposes. Among these crystals are quartz, Rochelle salt, and tourmaline. Rochelle salt is the most active piezoelectric substance because it generates the greatest amount of voltage for a given mechanical strain. However, the operation of a Rochelle salt crystal is affected to a large extent by heat, ageing, mechanical shock, and moisture.

Tourmaline is almost as good as quartz over a considerable frequency range, and is somewhat better than quartz between 3 and 90 MHz, but it has the disadvantage of being a semi-precious stone, and its cost is therefore prohibitive.

Quartz, although less active than Rochelle salt, is used universally for oscillator frequency control because it is inexpensive, mechanically rugged, and expands very little with heat. Crystals used in oscillator circuits are cut from natural or artificially grown quartz crystals that have the general form of a hexagonal prism (Fig. 6-18A), whose cross-section is shown in Fig. 6-18B.

The line joining the points at each end (or apex) of the crystal, is known as the *optical* or *Z-axis*. Stresses along this axis produce no piezoelectric effect. The X-axes pass through the hexagonal edges and the cross-sectional area at right-angles to the Z-axis. Such axes are known as the *electrical axes*, which are the directions of the greatest piezoelectric activity.

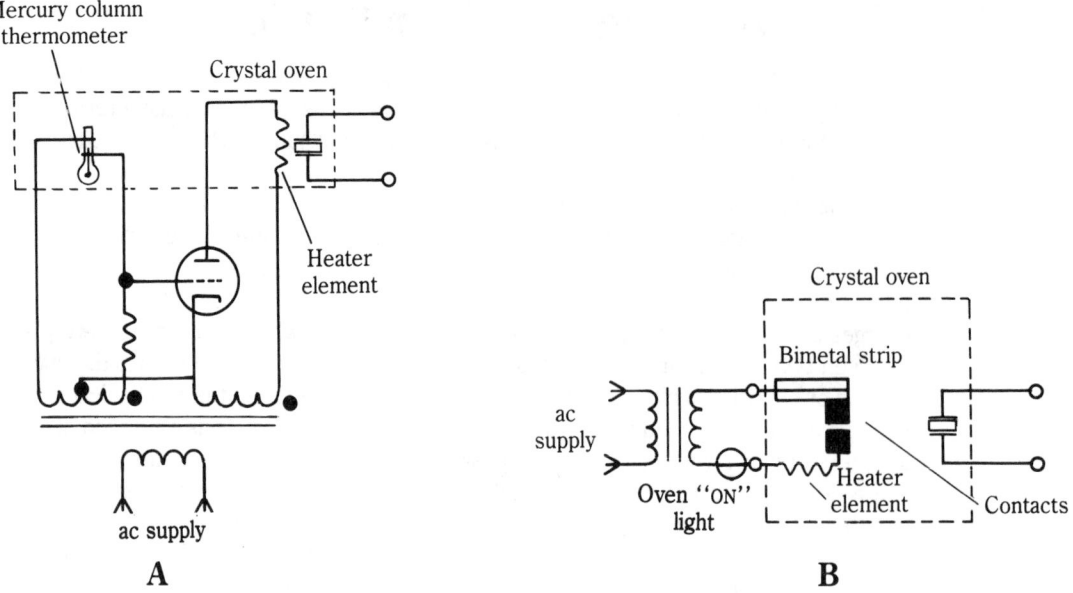

6-17   Thermostatically controlled crystal ovens.

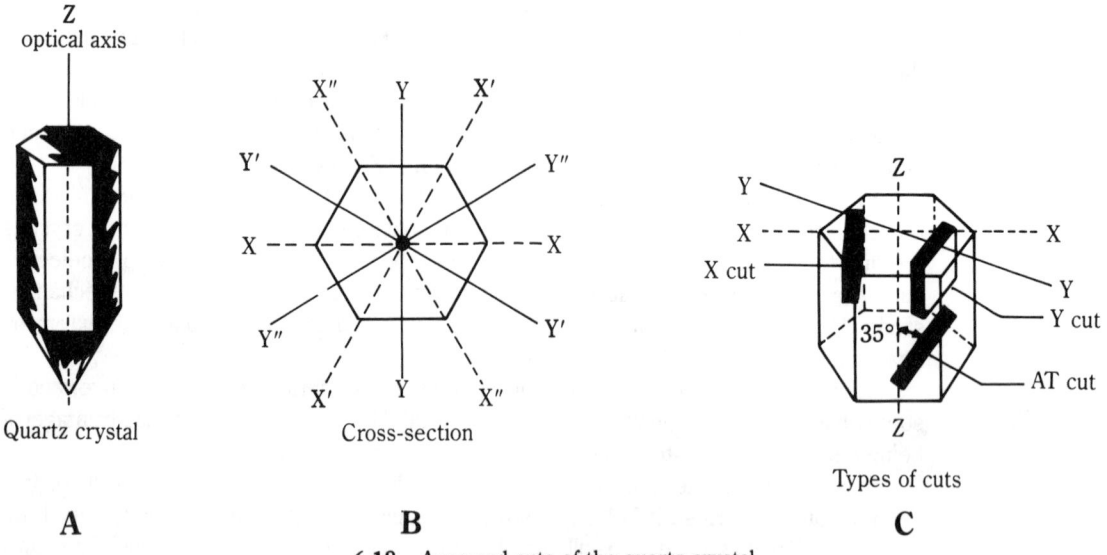

6-18   Axes and cuts of the quartz crystal.

The Y-axes, which are perpendicular to the faces of the crystal as well as the Z-axes, are called the *mechanical axes*. A mechanical stress in the direction of any Y-axis produces an electrostatic stress, or charge, in the direction of the Z-axis, which is perpendicular to the Y-axis involved. The polarity of the charge depends on whether the mechanical strain is a compression or a tension. An electrostatic stress, or voltage applied in the direction of any electrical axis produces a mechanical strain (either an expansion or a contraction) along that mechanical axis that is at right-angles to the electrical axis. For example, if a crystal is compressed along a Y-axis, a voltage will appear on the faces of the crystal along the X-axis. If a voltage is applied along the X-axis of a crystal, it will expand or contract in the direction of the Y-axis. This interconnection between the mechanical and the electrical properties is exhibited by practically all sections cut from a piezoelectric crystal.

Crystal wafers can be sliced from the mother crystal in a variety of directions along the axis. These are known as *cuts* and are identified by designations such as X, Y, AT, GT, etc. (Fig. 6-18C). Each has certain advantages, but, in general, we desire the crystal to operate at the intended frequency, have a single operating frequency, and possess minimum frequency variations due to temperature changes.

Both the X- and Y-cuts have unfavorable temperature characteristics, the FCC requires that such cuts are enclosed in a thermostatically-controlled oven with a maximum temperature variation of $\pm 0.1°C$. Better characteristics can be obtained by cutting wafers at different angles of rotation about the X-axis; the Y-cut serves as the zero degree reference, since it is lined up with both the X- and Z-axes and lies in a plane formed by those axes. By rotating this slice from its starting point, a number of different cuts can be formed.

Crystals used in oscillator circuits must be cut and ground to accurate dimensions. Crystals can also be cut in various shapes. Crystals in the lower-frequency range being square or rectangular and some of the crystals in the higher-frequency range being disc shaped, similar to a coin. The type of cut determines how active the crystal will be, and for any given crystal cut, the thinner the crystal, the higher its resonant frequency. For example, the thickness of a 25 MHz quartz crystal is only about 1 millimeter. This is the top limit for the fundamental frequency, since thinner crystals would tend to fracture easily.

The resonant frequency of quartz crystals is practically unaffected by changes in the load. Like most other materials, however, quartz expands slightly with an increase in temperature. This expansion affects the resonant frequency of the crystal.

The *temperature coefficient* of the crystal refers to the increase or decrease in the resonant frequency as a function of temperature. The temperature coefficients vary widely from one crystal cut to another. A positive temperature coefficient is assigned to those cuts that produce an increase in the frequency with an increase in the temperature. A negative coefficient refers to those crystals that decrease their natural resonant frequency when the temperature increases. One cut, the *GT*, has a practically-zero temperature coefficient over a wide range of temperature changes. The temperature coefficient also depends on the surrounding temperature at which it is measured. Heating of the crystal can be caused by external conditions, such as the high temperature of transmitter tubes. Heating can also be caused by excessive rf currents flowing through the crystal. The slow shift of the resonant frequency resulting from crystal heating is called *frequency drift*. This drift is avoided by the use of crystals with a nearly-zero temperature coefficient, and also by maintaining the crystal at a constant temperature.

The temperature coefficient of the crystal is measured in $\pm$ Hz/MHz/°C or parts per million per degree Celsius. The amount of the frequency drift is the product of the coefficient, the crystal's frequency in MHz, and the temperature change in degrees Celsius. This change is positive if the temperature increases but is negative if the temperature decreases. For example, if the temperature coefficient of a 2 MHz crystal is -10 Hz/MHz/°C and the temperature drops from 40° to 35°C, the total frequency shift = $(-10$ Hz/MHz/°C$) \times 2$ MHz $\times (40°C - 35°C) = +100$ Hz. Consequently the crystal's fundamental frequency has risen to 2 MHz + 100 Hz = 2.0001 MHz.

To maintain the extremely-close frequency tolerances required, the general practice is to construct the entire oscillator assembly in such a manner as to provide for nearly constant temperatures. This practice helps prevent excessive frequency drift. Power supply voltages are also kept as constant as possible by suitable voltage regulator circuits. In particular, the quartz crystal is operated within a constant temperature oven. The oven is electrically heated and temperature controlled by special thermostats. The entire assembly usually is constructed of an aluminum shell enclosed by thick layers of suitable material to insulate the assembly. For extreme stability, the entire compartment can be placed inside still another temperature controlled box. In this way, frequency stabilities as high as 1 part in 10,000,000 or better can be attained. Even with low temperature-coefficient crystals used in broadcast stations, the FCC requires that their temperature is maintained to within ± 1.0°C. The form of thermostat used in the oven can either be of the mercury thermometer type or the less sensitive, simpler, and cheaper thermocouple variety (Fig. 6-17A, and B). In Fig. 6-17A, any increase of the temperature above the tolerance limit will cause the bias arrangements to cut off the triode so that the heater element will carry no current. If the temperature in Fig. 6-17B falls below the tolerance limit, unequal contractions of the two dissimilar metals will cause the contacts of the bimetal strip to close and complete the circuit for the heater element.

At its resonant frequency, a crystal behaves like a series-tuned circuit, as far as the electrical circuits associated with it are concerned. The crystal and its holder (Fig. 6-19A) can be replaced by an equivalent circuit (Fig. 6-19B). C_H represents the capacitance of the mounting or holder with the crystal in place between the metallic electrodes. The series combination of L, R, and C, represents the electrical equivalent of the vibrational characteristics of the quartz crystal. The inductance, L, is the electrical equivalent of the crystal's mass effective in the vibration. The capacitance, C, is the electrical equivalent of the mechanical elasticity, and R represents the equivalent of the mechanical friction during vibration. The capacitance of the holder, C_H, is about 100 times as great as the vibration capacitance, C, of the crystal itself.

Typical values for L, C, and R are respectively of the order of henrys, picofarads, and kilohms. For example if $L = 1$ H, $C = 1$ pF and $R = 1$ kΩ, the value of Q is:

$$Q = \frac{1}{R} \times \sqrt{\frac{L}{C}}$$

$$= \frac{1}{10^3} \times \sqrt{\frac{1}{1 \times 10^{-12}}} = 1000$$

Crystals can have Q-factors of several thousand, which is large compared with the typical Q value (up to 300) of a conventional LCR circuit. As already discussed, a high-Q means superior frequency stability, and therefore crystal oscillators are much more sta-

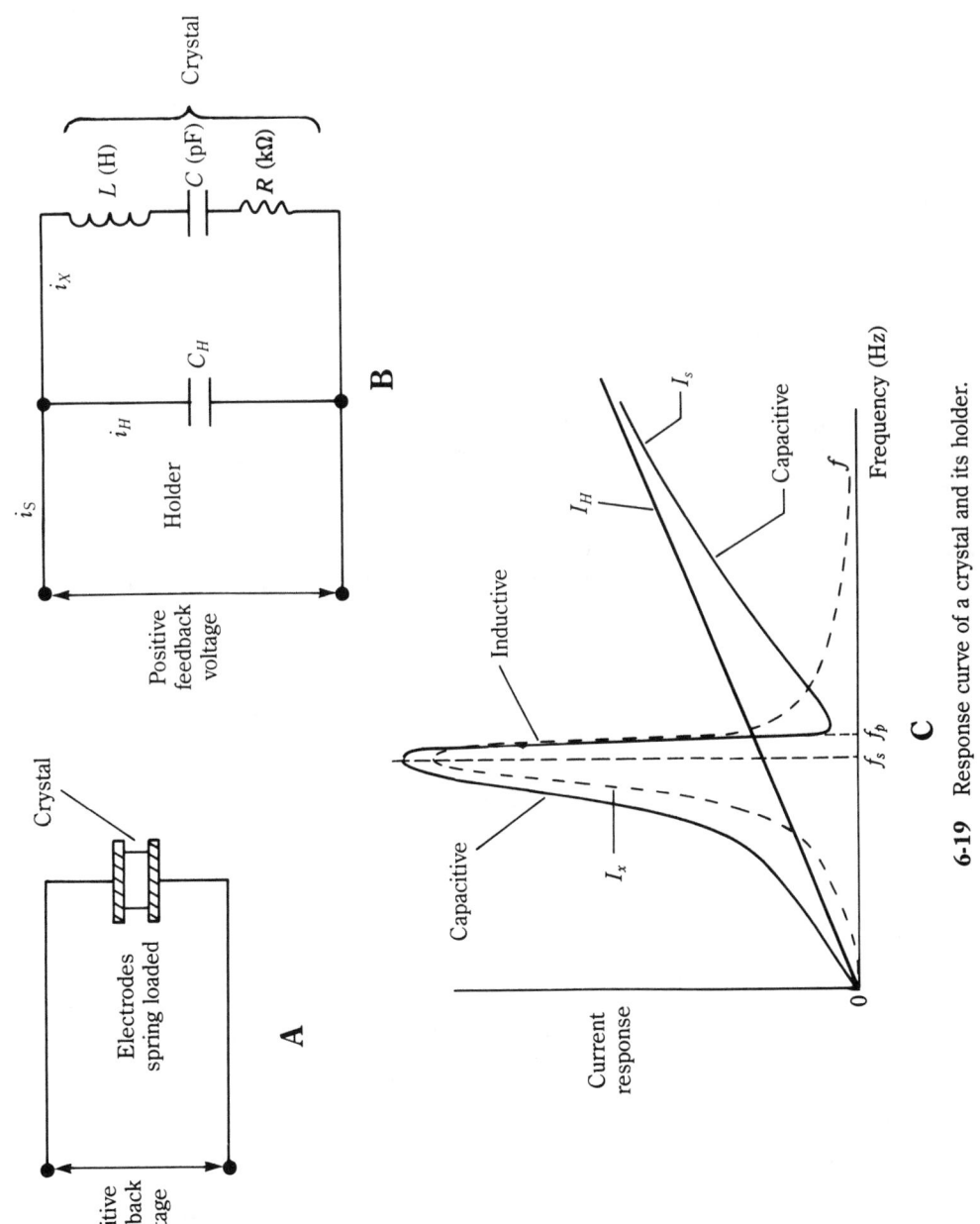

6-19 Response curve of a crystal and its holder.

ble than any of the variable frequency oscillators; as a secondary reason the temperature coefficient of the crystal is extremely low and can be virtually zero provided the crystal is maintained at the correct temperature.

To obtain the response curve of I_s versus f (Fig. 6-19C) combine the holder current, I_H, with the crystal current, I_x, taking into account their phase relationships. For a capacitor there is a linear relationship between its current and the frequency, whereas for the crystal, the current response has the familiar "bell" shape associated with a series LCR circuit. Below resonance the crystal behaves capacitively, while above resonance, it is inductive. After taking the phasor sum of I_x and I_H to produce I_s, it is revealed that the combination of the crystal and the holder has two resonant frequencies. There is a series resonant frequency, f_s, and a parallel resonant frequency, f_p, which, due to the crystal's high-Q, are very close together. As frequency increases, the combination is capacitive up to the point of series resonance (f_s) but behaves inductively over the narrow region between f_s and the point of parallel resonance, f_p. Above parallel resonance the combination is again capacitive. In some oscillator circuits the crystal and its holder can function in either the series mode or the parallel mode, while in other circuits the crystal and its holder operate at some position on the response curve between f_s and f_p and can therefore be used to replace an inductor in some of the VFO circuits already described.

So far only the crystal's fundamental frequency has been discussed. However the manner in which a crystal vibrates is complex and can be analyzed into the fundamental frequency together with overtone components. The *crystal overtones* are virtually harmonics of the fundamental frequency and it is the odd overtones that predominate. The presence of overtones allows the oscillator's output frequency to be much greater than the crystal's fundamental frequency. For example, if an overtone oscillator contains a 15 MHz crystal, the output frequency can be the third overtone of approximately $3 \times 15 =$ 45 MHz. It must be emphasized that a crystal oscillator is only capable of operating on certain fixed frequencies, namely the fundamental or the overtones, and cannot cover the complete frequency range such as a VFO. Sudden changes in output frequency can occur if the crystal is defective or dirty. To remove the dirt the crystal can be held by its edges and then cleaned with soap and water or alcohol.

Crystal oscillators

Remember the Colpitts oscillator employs a capacitive voltage divider for its positive feedback, is particularly suitable for the generation of high radio frequencies, and has an output waveform that is virtually devoid of harmonics. The inductor of the tank circuit can then be replaced by a crystal/holder combination which operates inductively between the conditions of series and parallel resonance. A transistorized common base version of this circuit is shown in Fig. 6-20. The trimmer capacitor, C_T, is capable of pulling the crystal frequency slightly; this is a common method of calibrating a crystal oscillator against a superior frequency standard. Other factors which affect the output frequency, are the type of crystal material, the size and nature of the cut as well as the value of the dc supply voltage (V_{cc}).

Failure to oscillate might be due to open circuits occurring in $L1$, $R1$, or a defective crystal which might possibly be fractured.

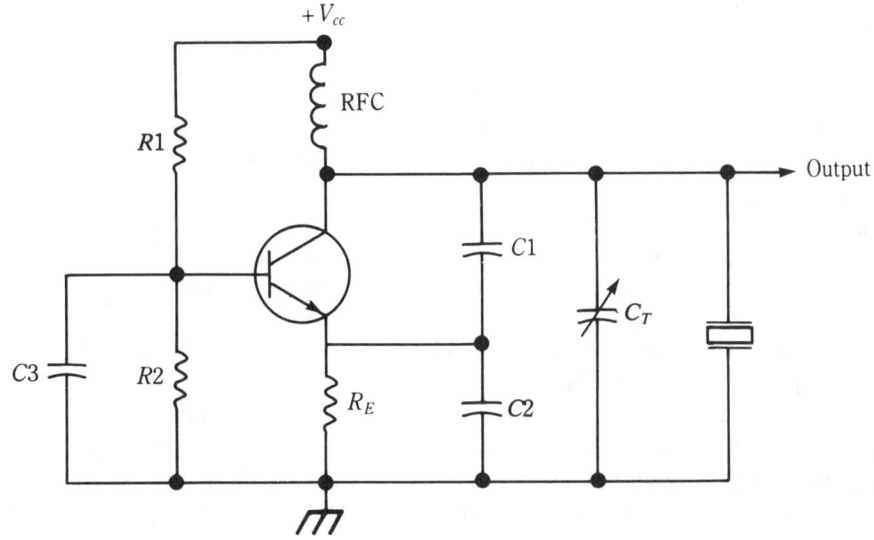

6-20 The crystal controlled Colpitts oscillator.

Miller oscillator

The JFET version of the Miller oscillator is shown in Fig. 6-21. You will recall that in order for the tuned-drain, tuned-gate circuit to oscillate, the gate tank circuit must be net inductive with respect to the crystal frequency. It is therefore possible to replace this

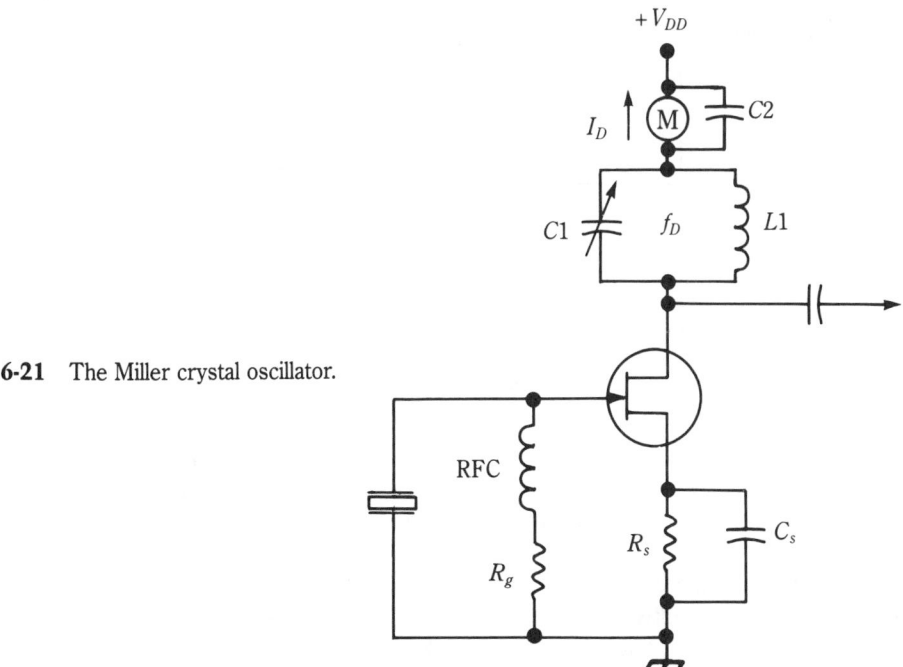

6-21 The Miller crystal oscillator.

circuit by a crystal/holder combination which will behave inductively between the series and parallel resonant frequencies. Positive feedback then occurs through the drain/gate capacitance.

Signal bias for the JFET is provided by the combination of R_g and the capacitance of the crystal/holder; the circuit also contains source bias provided by R_s and C_s. The radio frequency choke (RFC) is included to reduce the shunting effect of R across the crystal; such a shunting effect would tend to lower the crystal's Q and to increase the amount of frequency drift.

The circuit is set into oscillation by tuning the drain's tank circuit $L1$ $C1$. Initially $C1$ is adjusted to its maximum value so that the tank circuit behaves capacitively with respect to the crystal frequency. The feedback is then negative and no oscillation is possible. Consequently there is no signal bias generated, and there will be a relatively high dc level of drain current, I_D, which is recorded by the meter M. As $C1$ is gradually reduced in value, there will be no change in the reading of M until the resonant frequency, f_D, of the drain tank circuit slightly exceeds the crystal frequency, f_o. The tank circuit then represents a high value of inductive reactance, so there is considerable positive feedback and the circuit oscillates strongly. The large negative signal bias created will cause a sharp drop in the drain current. However, an increase in temperature can cause f_D to fall below f_o, and the oscillation will then cease. To prevent this from occurring, $C1$ is further decreased and consequently the drain tank circuit is less inductive. The oscillation and the bias are both reduced, and therefore the dc level of drain current, as recorded by the meter M, will slowly rise. The graph of I_D versus f_D illustrates the sharp fall and the slow rise in Fig. 6-22. If you continue to reduce the value of $C1$, additional sharp falls will occur for the overtones contained in the crystal's vibration. Note that $C2$ is an rf bypass capacitor, so the dc reading of M is not affected by the rf components of the drain current.

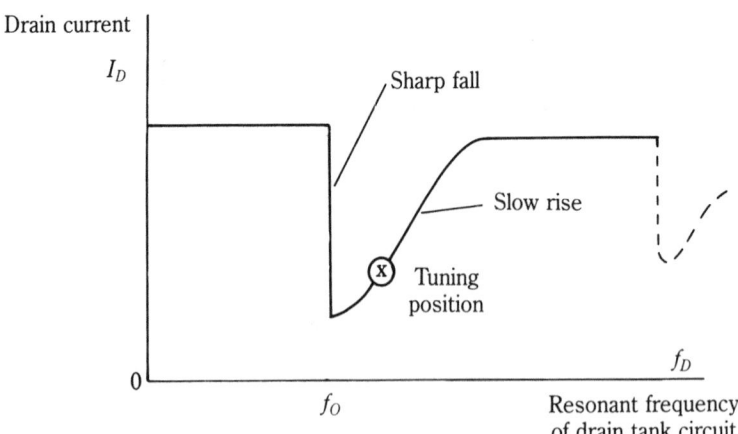

6-22   Tuning the Miller oscillator.

The Miller version is a popular crystal oscillator since the vibration is confined to the input circuit. Consequently, for a given excitation of the crystal, this type of oscillator will create the greatest power output because the feedback occurs from the drain to the gate and not through the crystal itself. The Miller oscillator is also reliable because the crystal is located in the gate circuit and is less subject to stresses and strains that might cause it to crack and fail.

The Pierce oscillator

The JFET version of the Pierce oscillator circuit is shown in Fig. 6-23. The crystal is now connected between the drain and the gate so it is more subject to mechanical stress than is the Miller circuit; consequently the main application of the Pierce oscillator is in low-power rf circuits. The crystal's holder in conjunction with the gate resistor, R_g, generates signal bias while C is a blocking capacitor used to reduce the amount of dc voltage developed across the crystal.

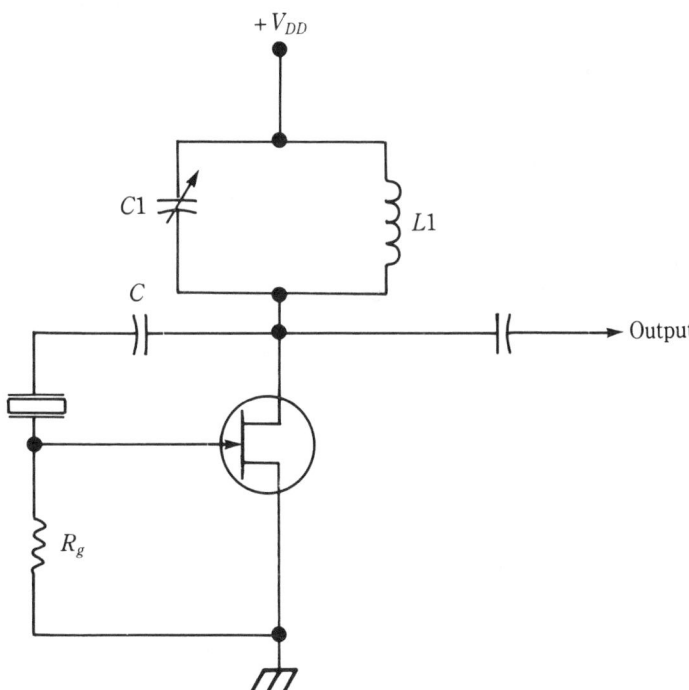

6-23 The Pierce crystal oscillator.

To provide positive feedback, the crystal/holder combination must behave inductively, and therefore oscillates at some position on the response curve between f_s and f_p. The indirect feedback path is through the gate-source capacitance and phasor analysis shows that the gate tank circuit must be *capacitive* for the circuit to oscillate. During the tuning process $C1$ is initially set to its minimum value and then its capacitance is gradually increased until a sharp dip occurs in the dc current meter, M. Due to the presence of the active device's output capacitance, the gate circuit is naturally capacitive; consequently the tank circuit is not essential, and can be replaced by either a resistor or a RFC. If this replacement is made, the Pierce oscillator is less subject to dynamic instability than the Miller circuit.

The advantages of the Pierce oscillator are its lack of need for an output tank circuit, and its ability to oscillate easily over a broad range of frequencies by using a number of different crystals. The disadvantage is the comparatively low-power operation, so it is mainly used in crystal calibrators, test equipment, receivers, and transmitters whose output power is only a few watts.

The Butler oscillator

The Butler oscillator is the most commonly used overtone oscillator with its advantages of simplicity, versatility, reliability, and good frequency stability. The circuit is least critical in its design and in the adjustments for its operation at frequencies up to the order of 100 MHz.

The tube version of the Butler oscillator is shown in Fig. 6-24. Basically the circuit consists of a grounded-grid amplifier ($V1$) which is coupled to a cathode follower ($V2$). Positive feedback between the stages is provided by the crystal, which is operated in the series mode (sometimes a small coil is placed in parallel with the crystal to tune out the holder capacitance at the series frequency).

In tracing the feedback loop, start at the cathode, K, of the tube $V1$. There is zero phase change between K_1 and the $V1$ plate, P_1, whose tank circuit $L1$, $C1$ is tuned to the required overtone (unlike the Pierce oscillator the Butler circuit *must* contain a tank circuit). A further zero shift occurs from P_1 to G_2 to K_2 since the output of a cathode follower is in phase with its grid input. The crystal completes the loop, and since it is in its series mode and behaving as a low-value resistor, there is no phase change from K_2 to

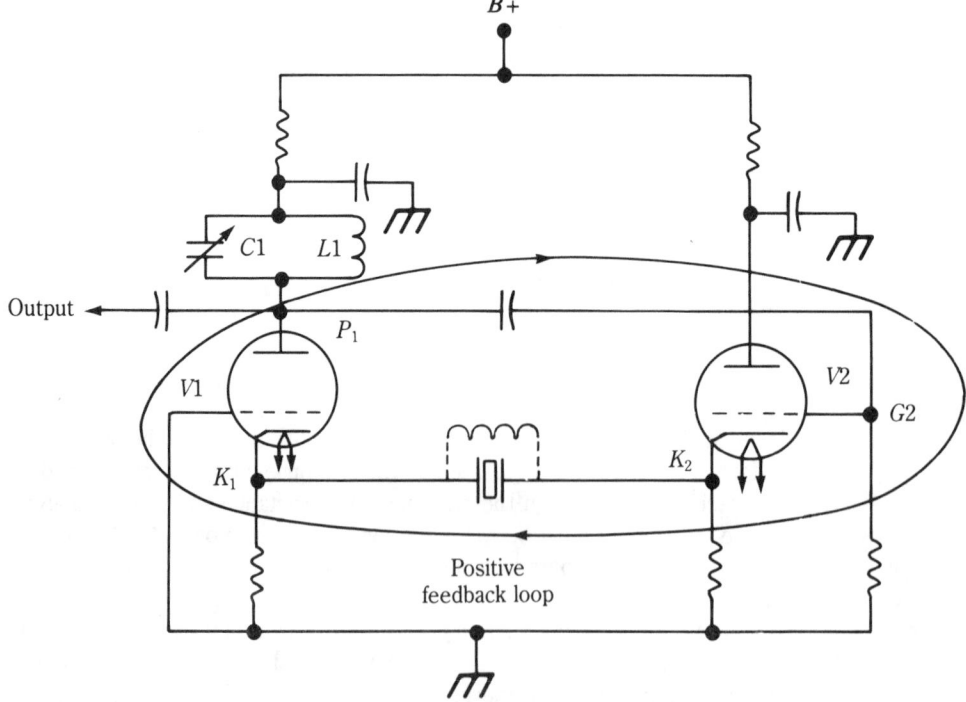

6-24   The Butler crystal oscillator.

K_1. Consequently, the total shift around the loop $K_1 \rightarrow P_1 \rightarrow G_2 \rightarrow K_2 \rightarrow K_1$ is zero and the feedback is positive. The circuit would oscillate if the crystal were replaced by an actual resistor but the generated frequency would then be unstable.

The disadvantages of the Butler circuit are its lower power output for a given crystal excitation compared with the Miller oscillator and its inability to operate over a broad range of frequencies without tuning (unlike the Pierce oscillator).

RC sine wave oscillators

In the oscillator circuits discussed so far, the frequency determining network has always been some form of LC circuit. However, it is possible to generate a low-frequency sine-wave output (up to 500 kHz) by using RC networks to provide the necessary phase shift in the positive feedback loop. Such oscillators are found in test equipment where their purpose is to generate a low power output with good frequency stability and excellent waveform. The most commonly used circuits of this type are the RC phase shift oscillator and the Wien bridge oscillator.

The RC phase shift oscillator

The phase shift oscillator (Fig. 6-25A, and B) is capable of generating a sine wave that is relatively free of harmonic distortion. Its frequency output can range from less than 1 Hz to a few hundred kHz.

The BJT circuit is used as a common emitter configuration so that there is 180° phase shift from the base to the collector; the feedback loop is completed by the RC phase shift network which contains a minimum number of three sections (as shown). For the feedback to be positive, the network must provide a further 180° shift (ignoring any effect of the transistor circuitry), and so it would appear at first glance that each RC section should contribute a 60° shift. But this simple approach ignores the shunting effect of one section on another, and an exact analysis gives the following results for the frequency of oscillation,

$$\text{Generated frequency, } f_o = \frac{1}{2\pi \sqrt{6}\, RC} \tag{6-9}$$

$$= \frac{0.159}{\sqrt{6}\, RC} \text{ Hz}$$

where R is measured in ohms and C in farads. The attenuation factor of the network is $\beta = V_o/V_i = 1/29$, and, if oscillations are to be sustained, the voltage gain of the common emitter amplifier must be greater than 29. Notice that the formula for f_o is inversely proportional to C (and not $\sqrt{C}$ as in an LC oscillator); consequently, if the capacitors, C, are variable and are mechanically ganged for tuning purposes, the oscillator can cover a complete frequency range for which the ratio of the maximum frequency to the minimum frequency is approximately 10:1 rather than 3:1.

If a four section RC network is used $f_o = 0.7/(2\pi RC)$ and the attenuation factor $\beta = 1/18.4$. The resistor, R_B, provides the starting bias and oscillation commences as the result of the transistor noise containing a component at the generated frequency, f_o.

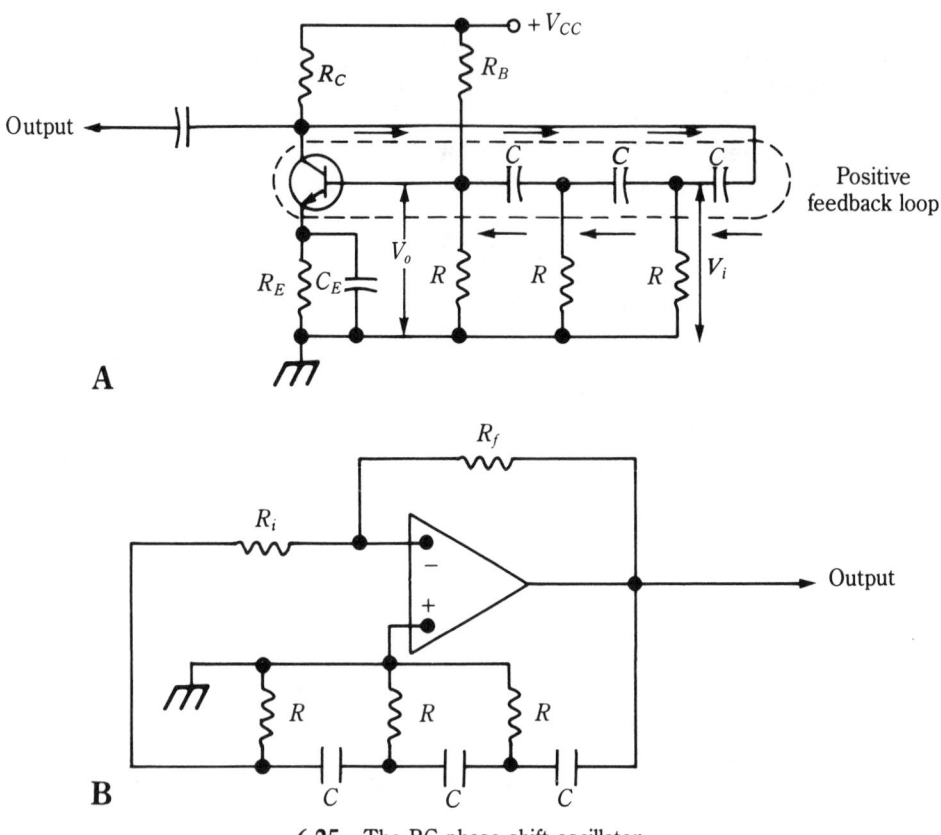

6-25 The RC phase-shift oscillator.

In the op amp version the resistors R_i and R_f must be set to provide a gain of greater than 29. The Barkhausen criterion will be satisfied and a continuous oscillation will result.

The Wien bridge oscillator

The Wien bridge oscillator (Fig. 6-26A) employs two common emitter stages so that there is theoretically zero phase shift between a signal voltage on the base of $Q1$ and the output voltage at the collector of $Q2$. The feedback loop is completed by the Wien filter consisting of $R1$, $R2$, $C1$, and $C2$. Hence, for the feedback to be positive, the input voltage, V_o, to the filter and the output voltage, V_i, from the filter must be in phase. This will occur at the frequency of the oscillation, f_o, which is given by

Generated frequency, $f_o = \dfrac{1}{2\pi\sqrt{R_1 R_2 C_1 C_2}}$

$$= \frac{1}{2\pi RC} \text{ Hz} \qquad (6\text{-}10)$$

if $R_1 = R_2 = R$ ohms and $C_1 = C_2 = C$ farads.

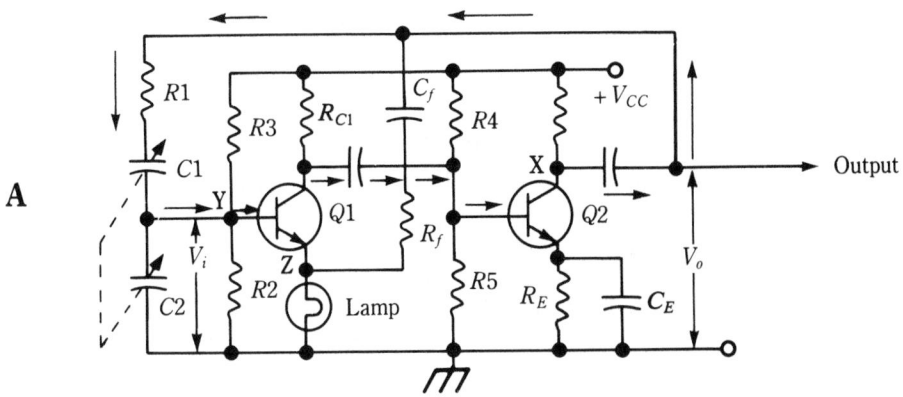

A

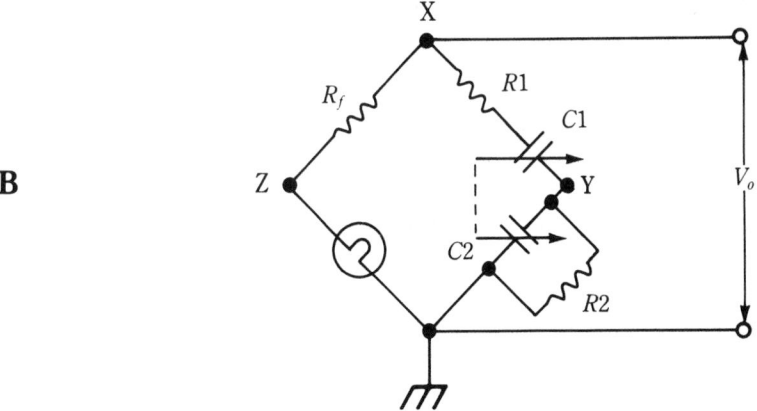

B

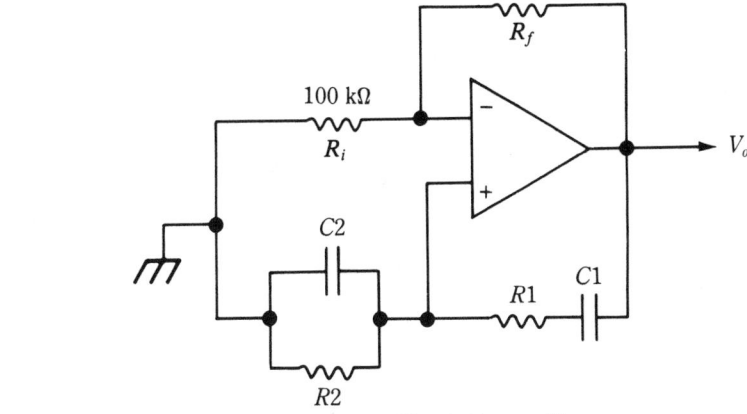

C

6-26 The Wien bridge oscillator.

At the frequency, f_o, the attenuation factor of the filter is $\beta = V_i/V_o = 1/3$. Consequently the combined voltage gain of $Q1$ and $Q2$ must be equal to 3 in order to fulfill the Barkhausen criterion. This is not practical. As a result, the oscillator circuit contains negative feedback provided by R_f and the lamp, which form the bridge circuit (Fig. 6-26B) in conjunction with $R1$, $C1$, $R2$, and $C2$. The combined gain of $Q1$ and $Q2$ with the negative feedback can then be high, but their gain with feedback will equal 3 under stable conditions, as determined by the operating resistance of the lamp.

In the op amp version the ratio of $R_f:R_i$ must be greater than 2 in order to sustain a continuous oscillation; this condition is satisfied by the values given.

Like the RC phase shift oscillator, the Wien bridge circuit is especially suitable for the generation of low-frequency sine waves with good stability and a lack of harmonic distortion.

Negative resistance oscillators

You have already seen that when a charged capacitor is connected across an inductor, the result is a damped oscillatory circuit. If an additional resistor is connected to the LC combination, the circuit losses are increased and the damping is more severe. The extra resistor represents *positive* resistance in the sense that as the voltage across the resistor decreases, the current decreases, so both changes are in the same direction. However, if you were able to add *negative* resistance across the LC circuit, the effect would be reversed and a continuous oscillation would be the result. But what is negative resistance? It is a region of some device's characteristic such that as the voltage across the device decreases, the current through the device increases; the two changes are now in the opposite direction. The following devices have characteristics which include negative resistance regions.

The tunnel (Esaki) diode

The characteristic of the tunnel diode is shown in Fig. 6-27A. The equivalent resistance derived from the slope of the negative region is typically about $-150 \, \Omega$. Provided the effect of this negative resistance is greater than that of the positive resistance contained in the LC combination, a tunnel-diode oscillator (Fig. 6-27B) will be capable of sustaining a continuous rf output which is typically in the microwatt range. This output is achieved at the expense of the power absorbed from the low dc voltage, E, which is necessary in order to bias the diode on its negative resistance region. The resistors, $R1$, and $R2$ form a voltage divider to provide the correct bias for the diode and, when power is first applied, $R1$ limits the initial surge current.

Since the charge carriers in a tunnel diode move much faster than in a conventional diode, a tunnel-diode oscillator can be used to generate microwave frequencies up to 100 GHz.

The following avalanche diodes have negative resistance characteristics similar to that of the tunnel diode but they are capable of greater microwave power outputs:

1. The Read or PNIN diode device has four layers, one of which is made from pure or intrinsic (I) semiconductor material. The combined tunneling and avalanche

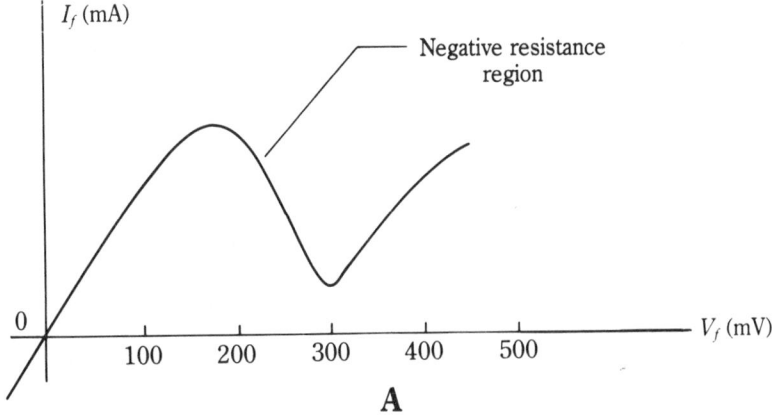

A

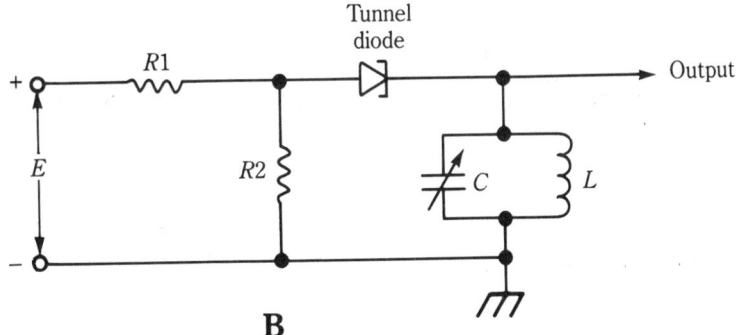

B

6-27 The tunnel-diode oscillator.

effects produce microwave powers of the order of hundreds of milliwatts in the 10 to 50 GHz range.

2. The Impatt (*Imp*act *a*valanche *t*ransit *t*ime) avalanche diode is basically equivalent to two Read diodes placed back to back with the doping so controlled that the device has a negative resistance characteristic in the avalanche condition. Such diodes can be stacked to produce microwave powers of hundreds of watts at frequencies of a few GHz.

Another method of producing a microwave output is to generate a low-frequency voltage which is used to overdrive a number of cascaded varactor diodes. These diodes have very sharp nonlinear characteristics and therefore generate strong harmonics that can reach into the microwave region. Normally such a unit would use integrated circuit construction and include microstrip thin-film elements.

An alternative to the varactor device is the step-recovery diode which allows a rapid release of forward-current stored charge when reverse bias is applied. This rapid transition creates strong harmonics so that the final output can extend into the GHz range.

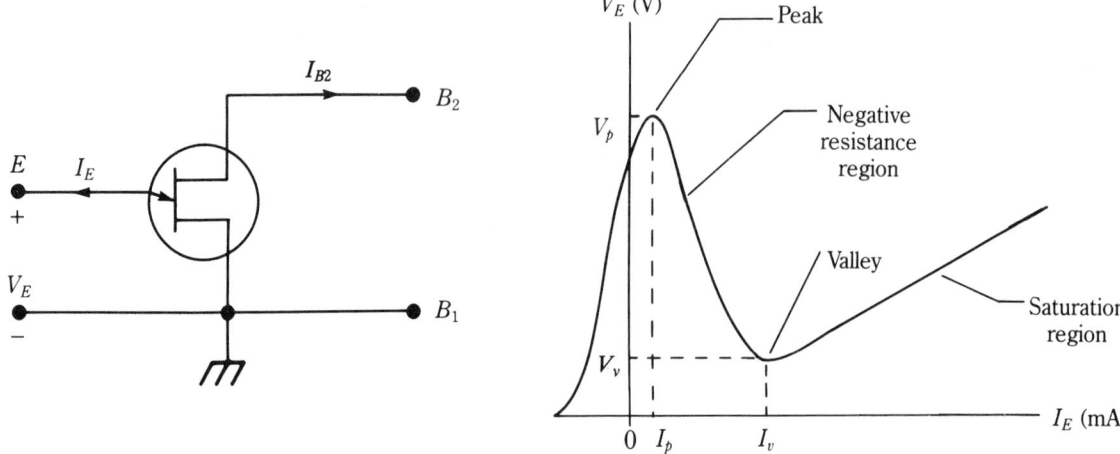

6-28 The unijunction transistor.

The unijunction transistor

The principles behind the unijunction transistor (UJT) are explained in Chapter 2. Basically the UJT consists of an emitter region (E) that forms a junction with a base region having connections B_1, and B_2 at its opposite ends (Fig. 6-28). A typical V_E/I_E emitter characteristic shows that there is a negative resistance characteristic between the peak voltage (V_p) and the valley voltage (V_v). The UJT can therefore be the active device in an rf oscillator circuit but it is more commonly used as a sawtooth generator or in a timing oscillator to trigger an SCR.

The tetrode

The emission of the slow moving secondary electrons from the plate results in the negative resistance region of a tetrode's I_p/E_p characteristic. This region was used in the now obsolete dynatron oscillator (Fig. 6-29) to create a continuous rf output.

Relaxation oscillators

During the cycle of relaxation oscillators there is a conducting state and a nonconducting state. These circuits then have the capability of producing a variety of waveforms such as the sawtooth for linear scanning, the square wave with its rich harmonic content, and the pulse for timing purposes.

The freely running sawtooth generator

A simple circuit for producing an approximate sawtooth is shown in Fig. 6-30A. Since the output will continue indefinitely the circuit is referred to as *freely running*.

The neon bulb contains "cold" electrodes and has a certain strike voltage at which the gas ionizes; this is the firing point when the bulb represents a low value of resistance. At a lower extinction voltage the gas ceases to ionize and the bulb is virtually an open circuit.

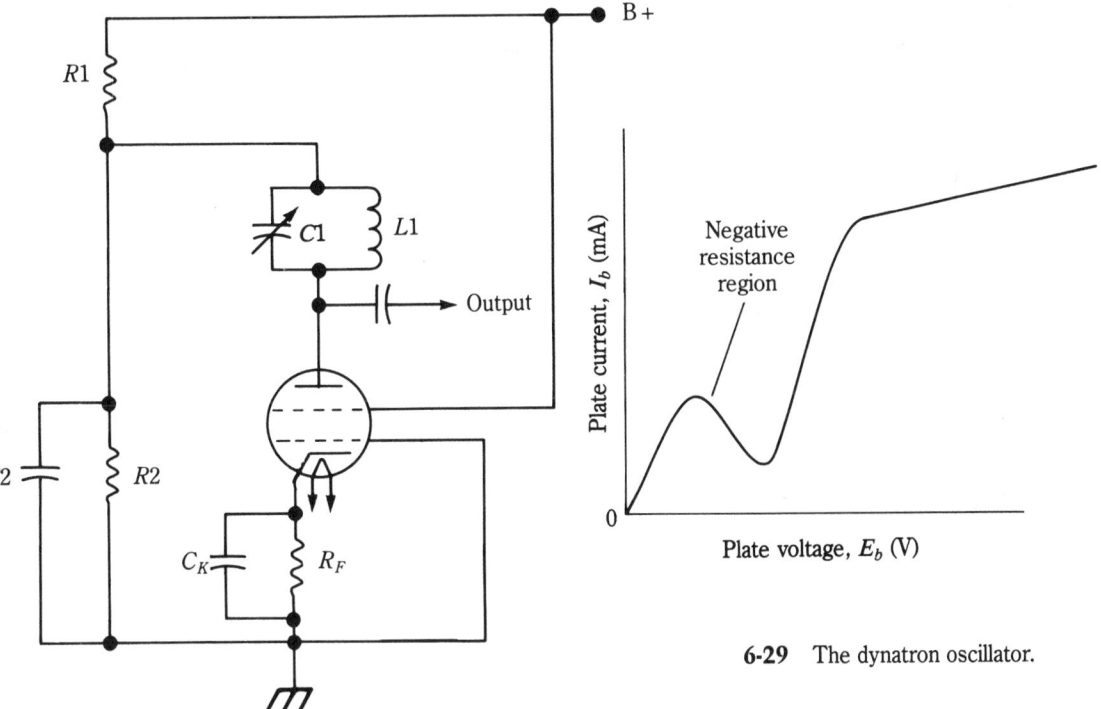

6-29   The dynatron oscillator.

When the switch, *S*, is closed, the bulb is not fired and the capacitor charges through the resistor towards the value of *E* with a time constant of *RC* seconds. However, the rise of the voltage across the capacitor is not linear (Fig. 6-30B) but follows an exponential growth curve. When the voltage, V_c, across the capacitor reaches the strike level (point X), the gas ionizes and the capacitor discharges through the bulb's low resistance. This rapid discharge, which is referred to as the flyback or retrace continues until V_c has fallen to the extinction level (point Y), at which point the gas ceases to ionize. The capacitor now recharges through the resistor to create the ramp of a sawtooth waveform whose peak-to-peak value is the difference between the strike and extinction levels. This peak-to-peak value is therefore fixed but the frequency can be changed by varying the value of *R* (or *C*).

The linearity of the ramp is improved by making the value of E many times greater than the bulb's strike voltage. The purpose of the capacitor, *C1*, is to block the dc level associated with V_c so that the final voltage across the load is an approximate sawtooth waveform with a mean level of zero volts.

The thyratron sawtooth generator

The peak-to-peak value of the sawtooth voltage can be changed if the neon bulb is replaced by a thyratron. If the bias on the grid is made more negative (Fig. 6-31A), the result is an increase in the level of the plate voltage at which the thyratron fires. This will raise the peak-to-peak value of the sawtooth output and, at the same time, the period increases so that the frequency is lowered; the peak-to-peak-value and the frequency are therefore not independent.

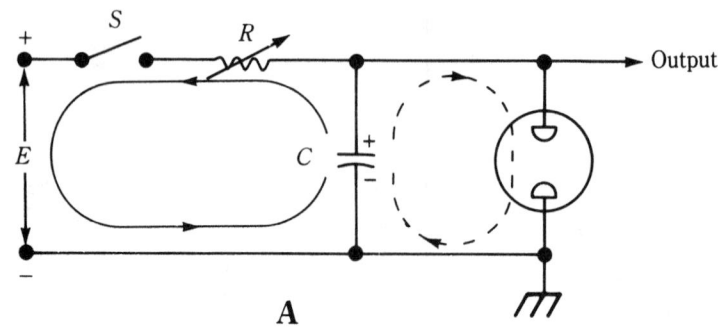

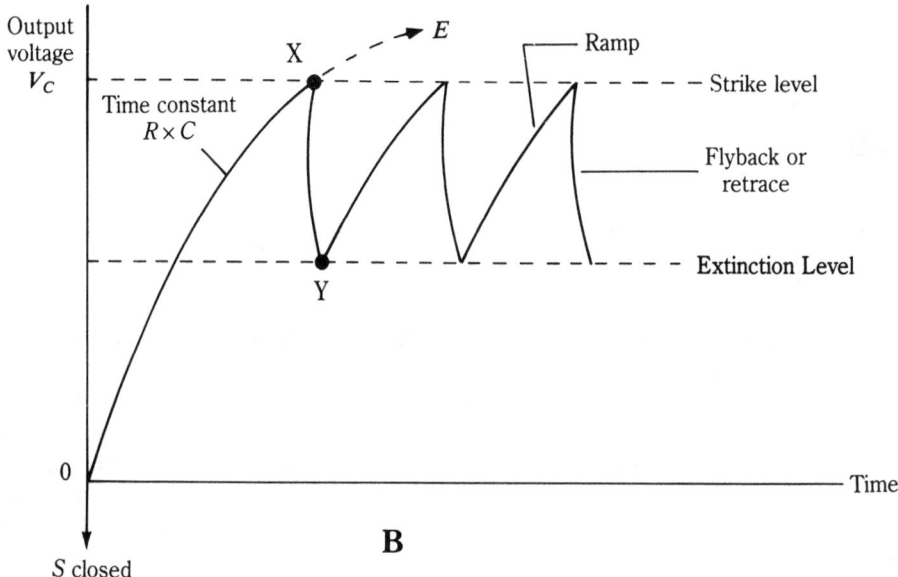

6-30 The freely running sawtooth generator.

A practical thyratron sawtooth generator is shown in Fig. 6-31B; such a circuit is capable of operating up to a few thousand hertz. The rheostat, $R1$, controls the time constant of the ramp while $R2$, and $R3$ form a voltage divider to provide the control bias. On occasions it is required to synchronize the point at which the thyratron fires; this can be done by feeding a positive trigger pulse to the thyratron's grid.

The solid state equivalent of the thyratron is the SCR, which can also be used in the generation of a sawtooth waveform.

The UJT sawtooth generator

The basic circuit of the UJT relaxation oscillator is shown in Fig. 6-32. The values of E_{BB}, $R1$, and $R2$ determine the UJT's peak voltage, so until V_E reaches this level, no emitter current flows and the lead to the emitter appears as an open circuit. Under these condi-

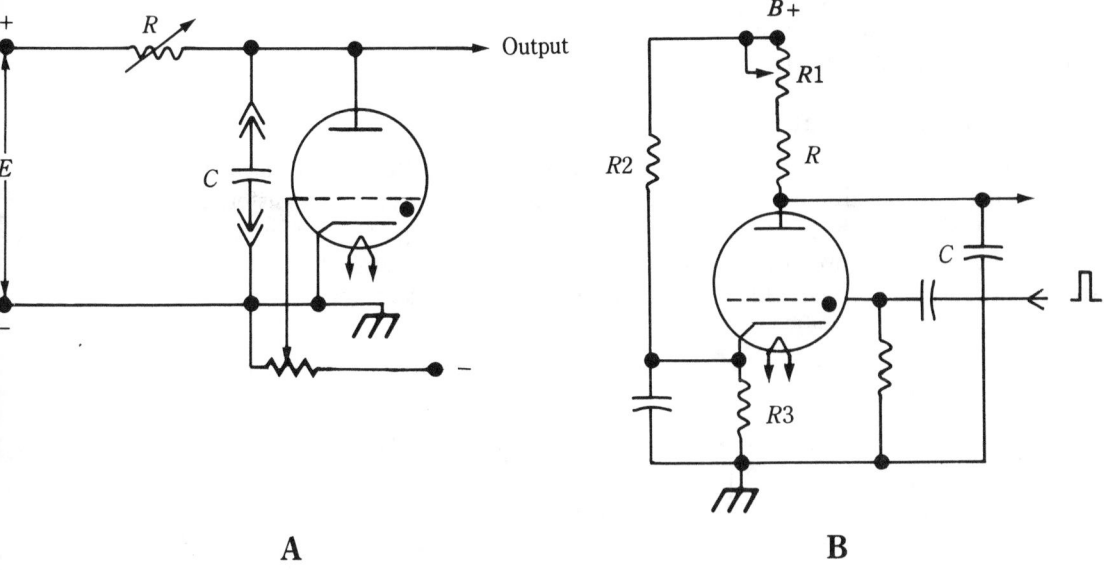

6-31   The thyratron sawtooth generators.

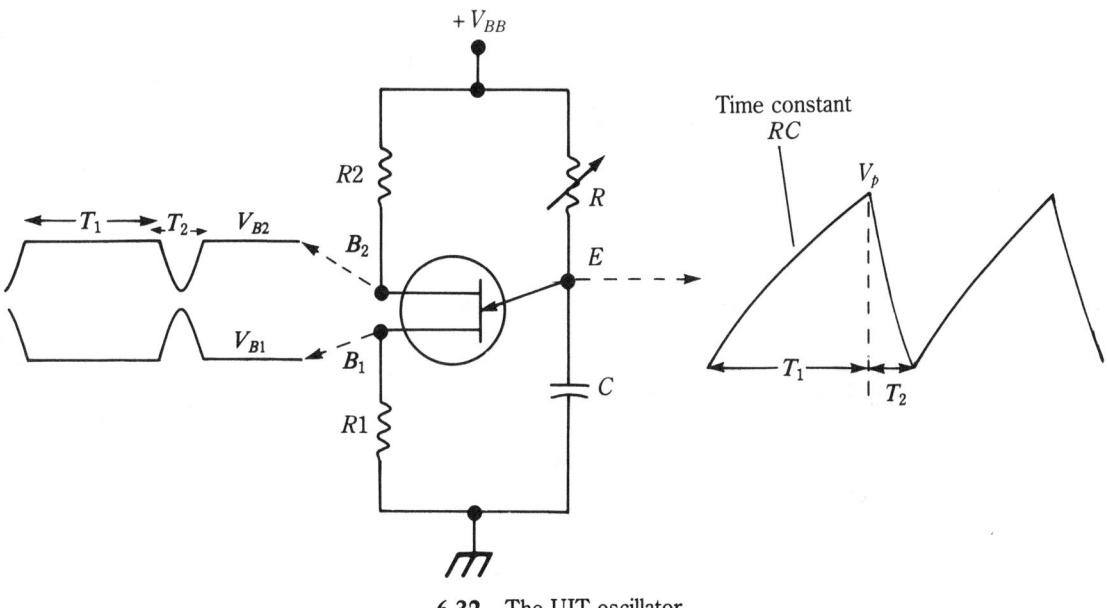

6-32   The UJT oscillator.

tions the capacitor, C, charges toward the value of the dc supply voltage, V_{BB}. The time constant involved is RC seconds so that the slope of the sawtooth can be controlled by varying the value of R. As soon as V_E reaches the peak point, the emitter current flows and the capacitor discharges rapidly to create the retrace. When V_E has fallen to the valley point, the emitter current ceases and the sequence is repeated at a frequency that can range from a few Hz up to 100 kHz.

For the short time during which the emitter current flows, the waveforms of V_{B1}, and V_{B2}, are respectively negative and positive-going pulses; one application of such pulses occurs in digital circuitry.

The astable multivibrator

The multivibrator circuit of Fig. 6-33 consists of two common-emitter stages which are cross-connected for positive feedback. The resultant instability causes the transistors to cut on and off alternatively so that approximate square-wave voltage outputs appear at the collectors; the base waveforms have a sawtooth appearance. Because the circuit has two unstable states and no stable state, it is astable and will therefore freely run.

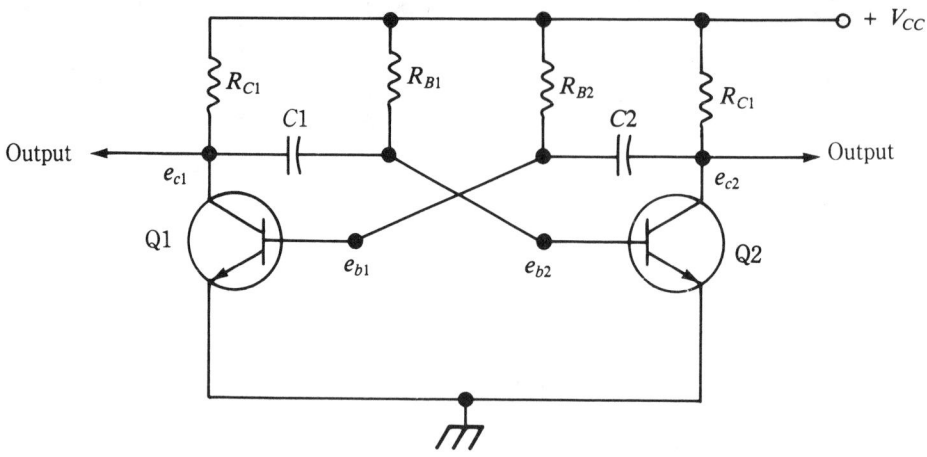

6-33 Transistorized astable multivibrator.

When the transistor Q1 has been driven to the cutoff condition as a result of the positive feedback action, the base potential E_{B1} is approximately equal to $-V_{CC}$. The capacitor $C2$ will then discharge through R_{B2} and Q2, so E_{B1} will rise toward $+V_{CC}$ with a time constant of approximately $C2R_{B2}$. When E_{B1} becomes slightly positive, Q1 will switch on, and the positive feedback action will drive Q2 to the cutoff condition. Since E_{B1} reaches approximately its halfway mark in rising from $-V_{CC}$ to a slightly positive potential on its way toward $+V_{CC}$, Q1 is cut off for a time interval of approximately $0.7C_2R_{B2}$.

$$\text{The multivibrator frequency} = \frac{1}{\text{total period}}$$

$$= \frac{1}{0.7\,(C_1R_{B1} + C_2R_{B2})}$$

$$= \frac{1}{1.4\,CR_B} \qquad \text{(6-11)}$$

if the multivibrator is symmetrical, with $C_1 = C_2 = C$, and $R_{B1} = R_{B2} = R_B$.

Multivibrator frequencies range from a few hertz to nearly 100 kHz. The square wave outputs from the collectors can be used to switch other electronic circuits ON and OFF; such square waves are also rich in harmonics and can be applied to an amplifier to test its frequency response.

Although the astable multivibrator is freely running, the changeover from one unstable state to the other can be timed to the arrival of a positive-going synchronizing pulse that is fed to the base of one of the transistors. That transistor will then start conducting earlier than it would have done so under freely running conditions. For this to occur the frequency of the synchronizing pulse must be higher than the natural frequency of the multivibrator. It is also possible for the synchronizing pulse frequency to be much higher than the multivibrator frequency; for example a multivibrator whose frequency is approximately 1 kHz, can be triggered by a 10 kHz pulse. Then nine out of the ten pulses would be inactive but the tenth would trigger the multivibrator. This represents a 10:1 "count down" factor and is known as *frequency division*, which results in the multivibrator's output frequency being a subharmonic of the synchronizing frequency.

The blocking oscillator

Blocking describes a condition in which a circuit oscillates for a period of time; this is followed by a quiescent interval during which no oscillation occurs. After the quiescent interval the circuit again oscillates, so the sequence is oscillation, no oscillation, oscillation, no oscillation, and so on indefinitely.

Oscillators employing signal bias tend to block provided there is a large amount of positive feedback and too high a time constant for the capacitor's discharge. The strong feedback oscillation causes the capacitor to charge, and drives the bias point far into the cutoff region of the active device. This reduces the device's gain to the point where the Barkhausen criterion is no longer fulfilled. As soon as the oscillation starts to decline, the bias under normal conditions rapidly decreases and the condition for continuous oscillation is restored. However, if the time constant of the capacitor's discharge is too high, the bias cannot decrease fast enough and the oscillation ceases. During the quiescent interval that follows, the capacitor discharges slowly until the active device again conducts to initiate a further burst of oscillation; the sequence is repeated indefinitely.

The blocking oscillator (Fig. 6-34A, and C) is an extreme example of the sequence described. A very high amount of positive feedback is ensured by the high degree of coupling between $L1$ and $L2$; this is achieved by the common soft iron core on which the two coils are wound. The components C and R create the high time constant for the signal bias but in the transistor version, one end of R is connected to $+V_{CC}$ in order to provide a starting bias. The diode $D1$ is included to prevent negative transient voltages from forward biasing the collector/base junction.

When the circuit is switched on, the initial plate current creates a magnetic flux which surrounds $L1$ and therefore cuts $L2$. This shocks into oscillation a tuned circuit primarily formed by $L2$ and its self-capacitance; the resonant frequency of this circuit is typically of the order of MHz. On the first positive swing of this oscillation the grid current charges C, and the bias point is driven well into the cutoff region. Oscillation then ceases and C starts to discharge slowly through R. Eventually the tube will rise above the cutoff level so that the plate current again flows and the sequence is repeated. An approximate sawtooth waveform is then present at the grid (Fig. 6-34B) while a narrow

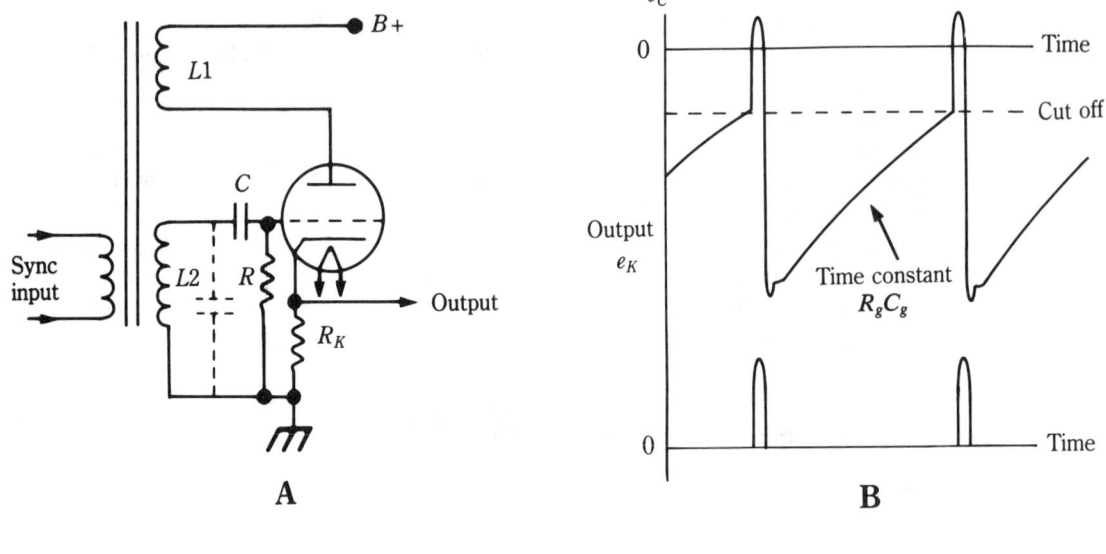

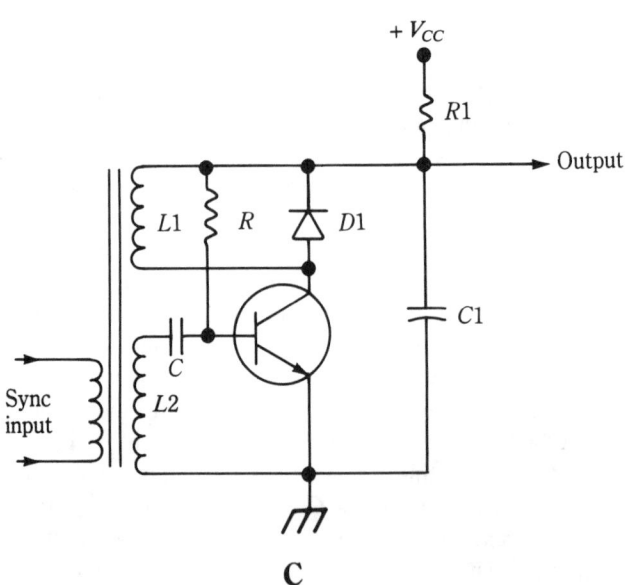

6-34 Examples of the blocking oscillator.

pulse is obtainable from the cathode. The natural frequency of these waveforms is primarily determined by the RC time constant.

Although the blocking oscillator is a freely running circuit, it can be triggered by a variety of synchronizing voltages such as a sine wave, pulse, etc. The synchronizing input causes the grid to rise above cutoff earlier than it would have done so under freely running conditions; it follows that the synchronizing frequency must be higher than the blocking oscillator's natural frequency.

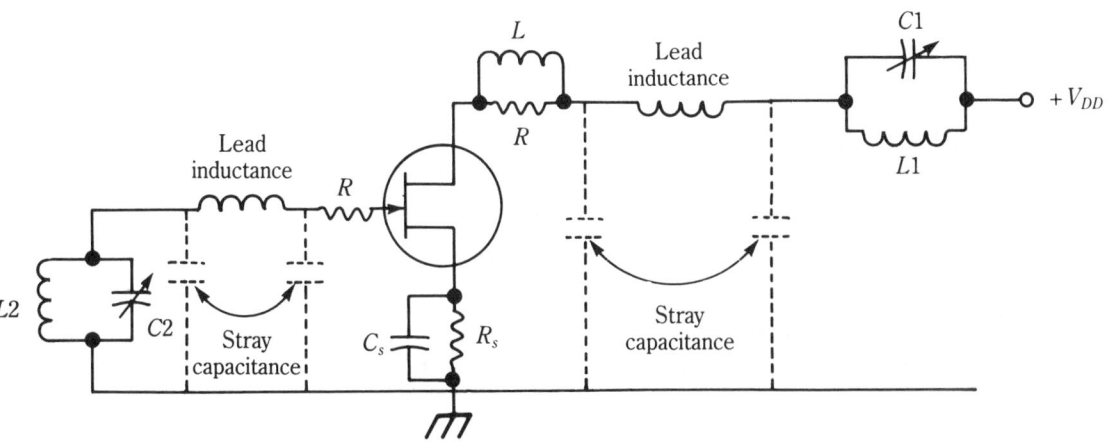

6-35 High-frequency parasitic oscillations.

The blocking oscillator can be used as the discharge circuit for a deflection generator (Fig. 6-34C) in a TV receiver and to form the pulse in a radar transmitter.

Parasitic oscillations

These unwanted oscillations occur at low frequencies or very high frequencies of the order of hundreds of MHz; in either case *parasitic oscillations* are in no way related to the design frequency. Such oscillations can appear in either rf amplifiers or oscillators and their presence can be detected by high or erratic readings in the dc meters located in the stage's input and output circuits. The importance of these oscillations lies in the additional power that is absorbed from the dc supply; this in turn will cause a greater power dissipation in the active device.

High-frequency parasitic oscillations are caused by stray inductance and capacitance associated with the leads, components, and active devices. These stray values can form some type of tuned-output, tuned-input oscillator, as illustrated in Fig. 6-35. The parasitic oscillation that the circuit generates can be eliminated by:

1. The inclusion of low value damping resistors, *R*, which are connected as close as possible to the active device. These lower the *Q* of the stray tank circuits (the visible tank circuits *L1C1*, and *L2C2* behave as short circuits at these very-high frequencies) so that the amount of positive feedback is no longer sufficient to sustain the unwanted oscillation.

2. Wrapping a few turns of copper wire around the damping resistor. At the high parasitic frequencies this forms an rf choke, *L*, which is connected in parallel with the resistor. The purpose of the rf choke is to introduce a phase shift into the feedback loop so that the Barkhausen criterion is no longer fulfilled for the parasitic oscillation. However at the design frequency the inductor, *L1*, would virtually short out the damping resistor.

The presence of *low* frequency parasitic oscillations is commonly due to the inclusion of rf chokes in the input and output circuits. The solution is either to replace one of the chokes by a resistor or to make the value of the output choke higher than that of the input choke.

Multiple-choice questions

1. In a crystal oscillator the frequency can be varied by a change in the temperature and also by:

   A. A change in the value of the dc supply voltage.
   B. A change in the value of the emitter resistor.
   C. A change in the value of the inductance in the plate tank circuit.
   D. A change in the value of the capacitance in the plate tank circuit.
   E. Changing the component values that provide the signal bias.

2. A test to determine whether an oscillator is generating its output, is to check:

   A. That drain current is flowing.
   B. The value of the signal bias on the gate.
   C. That source current is flowing.
   D. The dc value of the drain voltage.
   E. For a high value of the dc supply voltage.

3. Identify the circuit shown in Fig. 6-36.

   A. Audio frequency amplifier.
   B. Armstrong oscillator.
   C. Neutralized rf amplifier.
   D. Hartley oscillator.
   E. Colpitts oscillator.

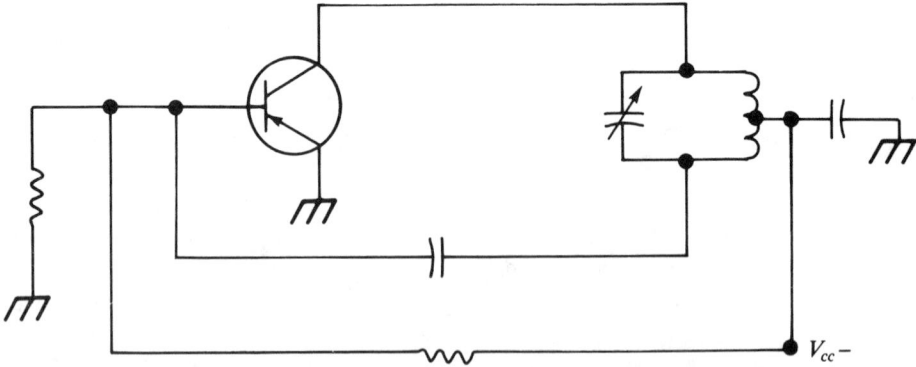

6-36   Circuit for Multiple-Choice question 3.

4. In an n-channel FET oscillator circuit, electron current would flow:

   A. From V_{DD} to the drain.
   B. From the drain to the source.

C. From gate to ground through the resistor providing signal bias.
D. From ground to gate through the resistor providing signal bias.
E. From the drain to the gate.

5. For successful operation a Hartley oscillator requires:

   A. Sufficient positive feedback to overcome the circuit losses.
   B. A very low-Q tank circuit.
   C. Indirect capacitive feedback from the output circuit to the input circuit.
   D. A tank coil that is grounded at its midpoint.
   E. A capacitive voltage divider.

6. In the circuit of Fig. 6-37, an approximate sawtooth waveform can be monitored between points:

   A. A and E.
   B. E and B.
   C. A and D.
   D. C and E.
   E. A and B.

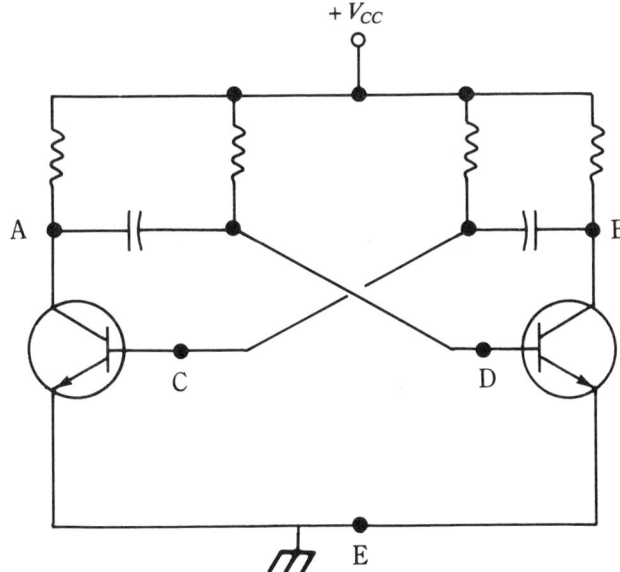

6-37 Circuit for Multiple-Choice question 6.

7. Which of the following factors will determine the output frequency of a crystal oscillator?

   A. The type of material from which the crystal is made.
   B. The size and cut of the crystal.
   C. The value of the dc supply voltage.

D. The temperature of the crystal.
E. All of the above are true.

8. What would cause oscillations to cease in the circuit of Fig. 6-38?

    A. A short circuit across the inductor *L2*.
    B. An open circuit in the inductor *L2*.
    C. A shorted capacitor, *C2*.
    D. *R1* open-circuited.
    E. A short circuit across the resistor *R2*.

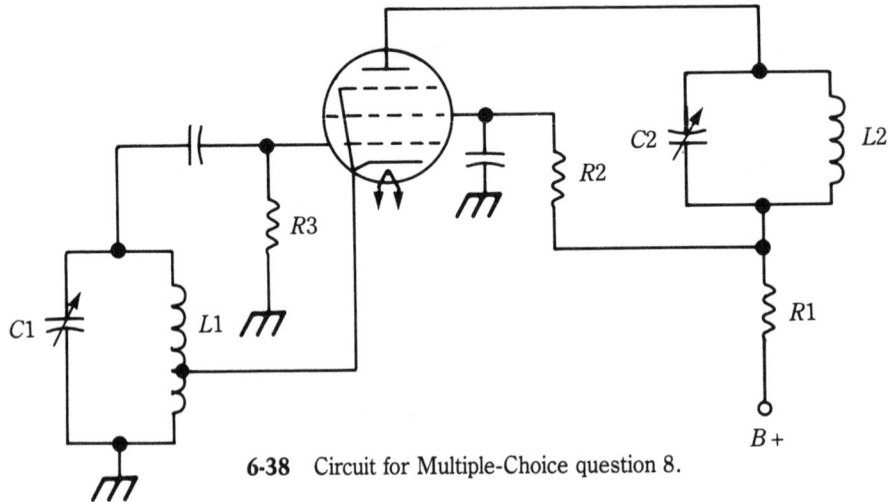

6-38 Circuit for Multiple-Choice question 8.

9. Which component is not required in the circuit of Fig. 6-39?

    A. L1
    B. C1
    C. R1
    D. C2
    E. All of the above are necessary.

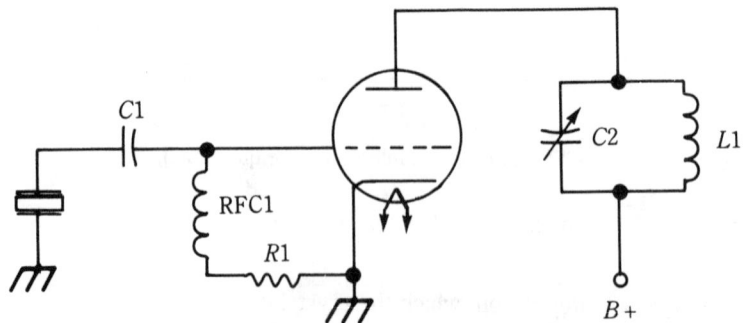

6-39 Circuit for Multiple-Choice question 9.

10. Signal bias in a JFET oscillator:

A. Is the result of the time constant provided by the resistor, capacitor combination in the gate, and source circuit.
B. Depends on the flow of source current to generate the bias.
C. Depends on the amplitude of the oscillation.
D. Prevents any flow of the drain current.
E. Both A and C.

11. Indirect positive feedback in a tuned-output – tuned-input oscillator occurs:

A. Through the capacitance that exists between the output and input circuits.
B. When the output and input tank circuits behave capacitively at the generated frequency.
C. When the output and input tank circuits are resonant at the generated frequency.
D. Through mutual coupling between the inductors in the input and output tank circuits.
E. Through a capacitive voltage divider.

12. What substance in its crystalline form can be used in oscillators?

A. Silicon.
B. Germanium.
C. Tourmaline.
D. Quartz.
E. Both C and D.

13. It is assumed that the circuit of Fig. 6-40 is oscillating. If the capacitor, C1 shorts:

A. There is no change in the meter readings.
B. The reading of the meter M1 is higher.
C. The reading of the meter M2 is lower.
D. The reading of the meter M2 is zero.
E. The reading of the meter M1 is lower.

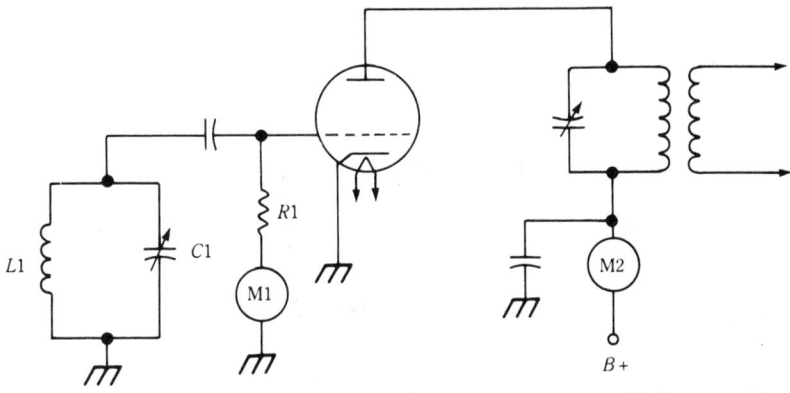

6-40 Circuit for Multiple-Choice question 13.

14. The frequency of a crystal oscillator is shifted slightly if:

    A. A resistor is added in series with the crystal.
    B. An inductor is added in series with the crystal.
    C. A capacitor is added in series with the crystal.
    D. An inductor is added in parallel with the crystal.
    E. A capacitor is added in parallel with the crystal.

15. What is the output waveform from the circuit in Fig. 6-41?

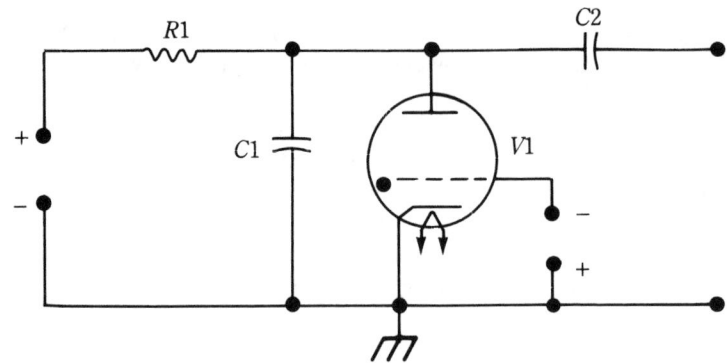

      6-41   Circuit for Multiple-Choice question 15.

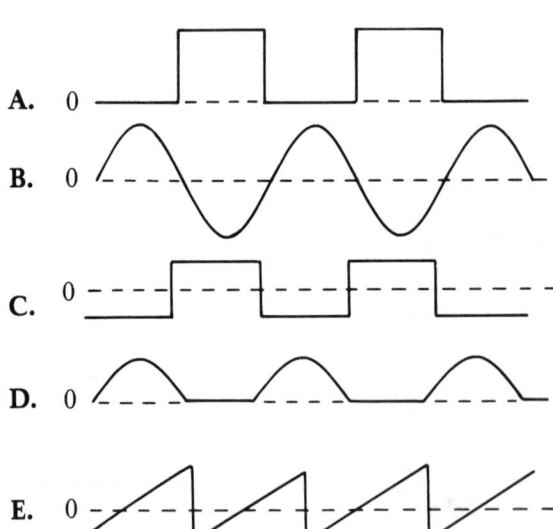

16. The frequency of a BJT multivibrator is determined primarily by:

    A. The collector supply voltage, V_{CC}.
    B. The junction capacitances of the transistors.
    C. The collector load resistors.

D. The time constants formed by the coupling capacitors and the base resistors.

E. The parameters of the transistors.

17. What happens to a shunt-fed JFET Hartley oscillator if there is a short circuit across the rf choke connected to the drain?

    A. There is no change in either the amplitude or the frequency of the oscillation.

    B. The amplitude of the oscillation increases.

    C. The circuit ceases to oscillate.

    D. The output frequency increases.

    E. The output frequency falls.

18. The circuit shown in Fig. 6-42 is a:

    A. Series-fed Armstrong oscillator.

    B. Shunt-fed Armstrong oscillator.

    C. Series-fed Hartley oscillator.

    D. Shunt-fed Hartley oscillator.

    E. JFET tuned-output oscillator.

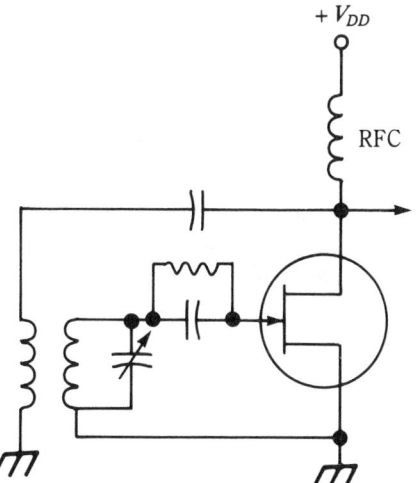

6-42   Circuit for Multiple-Choice question 18.

19. The Hartley oscillator:

    A. Uses indirect feedback.

    B. Is especially suitable for generating frequencies greater than 100 MHz.

    C. Has an output waveform rich in harmonics.

    D. Employs a capacitive voltage divider.

    E. Uses mutual coupling exclusively to provide the positive feedback.

20. The Clapp oscillator:

    A. Is a modification of the Hartley circuit.

    B. Uses a low-Q coil in its tank circuit.

C. Has improved frequency stability when compared with the Colpitts oscillator.

D. Uses low-value capacitors in its divider network to provide the positive feed-back.

E. Both B and C.

21. An electron-coupled oscillator:

   A. Uses the screen grid as the plate of its oscillator section.
   B. Requires a following buffer stage.
   C. Cannot be used for frequency multiplication.
   D. Uses a pentode's suppressor grid as the plate of its oscillator section.
   E. Is a form of relaxation oscillator.

22. A transmitter uses a 3 MHz crystal with a temperature coefficient of $+10$ Hz/MHz/°C. If the crystal oscillator is followed by three doubler stages, what is the transmitter's output frequency if the temperature falls by 20°C?

   A. 24 MHz
   B. 24.0016 MHz
   C. 23.9984 MHz
   D. 24.0048 MHz
   E. 23.9952 MHz

23. The RC phase shift oscillator:

   A. Suffers from considerable harmonic distortion.
   B. Generates a frequency that is inversely proportional to the square root of the capacitance in one of the RC sections.
   C. Uses a high-Q inductor to improve the stability of its tuned circuit.
   D. Generates frequencies of the order of several MHz.
   E. None of the above is true.

24. A tunnel-diode oscillator:

   A. Operates over the positive resistance section of its characteristic.
   B. Generates an rf output of the order of several watts.
   C. Can be operated successfully in the microwave region.
   D. Requires a bias of several volts.
   E. Is a form of relaxation oscillator.

25. When compared with a Pierce oscillator, the Miller oscillator:

   A. Has an output that is richer in overtones.
   B. Requires that its tank circuit behaves capacitively; the Pierce circuit requires no tuning.
   C. For a given crystal excitation, the Pierce circuit produces a greater power output.
   D. Uses indirect feedback through the active device's input capacitance.
   E. All of the above are false.

26. In Fig. 6-43 the output waveform at point X is a:

    A. Positive-going sawtooth.
    B. Negative-going sawtooth.
    C. Positive-going pulse.
    D. Negative-going pulse.
    E. Square wave.

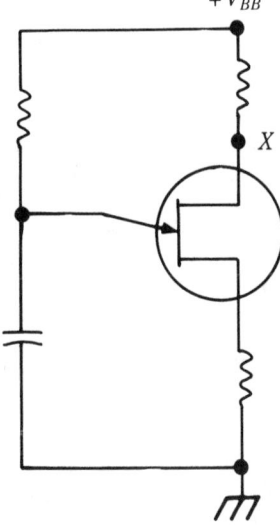

6-43   Circuit for Multiple-Choice question 26.

27. In Fig. 6-44 the waveform monitored at point X is a:

    A. Sawtooth.
    B. Square wave.
    C. Positive-going pulse.
    D. Negative-going pulse.
    E. Sine wave.

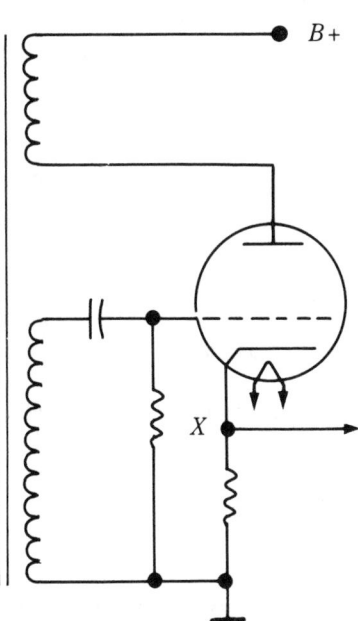

6-44   Circuit for Multiple-Choice question 27.

28. The Butler crystal oscillator:

  A. Generates an output at an overtone of the crystal's fundamental frequency.
  B. Operates with a crystal in its parallel mode.
  C. Operates with a crystal that behaves inductively at the fundamental frequency.
  D. Has a total phase shift of 180° around its positive feedback loop.
  E. Both A and C.

29. To reduce high-frequency parasitic oscillations:

  A. The circuit must be neutralized.
  B. Low-value damping resistors are connected to the active device.
  C. rf chokes are connected in series with the low-value damping resistors.
  D. An additional resistor is connected between the input and output circuits of the active device.
  E. Both B and C.

30. The Wien oscillator is used to generate a:

  A. Low-frequency sine wave with good stability.
  B. Linear sawtooth waveform.
  C. High-frequency, short duration pulse.
  D. Low-frequency square wave.
  E. Sine wave that has a rich harmonic content.

Basic problems

  1. An rf amplifier has a voltage gain without feedback (open-loop gain) of 20. If the amplifier now introduces 4% positive feedback, what is the value of the voltage gain with feedback (closed-loop gain)?

  2. An rf amplifier has a voltage gain of 50 without feedback. What percentage of positive feedback must be introduced to fulfill the Barkhausen criterion?

  3. In Fig. 6-7C, $L_1 = 125\ \mu H$, $L_2 = 25\ \mu H$ and $M = 15\ \mu H$. If the maximum and minimum values of the tuning capacitor are 250 pF and 25 pF, calculate the range of the Hartley oscillator.

  4. The electromechanical equivalent values of a crystal are $R = 500\ \Omega$, $L = 2\ H$ and $C = 0.5\ pF$. Calculate the values of the crystal's resonant frequency and its Q.

  5. A 3 MHz crystal, operating at 20°C, has a temperature coefficient of -12 Hz/MHz/ °C. At what frequency will the crystal resonate if the temperature falls to 15°C?

  6. In Fig. 6-25A, $C = 0.001\ \mu F$ and $R = 120\ k\Omega$. Neglecting any effects of the transistor circuitry, what is the value of the generated frequency?

7. In Fig. 6-9B, C_1 = 120 pF, C_2 = 330 pF and L = 50 μH. If the maximum and minimum values of the tuning capacitor are 150 pF and 15 pF, calculate the frequency range of the Colpitts oscillator.

8. In Fig. 6-33, C_1 = C_2 = 0.005 μF and R_{B1} = R_{B2} = 56 kΩ. What is the approximate frequency of the square wave generated by the multivibrator?

9. In Fig. 6-26C, R_1 = R_2 = 330 kΩ and C_1 = C_2 = 0.001 μF. What is the approximate frequency of the sine wave generated by the Wien bridge oscillator?

10. In the Clapp oscillator of Fig. 6-12, C_1 = C_2 = 2500 pF, C_3 = 40 pF and L = 800 μH. Calculate the value of the generated frequency.

Advanced problems

1. A Hartley oscillator covers the frequency range of 500 kHz to 1.5 MHz. If the maximum value of the tuning capacitor is 225 pF, calculate the required minimum capacitance.

2. In Fig. 6-16, C_2 = 1000 pF, C_1 = 220 pF, C_3 = 150 pF and L = 50 μH. If the oscillator is operated under class-C conditions and the tank circuit is tuned to the third harmonic, what is the oscillator's output frequency?

3. In Fig. 6-30A, E = 300 V, R = 100 kΩ, C = 500 pF. If the strike and extinction levels are respectively 180 V and 80 V, calculate the frequency of the relaxation oscillator (assume that the retrace time is 5% of the "linear" rise time).

4. A 4 MHz crystal at 25°C has a temperature coefficient of +10 Hz/MHz/°C. If the temperature rises to 35°C and the crystal oscillator is followed by three doubler stages, what is the value of the final output frequency?

5. In Fig. 6-33, C_1 = C_2 = 0.002 μF and R_{B1}, and R_{B2} each consists of a 33 kΩ resistor in series with a 0 – 50 kΩ rheostat. Calculate the approximate frequency range of the multivibrator.

6. In Fig. 6-26A, R_1 = R_2 = 120 kΩ and C_1, and C_2 are each variable from 30 to 300 pF. What is the frequency range of the oscillator? If the operating resistance of the lamp is 2 kΩ, suggest a suitable value for R_f?

7. A Miller crystal oscillator contains a 1 MHz crystal. The output tank circuit consists of a 50 μH inductor and a variable capacitor which is set to 200 pF. Will the circuit oscillate? If not, suggest a value to which the capacitor should be reset.

8. A tuned-plate oscillator contains a triode for which r_p = 10 kΩ and μ = 25. The plate tank circuit consists of an 80 μH inductor whose rf resistance is 15 Ω, and a 120 pF capacitor. Calculate the values of the generated frequency and the amount of

critical mutual inductance required to sustain oscillation. Suggest values for the grid-leak bias components.

9. A Butler oscillator contains a 15 MHz crystal which is being operated in its third overtone. If the output tank circuit contains a 3 μH coil, what is the required value for the capacitor?

10. An Armstrong oscillator is required to cover the frequency range of 990 kHz to 2065 kHz. The tuning capacitor is in parallel with a 100 μH inductor and a 30 pF fixed value capacitor. What are the maximum and minimum values of the tuning capacitor?

7
Digital technology

IT IS CONVENIENT TO SUBDIVIDE ELECTRONIC DEVICES AND EQUIPMENT INTO TWO categories—analog and digital. An analog device is thought of as having a continuous range of operations. A meter movement is one example. A linear amplifier for music reproduction is another. Such an amplifier is usually part of a complete analog sound system for recording, storing, and playing back programs of sound with the utmost fidelity. At each point in the system the sound is represented as an analogous electrical signal, magnetically recorded spots, etc. There are limits to the sound levels and frequencies that can be accommodated by this system, but the frequency and levels are continuously variable within these limits, subject only to the granularity of matter.

In direct contrast, digital devices, equipment, and systems are, by design, dependent on two possible states in the elementary unit of which they are composed. Any large quantity of information requires a correspondingly large number of such elementary units for its storage and processing.

This chapter is concerned with the principles and practices in the increasingly important digital world. The first topic to be treated will be the symbolic representation of numerical and other information appropriate for physical implementation in digital circuits. Binary numbers and the various common codes based on binary numbers will be considered.

The remainder of the chapter will feature the several classes of electronic devices and equipment in common use today. Starting with the elementary flip-flop, the logic circuits will be introduced and explained. These circuits will be combined to form shift registers, counters, and other digital devices which form even larger building blocks of systems such as the digital computer. It will also be necessary to form a basis for the logic that is implemented by digital circuits. Here you will encounter truth tables and Boolean algebra, which are specialized mathematical methods of expressing these logical relationships.

It is intended that this treatment be reasonably independent of the other chapters in this book. The purpose is to allow you the option of turning to this chapter as the need arises rather than to force you to cover all of the preceding chapters first.

This chapter consists of the following sections:

- Number systems and codes
- Logic concepts
- Combinatorial logic
- Flip-flops
- Counters
- Shift registers
- Arithmetic circuits
- Logic families
- Memories
- Microcomputers

Number systems and codes

The elementary electronic digital circuit has two possible normal states. Like an electrical switch, it can be "ON" or "OFF" and can contain a transistor which is saturated (fully ON) or nonconducting (fully OFF), for example. This suggests, if the condition can represent a number, that only two values are possible. Express these two values as 0 and 1. Larger quantities require a sufficient number of the same elementary binary numbers as will be seen shortly.

In the decimal number system you are able to count to nine—0, 1, 2, 3, 4, 5, 6, 7, 8, 9—with a single digit. Larger numbers require additional digits whose position value is a power of 10. Thus 743 is 700 + 40 + 3. The position values, or weights, are $100 = 10^2$, $10 = 10^1$ and $1 = 10^0$. (Note: Any non-zero finite number raised to the zero power is 1.) Ten is referred to as the base for the decimal number system.

By using the same reasoning, a number system with two counts, that is, 0 and 1, is a base two system. There is a binary digit, abbreviated *bit*, in each position. Additionally, each bit position has a weight value that is a power of two. Thus, the weight values for the binary number 1101 are $2^3 = 8$, $2^2 = 4$, $2^1 = 2$, and $2^0 = 1$. Adding the weights for the nonzero bits in 1101 gives the equivalent decimal number $8 + 4 + 0 + 1 = 13$.

Binary numbers appear to be strange for anyone accustomed to dealing with decimal numbers, since any quantity appears to be a string of 0s and 1s. More particularly, a larger number of bits is usually required than decimal digits. For example, ten bits are required to count to 1023. Actually, $1111111111_2 = 1023_{10}$, the subscripts indicate the number base for each.

In order to sidestep the human inability to grasp a long string of zeros and ones, still another number system is presently in common use. This number system is introduced in Table 7-1. The left column in the table lists the decimal number, the middle column

Decimal	Binary	Hexadecimal
00	0000	0
01	0001	1
02	0010	2
03	0011	3
04	0100	4
05	0101	5
06	0110	6
07	0111	7
08	1000	8
09	1001	9
10	1010	A
11	1011	B
12	1100	C
13	1101	D
14	1110	E
15	1111	F

Table 7-1. Number System Equivalents.

lists the four-bit binary number, and the right-hand column lists the new number representation. The digits for this new system use A through F of the alphabet in addition to 0 through 9 of the decimal system. Since you are able to count from zero to fifteen (sixteen counts) with one of the new digits, you are dealing with a base sixteen (hexadecimal) system. In particular, note that four bits correspond to one hexadecimal digit. Refer to the number $1111111111_2 = 0011\ 1111\ 1111_2$, this number is expressed in the hexadecimal number system as $3FF_{16}$. Take a closer look at it. The second form of 1111111111_2 is obtained by arranging the bits in groups of four starting on the right until all the ones are accounted for. This leaves two ones left over. Two zeros are attached to this last incomplete group to make 0011, whose value from Table 7-1 is 3_{16}. Of course the two groups of 1111 and 1111 are each recognized as F_{16} from the table. You should realize that the spaces placed in the binary number for purposes of separating groups are for convenience only and cause no numerical value change.

Table 7-2 supplies some additional number equivalents for anyone who does not understand that the primary reason for adopting the hexadecimal number system is to ease the mental and visual strains inherent in dealing with binary number strings. Converting from one number system to another is only required on occasion. Incidentally, the octal (base 8) number system was quite popular for the same reason the hexadecimal number system is used now. A conversion table for octal will require only eight entries as the digits in this number system are 0, 1, 2, 3, 4, 5, 6, and 7. Binary numbers are arranged in groups of three bits in preparing to convert to the octal system.

Binary numbers representing numerical information exist within digital systems where they are moved from point to point, are stored, undergo arithmetic and logic manipulation, and are subject to other forms of processing. For example, a set of binary

Table 7-2. Number Equivalents.

Decimal	Binary	Hexadecimal
2,479	1001 1010 1111	9AF
2,748	1010 1011 1100	ABC
3,414	1101 0101 0110	D56
50,658	1100 0101 1110 0010	C5E2
59,606	1110 1000 1101 0110	E8D6
5,280	0001 0100 1010 0000	14A0
27,540	0110 1011 1001 0100	6B94
63,718	1111 1000 1110 0110	F8E6
14,450	0011 1000 0111 0010	3872
39,258	1001 1001 0101 1010	995A
6,994	0001 1011 0101 0010	1B52
30,539	0111 0111 0100 1011	774B
41,735	1010 0011 0000 0111	A307
1,983	0000 0111 1011 1111	7BF
9,653	0010 0101 1011 0101	25B5

random numbers representing street addresses can be sorted into its correct numerical order by the logical process of comparing magnitudes.

In addition to bits signifying numerical information, bits can also form codes for non-numerical information, such as the letters of the alphabet. Table 7-3 lists the American Standard Code for Information Interchange (ASCII). The table contains the entire alphabet, both lowercase and capitalized, the decimal numbers, punctuation marks, and control codes are included. Such a code is important for two reasons:

1. It allows the exchange of information between points in a system or between systems using a standard language for the expression of the items covered.

2. It permits the processing of non-numerical information because each item is handled as a number. The conversion from one code to another is easily handled internally by a digital system using a process equivalent to table lookup.

Another alphanumeric code commonly encountered in applications of IBM equipment is the Extended Binary Code Decimal Interchange Code or EBCDIC (see Table 7-4). Another, Binary Coded Decimal, consists of four bits representing the decimal numbers 0 through 9.

The Gray Code contains a sufficient number of bit positions for the accuracy required. A four-bit example is shown in Table 7-5. It is useful for code wheels and similar devices that convert position to a digital form. It is an unusual code possessing the feature of having only one bit transition for a count increment or decrement. A code wheel is shown in Fig. 7-1 corresponding to the codes of Table 7-5. The code wheel can be implemented in various ways, for example, let the shaded area be conducting surfaces

Table 7-3. American Standard Code for Information Interchange (ASCII).

LSB		000	001	MSB 010	011	100	101	110	111
[0]	0000	NUL	DLE	SP	0	@	P	'	p
[1]	0001	SOH	DC$_1$	!	1	A	Q	a	q
[2]	0010	STX	DC$_2$	"	2	B	R	b	r
[3]	0011	ETX	DC$_3$	#	3	C	S	c	s
[4]	0100	EOT	DC$_4$	$	4	D	T	d	t
[5]	0101	ENQ	NAK	%	5	E	U	e	u
[6]	0110	ACK	SYN	&	6	F	V	f	v
[7]	0111	BEL	ETB	'	7	G	W	g	w
[8]	1000	BS	CAN	(	8	H	X	h	x
[9]	1001	HT	EM	)	9	I	Y	i	y
[A]	1010	LF	SUB	*	:	J	Z	j	z
[B]	1011	VI	ESC	+	;	K	[	k	{
[C]	1100	FF	FS	,	<	L	\	l	/
[D]	1101	CR	GS	-	=	M	]	m	}
[E]	1110	SO	RS	.	>	N	^	n	~
[F]	1111	SI	US	/	?	O	_	o	DEL

ACK	Acknowledge	FS	Form separator
BEL	Bell	GS	Group separator
BS	Backspace	HT	Horizontal tab
CAN	Cancel	LF	Line feed
CR	Carriage return	NAK	Negative acknowledge
DC	Direct control	NUL	Null
	(1-4)	RS	Record separator
DEL	Delete idle	SI	Shift in
DLE	Data link escape	SO	Shift out
EM	End of medium	SOH	Start of heading
ENQ	Enquiry	STX	Start text
EOT	End of xmission	SUB	Substitute
ESC	Escape	SYN	Synchronous idle
ETB	End of xmission	US	Unit separator
	block	VT	Vertical tab
ETX	End of text		
FF	Form feed		

with stationary electrical wipers bearing on them. The single transition feature avoids ambiguity problems that are inevitable where several transitions are to occur simultaneously for other codes with imperfect mechanical orientation of the wiper contacts. Many other codes are also used for special applications. Perhaps the most famous of these is the Morse code, which dates back to the invention of telegraphy. It is binary in the sense that two signals only can be transmitted, the short dot and the longer dash.

Communications involving the transmission and reception of digital information is an increasingly larger proportion of the total amount of communications traffic.

Table 7-4. Extended Binary-Coded Decimal Interchange Code (EBCDIC).

Positions 01 →	00				01				10				11			
Positions 23 →	00	01	10	11	00	01	10	11	00	01	10	11	00	01	10	11
Positions 4567 ↓																
0000	NUL		DS		SP	&	-									0
0001		SOS				/			a	j			A	J		1
0010			FS						b	k	s		B	K	S	2
0011		TM							c	l	t		C	L	T	3
0100	PF	RES	BYP	PN					d	m	u		D	M	U	4
0101	HT	NL	LF	RS					e	n	v		E	N	V	5
0110	LC	BS	EOB	UC					f	o	w		F	O	W	6
0111	DL	IL	PRE	EOT					g	p	x		G	P	X	7
1000									h	q	y		H	Q	Y	8
1001									i	r	z		I	R	Z	9
1010		CC	SM		¢	!		:								
1011					.	$	,	#								
1100					<	*	%	@								
1101					(	)	_	′								
1110					+	;	>	=								
1111	CU1	CU2	CU3		│	¬	?	"								

Code bit positions: b_0 b_1 b_2 b_3 b_4 b_5 b_6 b_7
Example: 0   1   1   1   1   0   1   1   =   #

BS	Backspace	NL	New line
BYP	Bypass	PF	Punch off
CC	Cursor control	PN	Punch on
CU1	Customer use	PRE	Prefix
CU2	Customer use	RES	Restore
CU3	Customer use	RS	Reader stop
DL	Delete	SM	Set mode
DS	Digit select	SP	Space
EOB	End of block	TM	Tape mark
EOT	End of xmission	UC	Upper case
FS	Field separator	│	Logical OR
HT	Horizontal tab	¬	Logical NOT
IL	Idle	-	Negative; hyphen
LC	Lower case	_	Underscore
LF	Line feed		

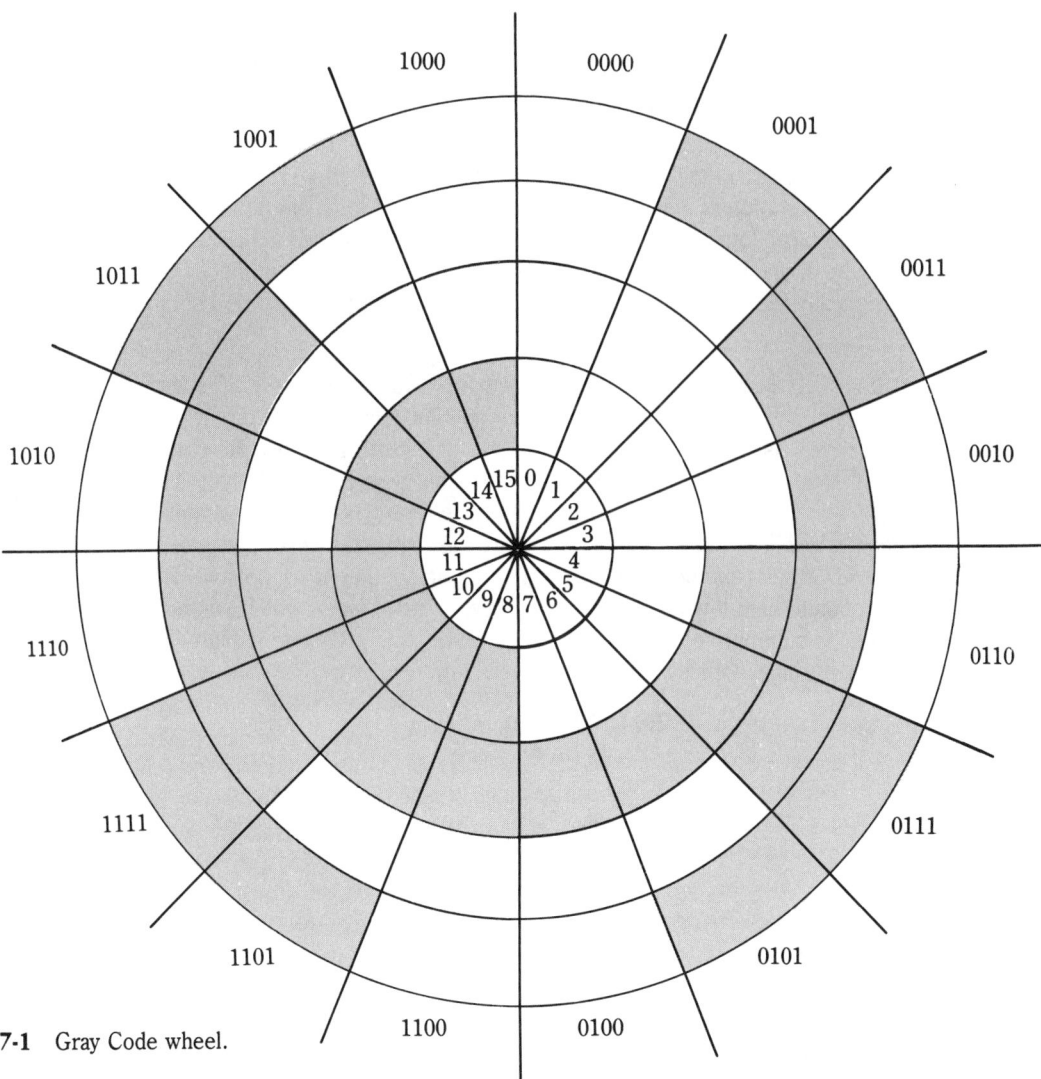

7-1 Gray Code wheel.

Table 7-5. Gray Code.

Decimal	Binary	Gray code	Decimal	Binary	Gray code
00	0000	0000	08	1000	1100
01	0001	0001	09	1001	1101
02	0010	0011	10	1010	1111
03	0011	0010	11	1011	1110
04	0100	0110	12	1100	1010
05	0101	0111	13	1101	1011
06	0110	0101	14	1110	1001
07	0111	0100	15	1111	1000

Logic concepts

It is fortunate that the emergence of digital technology, as we know it today, was preceded by a body of mathematical theory which is commonly referred to as Boolean algebra, named for its originator. Since the intent of George Boole was to apply mathematics to logic, a formal mode of reasoning, the word *logic* is still used. Moreover, this origin can be recognized by the *truth tables* which specify the operation of logic circuits.

To introduce the basic logic principles, consider a door having two separate locks. The door can only be opened if A (for Andrew) and B (for Benjamin) simultaneously use their keys. Symbolically the door opening is referred to as D using the common notation

$$D = A B$$

which means "If A AND B are both true, then D is true." You could use "T" for true and "F" for false, but the use of "1" for true and "0" for false is more directly connected to digital circuits as related to binary numbers. The truth table for the Boolean expression $D = A B$ is shown in Table 7-6. The truth table tells you that only if $A = 1$ and $B = 1$ will $D = 1$. Otherwise $D = 0$. Implement the AND truth table electronically using an AND gate. T and F or 1 and 0 are demonstrated as high and low voltage levels for both the inputs and outputs of the gate. Refer to this arrangement as positive logic. Adoption of the opposite high-low convention is named *negative logic*, which will not be considered further. To summarize, the output of a positive logic AND gate is high if, and only if, both inputs are high. Otherwise, the output is low.

Table 7-6. AND
Truth Table.

A	B	D
0	0	0
0	1	0
1	0	0
1	1	1

Now change to a different logic principle. Suppose that the door has one lock but A and B both have keys. In other words, if A OR B, OR the both of them, wish to unlock the door, it will be done. Only if neither wishes to open it will it remain locked. The notation for this logic is:

$$D = A + B$$

and the truth table appears in Table 7-7. The positive-logic OR gate has a high output if either input is high or both inputs are high. Otherwise, if both inputs are low, the output is low.

Table 7-7. OR **Truth
Table.**

A	B	D
0	0	0
0	1	1
1	0	1
1	1	1

Table 7-8. Exclusive
OR
(XOR) Truth Table.

A	B	XOR
0	0	0
0	1	1
1	0	1
1	1	0

Another basic logic circuit is the *inverter*, sometimes called a NOT gate. Very simply, a low input results in a high output. Likewise, a high input provides a low output.

There remains only one other basic logic gate—the exclusive-OR gate (XOR)—whose properties are precisely defined in Table 7-8.

Combinatorial logic

Circuits can be designed using the basic gates and inverter to satisfy any truth table specifications. Three forms then exist for the same logical combination:

1. The truth table
2. The Boolean equation
3. The gate circuit

To illustrate combinatorial logic a number of simple combinations will be explored. Consider first the truth table seen in Table 7-9. A Boolean equation that expresses the same relationship is:

$$D = A B + C$$

Corresponding to this equation, the circuit implementation is shown in Fig. 7-2.

Table 7-9. Truth Table
for AB + C = D.

Inputs			Output
A	B	C	D
0	0	0	0
0	0	1	1
0	1	0	0
0	1	1	1
1	0	0	0
1	0	1	1
1	1	0	1
1	1	1	1

Do not be concerned with the method of translating a truth table into a Boolean equation and the design process for obtaining a practical circuit that satisfies both the truth table and the equation. A logic designer who specializes in accomplishing the

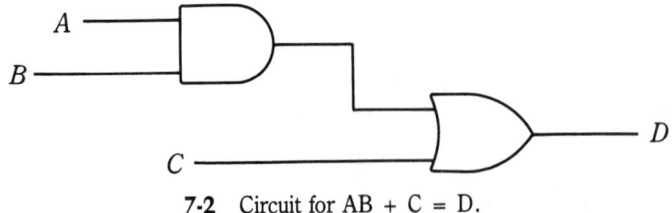

7-2   Circuit for AB + C = D.

implementation of such circuits has both the necessary training and experience. This specialized background is not required for the majority of technical personnel who deal with digital circuits.

Now consider a few more combinatorial logic circuits. Refer now to Table 7-10 for which the Boolean equation is:

$$D = (A + \bar{B})\,C$$

The bar over letter *B* in the equation indicates that the complement or opposite value of *B* is to be substituted. $\bar{B} = 1$ for $B = 0$ and $\bar{B} = 0$ for $B = 1$.

**Table 7-10. Truth Table
for $(A + \bar{B})\,C = D$.**

A	B	C	D
0	0	0	0
0	0	1	1
0	1	0	0
0	1	1	0
1	0	0	0
1	0	1	1
1	1	0	0
1	1	1	1

One circuit that will implement this equation is shown in Fig. 7-3. The inverter appears in the circuit as a means of obtaining the $\bar{B}$, that is, inverting *B*.

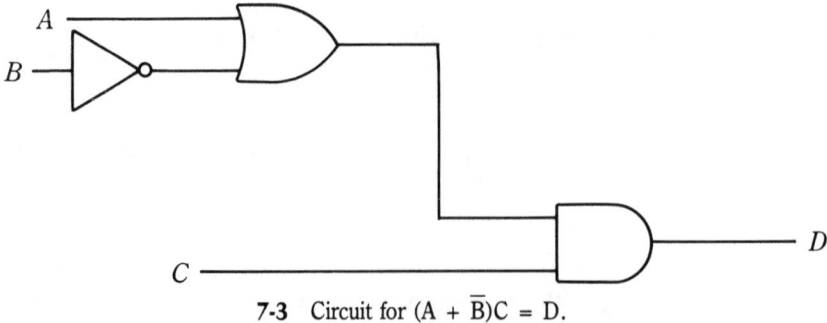

7-3   Circuit for (A + $\bar{B}$)C = D.

Refer to Table 7-8, the truth table for the exclusive-OR gate. It is interesting to consider how this truth table can be expressed as a Boolean equation, and how it can be implemented by using inverters and logic gates other than the XOR gate. This equation is:

$$\text{Output} = A\,\overline{B} + \overline{A}\,B$$

and the circuit is shown in Fig. 7-4.

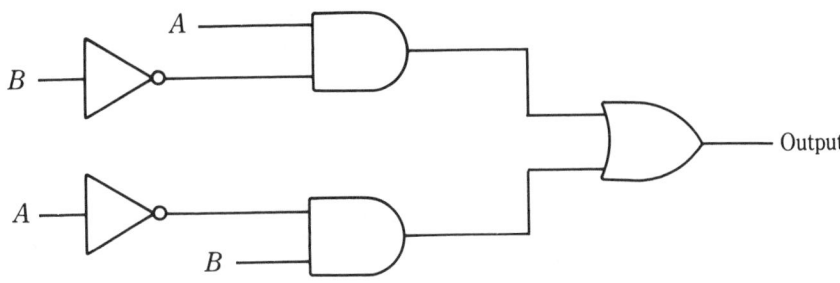

7-4   Circuit providing $\overline{AB} + \overline{AB}$ = OUTPUT.

Flip-flops

In all of the preceding logic circuits the output is solely a function of inputs at any moment. Any past history of electronic activity in the circuit has no bearing on the output now. Figure 7-5A is a circuit with NOR gates for which the outputs Q and $\overline{Q}$ are dependent on inputs S and R and the prior states of Q and $\overline{Q}$. The behavior of this circuit, a flip-flop, is summarized in Table 7-11. The truth table has one column for input S, one column for input R, and two columns for Q. Q_n represents the prior state, and Q_{n+1} represents the next state. No columns are assigned for $\overline{Q}_n$ and $\overline{Q}_{n+1}$ since they are always the complements of Q_n and Q_{n+1}. An additional column labeled *Mode* has been added to summarize the behavior under the given input conditions. For example, in the first two rows with both S and R zero, $Q_{n+1} = Q_n$, signifying no change, described in the table as *Hold*. In the next four rows either S or R, but not both, are 1, causing the flip-flop to be Set ($Q_{n+1} = 1$) or Reset ($Q_{n+1} = 0$) independently of Q_n. With S and R both equal to 1, the results are unpredictable, so this condition is prohibited for normal operation of the flip-flop. The SR flip-flop symbol appears in Fig. 7-5B.

The SR flip-flop and the various other flip-flops to be described have the capability of storing one bit as a memory element. They are the principal building blocks from which counters, shift registers, and the arithmetic and logic units of digital systems can be constructed.

Now examine other flip-flops using your knowledge of the SR flip-flop as a point of departure. Figure 7-6A shows how a D flip-flop can be constructed. The D flip-flop has a single input, 0 or 1, and Q_{n+1} will have the same binary value.

Another important member of the flip-flop family can be implemented using the SR flip-flop shown in Fig. 7-7. The truth table of Table 7-11 also applies to the JK flip-flop if several minor changes are made. Change S to J and R to K in the column headings. In

Table 7-11. Truth Table for SR Flip-Flop.

S	R	Q_n	Q_{n+1}	Mode
0	0	0	0	Hold
0	0	1	1	Hold
0	1	0	0	Reset
0	1	1	0	Reset
1	0	0	1	Set
1	0	1	1	Set
1	1	0		Prohibited
1	1	1		Prohibited

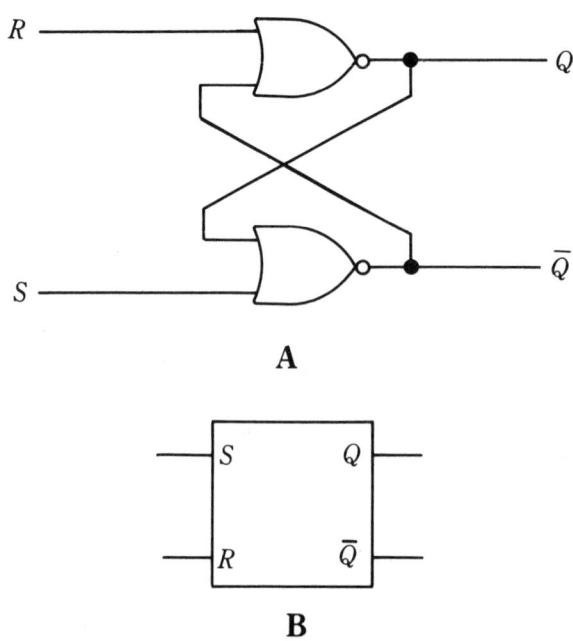

7-5 SR Flip-flop. A. Circuit using NOR gates. B. SR Flip-flop symbol.

the last two rows make Q_{n+1} be the complement of Q_n for which the mode designation is Toggle instead of being prohibited. The Toggle feature of the JK flip-flop is, by no means, prohibited and is extremely useful in counters and other applications.

Most flip-flops are designed to function in synchronism with other devices in a digital system. The master timing element of the system is referred to as the *clock* as are the periodic pulses provided by the clock. In order that the timing be more precise, the clocked flip-flop is permitted to change state only at the time of the clock edge. A circuit to accomplish this for the JK flip-flop is shown in Fig. 7-8, a modified version of the circuit in Fig. 7-7. Note that the three-input AND gate's behavior requires that all three inputs be high in order that the output be high. The RC combination in the figure is commonly referred to as a differentiator. It yields a sharp positive pulse for the positive-going

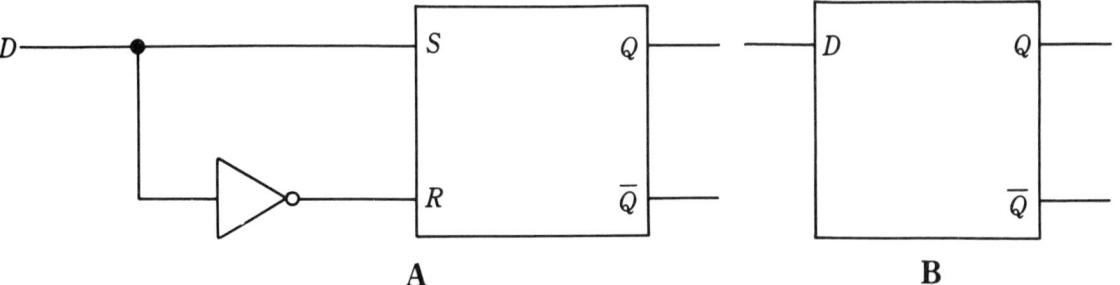

A   **B**

7-6   D Flip-flop. A. Using SR Flip-flop. B. Symbol.

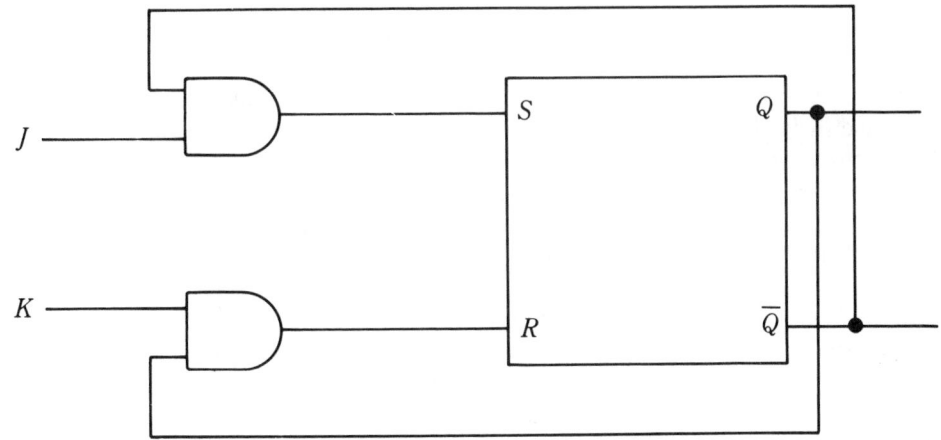

7-7   JK Flip-flop circuit using SR Flip-flop.

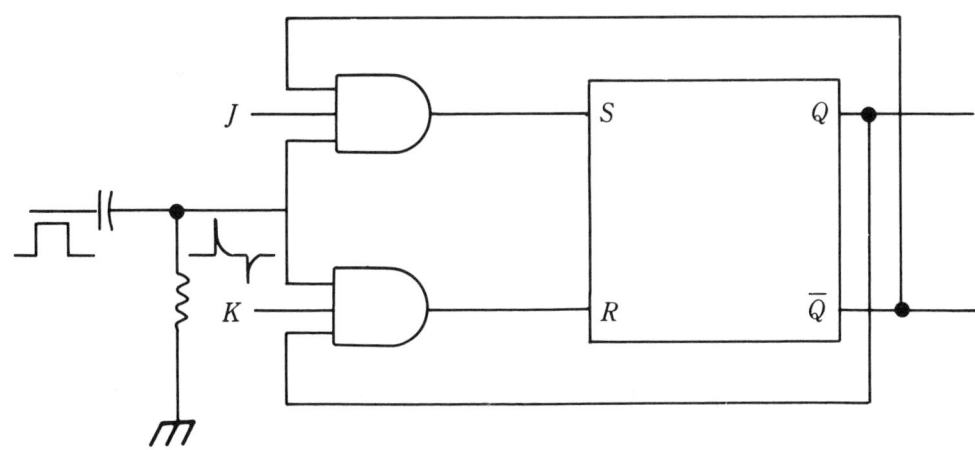

7-8   Clocked JK Flip-flop.

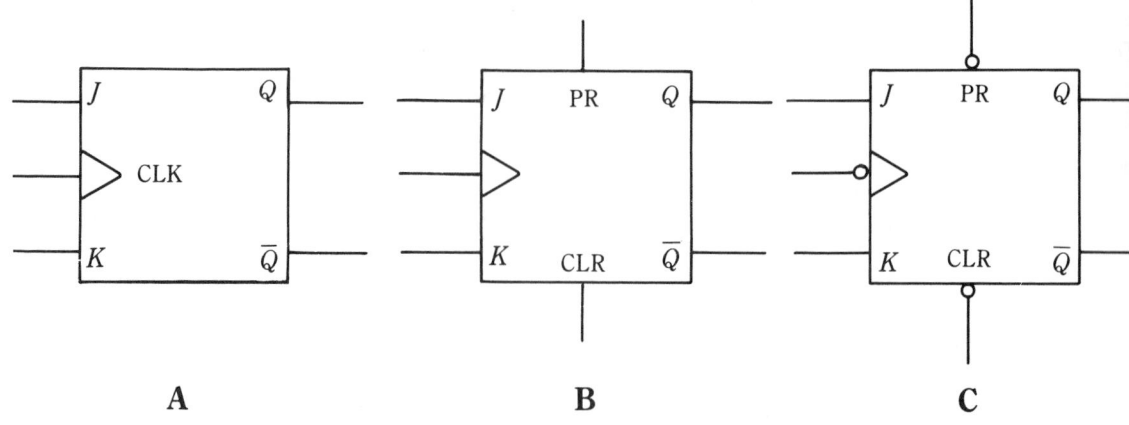

A   **B**   **C**

7-9   JK Flip-flop symbols.

edge, and a sharp negative pulse for the negative-going edge of the clock pulse. There-fore, the JK flip-flop state can only be changed during the positive pulse edge of the dif-ferentiator. The symbol for the positive-edge triggered flip-flop is shown in Fig. 7-9A.

Most flip-flops have nonsynchronous (not clocked) means of presetting or clearing the flip-flop. In Fig. 7-9B, if the *PR* (for preset) terminal is made to go high by some external circuitry, *Q* will go high and remain high. When PR is made to go low again, normal clocked operation is resumed with *Q* starting in the high state. A high CLR (for clear) terminal will make *Q* low in the same way. Note that *PR* and *CLR* dominate regardless of the inputs on *J* and *K*, and the existence of the clock pulse. *PR* and *CLR* are said to be *active high* for the operations just described.

In Fig. 7-9C *PR* and *CLR* are active low, i.e., *PR* and *CLR* are deliberately held high for normal synchronous operation of the flip-flop. The flip-flop can be set by bringing *PR* low and can be reset by bringing *CLR* low, which is indicated through the use of the bubble symbol (borrowed from the inverter). Note that *PR* and *CLR* cannot be active at the same time. Also, the negative edge of the clock pulse results in the triggering action.

Counters

A simple binary counter appears in Fig. 7-10. Its purpose is to count the number of input pulses to the clock input terminal of the JK flip-flop FF0. Table 7-12 summarizes the action, assuming that all flip-flops are clear before the arrival of the first pulse at the Input. (See the first row of the table.) Thereafter, FF0 will toggle on arrival of the input pulses. This is shown by the alternating zeros and ones in the rightmost column of the table.

FF1 also toggles as $\overline{Q}$ of FF0 goes high, i.e., *Q* goes low. This provides the pattern of zeros and ones for FF1 in the table. You see two zeros, then two ones, etc., moving down from row to row.

Finally, FF2 also toggles as $\overline{Q}$ of FF1 goes high, i.e., *Q* goes low. This provides the pattern of zeros and ones you see in the leftmost column of Table 7-12.

Viewing the overall result of the counter operation, you see binary numbers in suc-ceeding rows of 0, 1, 2, 3, 4, 5, 6, and 7 (in the next to the last row). The binary counter

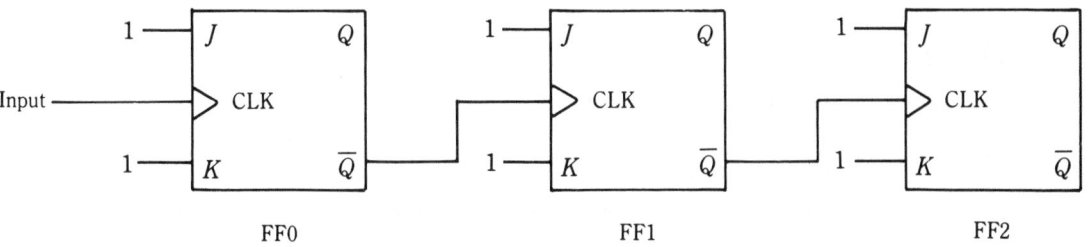

7-10 Simple binary counter.

cannot count to a number larger than 7 with three flip-flops. The very next count (last row in Table 7-12) causes the counter to recycle, so the counting cycle repeats as before. Higher count requirements can be satisfied with a larger number of flip-flops. With four flip-flops, the highest count will be 15. With five flip-flops, the highest count will be 31. With ten flip-flops, the highest count will be 1023.

The counter of Fig. 7-10 is referred to as a *ripple counter*. This name refers to the fact that each flip-flop requires some amount of time to change state from 0 to 1 or from 1 to 0. For example, in Table 7-12, next to the last row, on receipt of the next pulse to be counted, FF0 will change from 1 to 0 after a short time delay. Thereafter, FF1 will change from 1 to 0. Finally, FF2 will change from 1 to 0. Hence the term ripple as it refers to the counter.

Ripple counters have basic limitations in counting speed. Moreover, the flip-flop transitions do not occur simultaneously. Another, somewhat more complicated binary counter, responds in synchronism with the pulses being counted. This form of binary counter is called a *synchronous counter*.

The binary counters discussed all recycle at a count that is a power of two: 8, 16, 32, and 1024 for 3, 4, 5, and 10 flip-flops. These same counters are referred to as modulo 8, MOD8, MOD16, MOD32, and MOD1024 counters. Because you also live in a decimal world, it is convenient to also have a MOD10 or decade counter. Figure 7-11 illustrates

**Table 7-12. Counting
Action for Simple
Binary Counter.**

FF2	FF1	FF0
0	0	0
0	0	1
0	1	0
0	1	1
1	0	0
1	0	1
1	1	0
1	1	1
0	0	0

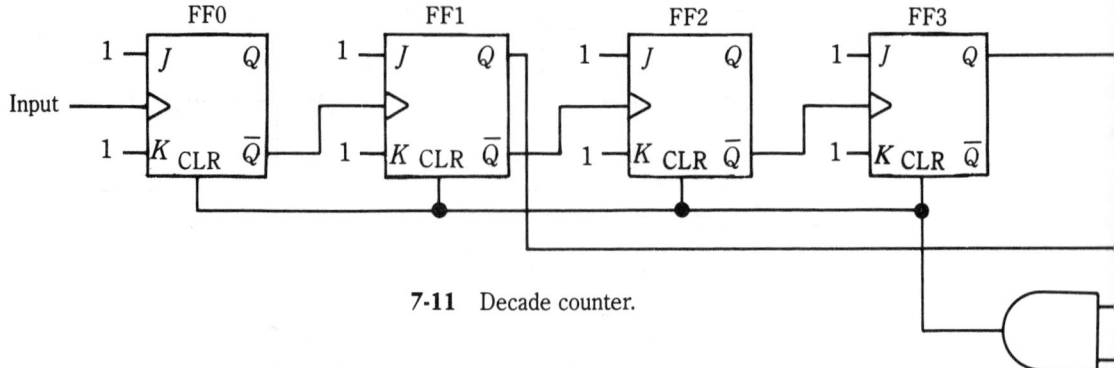

7-11   Decade counter.

how a MOD16 binary counter can be modified to become a MOD10 counter. Here you see a binary ripple counter and some additional logic circuitry connected to the CLR terminal of each flip-flop.

The counter works like this. It starts at 0000_2 and proceeds normally to $1001_2 = 9_{10}$. For each of these counts the output of the AND gate is low so that the CLR of each flip-flop is not active. However, with one more input pulse, the count becomes $1010_2 = 10_{10}$ and the output of the AND gate goes high, all flip-flops are almost immediately reset for a count of 0000_2, which brings the output of the AND gate low and the counting resumes. The logic circuit has, in essence, tricked the counter into recycling from 9_{10} to 0_{10}. The $1010_2 = 10_{10}$ occurs only momentarily and is not regarded as an indicated count.

Decade counters, as units, can be cascaded to give any number of significant decimal digits in the count. Many examples of such use can be found in modern test equipment including frequency counters, digital meters, etc.

Shift registers

A shift register permits a serial sequence of bits to be stored and, at some later time, to be shifted out in serial order. Other capabilities and applications will be identified shortly.

Figure 7-12 illustrates a basic, simple shift register constructed using clocked D flip-flops. Since the Input can be high or low at the time of the clock edge, Q will correspondingly either go high or low. That is, after the clock edge, the Q of FF3 will assume the same binary value as that of the Input at the time of the clock edge. Thereafter, at each

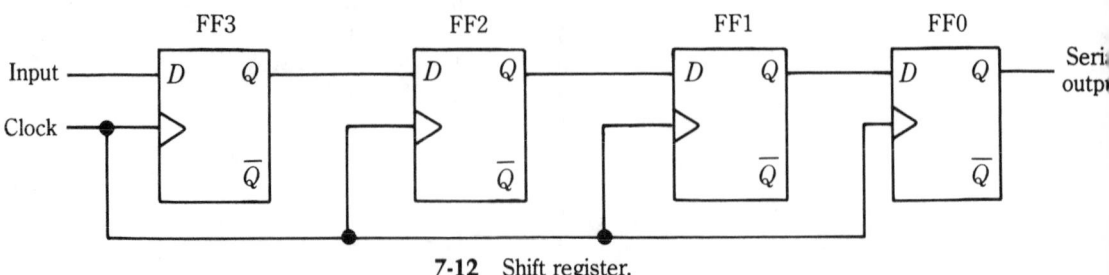

7-12   Shift register.

successive clock edge, the Q of FF3 will assume the value of the Input at that time and hold it until the next clock edge. The Input sequence of high or low levels is thus copied by FF3 at the clock edge times.

Likewise, FF2 will copy the Q of FF3. This means that the Q states of FF2, as would be seen using an oscilloscope, will be identical to those of FF3 except that there will be a time delay of one clock period. In the same way FF1 follows FF2, and FF0 follows FF1.

Examine the practical result of these actions. Visualize a quantity 1101_2 applied as Input with the rightmost bit arriving first and the others, in their proper order, in synchronism with the clock. The D of FF3 will become 1, then 0, then 1, and 1 for four clock pulses. At the end of four clock pulses the Qs for the flip-flops are:

FF3	FF2	FF1	FF0
1	1	0	1

If the CLOCK is now switched off, the binary quantity 1101_2 will be stored indefinitely as data.

At some time in the future the data might be required for some purpose. If the Clock is again started, the binary quantity will be shifted out as the Q of FF0 in the exact order and with the same timing arrangement as it was shifted in. Note that it takes exactly four clock periods to shift the number in and the same number to shift it out. The arriving data is in serial form and departing data is also in serial form. A single line is required for input or output data. Shift registers are commonly used at the sending and receiving ends of a serial data communications system.

A register composed of D flip-flops with preset (PRE) and clear (CLR) terminals allows the data to be loaded in parallel. That is, each flip-flop in the register can be selectively preset to contain a 1 or a 0. It is then possible to shift the data out serially. Likewise, it is possible to sense the state of all flip-flops in a loaded register simultaneously, i.e., in parallel.

The shift register is a memory device (usually temporary) and one element for parallel-to-serial and serial-to-parallel data conversion. It is also a possible part of an arithmetic digital circuit.

Arithmetic circuits

Addition is the basic binary arithmetic operation for digital equipment. When combined with complementation (inverting), shifting, and logic operations, including the numerical comparison of quantities, the other arithmetic operations such as subtraction, multiplication, and division, are possible.

Table 7-13 is the truth table for the binary addition of two bits designated X and Y. The sum is, Σ (Greek sigma), and the carry generated is C_o, for carry out. That is $X + Y = C_o$. For example, $1_2 + 1_2 = 10_2 = 2_{10}$, for which $\Sigma = 0$ and $C_o = 1$. A logic circuit to implement this addition is shown in Fig. 7-13. It is called a half-adder for reasons that will be clarified later.

The addition of multibit numbers requires the consideration of a carry in (C_{in}), which results from the addition of the next least significant bit pair. For example, if you wish to add $111_2 = 7_{10}$ to $011_2 = 3_{10}$.

Table 7-13. Truth Table for Adding Two Bits.

X	Y	C_o	Σ
0	0	0	0
0	1	0	1
1	0	0	1
1	1	1	0

$$
\begin{array}{c}
① ① ① \\
1 \; 1 \; 1 \\
+ \; 0 \; 1 \; 1 \\
\hline
1 \; 0 \; 1 \; 0
\end{array}
$$

Starting in the right-hand column $1 + 1 = 0$ with $C_o = 1$. This carry-out appears as a carry-in (encircled) located above the next column. The next addition is $1 + 1 + 1 =$ where $C_{in} = 1$. This gives $Σ = 1$ and $C_o = 1$. The addition carry in for each step is shown encircled above the proper column. The sum is $1010_2 = 10_{10}$. Obviously the original truth table for addition (Table 7-13) is inadequate. The enlarged truth table is shown in Table 7-14, which allows as inputs X, Y, and C_{in}. One implementation of a full adder is shown in Fig. 7-14.

The subtraction of one number from another is accomplished by taking the complement (in a particular way) of the number to be subtracted and adding the complement to the other number. The complement of a binary number is found by inverting each bit and then adding 1. For example, the complement, called the 2s-complement, of 0110 is 1010. It is seen that the subtraction can be accomplished by one adder for each bit and an inverter for each bit.

The process of multiplication is accomplished by adding and shifting. Each bit in the multiplier is examined one at a time starting on the right at the least significant bit. If a 1 is present, the number to be multiplied, the *multiplicand*, is added to the partial product. If, on the other hand, the number is 0, nothing is added to the partial product. In either

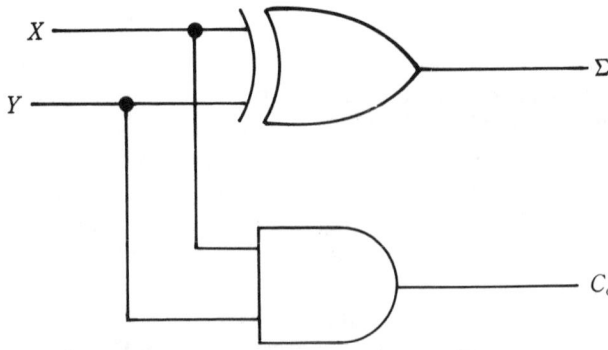

7-13 Half-adder using XOR, and AND gates.

**Table 7-14. Truth Table
for Full Adder.**

Inputs			Outputs	
X	Y	C_{IN}	Σ	C_o
0	0	0	0	0
0	0	1	1	0
0	1	0	1	0
0	1	1	0	1
1	0	0	1	0
1	0	1	0	1
1	1	0	0	1
1	1	1	1	1

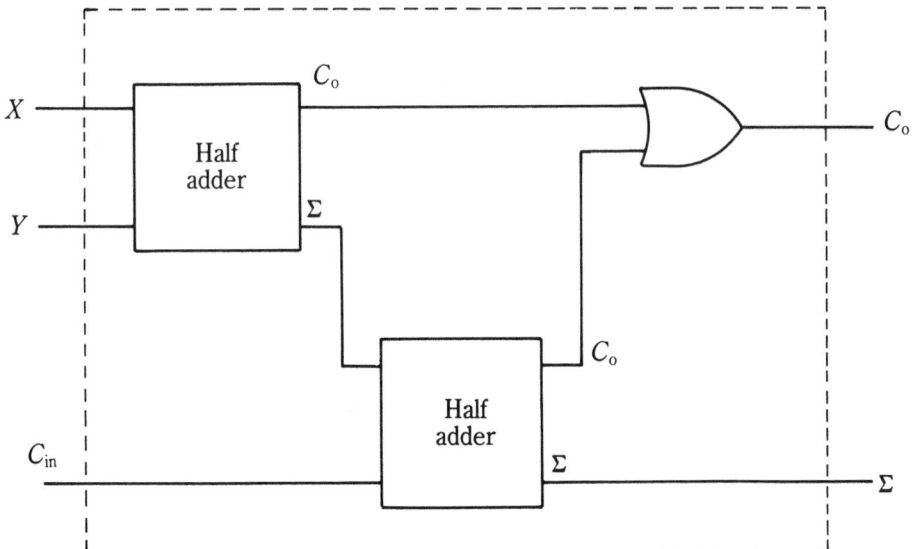

7-14   Full-adder constructed using Half-adders as components.

case the partial product is shifted one place to the right before the next bit of the multiplier is examined. This shifting is done until all bits of the multiplier have been considered. Multiplier equipment consists principally of a set of full adders and a shift register.

Division can be accomplished by an automated form of long division, much like you learned for decimal numbers in school. Note that the principal parts of the procedure involve multiplication, subtraction, and number value comparisons so that no new techniques apply.

Logic families

A logic family consists of compatible ICs which collectively implement some or all of the logic functions previously described. Within a family the members share common power supply levels, low input and output levels, high input and output levels, loading specifications, and noise immunity standards. Separate branches of large families will usually present different loads, load capabilities, and operating speeds. The most widely used families presently, TTL and CMOS, will be covered more thoroughly than the other families. Positive logic will be assumed unless there is specific mention to the contrary.

The *resistor-transistor logic* (RTL) family was the first to be introduced. Examination of Fig. 7-15 indicates that a high on *A* or *B* or both *A* and *B* will yield a low output at *C*, since either Q1, Q2, or both will be saturated. The logic is that of a NOR gate (an OR gate with inverted output). Similar circuits can be formed to become inverters, flip-flops, encoders, decoders, counters, shift registers, etc.

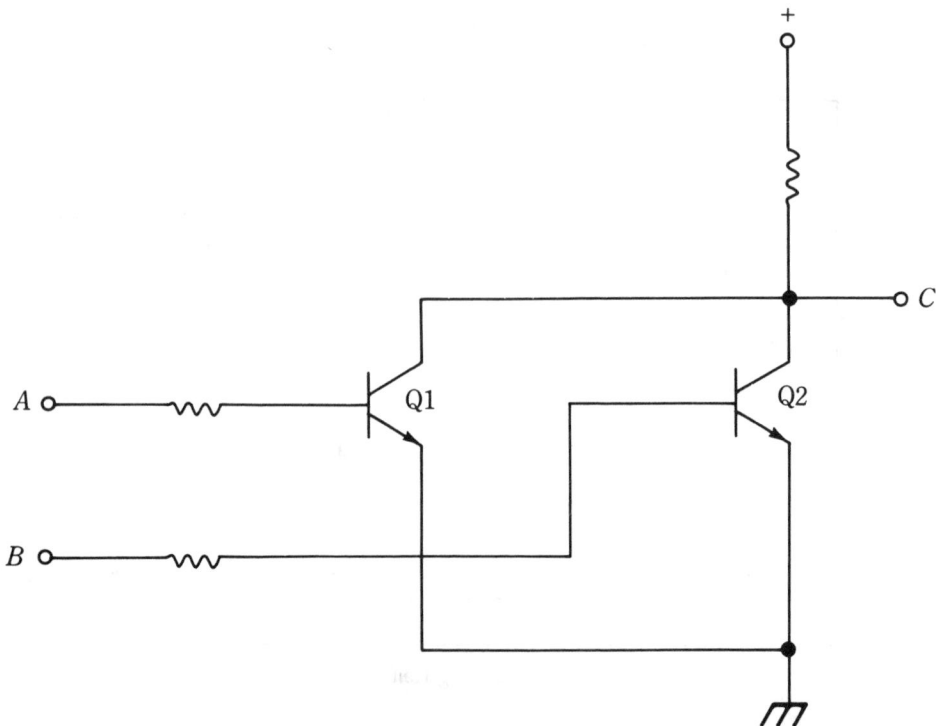

7-15   Resistor-Transistor Logic (RTL) NOR gate.

Figure 7-16 shows a *diode-transistor logic* (DTL) NAND gate (an AND gate with inverted output). All three inputs must be high in order that Q1, and hence Q2, conduct to give low output at *D*.

The next IC technology to evolve was *transistor-transistor logic* (TTL or T²L) characterized by the multiple-emitter transistor in the NAND gate of Fig. 7-17. It works as follows: If *A* and *B* are both high, Q1 will have no emitter current. Current will flow in the collector lead because the base-collector junction is forward-biased. Q2 will conduct, causing Q3 to be cut off and Q4 to conduct so that its collector output *C* is low. Q4 is capable of sinking current for an active load such as one or more input emitter leads of a second gate.

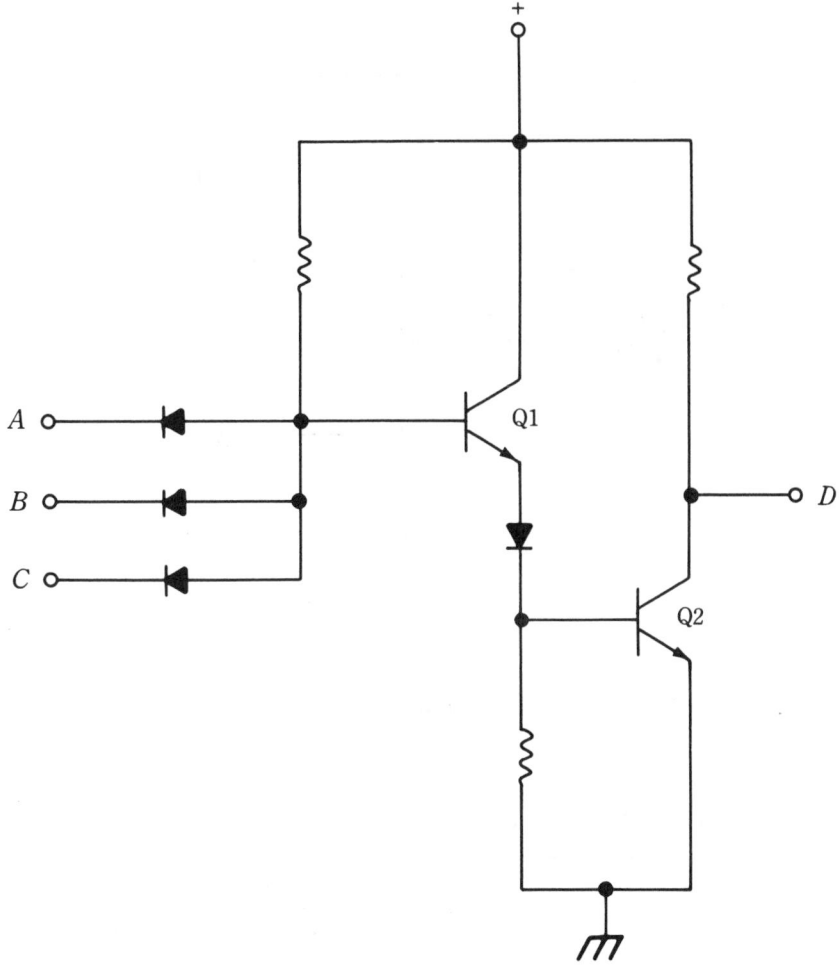

7-16   Diode-Transistor Logic (DTL) NAND gate.

If either *A* or *B* or both *A* and *B* are low, emitter current will flow because at least one base-emitter junction is forward-biased and the collector of Q1 will go low. This cuts off Q2, and hence Q4, while allowing Q3 to conduct and "source" current to the load. The Q3-Q4 arrangement is commonly referred to as a *totem pole*.

The IC packages in the TTL family have designations as exemplified in Table 7-15. Normally, there is a prefix indicating the manufacturer, like those shown in Table 7-16. A 54 or 74 designates the standard TTL family, followed by one or two letters indicating the subfamily, e.g., LS, and the next two or three digits the family member.

The 54 and 74 integrated circuits differ in the temperature ratings with the 54 having the greater temperature range and corresponding price tag.

Another logic family, *integrated-injection logic* (I^2L), usually exists in integrated circuits containing many thousands of gates. The mechanism employed causes the output of a current source to be steered away from the base of a bipolar transistor (for low input)

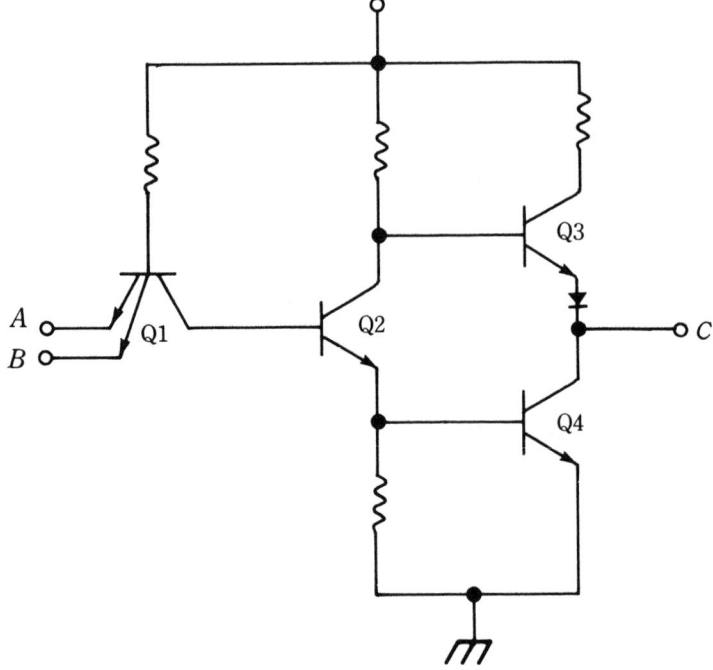

7-17   Transistor-Transistor Logic (TTL) NAND gate.

Table 7-15. Typical Characteristics of the TTL Family.

Designation	Series	Propagation delay (ns)	Power dissipation (mW)
74XX	Standard	10	10
74LXX	Low-power	30	1
74HXX	High-power	5	22
74SXX	Schottky-clamped	3	20
74LSXX	Low-power Schottky	8	2

or injected into the base (for high input). Gates are formed by directly connecting several inputs to the base of the transistor and connecting the collectors or several transistors for the outputs. The transistors usually have multiple collectors connected together in various combinations for efficient implementation of logical expressions.

The injector (current source) current for I^2L is controllable so that low current (hence low power) is achieveable at the expense of speed. This control feature is desirable in a digital watch for which speed is not required but low power consumption is vital. A second advantage is the ability to fabricate linear functions on the same I^2L chip. I^2L

**Table 7-16. Some Manufacturer's
Prefixes for TTL ICs.**

Prefix	Manufacturer
AM	Advanced Micro Devices
DS,DM	National
HD	Hitachi
M	Mitsubishi
MC	Motorola
MPY,TDC	TRW
N	Signetics
RC,RM	Raytheon
TD	Toshiba
ZN	Ferranti

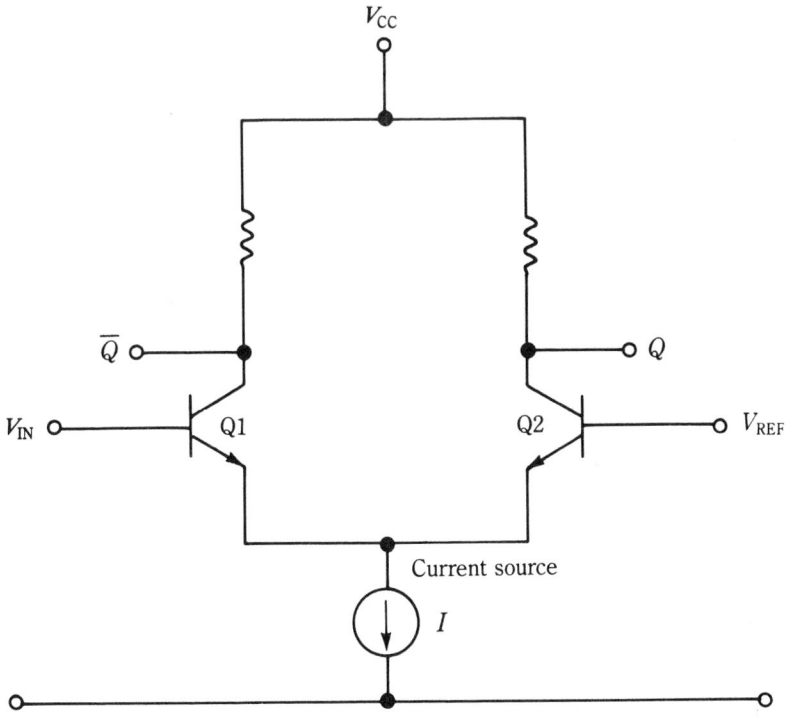

7-18   Emitter-Coupled Logic (ECL) gate element.

has great potential for many applications as exemplified by the Texas Instruments (SBP 9900) microcomputer family.

A basic *emitter-coupled logic* (ECL) switch is shown in Fig. 7-18. The configuration should be recognized as that of a differential amplifier with a common current source I. Transistors Q1 and Q2 control the path of this current. The base of Q2 is connected to a fixed reference voltage V_{REF}. If V_{IN} is greater than V_{REF}, I passes through Q1 so that output $\overline{Q}$ will be low and Q will be high. If V_{IN} is less than V_{REF}, the opposite is true. Additional resistors and transistors are necessary to form gates. ECL has the particular advantage of high-speed operation resulting from the design for low voltage levels and avoiding "storage time" delays by preventing saturation of the transistors.

Although p-channel MOS (PMOS) received earlier emphasis because of fabrication advantages, n-channel MOS (NMOS) is more widely used today. An enhancement-mode NMOS NOR gate is pictured in Fig. 7-19 with inputs A and B. The upper MOSFET is connected as a high resistance load. The other two are switches. If A and B are low, neither of the two MOSFETs conduct and the output is high. If either A or B is high (or both are high), conduction occurs and the output is low, because the effective resistance of the conducting MOSFET is lower, by orders of magnitude, than that of the upper MOSFET.

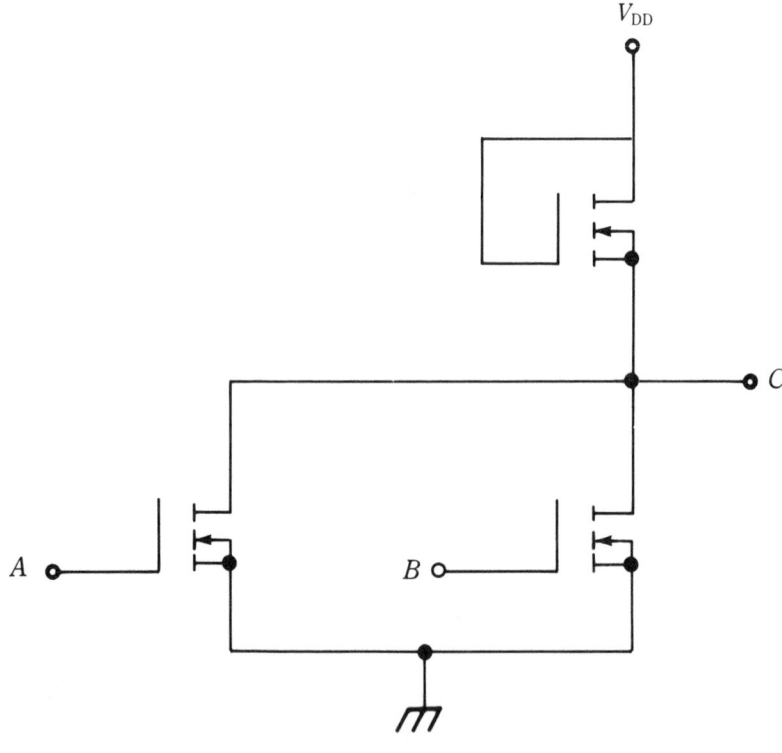

7-19 NMOS NOR gate.

The operation of the NAND gate in Fig. 7-20 is similar in nature. Viewing the two lower transistors as switches, both must be closed for the output to be low. Thus, both A and B must be high for conduction and low output. If either A or B is low, appreciable current will not flow and the output is high.

Complementary MOS (CMOS) involves both PMOS and NMOS transistors on the same chip and in the same gate. The NOR gate for CMOS is shown in Fig. 7-21. Suppose, for example, that A is high. The uppermost FET will be cut off whereas the left-

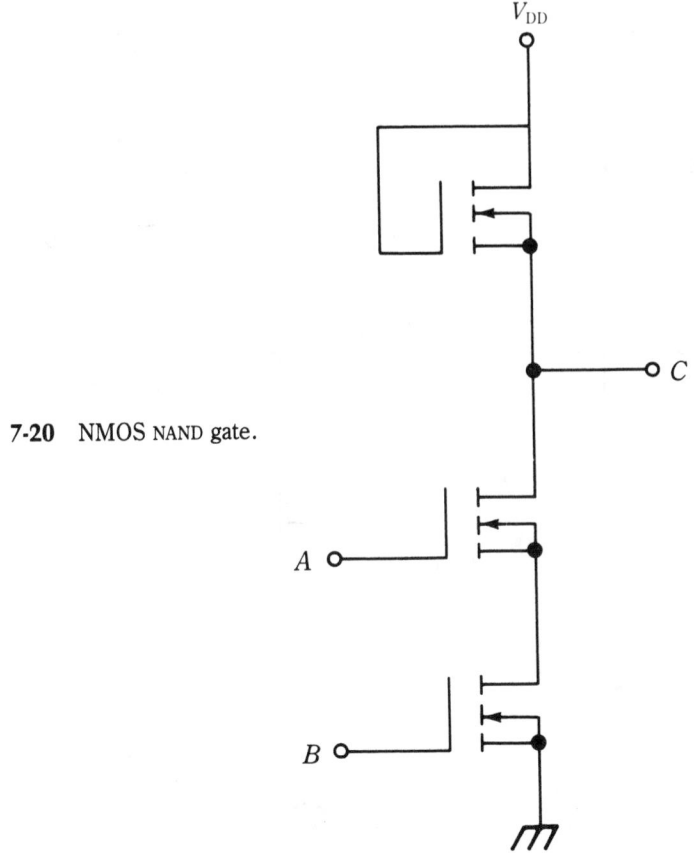

7-20 NMOS NAND gate.

most FET will conduct. The output is low. A similar argument holds if *B* is high. Both inputs can be high with the same result.

When both *A* and *B* are low, both of the upper series-connected FETs conduct, but the lower parallel-connected FETs are cut off. The output is high. Note that there is never a direct conducting path internal to the gate between V_{DD} and ground, except during a switching transition which cannot be avoided. Therefore, CMOS has very low power dissipation, a particular advantage where dense packing in equipment is necessary.

The circuit of a CMOS NAND gate is shown in Fig. 7-22. If inputs *A* and *B* are both high, neither of the upper two transistors conducts, whereas, both of the series-connected lower transistors conduct, yielding a low output. If either *A* or *B* is low, one of the lower series-connected transistors will be off, whereas, one of the upper parallel-connected transistors will conduct. The output will be high. As is the case for the NAND gate, no direct path exists within the gate for current flow, with a consequent saving in dissipated power.

Previously, one of the most significant disadvantages of CMOS was its speed. Current developments (the RCA QMOS series, for example) have made CMOS competitive with TTL speed-wise while retaining the significant power dissipation advantage.

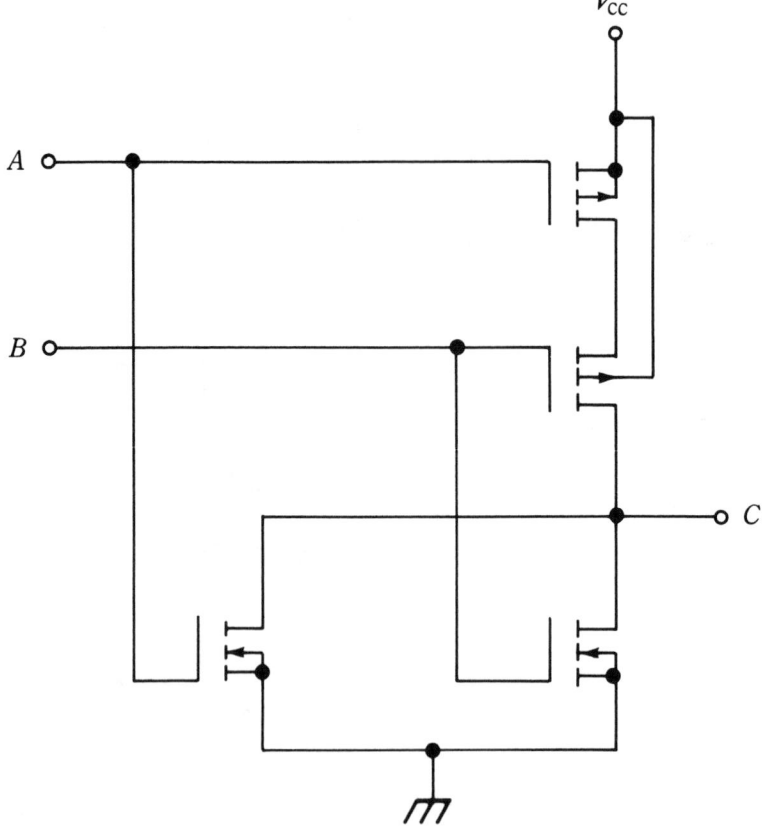

7-21   CMOS NOR gate.

Memories

Some of the logic devices previously introduced must store one or more bits in their normal operation. A flip-flop is the most elementary of these. A counter must be capable of storing the number of bits required to express the highest count, since the counting process requires that the previous count be known and incremented. Similarly, the state of each flip-flop in a shift register must be stored preparatory to shifting either to the right or left. Thus, many ongoing arithmetic and logic processes require the temporary storage of more bits. This storage requirement is an integral part of sequential logic in which the time dimension is added to the combinatorial logic originally presented.

Memory is more than a place of temporary storage as an ingredient of the arithmetic and logic processes taking place in computers and other digital devices. Memory means those devices and equipment that are employed in storing substantial amounts of information in locations which are systematically addressed for the purposes of placing the information in the memory and retrieving it. Computer programs are placed in the memory as are data that might be used in executing the programs. Intermediate and final results are also commonly placed in memory.

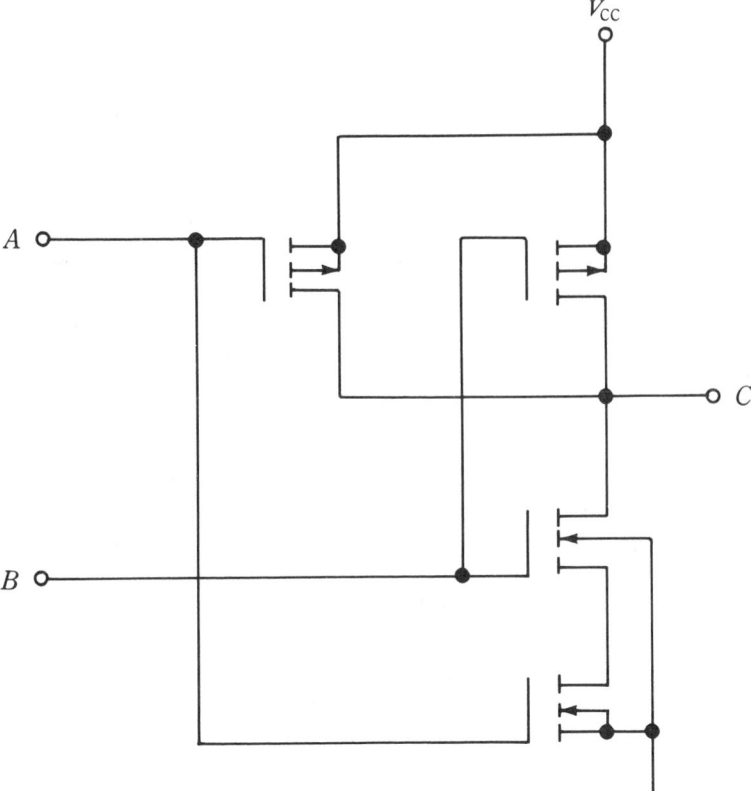

7-22 CMOS NAND gate.

In the years before semiconductor memories became available, punched cards and punched paper tape were widely used in computers and their predecessors. Magnetic recording on tape, cylindrical surfaces (drums), and disks has been, and continues to be, used in computer peripheral equipment. The access time for magnetic recording equipment is relatively high, as compared to the *r*andom *a*ccess *m*emories (RAM), and this equipment performs effectively where enormous amounts of information must be stored and the time to access it is not a serious problem in data processing. An example of this form of application is payroll preparation in which each employee's record is available on tape in alphabetical order.

Modern computers also require that RAM, as compared to the sequential access of magnetic tape, be available. The original form for the RAM was a grid of doughnut-shaped magnetic cores strung on intersecting arrays of wires allowing signals on these leads to pick one core for storing a bit or retrieving it. Stacking a number of core planes allows the whole word, usually consisting of 32 or more bits, to be stored or retrieved at one time. The magnetic core memory is still widely used, and is completely satisfactory. Semiconductor memories, serving the same function of random access for data storage and retrieval, are increasingly popular for all computers and are standard for all micro-computers.

Random-access memory

Functionally, semiconductor memories can be classified into two categories: random-access memories (RAMs) and read-only memories (ROMs). The method of classification distinguishes between the volatile RAM and the non-volatile ROM, which are employed in different ways in the microcomputer. It is the name RAM which is the source of confusion since technically, both are random-access memories. To be specific, a random-access memory is one in which the time to obtain data or to store data is independent of the address. (This feature is obviously not characteristic of magnetic tape, for example.) A RAM is capable of storing information during the execution of a program. A computer user can deliberately place program instructions and data in RAM. Moreover, memory locations in RAM can be used by the program being executed to store information. This information disappears when the computer is turned OFF, hence the term volatile.

In direct contrast, the user cannot store a program or data in the ROM in the same way that it can be stored in RAM. Execution of a program cannot cause information to be placed in ROM regardless of whether or not the memory locations presently contain information of any kind. Moreover, turning OFF the computer power does not cause the ROM contents to disappear. This accounts for the non-volatile term used in describing ROM. More discussion about the several types of ROM will be presented shortly.

There are two kinds of RAMs: *static* and *dynamic*. A dynamic RAM cell is formed in an IC from a single capacitor and a transistor as illustrated in Fig. 7-23. A 0 or 1 is stored in the cell depending on whether or not the capacitor holds a charge. The transistor is made to conduct by the voltage on the gate of the transistor so that, when the particular cell is selected, the capacitor can be charged to the data line level. This permits the original charging of the capacitor and the refreshing of the charge at regular intervals usually in milliseconds. The same transistor allows the charge on the capacitor to be sensed in the READ operation. The simplicity of the dynamic RAM cell is the reason that the IC will store 256 K in the new versions coming on the market. Note that K, as used in defining the memory capacity, means $2^{10} = 1024$ and not 1000.

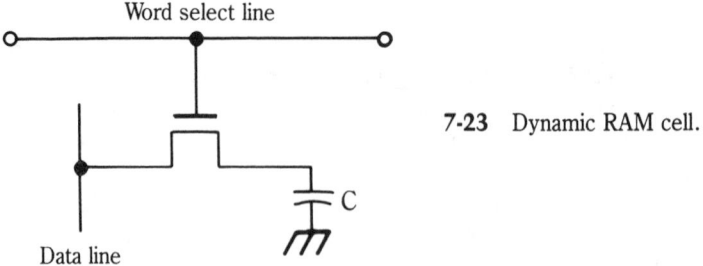

7-23   Dynamic RAM cell.

Static RAMs need no refreshing but are more complex, as is shown in Fig. 7-24. The static RAM cell requires six devices, four operating as transistors, with the two devices adjacent to the V_{DD} supply acting as resistances, in the flip-flop configuration. Because of the additional complexity, the static RAM IC has less storage capacity.

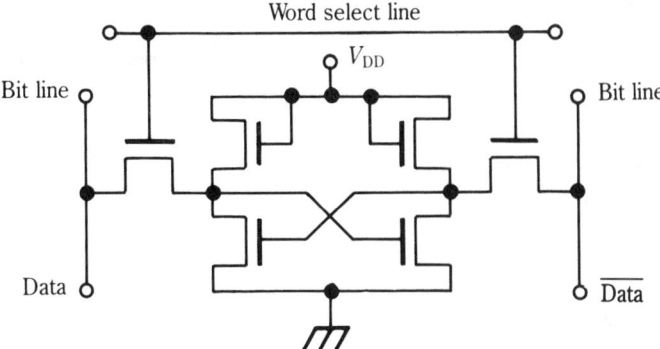

7-24 MOS static memory cell.

Read-only memory (ROM)

As indicated in Fig. 7-25, there are several types of semiconductor memories in the generic ROM category. Of these, only the mask ROM is a true read-only memory. It is programmed during the manufacturing process to contain the information specified by the customer. This is done by creating custom masks used in the lithographic steps for interconnecting the chip devices. This ROM cannot be altered by any means after the memory contents have been so created.

Programmable ROMs (PROMs) can have the memory contents inserted by the customer by a special procedure. This procedure consists of deliberately blowing tiny fuses that were left intact during manufacture. Special equipment providing sufficient current is required for this purpose. The fuses are selectively blown individually in sequence, in the cells requiring storage of this state. Note that a blown fuse could signify either a 0 or 1 by prior agreement. The contents of a PROM cannot be subsequently altered.

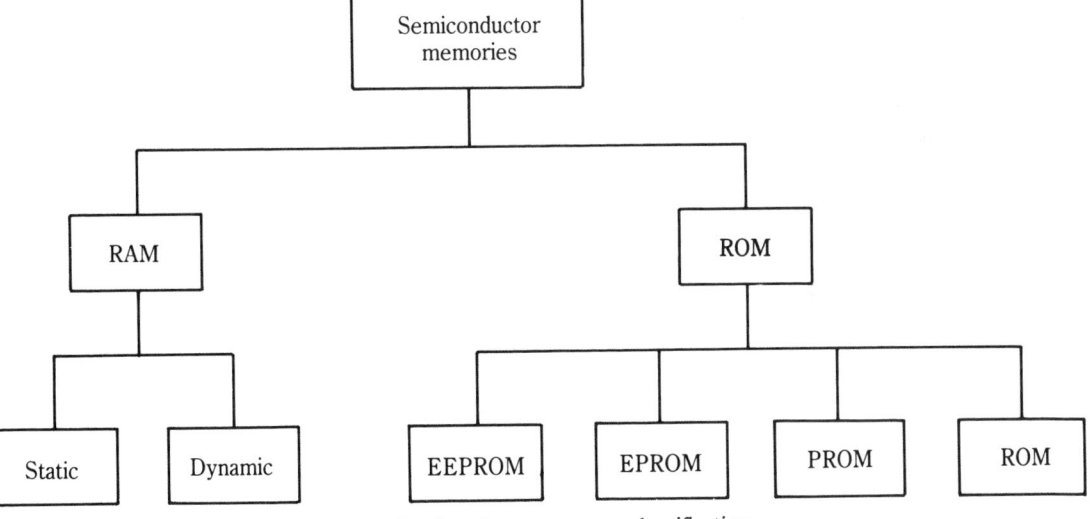

7-25 Semiconductor memory classification.

Next in order of increasing flexibility is the *e*raseable PROM (EPROM). Information is stored in EPROM cells by applying an electrical pulse which causes a charge to be trapped within the cell, insulated by a layer of silicon dioxide. The status of charge within the cell (in Fig. 7-26) is sensed during the READ operation. Unlike the PROM, whose information pattern is established permanently, the EPROM can be erased by the application of ultraviolet light, which causes the charge to escape from the cell. The EPROM can then be reprogrammed to store a completely different bit pattern.

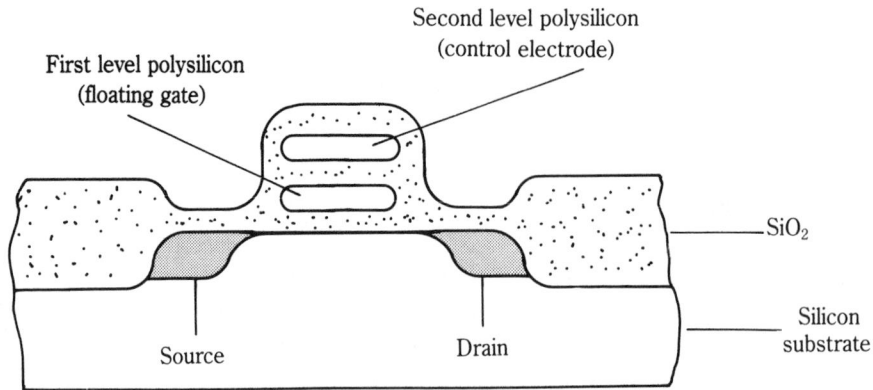

7-26   Erasable Programmable Read-only Memory (EPROM).

*E*lectronically *E*raseable PROMs (EEPROMs or E^{2}PROMs) are similar in construction to the EPROM. However, it is possible to modify the bit in each cell by electrical means. This is made possible (Fig. 7-27) by using a much thinner silicon dioxide electrical insulator, allowing erasure by electrical pulses.

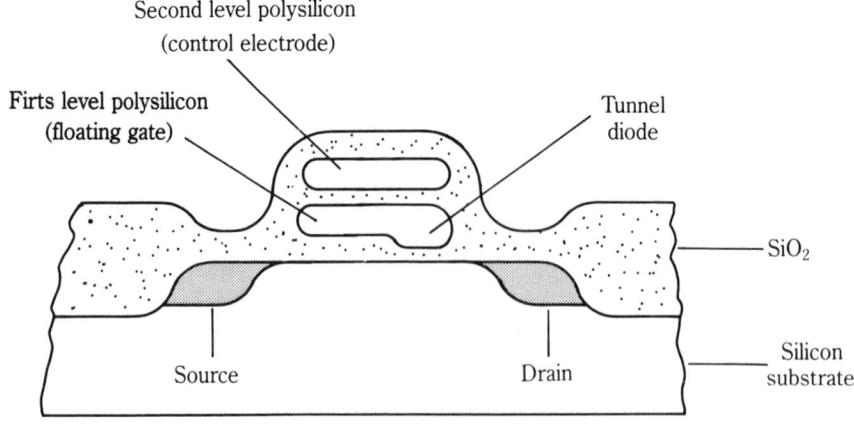

7-27   Electronically Erasable Programmable Read-Only Memory (EEPROM).

Microcomputers

Microcomputers are digital computers whose principal element is the microprocessor. The microprocessor is a comparatively recent IC development that incorporates most functions of a computer's central processing unit. The low price, small power consumption, and insignificant size of the microprocessor, coupled with its flexibility, has propelled it into a major role in virtually all categories of industrial, commercial, and consumer equipments for which electronic control is feasible. In particular, the microprocessor fits many categories of communications equipment including transmitters and receivers. A substantial part of the matching of digital technology to these control functions stems from the appearance of frequency synthesizers in virtually all equipment capable of operating on different carrier frequencies.

Microprocessors are now more popular in many applications than competing, hardwired (random) logic by reasons of cost. A microprocessor is mass-produced at low unit cost. It has many applications that are specifically determined by the additional supporting electronics and the software. Alternatively, the same applications can be implemented by a set of dedicated, interconnected electronic devices with no software. A different application calls for a different design, a different hardware configuration, and different software if the microprocessor approach is adopted.

In brief, the microprocessor and software can be substituted for large numbers of electronic IC parts, and, perhaps more importantly from the cost point of view, their method of interconnection via printed-circuit boards, harness, motherboards, etc. It must be concluded that the cost tradeoff has proven to be favorable for the microprocessor in many recent equipment applications.

In this section you will consider the general configuration and operation of a microcomputer; it is not typical of one which is to be found in a test instrument, a receiver or a transmitter. Indeed, this configuration has been specifically designed as an education tool which allows the entry of data and instructions at a keyboard and the display in alphanumeric form for the operator. Nevertheless, the main features of all microcomputers are present, along with the basic principles of operation.

Basic microcomputer system

A microcomputer system is configured to include a microprocessor and the other hardware elements necessary for the particular application. You will here be concerned with a basic microcomputer intended for direct use by a human operator. The differences between this basic microcomputer and one serving to automate some process, for example, will largely exist in the nature of the devices providing data to the computer, and those that must accept processed data, control signals, etc., from the computer. The basic computer is supplied with information from a keyboard as shown in Fig. 7-28. The results of the microcomputer processing actions are displayed to the operator. Both the keyboard and the display hardware are regarded as peripheral devices, not part of the microcomputer proper. These peripheral devices are interfaced with the rest of the microcomputer by means of hardware elements referred to respectively as the *Input port* and the *Output port*.

Perhaps the most prominent features shown in Fig. 7-28 are the buses labeled *Address bus*, *Data bus*, and *Control bus*. The Data bus, for example, consists of eight

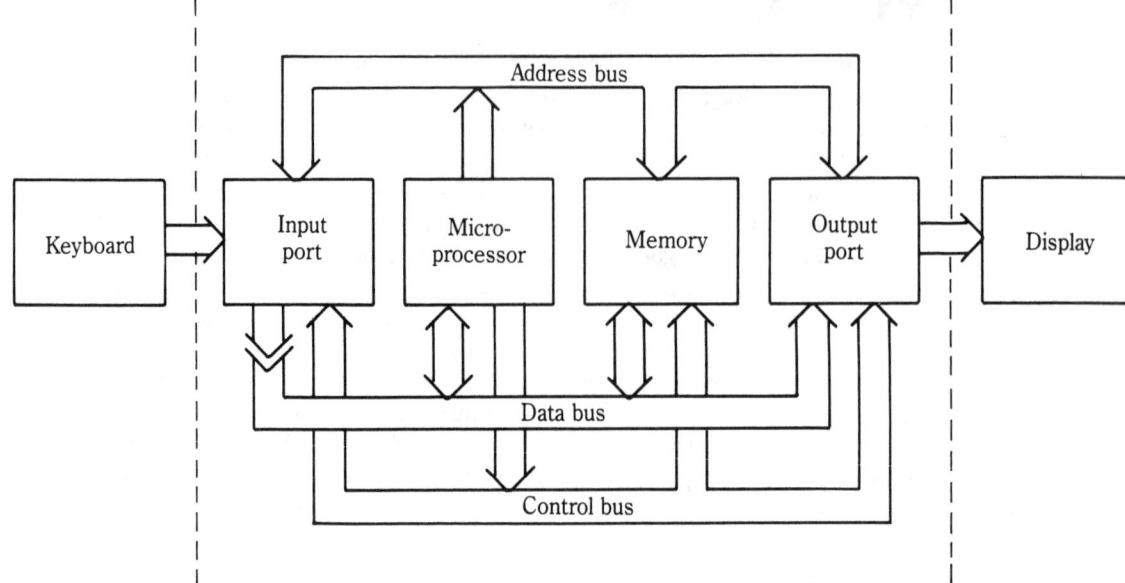

7-28   Basic microcomputer system.

parallel leads permanently connected to the ports and the other main hardware elements, namely, the *Microprocessor* and the *Memory*. At any one time, an electronic pathway is established so that one, and only one, of these can place data on the bus in order that another designated element can be the receiver. At another time the sender and the receiver might be different, so the bus is a general communications system between those elements for which data must flow. Eight bits, constituting one word, are transferred simultaneously over the data bus. However, sixteen or more bits can be included in a word for other microprocessors so that the bus must be correspondingly wide in the microcomputer.

This microcomputer has a total memory capacity of 2^{16} = 65,536 words. This means that a 16-bit address bus is needed. Addresses are transferred from the microprocessor to the memory for the purpose of designating the particular memory location from which a data word (8 bits) is to be extracted (read) or stored (written). The microprocessor can also designate the input port or the output port as participants in data transfer by specifying those unique addresses assigned to each by the micrcomputer designer. The address bus functions essentially the same way as the data bus in that one sender, always the microprocessor, controls the signal levels on the bus so that the designated receiver can obtain the 16-bit address.

The microcomputer is required to perform a number of specific actions in achieving an overall objective. These actions are prearranged through the preparation of a computer program which is stored in memory. The instructions in this program are interpreted by the microprocessor, which directs these actions by means of pulsed signals placed on one or more electrical leads known collectively as the *control bus*. The program can also provide addresses and data that are necessary in the execution of the program. The nature of a computer program will now be examined more thoroughly.

Computer programs

Computer programs are known as *software*, a term highlighting the comparison with the physical hardware of a computer system. Different types of computer programs are written for different purposes. However, they are all stored in memory, some in ROM and others in RAM, and the different instructions, of which the program is composed, are accessed in the proper sequence. Consider a simple program whose purpose is to demonstrate counting in a rather oversimplified application.

Table 7-17 lists the COUNTER PROGRAM. In the left column are placed the memory locations, i.e., the addresses in the program, starting at 0804 and ending at 080D. These addresses are each stated in the hexadecimal (base 16) number system. However, it must be understood that the microprocessor does not understand an address such as 0804, but rather 0000 1000 0000 0100, 0804's binary equivalent; the spaces have been placed here for emphasis and ease in reading. Each memory location is examined in sequence during program execution, except where a branch, or jump, is required. Such a situation occurs in the COUNTER PROGRAM as will be seen shortly.

Table 7-17. Counter Program Listing.

Address	Contents	Label	Instruction	Comments
0804	3E		MVI A,0	; Set A to zero
0805	00			
0806	32	LOOP:	STA 3000	; Output A to port
0807	00			
0808	30			
0809	00		NOP	; For later addition
080A	3C		INC A	; Increment A
080B	C3		JMP LOOP	; Loop back
080C	06			
080D	08			

In the second column indicated as "contents" are the numbers stored in the respective address locations. Thus, at 0804 you will find 3E, which, like the addresses, is expressed as a hexadecimal number. Note that 3E is the hexadecimal equivalent of an 8-bit binary number, eight being the width of the data bus. The significance of the words extracted from memory depend on the context of the program. The remaining three columns in the table are useful in explaining the program to a human reader and do not exist as far as the microcomputer is concerned.

You will now walk through the COUNTER PROGRAM and be introduced to some additional concepts along the way.

Starting again at 0804, the number 3E is obtained from the memory. It is interpreted by the microprocessor as the *operation code*, abbreviated as opcode, designating a specific instruction. This opcode causes a register, named the *accumulator* or A-register, to be set to contain a number given in the next memory location (0805). This number has previously been specified by the human programmer as 00. The action to be performed appears in the "instruction" column as MVI A, 0 which means "move immediate" the number 0 into A. This action is further described in the "comment" column.

Pause for a moment to ponder the three-letter group "MVI." It is your first example of *mnemonic* (first letter silent) *characters* selected as memory aids to suggest the nature of what the instruction is to accomplish.

At the third memory location, 0806, another opcode is stored, 32, whose mnemonic is STA, which means to store the accumulator, or rather, the contents of the accumulator. But where? The destination of the STA instruction is to be found as an address stored in the next memory locations, 0807, and 0808. Oddly enough, the destination of the data is given as an address in an inverted order. First 00 is found and then 30, whereas the address is properly written as 3000.

As indicated in the "comments" column, you have encountered an unusual memory reference. This is so because the address 3000 does not really specify one of many possible cells in the memory, but rather the output port. In other words, the instruction consists of 3E followed by 00 and 30 in turn, meaning to place the data, presently in the accumulator, into the output port.

Moving on to the next address in which the program is stored, namely 0809, several puzzling things are found. First the data at 0809 is 00, which is an opcode having the mnemonic NOP, meaning *no operation* or do nothing. While this might seem silly, the NOP can be used by a programmer to insert a space which can later, if necessary, be occupied by another, more vital, opcode instruction. The NOP can also be used to create a time delay of a few microseconds, the time required to execute the instruction. In your program nothing is accomplished by the NOP, so you must move on for more interesting future action.

At 080A opcode 3C is found, INR meaning (*increment* the accumulator contents, i.e., add 1). While this might seem strange, you should recall that you are examining a computer program whose purpose is to count (remember that this program might not be too useful. You are really using it to demonstrate a number of principles).

The last instruction in the program starts at 080B with opcode C3, meaning JMP for jump. The jump is to memory location 0806, specified as 06 in 080C and 08 in 080D, in inverted order as usual. The whole jump instruction requires three words (or bytes, the common designation of an 8-bit group) and causes an unconditional jump to 0806. This simply means that the next memory location accessed is to be 0806 in which is found 32 as before.

Actually, you have just traversed the loop for the first time in accomplishing nothing more than adding 1 to whatever number was in the accumulator before encountering 32 (STA).

But wait! What number was in the accumulator? It would appear that nothing was there since you started with 00 in A at the beginning and then transferred it to the output port. Actually, this is what happened, but a more accurate way of describing STA is the duplication of the amount stored in A, without destroying it or changing it. Therefore, in the first pass through the loop, 00 is incremented to 01. In the second pass, 01 will be incremented to 02, etc.

As a matter of fact, the accumulator will be incremented over and over again until A will hold FF (in hexadecimal) or 1111 1111 (in binary) after which one more incrementation makes it 00.

You must conclude that the COUNTER PROGRAM is not very useful. Nevertheless, it illustrates a number of principles, and hints at untapped power to be exploited by programming. Incidentally, you might wonder about the presence of the label *LOOP*. It exists only on paper for the benefit of whoever reads the program, to follow its operation.

Inside the microprocessor

Figure 7-29 shows a simplified block diagram of a typical microprocessor. The heart of the microprocessor is the *a*rithmetic and *l*ogic *u*nit (ALU) that performs all programmed logic operations and number processing, such as incrementing a number as specified in the instruction INR. You also see the accumulator, which is connected directly to the ALU and several other registers, most of which store eight bits, i.e., one data word. The flag register is actually five 1-bit registers that are set independently to indicate the condition resulting from an operation. Generally, the flag bits are used to cause conditional branching within a program, which lends great flexibility to its operation.

The program counter and the instruction register are used in fetching instructions from the memory, shown in Fig. 7-29 as the ROM. Refer to Table 7-17, which lists the COUNTER PROGRAM, the address 0804 is stored in the program counter which causes that location in the memory to be selected so that the opcode 3E appears on the data bus and is transferred to the instruction register.

The instruction register supplies the opcode to the instruction decoder as shown in Fig. 7-30. The instruction decoder, in turn, causes the control and timing logic to generate a sequence of control signals in synchronism with a sequence of clock pulses to cause the instruction execution actions. One of these actions is the incrementation of the program counter to contain the address 0805, from which is fetched 00, the data to be moved, via the data bus, to the accumulator.

You have just observed a series of activities within the microcomputer called a *fetch-execute cycle*. The execution of any computer program must include a number of such cycles. Programs having loops repeat the execution of instructions within the loops many times, usually until a certain condition is met and branching to the starting point in the loop is halted. In the interest of simplification, the COUNTER PROGRAM did not include such provisions, so that the counting process proceeds indefinitely.

Microcomputer hardware

The hardware in a microcomputer depends, to a large extent, on the application. For example, a general purpose, programmable computer application, such as is featured in this discussion, requires RAM memory for the storage of programs that are created and entered by the operator. A dedicated computer for the control of a test instrument, communication equipment, etc., will be delivered with a program in ROM, and a large memory capacity is not required.

The specific nature of hardware interfaces, the input and output ports, will also depend on the application. Hardware must be compatible with the sensor and transducer signal levels and impedances. Other than these requirements, the gates, flip-flops, and other electronic devices already presented in this chapter are appropriate. This is also largely true for the implementation of the control and timing logic within the microcomputer.

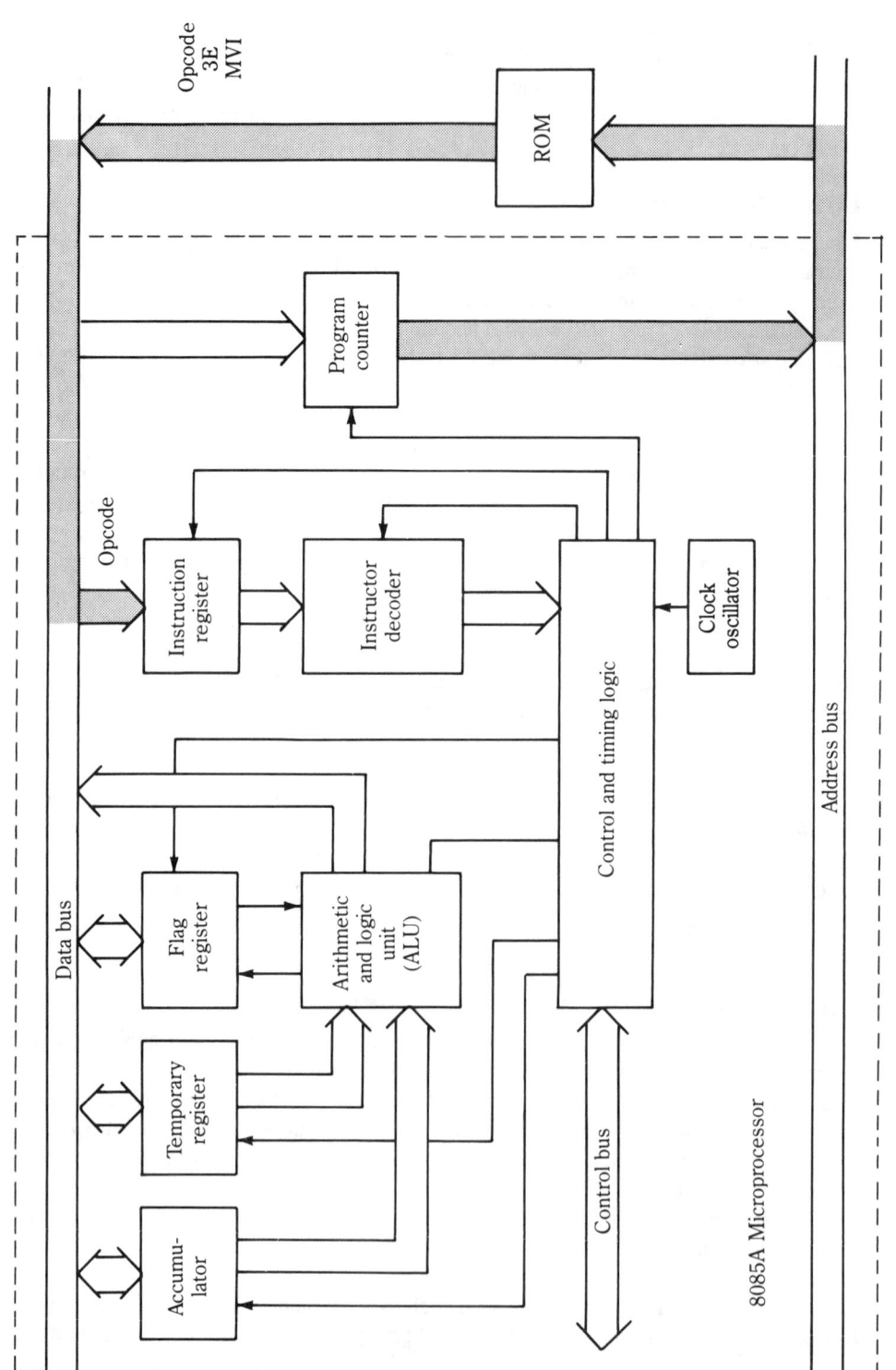

7-29 Reading the opcode from memory for a MVI A instruction.

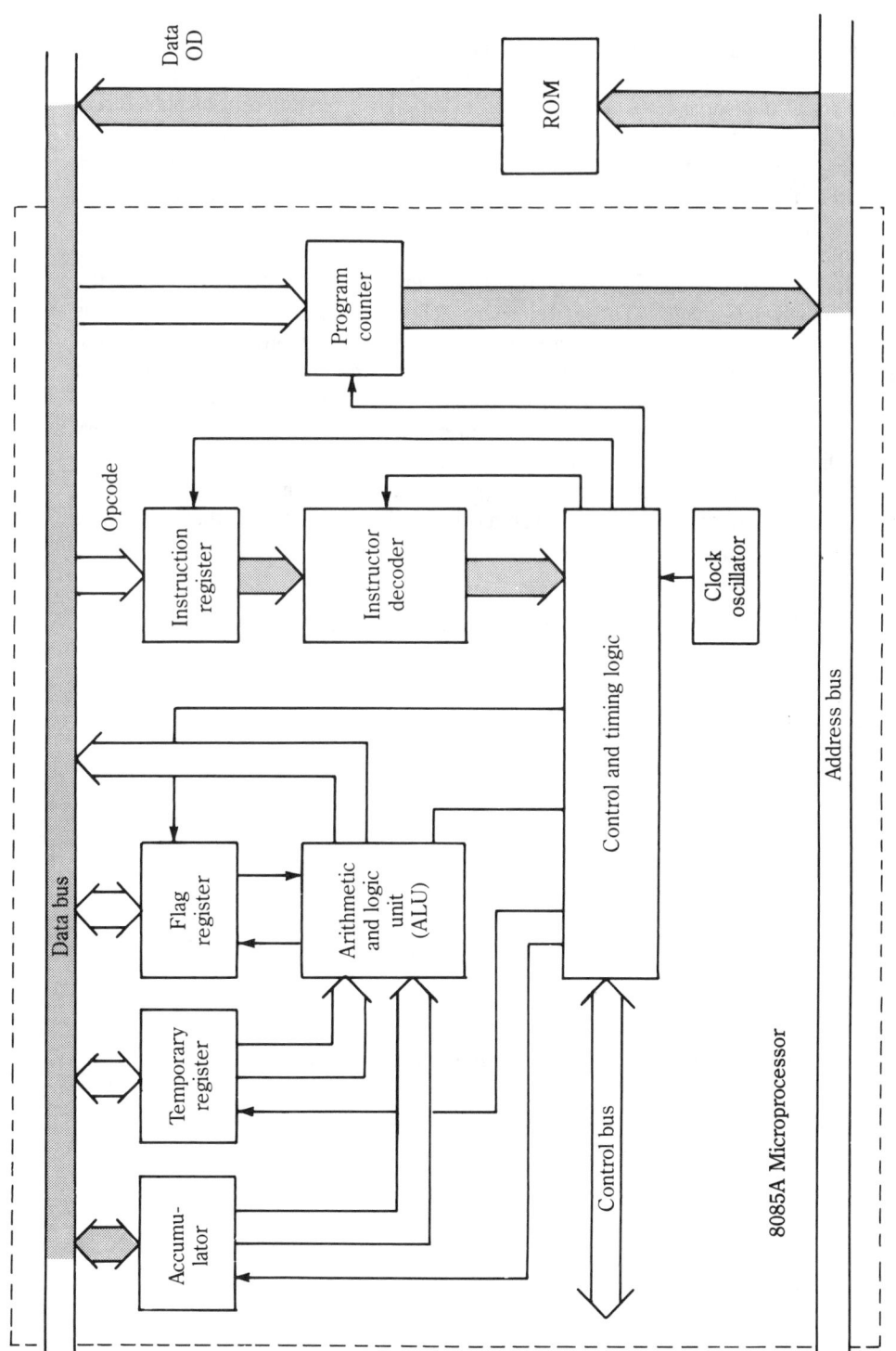

7-30 Reading the data for the MVI instruction.

It is necessary to recognize that many different microcomputers can be designed around one particular microprocessor chip. The microprocessor is a large scale integration (LSI) IC that contains the equivalent of thousands of logic gates. A microcomputer designer has a substantial number of microprocessor chips from different manufacturers to choose from. However, once chosen, no changes in the microprocessor are possible. In essence, a large majority of the internal microcomputer functions are frozen. Hardware, gates, flip-flops, ICs, and firmware, the program stored in ROM, are designed around the microprocessor. Briefly, the microprocessor is the only item of an unusual nature up to this point.

Microcomputer bus structure requires a different electronic item other than the ones previously introduced. These items are called *three-state drivers,* and one is required for each bit line on the bus for each microcomputer element that is intended to be a sender at any time. If data is being fetched or read from memory, each bit line must be independently driven high or low corresponding to a 1 or 0. These represent two of the three possible states. However, if data is not being read from memory, the memory driver is directed by a control signal to remain passive so as not to interfere with the driver in another unit that is properly in charge of placing the bit line high or low. Under these circumstances, the memory driver assumes the third state, which might be described as a high-impedance passive load.

Multiple-choice questions

1. The highest count possible with a single decimal digit is

  A. 0.
  B. 2.
  C. 7.
  D. 9.
  E. 10.

2. The number of different symbols required in the binary number system is:

  A. 0.
  B. 1.
  C. 2.
  D. 3.
  E. The answer depends on the number to be represented.

3. The binary equivalent of the decimal number 15 is:

  A. 1101.
  B. 1111.
  C. 1001.
  D. 0101.
  E. 1234.

4. The number of different symbols required for the hexadecimal number system is:

    A. 7.
    B. 9.
    C. 10.
    D. 15.
    E. 16.

5. Express the hexadecimal number B as the decimal equivalent:

    A. 1011.
    B. 15.
    C. 11.
    D. 0.
    E. 9.

6. The ASCII code designation 41 represents:

    A. A.
    B. B.
    C. C.
    D. D.
    E. No correct choice is given.

7. Express SYN using the ASCII code:

    A. 32.
    B. 61.
    C. 16.
    D. BA.
    E. F7.

8. The EBCDIC code designation 40 represents:

    A. PF.
    B. SP.
    C. 04.
    D. RES.
    E. 0100 0000.

9. Express 2 using its EBCDIC code:

    A. 61.
    B. 1111 0010.
    C. 23.
    D. 32.
    E. AB.

10. A Gray code wheel (disk) for the conversion of shaft angles to digital form has four separate, concentric, conducting surfaces and four fixed wipers. Express its resolution as the number of measureable positions in one revolution.

    A. 4.
    B. 8.
    C. 16.
    D. 32.
    E. 64.

11. Another name for the NOT gate is the:

    A. NAND gate.
    B. NOT AND gate.
    C. Inverter.
    D. Exclusive-OR gate.
    E. NOR gate.

12. There is one type of flip-flop for which the two inputs cannot simultaneously be high with predictable results. It has what designation?

    A. SR.
    B. JK.
    C. D.
    D. NOT.
    E. No correct answer is given.

13. The behavior of a single gate or interconnected gates is summarized by means of a:

    A. Boolean block diagram.
    B. Truth table.
    C. Boolean box.
    D. Black box.
    E. Clock circuit.

14. A particular kind of flip-flop can be caused to toggle by a certain combination of inputs. Name it.

    A. SR flip-flop.
    B. JK flip-flop.
    C. D flip-flop.
    D. XOR flip-flop.
    E. No correct answer is given.

15. Four JK flip-flops arranged as a binary counter will allow a maximum count of:

    A. 0.
    B. 4.
    C. 15.

D. 16.

E. None of the choices is true.

16. A binary counter consists of four JK flip-flops as suggested in Fig. 7-10. If all the flip-flops are cleared in advance, what will be the count after the arrival of 19 input pulses?

 A. 0.

 B. 1.

 C. 3.

 D. 9.

 E. 16.

17. A decade counter (see Fig. 7-11) is constructed from four JK flip-flops. Starting with a binary count equivalent to 7, what will be the binary count, expressed as a binary number, after the arrival of seven additional input pulses?

 A. 0000.

 B. 0010.

 C. 0100.

 D. 14.

 E. 1111.

18. A binary number, 1010, is to be shifted serially into a shift register consisting of four flip-flops holding 0000 initially. Note that the content of the shift register will be 1010 after four clock pulses. Identify the shift register content after exactly three clock pulses.

 A. 0000.

 B. 1000.

 C. 0100.

 D. 1010.

 E. 1111.

19. A half-adder can be constructed using a (an):

 A. Exclusive-OR gate.

 B. OR gate.

 C. AND gate.

 D. NAND gate.

 E. Both A and C.

20. Express 3 as a binary number using four bits, and then express the quantity -3, using four bit positions, in 2s-complement. The negative quantity in complement form is:

 A. 0011.

 B. 1110.

 C. 1001.

D. 1010.

E. 1101.

21. Identify one or more logic families:

   A. DTL.
   B. RTL.
   C. TTL.
   D. Both B and C.
   E. All of the choices are true.

22. Name a logic family that achieves speed by reducing storage time in the transistors:

   A. DTL.
   B. RTL.
   C. ECL.
   D. TTL.
   E. CMOS.

23. Name one or more logic families noted for low power consumption.

   A. TTL.
   B. ECL.
   C. I^2L.
   D. DTL.
   E. NMOS.

24. Identify a logic family, or families, for which there is particular danger to the device in handling:

   A. ECL.
   B. NMOS.
   C. DTL.
   D. CMOS.
   E. Both B and D.

25. One disadvantage of the dynamic RAM is a result of the need for:

   A. Refreshing.
   B. Complementing.
   C. High voltage.
   D. Blowing fuses.
   E. Ultraviolet light erasure.

26. One form of ROM can be erased for reuse through the application of ultraviolet light. It is identified as:

   A. E^2ROM.
   B. E^2RAM .

    C. ULROM.
    D. EPROM.
    E. ROMPROM.

27. In a microcomputer application much of the random logic of hardware is replaced with:

    A. PC boards.
    B. Ports.
    C. RAMs.
    D. Software.
    E. ROMs.

28. The hardware data entry and exit points of a microcomputer are known as:

    A. Sources.
    B. Sinks.
    C. Holes.
    D. Ports.
    E. Doorways.

29. A name for an instruction expressed like NOP or INR, for example, is known as a:

    A. Driver.
    B. Mnemonic.
    C. Program.
    D. Firmware.
    E. Three-state.

30. A NOP instruction is found at memory address 090A. The next instruction will be:

    A. 090B.
    B. NOP.
    C. INP.
    D. JMP.
    E. There is insufficient information given.

Basic problems

1. Name the simple logic gate for which the output is 1 when an odd number of inputs are 1, and is 0 otherwise.

2. Draw the circuit that will implement the following: $\overline{AB} + \overline{A}B$.

3. Draw the circuit that can be used to implement the SR flip-flop using NOR gates.

4. Briefly explain the fundamental difference between a half-adder and a full adder.

5. Demonstrate multiplication as a process of shift and add by multiplying 1010 by 0011.

6. Static electricity refers to the isolation of electrical charges. Why then must charge be restored in dynamic RAM but not in static RAM.

7. What ROM type, or types, would be used in an application in which a stored program might require subsequent modification?

8. Explain what parts of a microcomputer system will be most heavily influenced by the specific nature of the application.

9. Why does the address bus of your microcomputer have 16 bits while the data bus has only 8 bits?

10. Explain the significance of a jump instruction and the normal progression of the program in the absence of a jump instruction.

Advanced problems

1. Generate a table for a 5-bit Gray code disk using Table 7-5 as a guide.

2. Explain why the Gray code is superior to the binary code in the design of a device for converting shaft angle rotation into digital form.

3. Draw the circuit which will implement

$$\overline{A + B} \oplus AB$$

in which $\oplus$ is read as exclusive-OR.

4. Speculate as to why a binary ripple counter composed of flip-flops having significant switching time has counting speed limitations.

5. Demonstrate how a full adder can be constructed of half-adders.

6. Explain why CMOS logic has low power consumption as compared to either NMOS or PMOS.

7. Which ROM type would be selected for high-volume production in which the application requires a computer program to be permanently stored in memory?

8. Briefly describe the role of the microprocessor in the microcomputer.

9. What conclusions can be drawn about the number of memory locations required to store an instruction for the microcomputer illustrated in this chapter?

10. In the program listed in Table 7-17 a loop is included. Explain the number of times the instructions in the loop are executed and the conditions governing the number of times.

2
PART

Techniques, equipment and systems

8

Amplitude modulation

THE COMMUNICATION OF INFORMATION BY RADIO SYSTEMS USUALLY INVOLVES A spectrum of frequencies ranging from a few hertz to kilohertz (speech and music) or megahertz (data or TV). For a number of technical reasons it is necessary to superimpose the waveform representing the information on an ac radio frequency (rf) carrier. The more obvious advantages of this technique, called *carrier modulation*, are:

1. Multiple channels
2. Smaller antennas
3. Directional propagation

Figure 8-1 provides a block diagram of a simplified rf communication system. The main elements of this system are the information source, transmitter, channel, receiver, and information user. For the present purpose the key features are the modulation of the rf carrier by information in the transmitter, and the demodulation of the modulated carrier in the receiver. The objective of the entire system is to obtain, as the output of the demodulator, a faithful replica of the information waveform, free of any distortion and noise. This objective can never be fully realized.

Figure 8-2 concentrates on the transmitter to emphasize several points. First, there is an rf signal source whose frequency is accurately controlled and electronically protected by the buffer amplifier from *pulling*, an abnormal tendency of one circuit to cause another to slip into tune with it. The modulator causes the sinewave rf signal to be modified in some way. In this chapter the amplitude is made to follow the shape of the modulating signal waveform. In chapter 9 you will be introduced to frequency modulation (FM), in which the amplitude is maintained constant but the frequency is altered in accordance with the modulating signal. The transmitting system is completed with power amplification and coupling to the antenna.

Refer again to Fig. 8-2, the modulating signal originates in the audio frequency (af) source. You might immediately recognize that broadcast AM and other voice communi-

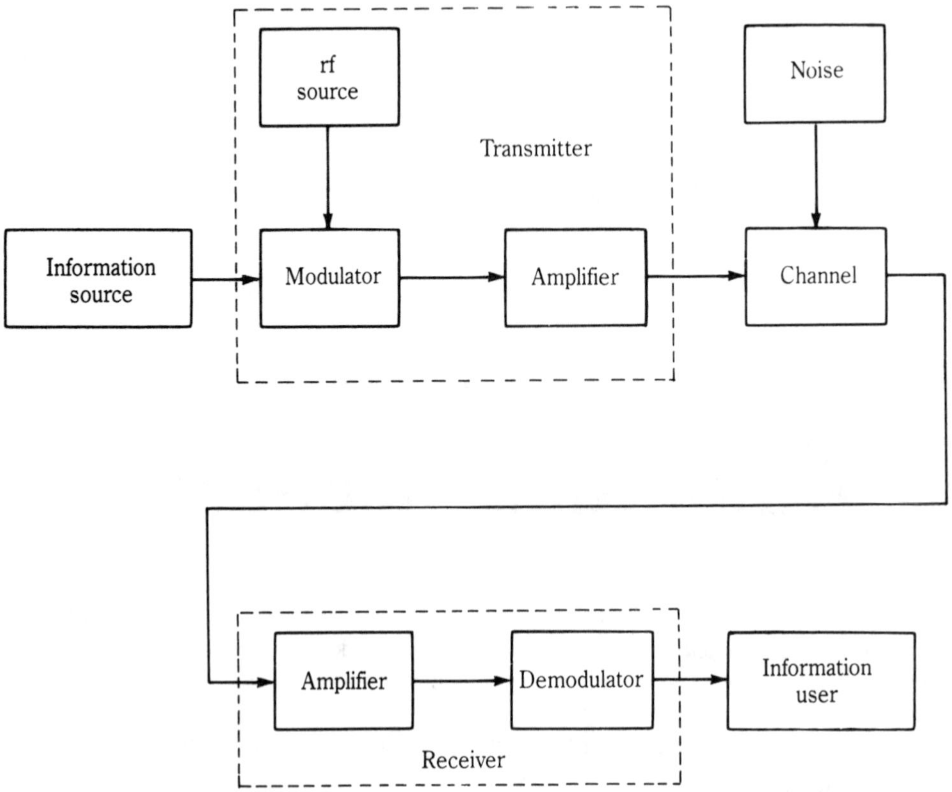

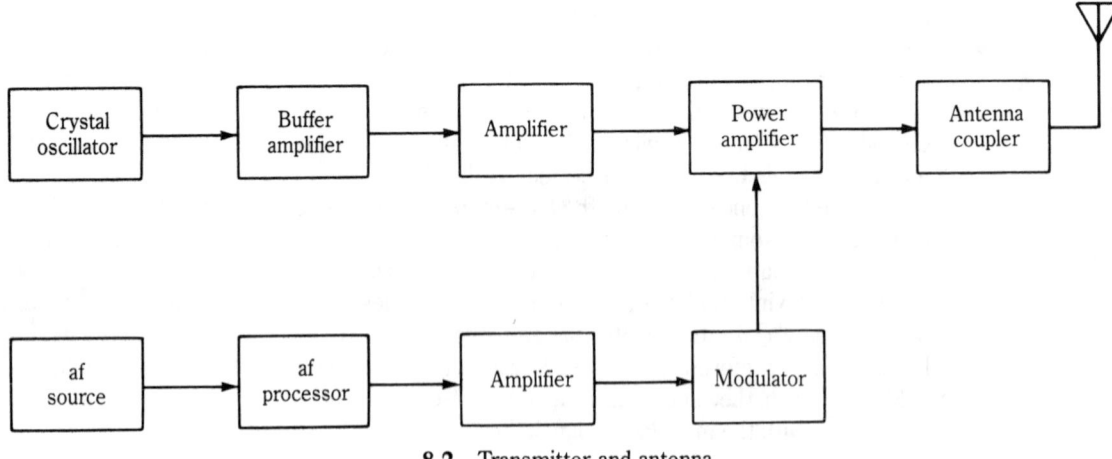

8-1 Simplified radio frequency communication system.

8-2 Transmitter and antenna.

cation radio systems fit the block diagram of Fig. 8-2. However, the basic ideas can easily be extended to other forms and frequency ranges for the information source.

Standard AM broadcast

Before launching into the technical specifics of AM waveforms (time domain), the spectra (frequency domain) and the equipment techniques for modulation and demodulation, an overview of "standard broadcast AM" will first be given. This emphasis is warranted by the sheer number of broadcast stations, the number of receivers, and the important position of standard broadcast AM as an element in public radio services within the United States.

The standard broadcast frequency band extends from 535 kHz to 1605 kHz with 107 assigned carrier frequencies of 540 kHz, 550 kHz...1600 kHz, with adjacent carrier frequencies being separated by 10 kHz. All stations are authorized by the Federal Communications Commission (FCC) to be operated on the assigned frequencies during particular hours of the day. Station licenses require strict adherence to technical standards of frequency tolerance, transmitter power, and service areas, in addition to a number of operation procedural rules, which collectively, are intended to reduce the interference of a received broadcast from one station caused by the transmission from another. The applicable regulations are quite complicated and, until recently, have remained relatively stable for a number of years.

This chapter is organized into the following sections:

- Basic amplitude modulation
- Sound conversion
- Sound processing and control
- The modulated carrier waveform
- Frequency components
- Pulse modulation
- Amplitude modulation techniques
- Single-sideband
- Envelope detection
- Product detection

Basic amplitude modulation

Once you have accepted the need for modulation as a practical means of superimposing information on an rf signal, the carrier, briefly examine the amplitude modulation technique.

Perhaps the most easily recognized property of a sine wave is its amplitude. The amplitude can be observed directly, and measured, with the cathode-ray oscilloscope (CRO). Other instruments, e.g., digital multimeters (DMMs), can be used for this purpose, subject to their frequency and other limitations. It should be understood that the instrument can be calibrated to directly measure the effective (rms) value, for example, so that a numerical conversion is required to obtain the *amplitude*, which is the peak value of the sine wave.

Now to focus on the communications aspects of modulation, it is possible to send as many different and distinct messages as there are achievable and measurable sine wave amplitudes for the system. In the interest of simplicity, the amplitude level is the message, except that the level represents a physical quantity such as sound intensity. To be more specific, if the sound, as in a recording studio, is converted to an electrical form and caused to amplitude-modulate an rf carrier, it is conceptually (and practically) possible to reverse the process at the receiver and recreate, with fair fidelity, the original sound. It was this ability to bring a musical, or other sound performance, to the millions of home radio receivers which laid the foundations for the present broadcast industry.

One comment is in order to elaborate on the technical significance of the preceding paragraph. On stage, a musician's product is an extremely complex sound. The sound level is not a constant, but more closely resembles a sine wave of one frequency (tone) with a large number of harmonics (overtones). Moreover, the musician mixes different

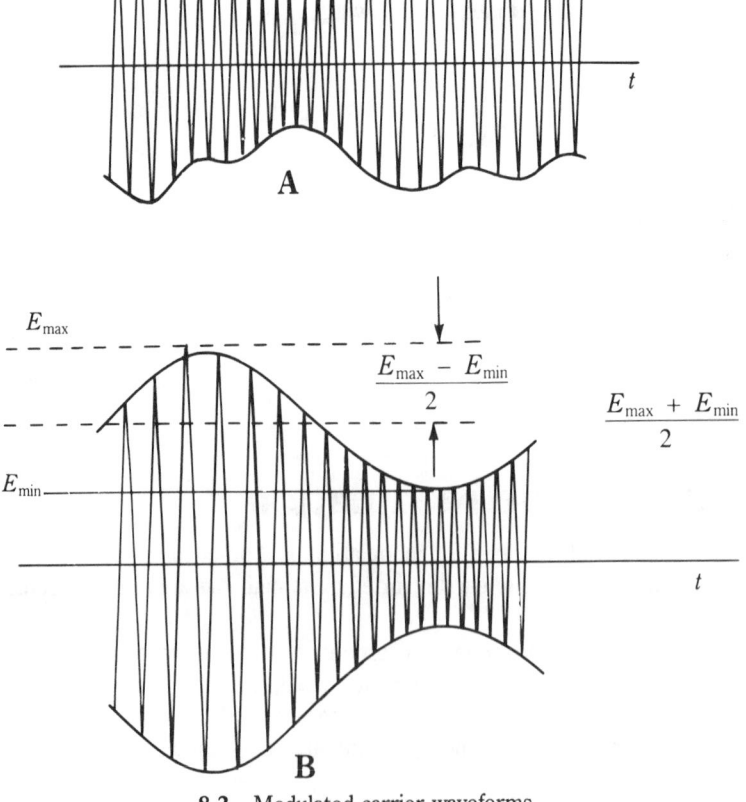

8-3   Modulated carrier waveforms.

tones, and changes them from moment to moment in creating the richness and beauty associated with fine music. The point to be made is that the instantaneous value of the modulating wave must change continuously to convey the musical sound. The amplitude of the rf carrier sine wave must be changed in the same manner, as illustrated in Fig. 8-3A. However, as will be emphasized later, the analysis of amplitude modulation is greatly simplified by considering modulation by a single sine wave alone (Fig. 8-3B). This allows certain conclusions to be drawn about the modulation process and its effects, conclusions that can be expanded in principle to cover modulation by the more complicated, but practical, waveforms.

Sound conversion

In order that sound can cause the modulation of an rf carrier, it must first be converted to an electrical form. Of course, the reverse process is necessary at the receiver in order that the original sound be reasonably well reproduced for the listener (see Fig. 8-4). Here you will be concerned with a brief description of the microphone (sound energy to electrical energy), of which there are several varieties in common use, and an even more brief description of the speaker (loudspeaker) device (electrical energy to sound energy).

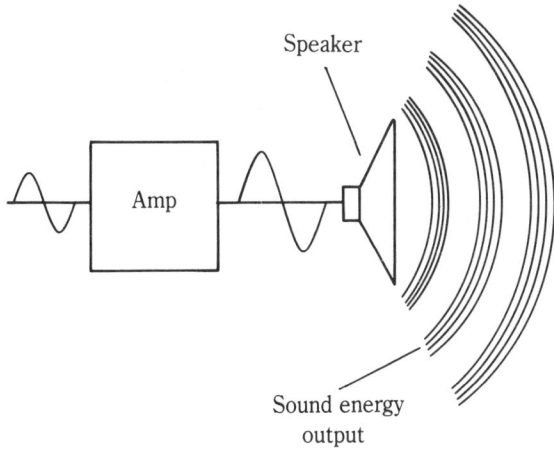

8-4 Conversion of electrical energy to sound energy.

A simplified sketch of a *dynamic microphone* appears in Fig. 8-5. The voice coil is fabricated of fine wire on a nonmagnetic form connected mechanically to the diaphragm. Differential air pressure on the opposite sides of the diaphragm cause it and the voice coil to move. The turns of the voice coil are in motion relative to a magnetic field in the air gap, so that an electromotive force (emf) is induced in the coil.

One domineering purpose of the radio communications system is to cause a diaphragm in the distant receiver speaker to duplicate the motion of the microphone diaphragm. The *ribbon microphone* of Fig. 8-6 is a simple variation of the dynamic microphone already described. It differs from the voice-coil microphone in the replace-

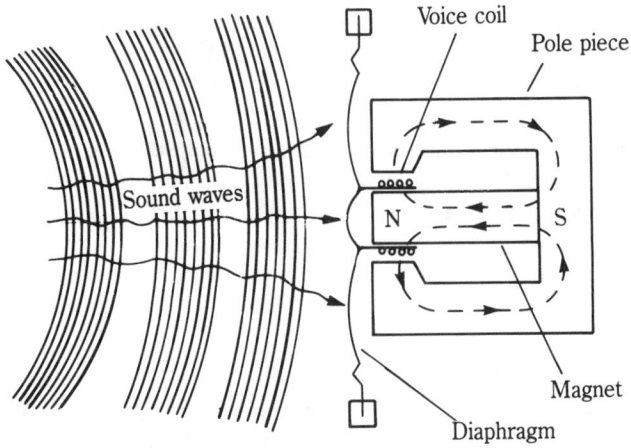

8-5 Dynamic microphone.

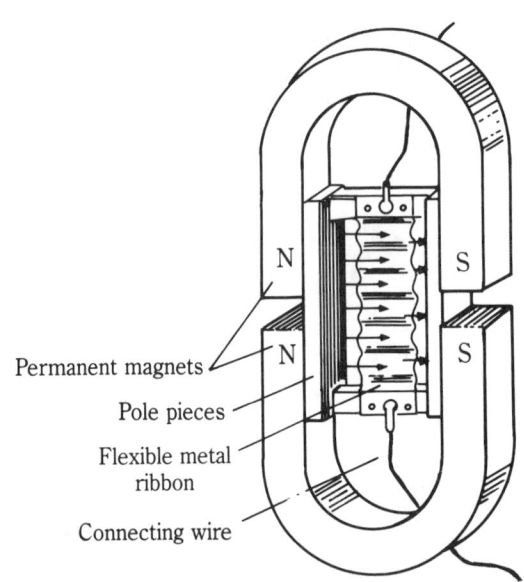

8-6 Ribbon microphone.

ment of the coil and diaphragm by a single, flexible metal ribbon that serves the purpose of both. Note that the ribbon, which is corrugated for structural strength, vibrates in its own magnetic field.

Microphones can also be constructed using the piezoelectric effect present in certain crystals. It will be recalled that the same phenomena is the basis for the crystal oscillator and the crystal filters. Mechanical deformation of the crystal causes voltages to appear across its faces, which are connected to electrodes.

The *condenser microphone* is also commonly used in commercial broadcasting. It works this way. Two parallel, conducting surfaces form a simple air capacitor. One of these surfaces is the diaphragm membrane which is made of a conducting material. A constant dc voltage is applied to the two plates of the capacitor. When the diaphragm vibrates in accordance with the ambient sound, the capacitance varies because of the changes in spacing between the plates. As a result, current in an external circuit connected to the plates must also flow to maintain the proper charge on the capacitor. This is the electrical signal desired from the conversion process. A condenser microphone requires a local dc supply voltage at the microphone. This can come from a battery at the microphone or be conveyed from remote electronic equipment by means of the microphone cable.

Dynamic microphones are inherently low-impedance devices so that the microphone cables are relatively insensitive to electrostatic pickup. However, they must be protected from electromagnetic fields such as those generated by ac power lines. Twisted pair (balanced) microphone cables greatly reduce this form of interference.

High input impedances, such as would match the impedance of condenser microphones, are susceptible to electrostatic pickup, so unbalanced lines—coaxial cable with a grounded outer conductor—are used.

Typical output levels for dynamic and ribbon microphones are of the order of -55 dBm and -40 dBm for condenser microphones. Because of these low levels, preamplifiers having low noise figures are necessary. Noise figure is an important design feature of radio receivers which will be explained in chapter 11.

Sound processing and control

The present treatment of this subject must necessarily be brief and oriented to its impact on modulation and communications effectiveness. You will first deal with processing equipment.

One item of processing equipment is the *equalizer*, which is a professional version of the consumer equalizer found in many stereo sound systems today. The studio equalizer consists basically of a parallel bank of band-pass filters with center frequencies separated by some fraction (typically one-third) of an octave. An octave refers to a frequency span for which the ratio of the highest to lowest frequency is two. The gain, or attenuation, of each filter is separately adjustable, with the result that the overall frequency response can be more-or-less continuously set across the entire audio frequency range. This allows compensation for the frequency response deficiencies of microphones, musical instruments, etc. It can also be used by the sound engineer for special effects, and as part of the process of blending the sounds from different sources such as microphones and tape playback equipment.

As will be seen later in the examination of the modulation process, the upper limit of modulation must be carefully observed. This presents a number of problems because a musical program, for example, exhibits a dynamic range of the order of 120 dB. If the gain of the amplifier driving the modulator is too low, the demodulated output of the receiver will be obscured by noise or other forms of interference in fringe areas of reception. Setting the gain level too high will cause distortion, if not illegal levels of overmodulation. Sound peaks are typically 8 to 14 dB above the rms (effective) level.

Broadcast equipment will usually include some form of *audio limiter*. Peak sound levels result in negative feedback similar to the automatic gain control (AGC) of receivers, so that the amplifier gain is reduced. This has the effect of compressing the peaks which reduces the possibility of overmodulation. The act of gain reduction, though automatic, requires some circuit reaction time, which might be referred to as *AGC attack time*. Abrupt peaks, for which the attack time is too great for effective peak compression, might be limited by the process of *clipping*, leveling off a signal peak at a predetermined level.

Compression, and the companion expansion techniques, are also employed in magnetic tape recording, which is the mainstay of program material for broadcast radio. In order to fit the 120 dB into the roughly 60 dB range between the noise level and saturation for magnetic tape recording, requires compression during recording, and expansion during playback. By way of contrast, the dynamic range of broadcast AM radio is of the order of 20 to 30 dB.

Noise reduction systems are intended to minimize one or more of the noise classes: hum, crosstalk between tape tracks, buzzes, tape and amplifier hiss, and modulation noise, also referred to as *asperity noise* which is due to irregularities in the coating of magnetic tape. These systems are much too complicated to be described in detail here. Dolby A, for example, provides 10 dB of noise reduction below 5 kHz, the amount increasing to a total of 15 dB at a frequency of 15 kHz. It is implemented solely in the studio.

Dolby B reduces noise 3 dB at 600 Hz, with the amount of reduction increasing to 10 dB at 5 kHz and higher frequencies. The system boosts the signal at higher frequencies before recording. A Dolby *decoder* is necessary in the receiver to restore the program to normal with consequent reduction in high frequency noise. Dolby B is used for FM broadcast, but the same principle could be applied to AM broadcast for which an equal improvement in program quality cannot be expected.

The purpose of the recording engineer console is to allow control of sound tone, volume, and the blend of separate sources such as microphones, electronic musical instruments, and tape playback equipment. Consoles consist of amplifiers, including combining amplifiers, attenuators, switches, patch cords, jacks, and plugs for flexible interconnection, etc. Prior to a live recording or broadcast session, the microphones must be physically placed relative to the performers, and all sources electrically connected to the console. During the performance, switching and continuous attenuation/amplification control are exercised by the operator in attempting to achieve the desired effects.

The modulated carrier waveform

Modulated carrier waveforms, resembling those which would be viewed on the screen of a cathode ray oscilloscope (CRO) are shown in Fig. 8-3. The one in Fig. 8-3A corresponds to an arbitrary modulating waveform whose shape is the same as the upper part of the *envelope* in the figure. An envelope is formed by joining the peaks by smooth curves. Note that the bottom of the envelope is the mirror image of the upper part with respect to the time axis.

Figure 8-3B corresponds to modulation by a single tone (frequency), which is convenient for the analysis to be presented. Two levels, E_{max} and E_{min} are marked on the figure. Midway between these marks, which is $(E_{max} + E_{min})/2$, seen on the right, is the level of the unmodulated carrier. The greatest departure from the carrier level corresponds to E_{max} and is $(E_{max} - E_{min})/2$. The ratio of this departure to the carrier level is the modulation index which is:

$$m = (E_{max} - E_{min}) \times 100 / (E_{max} + E_{min})$$

expressed as a percentage. In the extreme case, $E_{min} = 0$ and this occurs for $m = 100\%$, which is referred to as 100% modulation.

Any attempt to have a greater modulation index than 100% will result in two undesirable conditions:

1. Distortion of the demodulated signal in the receiver.

2. Splatter of the transmission beyond the assigned frequency band.

Modulation causes the total amount of transmitted power to be increased. The formula for the total power is given by:

$$P_T = P_C \left(\frac{1 + m^2}{2} \right)$$

where P_C is the power of the unmodulated carrier, and m is the modulation index expressed as a decimal fraction. For example, if the modulation index is 75%, $m = 0.75$.

For modulation by a number of tones having modulation indexes of m_1, m_2, m_3,..., the m^2 of the power formula becomes:

$$m^2 = m_1{}^2 + m_2{}^2 + m_3{}^2 + ... + m_n{}^2$$

This relationship holds whether or not the modulating frequencies are harmonically related.

Frequency components

The modulation of a carrier sine wave of frequency f_c by a single sine wave tone of frequency f_m results in three sinusoidal frequency components. One component has the amplitude of the original unmodulated carrier, A, with the frequency f_c. In other words, the component at the carrier frequency is exactly the same as the original unmodulated carrier. Therefore, the power of this component is exactly the same as the carrier power, P_C, which was expressed in equation form.

In addition to the carrier component at f_c, there are two other equal-amplitude components at $f_c + f_m$ and $f_c - f_m$ each having the amplitude $mA/2$ (see Fig. 8-7). Recall that power varies as the square of voltage, each of these side frequency components has a power level of $m^2 P_C/4$. The addition of all three power components yields P_T. It will be recalled that the modulated waveforms of Fig. 8-3A and 8-3B can be observed in what is called the time domain by using the CRO. Similarly, the spectral lines for the three frequency components can be observed and measured using a spectrum analyzer. Figure 8-8 is an illustration of the spectrum analyzer display showing the three components.

If the modulating signal consists of several sine waves, components of different frequencies and amplitudes, the same number of upper side frequency components will exist independently of each other. Similarly, lower side frequency components will exist as mirror images of the upper ones. Assume, for example, a modulating signal consisting

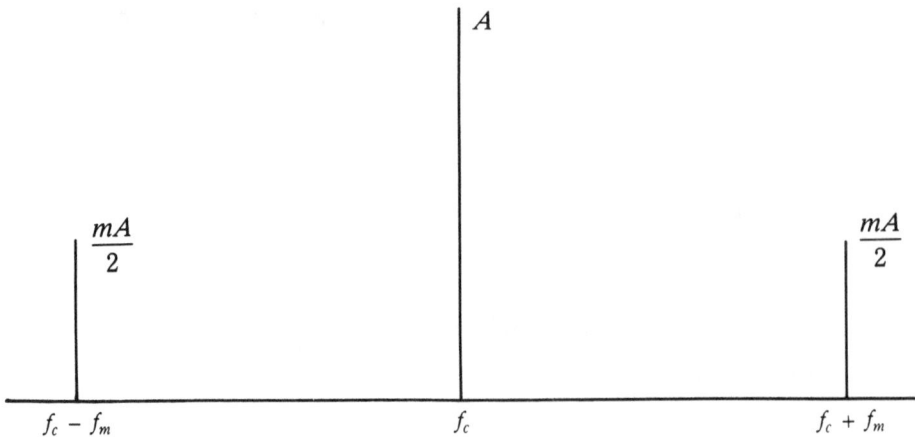

8-7 Frequency components for AM with a single sine wave.

of three components having frequencies f_1, f_2, and f_3, each capable of producing modulation indexes of m_1, m_2, and m_3 respectively when it alone is present in the modulating signal. The spectrum of the modulated waveform appears in Fig. 8-9. This figure illustrates a basic principle of amplitude modulation. The band assigned to the transmitter must be at least wide enough to accommodate all significant sideband components. A minimum bandwidth of $2f_3$ is necessary in this case with the understanding that f_3 is greater than either f_2 or f_1. One function of premodulation audio frequency processing is to ensure that the maximum frequency of the modulating signal components does not exceed one-half the effective bandwidth available. Note that the total bandwidth for a transmitting channel must also accommodate system defects such as frequency drift, and must, necessarily be greater than the effective bandwidth into which the sideband components are fitted.

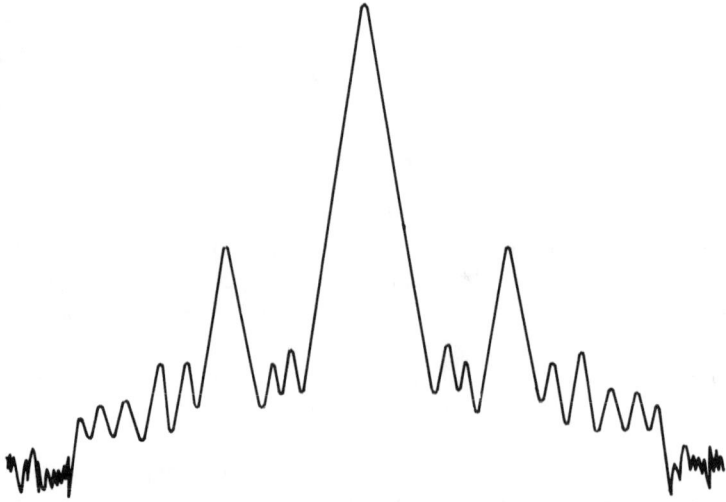

8-8 Spectrum analyzer display of an AM signal.

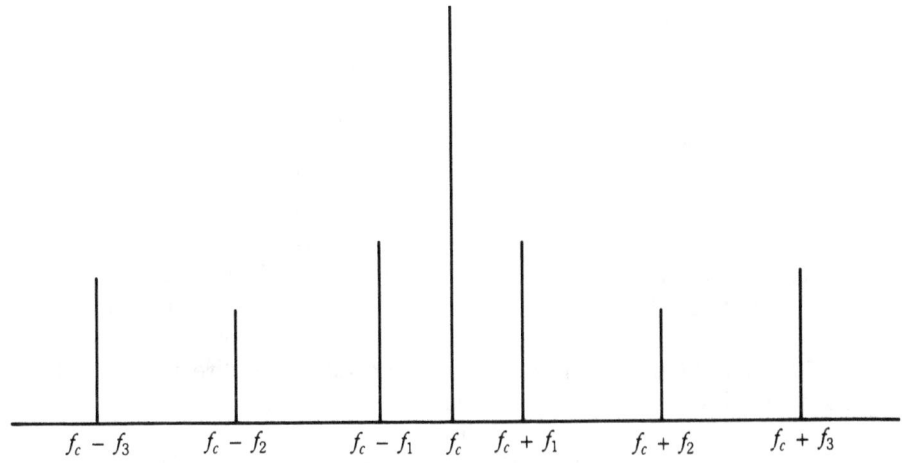

8-9   Frequency components for AM by three separate sine waves.

Pulse modulation

Morse (Baudot) code originated with telegraphy about 150 years ago. The telegraph system, in its most elementary form consisted of a series circuit with a battery voltage source. The normally-open circuit was completed at the sending end by manually closing a switch (key) causing current to flow through a receiving device some distance away. The key was closed for a short time to form a "dot" and a longer time for a "dash." The code consists of combinations of dots and dashes representing letters of the alphabet. Modulation, as we know it, was not involved.

In a similar way radiotelegraphy caused a transmitter, normally off, to be turned on for dots and dashes. This technique is still used. It is one form of *pulse modulation*. Another widely used pulse modulation method is *pulse code modulation* (PCM). A code group contains a fixed number of pulse positions. The existence of pulses in certain positions determines which alphabetical character, or other information element, is being transmitted. Codes in common use are described in chapter 7, Digital technology.

There are other pulse modulation techniques in which the pulse amplitude, pulse width, or pulse position in a pulse train are varied in proportion to the level of the modulating signal at any moment.

Some insight into pulse modulation can be obtained by assuming the existence of every other pulse. You might think of this as 100% amplitude modulation in which the modulating waveform is a square wave. "Fourier analysis" of a square wave of peak amplitude A reveals the following components:

Component amplitude	Component frequency
$A/2$	dc
$2A/\pi$	f
$2A/3\pi$	$3f$
$2A/15\pi$	$5f$
—	—
$2A/(n(n-2)\dots1)\pi$	nf

in which f is the frequency of the modulating square wave. It is seen that, in addition to the dc component, there is a fundamental component (frequency $= f$) and harmonic components at odd multiples of f. The amplitudes of the harmonics decrease with the order of the harmonic. For a nth order harmonic, i.e., the value of n, the amplitude of the harmonic is inversely proportional to $n(n - 2)(n - 4)...1$. For example, if $n = 7$, the value is $7 \times 5 \times 3 \times 1 = 105$, and if $n = 11$, the value is $11 \times 9 \times 7 \times 5 \times 3 \times 1 = 10{,}395$. Theoretically, a square wave is composed of a dc component, a sine wave of the fundamental frequency, and an infinite number of harmonics. Actually, the amplitudes of the harmonics decrease rapidly with large values of n so that their practical significance is negligible. Glancing at the example values presented in this paragraph, the 7th harmonic amplitude is about 1% of the fundamental amplitude. Moreover, the 11th is about 1% of the 7th or about 0.01% of the fundamental amplitude. A relatively small number of harmonic frequencies are sufficient to adequately represent the square wave.

The practical significance of this analysis follows directly. Modulation of the carrier by a square wave simply means that n sidebands are present on each side of the carrier in the spectrum of the modulated carrier waveform. Select the value of n to fit the accuracy required in the Fourier series representation, so that n might be 7, 9, or 11, but rarely more in any practical situation.

Basically the same reasoning applies when the modulating waveform is neither a square wave nor a sine wave but is one that can be represented by a fundamental and harmonic frequencies. The spectrum which must be accommodated will include all of the significant harmonic components.

Pulse modulation, as it is used for the transmission of digital data, does not require that the pulses be accurately reproduced in the receiver as is implied by the reception of higher-order harmonics. Indeed, the requirement is to distinguish between the presence or absence of a pulse. It can be shown that the bandwidth be such that the fundamental and not the harmonics be passed in the receiver. It is necessary to recognize that the transmission of digital information differs from the single square-wave modulating waveform in several respects, including the presence of noise which opens the two different noise-error possibilities, namely:

Interpreting the received signal as having a pulse where none is intended, and interpreting the received signal as not having a pulse where one was intended.

What should be clear is that the bandwidth required is greater as the pulse width is less. Thus, an increase in the information rate necessitates more pulses per second, decreases the period, and increases the fundamental frequency.

Closely related to these ideas is *time division multiplexing* (TDM), which concerns a number of low-rate information sources that are sampled in a regular and periodic order. Each sample in turn is converted into a code and transmitted, followed by the next. The total interleaved information rate is high, whereas that of each source is low. At the receiving end, the pulses in their individual groups must be separated and assigned to the appropriate information users.

Amplitude modulation techniques

Solid-state AM

A partial representation of a simple collector modulator circuit is shown in Fig. 8-10. It closely resembles a class-C amplifier, differing principally in the injection of the modulating signal into the collector circuit of QB by means of the transformer *T*. The driver stage with QA has a parallel tuned circuit (*L1C1*) and a series resonant circuit (*L2C3*) in the base circuit of QB for impedance-matching the low base resistance. Capacitors *C2* and *C4* provide the rf connections to ground. A π-matching load impedance network (*C6L5C7*) is inserted in the output circuit. Inductor *L4* provides shunt feed to the collector which includes $-V_{CC}$ upon which is superimposed the modulating signal.

Base modulation requires less modulating power as the amount of base drive is less than that necessary for the collector of Fig. 8-10. The circuits of Fig. 8-11 are an example of base-modulation circuits.

In order to achieve 100% modulation, it is usually necessary to combine both the base and the collector modulation. This is illustrated in Fig. 8-12. Note that a portion of the transformer *T* secondary voltage is placed in series with the $+V_{CC}$ supply of QA so that the base voltage of QB rises and falls in unison with the collector voltage. The result is a greater variation of collector current and percentage modulation than would be the case for Fig. 8-10 or Fig. 8-11.

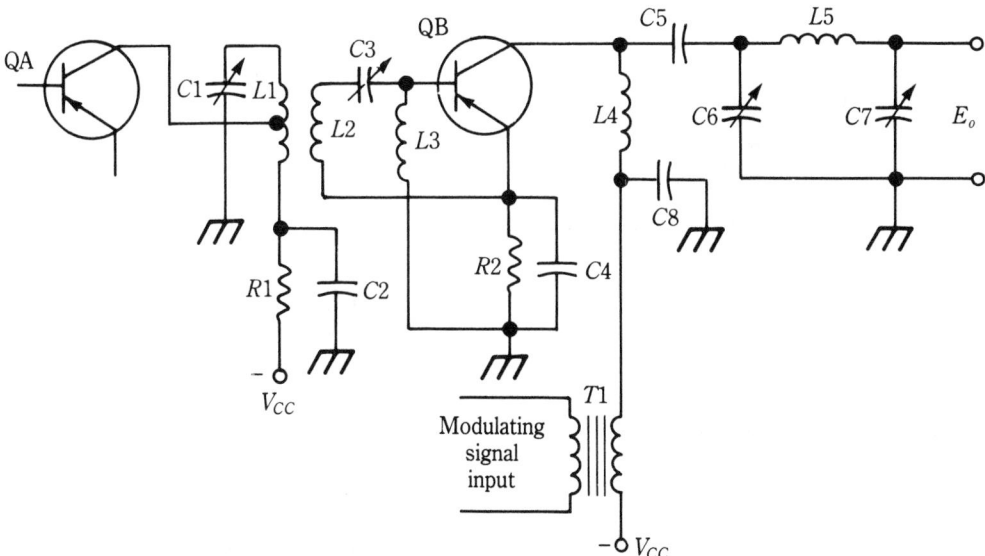

8-10 A collector modulated circuit.

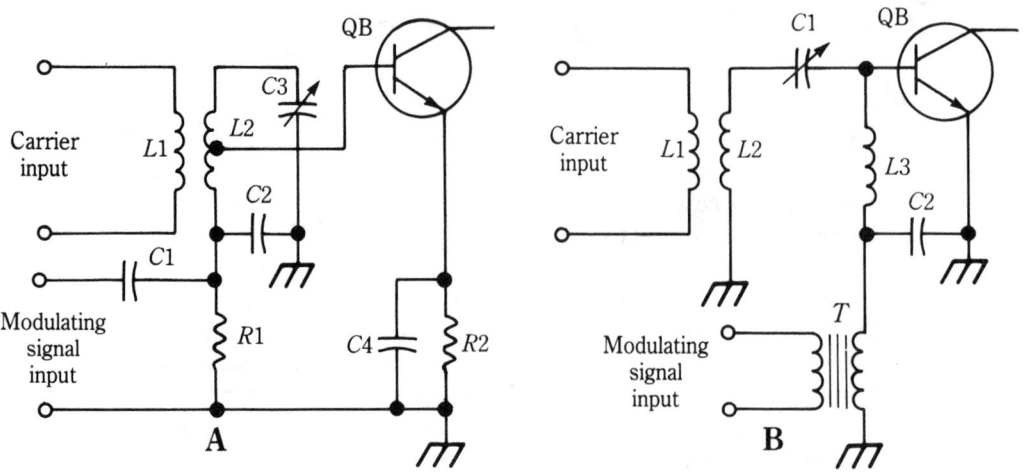

8-11   Base modulation circuits A and B.

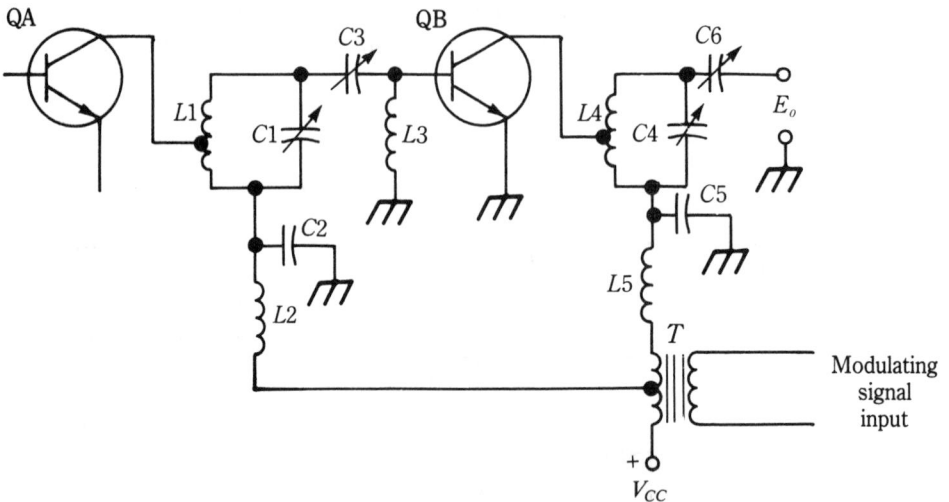

8-12   Combined base and collector modulation for high-modulation levels.

Vacuum-tube AM

An rf carrier can be amplitude modulated at various locations in the rf amplifier section of the transmitter. The method of modulation refers to the electrode or element of the rf amplifier to which the modulating voltage is applied.

In plate modulation, the most commonly used method, the modulating voltage is impressed on the dc supply voltage for the plate of the final rf amplifier in the transmitter. The output of the modulated stage is varied by adding the modulating signal to the plate supply voltage.

Application of the modulating voltage to the grid of an rf amplifier tube is called *grid modulation. Cathode modulation* is a method whereby the modulating voltage is applied to the cathode circuit of the modulated stage.

Pentode type power amplifiers can be modulated by applying the modulating voltage to the suppressor grid for suppressor modulation or to the screen grid for screen-grid modulation. Screen-grid modulation can also be used with tetrode power amplifier tubes. In many cases, screen-grid and plate modulation are used in combination.

Modulation is also identified as to level. High-level modulation occurs when the modulating voltage is applied to the plate circuit of the final power amplifier. With high-level modulation, the final stage is operated class C and the preceding stages are also operated class C. The overall efficiency of such an amplifier is very high. A disadvantage of high-level modulation is that comparatively high audio power is needed, and several stages of voltage and power amplification might be required in the audio amplifier and modulator circuits.

Low-level modulation takes place in the control grid or the cathode of the final power amplifier or in a previous stage. It is difficult to obtain any large degree of modulation using this method.

The modulation of a carrier signal by the modulating signal can be accomplished in any nonlinear device. However, since high power output is a requirement of the transmitter, modulation usually takes place in a transistor or vacuum-tube stage.

Figure 8-13 illustrates a plate-modulated rf power amplifier. The $V1$ driver circuit is an audio frequency voltage amplifier whose output is transformer coupled to the $V2$ modulator stage. The modulator is an audio frequency power amplifier whose output is transformer-coupled to the rf power amplifier plate circuit. The af driver and modulator circuits must be biased class A to minimize audio frequency signal distortion. The rf power amplifier is usually operated class C to obtain high efficiency.

The unmodulated rf input is applied to the $V3$ grid through the coupling capacitor, C_C, and developed across RFC1 and $R2$. $V3$ is biased with a combination of the grid-leak bias across R2 and the fixed bias, E_{CC}. C2 couples the $V3$ plate current pulses to the tank circuit while blocking the dc plate supply voltage. The modulated rf output of $V3$ is developed in the $C3$, $T3$ primary tank circuit and coupled through transformer $T3$ to the following rf amplifier or the antenna circuit. RFC2 presents a high impedance to the rf signal and prevents the rf signal from feeding back into the power supply.

The $C3$, $T3$ primary tank circuit is tuned to the same frequency as the rf input to $V3$. With no modulating input, the rf output will be an amplified version of the input. Even though $V3$ is operated class C, the pulses of current through the tank circuit will trigger the flywheel action and the missing portions of the waveform will be restored.

Assume a polarity reversal across the transformers in the circuit. When the af input on the $V2$ grid is on the negative alternation, the $V2$ plate current will decrease, and plate voltage will increase. This results in a positive alternation at the top of the $T2$ primary. The voltage induced in the $T2$ secondary will oppose the $V3$ power supply voltage and the $V3$ plate current pulse amplitude will decrease. The signal being developed in the tank circuit can follow the amplitude change of the modulating signal due to the low Q of the tank circuit. The low Q is primarily the result of heavy loading of the tank circuit by the antenna or other secondary load.

When the audio signal to the $V2$ grid swings positive, the tube will conduct harder,

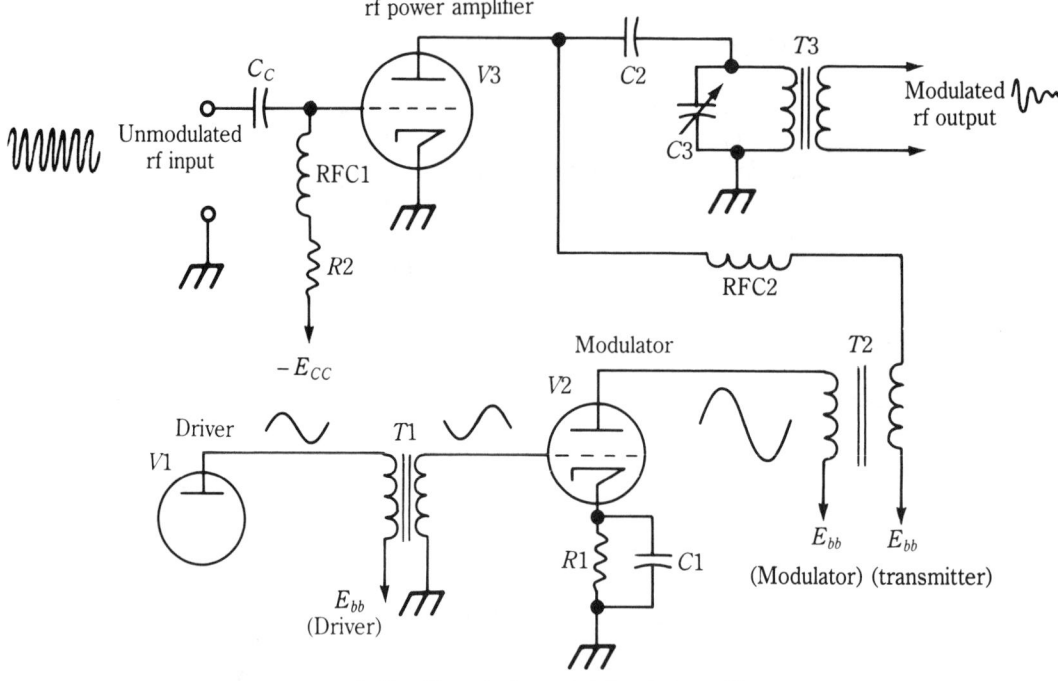

8-13   Class C plate-modulated rf amplifier.

and plate voltage will decrease. The voltage induced in the $T2$ secondary now aids the $V3$ plate supply voltage. The $V3$ plate current pulses increase in amplitude, and the signal developed in the tank circuit increases in amplitude accordingly.

High-power modulation applications use plate and screen-grid modulation with tetrode and pentode tubes. The plate and screen-grid voltages are both varied to achieve modulation. The modulation power required is equal to one-half of the rf carrier plate and screen power to produce 100% modulation. Efficiency is extremely high and the circuit will produce maximum power output for the tube used.

A simplified schematic of a basic plate and screen-grid modulated pentode is shown in Fig. 8-14. $V1$ is operated class C with a combination of grid-leak and fixed bias. If the rf input power is lost, the fixed bias provides enough protective bias to prevent tube damage.

The unmodulated rf input is applied to the $V1$ grid through a coupling capacitor and is developed across RFC1 and R_g. $C1$ decouples the fixed grid bias supply E_{CC1}. The suppressor grid is shown connected to a small fixed positive bias E_{CC3} and placed at rf ground by capacitor $C2$. In other types of pentodes, the suppressor grid might be connected directly to the cathode, either internally or externally. In tetrodes, of course, the suppressor grid is not present. The positive bias, as used in the schematic, improves the shielding effect of the suppressor grid at low plate voltages. When the plate is almost at zero potential, the suppressor grid will intercept electrons and provide a return path to

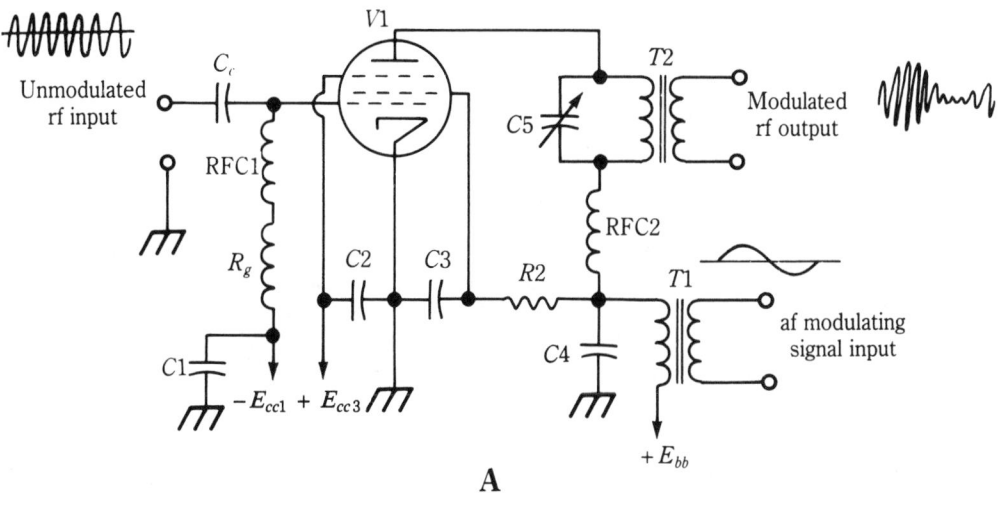

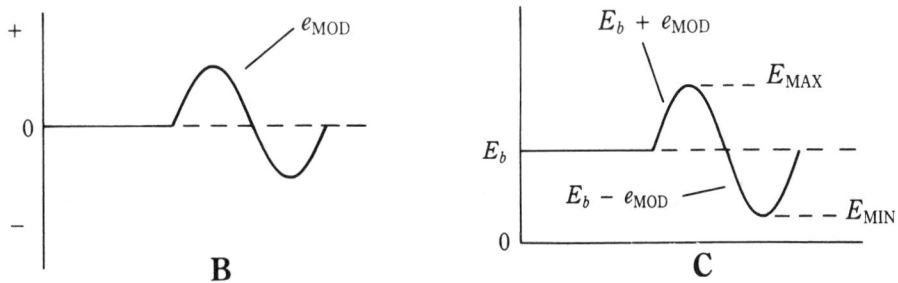

8-14 Plate and screen modulated pentode: A. Circuit. B. Modulating signal. C. Instantaneous effective plate voltage.

ground. Thus, zero plate current is possible with a sharper cutoff than would be possible if the suppressor were connected directly to the cathode.

The screen-grid voltage is obtained from the plate supply through dropping resistor $R2$ and is decoupled by capacitors $C3$, and $C4$. $C3$ places the screen grid at rf ground potential and $C4$ prevents rf feedback from entering the audio circuits through $T1$.

The secondary of $T1$ is connected between the E_{bb} supply and the common junction of the screen and plate supply. The audio modulating signal from the modulator is induced in the $T1$ secondary and produces modulation by varying the instantaneous effective plate and screen-grid voltages. RFC2 isolates $T1$ and the audio circuits from the tank circuit rf energy variations. The plate tank circuit is comprised of $C5$ and the $T2$ primary. $T2$ inductively couples the modulated rf to the output load.

A cycle of operation will now be analyzed in detail. With no modulating signal input, the plate and screen voltages applied are dc, and the output is an amplified, unmodulated carrier. When the signal applied to the control grid swings far enough positive to overcome the class-C bias, the tube will conduct until the signal again drops the bias below cutoff. This pulse of current excites oscillations in the plate tank circuit, which is tuned to the frequency of the input. The oscillations in the tank circuit restore the missing portions of the tube current pulse for a sinusoidal output.

When the applied modulating signal is going positive at the top of the secondary of $T1$, the induced voltage in the secondary will add to E_{bb} to increase the plate and screen-grid voltages. When the signal is reversed, the voltage at the secondary (top) of $T1$ will subtract from E_{bb} and decrease the plate and screen-grid voltages. Varying either the plate or screen-grid voltage alone will vary plate current, resulting in modulation. However, the variation is usually not great enough to produce 100% modulation in tetrode or pentode electron tubes. Changing both the plate and screen-grid voltages simultaneously will result in a change in plate current large enough to produce 100% modulation.

When plate and screen-grid modulation is to be obtained using a separate screen supply, the screen voltage must be varied in some manner. One method is to apply the modulating signal to the plate only, but as the tube current varies, a large choke in the screen circuit will develop the modulating signal on the screen grid as well. Another method would be to inductively couple the modulating signal to both the plate and screen-grid circuits through a transformer with separate secondary windings.

Single sideband

It has been shown previously that modulation of a carrier signal (f_c) by a single-frequency waveform (f_m) creates three separate and distinct output frequency components: $f_c - f_m, f_c$, and $f_c + f_m$. The carrier frequency component amplitude, A, is exactly that of the unmodulated carrier, while those of the lower and upper side frequency components are the same and are equal to $mA/2$. Recall that m is the modulation index expressed as a decimal fraction. Also, the carrier power, P_C, is unchanged by the process of modulation but each side frequency component has a power of $P_C m^2/4$. In the normal situation for which modulation is achieved by complex waveforms having a continuous information signal spectrum, e.g., musical program, continuous sidebands exist, but the power relations are unchanged. In this event it is meaningful to speak of the rms or effective modulating signal, but the peak value is usually not known and is of interest only because of overmodulation possibilities.

Each sideband contains all of the information of the modulating signal so that it can be reconstructed in the receiver by the proper demodulation process. This clearly means that one sideband alone is sufficient, and the other is redundant. The carrier frequency component has no direct information value. The single-sideband suppressed-carrier (SSBSC) radio communication system is based on the suppression of one sideband together with the carrier and the transmission of the other sideband alone. In addition to the spectrum savings, which allows the number of available rf channels to be doubled in principle, there is a significant savings in the amount of power that is transmitted. In summary:

$$\text{Power transmitted (SSBSC)} = m^2 P_C / 4$$

$$\text{Power savings (SSBSC)} = P_C \left[(1 + m^2)/4 \right]$$

in which the comparison is made with the power of the AM carrier component P_C.

You should be concerned with the techniques for single-sideband modulation and demodulation. It will quickly be recognized that these processes are more complicated than those for standard broadcast AM modulation and demodulation. Starting with single sideband generation, the balanced modulator will first be introduced.

Balanced modulator

The balanced modulator is a key circuit employed in single-sideband modulation. Its salient feature is the elimination of the carrier as part of the amplitude modulation process. One balanced modulator circuit is shown in Fig. 8-15. It resembles a push-pull amplifier for the modulating signal. The carrier is injected in such a way that it produces no output in the absence of the modulating signal. When both the modulating signal and the carrier are present as inputs, the result is more difficult to analyze.

Consider first the composite signals at the gates of Q1 and Q2. Each is the sum of the two input signals. Because of the nonlinear FET characteristics the drain currents will include components of the original waveforms and their harmonics together with other components of all sum and difference frequencies. The modulating frequencies and their harmonics are of little interest because they are much lower than the carrier rf. The fundamental carrier rf component together with the sum and difference modulating

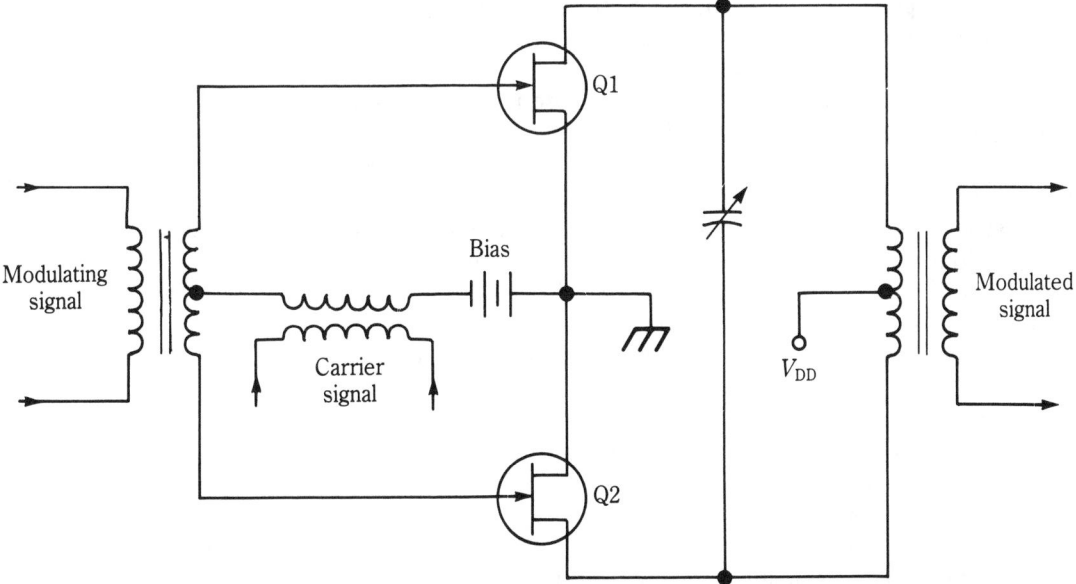

8-15 Push-pull balanced modulator.

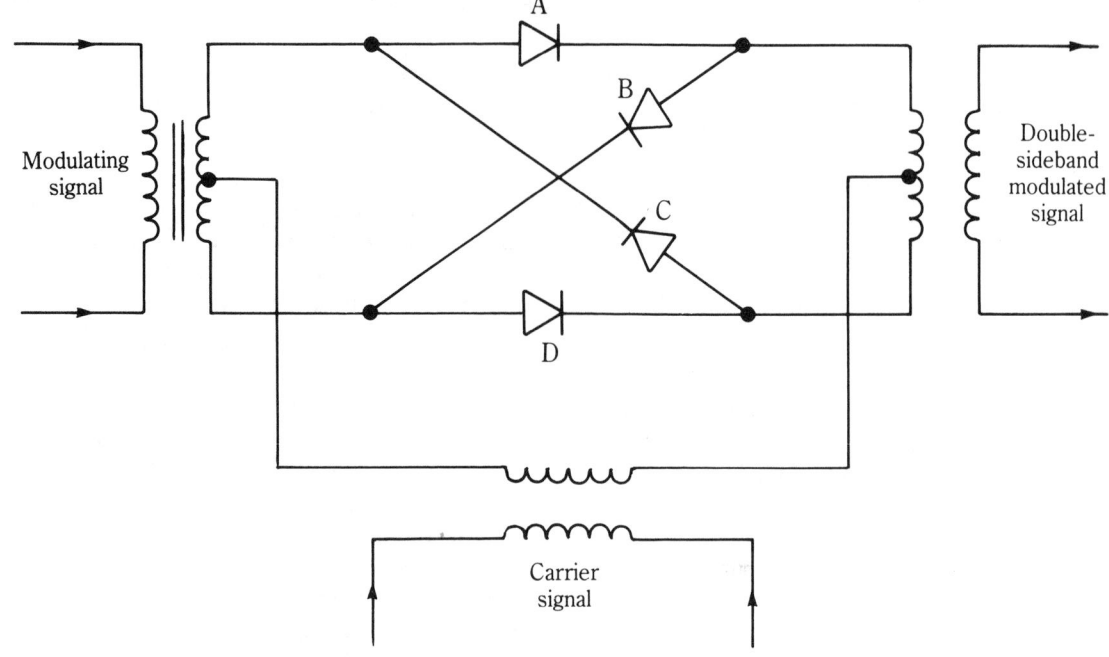

8-16 Balanced ring modulator.

frequencies exist, and together are similar to the standard double-sideband AM spectrum. However, the carrier rf component cancels because of the balanced push-pull arrangement. This leaves only the sidebands, which are wanted, and the frequency components at multiples of the carrier rf, which are caused to disappear during amplification by subsequent tuned stages. The practical result is the suppression of the carrier rf which is the purpose of the balanced modulator.

A double-balanced diode ring modulator, whose name appears not to be standardized, is shown in Fig. 8-16. The circuit is sufficiently complicated to be approached with caution and respect. Suppose that no modulating signal is applied and a rather large carrier rf voltage is placed across the primary of the lower transformer. It is clearly seen then that for one alternation diodes A and D conduct completing parallel conduction paths from the center tap of the left-hand transformer to the center tap of the right-hand transformer. Moreover, parallel paths exist through diodes B and C for the other alternation of the rf ac cycle. However, with a perfectly balanced circuit, there will exist no output in the secondary winding of the right-hand transformer. Note that the level of rf signal is sufficient to cause full conduction of the diodes. The currents that flow are essentially square in shape.

Now modify the conditions just described by also applying a modulating signal so that the previous balanced condition no longer prevails. The relatively low-frequency modulation waveform is now superimposed on the square-wave currents so that the net current in the center-tapped primary of the output transformer is not zero, and a voltage is induced in the output. In an argument similar to that used previously for the FET balanced modulator, only the sidebands, but not the rf carrier, will survive in the frequency range that will be passed by the following amplifier stages. It is of interest to notice that this modulator is balanced for both the carrier and modulating frequencies so that neither appear in the output. The "double-balanced" descriptor in the name derives from this feature.

Carrier suppression is highly dependent on the ability to produce nearly perfect balanced-modulators. This suggests that the natural advantages of integrated circuits should be exploited. The LM1596/LM1496 balanced modulator-demodulator is a good example of a double-balanced circuit, its schematic is reproduced in Fig. 8-17. Here the basic differential amplifier configuration prevalent in IC op amps is expanded to yield the product of two separated differential waveforms. The output, which contains essentially the same components as the other balanced-modulator circuits explained earlier in this section, must be appropriately filtered to select the sidebands and reject the many harmonics that are not desired in this application of the IC. The carrier suppression for this

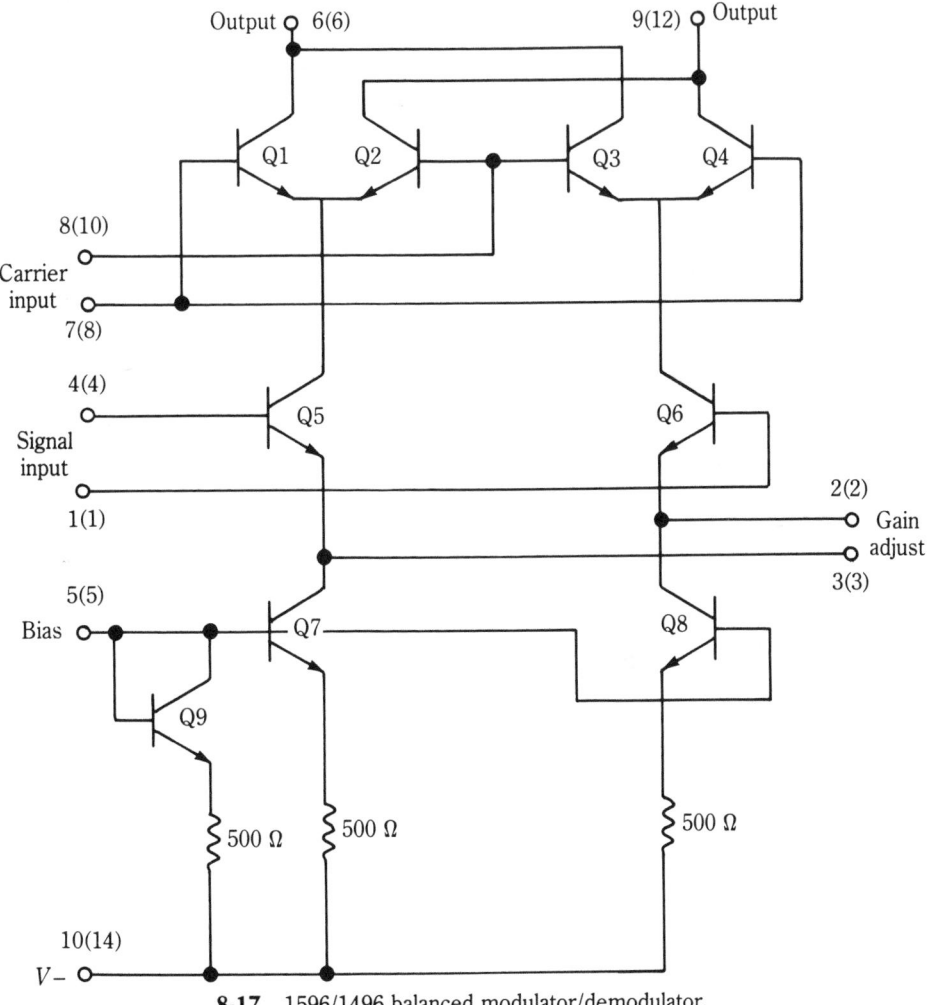

8-17 1596/1496 balanced modulator/demodulator.

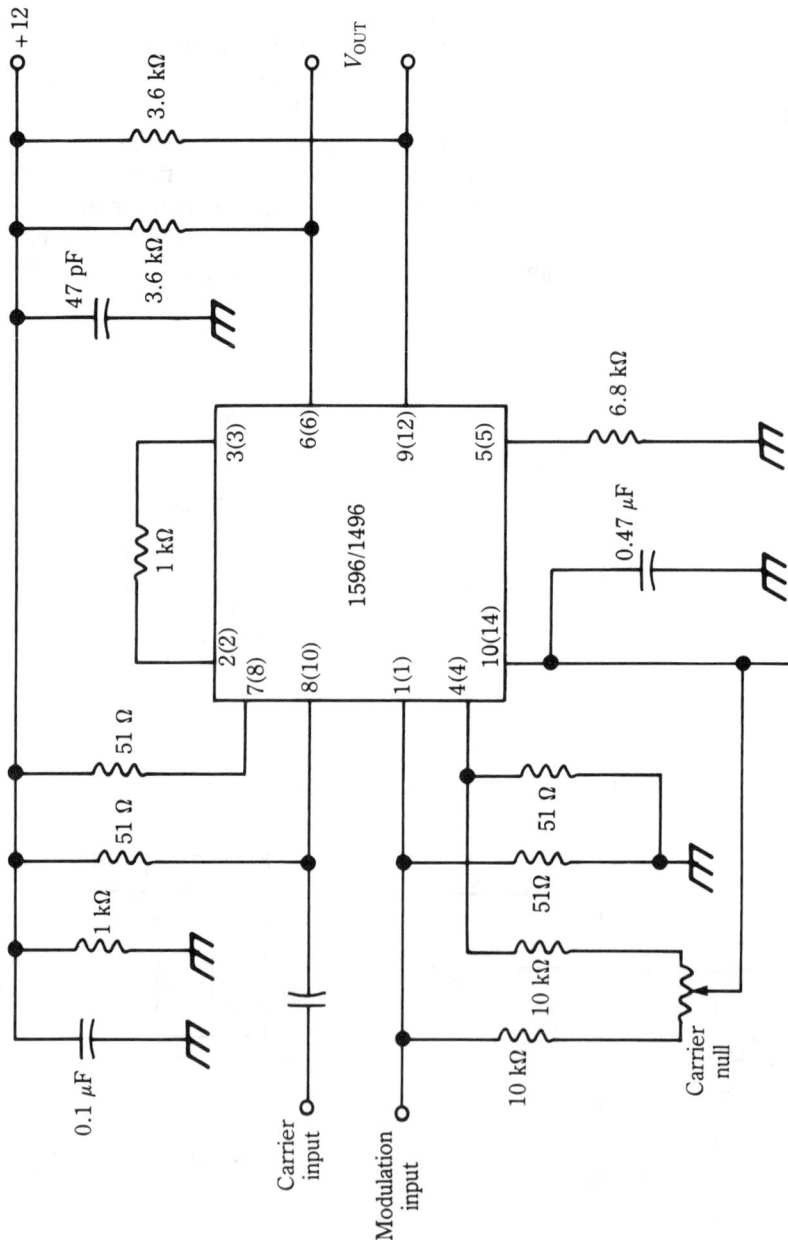

8-18 Typical application of 1596/1496 balanced modulator/demodulator.

device is typically 65 dB at 0.5 MHz, and 50 dB at 10 MHz. It should be noted that the simplified schematic of Fig. 8-17 is incomplete. A typical circuit for an application of the IC is shown in Fig. 8-18.

In line with the traditional methods of treating this subject, three techniques for eliminating the unwanted sidebands will be considered.

The filter method

The *filter method* is straightforward but technically challenging. It consists of inserting a bandpass filter in cascade with the balanced modulator for the sole purpose of eliminating the unwanted sideband. The technical problem stems from the narrow separation of about twice the lowest modulating frequency between the upper end of the lower band and the lower end of the upper band. This requires a filter having extremely steep skirts. Assume that the lower sideband is to be selected. Figure 8-19 illustrates the problem in which the separation is about 200 Hz at most. The Q of passive filters, which affects the skirt steepness, is not satisfactory for this purpose. The practical solution provides for filtering at comparatively low frequencies (much lower than the rf) and subsequent frequency translation to achieve the desired rf, together with selection of the proper bandpass filter. The filter itself is usually one of those based on mechanical vibration: mechanical filter, ceramic filter, or crystal filter.

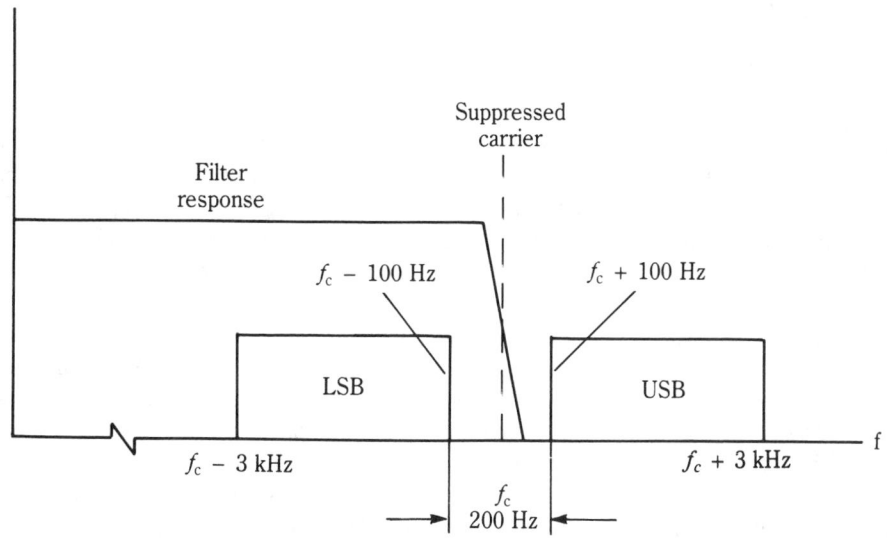

8-19 Upper sideband suppression by filter.

As stated previously, because of the filtering problem, the carrier frequency at the filtering stage is deliberately selected to be low—typically of the order of 100 kHz. Therefore, it is necessary to subsequently translate this frequency to the actual transmission frequency. The balanced modulator is also very effective for this frequency translation purpose. As shown in Fig. 8-20, a second crystal oscillator and a second balanced modulator appear in the SSB transmitter block diagram. The balanced modulator must be followed by a filter, not shown in the figure, to eliminate the difference-fre-

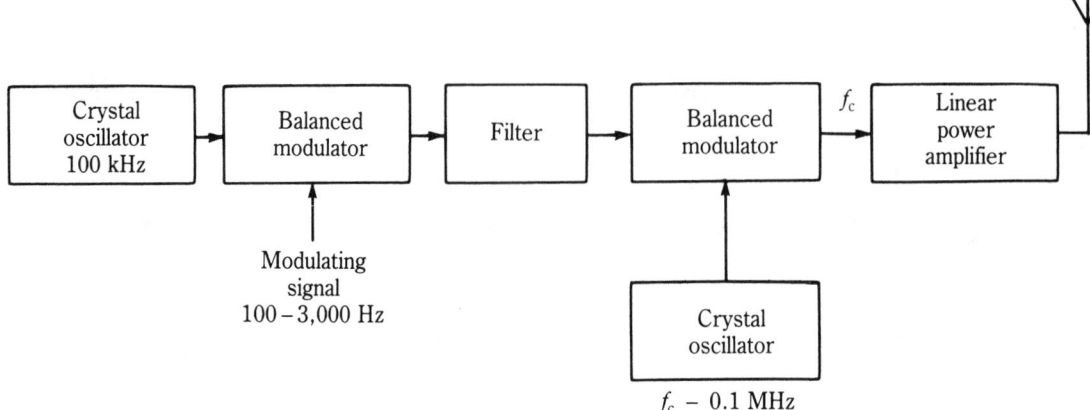

8-20 Single sideband transmitter.

quency components from the second balanced modulator. This filtering process presents no difficulties as a 200 kHz separation between the desired and undesired components now exists. Finally, a linear power amplifier is required for the transmitter as contrasted with the usual class-C power amplifier for AM.

Phasing method

A block diagram of a circuit implementing the *phasing method* of unwanted sideband elimination appears in Fig. 8-21. In this figure the phase shifts are indicated so that the disappearance of the undesired sideband can be explained. Starting from the left the modulating signal is assumed to be a sine wave of frequency f_m so that its angle is $\omega_m t$. In the one path, the angle becomes $\omega_m t + 90°$ after passing through the phase shifter. The oscillator produces another sine wave of frequency f_c which also undergoes a 90° phase shift before encountering the upper balanced modulator.

Recall that the outputs of the balanced modulator that are of interest to you are the sum and difference frequencies. These are marked as the upper inputs to the summing amplifier on the right. A similar, but critically different, set of outputs are generated by the lower balanced modulator. Notice that one of the 90° phase angles now has a negative sign because of the location of the phase shifter in the signal path. Because of this negative sign, the summing amplifier causes the lower sideband components to cancel and the upper sideband components to add directly. This accomplishes the intended purpose.

The principal difficulty in implementing the phasing method arises from the need to shift all modulating frequencies by exactly 90°. This difficulty is reduced through the adoption of the Weaver method.

The Weaver method

The *Weaver method*, called after its originator, is implemented by the block-diagram system of Fig. 8-22. Two carrier oscillators of frequencies f_1 and f_2 are used in this system. Here f_1 is considerably less than f_2, this last frequency being in the rf range. It will also

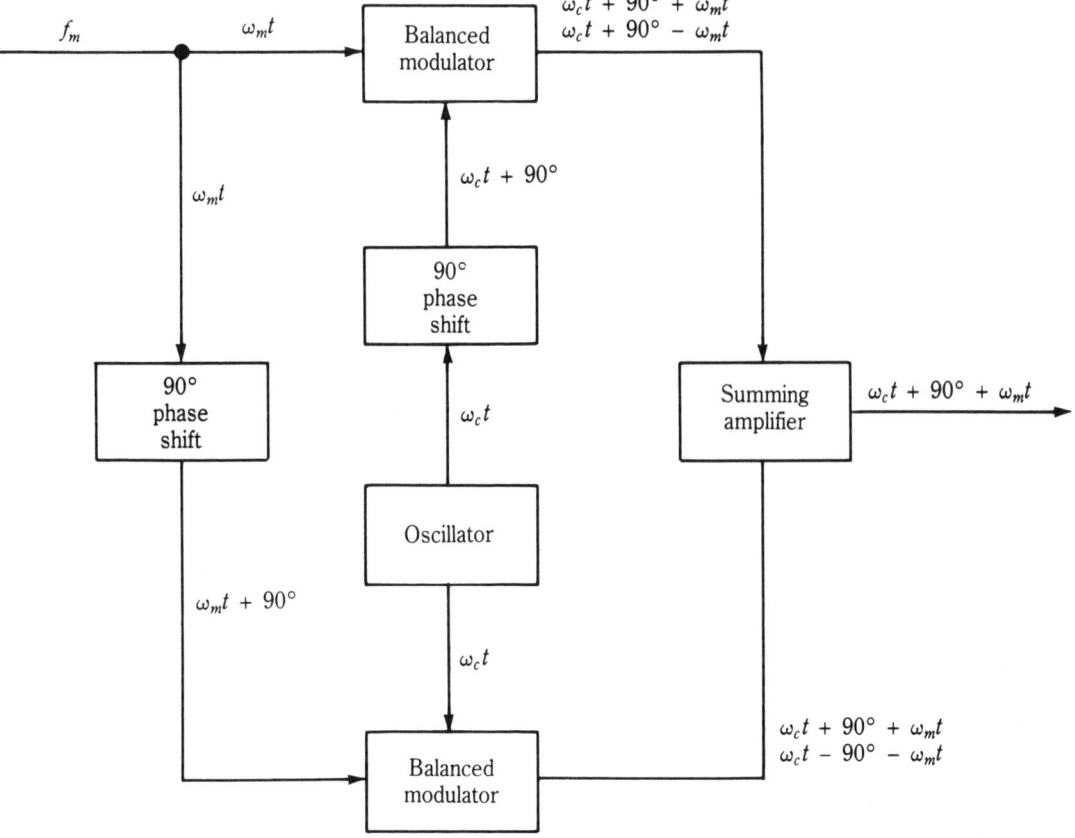

8-21 Phasing method for sideband suppression.

be seen that two low-frequency filters are necessary, but that there is no 90° shifting of the modulation frequency components. There is, however, phase shifting of both carrier frequencies which presents no significant difficulties. At the output of the summing amplifier, the effective carrier frequency is $f_1 + f_2$ and only the lower sideband is present. Note that cancellation of one sideband, the upper, takes place in the summing amplifier, because one input has a $+90°$ phase angle, whereas the other input has a $-90°$ phase angle.

AM designations

Amplitude modulation bears the FCC designation *A3* which signifies that the carrier and both sidebands are present. If one sideband alone is removed, leaving the carrier and the other sideband, the modulation is named *A3H*. Emphasis in this chapter has been placed on A3J, SSB with suppressed carrier. A variation of A3J, in which a small fraction of the carrier power is reinserted, is labeled *A3A*, SSB with reduced carrier. A3A allows the carrier to be received, amplified, and used in the demodulation process. Table 8-1 supplies this information in capsule form.

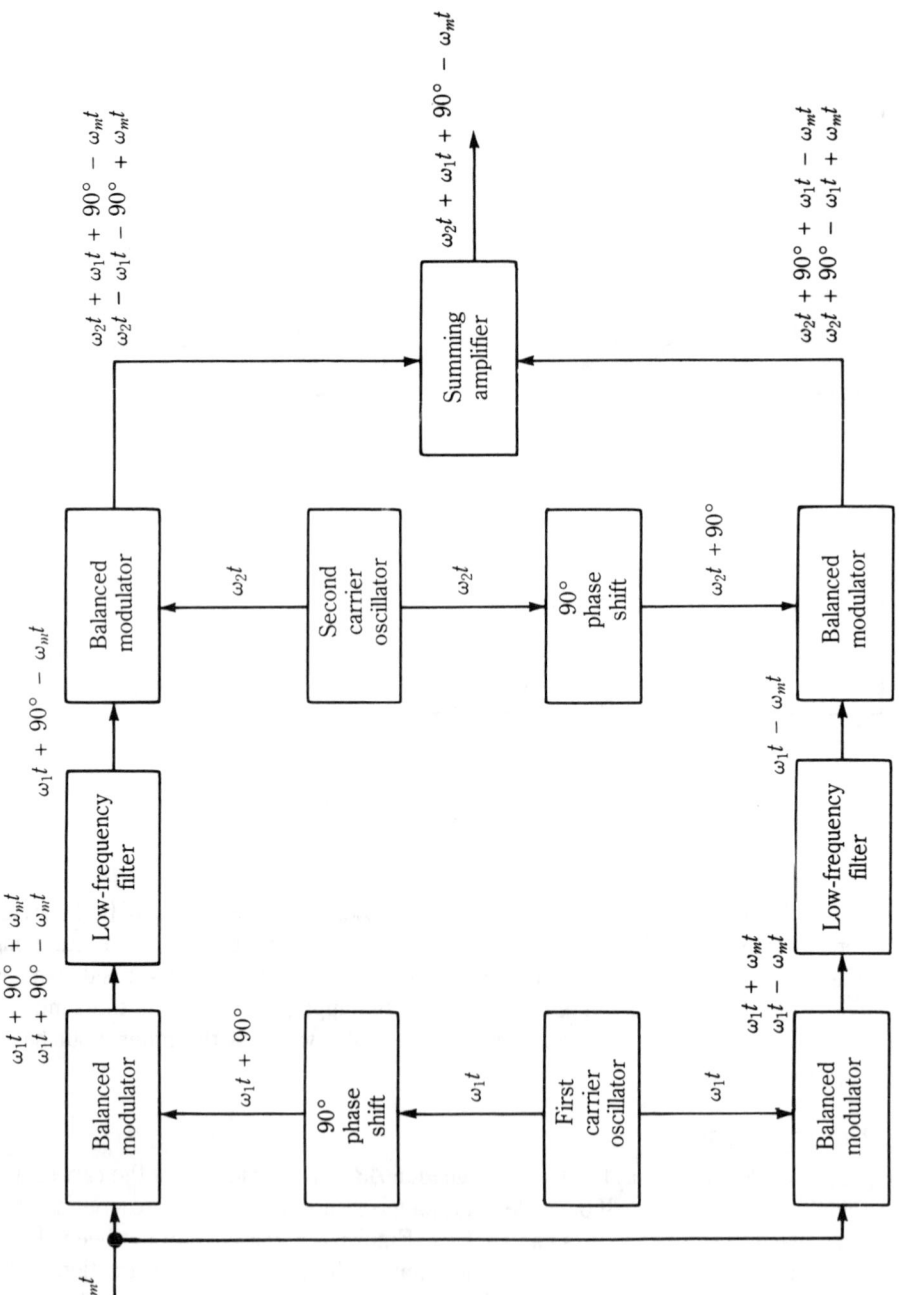

8-22 The Weaver method of sideband suppression.

**Table 8-1. Amplitude Modulation
Designations.**

Designation	Description
A3	AM with carrier and sidebands
A3A	SSB with reduced carrier
A3H	SSB with full carrier
A3J	SSB with suppressed carrier

Envelope detection

Demodulation of an AM carrier is most often accomplished in inexpensive broadcast receivers by a combination of rectification and low-pass filtering (smoothing). Consider Fig. 8-23A representing a sine wave carrier which is amplitude-modulated 100% by a much lower frequency sine wave. For reasons which will be more apparent in chapter 11, E_{IN}, the modulated waveform is coupled by means of a transformer, $T1$, to the demodu-

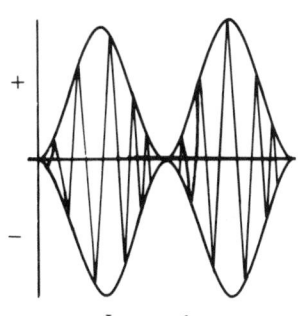

Input voltage
waveform, E_{IN}

A

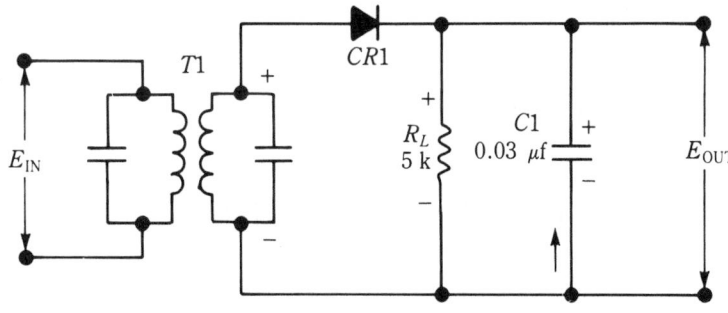

B

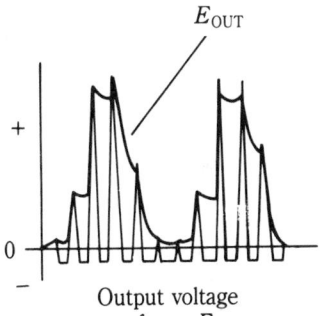

Output voltage
waveform, E_{OUT}

C

8-23 Envelope detection with rectifier and filter.

lation circuit, still referred to as the detector circuit after many decades of usage. The fact that the primary and secondary of the transformer are tuned to the carrier frequency relates to other receiver functions and not to the process of demodulation.

From the figure it is seen that a semiconductor diode is placed in series with the transformer secondary voltage. Current will then flow in the parallel combination of R and C (Fig. 8-23B) during alternations of one polarity only because of the rectifying action of the diode. If C is removed from the circuit, the voltage waveform across R will be exactly the same shape as that of the positive alternations in Fig. 8-23A. With C added in parallel with R, the capacitor charges to the alternation peaks and partially discharges between peaks. The net effect is to fill in (incompletely) the gaps between the peaks. This produces a scalloped near-replica of the modulating waveform shown in Fig. 8-23C.

The time constant, RC, is significant in establishing the fidelity of recovered modulating waveforms. When the time constant is short, excessive discharge occurs between peaks as shown in Fig. 8-24. With an unduly long time constant, the capacitor voltage cannot track the variations in modulating waveform and the resulting distortion is attributed to *diagonal clipping* (see Fig. 8-25).

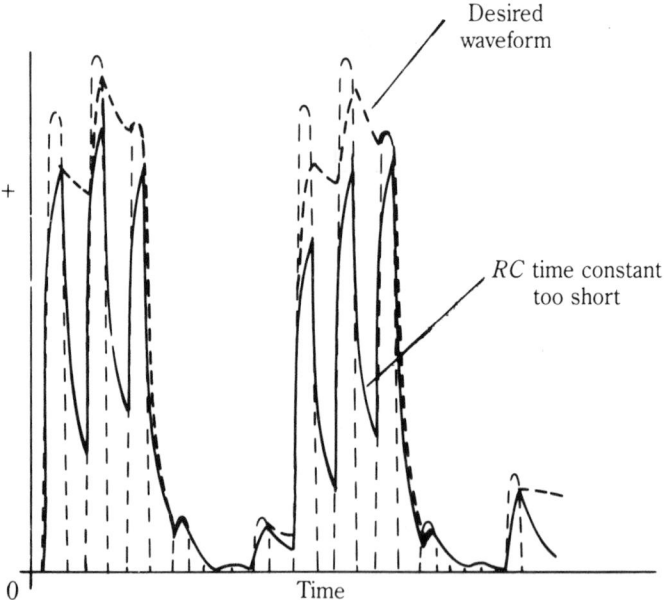

8-24 Envelope detection with excessively short time constant.

The envelope detector output is amplified by audio amplifier stages before being introduced into headphones or speakers. Further low-pass filtering of the detector output yields a slowly varying, much lower than any audio frequency, voltage which is used for another receiver function (AGC) to be explained later.

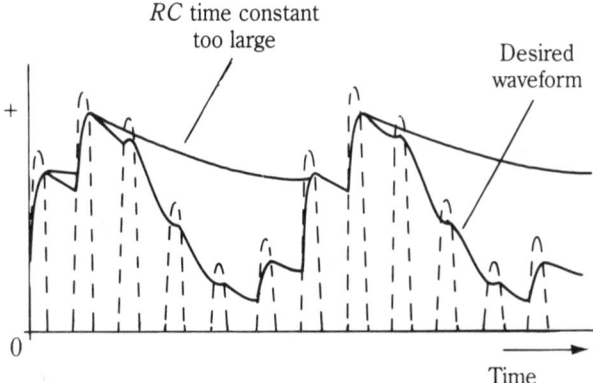

8-25 Envelope detection with excessively long time constant.

Product detection

Product detection is the process of mixing (heterodyning) a modulated waveform with a sine wave of the carrier frequency. The carrier copy is *coherent* if the phase angle is identical with that of the unmodulated carrier, in which case it is called *synchronous detection*, *coherent detection* or *heterodyne detection*, depending on the method employed. Detection (demodulation) is accomplished by taking the product of the two signals which yields the *baseband* information signal (modulating signal waveform) and an unwanted spectra near the second harmonic of the carrier frequency, which is easily disposed of by filtering. See Fig. 8-26 for a simplified block diagram of the product detection process.

A coherent signal for product detection can be obtained in the receiver using a *phase-locked loop*. The phase-locked loop is a servomechanism which will track the carrier component of a modulated signal by comparing its phase angle with that of a locally generated sine wave and adjusting the frequency of the local oscillator to reduce the phase error. Figure 8-27 shows the block diagram of the phase-locked loop. The input signal is compared with the output of the VCO in the phase detector. If the phases do not match exactly, an error voltage is produced. It is filtered to eliminate any high-frequency error components, such as exist because of the presence of sidebands, and then applied to the VCO which causes the oscillator to increase or decrease until the phase error is reduced. If the "free-running" frequency of the VCO exactly matches the carrier fre-

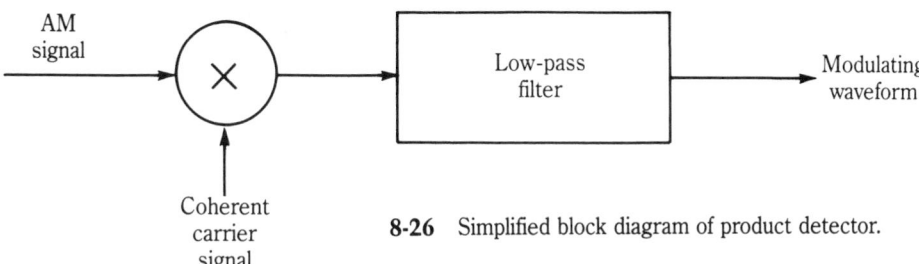

8-26 Simplified block diagram of product detector.

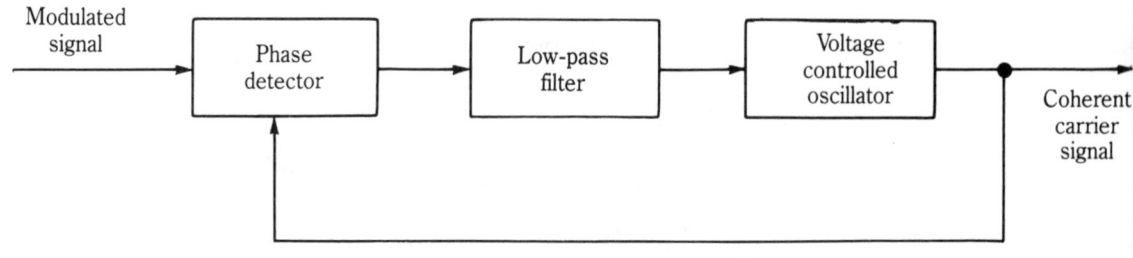

8-27   Phase-locked loop.

quency, the error voltage is reduced to zero. Otherwise, a small error voltage is required to match the frequencies with a nearly-perfect match of the phase angles. Any subsequent slow drift of the transmitter frequency is followed by the VCO in providing the coherent replica of the carrier.

The phase-locked loop is particularly effective in narrow bandwidth, low-information-rate communication systems, such as are employed in deep space exploration, since low-pass, as contrasted with band-pass filters determine the bandwidth.

The phase-locked loop, formerly quite expensive when constructed of discrete components, is now delightfully affordable in IC form. It is also a part of the ICs incorporating some or all of the receiver functions to be described later.

Coherent detection is not a required feature of product detection. If an oscillator can produce a reasonable approximation of the original carrier signal, product detection is possible. You should realize that the output is then only an approximation of the modulating waveform, being in error by a component whose frequency is that of the local oscillator frequency error. Intelligibility, or esthetic criteria, will determine the allowable error.

Multiple-choice questions

1. Modulation of an rf carrier results in:

   A. Multiple channels.
   B. Smaller antennas.
   C. Directional propagation.
   D. All of the preceding choices.
   E. None of the preceding choices.

2. A process which occurs in the transmitter is:

   A. Correlation.
   B. Modulation.
   C. Sublimation.
   D. Demodulation.
   E. None of the preceding choices.

3. A process which occurs in the receiver is:

   A. Correlation.

 B. Modulation.
 C. Sublimation.
 D. Demodulation.
 E. None of the preceding choices.

4. One part of the transmitter protects the crystal oscillator from ''pulling.'' It is the:

 A. Buffer amplifier.
 B. Modulator.
 C. Power amplifier.
 D. Antenna coupler.
 E. af processor.

5. What aspect of the carrier is changed by modulation?

 A. Power.
 B. Frequency.
 C. Phase.
 D. Amplitude.
 E. This question cannot be answered without knowing the type of modulation.

6. The lowest assigned carrier frequency for standard AM broadcast is:

 A. 107 kHz.
 B. 535 kHz.
 C. 540 kHz.
 D. 1600 kHz.
 E. 1605 kHz.

7. The highest assigned carrier frequency for standard AM broadcast is:

 A. 107 kHz.
 B. 535 kHz.
 C. 540 kHz.
 D. 1600 kHz.
 E. 1605 kHz.

8. In standard AM broadcast the frequency separation of assigned carrier frequencies is:

 A. 5 kHz.
 B. 10 kHz.
 C. 20 kHz.
 D. 107 kHz.
 E. 540 kHz.

9. The amplitude of a sine wave which is modulated by a musical program will:

 A. Be complex.

B. Contain fundamental frequencies.
C. Contain harmonic frequencies.
D. Contain a mixture of tones.
E. All of the preceding choices.

10. A dynamic microphone will likely have:

A. A diaphragm.
B. A moveable coil.
C. A FET amplifier.
D. A static magnetic field.
E. All of the choices except C.

11. A condenser microphone will:

A. Cause an emf to be induced.
B. Require a dc supply.
C. Operate on the same principle as the ribbon microphone.
D. Be useless when the humidity is high.
E. All of the preceding choices are incorrect.

12. The output level of a dynamic microphone is of the order of:

A. 1 μV.
B. 0 dB.
C. -40 dBm.
D. -55 dBm.
E. -100 dBm.

13. What will be the result of the gain level being too high for signals entering the modulator?

A. Receiver noise.
B. Excessive volume of receiver output.
C. Oscillator disturbance.
D. Fine levied by the FCC on the receiver owner.
E. Distortion and splatter.

14. Amplitude modulation causes the amount of transmitter power to:

A. Increase.
B. Decrease.
C. Remain the same.
D. Double.
E. Halve.

15. When a carrier is modulated 100%, the total power increases by what percentage over that of the carrier alone?

A. 25%.
B. 50%.
C. 75%.
D. 100%.
E. 150%.

16. When the amplitude of the modulating voltage is increased for AM, the antenna current will:

A. Increase.
B. Decrease.
C. Remain constant.
D. Decrease exponentially.
E. The answer to this question cannot be answered without knowing the antenna impedance.

17. An increase in transmitter power from 25 W to 30 W will cause the antenna current to increase from 700 mA to:

A. 500 mA.
B. 683 mA.
C. 767 mA.
D. 840 mA.
E. 1000 mA.

18. A second modulating tone having the same amplitude, but a different frequency, is added to the first at the input to the modulator. The modulation index will be increased by a factor of:

A. 1.
B. $\sqrt{2}$.
C. 2.
D. $2\sqrt{2}$.
E. 4.

19. A 1000 kHz carrier is modulated by a 2500 Hz tone. One frequency component of the modulated signal is:

A. 997.5 kHz.
B. 1000 kHz.
C. 1002.5 kHz.
D. 2500 Hz.
E. Choices A, B, and C.

20. Assume that the amplitude of the unmodulated carrier in Question 19 is 100 mV. The amplitude of each sideband for 100% AM will be:

A. 0.
B. 25 mV.

C. 50 mV.
D. 100 mV.
E. 200 mV.

21. A 1200 kHz carrier is amplitude-modulated by two tones of 500 Hz and 700 Hz. Identify a frequency component of the modulated wave.

A. 1195 kHz.
B. 1199.3 kHz.
C. 1199.7 kHz.
D. 1205 kHz.
E. 1207 kHz.

22. Identify a modulation method, or methods, in use for a common-emitter configuration.

A. Base modulation.
B. Emitter modulation.
C. Collector modulation.
D. Both A and C.
E. Both B and C.

23. The rf signal injected into a balanced modulator is 10 MHz and the modulating frequency is 1 kHz. Which frequency, or frequencies, will not appear in the output?

A. 9.999 MHz.
B. 10 MHz.
C. 10.001 MHz.
D. Both A and B.
E. Both A and C.

24. Unwanted sidebands in SSB equipment can be suppressed by one or more of the following methods:

A. Phasing method.
B. Filter method.
C. Decoder method.
D. Both A and B.
E. None of the preceding choices.

25. Envelope detection is concerned with the process of:

A. Mixing.
B. Heterodyning.
C. Modulation.
D. Rectification.
E. Band splitting.

26. Diagonal clipping in envelope detection will result in:

    A. Distortion.
    B. Phase reversal.
    C. Reduced sensitivity.
    D. Amplitude damage.
    E. Adjacent channel interference.

27. Product detection requires the process of:

    A. Rectification.
    B. Heterodyning.
    C. Decoding.
    D. Phase shifting.
    E. Both A and D.

28. A sine wave which is coherent with the carrier has identical:

    A. Amplitude.
    B. Frequency.
    C. Phase angle.
    D. Both B and C.
    E. Choices A, B, and C.

29. The FCC designation for amplitude modulation is:

    A. A3.
    B. A3H.
    C. A3J.
    D. AA.
    E. AM.

30. The FCC designation for single sideband, suppressed-carrier is:

    A. A3.
    B. A3H.
    C. A3J.
    D. SSB.
    E. SSBSC.

Basic problems

1. Name at least three basic elements of an electronic communications system.

2. Name at least three parts of a transmitter.

3. Briefly describe the standard broadcast AM frequencies.

4. Briefly state the compression problem arising from the recording of a musical program.

5. A sine wave carrier is modulated by a single tone so that E_{max} = 5 mV and E_{min} = 1 mV. Calculate the modulation index.

6. An unmodulated transmitter has an output of 600 W. Determine the amount of power when the same transmitter is modulated at the 100% level.

7. Find the carrier power component when m = 90% and P_T = 35 kW.

8. A spectrum analyzer for an AM wave reveals three equally-spaced components having amplitudes of 500 μV, 1500 μV, and 500 μV respectively. What is the modulation index?

9. In Basic Problem 8, the frequency separation of two adjacent components is 1200 Hz. What is the modulation frequency?

10. In product detection, the oscillator in the receiver is in error by 5 Hz. What effect will this have in the output of the detector?

Advanced problems

1. Assuming a 150 Ω impedance level, a power level of -60 dBm requires what rms voltage level?

2. The carrier wave has an amplitude of 20 V before modulation and reaches a maximum amplitude of 30 V during modulation. Determine the modulation index.

3. What modulation index will cause the total transmitted power to increase 25% over that of the unmodulated carrier?

4. With a modulation index of 75% the total power transmitted is 1250 W. What will be the total power when the modulation index is decreased to 50%?

5. At 100% modulation, what percentage of the total power of an AM transmission resides in one set of sidebands?

6. Anticipating single-sideband AM, what percentage of the total power is saved by suppressing the carrier and one set of sidebands while transmitting the other set of sidebands only?

7. Three tones are used in the amplitude modulation of a 1000 W carrier. These are capable of individually establishing modulation indexes of 40%, 50%, and 60% respectively.

   A. Calculate the total power.

B. Calculate the power in the carrier and each set of sidebands.

8. A practical IC balanced modulator can have a carrier level 40 dB (or more) below the output sideband levels. Assuming a sideband level of 1 V, what voltage level can be expected for the carrier?

9. A 455 kHz i-f amplifier feeds an envelope detector. Assuming original modulation by a 1 kHz tone, what frequency components are deliberately attenuated by filtering in the detector?

10. Briefly explain how a phase-locked loop can be applied to supply a coherent signal for product detection.

9

Frequency modulation

IN ADDITION TO AMPLITUDE MODULATION THERE ARE TWO OTHER CLOSELY RELATED modulation techniques—*frequency modulation* (FM) and *phase modulation* (PM)—in common use. These two are collectively known as *angle modulation*, since the phase angle associated with the carrier is deliberately changed in accordance with the modulating signal while the amplitude is held constant.

You are here concerned with the different forms of angle modulation, the practical means for achieving modulation and demodulation and both the advantages and disadvantages of angle modulation. FM will be emphasized because of its established role in radio communication.

This chapter is divided into the following sections:

- The FM signal
- The FM spectra
- Noise
- Direct FM
- Indirect FM
- FM stereo
- Demodulation using tuned circuits
- Other FM demodulation methods

The FM signal

As a means of visualizing frequency modulation (FM) imagine a manually-controlled oscillator set to the carrier frequency f_C. This frequency corresponds to a zero modulating signal level. The operator (knob twister) is obliged to monitor the modulating signal level and to increase (or decrease) the oscillator frequency so as to cause a change in frequency that is directly proportional to that level. Positive signal levels result in an increase in frequency, and negative levels result in a decrease in frequency. The change in frequency is named the *frequency deviation* and the maximum change in frequency for which the system is designed is the *maximum frequency deviation*. If sound levels are the source of the electrical modulating signal, the loudest sound will cause the largest frequency deviation.

As in AM, it is useful to consider the case of a sinewave modulation waveform. Study Fig. 9-1A which represents one cycle of a sound wave that has a sinusoidal shape. Figure 9-1B shows the modulated carrier. Note that it appears to be bunched up at the region on the time scale for which the amplitude of the modulating signal is maximum. The opposite is true when the amplitude is reaching its greatest negative value.

Figure 9-1C represents the situation differently and shows how the frequency varies with time. The relationships of these three waveforms is most important and should be understood before pressing on.

To provide a contrast with the previous figures, Fig. 9-1D shows how the same sound wave would amplitude-modulate the carrier.

One aspect of FM has been deliberately concealed up to this point. This relates to the frequency of the modulating waveform. This frequency has absolutely no influence on the frequency deviation of the modulated carrier. If, for example, the modulating signal frequency is doubled in Fig. 9-1 without changing the time scale, two cycles will be seen in Fig. 9-1A and 9-1C and two bunchings and two "expansions" will be seen in Fig. 9-1B.

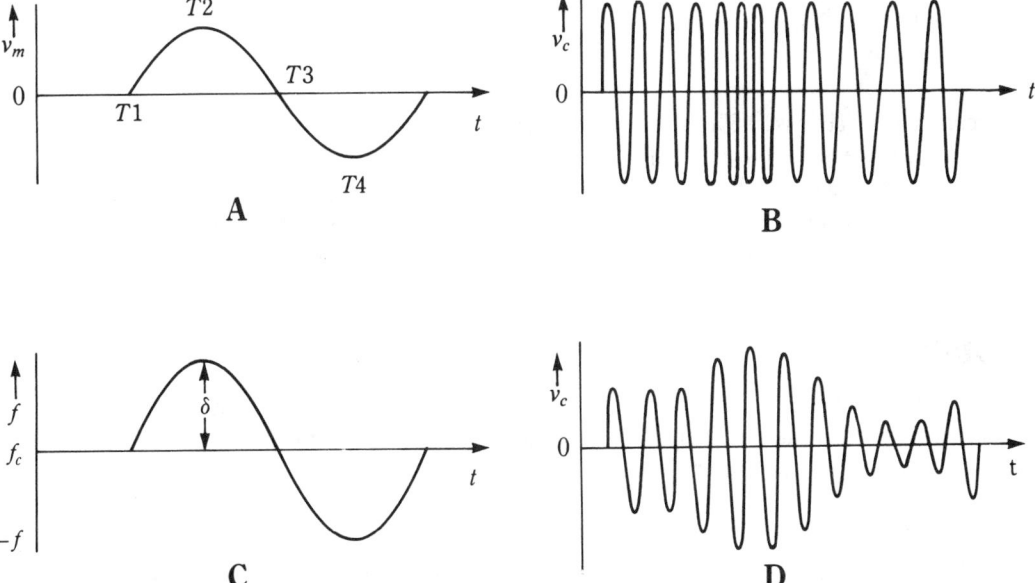

9-1 Modulation methods comparison: A. Modulating waveform. B. Frequency-modulated carrier. C. Frequency variation with time for FM. D. AM carrier.

The FM spectra

It will be recalled from the previous study of AM that frequency spectra referred to the distribution of the modulated carrier frequency components in the frequency domain. This concept is relatively easy to absorb for AM in which there is exactly one frequency component in the upper sideband and its mirror image in the lower sideband for each frequency component of the modulating signal. The situation for FM is dramatically different. If, for example, the modulating waveform is sinusoidal with a frequency of 1 kHz, the modulated carrier spectra will have lines at 1, 2, 3, 4,.... kHz above f_C and the mirror image lines below f_C. See Fig. 9-2 in which the spacing between spectral lines is 1 kHz.

A closer examination of Fig. 9-2 confirms that a number of spectral lines exist with fixed spacing, but that their relative amplitudes appear to vary in accordance with some quantity labeled m_f, which is the modulation index for FM. It should be pointed out that there is no basic limitation on the maximum frequency deviation, as compared with the fundamental inability to reduce the carrier amplitude below zero in AM. Therefore, if the maximum frequency deviation, to which we assign the Greek letter delta (δ), is not constrained, the modulation index, m_f, likewise has no fundamental constraint:

$$m_f = \delta/f_m$$

where f_m is the modulating frequency. (The maximum frequency deviation is restricted in the FCC station authorization, which is not an issue at this point.)

As shown in Fig. 9-2, with a given modulating frequency f_m, the overall span of the spectra expands with both δ and m_f. Theoretically, the extent of the spectrum for FM is infinite. Practically, it is characterized by Carson's rule which states that the required bandwidth is given by:

$$BW = 2(1 + m_f)f_m$$

to accommodate at least 98% of the total power emission. The immediate practical impact of this rule is not favorable. As compared to AM for which $BW = 2f_m$, you pay a penalty for larger values of the modulation index, m_f, in terms of larger bandwidth. While the underlying reasoning is somewhat complicated, the sacrifice of spectrum in the adoption of FM has compensating advantages which will be explored in the next section.

Noise

All radio communications must contend with extraneous, undesirable signal variations that contaminate the modulated carrier. Here you will be concerned with external noise. In chapter 11 the subject of noise originating within the receiving equipment will be covered.

External noise, classified in accordance with its source, is atmospheric, galactic, and man-made. *Atmospheric noise* is caused principally by lightning discharges in the atmosphere. Its magnitude varies inversely with frequency, and with other factors governing thunderstorm activity. *Galactic noise* originates outside the atmosphere and, to a large extent, outside the boundaries of the solar system. *Man-made noise* is particularly troublesome in an industrial environment, but can present problems in any location hav-

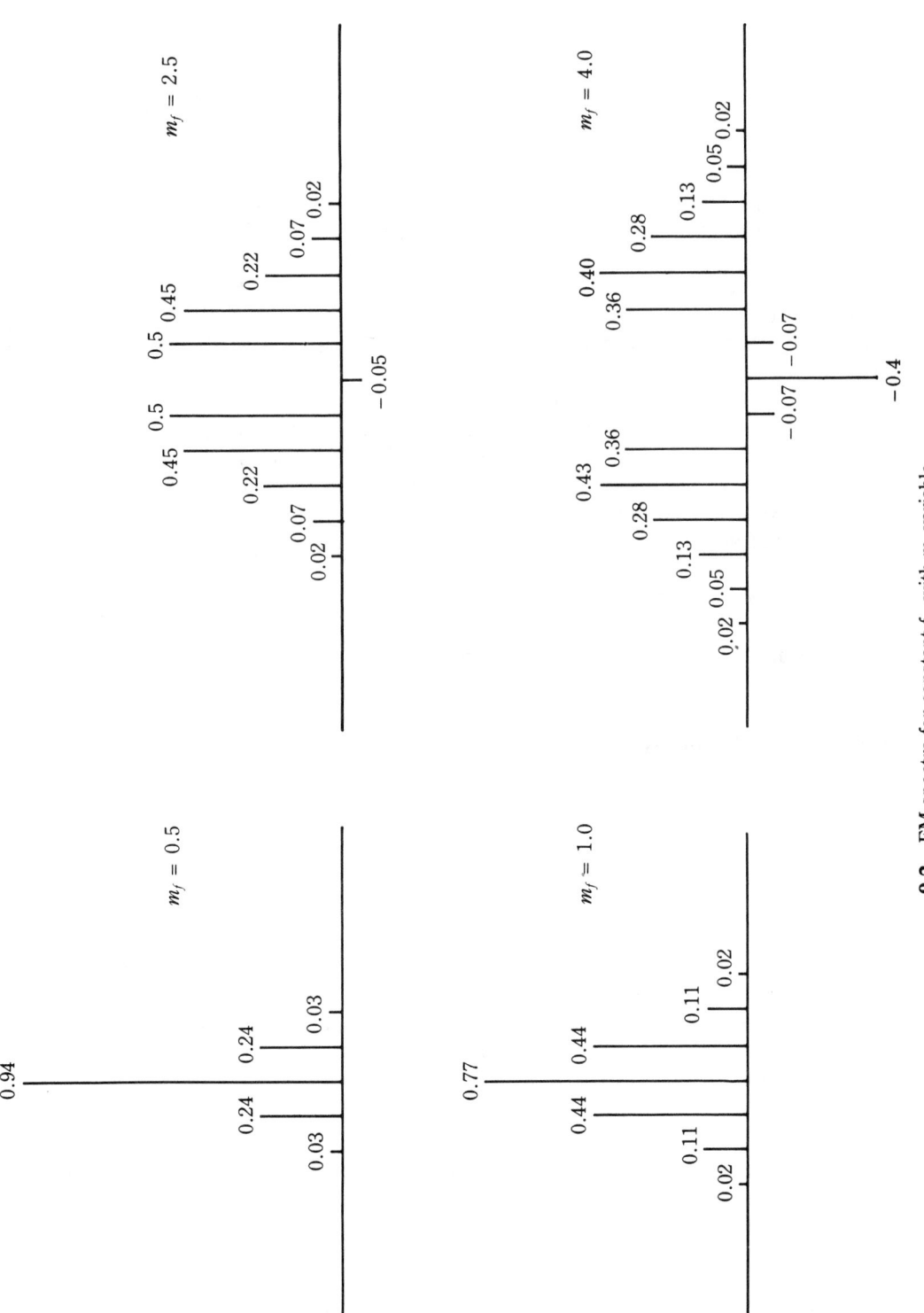

9-2 FM spectra for constant f_m with m_f variable.

ing high-voltage power lines and wherever internal combustion engine ignition systems are operating. It tends to decrease with frequency, but might be unevenly distributed depending on its specific source. Noise components can be propagated long distances on the earth's surface under the same conditions as are the propagation of deliberate radio transmissions.

The central feature of noise is the inability to predict its instantaneous value. You can usually specify its bulk properties such as average power in much the same way as life expectancy is known about a human population but not for any particular individual in the population. If noise could be predicted in detail, equipment could be devised to effect its cancellation. Noise power increases with the channel bandwidth. Because the information rate of the channel also increases with bandwidth, a tradeoff exists between the two. This explains why deep space communications often have a limited bandwidth, hence limited information rates.

The effect of noise on an AM waveform is to superimpose a random noise component on the envelope. A lightning discharge, for example, will impose a spike on the envelope variations that convey the modulating waveform shape. This form of interference is not so serious with FM because the spikes are deliberately limited by circuitry designed into the receiver. The strategy of limiting to eliminate spikes cannot be used in the AM receiver without modifying the envelope with the consequent distortion of the output.

Some forms of noise are uniformly distributed throughout the pass band of the filters incorporated in the receiving equipment. Thus the amount of noise power is proportional to the product of this noise spectral density, n, which is commonly given as watts per hertz, and the receiver bandwidth which will be expressed as $2B$. (The largest component frequency of modulating signal for such a receiver is no larger than B.) With 100% amplitude modulation, and carrier power, P_C, the receiver output signal-to-noise power ratio is:

$$(S/N)_{AM} = \frac{P_C}{2Bn}$$

In comparison, the signal-to-noise ratio for single-sideband-suppressed-carrier operation is:

$$(S/N)_{SSB} = 3(S/N)_{AM}$$

which is valid only for 100% modulation.

With FM, the signal-to-noise ratio depends on the maximum frequency deviation, δ. It is given by:

$$(S/N)_{FM} = 3\left(\frac{\delta}{B}\right)^2 (S/N)_{AM}$$

It is clear from the examination of the preceding equation and Carson's law that improved signal-to-noise ratio can be gained at the expense of rf spectrum. This is a technical trade-off. However, you should recognize that the improvement, as expressed by the equation, is true when the carrier power is significantly greater than the noise power.

The preceding analysis does not recognize the different effect of noise on the several components of an information signal. It comes about for the following reason: In Fig.

9-3 is a simple diagram of the carrier phasor and a noise phasor corresponding to the noise voltage within a narrow span in the receiver pass band *B*. The existence of the noise component, which is randomly variable, causes a phase shift angle as shown in the figure, labeled ϕ_N. The noise phasor can be considered as fixed in length, which it is not, and rotating with respect to the carrier voltage phasor. Its relative rate of rotation is exactly the same as the component position in the spectrum relative to the carrier frequency. In other words, the farther from the carrier frequency the noise component, the more rapidly will its phasor rotate with respect to the carrier phasor, and the greater will be the FM noise due to that single component.

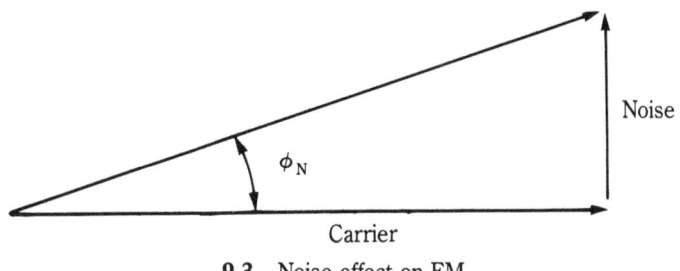

Noise

ϕ_N

Carrier

9-3   Noise effect on FM.

The preceding discussion leads to the noise triangle concept, which is a way of illustrating the overall effect of all noise components for FM. The noise triangle is illustrated in Fig. 9-4. Even with the random noise distributed uniformly across the spectrum, the noise will exert the maximum interfering effect on higher-frequency components of the modulating waveform and very little effect for the lowest-frequency components of the modulating signal.

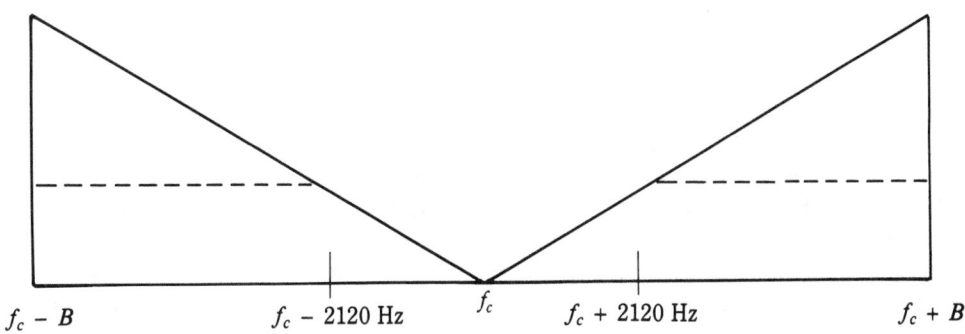

$f_c - B$ $f_c - 2120$ Hz f_c $f_c + 2120$ Hz $f_c + B$

9-4   Noise triangle.

The degradation of the higher audio frequency components for broadcast FM, for example, is not acceptable. A method of reducing this noise effect in FM systems will now be explained. The basic strategy is to boost the higher-frequency components of the modulation waveform prior to the actual modulation process, so as to increase these more vulnerable elements relative to the noise at the upper end of the band. This is known as *preemphasis*. It is accomplished in broadcast FM by inserting a 75 microsecond time constant circuit in one of the amplifier stages so that a simple high-pass filter is formed with the gain increased by 3 dB at 2120 Hz (Note that $1/(2\pi \times 75 \times 10^{-6}) = 2120$). The preemphasis curve is shown in Fig. 9-5.

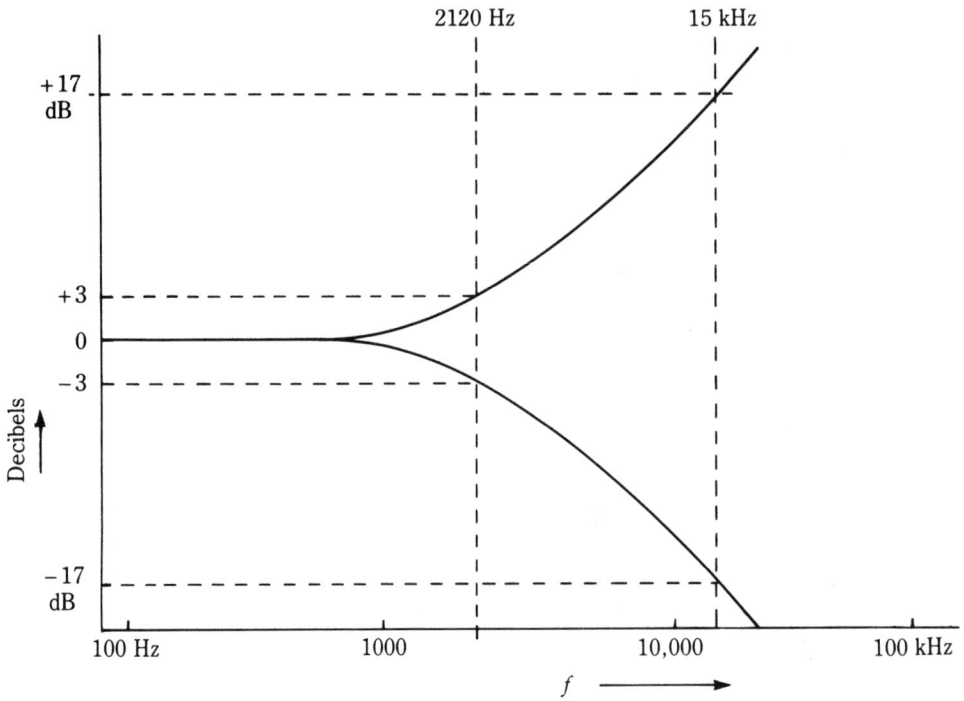

9-5 Emphasis and deemphasis curves.

Preemphasizing the higher-frequency signal components has an accompanying disadvantage. Music and voice quality is sacrificed because of the distortion introduced by amplifying high frequencies as compared with low frequencies. A similar effect can be achieved with the home high-fidelity amplifier through settings of the equalizer controls.

The remedy to the distortion introduced by preemphasis is to impose a reverse process in the receiving equipment. In other words, a low-pass filter can be introduced so that the lower frequencies are favored over the higher frequencies. While this would seem to cure the distortion problem, it would appear that the desired advantages of preemphasis have been negated through the introduction of deemphasis. This is not so. Preemphasis occurs before the noise in the channel is mixed with the signal. *Deemphasis*, the opposite of preemphasis occurs after the noise is mixed with the signal. Therefore, the troublesome high-frequency noise components are deemphasized along with the higher-frequency signal components, but without having had the earlier preemphasis. Therefore, the high frequency noise is lowered relative to the higher frequencies of the modulation waveform with a significant further improvement of the FM broadcast signal-to-noise ratio. The noise triangle of Fig. 9-4 has been altered using horizontal dotted lines to reflect the change caused by preemphasis and deemphasis.

Direct FM

Direct FM refers to the changing of the resonant frequency of a tuned circuit by the deliberate alteration of an effective reactive element.

Varactor diode modulator

In Fig. 9-6 $D1$ is a varactor diode that is reverse biased by voltage source E, which is placed in series with a modulating signal source e_m. C_C is a blocking capacitor. As e_m rises, increasing the total bias on $D1$, the junction capacitance decreases, so the resonant frequency increases. When e_m has the opposite polarity, the capacitance increases, so the resonant frequency decreases. Therefore, the frequency modulation is achieved through the application of the modulating signal to the varactor diode.

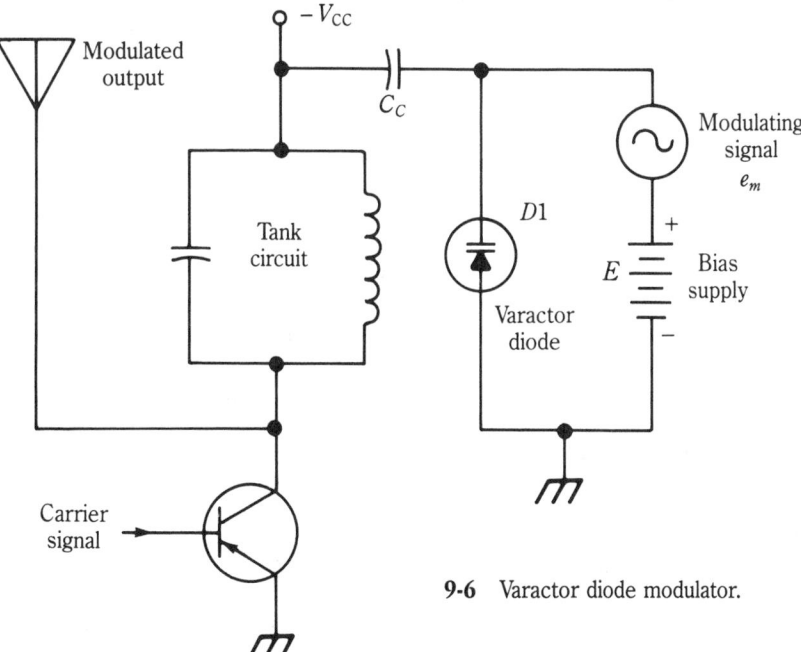

9-6   Varactor diode modulator.

Reactance modulator

In the circuit of Fig. 9-7, $R1$ and $C2$ perform as a series circuit connected across the $L1C1$ parallel tuned circuit. This assumes a relatively small base current and relatively small currents in $R2$ and $R3$. If the reactance of $C2$ is much larger than that of $R1$, the current through them leads the total tank voltage by 90° and the base voltage will also lead by 90°. Therefore, the collector current will lead the total voltage so that the transistor behaves as a capacitor.

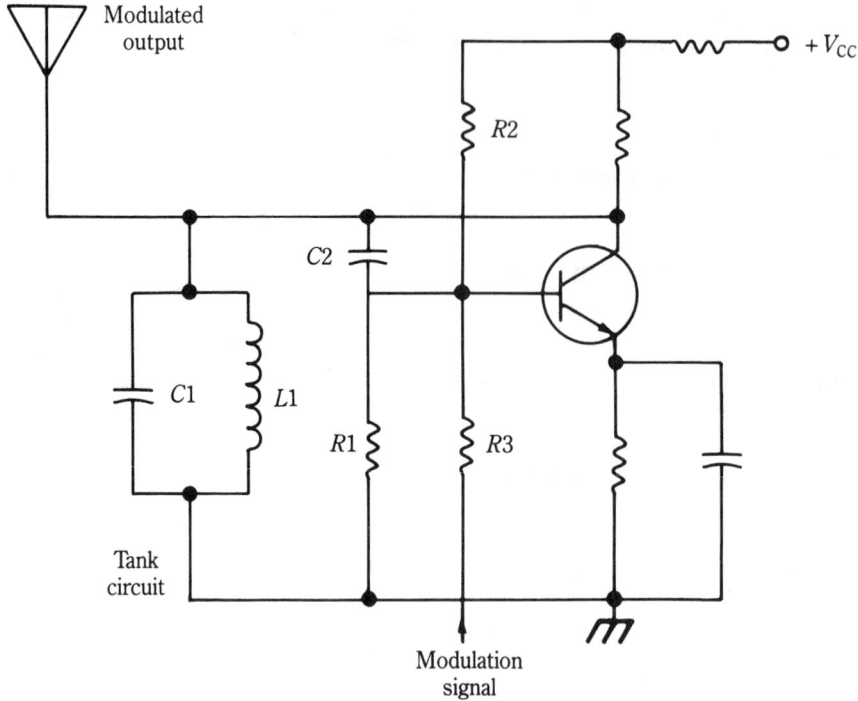

9-7   Reactance modulator.

When a much lower-frequency modulating signal, as compared to the resonant frequency of the tank circuit, is applied to the base circuit, the base bias point will also change position at this frequency by an amount depending on the amplitude of the modulation drive. This in turn will cause the transconductance, g_m, to be varied so that the amount of collector quadrature current also varies. The transistor now behaves as a variable capacitor controlled by the modulating waveform and the frequency is so modulated as to produce FM.

Voltage controlled oscillator

The NE/SE 566 function generator IC is a general-purpose voltage controlled oscillator. On the chip a multivibrator-type oscillator has a center frequency between about 10 Hz and 100 kHz established by an externally-connected resistor and capacitor. The frequency can then be varied linearly over approximately a 5:1 range by means of a control voltage. Both triangular and square waves are available as outputs.

Indirect FM

Phase modulation and FM are often considered to be the same. You are first reminded that they are related, but definitely not identical. In phase modulation the phase deviation is proportional to the instantaneous value of the modulating signal. In FM there is a frequency deviation proportionality.

However, it is impossible to have a changing phase deviation without also having a frequency deviation. Conversely, it is also impossible to have a frequency deviation without a changing phase deviation. This means that, with the proper processing of the modulating signal prior to the modulation step proper, phase modulation can be used to achieve frequency modulation and vice versa. Indeed, one specific disadvantage of direct FM, that of the unstabilized carrier frequency resulting from the direct control of frequency, can be overcome by phase modulation of a crystal-stabilized carrier frequency to achieve FM. To understand how this is accomplished, you must first understand basic phase modulation.

Consider the circuit of Fig. 9-8. The crystal oscillator provides a stable sine wave whose frequency is a submultiple of the carrier, that is, f_C/N, where f_C is the carrier frequency and N is a whole number. The reactance of C combined with the effective resistance of the FET, which is controlled by the applied gate voltage bias, the modulation input, causes the output to vary both in amplitude and phase. The amplitude variation must later be removed in order that phase modulation with an accurate, stabilized carrier frequency is achieved. You immediately recognize that the maximum phase variation possible is limited to an absolute maximum of 90° and to a fraction thereof with reasonable preservation of linearity. A similar phase modulator can be constructed using a varactor diode whose capacitance is controlled by the modulating signal.

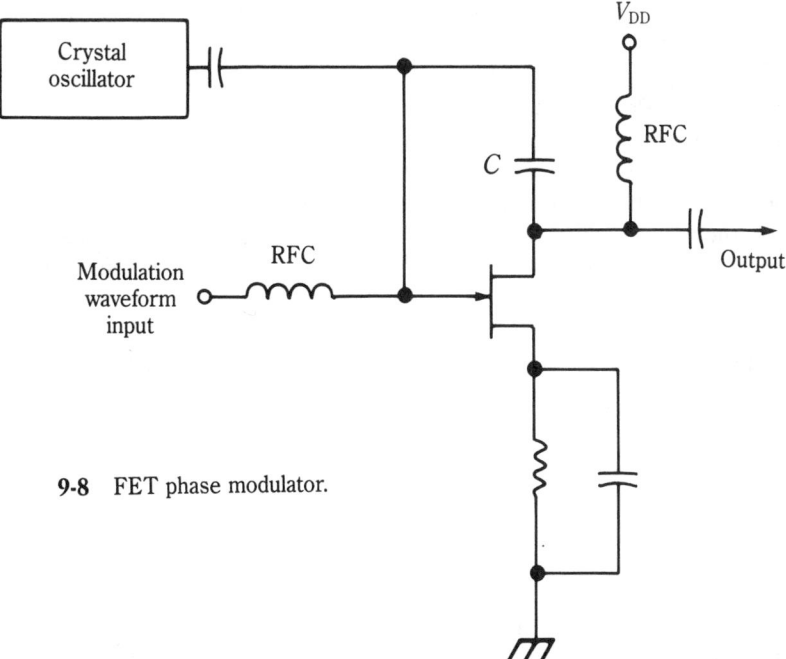

9-8 FET phase modulator.

Phase modulation can be converted to FM by a process that is referred to as *indirect FM*. It requires that the modulating signal frequency components first be attenuated in direct proportion to their respective frequencies. Alternately, these components can be amplified amounts that are inversely proportional to their respective frequencies prior to the phase modulation process. For example, let the modulating signal consist only of two equal-amplitude components of 300 Hz and 1800 Hz. However, after being selectively amplified and/or attenuated and prior to being applied to the modulating circuit, the amplitude of the 300 Hz component divided by that of the 1800 Hz component is 6. After phase modulation is completed using these two pre-processed signal components, the peak phase deviation for the lower-frequency component will be six times that of the higher. A little thought reveals that the peak frequency deviation caused by the two different frequency components is exactly the same, which is the result desired in adopting the indirect FM method.

Frequency translation and multiplication

There might be several reasons why the carrier frequency of a transmitter (or receiver) is inconveniently high for amplification and other processes, including modulation, which must take place in the electronics equipment. If, for example, the linear phase deviation is restricted to $1/2$ radian (approximately 28.6°) for the phase modulator of Fig. 9-8, it might be necessary to multiply the frequency deviation by a certain amount to achieve the nominal 75 kHz maximum frequency deviation of broadcast FM. Multiplication of a center frequency by some whole number, M, causes the phase deviation and the frequency deviation both to also be multiplied by the same factor.

Continuing the example, the frequency deviation at the output of the phase modulator for a highest modulating frequency of 15 kHz is $(0.5 \times 15,000)/2\pi = 1194$ Hz. To achieve 75 kHz peak frequency deviation, the value of N must be $75,000/1194 = 63$. The method whereby frequency multiplication takes place, in which 63 is the frequency multiplication factor in the present example, will be presented in chapter 10.

Frequency translation is the process of moving the entire spectrum of a modulated waveform without changing the modulation index. Suppose that the crystal oscillator frequency of the previous example was 100 kHz. When multiplied it becomes $63 \times 100 = 6300$ kHz = 6.3 MHz. Now, to achieve a center frequency of 100 MHz, a translation of $100 - 6.3 = 97.3$ MHz additionally is required.

Translation can be utilized for either AM or angle modulation. Multiplication causes distortion with AM and changes the modulation index for angle modulation.

FM stereo

A stereo FM broadcast system requires that two separate channels, or their equivalent, be provided. Moreover, the system must be compatible for receivers not having stereo capability. This means that such a receiver will be able to receive a stereophonic broadcast automatically as though it was not stereophonic. The receiver operator will not be aware of whether or not there is a stereo broadcast program with compatibility.

The method adopted for stereo broadcasting starts with two microphones labeled ''L'' for left and ''R'' for right, located at the origin of the program. After the L and R audio signals pass through separate preamplifier circuits they are added to produce the L + R and subtracted for the L − R audio signal. See the matrix network of Fig. 9-9.

The L − R signal amplitude modulates a 38 kHz subcarrier by means of a balanced modulator, which causes the carrier to be suppressed. Three signals, the 19 kHz pilot

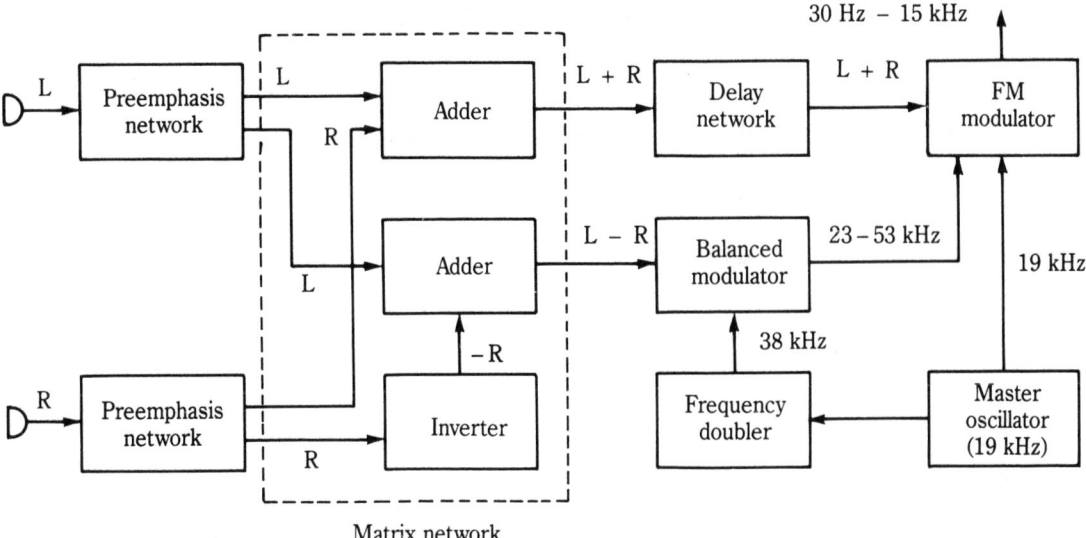

9-9 Stereo processing and modulation for FM transmitter.

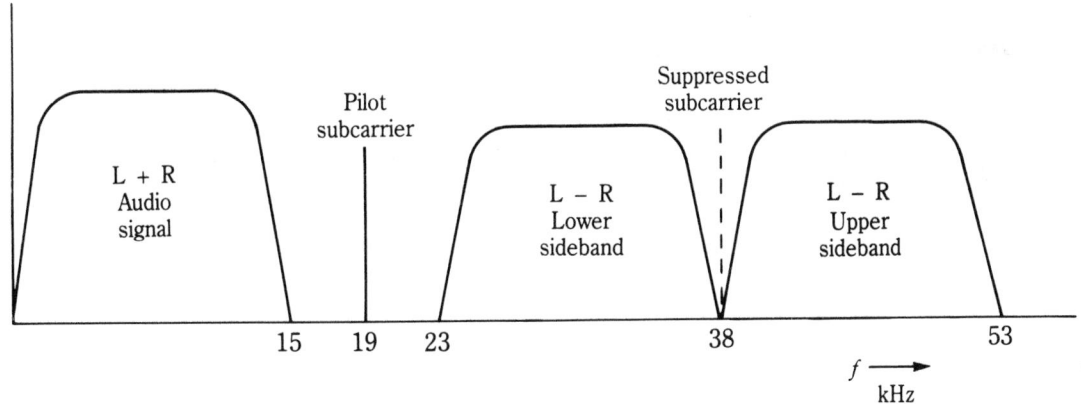

9-10 Composite modulation spectrum for FM stereo.

subcarrier, from which the 38 kHz subcarrier was derived, the AM double sideband suppressed carrier (DSBSC) signal, and the L + R (30 Hz to 15 kHz) signal are now combined and enter the FM modulator. The composite modulating spectrum is shown in Fig. 9-10. Note that the L − R spectrum ranges from 38 − 15 = 23 kHz to 38 + 15 = 53 kHz. This last part of the modulation arrangement is shown in the right half of Fig. 9-9. A delay network is placed in the L + R path to match the delay presented by the balanced modulator in the L − R path. The role of the 19 kHz pilot subcarrier will be explained later in chapter 11.

Demodulation using tuned circuits

A tuned circuit can be made to convert FM to AM by adjusting its resonant frequency, f_R, so that the center (carrier) frequency, f_C, falls on the flank of the response curve (see Fig. 9-11). Except for emergency use, *slope detection*, as this technique is known, is unsatisfactory because of the nonlinearity of the response and its vulnerability to unwanted amplitude modulation of the received signal. Nevertheless, the principle of operation is easily understood and is often introduced as a preliminary to the other, more acceptable, FM demodulation methods.

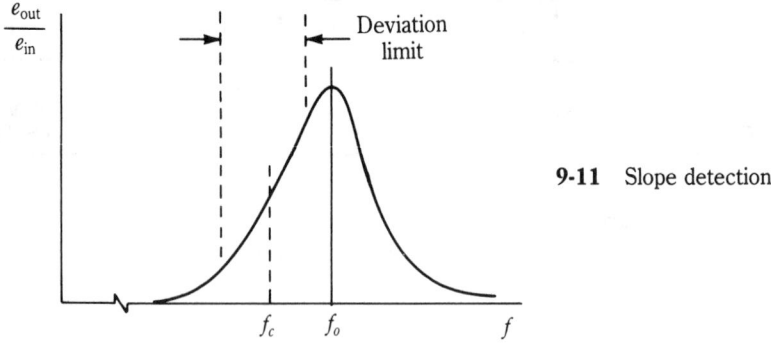

9-11   Slope detection.

Double (stagger) tuned discriminator

Improved performance, as compared with the slope detector, can be achieved using two tuned circuits as shown in Fig. 9-12. The upper parallel LC circuit is tuned to a frequency a fixed amount above the center frequency of the modulated carrier, and the lower is tuned to a frequency the same fixed amount below. As a result the linearity of the composite circuit is greatly improved. This demodulator has never enjoyed the popularity of the other demodulators to be described.

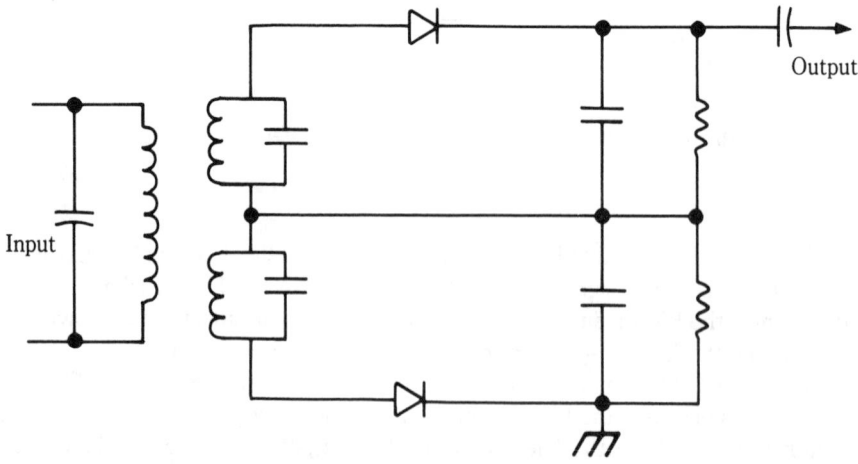

9-12   Double-tuned slope detection.

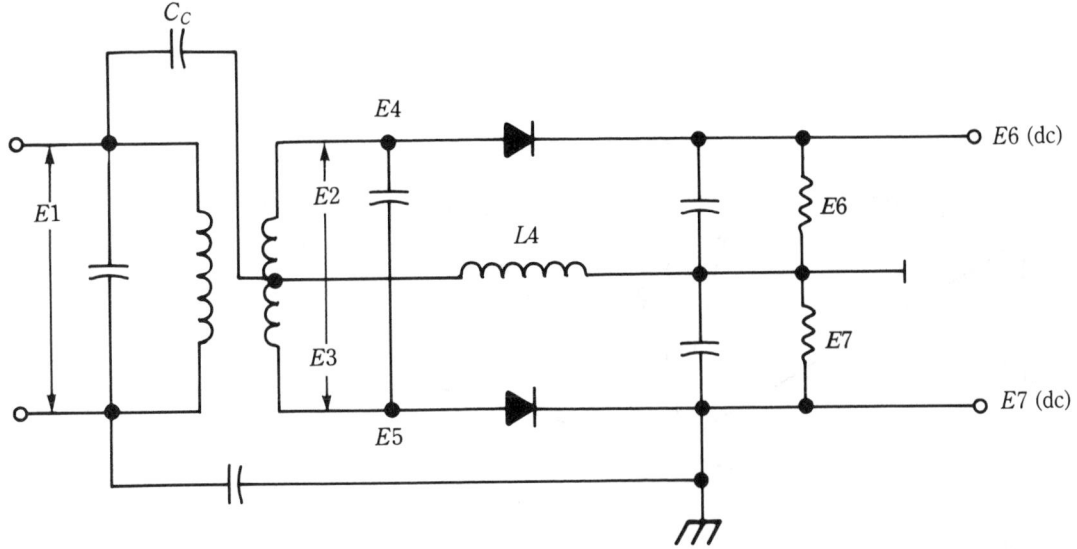

9-13   Foster-Seeley discriminator.

Foster-Seeley discriminator

The Foster-Seeley discriminator circuit for FM demodulation has been in use for several decades. A typical circuit is shown in Fig. 9-13. In it the secondary circuit of the input transformer is adjusted so that the audio output is zero at the center frequency. Phasor diagram Fig. 9-14A applies. Note that in Fig. 9-13 $E1$ is coupled to the transformer secondary center tap by C_C (low reactance) and appears across $L4$ (high reactance) providing equal magnitude $E4$ and $E5$ in the respective rectifier circuits. Therefore, for the conditions of this figure, dc outputs $E6$ and $E7$ are equal but of opposite polarities, which explains why the discriminator output is zero.

If the frequency of $E1$ is now raised, I_S will lag the induced secondary voltage E_S because the tuned circuit is now inductive, causing $E2$ and $E3$ to be retarded relative to their former positions (see Fig. 9-14B). Since $E4$ is now larger than before, and $E5$ is smaller, the output voltage is positive.

Likewise, as presented in Fig. 9-14C, if the frequency is reduced, I_S will lead E_S because the circuit is capacitive, and $E4$ and $E5$ will again be unbalanced, but in the opposite direction, to provide a negative discriminator output voltage.

One major disadvantage of the Foster-Seeley discriminator is its sensitivity to possible amplitude variations of the frequency-modulated waveform.

Ratio detector

The ratio detector of Fig. 9-15 closely resembles the Foster-Seeley discriminator except for its output circuit. The voltages appearing across $C3$ and $C4$ will be equal for $f = f_C$ so that no signal voltage appears across the output terminals X, and Y, and the voltage across large capacitor $C5$ is zero. If the frequency is increased, the voltage across $C3$ will exceed that across $C4$. However, since the frequency change will be occurring at an

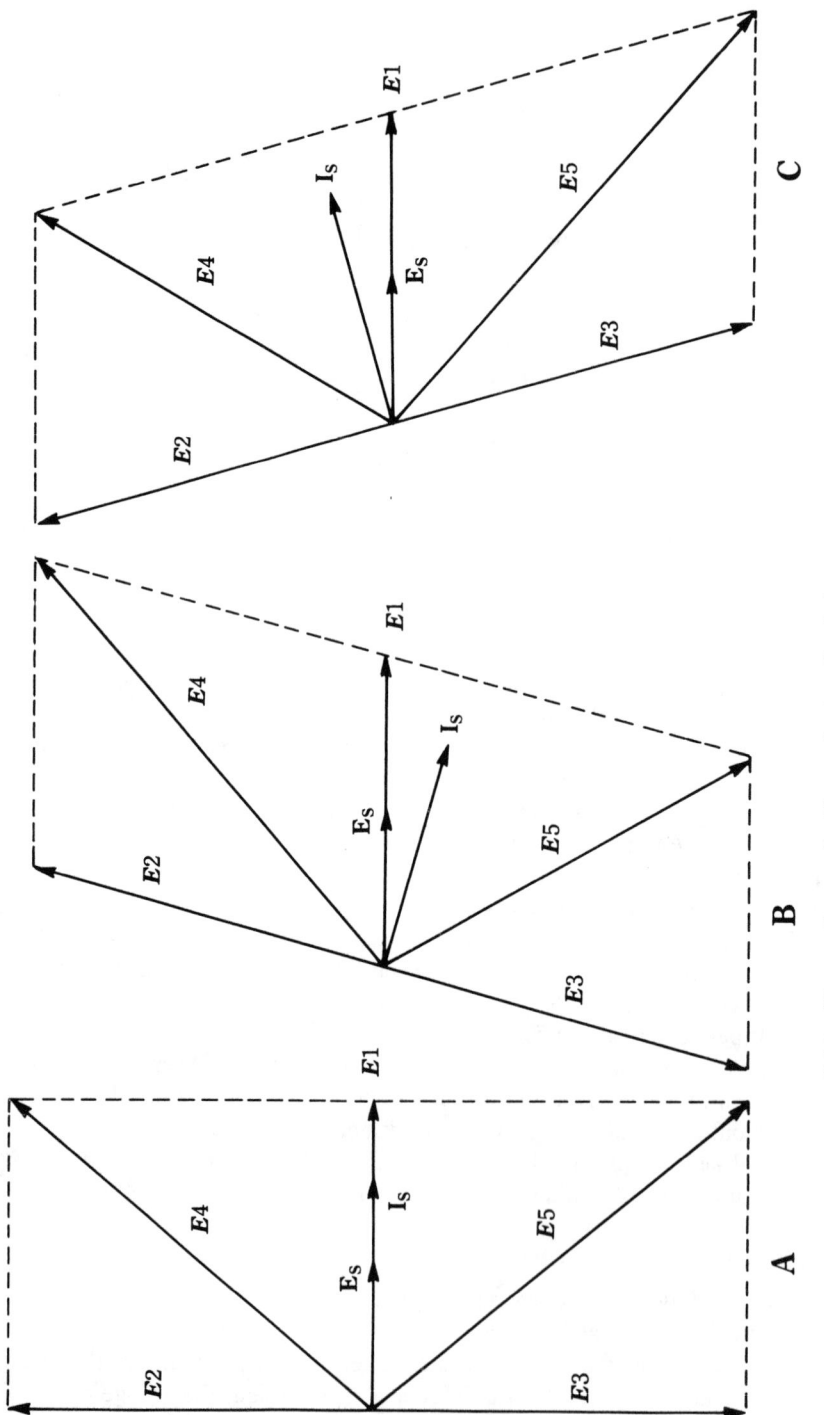

9-14 Phasor diagram for Foster-Seeley discriminator shown in A, B, and C.

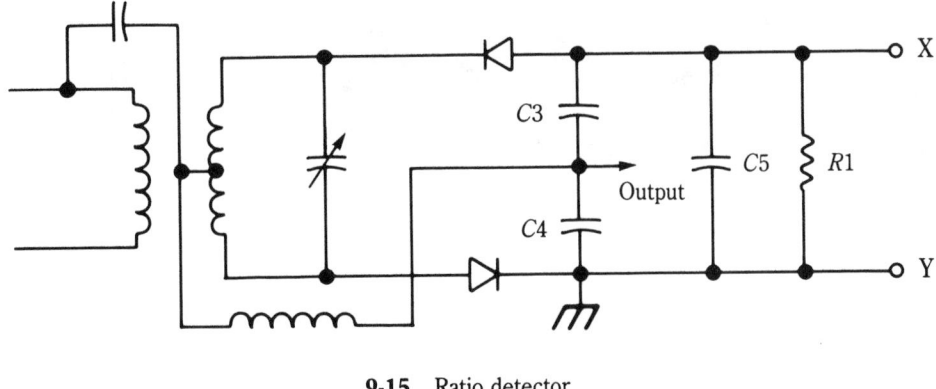

9-15 Ratio detector.

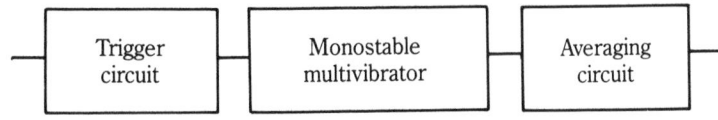

9-16 Pulse-averaging FM demodulator.

audio frequency rate, the voltage across large capacitor $C5$ will not be able to change and the voltage at the output Y will be positive. Similarly, if the frequency is decreased, the output will be negative.

Any rapid amplitude variations of the input frequency-modulated signal cannot change the voltage across $C5$, hence $C3$ and $C4$, and will not cause undesired output variations.

Other FM demodulation methods

The previous demodulation circuits have largely disappeared from modern equipment because of the availability of inexpensive ICs to replace discrete elements and the application of techniques not requiring tuned circuits having relatively expensive inductors and transformers.

Pulse averaging demodulator

A completely different approach to FM demodulation appears in Fig. 9-16. The trigger circuit produces a short pulse each time the frequency-modulated waveform crosses the zero axis. Each of these pulses triggers the monostable multivibrator, which is timed to remain in its unstable state for approximately one-half the average input period. The multivibrator then reverts to its stable state and awaits the next trigger pulse.

The output of the multivibrator is a string of constant-width pulses separated by smaller time gaps for positive swings of frequency, and larger time gaps for negative frequency swings (Fig. 9-17A). It follows that the average level for the pulse train is larger in the former case (see Fig. 9-17B).

A. Input

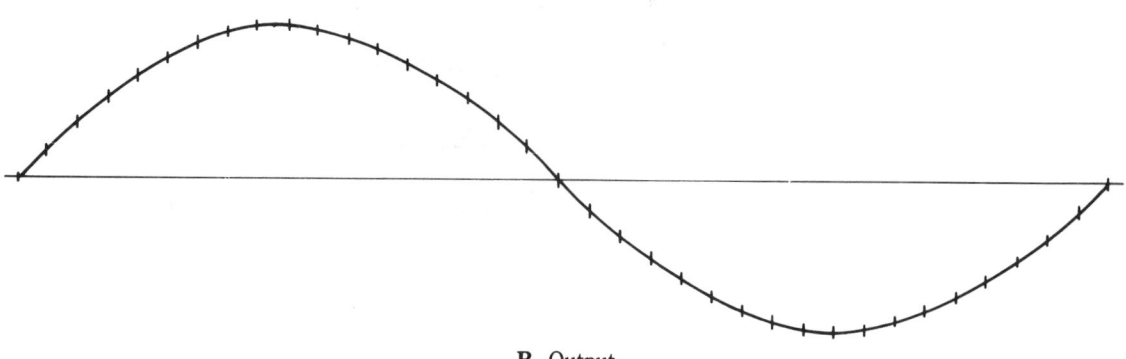

B. Output

9-17　Demodulation of FM by averaging circuit.

Quadrature detector

The quadrature detector supplies an output that is proportional to the phase difference of two coherent inputs. Its action is exemplified by the LM3089 FM receiver i-f system whose block diagram appears in Fig. 9-18. The quadrature detector proper is only a part of the IC.

Phase-locked loop FM demodulation

Operation of the phase-locked loop will be described more completely in chapter 10. For the present purpose the block diagram appears drawn in slightly different form in Fig. 9-19 to emphasize the intended function. However, it is basically no different than any other form of this versatile device. The explanation of the operation is simple. The servomechanism, which is the phase-locked loop, exists with the objective of matching the phase of the voltage-controlled oscillator (VCO) and that of the input signal, which here is the output of the i-f amplifier. Assuming that it meets the objective reasonably well so that the phase difference error is small at all times, you must then recognize that the frequency of the VCO indeed matches that of the frequency-modulated i-f carrier with correspondingly small errors at all times. You must then conclude that the VCO input stimulus accurately reflects the frequency deviation, hence the waveform of the modulating signal.

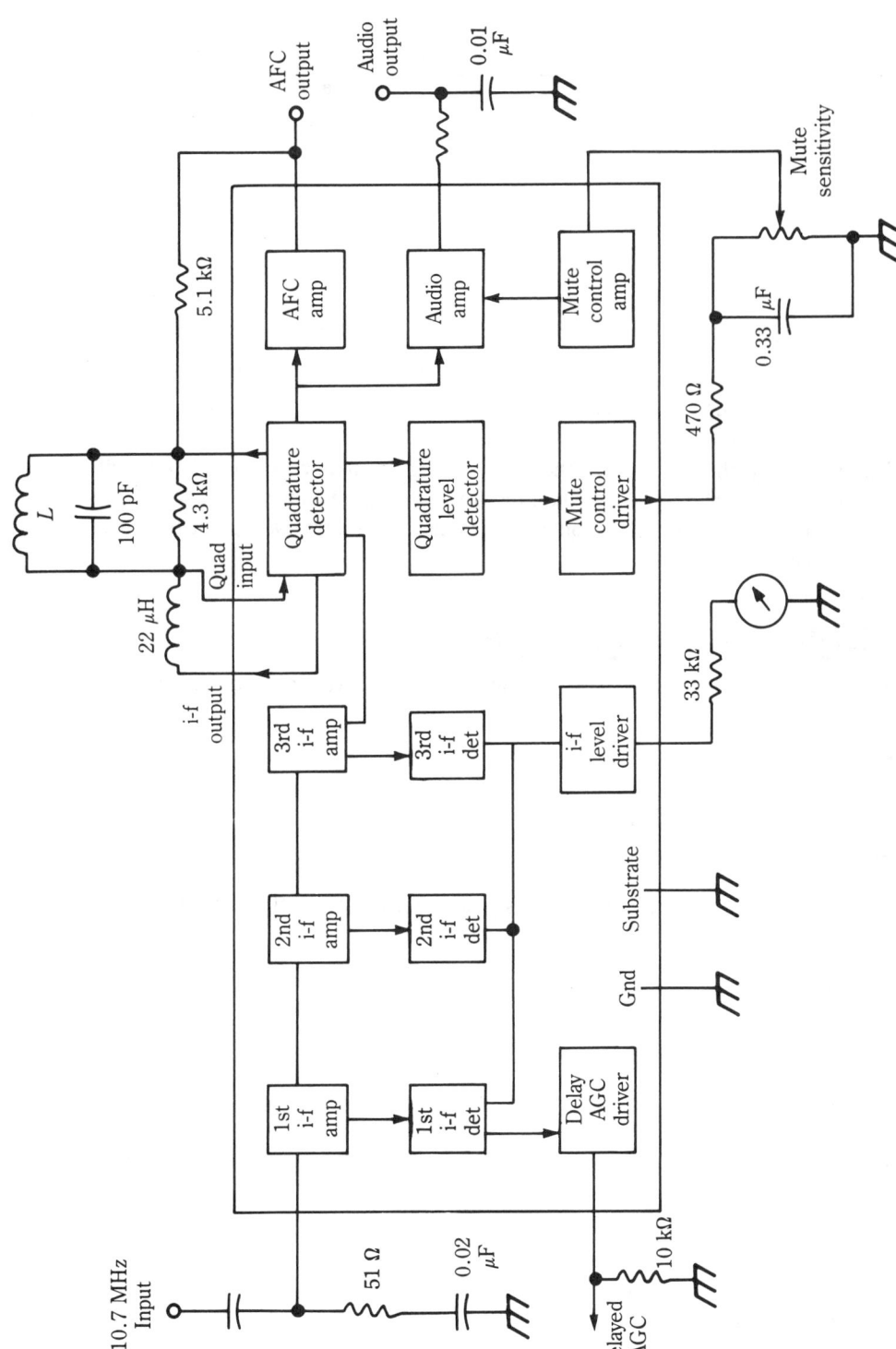

9-18 Simplified diagram for 3089 application.

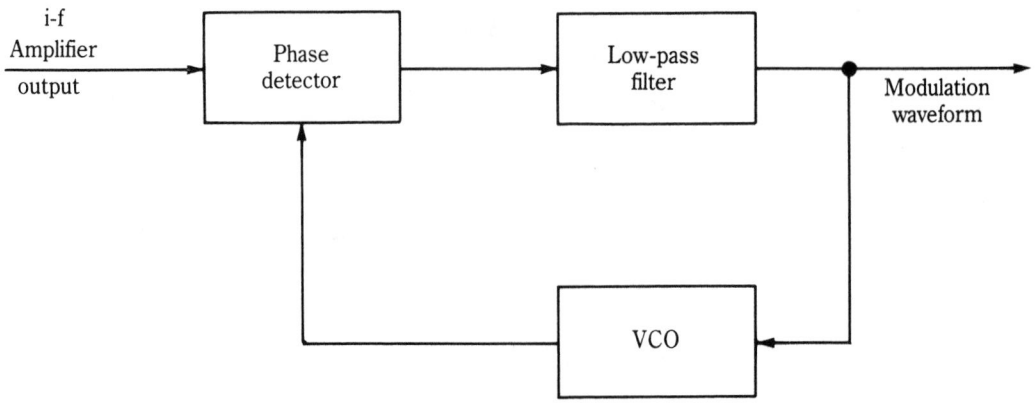

9-19   Application of phase-locked loop for FM demodulation.

This is exactly the case when the free-running frequency of the VCO exactly matches the i-f (carrier) frequency. When the free-running frequency of the VCO does not exactly match the i-f, a dc component must exist in the error voltage at the phase detector output.

Multiple-choice questions

1. Frequency modulation and phase modulation are collectively referred to as:

   A. Stereo.
   B. Angle modulation.
   C. FCC modulation.
   D. High fidelity modulation.
   E. Noiseless modulation.

2. In FM the change in carrier frequency is proportional to what attribute of the modulating signal?

   A. Angle.
   B. Frequency.
   C. Amplitude.
   D. Tone.
   E. Quality.

3. A louder sound, when generating the modulating waveform for FM, will cause a greater:

   A. Carrier amplitude.
   B. Angle amplitude.
   C. Distortion at the receiver.
   D. Frequency deviation.
   E. Both A and B.

4. If a positive change in modulation signal level of 200 mV will cause a positive frequency deviation of 10 kHz, what will be the frequency deviation for a negative change of 100 mV in the level of the modulating signal?

   A. 0.
   B. −5 kHz.
   C. +5 kHz.
   D. +10 kHz.
   E. This is not a valid question.

5. Referring to Question 4, if the first modulating signal has a frequency of 500 Hz and the second modulating signal frequency is 1000 Hz, the answer to the question will be:

   A. −10 kHz.
   B. −5 kHz.
   C. +5 kHz.
   D. +10 kHz.
   E. This is not a valid question.

6. A particular 15 kHz modulation tone results in a peak frequency deviation of 75 kHz. The modulation index is:

   A. 5.
   B. 15.
   C. 75.
   D. Both A and B.
   E. Insufficient information has been supplied.

7. A 15 kHz sine wave frequency-modulates an 88 MHz carrier. A sideband frequency will be found at:

   A. 87.970 MHz.
   B. 87.985 MHz.
   C. 88.015 MHz.
   D. Both B and C.
   E. All of the preceding choices.

8. For a carrier of 94 MHz, sideband frequencies of 94.005 MHz, 94.010 MHz, and 94.020 MHz are known to exist. Identify any other sideband frequencies.

   A. 93.980 MHz.
   B. 93.985 MHz.
   C. 93.990 MHz.
   D. Both A and C.
   E. All of the preceding choices.

9. Name one or more sources of noise bearing on electronic communications. (Make one choice only.)

A. Steam boiler.
B. Galaxies.
C. Both A and B.
D. Internal combustion engines.
E. Both B and D.

10. A central feature of noise is:

   A. Predictability.
   B. Unpredictability.
   C. Usefulness.
   D. Information content.
   E. Both B and C.

11. With FM, noise has a larger effect in corrupting certain ranges of modulating frequencies. They are:

   A. Low.
   B. Intermediate.
   C. High.
   D. All ranges.
   E. There is no dependence on frequency.

12. The transmitter technique adopted to reduce the noise effect of Question 11 is called:

   A. Noise masking.
   B. Antinoise.
   C. Noise killing.
   D. Derandomize.
   E. Preemphasis.

13. Deemphasis in the receiver in effect attenuates modulating signal components and noise in what frequency range?

   A. dc.
   B. Low.
   C. Intermediate.
   D. High.
   E. All ranges equally.

14. Name a device whose capacitance is deliberately made to be a function of the applied voltage.

   A. Varactor diode.
   B. UJT.
   C. SAW.
   D. Variable capacitor.
   E. Electrolytic capacitor.

15. A reactance modulator is one method of obtaining:

A. Indirect FM.
B. Direct FM.
C. Demodulation.
D. Low-frequency filtering.
E. Automatic gain control.

16. A device, now available in IC form, is useful for direct FM and as one element in the phase-locked loop. It is the:

A. AFC.
B. AGC.
C. Frequency counter.
D. VCO.
E. Reactance modulator.

17. A frequency change process, whereby the phase deviation and frequency deviation are multiplied by some fixed constant, is frequency:

A. Translation.
B. Multiplication.
C. Division.
D. Addition.
E. Subtraction.

18. In stereo FM a 38 kHz subcarrier is used for the L − R signal. It is derived from a pilot subcarrier of:

A. 9.5 kHz.
B. 19 kHz.
C. 38 kHz.
D. 76 kHz.
E. No correct choice is given.

19. This circuit has the function of demodulating the frequency-modulated signal. It is a:

A. AGC.
B. AFC.
C. Envelope detector.
D. Decoder.
E. Foster-Seeley discriminator.

20. Slope detection for FM has the disadvantage of response to:

A. Phase modulation.
B. AM.
C. Charge storage.
D. Overheating.

  E. Diode breakdown.

21. The ratio detector is superior to the slope detector because:

  A. It is less sensitive to phase modulation.
  B. It is less sensitive to noise spikes.
  C. It is less sensitive to interference-causing AM.
  D. No power supply is required.
  E. Both B and C.

22. One implementation of a pulse-averaging discriminator has:

  A. A free-running multivibrator.
  B. A crystal-controlled oscillator.
  C. A quartz crystal filter.
  D. A triggered multivibrator.
  E. Both B and D.

23. A 10% increase in the frequency of a constant-width pulse train should cause what change in its average value?

  A. −10%.
  B. −1%.
  C. 0.
  D. +1%.
  E. +10%.

24. Two different signals can be coherent if they:

  A. Have the same amplitude.
  B. Are both sine waves of different frequencies.
  C. Originate in the same physical equipment simultaneously.
  D. Have the same frequency.
  E. Both B and C.

25. A quadrature detector requires that:

  A. Four gates be provided.
  B. The inputs are coherent.
  C. The inputs are incoherent.
  D. The inputs are identical.
  E. Both B and D.

26. In a phase-locked loop, the VCO is the abbreviation for:

  A. Variable coherent output.
  B. VHF communication oscillator.
  C. Voltage-controlled oscillator.
  D. Vien-count oscillator (neutralized).

E. All of the preceding choices are nonsense and are incorrect.

27. The bandwidth of the phase-locked loop in Fig. 9-19 is controlled by:

    A. The phase detector.
    B. The low-pass filter.
    C. The VCO.
    D. The noise level.
    E. A microcomputer which does not appear in this simplified figure.

28. Select a cutoff frequency for the low-pass filter of Fig. 9-19 that is reasonably compatible with a standard FM broadcast application.

    A. 2 Hz.
    B. 20 Hz.
    C. 200 Hz.
    D. 2000 Hz.
    E. 20,000 Hz.

29. A deemphasis circuit is to be added to the circuit of Fig. 9-19. Where should it be placed?

    A. Between the phase detector and the low-pass filter.
    B. Immediately following the low-pass filter.
    C. At the point labeled: Modulation waveform.
    D. At the VCO input.
    E. Between the VCO and the phase detector.

30. Identify an advantage, or advantages, of a properly designed FM system.

    A. Relative immunity to atmospheric noise (lightning).
    B. Reduced bandwidth required.
    C. No noise of any kind.
    D. The noise figure is inversely proportional to the modulation index.
    E. Does not respond to galactic noise.

Basic problems

1. When the modulating voltage changes from 200 mV to 300 mV, the frequency deviation will change from 25 kHz to what new value?

2. Estimate the number of significant sidebands for $m_f = 5$ using Carson's rule.

3. Briefly describe the effect of noise on an AM waveform.

4. One source of noise is particularly ineffective against FM transmission. Why?

5. What is the significance of the noise triangle to FM?

6. State exactly the preprocessing conditions that allow indirect FM.

7. Illustrate how direct and indirect FM are related by calculating the frequency deviation equivalent to a 1 radian phase deviation at 1000 Hz.

8. The primary of the transformer of a Foster-Seeley discriminator has a leakage inductance of 10 μH. Determine the required parallel capacitance when the receiver is tuned to 88 MHz.

9. A deemphasis circuit in a receiver has a 1000 ohm resistor in series with a capacitor. If the time constant is 75 microseconds, how much capacitance is required?

10. The circuit of Basic Problem 9 is redesigned to employ the same resistor and an inductor. Calculate the value of inductance required for the same time constant.

Advanced problems

1. A 5 kHz modulating signal creates a maximum frequency deviation of 37.5 kHz for an FM system. If the frequency of the modulating signal is changed to 10 kHz and the amplitude is halved, what is the new frequency deviation?

2. The maximum frequency deviation authorized for FM broadcast radio is 75 kHz. Disregard any audio frequency bandwidth limitations imposed by current broadcast transmitter and receiver equipment and establish a theoretical upper limit on the modulation frequency.

3. A bandwidth of 100 kHz is required for a particular FM transmission in which the modulation index is 4. Assuming only enough bandwidth to accommodate the significant sidebands, what is the modulating frequency?

4. In Advanced Problem 3, what will be the required bandwidth if no change is made other than to make the modulating frequency equal to 5 kHz?

5. What is the difference between direct and indirect FM?

6. In the section, Indirect FM, the roles of frequency multiplication and translation are demonstrated. Repeat the calculations assuming a maximum phase deviation of 0.4 radians is obtainable.

7. A VCO changes frequency at the rate of 1 kHz/mV. What will be the constant error voltage for a tracking phase-locked loop if the 100 MHz transmitter frequency is in error by 0.0001%?

8. If the phase-locked loop of Advanced Problem 7 is used for FM demodulation, determine the instantaneous input to the VCO corresponding to a maximum frequency deviation of 75 kHz.

9. A preemphasis circuit has a time constant of 75 microseconds. Assume that the output of the preemphasis circuit is taken across an inductor and resistor placed in series with a current source. At what frequency will the output be 3 dB above the response at the low frequency?

10. A triggered multivibrator running at 750 kHz provides a train of 0.5 microsecond, 1 V amplitude pulses. Determine the changes in average level of the pulse train for:

    A. A negative frequency deviation of 75 kHz.
    B. A positive frequency deviation of 75 kHz.

10

Transmitters

IN THIS CHAPTER THE TRANSMITTER IS VISUALIZED AS A SYSTEM CONTAINING the reference carrier frequency source, the means for frequency multiplication and translation, the modulator, and the monitoring and control, in addition to the low-level and buffer amplifiers, filters, and the power amplifier. These parts of the transmitter must be compatible in the sense of being capable of being interconnected. Moreover, the selection of frequency multiplication and translation amounts is not only related to the frequency of an accurate and stable source, but also to the modulation method for FM. In keeping with current equipment development trends, the frequency synthesizer source is assumed.

The following sections are included in this chapter:

- Frequency conversions
- Frequency synthesis
- CW transmitters
- FM transmitters
- AM transmitters
- Broadcast station operations

Frequency conversions

Frequency translation is achieved when the spectrum is shifted without any other change in the spectrum. This means that the relative spectral frequency component amplitudes and phases are not altered by the shift. A balanced modulator, followed by a filter to pass the desired sideband and to eliminate the undesired sideband, together with other spurious products, performs frequency translation in the sense intended here. Figure 10-1 illustrates the phenomenon of frequency translation for the baseband, i.e., the totality of

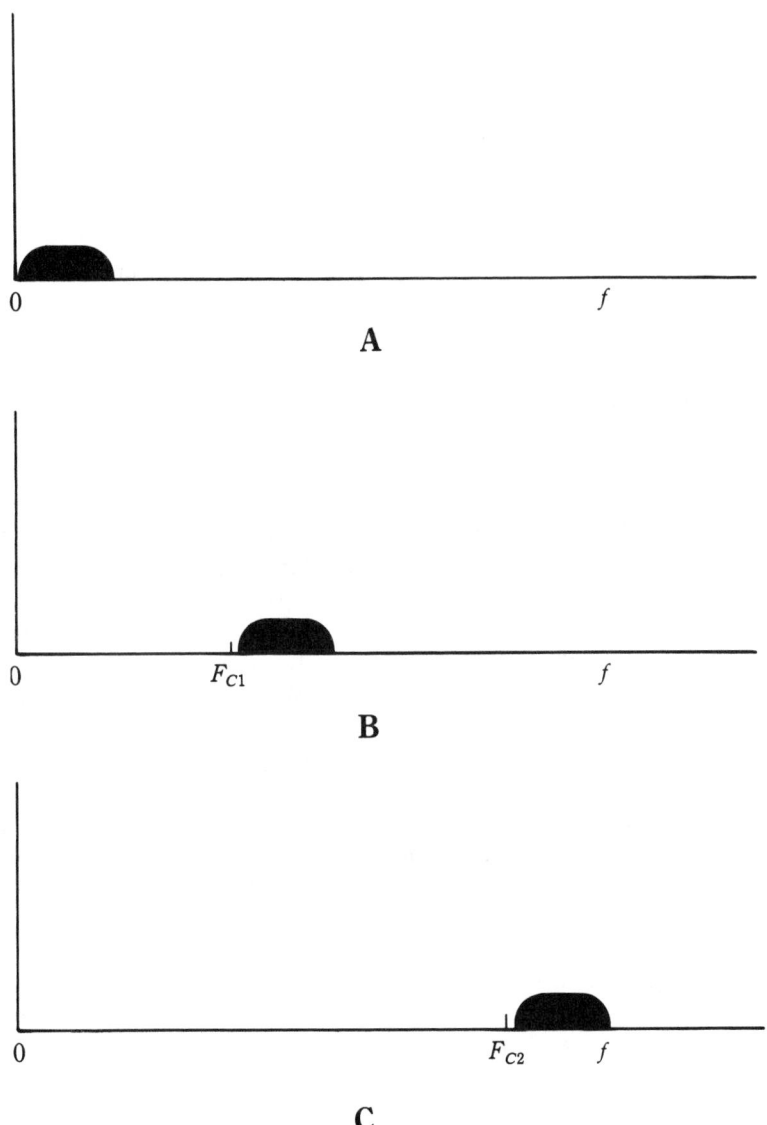

10-1 Frequency translation.

modulating frequency components shown as the spectrum in Fig. 10-1A. In Fig. 10-1B this same spectrum has been translated an amount f_{C1}, and an amount f_{C2} in Fig. 10-1C. The frequency translation from Fig. 10-1A to Fig. 10-1B or to Fig. 10-1C is referred to as *up-conversion* as is the translation from Fig. 10-1B to Fig. 10-1C. Frequency translation causing movement of the spectrum in the opposite direction is *down-conversion*.

Mathematically, frequency translation is easily illustrated by taking the product of the spectral components to be translated and a sine wave having a frequency equal to the amount of translation desired. The lower sideband produced by this process must be removed by filtering in order that the translation be completed as illustrated in Fig. 10-1. Other translation circuits, such as balanced modulators, can employ switching at the rate of the single translation frequency so that the spectral components to be removed by filtering will include more than the one sideband. This occurs because switching is equivalent to multiplication by a sine wave of the switching frequency and all of its odd harmonics.

While down-conversion is of interest as a means of demodulation, you are principally interested in translation as one process in a transmitting system. To repeat, a key feature of frequency translation is the preservation of amplitude modulation and angle modulation.

Frequency multiplication

Frequency multiplication, unlike the ideal function multiplication just described, requires a nonlinear circuit element. A sine wave of a particular frequency applied to such a device causes the generation of components having harmonics of the original frequency. Usually, the nonlinearity is chosen to emphasize one harmonic. For example, a square-law nonlinear element has a strong second harmonic output. It is necessary to filter the output of the other, undesired frequency components for the purpose of eliminating them.

Before examining some practical frequency multipliers in more detail, it is important to recognize their impact on a modulated carrier. In knowing that any waveform passing through a nonlinear circuit element is distorted, you can immediately conclude that amplitude modulation can also undergo distortion with frequency multiplication. However, the effect on angle modulation is both simple and useful. Both the effective phase and frequency deviation are multiplied by the same integer multiplication factor as is the frequency.

A class-A amplifier constructed with a BJT can deliver a significant harmonic component to a resonant load tuned to that harmonic frequency. Without changing the operating point, the amplifier can be overdriven into cutoff and saturation to establish a large harmonic content in the output current. FETs having square-law transfer characteristics can be the basis of frequency doublers.

Class-C amplifiers have strong harmonics, especially for small conduction angles, which can be adjusted to maximize the harmonic content for the frequency desired. A simple class-C frequency multiplier is shown in Fig. 10-2. The dc base voltage of the

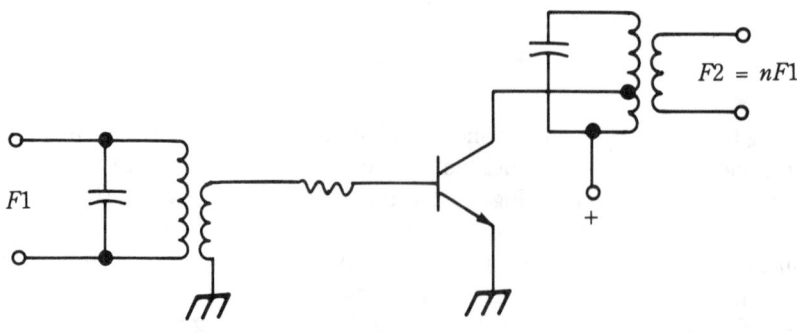

10-2 Frequency multiplier stage.

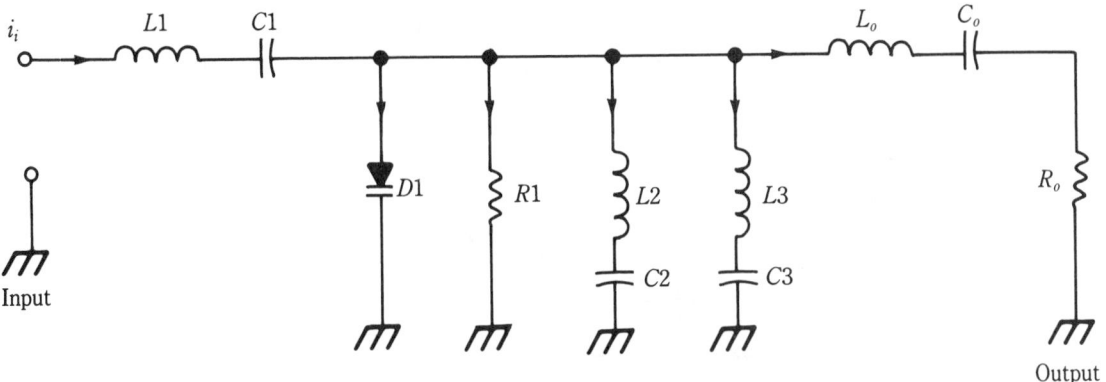

10-3 Frequency multiplier with varactor diode.

transistor is zero so that base current conduction, limited by the series resistor, will occur for somewhat less than one-half cycle. The output tank circuit is tuned to an integral multiple of the input frequency.

Diodes can be used effectively to multiply the frequency because of their inherent nonlinearities. Figure 10-3 illustrates a multiplier using a varactor diode. Series resonance is employed for the input circuit ($L1C1$), the output circuit L_oC_o and to the shunt circuits $L2C2$ and $L3C3$, which are respectively tuned to strong harmonics not wanted in the output. In this application a varactor diode is used.

The circuit configuration influences which harmonics are most effectively produced for a frequency multiplier. For example, a class-B push-pull amplifier is normally intended as a linear amplifier with a minimum of harmonic content in the output, so it would be a disastrous choice for a frequency multiplier. The class-C amplifier drives relatively short pulses through an output resonant circuit. These pulses are rich in the fundamental frequency component and the harmonics, which suggests that a tripler design, for example, will be very effective.

Recognizing the advantage of having a pair of pulses produced for each fundamental cycle, early designers settled on the "push-pull" frequency doubler illustrated in Fig. 10-4. You must look at the circuit, as well as its name, to avoid confusing it with the more familiar push-pull amplifier. In the figure you should recognize that the vacuum-tube plates are connected together. The grid excitation then causes conduction of the tubes to alternate causing two current pulses to pass through the output tank circuit for each cycle of the input signal. Such a pulse train has a large second harmonic content, to which the tank is tuned, and odd multiples of the second harmonic, i.e., $2f$, $6f$, $10f$, etc., where f is the input excitation frequency.

Frequency division

Frequency division, unlike frequency translation and frequency multiplication, is not used directly in transmitters to attain a specified modulation index for the modulated carrier signal at the actual rf transmitted. It is, however, one technique used in frequency synthesizers which are commonly found today in the equipment for the generation or measurement of the carrier frequency.

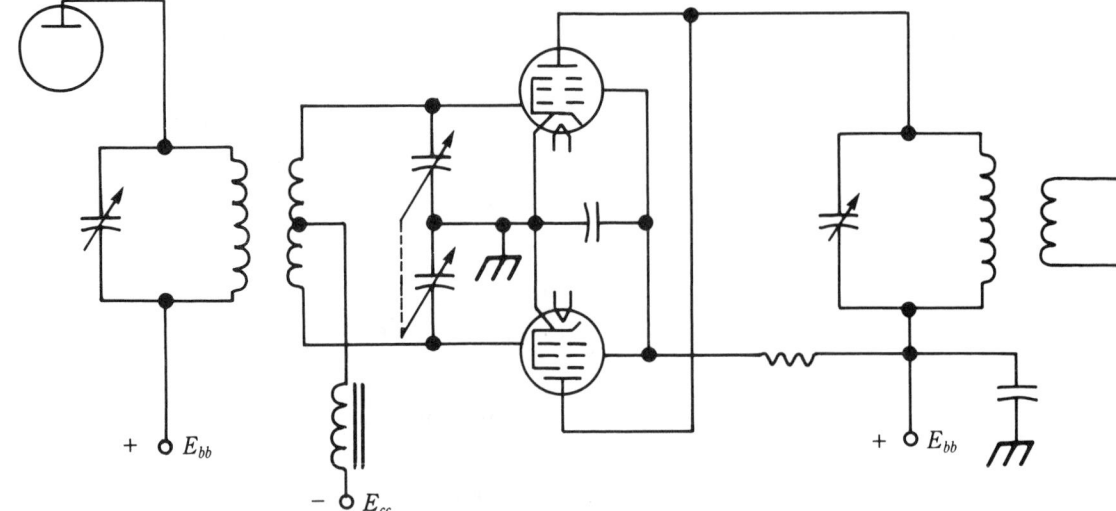

10-4 A push-pull frequency doubler.

As the name implies, frequency division causes a signal of one given frequency to produce another whose frequency is that of the first divided by an integer (whole number, N, i.e., f/N). Frequency division is easily accomplished with digital counters. The subject of digital counters is covered more completely in chapter 7, but a brief explanation will be offered here. First consider a decade counter such as is incorporated in the units position of an automobile odometer (mileage meter). As the automobile travels, the indicated mileage will change from 0 to 1, then 2, 3, 4, 5, 6, 7, 8, 9, and will then return to 0 as the count in the next (tens) position is incremented by 1. An electronic decade counter behaves in a similar way in counting pulses. It cycles from 0 to a count of 9 and returns to 0. It can be made to produce one output pulse for each ten input pulses. Therefore, it is a frequency divider with $N = 10$. The decade counter is also termed a *modulo ten counter.*

Binary counters are modulo 2 for one flip-flop, modulo 4 for two flip-flops, modulo 8 for three flip-flops. More generally, the binary counter is modulo N with $N = 2^F$ where F is the number of flip-flops. The addition of feedback through logical circuit elements can cause N to be less than 2^F but usually greater than 2^{F-1}. A modulo N counter can be the basis of an f/N frequency divider in which N is any positive integer.

Frequency synthesis

Frequency synthesis is the process of generating a sine wave having one of a large number of controllable frequencies based on those generated by one, or a relatively small number of fixed-frequency reference oscillators. Two classes of frequency synthesizers are recognized. A direct synthesizer uses mixing, multiplication, and filtering to obtain the desired frequency. An indirect synthesizer incorporates phase-locked loops containing VCOs from which the sine wave of the desired frequency is obtained.

Direct frequency synthesis

A particularly simple form of frequency synthesizer is illustrated in Fig. 10-5. A switchable group of ten crystals having frequencies, for example, of 20.000 MHz to 20.450 MHz, in steps of 50 kHz controls the output frequency of oscillator 1. Additionally, a second group also has ten crystals spaced at 5 kHz intervals from 25.000 to 25.045 MHz controls the output frequency of oscillator 2. Following the mixing of the oscillator outputs, a filter selects the sum frequency and rejects the other mixer products.

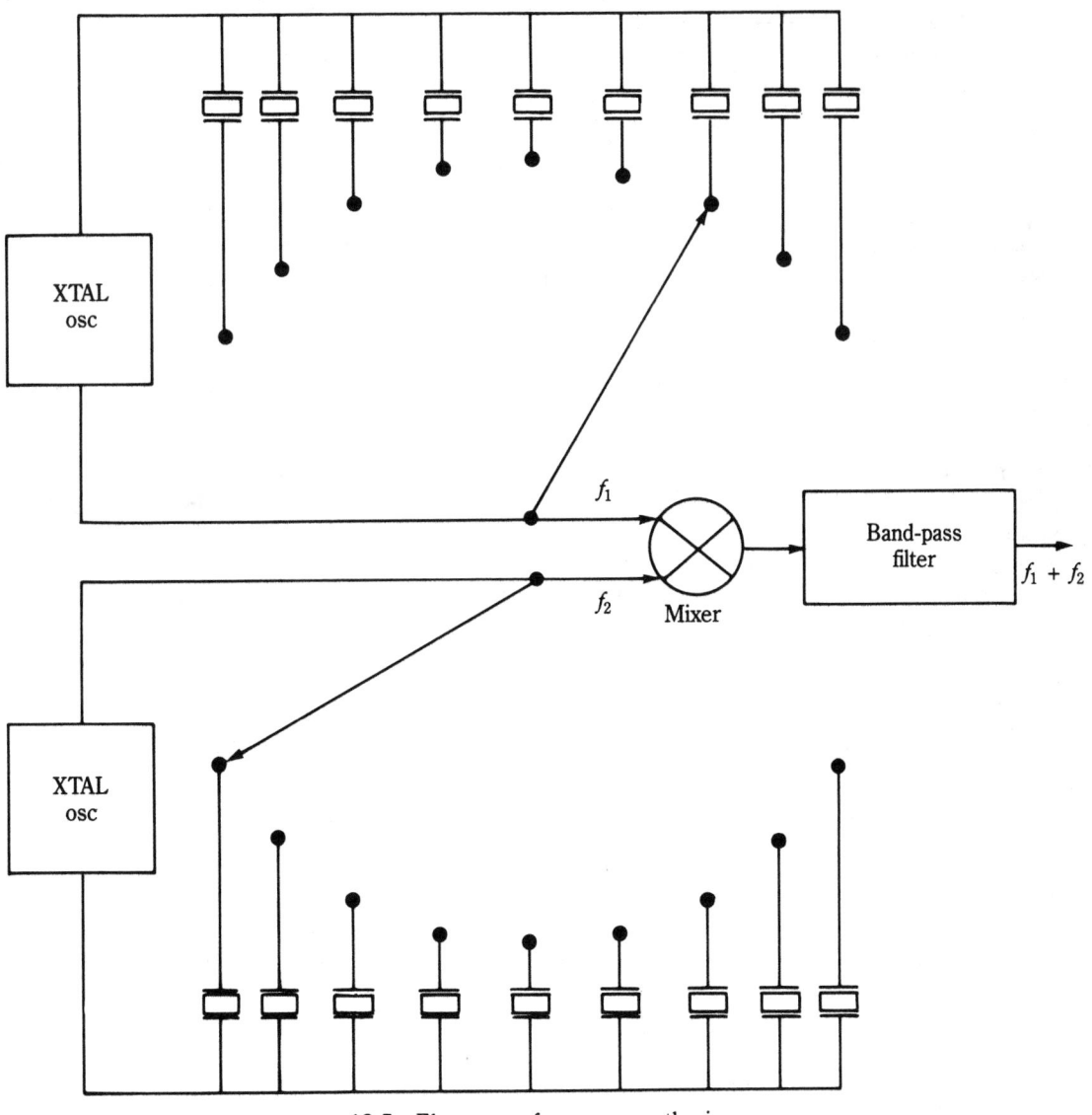

10-5 Elementary frequency synthesizer.

The phase-locked loop

Before launching into the indirect frequency synthesizer, a subject which can occupy volumes, it is first necessary to describe the nature of the phase-locked loop and its basic constituent parts, another subject of considerable breadth and depth. Since a great deal written about the phase-locked loop is misleading, if not erroneous, some care will be taken in emphasizing not only the correct points, but also some warnings about the limitations of this necessarily brief treatment.

A rudimentary phase-locked loop circuit is presented in Fig. 10-6. A reference V_R of frequency f_R is compared with the output of the VCO designated as V_O with frequency f_O. After lock has been achieved, any fixed relative change of phase of the two inputs will cause an error voltage to be generated by the phase detector. This error voltage will, in turn, cause an adjustment of f_O, which will reduce the error to zero after which there will be an exact match of f_O and f_R. Any drift of either oscillator frequency caused by temperature changes, aging, etc., will result in a fixed phase error after the two frequencies have been exactly matched.

Stated another way, with an ideal condition in which f_R and f_O are exactly stable and exactly matched with the VCO free-running, the phases will also be exactly matched with zero error voltage. For a more realistic assumption of a stable situation with matched frequencies, there will exist a phase error and error voltage. Both can be small by design. Actually, the frequencies and phases will be exactly matched only momentarily in a practical, dynamic application, so the error voltage will be constantly changing in supplying directions to the VCO on what to do to reduce the errors.

Two basic varieties of phase detectors are used for the PLL: Linear (analog) and digital. This last designation is a common one, but questionable. A double-balanced modulator, such as the MC1496 IC, can be used as a linear phase detector. It produces an error voltage which is a sine function of the phase error. An exclusive-OR gate will form a phase detector when V_R and V_O are converted to square waves. With no phase error, the gate output is zero; and with 180° (pi radians) phase error, the error voltage is maximum as shown in Fig. 10-7. The limited, linear range, π radians, and the repeatable pattern of the output error voltage of this phase detector causes some problems including the locking at undesirable VCO frequencies in the frequency synthesizer applications not yet fully introduced. A more complex phase detector, composed solely of logic will have filtered output in the form graphed in Fig. 10-8. The polarity of error voltage must correspond to that of the phase error so that it is impossible to achieve lock for other than correctly matching frequencies.

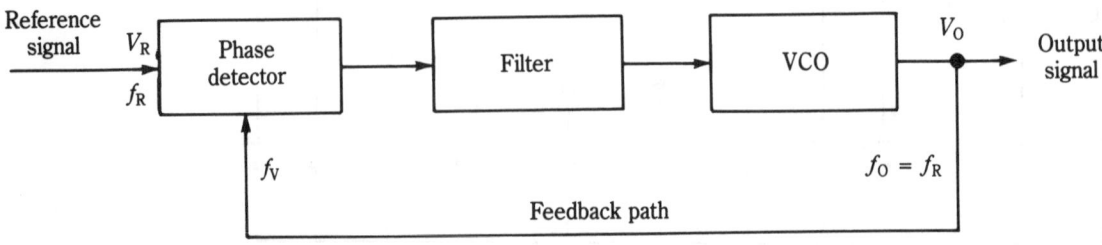

10-6 Block diagram of phase-locked loop.

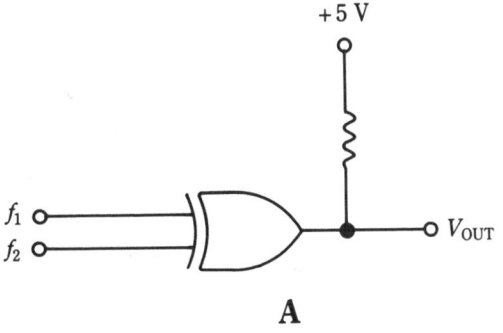

A

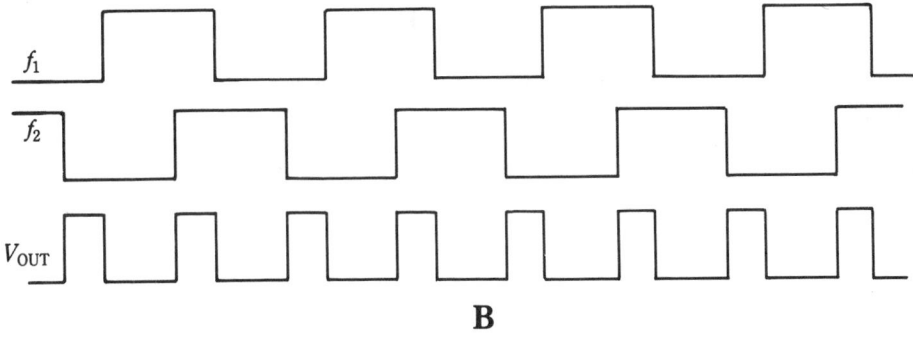

B

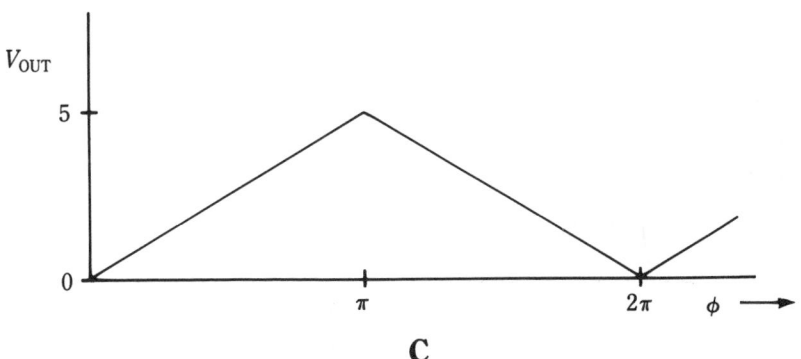

C

10-7 A. Exclusive-OR gate phase detector. B. Timing diagrams. C. Average dc output.

A low-pass filter following the phase detector removes the harmonic components of f_R resulting from the pulse train output (see Fig. 10-9). It also contributes to the spectral purity of the VCO, i.e., the VCO output is less contaminated with phase noise and other noise corruptions that cause the waveform to depart from an exactly periodic wave.

A multivibrator-type VCO will provide the correct waveform for the digital phase detector. The MC4024 IC will function over a number of frequency ranges, as deter-

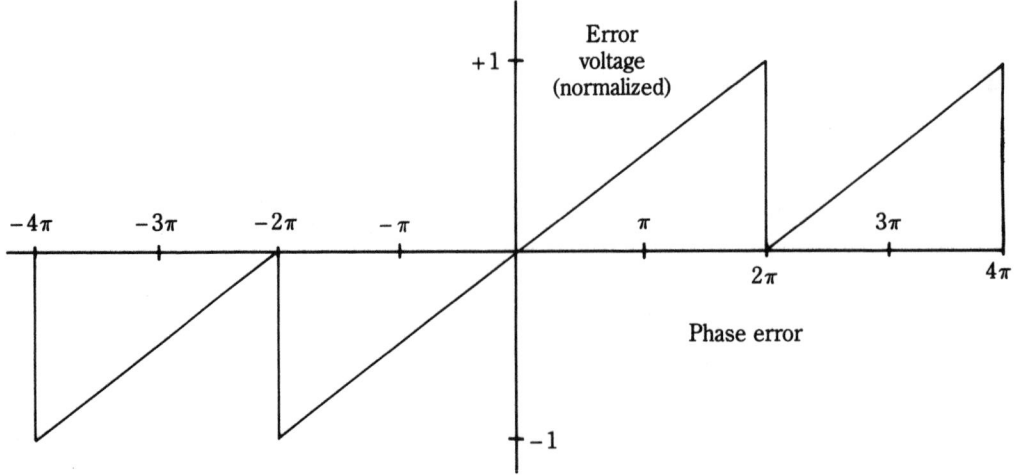

10-8   Digital phase-frequency detector.

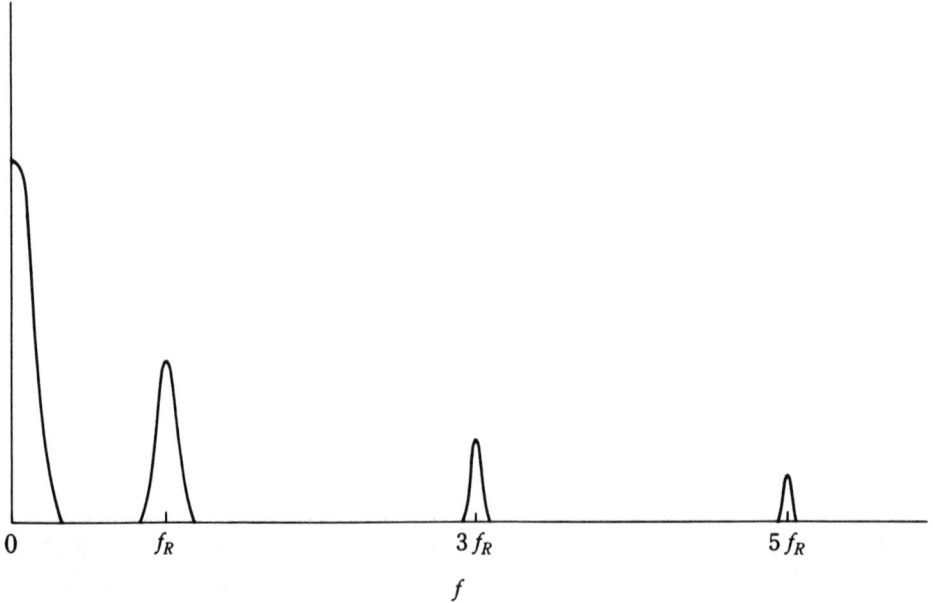

10-9   Phase detector output spectrum.

mined by the external timing resistance and capacitance combination from about one hertz to about 30 MHz and has better than a 2:1 linear frequency variation controlled by the input voltage for any range selection made in the design.

Figure 10-10 shows a PLL in which a programmable counter has been inserted in the feedback path so that $f_O = Nf_R$. Since the N can be changed by programming, a large number of output frequencies, with f_R adjacent frequency spacing, are available in what

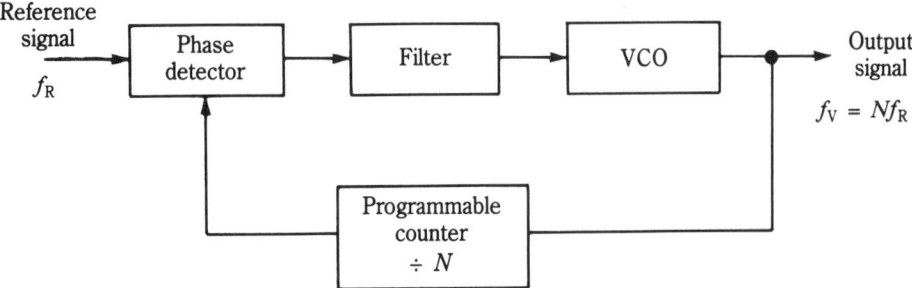

10-10 Exact frequency multiplier using PLL.

you can now call a *frequency synthesizer*. For maximum flexibility, f_R should be relatively small, perhaps 1 kHz to 100 kHz, depending on the application, while N is large. It should be emphasized that the accuracy of f_O corresponds exactly to that of f_R which, in turn, depends on the accuracy of the crystal oscillator from which it is derived.

Figure 10-11 presents all of the major elements of an indirect PLL frequency synthesizer. A crystal oscillator, typically operating at a frequency of the order of 10 MHz, establishes the frequency accuracy of the synthesizer. The reference frequency counter in the figure reduces, by division, f_O to f_R. This establishes f_O as being an integral multiple of a convenient value for f_R/M.

There are a number of ICs specifically developed for crystal oscillator applications. Table 10-1 lists oscillator and other ICs available for phase-locked loop applications. The dual-modulus prescalers are counters which, typically, divide by two values P and $P+1$. When combined with two programmable counters performing division by Q and R respectively, with R less than Q, the effective division for the combination is $N = PQ + R$.

CW transmitters

The CW transmitter incorporates most of the basic principles of stable carrier frequency generation, frequency multiplication, power amplification, modulation, and power trans-

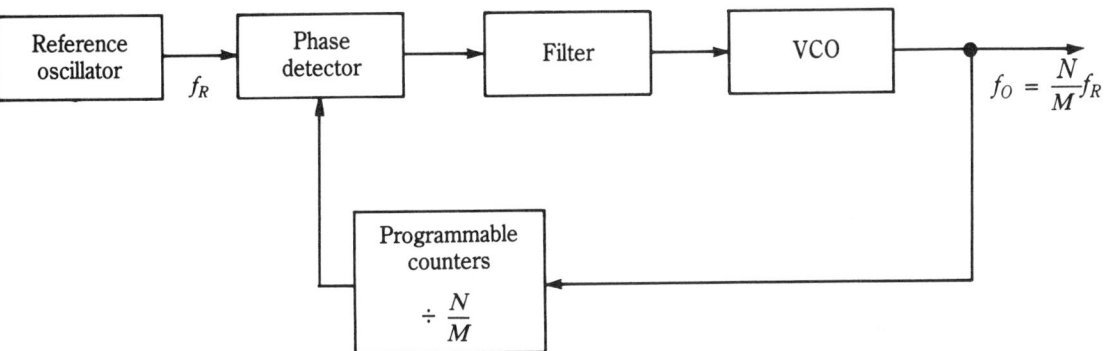

10-11 Principal features of frequency synthesizer using PLL.

Table 10-1. Components for Phase-locked Loop Applications.

Functions	Family	Frequency MHz (typ)	Power dissipation mW (typ)
Combination			
Digital Mixer Translator	MECL	250	470
Phase-Locked Loop	linear	0.5	825
Oscillators			
Crystal Oscillator	MECL	2 to 20	210
Crystal Oscillator	MECL	0.1 to 2	175
Voltage-Controlled Osc.	MECL	225	150
Volt.-Cont. Multivib.	MECL	150	150
Dual Volt.-Cont. Multivib.	MTTL	30	150
Phase Detectors			
Digital			
Phase-Frequency Det.	MECL	70	520
Phase-Frequency Det.	MTTL	8.0	85
Analog			
Mixer, Double Balanced	MECL	100	60
Modulator/Demodulator	linear	10	575
Control Functions			
Counter Control Logic	MECL	25	150
Prescalars			
High-Speed (/256, /64)	MECL	950	350
/4 Counter	MECL	1100	322
/4 Counter	MECL	1100	322
Two-Modulus (/5, /6)	MECL	500	350
Two-Modulus (/8, /9)	MECL	550	350
Two-Modulus (/10, /11)	MECL	600	350
uhf Type D Prescalar (/2)	MECL	500	-
Two-Modulus (/32, /33)	MECL	225	7 mA
Two-Modulus (/40, /41)	MECL	225	7 mA
Two-Modulus (/64, /65)	MECL	225	7 mA
Two-Modulus (/2, /(5/6), /(10/11), /(10/12)	MECL	200	500
Dual Type D	LS TTL	45	20

fer to the antenna system. A simple CW transmitter configuration is shown in Fig. 10-12. It is referred to as the *master oscillator power amplifier* (MOPA) transmitter. The circuitry for a vacuum-tube MOPA CW transmitter appears in Fig. 10-13. You should recognize that such a transmitter is still to be found in use, though obsolescent. The basic tuning principles are largely applicable, or adaptable, to modern CW transmitters having other master oscillator circuits and semiconductors replacing one or more vacuum tubes. A recommended sequence of steps for the tuning operation is as follows:

1. Turn off the power and remove the $B+$ lead from $V2$ so that it will not receive plate current when the power is again turned on, which is the next step. Detune the antenna circuit by reducing $C13$ to its minimum value. Also reduce the cou-

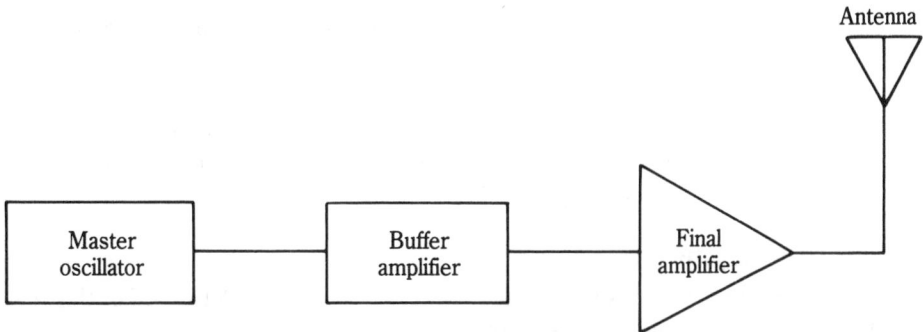

10-12 *Master oscillator power amplifier* (MOPA) transmitter.

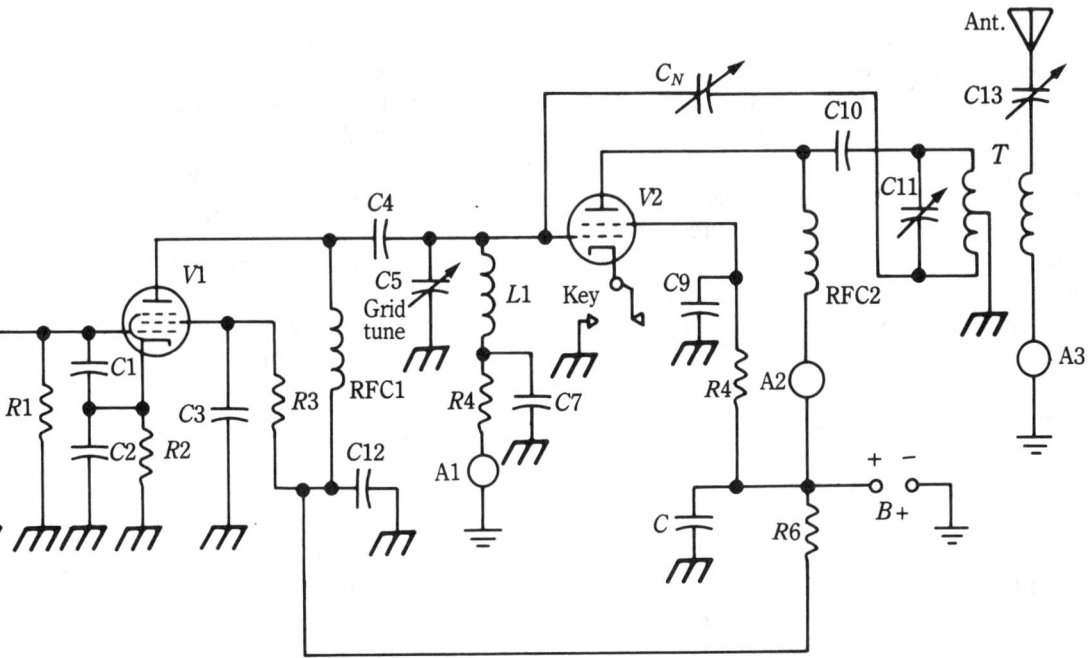

10-13 A simple MOPA (CW) transmitter.

pling (mutual inductance) between the windings of transformer T to a minimum value.

2. Turn the power on again and adjust $C5$ for a maximum reading of $A1$. Note that the crystal oscillator of $V1$ must be functioning for a nonzero reading of A1.

3. Couple a frequency meter or receiver to the primary of T after ensuring that the instrument is adjusted to the oscillator frequency. Neutralize the amplifier of $V2$ by adjusting C_N for a minimum reading of the instrument.

4. Turn off the power a second time, reconnect the $B+$ plate lead of $V2$ and turn the power on again. Now, adjust $C13$ for maximum rf antenna current as read by A3.

5. Increase the antenna coupling observing that the readings of A2 and A3 both increase.

6. Readjust C5 for a maximum reading of A1, C11 for greatest dip of A2, and C13 for maximum readings of A2 and A3. In this final sequence of adjustments, the antenna coupling should be varied to obtain the rated, or the desired value below the rating, for the plate current A2.

Transmitter keying

On-off (CW) keying is perhaps the simplest form of amplitude modulation. For this reason the problems associated with effective CW keying might possibly be overlooked, with a consequent reduction in communications efficiency.

There are at least two possible reasons why a CW transmitter might exceed the FCC limits for spurious emissions: The most obvious reason, parasitic oscillations in any transmitter stage, is not confined to CW modulation. A second reason results from harmonics of keying frequencies when the modulation waveform envelope is not properly shaped. Stated differently, the too rapid make and break of power in the keyed circuit widens the emission bandwidth with possible interference to adjacent channels. If the pulse rise time is too short, clicks can sometimes be heard from receivers tuned to other frequencies as the transmitter is keyed. Careful design to achieve an acceptable pulse rise (and fall) time in the keying circuit can be rendered ineffective by later amplifier stages in the chain that saturate. Excessive rise time, on the other hand, can produce an objectionable soft sound for the intended receiver.

Two other adverse effects might be experienced. If the oscillator is "pulled" by the keying action, the small, objectionable frequency variations are known as *chirps*. Inadequate power supplies can also result in unexpected variations in the envelope shape when the transmitter is keyed.

FM transmitters

Two basically different methods of FM were introduced in chapter 9: Indirect FM and direct FM. Each of these can be accomplished by several different techniques which will not be emphasized here. You will, however, examine the two methods in the context of the FM broadcast transmitter for which considerations other than modulation alone apply.

Indirect FM transmitter

Figure 10-14 summarizes one design for a broadcast FM transmitter with indirect modulation. It introduces some of the salient features for such a transmitter without pretending that the design is an optimum one.

The first block in the diagram is a phase modulator. A 100 kHz sine wave, derived from a precision crystal oscillator is phase modulated by the 30 Hz – 15 kHz information waveform which has been preprocessed by a filter whose response is inversely proportional to the frequency. Any modulating frequency component will then result in the same frequency deviation as any other component of a different frequency, but with the same amplitude—one requirement for frequency modulation.

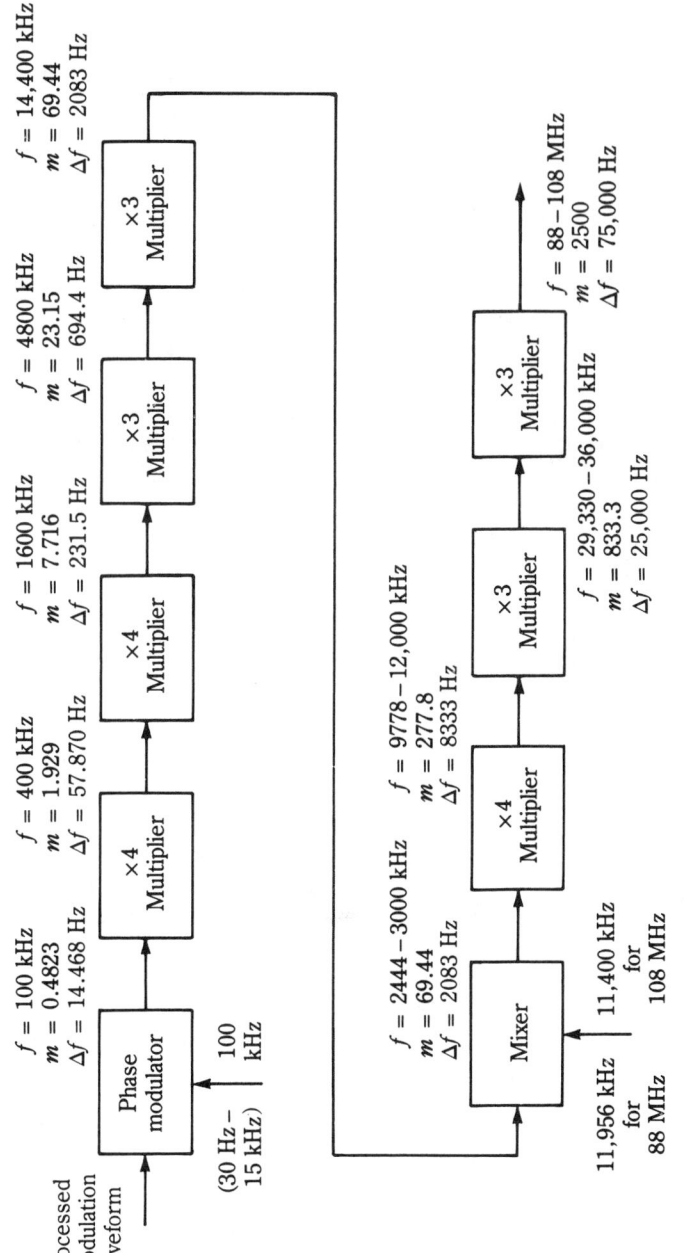

10-14 Design for broadcast indirect FM transmitter (shown for $f_m = 30$ Hz).

The maximum phase modulation is restricted to about 0.5 radian to preserve linearity. Those stages following the phase modulator perform two basic functions:

1. Multiplication, which increases the phase deviation, frequency deviation, and (FM) modulation index by the amount of the multiplication factor

2. Frequency translation to place the carrier at the correct point in the assigned rf frequency range. The reasons for parameter selection in this design will be more apparent as the functions of the parts are described.

Starting with the output of the phase modulator, the center frequency is multiplied successively from 100 kHz to 400 kHz, 1600 kHz, 4800 kHz, and 14,400 kHz. It is then mixed with a sine wave and translated to one of twenty different frequencies, which, after further multiplication, becomes an assigned vhf channel frequency. One frequency, corresponding to a final rf frequency of 88 MHz, is 11,956 kHz, yielding a difference of 2,444.4 kHz. The other frequency is 11,400 kHz yielding a difference frequency of 3,000 kHz, which, after multiplication by thirty-six, becomes 108 MHz. Figure 10-14 is incomplete in that a number of filters are required to fulfill the transmitter multiplication and translation functions, but are not shown. A filter is required after each multiplication and translation to eliminate the undesired products of the process. These band-pass filters must have sufficient bandwidths to pass not only the center frequency, but the side frequencies (sidebands, in general) corresponding to the modulating frequency and the modulation index.

Direct FM transmitter

In contrast to the indirect FM transmitter, the direct FM transmitter requires very little frequency multiplication. A second distinction in this comparison is the frequency modulation of a free-running oscillator and the necessity of having an *automatic frequency control* (afc) loop to maintain the carrier frequency within tolerance. For the most part, the technical principles involved have already been introduced so that it will not be necessary to cover all of them in detail.

Modification of the resonant frequency of an oscillator was formerly accomplished using a *reactance-tube modulator*. Briefly, the reactance tube plate and cathode terminals are placed across the parallel resonant circuit of the oscillator. The tube current is made to be in quadrature with its plate-cathode voltage because of a reactive element in its control-grid feedback circuit. The modulating signal, also applied to the grid, controls the tube transconductance and the plate current. Therefore, the reactance-tube modulator behaves as a reactive element whose value varies with the modulating signal. The net result is a corresponding variation of the resonant frequency with changes of the modulating signal level.

A more common method of direct FM today involves the use of a voltage-variable diode capacitor, more popularly known as a *varicap* or *varactor*. It will be recalled from chapter 9 that this diode is operated with reverse bias, the level of which affects the thickness of the depletion layer, hence the capacitance across the junction. Figure 10-15 illustrates how the frequency of a Colpitts oscillator is controlled for direct FM with the diodes. With no modulating signal input, the diodes are subjected to fixed reverse bias. A negative input signal will reduce the bias, decrease the width of the depletion layer, and increase the capacitance of each diode so that the resonant frequency will be decreased.

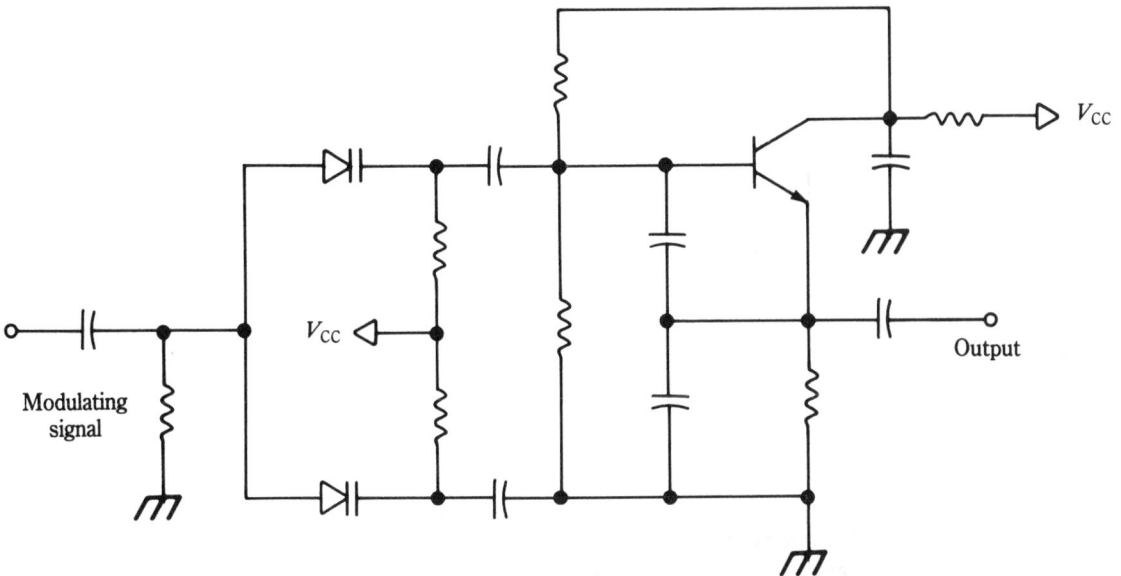

10-15 Direct frequency modulation of Colpitts oscillator.

A positive modulating signal input will increase the resonant frequency, by similar reasoning.

Figure 10-16 is a block diagram of an afc loop for a direct FM exciter. Frequency control is based on a TXCO (temperature-compensated crystal oscillator) operating at one frequency between 11 MHz and 13.5 MHz, this frequency being one-eighth of the assigned rf carrier frequency. This frequency is divided by eight before the reference signal is introduced to the phase detector.

The FM oscillator provides a frequency-modulated output with center frequency f_O in the range 22 MHz to 27 MHz. After division by sixteen in the feedback loop, its output is also introduced into the phase detector.

The output of the phase detector includes an error voltage and a number of other products that are removed by the low-pass filter. The error voltage is applied to the FM oscillator where it is instrumental in changing the center frequency in such a direction as to reduce the phase error. In addition, the modulating signal is also superimposed on the low-frequency error signal to establish the frequency modulation. Finally, multiplication by a factor of four brings the center frequency to the nominal rf value.

Broadcast FM transmitters

Table 10-2 summarizes specifications for a number of radio transmitters for the 88 – 108 MHz band. The rf output impedance is typically 50 ohms. Values given are for monophonic aural transmission.

Mobile narrow band FM

There are a number of frequency bands for which land mobile assignments are made. These include vhf near 150 MHz and uhf near 450 MHz and 950 MHz. Channel spacing

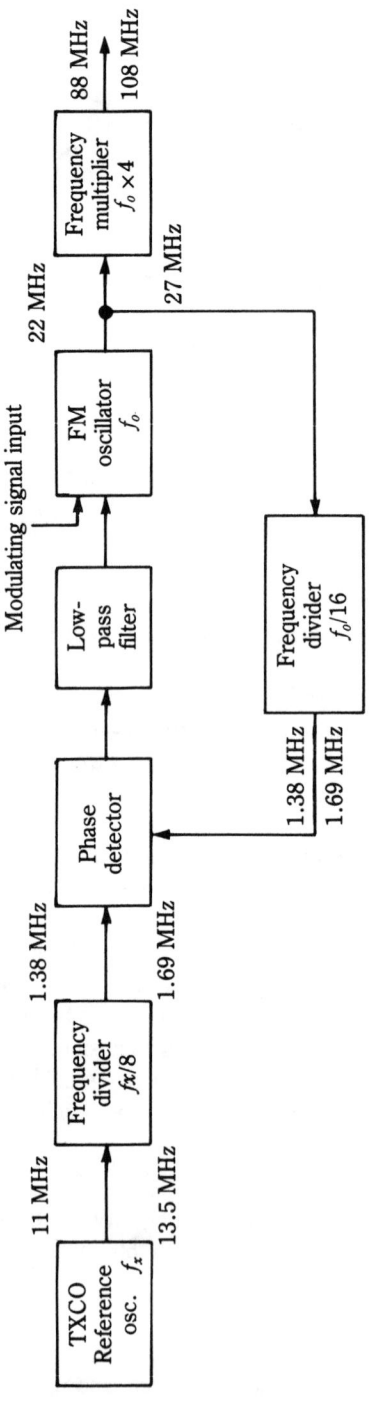

10-16 Simplified block diagram of afc loop for broadcast FM exciter.

Table 10-2. FM Broadcast Transmitter Specifications.

This equipment includes transmission systems for operation in the 88-108 MHz (CCIR band II) spectrum with frequency modulation. Intermediate power amplifiers are solid-state unless the model number is marked with a diamond (◆). All models are capable of operating in FM stereo service, with each manufacturing providing appropriate exciters and generators. All include an auto-recycle feature.

Manufacturer Model Series	AEG/Bayly S-3152 (S-3168)	Broadcast Electronics			Cemtys
		FM-10A	FM 35A	FM-3.5A	EF-5KX/80 (10KX/20KX)
RF output power	10 kW (3kW)	10kW	35kW	3.5kW	5kW (10kW/20kW)
Output connector (in.)	.....	3 1/8	3 1/8	1 5/8	1 5/8 (1 5/8, 3 1/8)
Modulation capability	<100kHz	±200kHz	±200kHz	±200kHz	±150kHz
Final amp device	Tetrode	4CX7500A tetrode	4CX20000C tetrode	4CX35000A tetrode	YL-1631 cer. tetrode
AF input connection	Unbalanced	Bal 600Ω	Bal 600Ω	Bal/unbal resistive	Bal/unbal 600Ω, 1kΩ
AF response	±0.2dB 40Hz-43kHz	±0.5dB, 30Hz-15kHz	±0.5dB, 30Hz-15kHz	±0.5dB, 30Hz-15kHz	±0.5dB 40Hz-15kHz
THD distortion	<0.4%	0.08%	<0.08% THD/IMD	0.08%, 100% mod	<0.25%
Pre emphasis	Any std value	Switch selected	Switch selected	75μs	25/50/75μs
AM noise figure	<-50dB unwtd	-45dB	-55dB	-40dB	<-55dB
FM noise figure (mono)	<-62dB unwtd	-72dB 75μs deemp	-72dB 400Hz	-75dB 75μs deemp	-70dB unwtd
Floor space (in.)	23.6W×39.4D	33.7×37.2	56.5×31.5	34.5×37.25	48.7×35
Separate power rack	No	Yes	Yes	No	No
Input ac power	380Vac 3φ (110/220)	240V 3φ	240V 3φ	240V 1φ	220-380Vac 3φ
Typical power factor	.....	>0.90	>0.94	>0.9	>0.93
Power consumption	20kVA (5.5kVA)	17.2kW	51kW	6.8kW	.....
System efficiency	.....	.....	68%	.....	.....
Fault logic architecture	.....	μP system	μP control	μP system	CMOS
μP diagnostics	.....	Yes	Yes	No	.....
	Other models 50W-1kW				

Manufacturer Model Series	Varian Continental TVT				
	816-2A◆ (817R-2A)*	817A	816R-5 (817R-5)	814R-1 (814B/815A)	814C solid-state
RF output power	20kW (40)kW	60kW	35kW (70kW)	2.4kW (4.3/5kW)	3.8kW
Output connector (in.)	3 1/8, 6 1/8	6 1/8 EIA male	3 1/8 (6 1/8)	1 5/8 EIA male	1 5/8 EIA
Modulation capability	±150kHz	±150kHz	±150kHz	±150kHz	±150kHz
Final amp device	4CX15,000A	4CX40,000G	9019/YC130 (2 tubes)	5CX1500B (4CX3500A)	Solid-state
AF input connection	Active bal	Active bal	Active bal	Active bal	Active bal
AF response	0.5dB, 20Hz-15kHz	0.5dB, 20Hz-15kHz	0.5dB, 20Hz-15kHz	0.5dB, 20Hz-15kHz	0.5dB, 20Hz-15kHz
THD distortion	0.08%	0.2%	≤0.1%	0.08%	.....
Pre emphasis	25/50/75μs	25/50/75μs	25/50/75μs	25/50/75μs	25/50/75μs
AM noise figure	-55dB	-55dB	-55dB	-40dB (-50dB)	.....
FM noise figure (mono)	-75dB unwtd (-72dB)	-72dB unwtd	-72dB (-75dB)	-75dB	.....
Floor space (in.)	72(160)W×28D	128W×40D	72(160)W×28D	35W×24D (×34D)	35W×34D
Separate power rack	No	No	Yes	No	No
Input ac power	200-250Vac 3φ	200-250Vac 3φ	200-250Vac 3φ	200-250Vac 1φ	200-250Vac 1φ
Typical power factor	>0.92	>0.95	>0.92	>0.94 (>0.92/...)	.....

Table 10-2. Continued.

Power consumption	33.5 (67.3)kVA	99kVA	57kVA (114kVA)	4.9kVA (7.6kVA/9.7kVA)
System efficiency	58% (57%)	64%	66%	57% (55%/...)
Fault logic architecture	Relay	μP based	Relay	Relay
μP diagnostics	No	No	No	Solid-state
				No

◆816-3 rated 25kW, 816-4 rated 27.5kW, 817R-1 rated 50kW, 817R-4 rated 55kW

Manufacturer	CSI Electronics		Elcom-Bauer		
Model Series	T-25-F◆	◆610A (625A/610C)	6100 (6300)	61000	605-FAT ◆(605C/603)
RF output power	27.5kW	10kW (25/10kW)	100W (300W)	1kW	5kW (5/3kW)
Output connector (in.)	3 1/8 EIA female	3 1/8	Type N	7/8	1 5/8
Modulation capability	±150kHz	±150kHz	±200kHz	±200kHz	±200kHz (±150kHz)
Final amp device	3CX15000A7	3CX10000A7 (15000A7)‡	Solid-state	Solid-state modules	4CX3500A (3CX3000A7)†
AF input connection	Balanced	Bal. 600Ω, unbal 10kΩ	Bal, unbal, 10kΩ	Bal, unbal, 10kΩ	Bal 600Ω, unbal 10kΩ
AF response	±0.5dB 30Hz-15kHz	±0.1dB 30Hz-100kHz	±0.1dB 30Hz-100kHz	±0.1dB 30Hz-100kHz	±0.1dB 30Hz-100kHz
THD distortion	0.5%	0.25%	0.08%, 0.1% IMD	0.08%, 0.1% IMD	0.08% (0.1% IMD)
Pre emphasis	75μs std	75μs	25, 50, 75μs	75μs	25, 50, 75μs
AM noise figure	-55dB	-50dB	-50dB sync, -70dB sync	-60dB	-40dB sync, -50dB nsync
FM noise figure (mono)	-65dB	<-65dB	-72dB	<-72dB	<-75dB
Floor space (in.)	76.5W×37.8D	70W×30D	Rack mount	23W×30D	23W×34D (35W×30D)
Separate power rack	No	(34W×26D 625 only)	No	No	No
Input ac power	190-460Vac 3φ	208-240Vac 3φ	110-220Vac 1φ	110-220Vac 1φ	440 or (220)Vac 3φ
Typical power factor	....	0.9	>0.9	0.9	0.9
Power consumption	36kW	21kW (38kW)	250W (600W)	2.7kVA	8.2kW (8.9/7.1kW)
System efficiency	....	80% final amp	Solid-state	....	....
Fault logic architecture	Relay/resistor	Relay	Solid-state	CMOS/TTL	Solid-state
μP diagnostics	No	No	....	No	DTMF rmt ctl (No)
	Models for 1kW, 10kW	‡4CX7500A for 610C	....	Fluorinert coolant	†4CX3500A for 603
					601A, 1kW, 602A, 2kW

Manufacturer	Energy/Onix			Harris Broadcast	
Model Series	MK 15	MK 1.5	MK 30	FM 5K1	FM 25K1◆
RF output power	15kW	1.5kW	30kW	5kW	25kW
Output connector (in.)	1 5/8	1 5/8	3 1/8	1 5/8	3 1/8
Modulation capability	±200kHz	±200kHz	±200kHz	±133%	±75kHz
Final amp device	3CX10000A7 triode	3CX3000A7 triode	3CX20000A7 triode	4CX35000A	4CX20000A
AF input connection	Balanced	Balanced	Balanced	Balanced	Balanced
AF response	0.2dB, 30Hz-100kHz	0.2dB, 30Hz-100kHz	0.2dB, 30Hz-100kHz	0.5dB, 30Hz-15kHz	0.5dB, 30Hz-15kHz
THD distortion	0.2%	0.2%	0.2%	0.2%	0.2%
Pre emphasis	75μs	75μs	75μs	25/50μs	25/50μs
AM noise figure	-60dB	-60dB	-60dB	-50dB	-50dB
FM noise figure (mono)	-70dB	-70dB	-70dB	-80dB	-80dB
Floor space (in.)	58×32.5	38.75×32.5	58×32.5	33W×34D	33W×30D
Separate power rack	Yes	No	Yes	No	Yes

Input ac power	230V 3ϕ	230V 1ϕ	230V 3ϕ	240Vac 3ϕ	240Vac 3ϕ
Typical power factor	. . . .	. . . .	. . . .	. . . .	0.95
Power consumption	24kW	3.1kW	48kW	10kW	40kW
System efficiency	. . . .	. . . .	. . . .	. . . .	. . . .
Fault logic architecture	μP	μP	μP	μP	μP
μP diagnostics	No	No	No	No	. . . .

Manufacturer	ITAME		ITELCO	LDL Comm./Larcan	Marconi Comm.
Model Series	Tauro 30 kW	Europa 600BT (100/300)	T-294 (T-234)	◆FMT25L	B6525 (B6526)
Rf output power	30kW	600W (100/300W)	55kW (12.5kW)	25kW	10kW (20kW)
Output connector (in.)	$3^{1}/_{8}$	Type N	. . . .	$3^{1}/_{8}$	$3^{1}/_{8}$ EIA
Modulation capability	> ±75kHz	> ±75kHz	±75kHz	±200kHz	±75kHz
Final amp device	Tetrode	Solid-state	Solid-state modules	9011 tetrode	8986 (9011 tetrode)
AF input connection	Balanced	Unbalanced	600Ω balanced	Balanced	Xfmr bal
AF response	0.5dB, 40Hz-15kHz	50dB, 40Hz-15kHz	0.2dB 20Hz-15kHz	±0.1dB, 30Hz-15kHz	±0.5dB 40Hz-15kHz
THD distortion	0.5%	0.5%	<0.05% ±75kHz dev.	0.1%	<0.4% 100% mod
Pre emphasis	50μs	50μs	50/75μ jumper	75μs	0/50μs switchable
AM noise figure	< −65dB	< −65dB	< −70dB	−55dB	−50dB
FM noise figure (mono)	< −70dB	< −70dB	< −84dB CCIR wtd	−75dB	−70dB de-emph unwtd
Floor space (in.)	68 × 39	Rack-mount	86W × 32D (43W × 32D)	47.75W × 32.5D	24W × 40D (42W × 40D)
Separate power rack	No	Integral circuit	No	No	No
Input ac power	220-380Vac 3ϕ	220Vac 1ϕ	208/380Vac 3ϕ	240Vac 3ϕ	380/415V 3ϕ
Typical power factor	. . . .	. . . .	. . . .	. . . .	>0.9
Power consumption	43kW	1.2kW (220W, 650W)	96kVA (19kVA)	37kVA	<19kVA (<37kVA)
System efficiency	. . . .	. . . .	. . . .	. . . .	60% (61%)
Fault logic architecture	μP	TTL, solid-state	μP	μP	CMOS/TTL & relay
μP diagnostics	Yes	. . . .	. . . .	Yes	No
			Models 50W-27.5kW		Models 500W-4kW

Manufacturer	QEI Corporation			Rhode & Schwarz	Thomson-LGT
Model Series	695T30 kW	FMQ-3500	FMQ-10000	NU-421	EVHF-5000
Rf output power	30kW	3.5kW	10kW	20kW	5kW
Output connector (in.)	$3^{1}/_{8}$	$1^{5}/_{8}$	$1^{5}/_{8}$	Air line 25/58	EIA type
Modulation capability	. . . .	. . . .	. . . .	. . . .	> ±75kHz
Final amp device	3CX15000A7 triode	3CX3000A7 triode	YU148 triode	Tetrode	Solid-state
Af input connection	. . . .	Balanced	Balanced?. . . .	5kΩ bal/unbal	Balanced
Af response	. . . .	. . . .	. . . .	±0.1dB, 40-43k	±0.2dB, 40Hz-15kHz
THD distortion	. . . .	0.025%	0.025%	<0.4% ±40kHz dev	<0.5%
Pre emphasis	0.025%	. . . .	. . . .	50, 75	50 (75 opt)
AM noise figure	. . . .	−55dB	−55dB	. . . .	< −52dB
FM noise figure (mono)	−55dB	−75dB	−75dB	68dB ±40kHz dev	< −60dB
Floor space (in.)	772F generator	24 × 30	24 × 30	70W × 29D	23W × 30D
Separate power rack	48 × 30	No	. . . .	No	No
Input ac power	240V 3ϕ	240V 1ϕ	240V 1ϕ	220-380V, 3ϕ	120/240Vac
Typical power factor	. . . .	. . . .	. . . .	>0.95	. . . .

Table 10-2. Continued.

Power consumption	50kW	6.4kW	17kW	42kVA	6.5kW
System efficiency	. . . .	. . . .	. . . .	. . . .	. . . .
Fault logic architecture	. . . .	CMOS/TTL	CMOS/TTL	Yes	CMOS/TTL
μP diagnostics	No	No	No	. . . .	No
				NU351, 5kW; NU411, 10kW	

Manufacturer Model Series	TTC/Wilkinson Div ◆FM-5000	Varian Continental TVT	
		LDM-1233 (−1236)	LDM-1239 (−1238)
Rf output power	5kW	10kW (27.5kW)	1kW (0.5kW)
Output connector (in.)	1 5/8	1 5/8 (3 1/8)	Type N (1 5/8)
Modulation capability	75kHz deviation	±100kHz	±100kHz
Final amp device	Tube	TH-339 (TH-374)	Solid-state
Af input connection	Xfmr balanced	Bal, unbal	Bal, unbal
Af response	±0.5dB, 25Hz-100kHz	±0.5dB, 30Hz-15kHz	±0.5dB, 30Hz-15kHz
THD distortion	0.5% at 100% mod.	0.3%, ±75kHz dev	0.3%, ±75kHz
Pre emphasis	75μS	Any std value	Switch selectable
Am noise figure	−58dB	−65dB	−65dB
FM noise figure (mono)	−80dB, 75μs deemp	−75dB	−75dB
Floor space (in.)	24.5×38	44W×23D (43W×37D)	22W×29D
Separate power rack	. . . .	Yes	No
Input ac power	240V 3φ	220Vac 3φ	220V 1φ
Typical power factor	. . . .	>0.9	>0.9
Power consumption	9.5kHz	18.5kVA (50kVA)	2.1kVA (1.3kVA)
System efficiency	. . . .	55% (55.5%)	48%
Fault logic architecture	Relay	CMOS/TTL	CMOS/TTL
μP diagnostics	No	No	No
	FM-10000J, 10kW FM-25000J-25kW		Models for 0.3/0.1kW

is typically 15 kHz with frequency tolerances of 1.5 to 5 parts per million (ppm) depending on the frequency band and whether a fixed or mobile transmitter is specified. The maximum frequency deviation permitted is usually 5 kHz as compared with 75 kHz for broadcast FM.

Direct FM, in which the varactor is placed across the crystal to "pull" the frequency the necessary amount for the deviation, is often used in the transmitter. This has the advantage of requiring no afc loop to maintain center frequency stability. Audio response is usually confined to the 300 – 3000 Hz range.

Mobile transmitters often have a power output of 50 W or less. The low power, combined with line-of-sight propagation limitations, will limit the operating range to only a few miles between mobile units in adverse terrain with hills, buildings, etc. For this reason repeater stations with antennas are placed at high points, e.g., tops of tall buildings, in the geographical area served. Mobile stations then transmit on the same frequency among themselves or with the base station via the repeater. A link between the base station and the repeater is established on another frequency. It will be recognized immediately that many channels and more sophisticated control are essential in a crowded communications environment. Some idea of how this is accomplished will be discussed for mobile telephone systems.

Mobile telephone

The Improved Mobile Telephone System (IMTS) provides fully automated service, as contrasted with MTS requiring the intervention of operators, in which the subscriber has essentially the same telephone capabilities as a fixed subscriber. Each mobile unit has an assigned telephone number, and is able to dial and communicate with any conventional telephone or other mobile unit (see Fig. 10-17). Since a limited number of channel

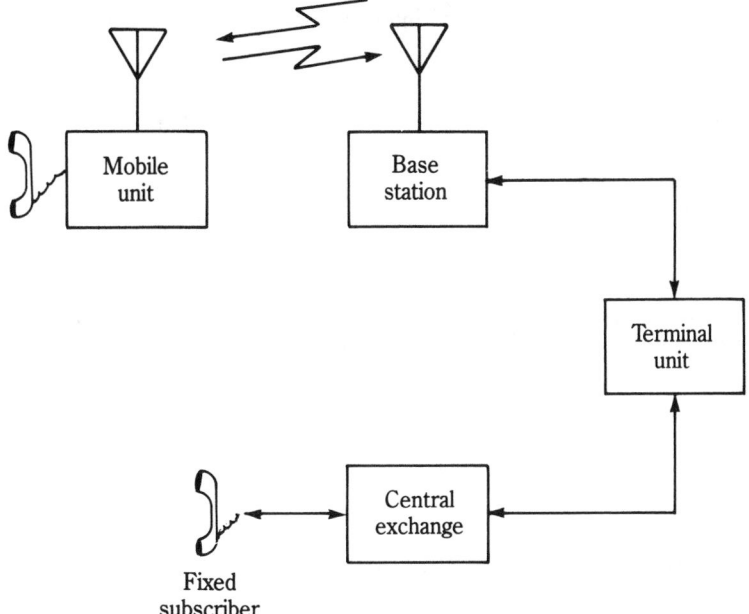

10-17 MTS system block diagram shown for typical operation.

frequencies are available, each mobile unit must have the capability of transmitting and receiving with a pair of the frequencies assigned to the overall system. A sophisticated control system is necessary to accomplish this.

The terminal unit must know the status of each channel at all times. If all channels are busy, the mobile units must be aware of this so that none will institute new calls. When one or more channels are free, each mobile unit must have a channel assignment for inaugurating calls. Such assignments are shared by a number of mobile units so that any call placed by one will change the system and, possibly, some channel assignments. Likewise, a terminated call will release a channel to be made available for reassignment. At no time can more active mobile units be connected to a base station than the total number of channels assigned to the system. Also, conforming to the standard telephone, *full-duplex* operation is needed. This simply means that simultaneous two-way communications is possible, so that two frequencies—one for transmission *from* the base station, and one for transmission *to* the base station are necessary for each assigned channel.

System control means automatic signaling between the base station and concerned mobile units. A *protocol* or procedure has been established in hardware (or software) for every conceivable sequence of events in system operation. A popular form of signaling consists of transmitting codes using *frequency shift keying* (FSK). This means that, at any one moment, the carrier can be frequency-modulated by one of a number of preassigned audio frequencies. The audio tones are in the neighborhood of 2000 Hz, with sufficient spacing for easy distinguishability. A sequence of frequency shifts constitutes a code. Both the base station and each mobile unit must be capable of transmitting and receiving the codes necessary for inaugurating calls, assigning frequencies, acknowledgments, ringing, etc.

IMTS was established in 1964. However, IMTS was allocated only 33 frequency channels by the FCC. Since the same frequencies cannot be used by cities in the same geographical region because of mutual interference, the channels available in heavily populated areas are severely limited. Moreover, each base station with a single antenna must serve a relatively large area, so that service in the fringe areas is vulnerable to noise and other forms of interference. The cellular mobile system to be introduced shortly is one answer to these problems.

Cellular mobile communications

Because of the channel availability limitations of current mobile radio communications, the cellular system has been developed. Basically, the geographical area is subdivided into a number of cells, each containing a base station with a relatively low-power transmitter (see Fig. 10-18 in which the cells are represented as being hexagonal in shape). In a cluster (not shown in the figure) of N cells, the frequencies assigned to each cell are different so that the same frequencies can be assigned to other cells in other clusters. The trick is to make the frequency assignments such that any one cell with particular frequencies has sufficient spatial separation from another cell with the same frequencies, so the probability of interference is statistically acceptable. The number of channels made available in a large urban area is greatly multiplied through the reuse of channel frequencies. A cell might cover an area of 200 square miles (for example), whereas many ''greater metropolitan areas'' cover many thousands of square miles.

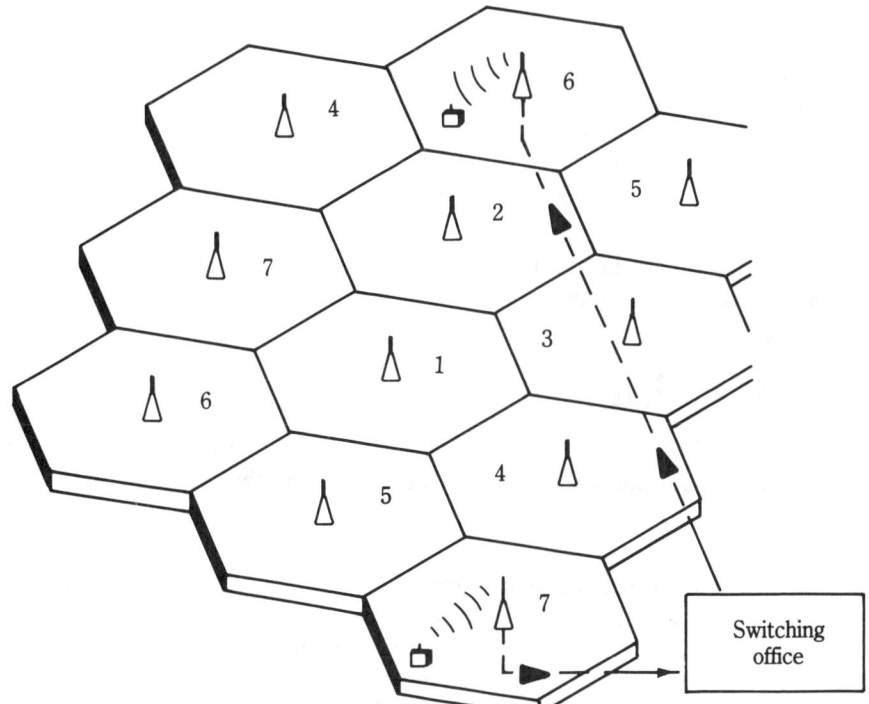

10-18 Cellular mobile communication system.

Each cell base station consists of antennas, transmitters, receivers, and computer controllers. These, in turn, are connected to a switching office by land line. Consider briefly how the system works for voice communications between two mobile units by tracing the steps required in placing the call and during its duration.

1. The calling unit causes coded digital source and destination data to be transmitted on an assigned cell frequency.

2. The nearest cell base station receives the transmission and forwards the data to the switching office.

3. The switching office pages each cell base station which, in turn, attempts to locate the called unit in its cell. When the called unit responds, this information is forwarded to the switching office.

4. Channel frequencies, from within the respective cell assignments, are now assigned since the switching station knows the location of each mobile unit. When each unit receives its channel assignment, the transmitters and receivers are automatically tuned to the correct frequencies and the link is completed.

5. Data bursts are transmitted from the mobile unit by momentarily muting the voice so as not to cause interference. This allows the channel status of each mobile unit to be reported.

6. When a mobile unit crosses the cell boundary during the communications link-up, handover, controlled by the switching, is accomplished.

Advanced Mobile Phone Service (AMPS), developed by Bell Telephone Laboratories, was first tested in Chicago in 1979. Some information about AMPS is useful in understanding the cellular approach. The mobile transceiver is a solid-state, narrow-band, FM unit. It transmits in the frequency range 825 – 845 MHz band (12W) and receives in the 870 – 890 MHz band. Control of the transceiver is exercised by a microprocessor. An area of about 2000 square miles with no more than 2500 mobile units and 10 cells is covered in the Chicago tests.

In looking toward the future, the need for and growth potential of cellular mobile communications is great. However, other modulation techniques such as SSB or *spread spectrum* might replace narrow-band FM.

AM transmitters

Circuitry for the achievement of AM has received considerable attention in chapter 8. For this reason the present discussion will concentrate on only a few other points.

Solid state

Transistors and other solid-state devices are found in all modern AM transmitters. Vacuum tubes still appear in the final modulator and power amplifier stages of high power, medium frequency AM transmitters and in the final amplifier stages of vhf and uhf broadcast transmitters. Figure 10-19 reproduces the block diagram and specifications for the Harris Corporation SX-5 5000 W, all-solid-state AM broadcast transmitter. This transmitter has MOSFET power transistors in the final power amplifier stages. It also has polyphase pulse duration modulation in which the audio modulating signal is sampled four times during each PDM cycle. The SX-5 contains four rf power amplifiers grouped in a quad configuration whose outputs are combined in a low-loss, series-configuration ferrite combiner. An overall power amplifier efficiency exceeding 85% is achieved. Should there be a malfunction in one phase of the system, operation can continue with a slight reduction in performance until a convenient maintenance action can be scheduled.

Vacuum-tube transmitters

Table 10-3 presents consolidated information on AM broadcast transmitters. The equipment listed is for the generation, modulation, and amplification of AM signals from audio signals. The rf output impedance is 50 ohms. The level of audio signals applied to the transmitter must be about +10 dBm for 100% modulation. It will be noted that the modulator final and final amplifier stages are designated as solid-state or that the specific vacuum-tube designation is given.

Broadcast station operations

You have previously considered the AM signal for a radio transmitter. To review briefly, the sound amplitude and pitch vary continuously to yield an irregular waveform such as is represented in Fig. 10-20. In this figure the dashed line depicts the response of a monitoring meter, such as a VU (volume unit) meter which is unable to follow the audio wave-

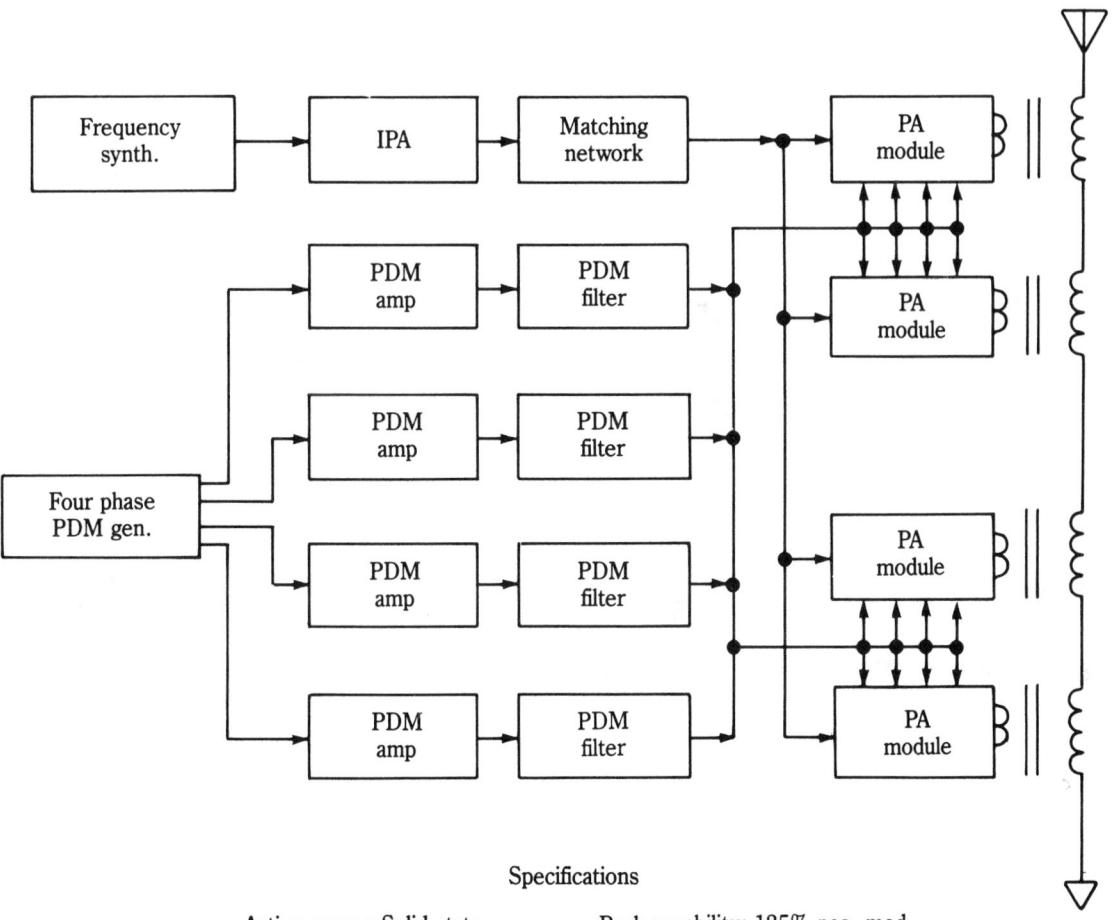

Specifications

Active comp.: Solid state	Peak capability: 125% pos. mod
Audio freq. resp.: 20 – 12500 Hz	Power consumption: 11.6 kW (100%)
Audio input: – 10 to + 10 dBm	Power output: 5.6 kW (capable)
Freq. stability: 10 Hz	Size: 72″ × 28″ × 30″
Overall eff.: 65% or better	Weight: 500 pounds

10-19 5-kW AM broadcast transmitter block diagram and specifications.

form in detail, and whose indication more closely resembles the average of the instantaneous sound level.

In the station the audio signal is controlled both automatically and manually by means of channel gain controls and the master gain control (see Fig. 10-21). The automatic controls are referred to as automatic gain control (AGC), limiter, compression amplifier, and the loudness controller or peak control. Their basic purpose is to maintain a high average audio level but prevent excessive modulation such as is demonstrated in Fig. 10-22 at point (a). A high average level is not governed by rules and is largely dependent on the program material. Expressed as the *modulation index*, it might reasonably be expected to be between 25% and 75%.

Table 10-3. AM Broadcast Transmitter Specifications.

| Manufacturer | Cemtys | | | Varian Continental TVT | |
Model Series	◆EB-KV/70 (EB-2VK/70†)	◆EB-VD/70 (EB-DT/70)	315-R1	419F-2	◆420B
Service band	Broadcast	Broadcast	Broadcast	Shortwave	Shortwave
Rf output rating	10kW (20kW)	5kW (2kW)	5kW	250kW	500kW
Rf connector (in.)	1⅝ 50Ω	50Ω unbal	1⅝	9 EIA, 75/300Ω	9 EIA 75/300Ω
Carrier stability	±5Hz	±5Hz	±5Hz	±1×10⁶Hz	±1×10⁶Hz
Carrier shift	....	....	2%, 100% mod	<5%, 100% mod	<5%, 100% mod
Modulation (max)	125%	125%	125%	100%	100%
Af input connection	Bal, 600Ω	Bal, 600Ω	Active bal	Bal, unbal	Bal, unbal
Af response	±1dB 40Hz-8.5kHz	±1dB 100Hz-9kHz	1 dB, 20Hz-20kHz, 95% mod	1dB, 50Hz-7.5kHz	1dB, 50Hz-7.5kHz
THD distortion	<1.5% 100% mod	<2% 95% mod	<2%, 100% mod	<3%, 90% mod	<4%, 95% mod
AM noise figure	<-65dB	-60dB	< -60dB unwtd	< -55dB	< -55dB
Final amp device	2 - 3CX2500F3 (2 prs)	3CX2500F3	3CX3000F7	4CV250,000B	4CM400,000
Modulator final device	2 - 4CX3000A (2 prs)	2 5CX1500A	3CX3000F7	2-4CV250,000B	4CM400,000
Floor space required	16'W×37D	67W×34D (63W×31D)	35W×34D	34m²	42m²
Separate power rack	No	No	No	Yes	Yes
Input ac power required	220Vac 3φ	220Vac 3φ	208-250Vac 3φ	480V, 3φ	4160V, 3φ
Power factor	>0.94	>0.92	0.95	0.9	0.9
Power consumption	54kVa for 20kW	15.6kVa (6kVA)	12.8kVA	625kVA	1075kVA
System efficiency	80% final/40% modulator	... (80% final)	>55%	60%	70% typical
Fault logic architecture	Solid-state, relay	Solid-state, relay	Relay	Relay	μP system
μP diagnostics	No †2-10kW w/combiner	....	No	No	Yes

| Manufacturer | Varian Continental TVT | | | | CSI Electronics |
Model Series	◆418D-2	316F-1	317C-3	314F solid-state	◆T-100-A1
Service band	Shortwave	Broadcast	Broadcast	Broadcast	Broadcast, shortwave
Rf output rating	100kW	10kW	50kW	1kW	110kW
Rf connector (in.)	6 EIA, 75/300Ω	1⅝	3⅛ EIA female	L/C, type N	6⅛
Carrier stability	±1×10⁶Hz	± Hz	± 5 Hz	±5Hz	±5Hz
Carrier shift	<5%, 100% mod	<2%, 100% mod	<2%	0.5% @ 100% mod	3% 100% mod, 400Hz
Modulation (max)	100%	125%	125%	125%	125%
Af input connection	Bal, unbal	Active bal	Active bal	Active bal	Xfmr bal
Af response	1dB 50Hz-7.5kHz	1.5, 30Hz-15kHz	0.5dB, 10H-7.5kHz	±1dB, 20Hz-12kHz	±1.5dB, 50Hz-10kHz
THD distortion	<3%, 90% mod	<3%, 100%	<2%, 100% mod	<2%, 100% mod	3% 95% mod
AM noise figure	<-60dB below 100% mod	< -60dB unwtd	> -60dB unwtd	< -60dB unwtd	-55dB unwtd
Final amp device	4CV100,000C	4CX35000A	4CX35000C	Solid-state	4CV100000E
Modulator final device	2-4CV100,000C	Solid-state	3CX000A	Solid-state	4CV50000E
Floor space required	20m²	67W×26D	144W×54D	22W×26D	237W×55D
Separate power rack	Yes	No	24W×46D	No	Yes
Input ac power required	480Vac, 3φ	208-230Vac 3φ	460Vac, 3φ	180-250Vac 1φ	460Vac 3φ
Power factor	0.9	0.93	0.95	....	0.89
Power consumption	275kVA	28.4kW 100% mod	118kVA	2.6kW, 100% mod	295kW

(continued from previous page)

System efficiency	55%	53%	60% 100% mod	>57%, 100% mod	64%
Fault logic architecture	Relay	Relay	Relay	Digital relay	Relay
µP diagnostics	No	No	No	Option	No
					Also 27.5kW, 55kW

Manufacturer	CSI Electronics		Elcom Bauer		Harris Broadcast
Model Series	**T-10-A1†**	**725 (715C★)**	**710C**	**SX-5A**	**SX1A**
Service band	Broadcast, shortwave	Broadcast, shortwave	Broadcast, shortwave	Broadcast	Broadcast
Rf output rating	12kW	25kW (15kW)	10kW	5kW	1kW
Rf connector (in.)	1 5/8	1 5/8	1 5/8	7/8	Type N
Carrier stability	±5Hz	±5Hz	±5Hz	±20Hz	±20Hz
Carrier shift	3%, 100% mod 400Hz	3%, 100% mod, 400Hz	3%	2%	2%
Modulation (max)	125%	125%	125%	125%	·····
Af input connection	Xfmr bal	Xfmr bal	Xfmr bal	Balanced	Balanced
Af response	±1.5dB, 50Hz-10kHz	±1dB, 50Hz-7.5 kHz	±1.5dB, 50Hz-10kHz	±0.5dB, 20Hz-12.5kHz	±0.5dB, 20Hz-12.5kHz
THD distortion	2.5% 95% mod	<2.5%	<2% 95% mod	1%	1%
AM noise figure	−55dB unwtd	−60dB unwtd	−60dB unwtd	−60dB	−60dB
Final amp device	3CX10000A3	4CX20000B/8990	4CX15000	Solid-state	Solid-state
Modulator final device	4CX3000A	4CX20000B/8990	4CX5000A	Solid-state	Solid-state
Floor space required	76.5W×34D	140W×30D	70W×31D	28W×30D	28W×30D
Separate power rack	Yes	Yes	No	No	No
Input ac power required	240Vac 3φ	380-450Vac 3φ (220Vac)	208-240Vac 3φ	200Vac 1φ	220Vac 1φ
Power factor	0.89	0.9	0.9	·····	·····
Power consumption	27kW	56kW 100% mod (28kW)	25kW	11.6kW	1.5kW
System efficiency	56%	54%	54%	65%	66%
Fault logic architecture	Relay	Relay	Relay	µP	µP
µP diagnostics	No	No	No	No	No
Other models	†1.1kW/3kW/6kW	●uses 4CX15000A final 4CX5000A for mod	701B, 1kW 705C, 5kW		

Manufacturer	Harris Broadcast	ITAME	LPB Inc	Marconi Comm.	
Model Series	**DX-10**	**Tauro-S series**	**AM series**	**B6127 (B6126)**	**B6034**
Service band	Broadcast	Broadcast	Broadcast	Shortwave	Broadcast
Rf output rating	10kW	10kW (5/2kW)	25/30/50/100/150W	500kW (300kW)	1kW
Rf connector (in.)	1 5/8	1 5/8	SO-230 UHF	300Ω bal/50-75Ω unbal	7/16
Carrier stability	±10Hz	±5Hz	±10Hz	<±10Hz	±10Hz
Carrier shift	2%	3%	2% 100% mod	<5% 100% mod 400Hz	<5%
Modulation (max)	·····	125%	110-120%	100%	100%
Af input connection	Balanced	Balanced	Xfmr bal	600Ω bal	600Ω bal
Af response	±0.5dB, 50Hz-12.5kHz	1dB, 20Hz-10kHz	±1dB, 20Hz-15kHz	±1dB, 50Hz-6kHz	±2dB, 40Hz-10kHz
THD distortion	1%	2%	·····	<4% 90% mod	<3%, 90% mod
AM noise figure	−60dB	<−60dB	−55dB	<−56dB	−60dB unwtd
Final amp device	Solid-state	3CX10000A†	Solid-state	TH558 Hypervapotron	Solid-state modules
Modulator final device	Solid-state	4CX15000A	Solid-state	4CW25000A water cooled	Solid-state
Floor space required	72W×30D	45W×37D	Rack, wall mount	181W×189D	32W×22D
Separate power rack	No	Yes	No, 25/30W	314W×433D	No

Table 10-3. Continued.

Input ac power required	220Vac 3φ	220-380Vac 3φ	110Vac 1φ	308-415Vac 3φ	230Vac 3φ
Power factor	....	....	....	>0.9	>0.9
Power consumption	19kW	23.5kVA (13.5, 5.5)	0.7kVA for AM-50	....	3.9kVA
System efficiency	75%	....	25%	62%	48%
Fault logic architecture	μP system	....	Relay	MOS/TTL, relay	CMOS/TTL
μP diagnostics	No	†3000F7/1200A7 for 5/2kW mod and final	No	Yes	No

Manufacturer	Marconi Communications		Nautel		Radio Systems
Model Series	B6034†	B6040	AMPFET-25 (−50)	AMPFET-10 (−5, P400)	TR-20
Service band	Broadcast	Broadcast	Broadcast	Broadcast	Broadcast
Rf output rating	50kW	10kW	25kW (50kW)	10kW (5kW, 400W)	20W
Rf connector (in.)	3 1/8	Coaxial cable	3 1/8	1 5/8 (1 5/8, Type N)	UHF SO-239
Carrier stability	±10Hz (hi stab opt)	±10Hz (hi stab opt)	±5Hz	±5Hz	±0.002%
Carrier shift	<4%, 100% mod	<5%	1%, 100% mod, 400Hz	<3%	2%
Modulation (max)	125%	125%	125%	125%	100%
Af input connection	Xfmr bal	Active bal	Xfmr bal	Xfmr bal	Unbal
Af response	±1dB, 30Hz-6kHz	±1dB, 50Hz-6kHz	±0.5dB, 30Hz-10kHz	±0.5dB, 50Hz-10kHz	1dB, 20Hz-15kHz
THD distortion	<3%, 90% mod	<3%, 90% mod	<2%, 90% mod	<2%, 95% mod	2%
AM noise figure	−58dB unwtd	−58dB unwtd	−60dB unwtd	−60dB unwtd	−60dB
Final amp device	2 4CX35000C tetrode	2 4CX1500A tetrode	Solid-state modules	Solid-state	Solid-state
Modulator final device	2 4CX1500B tetrode	Solid-state	Solid-state	Solid-state	Solid-state
Floor space required	144W×61D	65W×33D	90W×24D (135W×30D)	25W×24D	Rack-mount
Separate power rack	No	No	Yes	Yes	No
Input ac power required	380-415Vac 3φ	380-415Vac 3φ	208-480Vac 3φ	208-480Vac 3φ‡	110Vac 1φ
Power factor	>0.9	>0.9	0.98	0.95	....
Power consumption	<129kVA	<33kVA	50kW (100kW)	20kVA (<10kVA, 0.9kVA)	0.1kVA
System efficiency	52% at carrier	42%	>75%	>75% (>75%, >65%)	....
Fault logic architecture	CMOS, TTL relay	CMOS, TTL relay	CMOS, TTL	CMOS/TTL	....
μP diagnostics	No	No	No	No	No
Other models	†Also 100kW, 200kW			Also 1kW, 2.5kW ‡P400, 110Vac 1φ	

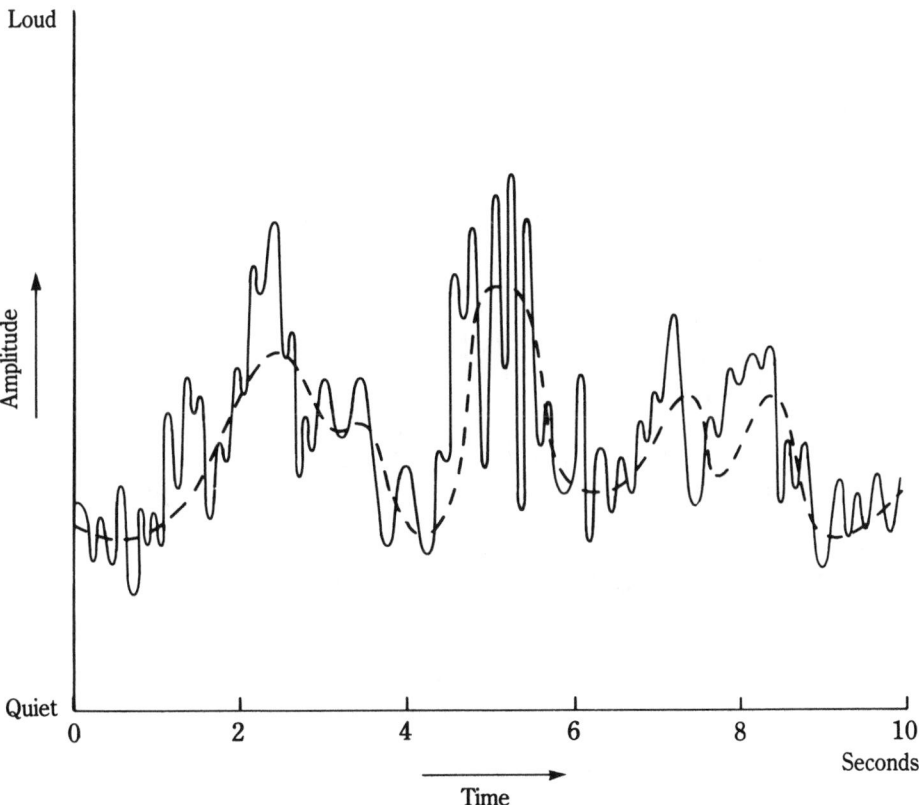

10-20   Representation of an audio signal.

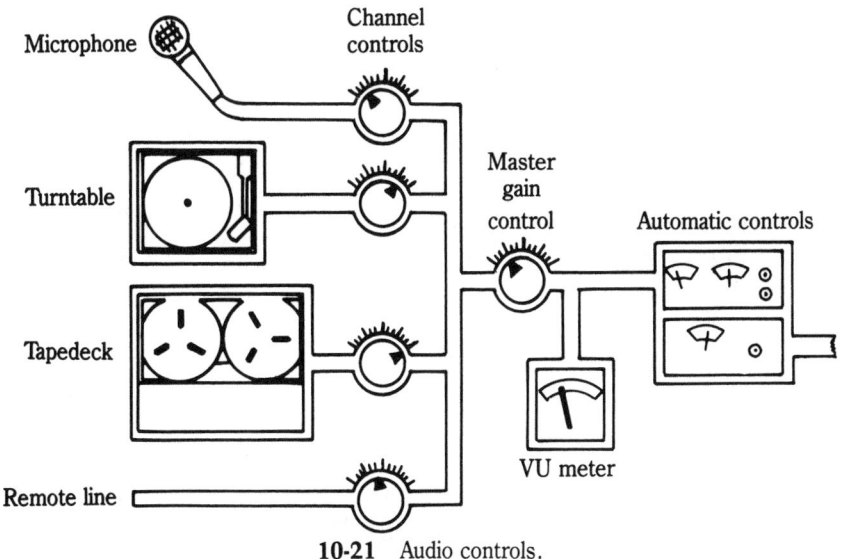

10-21   Audio controls.

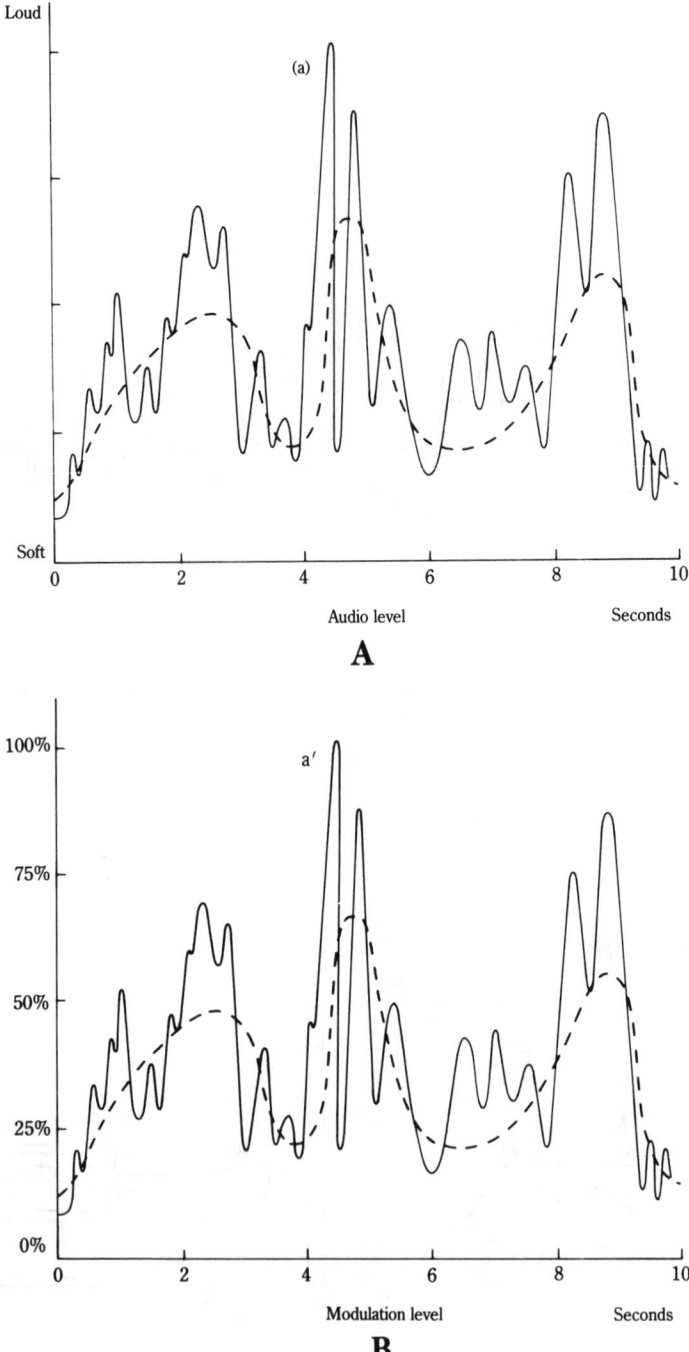

10-22 Comparison of audio and modulation levels: A. Represents an audio signal that has modulated a transmitter. B. Represents the resulting modulation (the dashed line is the average level).

At any radio broadcasting station the modulation must not exceed the following limits of *frequent occurrence*, i.e., averaging more than once per second.

Modulation	AM	FM
Positive	125%	100%
Negative	100%	100%

The *modulation monitor* is a separate piece of equipment that measures and indicates the modulation level on a meter (see Fig. 10-23) and a peak indicator light. The circuitry is arranged to follow the modulation envelope in the positive or negative direction, as selected, with slow decays following the peaks so that the meter indication fluctuates more-or-less continuously.

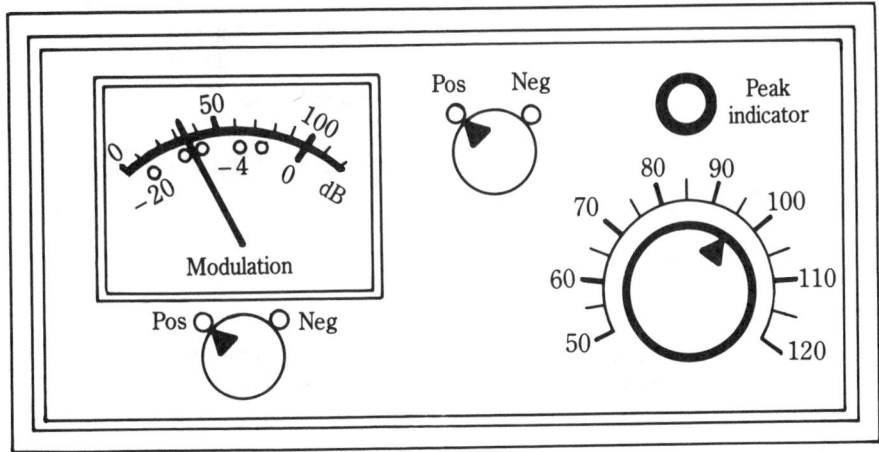

10-23   Modulation monitor.

Peaks exceeding a modulation percentage level, preset on the front of the panel, cause the peak indicator to light.

Because of the different prescribed ballistic constants (time constants) of the VU meter, the modulation monitor, and standard voltmeters, there is no exact equivalence in their indications for broadcast material. Experience and skill are required on the part of the operator in meeting objectives of high average modulation level without exceeding the allowable peak modulation limits.

Operating power

The most straightforward and usual method for determining operating power is by the measurement of the power or current flowing into the antenna. This is called the *direct method*, which is required for broadcast AM stations. Measurements made at other points in the system require that power be calculated on the basis of some assumptions; this is the *indirect method* most commonly used for FM broadcast stations.

Figure 10-24 illustrates the arrangement for a nondirectional AM station in which antenna current is obtained, with the transmitter unmodulated, from a meter installed at the tower base and whose reading is displayed remotely at the station for the operator.

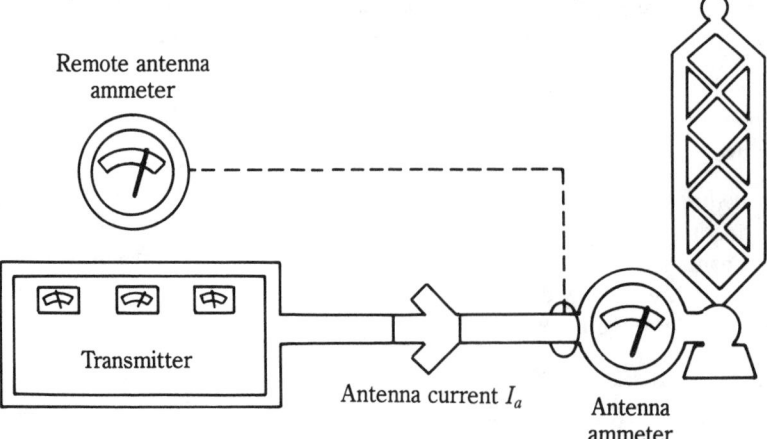

10-24 Nondirectional AM antenna.

Using the measured value of antenna resistance R_a, the operating power is:

$$P = I_a^2 R_a$$

For example, if $R_a = 14.9$ ohms and $I_a = 8.2$ A,

$$P = (8.2)^2 \times 14.9 = 1002 \text{ W}$$

In Fig. 10-25, the measurement of the total antenna current for a directional antenna array is shown. The calculation of power remains the same with the understanding that the resistance is for the combination of all antennas as measured at the common point. (Note that the resistance is not the same as that of all the antennas placed directly in parallel.)

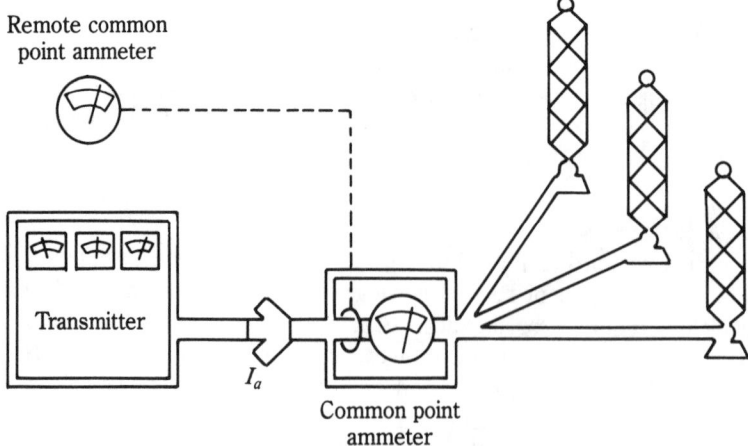

10-25 Measurement of the total antenna current for a directional array.

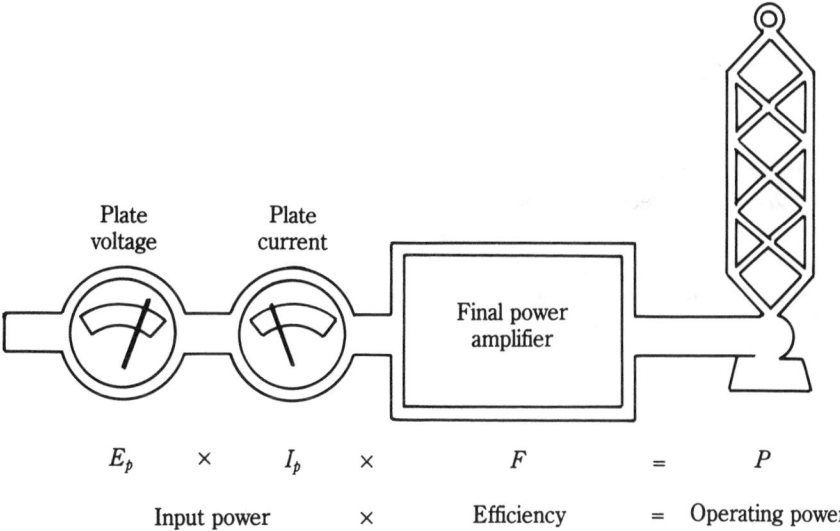

Plate voltage   Plate current

Final power amplifier

E_p × I_p × F = P

Input power × Efficiency = Operating power

10-26 Operating power calculation by the indirect method.

Figure 10-26 summarizes the calculations of operating power by the indirect method in which the formula used is:

$$P = E_P \times I_P \times \eta$$

in which E_P is the dc plate voltage, I_P is the dc plate current, and eta (η) is the final amplifier efficiency expressed as a fraction (usually decimal fraction). As an example, let E_P = 3009 V, I_P = 0.382 A, and the efficiency is assumed to be 71.5% = 0.715.

$$P = 3009 \times 0.382 \times 0.715 = 822 \text{ W}$$

Directional AM stations

A directional AM station is one that has a directional antenna array that causes the antenna to increase the received signal for receivers located in certain directions from the antenna, while reducing the received signal for receivers located in other directions. The actual antenna pattern is influenced by factors over which little or no control is possible:

1. Surrounding terrain

2. Ground moisture

3. Structures in the vicinity of the antenna

In contrast, the antenna pattern can be controlled through the selection, by design, of the:

1. Size and shape of towers

2. Number of towers

3. Relative location of towers

4. Relative phase of rf current fed to each tower

5. Amplitude of rf current fed to each tower

These last two factors, phase and amplitude, are properly the concern of the station operating personnel.

Figure 10-27 represents a station having a directional antenna system fed by a device called a *phasor*. This piece of equipment is adjusted to regulate both the amount and relative phase angle of the individual tower base currents. It is the combination of the tower placements, that is, the spacing and directional orientation, the base current amplitudes, and the relative base current phases that establish the directional antenna pattern for a given antenna site.

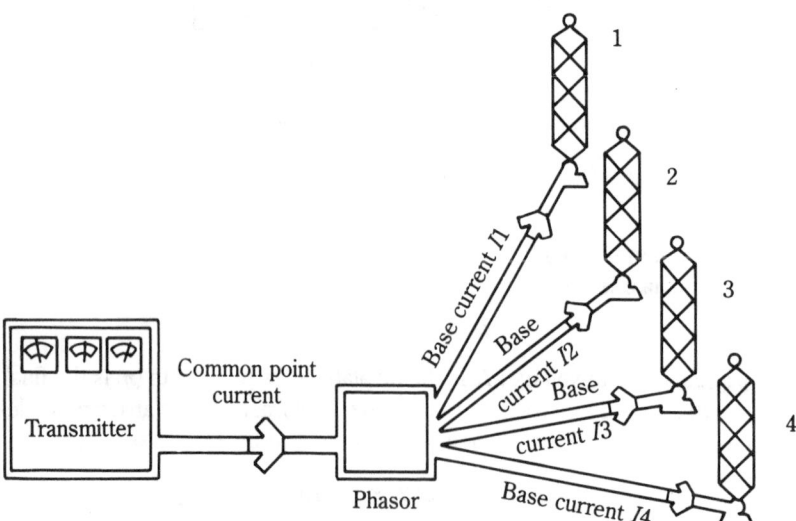

10-27   Division and distribution of the antenna current.

It is standard practice to choose one antenna tower as the reference for the purpose of expressing measured tower currents. Base tower currents for the other towers in the array are then expressed as current ratios, relative to that of the reference. Likewise, other individual tower current phase angles are referred to the reference tower current phase angle. For example, if the currents for Fig. 10-27 are:

Tower	1	2	3	4
Base current	2.45 A	5.9 A	6.6 A	4.7 A

and tower 3 is chosen as the reference, the current ratios are then

Tower	1	2	3	4
Current ratio	0.37	0.89	1.0	0.71

These last numbers are derived by dividing the respective tower base currents, 2.45 A, 5.9 A, 6.6 A, and 4.7 A, by the base current for tower 3, i.e., 6.6 A. For example, the first ratio is 2.45/6.6 = 0.37.

Similarly, the phase angles for the tower base currents are referred to that of tower 3 base current.

Tower	1	2	3	4
Phase	18°	−144°	0°	105°

Here, positive phase angles simple mean that the current leads the reference by the stated amount. Negative phase angles mean that the current in question lags the reference tower base current. Thus, the base current for tower 1 leads that of tower 3 by 18°, whereas, the base current of tower 2 lags that of tower 3 by 144°. A typical *antenna monitor* appears in Fig. 10-28. Its purpose is to determine by measurement the actual current ratios and relative phase angles. The operator is then able to compare the measurement, as read from the antenna monitor, with the authorized values to establish whether or not the station antenna array is performing properly.

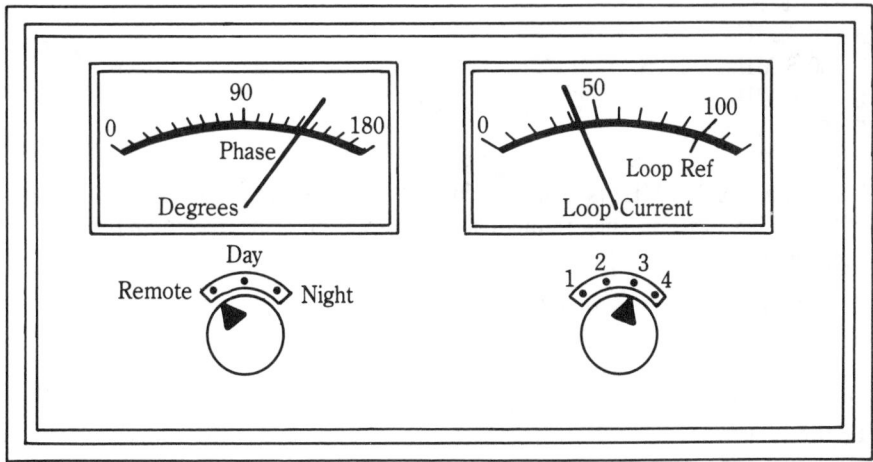

10-28   Typical antenna monitor.

Multiple-choice questions

1. If the spectrum is shifted in frequency with no other changes, this is known as:

    A. Frequency multiplication.

    B. Sideband movement.

    C. Baseband reorientation.

    D. Frequency translation.

    E. None of the preceding choices.

2. A device which is capable of causing frequency translation is the:

    A. High-Q tank circuit.

    B. Balanced modulator.

    C. Low-Q tank circuit.

D. i-f strip.

E. Op amp comparator.

3. If the frequency of each component in a signal spectrum is increased by the same fixed amount, this is known as:

   A. Modulation.

   B. Demodulation.

   C. Frequency translation.

   D. Up conversion.

   E. Both C and D.

4. A particular amplifier is designed to be a frequency doubler. If the input signal frequency is 15.4 MHz, a circuit in the output will be tuned to:

   A. 7.7 MHz.

   B. 15.4 MHz.

   C. 30.8 MHz.

   D. 61.6 MHz.

   E. Both A and C.

5. A sine wave of 293 MHz is phase-modulated to achieve a maximum phase deviation of 0.2 radian. After passing through a frequency tripler, the maximum phase deviation will be:

   A. 0.0667 radian.

   B. 0.2 radian.

   C. 0.3 radian.

   D. 0.4 radian.

   E. 0.6 radian.

6. Any device to be used as a frequency multiplier must be:

   A. Active.

   B. Passive.

   C. Linear.

   D. Nonlinear.

   E. Digital.

7. A particular amplifier circuit used for frequency doubling is known as:

   A. Push-push.

   B. Push-pull.

   C. Pull-push.

   d. Pull-pull.

   E. All of the preceding choices are nonsense.

8. Frequency division is useful in the implementation of a:

   A. AM demodulator.

B. Frequency synthesizer.

C. AGC circuit.

D. FM demodulator.

E. Both A and D.

9. A particular digital device can be used effectively for frequency division. It is the:

A. AND gate.

B. OR gate.

C. Exclusive-OR gate.

D. Shift register.

E. None of the choices is true.

10. Frequency division by 12 will require how many flip-flops in the counter?

A. 3.

B. 4.

C. 6.

D. 12.

E. No correct choice is given.

11. Indirect frequency synthesizers will include:

A. Phase-locked loops.

B. Voltage-controlled oscillators.

C. Multiple banks of crystals.

D. Both A and B.

E. All of the choices are true.

12. Identify an electronic device, not specifically designed for the purpose, which can be used as a phase detector:

A. Wien bridge.

B. Colpitts oscillator.

C. Balanced modulator.

D. Butterworth filter.

E. Envelope detector.

13. A particular frequency synthesizer contains only a single crystal. What words describe this synthesizer?

A. Crystal modulated.

B. Inexact.

C. Indirect.

D. Deficient.

E. Phase-unlocked.

14. A recognizable feature of a CW transmitter is:

A. Keyed transmitter.

B. Power amplification.
C. Frequency generation.
D. Frequency multiplication.
E. All of the choices are true.

15. The term "pulling" refers to:

    A. The change of the crystal oscillator frequency by loading.
    B. One half-cycle operation of a push-pull amplifier.
    C. Loading on the transmitter caused by the antenna connection.
    D. Reduction of the power supply terminal voltage as the transmitter is keyed.
    E. Both B and C.

16. When frequency modulation is achieved by initial phase modulation, this is called:

    A. Angular modulation.
    B. Direct FM.
    C. Indirect FM.
    D. Indirect synthesis.
    E. Direct synthesis.

17. A disadvantage of direct FM is the need for

    A. AGC.
    B. AFC.
    C. A frequency synthesizer.
    D. Phase modulation.
    E. Both B and D.

18. Direct FM can be achieved by:

    A. A reactance tube modulator.
    B. A varactor diode.
    C. An AGC circuit.
    D. A phase modulator.
    E. Both A and B.

19. In broadcast FM the maximum frequency deviation is:

    A. 2500 Hz.
    B. 0 – 2 kHz.
    C. 75 kHz.
    D. 88 MHz.
    E. The answer depends on the frequency of the modulating signal.

20. The frequency deviation permitted for narrow-band FM is usually:

    A. 1 kHz.
    B. 5 kHz.

C. 25 kHz.

D. 75 kHz.

E. None of the choices is true.

21. Audio frequency response for narrow-band FM is usually confined to:

    A. 30 – 20,000 Hz.
    B. 30 – 300 Hz.
    C. 300 – 3000 Hz.
    D. 3000 – 30,000 Hz.
    E. Both B and C.

22. In a mobile radio communications system will be found:

    A. A base station.
    B. Repeaters.
    C. Links.
    D. Mobile transceivers.
    E. All of the choices.

23. The principal reason for adopting the cellular mobile communications concept is to:

    A. Increase privacy.
    B. Allow digitized services.
    C. Overcome limitations on the number of available channels.
    D. Increase neighborhood and community spirit.
    E. Subdivide congressional districts for campaign equity.

24. This is a question requiring reference to Table 10-2 for vhf FM broadcast transmitters. The final power amplifiers for transmitters of multi-kilowatt rating consist of:

    A. Transistors.
    B. ICs.
    C. Magnetrons.
    D. Vacuum tubes.
    E. Varactor diodes.

25. Refer to Table 10-2. For vhf FM broadcast transmitters in the multi-kilowatt operating range the driver stages consist of:

    A. Solid-state devices.
    B. ICs.
    C. Magnetrons.
    D. Vacuum tubes.
    E. Both A and D.

26. An AM broadcast transmitter in the multikilowatt operating range will have what form of final amplifier?

    A. Solid-state devices.

   B. Vacuum tubes.
   C. Traveling-wave tubes.
   D. Both A and B.
   E. All of the choices are true.

27. In a broadcast station, the AGC is referred to as:

   A. Automatic gain control limiter.
   B. Compression amplifier.
   C. Loudness controller.
   D. Peak control.
   E. All of the choices are true.

28. The operating power of a transmitter is to be calculated by the indirect method. Indicate which of the following quantities is not required.

   A. Plate voltage.
   B. Plate current.
   C. Final amplifier efficiency.
   D. Both A and C.
   E. None of the choices is correct.

29. For a given antenna site, which factor or factors influence the antenna pattern?

   A. Tower placement and orientation.
   B. Tower base current amplitudes.
   C. Tower base current phases.
   D. All of the preceding choices are true.
   E. Only choices C and D are not true.

30. The purpose of the antenna monitor is to determine by measurement:

   A. Total power input to the transmitter final stage.
   B. Antenna pattern as influenced by ground moisture content.
   C. Electric field strength in the direction of greatest area coverage.
   D. Current ratios and relative phase angles.
   E. Both A and B.

Basic problems

1. A sine wave, amplitude-modulated to yield frequency components at 1.995 MHz, 2.000 MHz, and 2.005 MHz, undergoes an up-conversion frequency translation of exactly 10 MHz. Calculate the new frequencies of the components.

2. A class-C amplifier has a conduction angle of 30°. Calculate the frequency multiplication possible.

3. An indirect frequency synthesizer has a 10 MHz, temperature-controlled crystal with a short-term accuracy of 1 part in 10^{10}. If the maximum synthesizer frequency available is 100 MHz, estimate the frequency error for the synthesizer at that frequency.

4. Explain the adverse results of improper pulse shaping in a CW transmitter.

5. In Fig. 10-14, which details the design of an FM transmitter, a number of existing filters are not shown. What is the purpose of these filters?

6. Assuming that a peak value of 100 mV at a frequency of 10 kHz yields a maximum permissible frequency deviation for broadcast FM, what will be the peak frequency deviation if the peak value of the modulating signal is reduced to 25 mV without changing the frequency of this modulating signal?

7. Comment on the implications of low transmitter power and line-of-sight limitations for narrow-band FM.

8. Calculate the power consumption of a Marconi B6127, 500 kW shortwave transmitter using data from Table 10-3.

9. The antenna current for a transmitter is measured to be 25.3 A and the antenna resistance is 19.5 ohms. Calculate the operating power of the station.

10. Discuss those factors which can be controlled and cannot be controlled in establishing the antenna pattern of an AM broadcast station.

Advanced problems

1. A tripler has a parallel resonant circuit placed in series with the collector of a common-emitter amplifier. If L = 12 μH, and C = 50 pF, calculate the input driving frequency for the base circuit.

2. A counter is to be used to provide frequency division by N = 19. Determine the maximum count number to achieve this division.

3. Explain the circumstances for which the voltage applied to the VCO of a phase-locked loop is absolutely constant.

4. In Fig. 10-14 we see examples of frequency multiplication and translation. Briefly explain the result of using only frequency multiplication to change from 100 kHz to an assumed transmitter carrier frequency of 88 MHz.

5. Explain what determines the accuracy of the center frequency for direct FM.

6. Referring to Problem 6 in the Basic problems, let the modulating signal frequency also be changed from 1 kHz to 5 kHz. Calculate the modulation index for the initial and final conditions.

7. Summarize the salient features of a cellular mobile communications system.

8. Refer to Table 10-3 for standard AM broadcast transmitters. Calculate the power consumption of the Varian Continental TVT 317C-3, 50 kW transmitter.

9. The operating power of a station is 50 kW. If the antenna resistance is 33.4 Ω, what is the antenna current?

10. In a given directional antenna having the currents given in Broadcast Station Operations, select tower 4 to be the reference and determine the new current ratios.

11

Receivers

A RADIO RECEIVER MUST PROVIDE MOST OF THE TRANSMITTER FUNCTIONS, BUT IN reverse order. It must be capable of selecting one from among many rf signals, amplify it, demodulate it, and process it to recover the original intelligence in a proper form. The receiver must also cope with noise, interference sources, and a variety of other problems arising from the environment, the mode of operation, and so forth. The span of different capabilities, price range, and applications for receivers is impressive.

The following sections are included in this chapter:

- The superheterodyne receiver
- General performance specifications
- The receiver front end
- i-f amplifier
- Integrated circuit receivers
- Control circuitry
- The communications receiver

The superheterodyne receiver

All of the basic requirements of a radio receiver for voice communications are met by the equipment whose block diagram appears in Fig. 11-1. A weak rf signal from the antenna is amplified and demodulated. The recovered modulation waveform is further amplified and then applied to the speaker. Tuning to a different radio frequency requires only that the rf amplifier be tuned. However, since the rf amplifier would necessarily require two or three stages of amplification, each must be individually tuned.

In the early days of radio broadcasting, tuning was accomplished by mechanically adjusting the capacitance or inductance in resonant circuits associated with each stage. This process required both skill and patience, which prompted many ingenious schemes for mechanically coupling the tuning element shafts.

Antenna

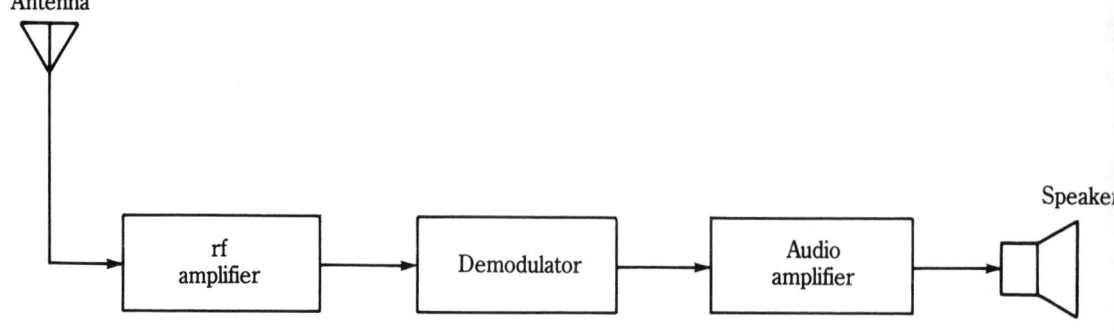

11-1   Block diagram of a simplified voice channel radio receiver.

Aside from the tuning coordination problem inherent in the AM broadcast tuned radio frequency (TRF) receiver, the change in bandwidth over the tuning range is undesirable for a number of reasons. The relatively high gain required for the rf signal also poses problems in stability because of the possibility of a small amount of positive feedback resulting in oscillation of the receiver.

The superheterodyne receiver solves many of the TRF technical shortcomings and, except for special communications applications in which a fixed carrier frequency is used, has almost completely replaced it. As is usually the case, the superheterodyne receiver, while overcoming many of the TRF receiver shortcomings, introduced several which are uniquely its own. These will be considered later.

Figure 11-2 is the block diagram of a simple superheterodyne receiver. It has one rf amplifier stage and an intermediate frequency (i-f) amplifier which normally consists of at least two stages. Although the rf amplifier stage is tuned to whichever rf carrier frequency is desired, the i-f amplifier operates at the fixed i-f. The carrier frequency signal from the rf amplifier is converted to the i-f in the mixer. The mixer is a nonlinear device which produces the i-f by combining the carrier frequency, f_C, with the local oscillator frequency, f_{LO}. The difference frequency is the i-f:

$$f_{IF} = f_{LO} - f_C$$

which assumes that f_{LO} is greater than f_C. However, additional frequency products, stemming from f_C and f_{LO} originate in the mixer. These include:

1. The original frequencies f_C and f_{LO}
2. The sum frequency $f_{LO} + f_C$ in addition to f_{IF}
3. Harmonics of the frequencies in 1 and 2 above

Only the difference frequency is accepted by the i-f amplifier and the other mixer products are rejected. This feature of the i-f amplifier makes the superheterodyne receiver practical. It also creates new difficulties in that additional, undesirable components at, or near, the i-f are sometimes produced in the mixer as the result of particular interfering frequencies that enter the receiver through the *front end*, i.e., from the antenna. These image and spurious frequencies will be analyzed later.

It will be recalled that sidebands are generated by the modulation process which are essential in the conveyance of information. These sidebands occupy the spectrum on one

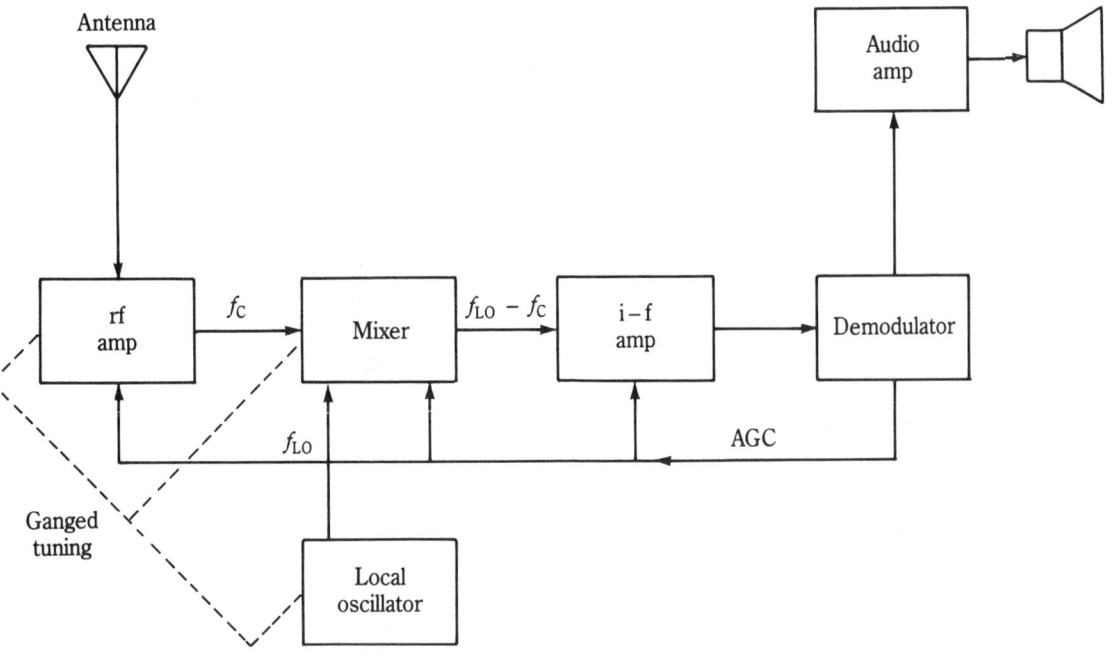

11-2 Voice channel superheterodyne receiver block diagram.

side of the carrier for SSB modulation, and on both sides for AM or FM. During frequency translation, accomplished by the mixer, each sideband frequency component, together with the carrier, is moved by exactly the same amount. Therefore, the spectra associated with the i-f signal have exactly the same shape as that transmitted with the rf carrier.

In order that the superheterodyne receiver functions properly, f_{LO}, as produced by the local oscillator, must be positioned so that it is greater than f_C by the same amount for any f_C to which the receiver is tuned. This means that the tuning of all circuits preceding the mixer, including those in the rf amplifier, and those circuits of the local oscillator, be coordinated to maintain this frequency difference. Perfect tracking is not possible, but it is necessary that the departure from this ideal performance shall not cause unacceptable degradation of the audio quality for broadcast and voice communication receivers.

Following amplification by the i-f amplifier stages, demodulation occurs. Of course, the method for demodulation will depend on the form of modulation employed. Modulation and demodulation techniques were covered in chapters 8 and 9. Additional information about modulation, with particular attention to the integrated semiconductor devices presently being used, will be postponed until later in this chapter. If your background extends to the vacuum tube and discrete semiconductor era you will recognize a wider variety of techniques in the current methods of demodulation, and a greater degree of sophistication.

Before leaving this introduction to the superheterodyne receiver, one aspect of Fig. 11-2 requires further explanation. The rf amplifier in this figure is not an essential feature of superheterodyne receivers in general. Indeed, AM broadcast receivers seldom have an rf amplifier stage for reasons which will be explained shortly. More specifically, the necessary overall receiver gain can be obtained in the i-f amplifier and audio amplifier following demodulation. The presence of the rf amplifier is the proper design decision under some circumstances which have not yet been examined.

A second point requires clarification. It was stated that the local oscillator frequency, f_{LO}, is greater than f_C. This is not necessarily so. Indeed there are many receivers for which f_{LO} is less than the frequency to be translated to the intermediate frequency. However, in simple superheterodyne receivers having one mixer, one local oscillator and one i-f amplifier, there is a pronounced technical advantage in positioning the local oscillator frequency above the carrier frequency. This occurs because the ratio of the highest to the lowest local oscillator frequencies required over the receiver tuning range is less than would otherwise be the case. Similarly, the relative range of tuning capacitance for the local oscillator is less—a decided advantage.

General performance specifications

The purpose of this section is to provide a general overview of the basic receiver functions expressed in terms of performance. As the organization of the superheterodyne receiver system and the functioning of its individual subsystems are subsequently analyzed, additional performance specifications will be recognized and discussed.

Selectivity is a measure of the receiver's ability to select the desired, modulated rf signal to the exclusion of undesired emissions in the same general frequency range. As a familiar example, a standard AM broadcast receiver will be tuned to a specific carrier frequency in the 540 Hz to 1600 kHz range. Assume that this frequency is 1000 kHz to be specific. Elsewhere in the United States there might be one or more stations broadcasting on 990 kHz and others on 1010 kHz. Your receiver must be capable of rejecting these neighboring transmissions without significant interference to the selected channel centered at 1000 kHz.

Continuing with this example, you will see that the technical solution to the selectivity problem is not quite so simple. In addition to the carrier frequency alone, each transmission must include sidebands so that its spectra width is twice the amount of the highest significant modulating frequency, i.e., not more than 5 kHz. Figure 11-3 illustrates a simplified interpretation of this situation. Ideally, the receiver tuned to 1000 kHz should accept the carrier and all the frequency components in the sidebands between 995 kHz and 1005 kHz equally, while absolutely rejecting frequencies outside this range.

You should recognize that the receiver challenge in Fig. 11-3 is exaggerated. Station assignments are staggered so that frequency assignments, service areas, and operating hours reduce the amount of potential adjacent channel interference.

Sensitivity is the ability of the receiver to correctly receive weak signals. This would appear to be a question of designing sufficient gain into the receiver to amplify the received signal from any transmitter, however distant it might be. However, the cost and difficulty of adding amplifier stages and achieving gain is no longer a formidable problem when compared with the true limit to sensitivity, namely noise. Noise from external sources will always be received and cannot be rejected or reduced without similar rejection or reduction of signal components in the same frequency bandwidth.

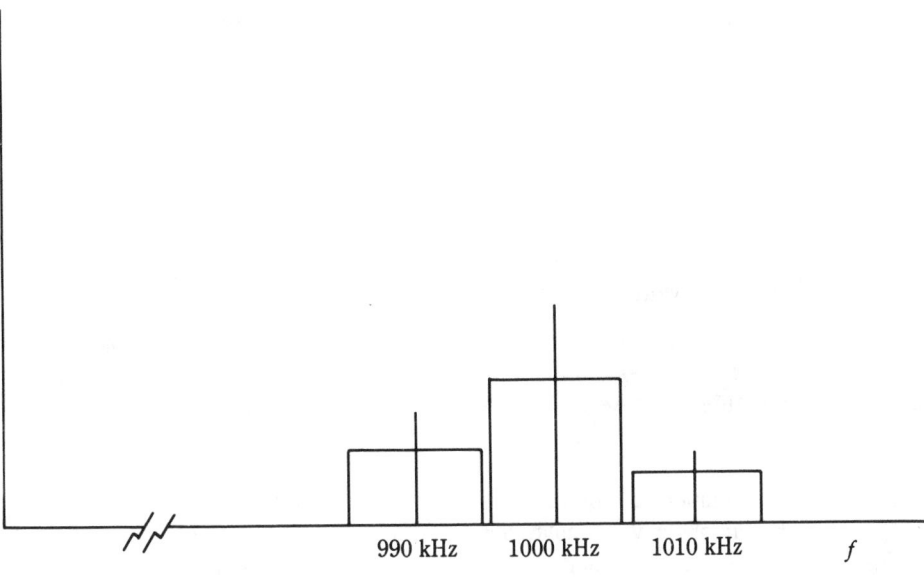

990 kHz   1000 kHz   1010 kHz   *f*

11-3   AM broadcast selectivity problem.

The key to achieving the best sensitivity possible is the reduction of the internally-generated receiver noise to the lowest amount possible. Noise performance of a receiver, inasmuch as it is influenced by design, is established by those parts of the receiver preceding the i-f amplifier.

Practical radio receivers, and especially communications-type receivers, embody a substantial number of control and other functions that improve their operations in some way. These additional functions, together with refinements in the consideration of the basic receiver, require the definition of more performance specifications which will be encountered later in the chapter.

The receiver front end

An ideal amplifier will reproduce at its output an amplified, but otherwise identical, version of the waveform at the input terminals. There should be no additional noise in the output. There should be no distortion of the waveform. Additionally, there should be no interaction of different frequency signals caused by amplifier nonlinearities so that different and unwanted products appear in the output. Every practical amplifier possesses these deficiencies to some degree.

rf amplifier noise

To place the matter of rf amplifiers in perspective, you should recognize that external sources of noise, such as atmospheric noise and man-made noise, dominate at lower rf frequencies. At vhf and uhf frequencies, the receiver noise significantly degrades the signal-to-noise ratio (S/N). The present discussion is concerned with this problem.

Every electrical network having resistance produces noise, notably thermal noise.

The available thermal noise power from a resistor, or an equivalent resistance, is given by:

$$P_N = kTB$$

in which,

> P_N is in watts
> k is Boltzman's constant = 1.38×10^{-23} watts per kelvin
> T is the absolute temperature in kelvins
> B is the bandwidth in hertz.

This formula suggests why receivers for deep-space communications are cooled to near absolute zero (a few kelvins) and the information bandwidth is severely restricted. The absolute temperature, T, in kelvins, formerly degrees kelvin, is equal to:

$$T = C + 273$$

with C being the temperature on the Celsius scale.

A perfect rf amplifier will amplify any noise introduced into its input terminals. In addition, a practical amplifier will generate excess noise internally, which has the equivalent effect of increasing the effective noise temperature of the input noise source. We account for the actual available noise, as compared to that of the ideal amplifier by assigning a noise figure, F, whose numerical value is greater than one and is often expressed in dB since a power ratio (actual noise power/ideal noise power) is implied. Thus an amplifier with a noise figure $F = 1.26$ is also said to have a $10 \log_{10} 1.26 = 1$ dB noise figure. An ideal amplifier system heated to $273 \times 1.26 = 344$ K will have the same available noise power so that the excess noise corresponds to an additional source of $344 - 273 = 71$ K.

The sinister effect of noise figure appears vividly in examining the actual amplifier output signal-to-noise ratio S_O/N_O, which is degraded by internally-originated amplifier noise, as compared with the input signal-to-noise ratio S_I/N_I.

$$F = \frac{S_I/N_I}{S_O/N_O}$$

When amplifier stages are cascaded, the first stage noise figure is the key to the overall noise performance. Let F_1 be the noise figure of the first amplifier stage and F_2 be the noise figure of the second amplifier stage of a multistage amplifier. For F_T, the noise figure of the combination, you have:

$$F_T = F_1 + (F_2 - 1)/A_{P1}$$

in which A_{P1} is the power gain of the first stage. For example, let the gain of the first stage be 8 dB and the noise figures of each stage be 3 dB. Then:

$$F_T = 2 + (2-1)/6.31 = 2.16 = 3.34 \ dB$$

Image frequency

It was previously stated that the local oscillator frequency, f_{LO}, is greater than the carrier frequency, f_C, by the amount of the intermediate frequency, f_{IF}. In principle, f_{LO} could actually be less than f_C by the same amount since the difference between the two is the desired product of the (nonlinear) mixer. The reason for selecting the larger of the two possible, workable values for f_{LO} is a very practical one as was explained earlier. Assume that the local oscillator is tuned, so as to track f_C with a fixed difference frequency, by changing the capacitance of a resonant LC circuit. The actual variation of capacitance can be achieved mechanically with the conventional interleaved-plate variable capacitor (Fig. 11-4) or electronically by adjusting the reverse bias on a varactor diode in the circuit.

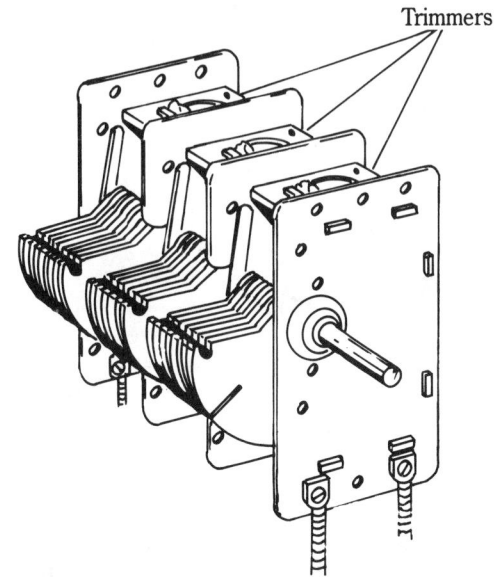

Trimmers

11-4   Variable ganged capacitor.

Whichever the actual method employed, the capacitance must be increased as the local oscillator frequency is decreased. The percentage change of capacitance required is less in tuning from the highest f_C to the lowest f_C in the band with the normal selection of f_{LO} greater than f_C.

It should be recognized that one additional frequency, called the image frequency, f_I, meets the requirement for conversion to f_{IF}. This particular frequency will be greater than f_{LO} by the amount of f_{IF}. For example, consider a citizens band receiver having f_{IF} = 11.275 MHz, which is tuned to f_C = 27.005 MHz. f_{LO} = 27.005 + 11.275 = 38.280 MHz. The image frequency f_I is 38.280 + 11.275 = 49.555 MHz.

You must bear in mind that input signals of equal strength entering the mixer of 27.005 MHz and 49.555 MHz will provide equal level inputs to the i-f amplifier. However, one or more tuned circuits precede the mixer, so the actual response to the image frequency will be much less than that of the carrier. This is illustrated in Fig. 11-5 in which no effort has been made to provide realistic scales.

The image rejection ratio, which will here be abbreviated IRR for want of a commonly-accepted symbol, is the ratio of the two responses, namely, that at f_C to that at f_I. It is given by:

$$IRR = \sqrt{1 + (FR \times Q)^2}$$

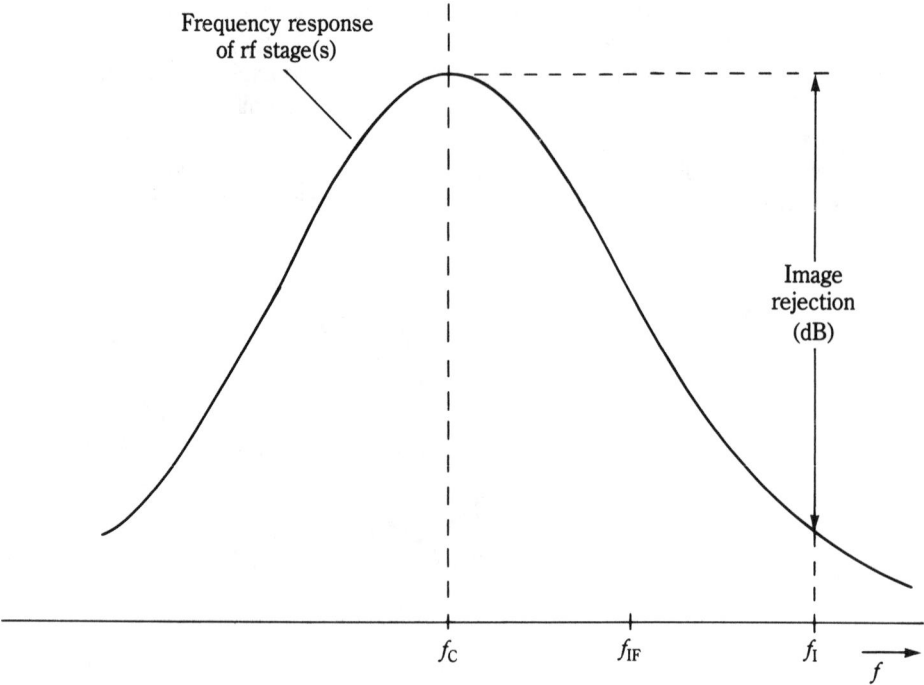

11-5   Image frequency rejection.

where

$$FR = \frac{f_I}{f_C} - \frac{f_C}{f_I}$$

and Q is that of the tuned circuit.

Using the previous CB frequency data and assuming $Q = 100$ for a single tuned circuit gives:

$$FR = \frac{49.555}{27.005} - \frac{27.005}{49.555}$$

$$= 1.29$$

$$IRR = \sqrt{1 + (1.29 \times 100)^2}$$

$$= 129$$

$$= 42dB$$

For this example, any interference source for which the received signal level is about 40 dB greater than that of the desired signal will cause severe interference. One solution to this problem is the addition of a second tuned circuit as might be accomplished by preceding the mixer with an rf amplifier. With a $Q = 100$ as before an additional 42 dB of image rejection is realized for a total of 84 dB.

Another phenomenon, related to the possible reception of interference at the image frequency, but otherwise quite different, occurs in the operation of superheterodyne receivers. In order to visualize it, return briefly to the image frequency situation of the previous example. The frequencies involved are presented in Fig. 11-6A.

Now proceed to change the dial setting in selecting a lower frequency by tuning until the local oscillator frequency is positioned at $f_C - f_{IF} = 27.005 - 11.275 = 15.730$ MHz. That is, the local oscillator frequency has been reduced by twice the i-f ($2f_{IF} = 2 \times 11.275 = 22.550$ MHz). Actually, f_{LO} has tracked the dial setting during this change from an indicated frequency $f_C = 27.005$ to $f_C = 27.005 - 22,550 = 4.455$ MHz as shown in Fig. 11-6B. The appearance of the same transmitted frequency at two points on the dial is called *double spotting*.

Spurious responses

Nonlinearities inherent in the rf amplifier and mixer lead to potential interference from radiating sources at other frequencies. It can be shown that the fundamental frequency of the local oscillator can combine with particular harmonics of an interfering signal at particular frequencies to produce products which will pass through the i-f amplifier. The same is generally true of harmonics of the local oscillator and the interfering signal. It is only necessary that the sum or difference of the fundamental and/or harmonics equal f_{IF}.

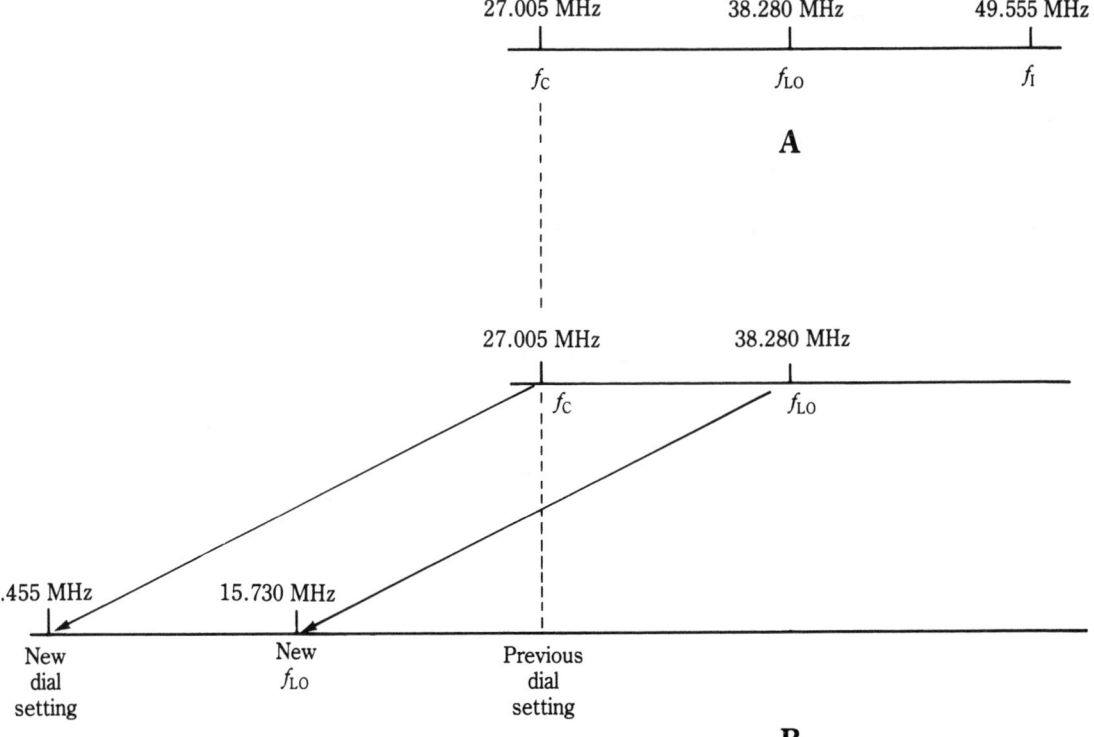

11-6   Illustrating: A. Image rejection. B. Double spotting.

To demonstrate how a spurious response is possible, consider that a receiver having $f_{IF} = 455$ kHz is tuned to $f_C = 1100$ kHz. In this case $f_{LO} = f_C + f_{IF} = 1100 + 455 = 1555$ kHz. Now, let a strong source exist at 1005 kHz for which the second harmonic generated in the mixer is 2010 kHz. The difference, $2010 - 1555 = 455$ kHz, the intermediate frequency.

In general, spurious responses are possible for the following:

$$f_{IF} = |mf_C \pm nf_{SP}|$$

in which f_{SP} is the frequency of a source giving rise to the spurious response, and both m and n are independent, positive integer values—0, 1, 2, 3, etc. The vertical lines on the right-hand side of the equation indicate that the sign of a negative difference is disregarded in testing for vulnerability to interference by spurious frequency sources. This interference is called *intermodulation distortion* (IMD). Increased levels of signal applied to the mixer result in a greater number of higher-order (larger values of m, n, or both m and n) IMD products.

In addition to IMD, there is *cross-modulation*, in which the modulation of an interfering signal passes through the receiver. The nonlinear nature of the mixer also causes its output to be reduced (compressed) in the presence of high input levels.

The *dynamic range*, the ratio of the faintest to the sounds reproduced without distortion, of a receiver is limited at the lower end by noise and at the higher end by intermodulation distortion, cross-modulation, and *gain compression*. This range can be of the order of 100 dB, depending on the receiver design, frequency range, etc. Extending the range at the lower end requires that the rf amplifier gain be sufficient to overcome mixer noise, that is, to essentially establish the receiver noise figure prior to the mixer, but not so great as to unnecessarily restrict input signal levels at the higher end.

The i-f amplifier

The i-f amplifier has three basic functions in the superheterodyne receiver.

1. It provides the major part of the overall receiver gain preceding the demodulator
2. It establishes the receiver bandwidth and selectivity
3. It incorporates the means for variable gain control, and automatic gain control (AGC), to accommodate the necessary dynamic range of the receiver

Most receivers are currently constructed using ICs for the i-f amplifier and other superheterodyne functions. Previously, receivers were made of discrete elements. Many of these receivers continue to be used, and their analysis demonstrates many important features of the i-f amplifier.

Figure 11-7 is the schematic diagram for a typical transistor i-f amplifier for which the means of controlling the gain, the AGC, is deliberately not shown. In this figure, there is a tuned input circuit, the output tank circuit of the mixer, which is formed by the primary winding self-inductance of transformer T1 in parallel with C1. Tuning of the T1 resonant circuit is accomplished by the adjustment of a powdered-iron core position (called *permeability tuning*). The secondary circuit of transformer T1 is not tuned. The overall objective in establishing the turns ratio and the primary tap position is the matching of the mixer output impedance to the i-f amplifier stage input impedance.

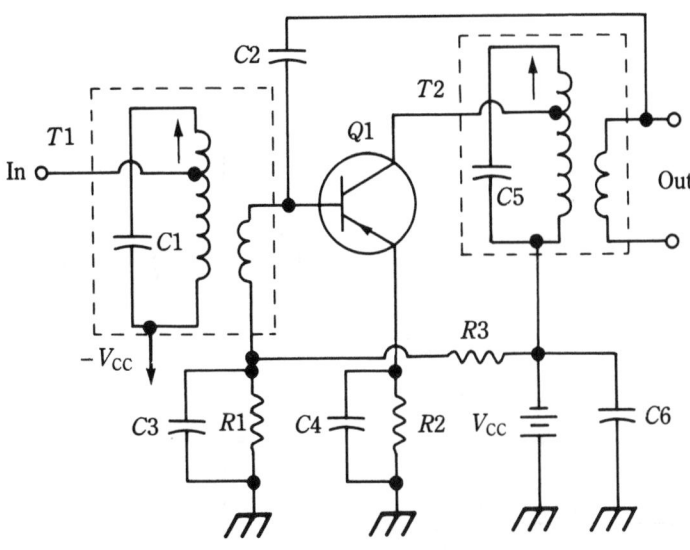

11-7 Transistor i-f amplifier.

Bias for class-A operation of the common-emitter amplifier configuration is determined by the voltage divider consisting of resistors R1, R2, and R3. Capacitors C3, C4, and C6 are bypass capacitors having low reactance values at the intermediate frequency.

The primary circuit of transformer T2 is also tuned. A tap on the primary connected to the collector of Q1 causes a larger output resistance to be reflected into the tuned circuit for the purpose of maintaining a high value of Q.

Gain control

Variable gain control is one essential feature of the i-f amplifier stage or stages. This can be accomplished for the bipolar transistor of Fig. 11-7 by disconnecting resistor R3 and applying the bias control voltage to the base. This way the dc base current can be adjusted with the consequent variation of collector current and transconductance, g_m.

Field effect transistors are frequently used in i-f amplifier stages. Adaptation of Fig. 11-7 will permit the use of a FET. Another popular circuit in Fig. 11-8 uses a dual-gate MOSFET. Note that the control voltage is applied to one gate.

In ICs gain control is usually more elaborate. For example, in Fig. 11-9 the control voltage is applied to Q2, which can alter the proportions of the Q1 current that pass through Q2 and Q3. For maximum gain the current in Q2 will be cut off. As the control voltage is increased, current is "robbed" by Q2 causing more of the Q1 current to be diverted with a consequent loss of gain.

i-f transformers

The characteristics and physical construction of transformers used as coupling devices at intermediate frequencies are quite different from those used at power and audio frequencies. In particular, i-f transformers are very small, having powdered-iron cores for high inductance, low losses, and high Q. As stated before, the core position is adjustable for permeability tuning.

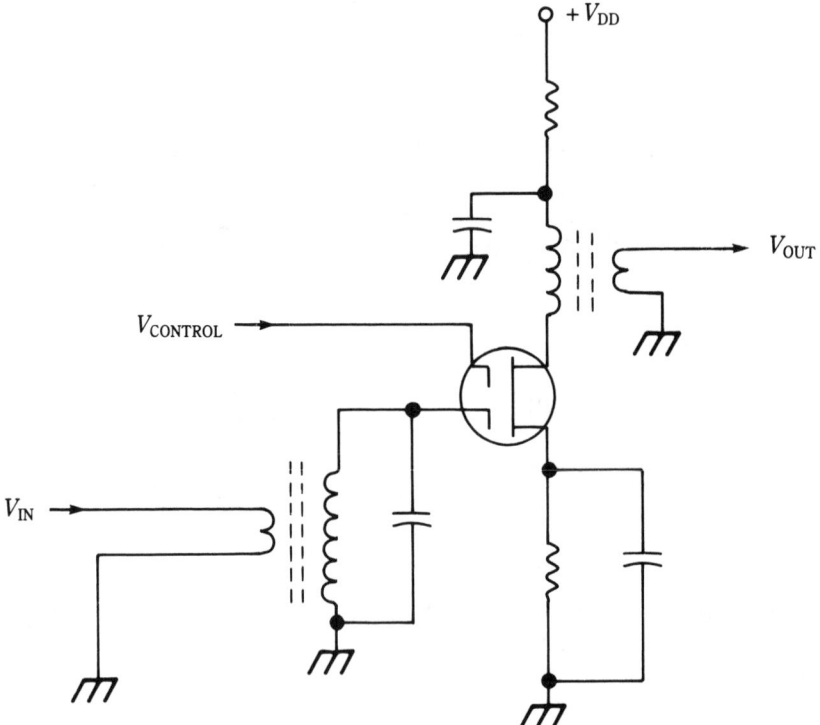

11-8 An i-f amplifier stage with dual-gate MOSFET.

One of the functions of an interstage transformer is to match the relatively low input resistance, R_I, to the relatively large output resistance, R_O (see Fig. 11-10). When, with the powdered-iron core, the coupling between the primary and secondary windings is sufficiently close to unity, the primary to secondary turns ratio is:

$$\frac{N_P}{N_S} = \sqrt{\frac{R_O}{R_I}}$$

For example, if $R_I = 1000 \ \Omega$ and $R_O = 20 \ k\Omega$

$$\frac{N_P}{N_S} = \sqrt{\frac{20}{1}}$$

$$= 4.47$$

In most applications the bandwidth requirements are such that double tuning, that is, tuning of both the primary and secondary circuits, is required. The actual characteristics of a double-tuned transformer depend on the Qs of the primary and secondary circuits and on the coefficient of coupling, k. Figure 11-11 illustrates the manner in which the values of k affect the response curves for the double-tuned transformer. Each response

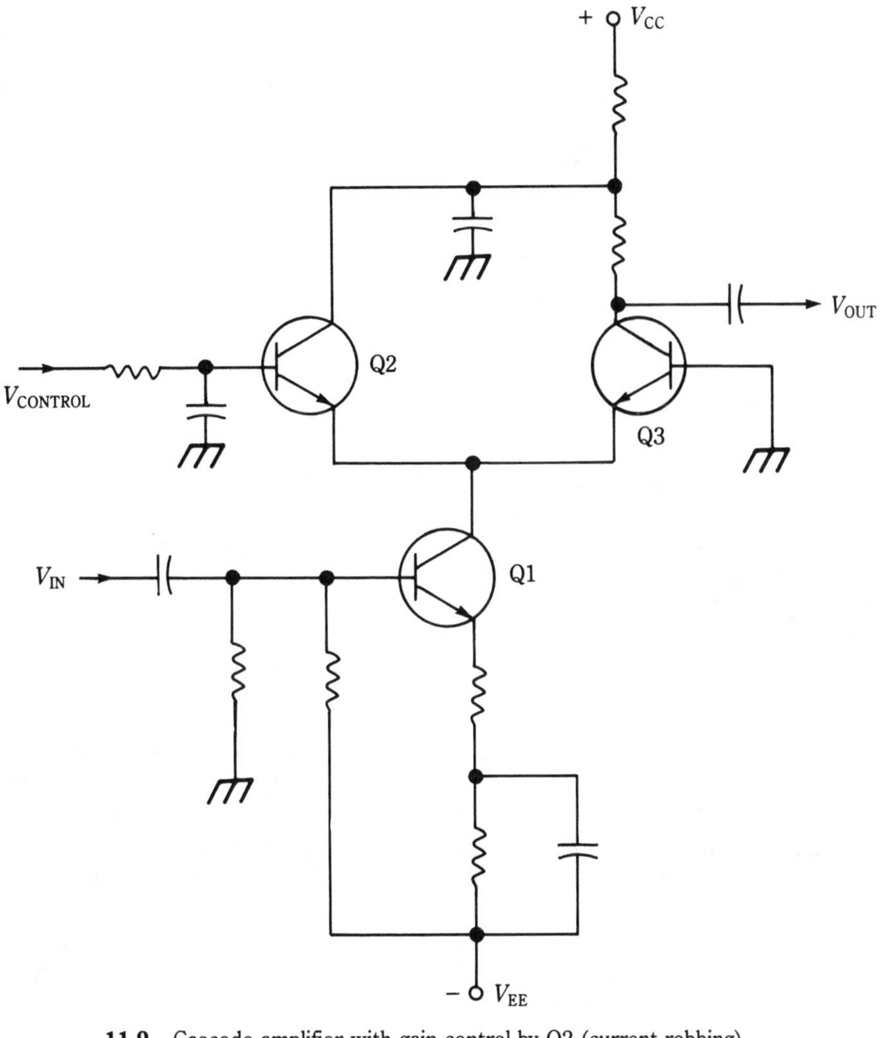

11-9 Cascode amplifier with gain control by Q2 (current robbing).

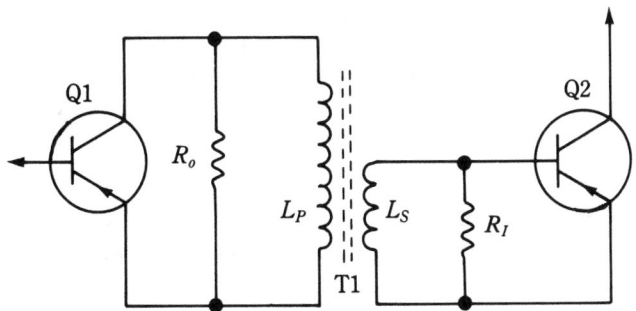

11-10 An i-f amplifier with transformer interstage coupling.

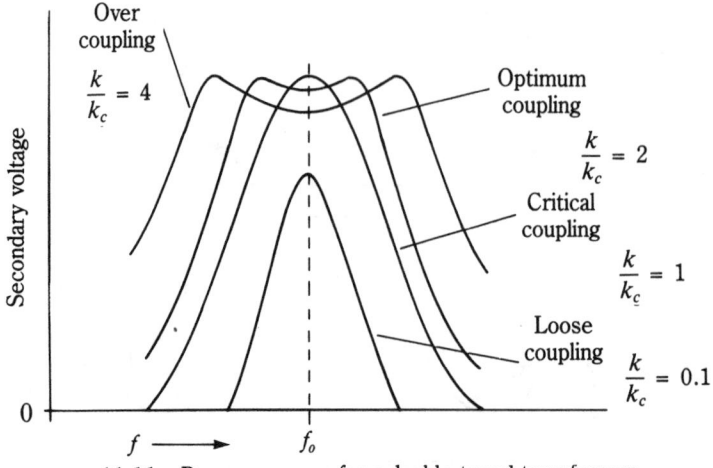

11-11 Response curve for a double-tuned transformer.

curve in the figure is labeled with the ratio of the actual coupling coefficient to the *critical coupling coefficient*, k_c, for which:

$$k_c = \frac{1}{\sqrt{Q_P Q_S}}$$

in which Q_P and Q_S are the primary and secondary Qs.

A double-tuned i-f transformer response curve more closely resembles the ideal flat-top, vertical-side ideal response than does that of a single-tuned i-f transformer. In order that the carrier and sidebands be amplified without distortion the response should be reasonably flat near the carrier frequency. However, the selectivity is improved for steep skirts of the response curve. *Shape factor* is a convenient measure of this property. It is the ratio of the response curve width at the −60 dB points divided by the width at the −6 dB points, so that the shape factor is ideally 1. The shape factor is improved for cascaded i-f amplifier stages coupled by double-tuned transformers. Table 11-1 demonstrates this effect for both single-tuned and double-tuned transformers.

Interstage filter coupling

Most i-f transformers are relatively bulky for i-f amplifiers having ICs. Moreover, the shape factor presents severe limitations in using passive electrical components, and, especially, for SSB receivers. For these reasons, alternate filters have undergone intense development for a number of years. The various forms of filters based on mechanical vibration and the SAW filter were introduced in chapter 1.

- Mechanical filters have an equivalent Q in the range 10 to 1000. They are useful for frequencies to about 500 kHz.
- Ceramic filters, based on the piezoelectric effect, have Qs to about 2000 and are useful to about 10 MHz.
- Piezoelectric quartz crystals have even higher values of Q and can be used for frequencies in excess of 100 MHz.

**Table 11-1. Influence of
the Number of Stages on
the Shape Factor.**

Number of stages	Shape factor	
	Single-tuned	*Double-tuned*
1	577	23.9
2	33	5.63
3	13	3.59
4	8.6	2.94

To achieve the desired shape factor, multiple piezoelectric quartz crystals are used in various configurations which take advantage of the crystal's various resonant modes. The crystal gate and crystal lattice filters of Fig. 11-12 have been designed into SSB receivers for some time.

Three-electrode resonators are constructed by dividing one electrode of the ceramic or quartz crystal filter element as suggested in Fig. 11-13 which also include approximate equivalent circuits including resistances to account for losses. Figure 11-14 shows the response of a three-electrode crystal resonator with a center frequency of 75 MHz.

Integrated circuit receivers

The present trend in broadcast receivers is toward the complete integration of most, if not all, receiver functions. Your present objective will be more modest in examining the application of an IC for the replacement of the active devices in a simple receiver, confined to the functions preceding demodulation.

Figure 11-15 is a schematic diagram of the National Semiconductor LM1820 AM/FM radio system which was designed primarily for AM superheterodyne applications utilizing an rf amplifier stage preceding the mixer-oscillator. However, the present application is for a low-cost AM receiver without the rf amplifier as seen in the block diagram of Fig. 11-16. The schematic diagram (Fig. 11-17) for the receiver shows the circuitry external to the IC chip including the tuning capacitor with sections that tune the antenna coupler and the local oscillator. The AGC function, to be explained presently, is implemented by the LM1820.

In Fig. 11-17 the small radio receiver is implemented with the LM1820 and a second IC (IC2), the LM386, operating an audio amplifier.

The application information just described for the LM1820 was published in 1975, which establishes a time reference for the level of sophistication in this application.

Control circuitry

The present emphasis will be on the more conventional automatic control of gain which is necessarily embodied in all practical AM and FM receivers. Additionally, squelch (muting) will be briefly covered.

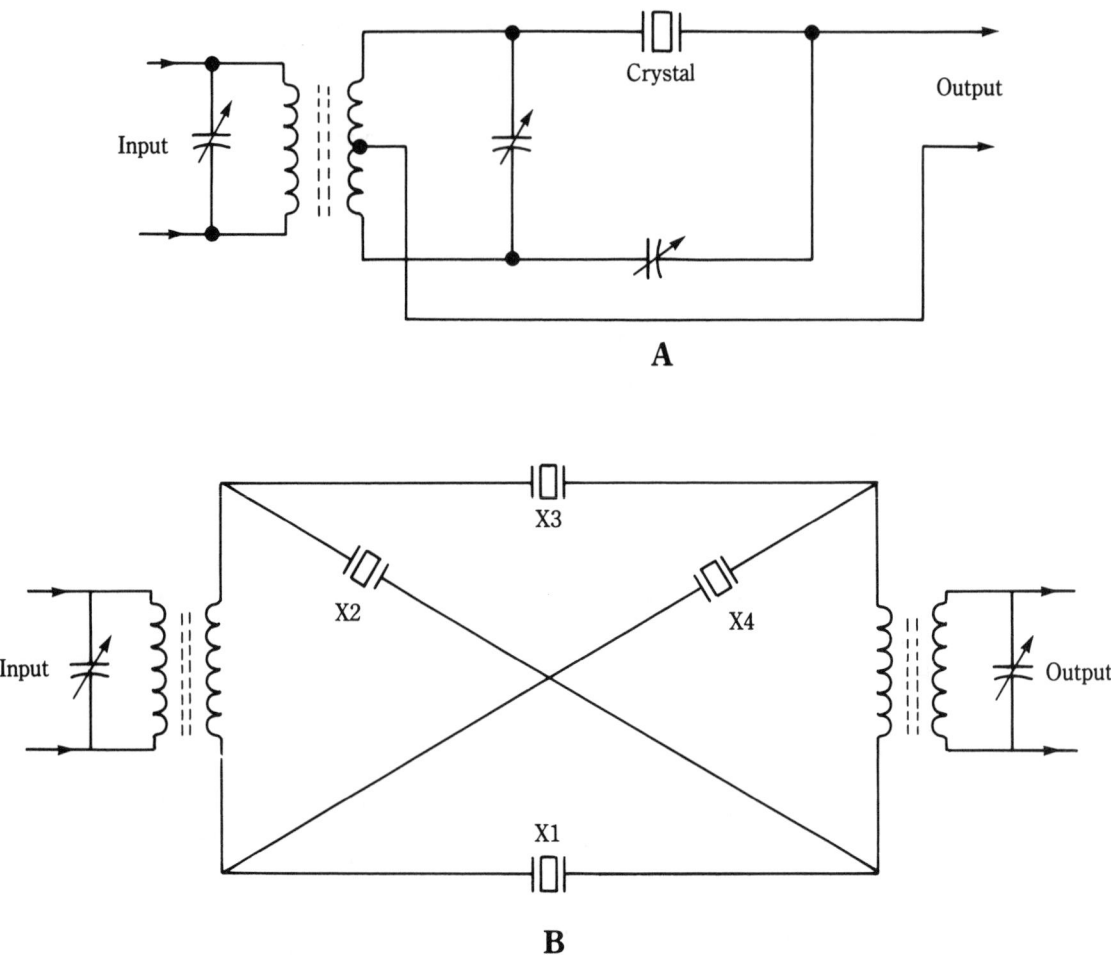

11-12 Piezoelectric crystal filters: A. Crystal gate. B. Crystal lattice.

A large number of sophisticated control functions are presently designed into receivers which simply did not exist a few years ago. For example, the hand-off of mobile transceivers from one cell to another in cellular radio systems is very complex. Of course, the principal control functions are directed by fixed land installations. Nevertheless, a significant number of control functions are necessarily a part of the mobile equipment, both receiver and transmitter. Further treatment of the cellular system operation is beyond the scope of this book.

Automatic gain control (AGC) causes the gain of one, or several, amplifier stages to be modified in accordance with the received signal strength. More accurately, negative feedback contrives to maintain the average dc level, obtained from smoothing the demodulator output, constant. A stronger received signal will provide a stronger output from the demodulator in the absence of negative feedback. Using negative feedback, the bias of the amplifier stages is automatically shifted to reduce the overall gain. Conversely,

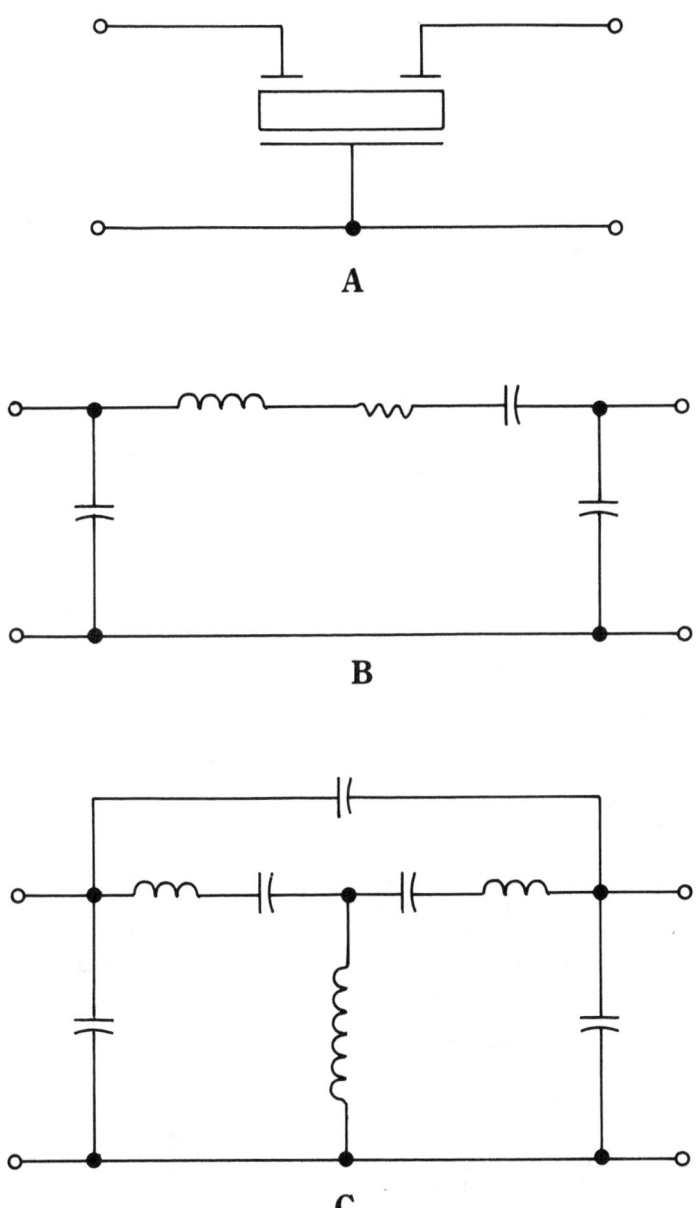

11-13 Three-electrode resonator: A. Symbol. B. Approximate equivalent circuit for ceramic filter. C. Equivalent circuit for monolithic crystal filter.

a weaker received signal will cause the bias, as determined by the AGC system, to increase the amplifier gain. The net effect is to maintain the volume level at the receiver substantially constant for a large field strength range of variations at the antenna.

Delayed AGC is quite similar, except that no gain reduction occurs until a minimum threshold signal is received. For larger input signal levels, the gain is reduced. There-

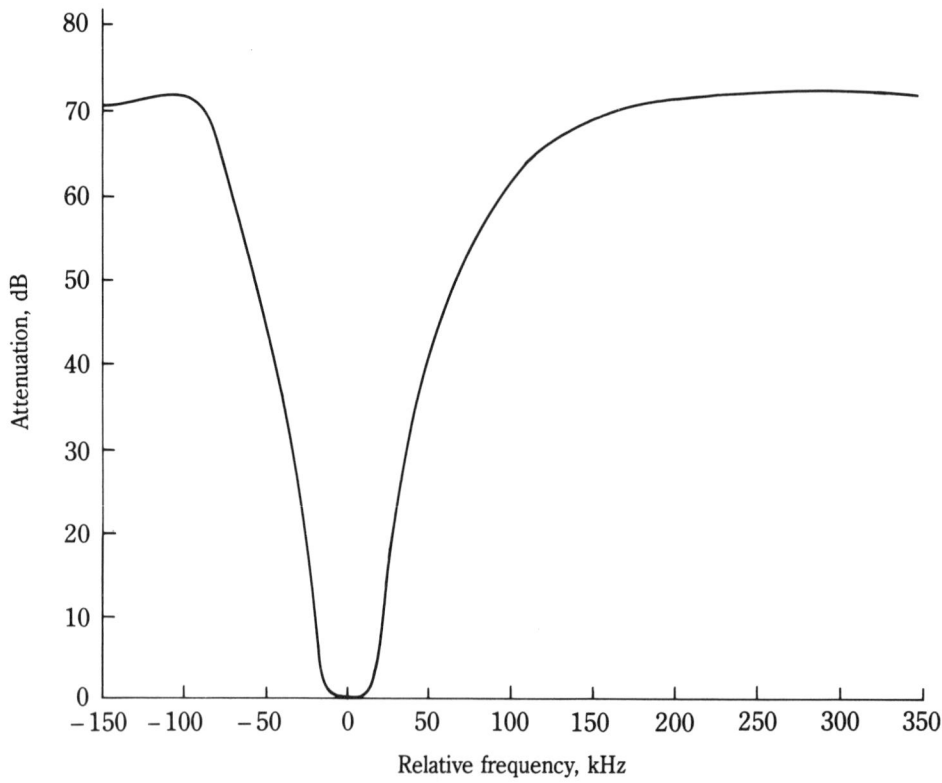

11-14   Frequency response of four-pole monolithic crystal filter (f_0 = 75 MHz)

fore, the AGC does not interfere with the reception under the difficult circumstances for which all possible gain might be necessary.

Squelch (muting), while related to AGC indirectly, is quite different. Consider the action of a receiver for which the received signal is decreasing. With AGC, the gain will automatically be increased, until the noise level completely masks and makes the received information completely unintelligible.

There is no recourse for improving the conveyance of information for this condition. However, with an extremely weak received signal, or none at all, the operator must listen to noise unless the receiver output is turned off. Squelch, a descriptive name assigned to this function, does indeed turn the output off in the absence of an rf carrier above an assigned threshold level. Squelch is especially useful when strong transmissions are sporadic and brief, as in mobile radio communications. To the listener, the receiver is silent unless a message is being received.

The communications receiver

A typical communications receiver, the Kenwood R-1000, has been chosen to illustrate some of the features usually found in this type of equipment. The specifications for the R-1000 are presented in Table 11-2. A preliminary look at these specifications reveals

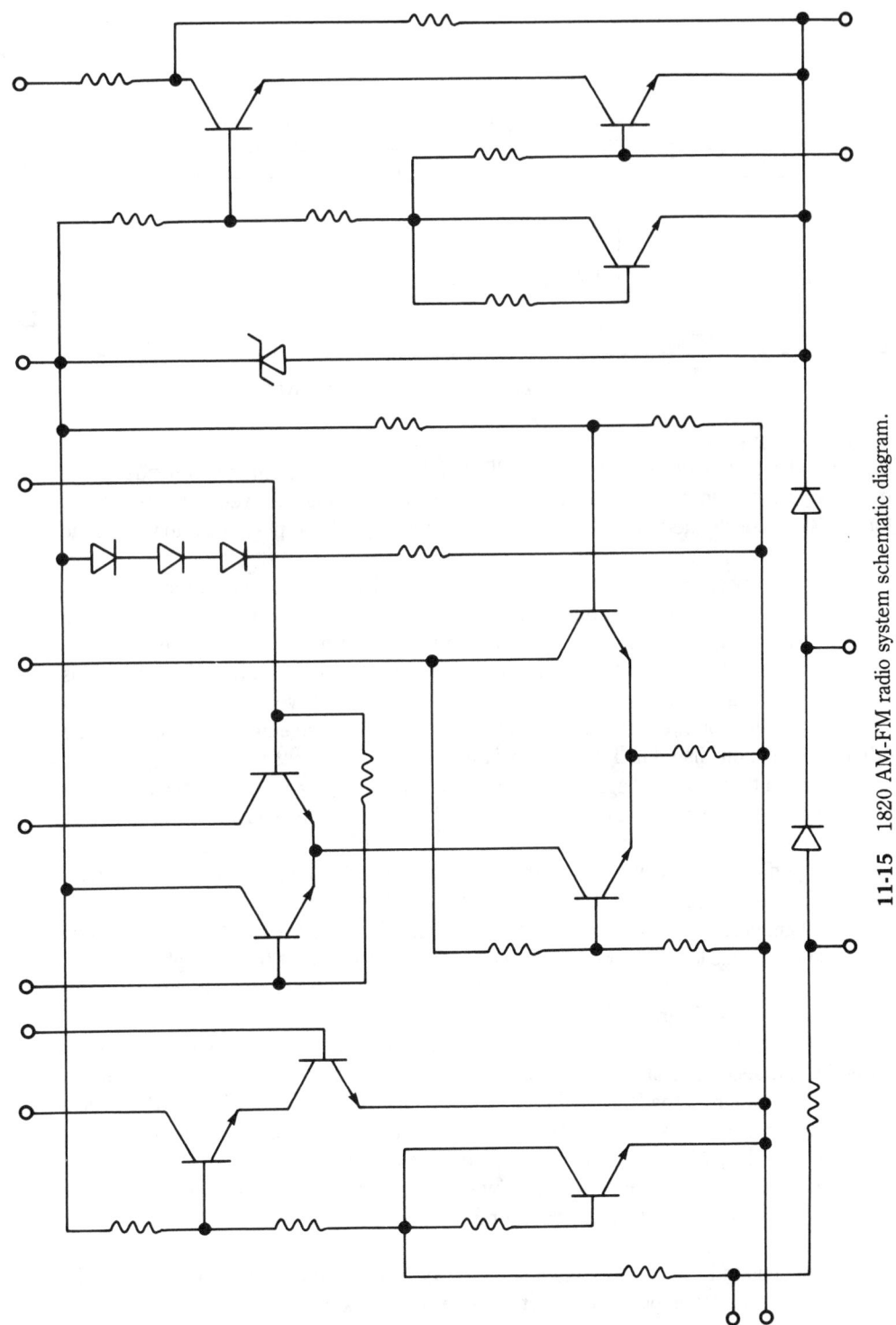

11-15 1820 AM-FM radio system schematic diagram.

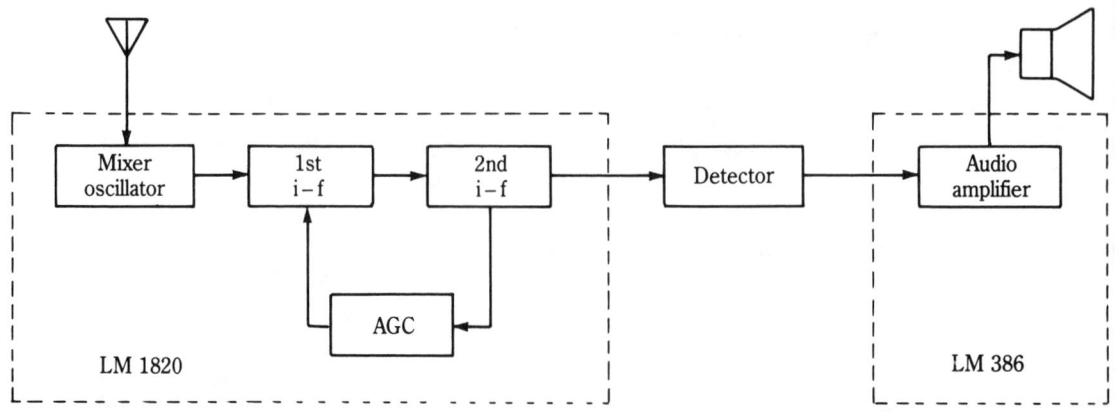

11-16   Low cost AM receiver.

that the rf frequency range is 200 kHz to 30.0 MHz. Receiver sensitivity depends on the frequency range and whether or not SSB modulation is being received. The sensitivity for the R-1000 is defined to be the rf input level yielding a signal-plus-noise to noise ratio of 10 dB.

From the specifications it is seen that the selectivity is expressed as the shape factor or an equivalent measure of the frequency response characteristic. For AM reception wide or narrow bandwidth options are available through front panel switch selection.

Two antenna terminals have been placed on the back of the receiver chassis. One is for MW (medium frequency) and the other for SW (short wave, i.e., high frequency) operation. Either of two antenna impedance levels can be selected by a switch. The other specifications listed in Table 11-2 will not receive comment here.

Now turn to the block diagram of the receiver shown in Fig. 11-18. At the upper left corner, the inputs from the antennas are coupled to a variable attenuator permitting 0 to 60 dB to be set, and the output of the attenuator is fed to one of six bandpass filters, depending on the frequency to which the receiver is tuned. The bands are 0.2 – 1 MHz, 1 – 2 MHz, 2 – 4 MHz, 4 – 8 MHz, 8 – 16 MHz, and 16 – 30 MHz.

The bandpass filters are followed by two rf amplifier stages and the first balanced mixer. The rf signal is there mixed with the local oscillator output from the phase-locked loop unit (PLL UNIT) to produce an i-f of 48.055 MHz. See the lower left corner of Fig. 11-18 for the block diagram of the PLL UNIT.

The first i-f strip consists of two monolithic filters and associated circuitry and is followed by the second balanced mixer which also receives a 47.6 MHz output from the local oscillator source, the PLL UNIT. Next in the chain is the NOISE BLANKER which senses the presence of a noise pulse and silences the receiver during its duration. This circuit is especially effective in reducing the effects of impulse noise.

The NOISE BLANKER is followed by one of three 455 kHz ceramic filters. The narrow AM (6 kHz), wide AM (12 kHz), or SSB (2.7 kHz) filter selection is made by the front panel MODE switch.

Two stages of i-f precede demodulation of the AM or continuous wave (CW) SSB. Amplification then takes place sufficient to drive the speaker or head set. The output of

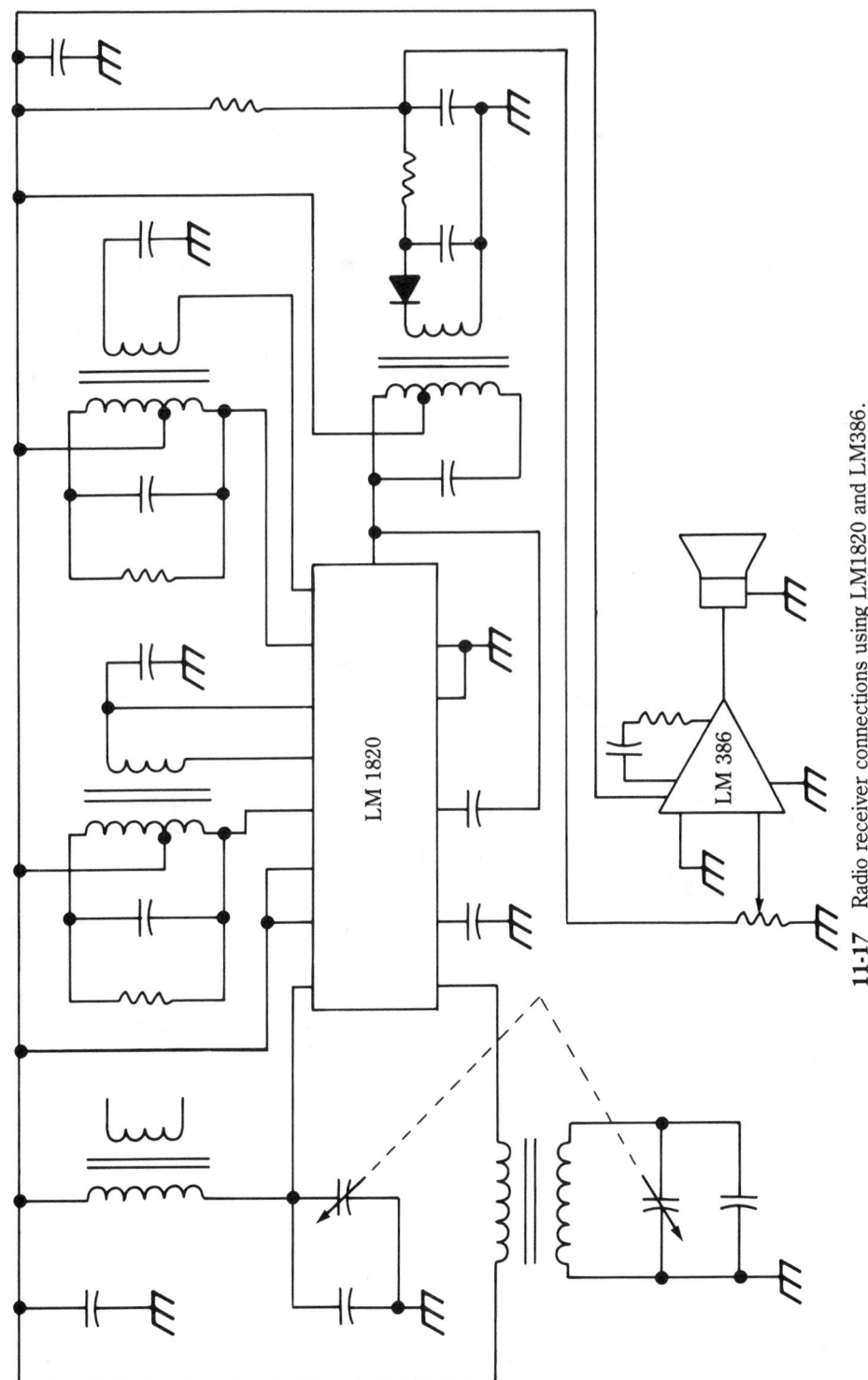

11-17 Radio receiver connections using LM1820 and LM386.

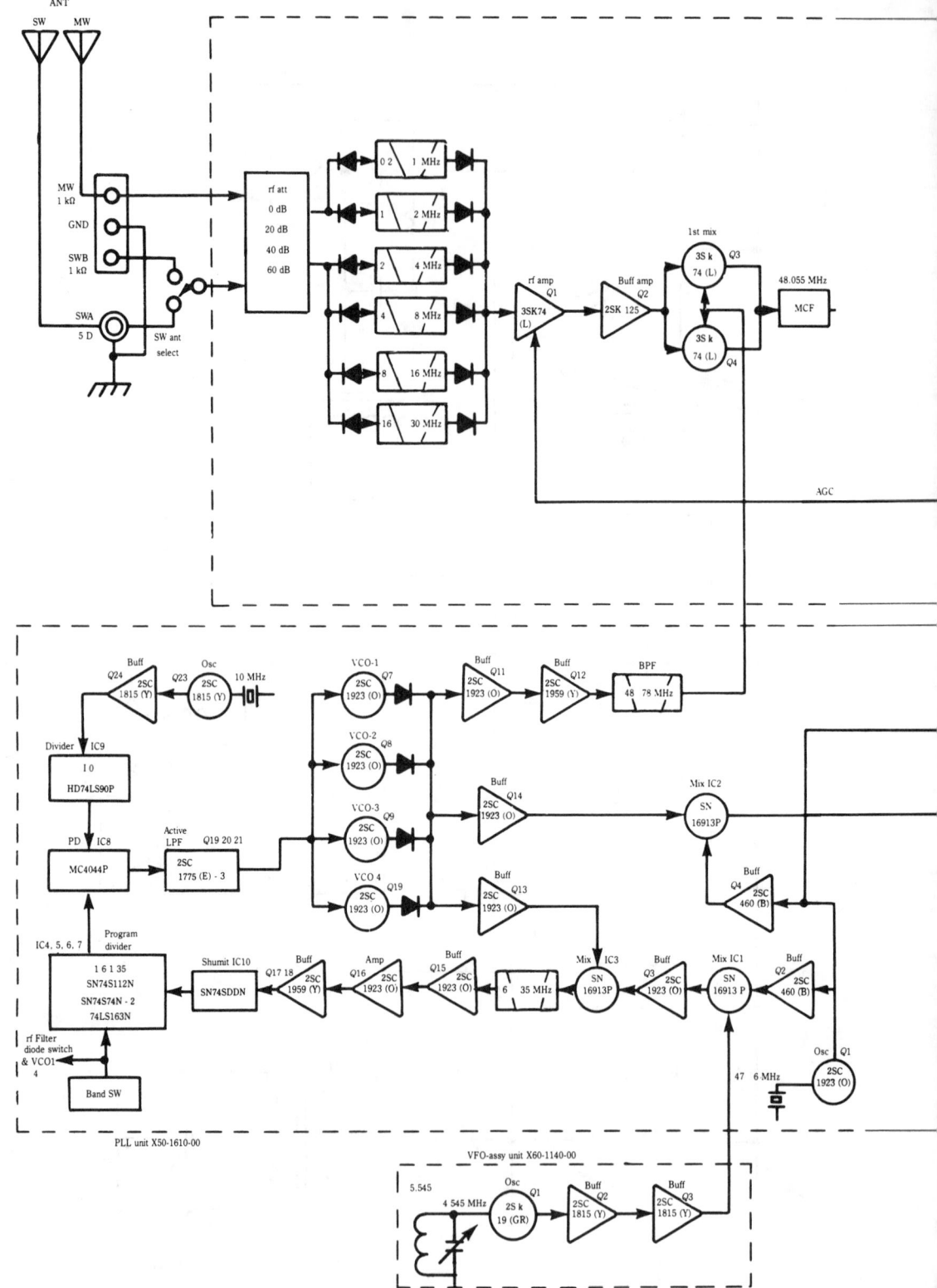

11-18 Block diagram of Kenwood R-1000 communications receiver.

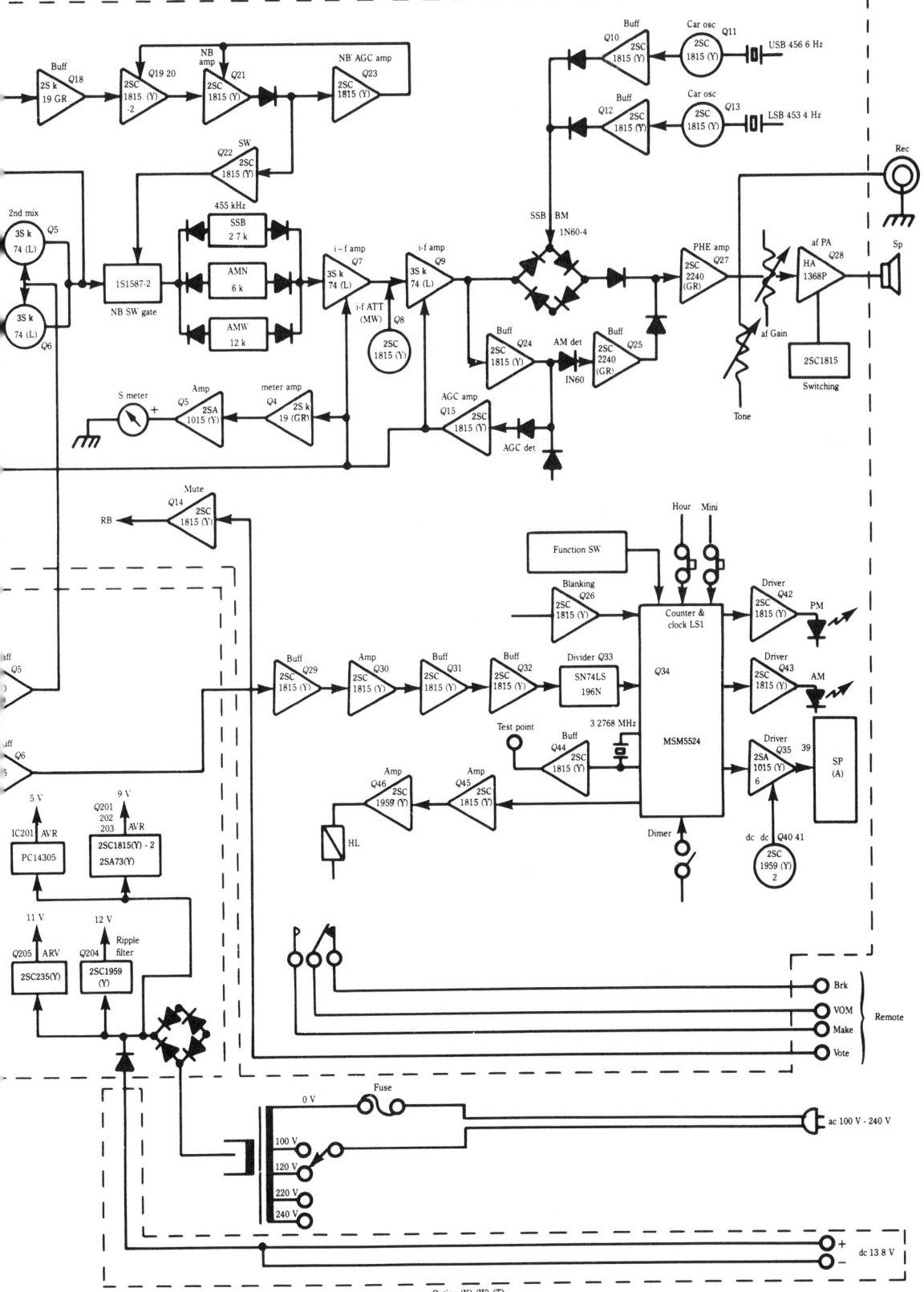

Option (K) (W) (T)

Table 11-2. Kenwood R-1000 Communications Receiver Specifications.

Frequency Range	200 kHz-30.0 MHz	
Mode	AM, SSB, CW	
Sensitivity (S + N/N 10 dB or more):		
	SSB	AM
200 kHz-2 MHz....................	5μV	50μV
2 MHz-30 MHz....................	0.5 μV	5 μV
Image ratio	More than 60 dB	
i-f rejection	More than 70 dB	
Selectivity:		
AM (wide)	12 kHz at −6 dB, 25 kHz at −50 dB	
AM (narrow)	6 kHz at −6 dB, 18 kHz at −50 dB	
SSB/CW.........................	2.7 kHz at −6 dB, 5 kHz at −60 dB	
Frequency stability:		
±2 kHz max. from 1 to 60 minutes after power on		
±300 Hz max. in every subsequent 30 minutes		
Antenna impedance..................	MW 200 kHz-2 MHz, 1 kΩ (unbalanced)	
	SWA 2 MHz-30 MHz, 50 Ω (unbalanced)	
	SWB 2 MHz-30 MHz, 1 kΩ (unbalanced)	
Audio output	1.5W min. (8Ω load, 10% distortion)	
Audio load impedance...............	4 − 16Ω, external speaker or headphone	
Power consumption..................	20W	
Power requirements	100, 120, 220, 240 V AC, 50/60 Hz	
Semiconductors.....................	40 ICs, 11 FETs, 63 transistors, 71 diodes,	
	1 display tube	
Dimensions	W 300 mm (12-3/4 inch)	
	H 115 mm (4-1/2 inch)	
	D 218 mm (8-9/16 inch)	
Weight	5.5 kg (12.1 lbs)	

(Courtesy Kenwood U.S.A. Corporation)

the demodulator is also filtered and amplified to provide automatic gain control. Briefly the function of the AGC is to sense the dc level in the demodulator output and to control the gain of the preceding rf and i-f amplifier stages by means of a negative feedback loop, so as to maintain this level nearly constant for a fairly wide range of rf input signal levels.

The significance of the PLL UNIT must be acknowledged. Its configuration is shown in Fig. 11-19. Within the PLL UNIT are four VCOs which, together with a VFO UNIT, supply frequencies from 48.255 to 78.055 MHz for the first mixer in the main receiver unit. It must also supply the correct frequencies for the second mixer. The exact local oscillator frequencies are obtained using synthesizer techniques with a 10 MHz crystal oscillator reference.

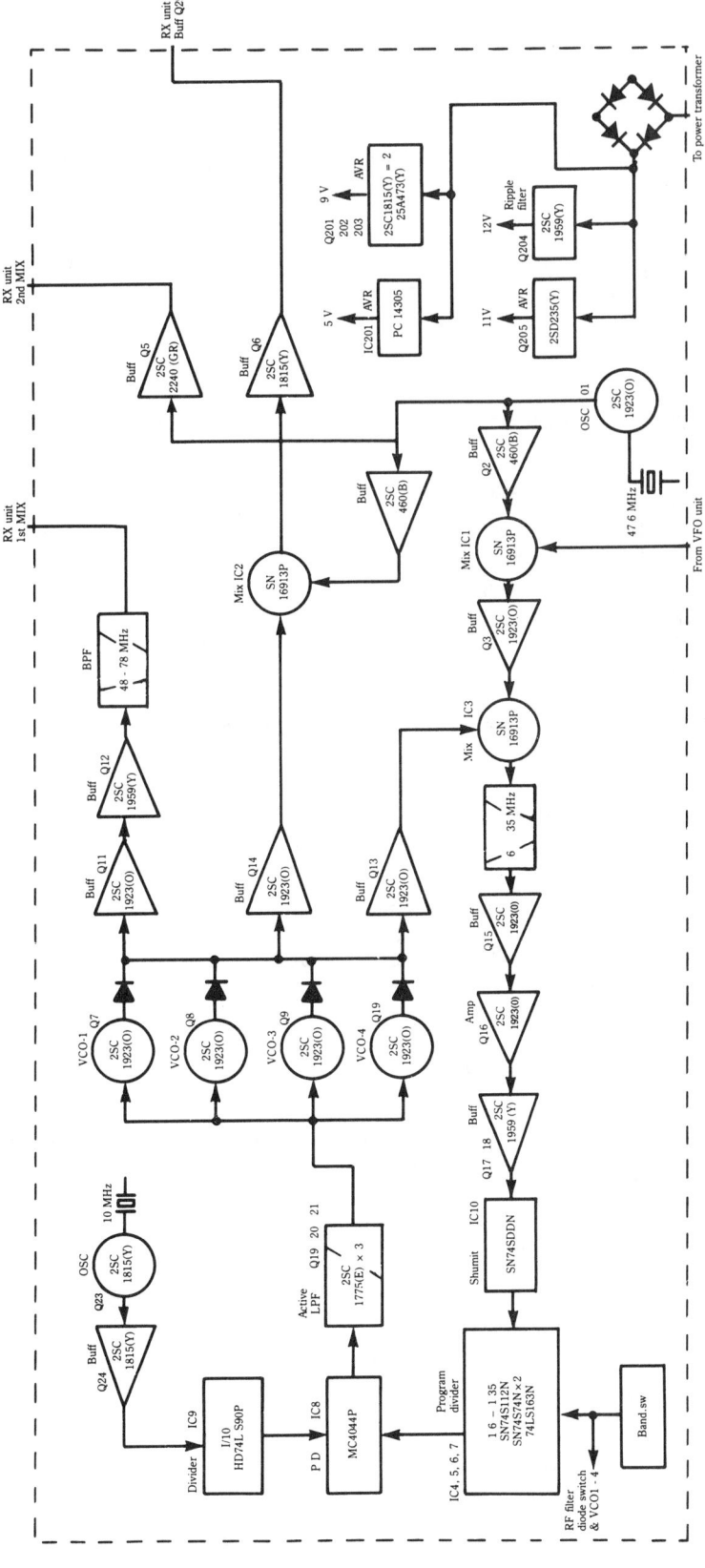

11-19 PLL unit for Kenwood R-1000 receiver.

Multiple-choice questions

1. A receiver in which all rf amplifier stages require manual tuning to the desired rf is called:

A. Superheterodyne.
B. Autodyne.
C. TRF.
D. AFC.
E. AGC.

2. Why is it often necessary to precede the demodulator by amplifier stages in a receiver?

A. To improve fidelity.
B. To reduce receiver noise.
C. To eliminate image response.
D. Weak antenna signals.
E. Because of spurious oscillations.

3. A serious disadvantage of the TRF receiver is:

A. Bandwidth variations over the tuning range.
B. The weight and cost.
C. The requirement for a closely regulated power supply.
D. The requirement for a half-wave antenna.
E. Expensive shielding requirements in export models.

4. Identify any major part of a superheterodyne receiver:

A. Local oscillator.
B. Modulator.
C. i-f amplifier.
D. Demodulator.
E. Choices A, C, and D.

5. Which major element will not be found in every superheterodyne receiver?

A. rf amplifier.
B. Mixer.
C. Local oscillator.
D. i-f amplifier.
E. Demodulator.

6. Which major element of a superheterodyne receiver must be nonlinear?

A. rf amplifier.
B. Mixer.
C. Local oscillator.

D. i-f amplifier.
E. Demodulator.

7. An FM receiver is tuned to 93.5 MHz. Assuming an i-f of 10.7 MHz, which of the following frequencies will not appear at the mixer output?

    A. 5.35 MHz.
    B. 10.7 MHz.
    C. 93.5 MHz.
    D. 104.2 MHz.
    E. 197.0 MHz.

8. The change of the modulated carrier frequency from the original rf to the i-f of the superheterodyne receiver is known as:

    A. Frequency multiplication.
    B. Frequency allocation.
    C. Frequency substitution.
    D. Frequency translation.
    E. All of the choices are appropriate.

9. A receiver must be able to select one carrier frequency while disregarding (ideally) the existence of other carrier frequencies in the same overall frequency range. This receiver quality bears the name:

    A. Sensitivity.
    B. Selectivity.
    C. Image rejection.
    D. Noise rejection.
    E. Tuning range.

10. The key to achieving receiver sensitivity is the reduction of:

    A. Image response.
    B. Mixer harmonic products.
    C. Spurious frequency response.
    D. Mixer nonlinearity.
    E. Internal noise.

11. Which of the following receiver design objectives is impossible?

    A. Elimination of galactic noise.
    B. Elimination of atmospheric noise.
    C. Elimination of man-made noise.
    D. Reduction of receiver internal noise.
    E. A, B, and C represent impossible design objectives.

12. In comparing the (S/N) ratio for the input to the receiver with the (S/N) ratio for the output, you find the latter to be:

A. Less.
B. The same.
C. Greater.
D. Infinite.
E. There is insufficient information to answer the question.

13. At 5 K (five degrees on the Kelvin scale) the temperature on the Celsius scale is:

A. −293°C.
B. −268°C.
C. −20°C.
D. +20°C.
E. +293°C.

14. Name the characteristic of a receiver which specifies the self-generated noise:

A. Noise immunity.
B. Noise factor.
C. Noise figure.
D. Muting factor.
E. Quietness coefficient.

15. An FM receiver with an i-f of 10.7 MHz is tuned to 98.7 MHz. What is the numerical value of the image frequency?

A. 77.3 MHz.
B. 88.0 MHz.
C. 109.4 MHz.
D. 120.1 MHz.
E. None of the choices is true.

16. A source of rf interference exists at 109.9 MHz. For which frequency in the FM broadcast band will this be the image frequency?

A. 21.4 MHz.
B. 88.5 MHz.
C. 99.2 MHz.
D. 110.7 MHz.
E. 121.4 MHz.

17. The ratio of the superheterodyne receiver response at the desired carrier frequency to that at the image frequency is called:

A. The sensitivity.
B. The selectivity.
C. The image frequency.
D. The image rejection ratio.
E. Both B and D.

18. In order that the same station be received at a different point on the dial by double-spotting:

    A. There must be no image frequency.
    B. The receiver must be a superheterodyne.
    C. There must be two mixers.
    D. The local oscillator frequency must be less than the carrier frequency.
    E. Both B and D.

19. Intermodulation distortion (IMD) interference is related to:

    A. Galactic noise.
    B. Inadequate stereo separation.
    C. Delayed AGC.
    D. Audio amplifier nonlinearity.
    E. Spurious responses.

20. When the modulation of an interfering signal passes through the receiver, this is known as:

    A. Intermodulation interference.
    B. Spurious modulation.
    C. Interfering modulation.
    D. Cross-modulation.
    E. Hash.

21. The part of the receiver that provides the major share of the overall gain is the:

    A. rf amplifier.
    B. Mixer.
    C. Local oscillator.
    D. i-f amplifier.
    E. Audio amplifier.

22. Receiver bandwidth is established by the:

    A. rf amplifier.
    B. Mixer.
    C. Local oscillator.
    D. i-f amplifier.
    E. Audio amplifier.

23. An essential feature of i-f amplifier stages is:

    A. Zero phase shift.
    B. Variable gain.
    C. Single sideband operation.
    D. dc gain.
    E. Dual channels for stereo capability.

24. The core of an i-f transformer usually contains:

   A. Teflon.
   B. High vacuum.
   C. Computer nylon.
   D. Powdered iron.
   E. Laminated steel.

25. Shape factor is a measure of:

   A. Bandwidth.
   B. Skirt steepness.
   C. Coupling coefficient.
   D. Critical coupling.
   E. Peak-to-valley ratio.

26. Referring to the block diagram of a particular IC for radio receivers, the LM1820, its functions include:

   A. All transistors and circuitry for the mixer and local oscillator.
   B. All transistors and circuitry for the i-f amplifier.
   C. All transistors and circuitry for the demodulator.
   D. A and B only.
   E. None of these choices is correct; external circuitry is required.

27. The function which tends to maintain the sound volume level of a voice receiver nearly constant for a large signal strength range is called:

   A. Squelch.
   B. Muting.
   C. AGC.
   D. AFC.
   E. Both A and B.

28. The function which tends to silence the receiver in the absence of a transmitted carrier is called:

   A. Squelch.
   B. Muting.
   C. AGC.
   D. AFC.
   E. Both A and B.

29. A device incorporated in a communications receiver to reduce impulse noise is called the:

   A. Front-end cooler.
   B. Squelch.
   C. AGC.

    D. Noise blanker.

    E. Image rejector.

30. In the Kenwood R-1000 receiver, the local oscillator frequencies are derived from:

    A. The first mixer.

    B. The second mixer.

    C. The IFF UNIT.

    D. The PLL UNIT.

    E. The NOISE BLANKER.

Basic problems

1. What must be the tuning range of the local oscillator of an AM broadcast receiver for which the rf range is 540 kHz to 1600 kHz?

2. Determine the change in noise power when the receiver bandwidth is doubled.

3. When the receiver bandwidth is doubled, what is the effect on the noise voltage?

4. Calculate the image frequency for an FM station at 101.1 MHz assuming an i-f of 10.7 MHz.

5. If the receiver of Basic Problem 4 has sufficient tuning range, at what point on the dial will the same station be received by double spotting?

6. Calculate the turns ratio of a tightly-coupled i-f transformer which matches a 10-kΩ source to a 500-Ω load.

7. A double-tuned i-f amplifier has two stages and a shape factor of 5.65. What shape factor is to be expected for three stages?

8. A receiver has a noise figure, F, of 3 dB. If the (S/N) ratio of the output is 10 dB, what is the input (S/N) ratio?

9. The noise figure for each amplifier stage of a two-stage amplifier is 3 dB, and the gain of the first stage is 10 dB. Find the overall noise figure, F_T.

10. For what reason would an adjustable attenuator be placed between the antenna and the first rf amplifier of a communications receiver?

Advanced problems

1. Consider the very real possibility of the local oscillator frequency being less than the carrier frequency, as contrasted with the normal arrangement. Use the data of Basic Problem 1 and examine the practicality of this proposal.

2. A communications receiver has a 2 kHz bandwidth and an input resistance of 50 Ω. Calculate the available noise power at standard room temperature of 20°C.

3. Determine the noise voltage corresponding to the noise power of Advanced Problem 2.

4. The (S/N) ratio at the output of a receiver is 10 dB, whereas the (S/N) ratio at the input is 14 dB. What is the noise figure of the receiver?

5. In the receiver of Advanced Problem 4, assume that all of the receiver noise originates in the first two rf stages, the first stage has a gain of 9 dB, and both stages have equal noise figures which are to be calculated.

6. Determine the image rejection ratio for an FM receiver tuned to 88.1 MHz assuming no rf amplifier stage and a *Q* of 100 for the antenna coupler. Use 10.7 MHz for the i-f.

7. Another amplifier stage is to be added to the receiver of Advanced Problem 6 for the purpose of making the total IRR equal to 60 dB. What value of *Q* is required of the tuned circuit to be added?

8. A mechanical filter response has a width of 4.2 kHz at −60 dB and a width of 3.5 kHz at −6 dB. Calculate the shape factor.

9. A particular receiver has an input (S/N) ratio of 8 dB and an output (S/N) ratio of 7 dB. Determine the noise figure of the receiver as a numerical ratio and also express it in decibels.

10. Refer to the Kenwood R-1000 in the communications receiver section. Determine the frequencies supplied by the PLL UNIT to

   A. the first mixer

   B. the second mixer for an rf carrier frequency of 29.000 MHz.

12

Television

FACSIMILE AND TELEVISION ALLOW VISUAL IMAGES TO BE TRANSFERRED BETWEEN geographically-separated points by electronic means. Facsimile most often causes a single page to be reproduced at a distant location; a printed record, weather map, or photograph are typical examples. Television resembles motion pictures in that frames, usually originating from live scenes or from recordings, are handled in rapid succession to give the illusion of continuous motion. An overwhelming percentage of television receivers display the reproduced images on cathode ray oscilloscopes in black and white (monochrome) or in color.

Despite the pervasive influence of television in modern life, few people have any concept of color television complexity. Indeed, the color TV receiver is probably the most intricate electronic equipment being mass-produced today. However, the widespread use of ICs has reduced the parts count to the point that the visible circuitry is no longer impressive.

You will first examine the facsimile basics because this system continues to be an important communications technique, and also it is a simple introduction to the scanning of images employed by television.

Next, you will deal with the TV video waveform, which includes the specialized pulses used to synchronize the electron beam motions in the CRTs at the sending and receiving points. You will see that the video information is multiplexed (in time) with the synchronizing and blanking pulses.

Having examined the TV video signal in the time domain, you will look at the modulation methods for the video information and the separate modulation system for the sound. The frequency spectrum will then be analyzed for monochrome TV.

Next, the subject of color TV will be introduced, to be followed by sections on transmitters, receivers, and other equipment used in TV broadcasting.

The following sections are included in this chapter:

- Facsimile
- TV scanning and synchronization

- TV camera tubes
- Modulation and spectrum
- Color television
- Transmitting systems
- Monochrome television receiving systems
- Color television receiving systems
- Digital signal processing for TV

Facsimile

The original facsimile, which preceded the growth of the TV broadcast industry by a number of years, was analog in nature and very simple. A sheet of paper containing the hard-copy image is fastened to the outside of a cylinder that is mounted on a lead screw (see Fig. 12-1). An optical system containing both a lamp light source and photoelectric cell is focused at one point on the cylindrical surface. As the cylinder is turned by the motor, the image on the paper is scanned so that a varying electrical signal is produced whose amplitude depends on the light reflected from the image in the scanning path. The lead screw causes the cylinder to move longitudinally to form a helical scan of the cylinder surface.

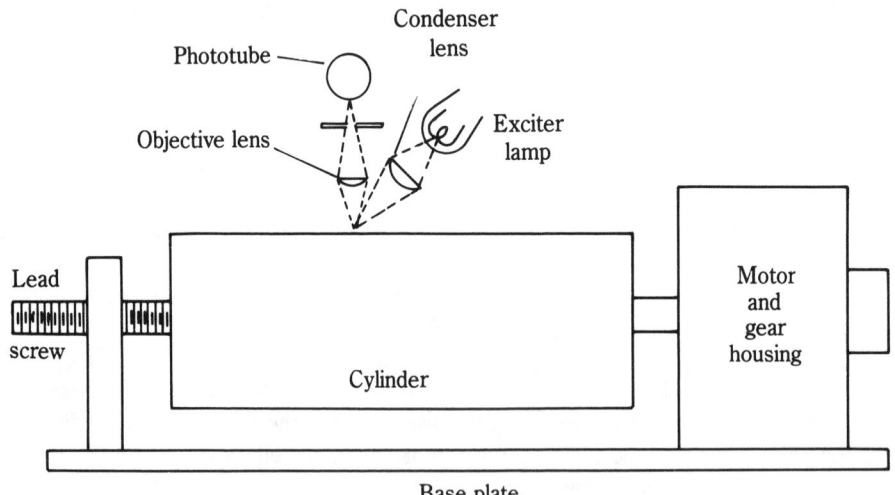

12-1   Early facsimile drum scanner.

At the receiving end a similar device has a cylinder rotating in synchronism with the sender. An optical system contains a light source whose intensity varies with the received signal to reproduce the original image on photosensitive paper.

Of course there are many different variations of the basic configuration, including stationary cylinders, rotating mirrors, laser light sources, etc. A key feature of facsimile is the need to synchronize the motions of the two cylinders. If the sending-end motor and the receiving-end motors are started at the same time with both cylinders in the same position, satisfactory synchronization is accomplished using common 60 Hz utility power for both.

TV scanning and synchronization

In the National Television System Committee (NTSC) television system the image is scanned line-by-line from left to right, starting at the top and moving to the bottom (see

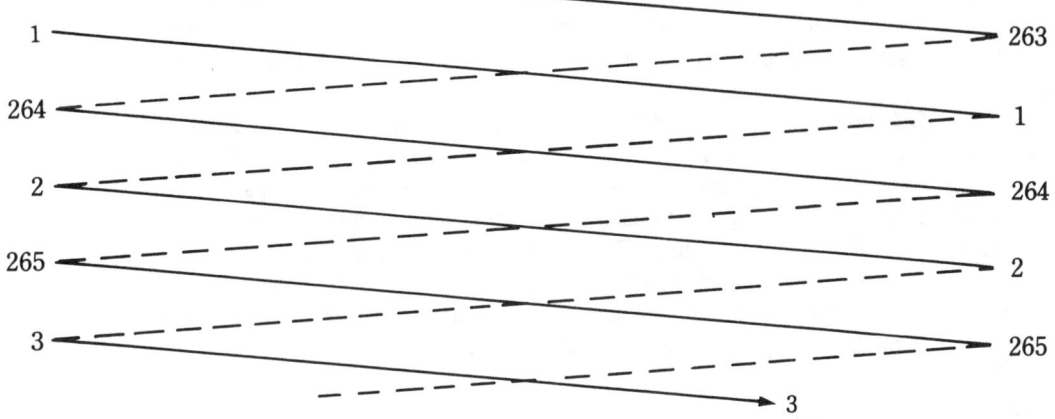

Lines 4 through 238 and lines 266 through 501 are not shown in this figure.

Lines 243 through 262 and lines 505 through 525 are not visible because of blanking during the vertical retrace time interval.

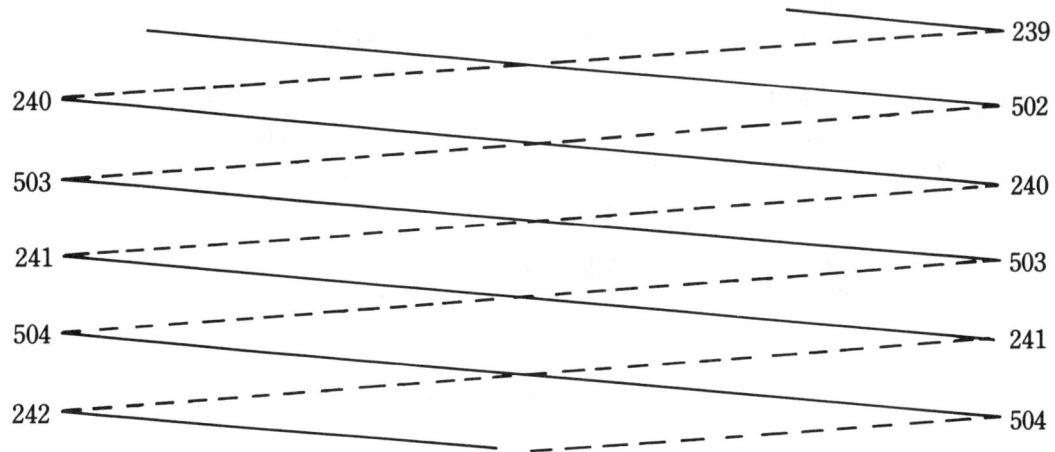

12-2 Details of 525-line frame and interleaved fields.

Fig. 12-2). One complete scanning cycle is referred to as a *frame* and consists of 525 lines. The frame frequency is approximately 30 Hz. However, to eliminate flicker, as perceived by a human viewer, every other line is scanned in one field, followed by the interlaced scanning of the remaining lines of another field. See the even and odd-numbered lines of Fig. 12-2. Thus there are 60 fields of $262^{1/2}$ lines each second. Less than 500 lines are available for picture information in each frame.

At the transmitter, horizontal synchronizing pulses are generated and superimposed on the video information (see Fig. 12-3). The wider part of the blanking pulses, including both the smaller front porch (approximately 1 μs) and the larger back porch, blank the electron beam as it retraces from right to left after scanning a line. Sync pulses occur at the line rate, f_H, of $30 \times 525 = 15.75$ kHz for which the period $H = 63.49$ μs. Color TV transmissions also superimpose a color burst on the back porch. Its use will be explained later.

Figure 12-4 shows the vertical blanking pulse whose duration is 833 to 1333 μs. Riding on the vertical blanking pulses are three groups of $2 \times 15.75 = 31.5$ kHz frequency pulses each spanning a nominal time duration of 3H. The first and third groups have fairly narrow (approximately 2 μs) equalizing pulses while the second group, having a serrated appearance, has wider (approximately 27 μs) vertical sync pulses. It should be recognized that the vertical blanking pulse and its associated riders must be positioned differently for the field that starts at the beginning of line 1, and midway across the tube face for line 263 at the beginning of the second field of the frame.

TV camera tubes

The first commercially available monochrome cameras used photoemissive iconoscopes and image orthicons for pickup tubes. Later tubes are photoconductive, bearing names such as *vidicon* and *plumbicon*. The vidicon will be described first. Figure 12-5 shows a simplified drawing of the vidicon camera tube. It has an electron gun (electron source, cathode), electrodes for accelerating and decelerating the electron beam, and coils surrounding the tube for focusing and deflection.

The electron target of the vidicon is a photoconductive mosaic layer separated from a transparent conductive film by a semiconductor photoresistive layer. When the target is dark, the electrons are stored on the inner target surface making the target potential roughly the same as that of the cathode. A point of light, corresponding to a picture element, will cause electrons to leak from the target so that the corresponding point on the target surface will become more positive. These electrons will be replaced by the electron beam on its next scan; this causes a pulse whose amplitude is a function of the picture element spot light intensity.

The Sulfide vidicon has chemicals in the photoconductive layer that have increased sensitivity, and improved light dispersion and light flare properties. The Saticon vidicon also has improved properties, including spectral sensitivity, which is higher in the blue and negligible in the infrared regions making it especially suitable for live-color TV pickup. Silicon-Intensifier Target (SIT) and Intensifier-Silicon-Intensifier Target (ISIT) camera tubes have scanning and read-out methods similar to those of the vidicon. A schematic representation of a SIT tube appears in Fig. 12-6. The photoelectrons are focused onto a special target which provides relatively high gain before the scanning

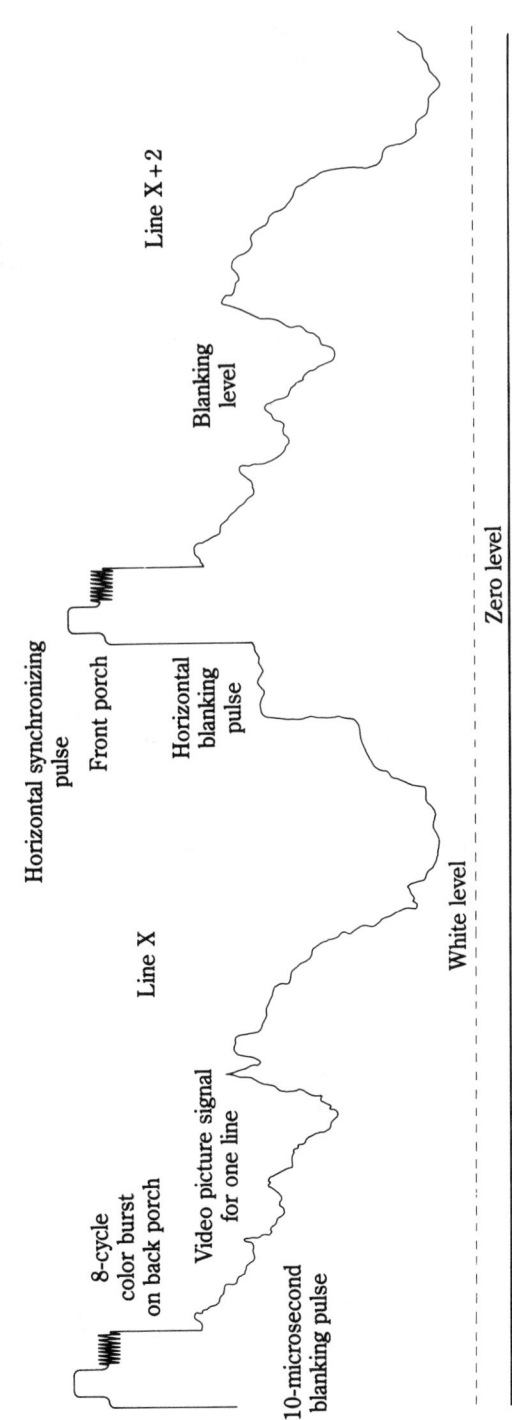

12-3 Horizontal synchronizing pulses.

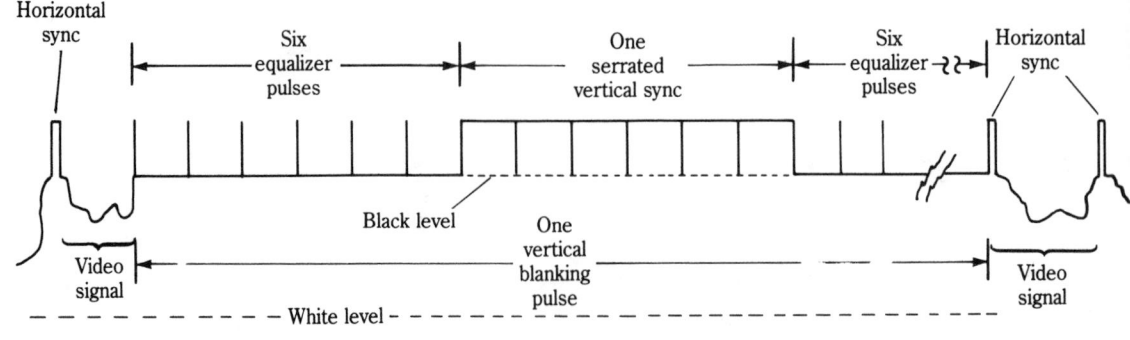

12-4 Vertical blanking pulse details.

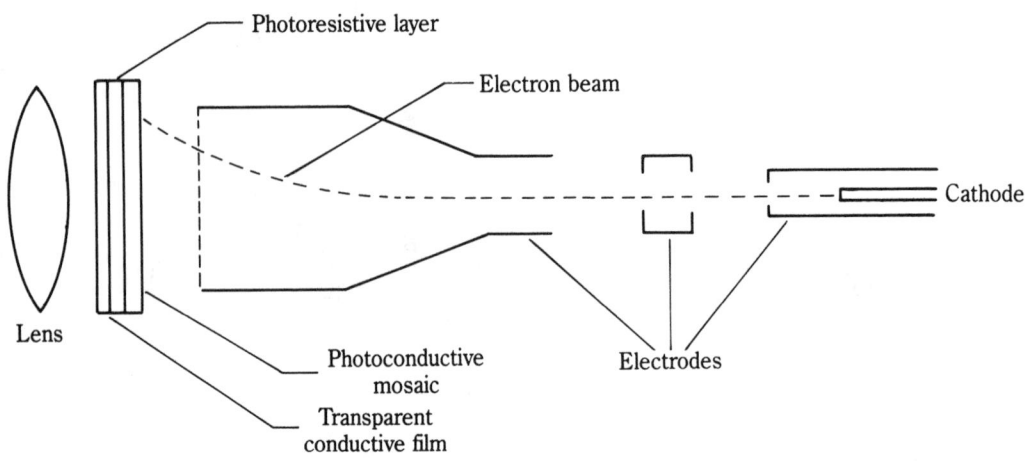

12-5 Vidicon camera tube.

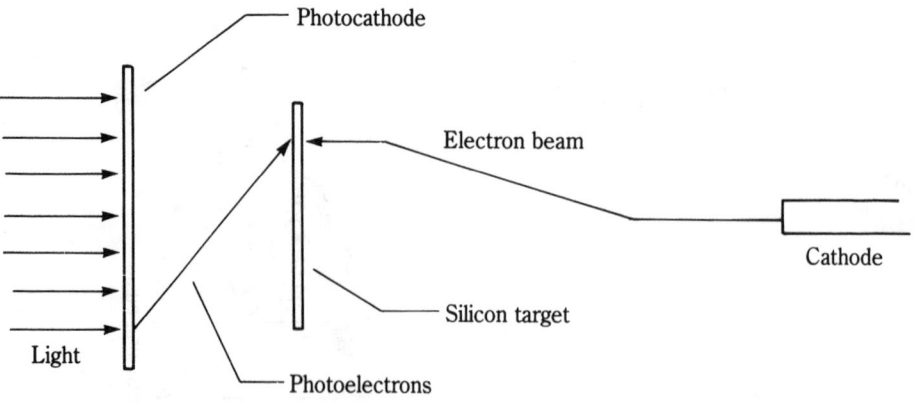

12-6 Schematic representation of a *silicon-intensifier target* (SIT) tube.

operation commences. The target is a very thin silicon wafer on which a tightly-spaced matrix of p-n junction diodes is formed. Gain is provided when a primary photoelectron, accelerated to perhaps 10 keV (this refers to the energy of an electron attained after it has been accelerated by a potential difference of 10,000 V), impinges on the target and causes multiple dissociations of electron-hole pairs. Gains of 1000 or more can be achieved. The holes are collected at the p-side of the diode where the charge is neutralized by the scanning beam. The signal is read out on the backplate of the target. SIT tubes have been used extensively for low-light-level pickup.

Image orthicons are television camera tubes employing a photocathode as a light sensor (Fig. 12-7). They are high sensitivity devices designed for high resolution TV systems where low light levels, low lag, and exceptional dynamic range are important. In the figure, when the electron scanning beam approaches the target, several events occur: Electrons might enter the target and neutralize a positive charge; electrons might fail to land and be electrostatically reflected; or electrons might land on the target and be scattered back at various angles. The scattered electrons constitute the signal in the return beam. The reflected electrons are eliminated by means of a baffle as shown in the figure.

The *image orthicon* is primarily a replacement type of camera tube. For many years it was used almost exclusively for live pickup in studio and outdoor broadcast cameras. Figure 12-7 shows that the tube is more intricate than the vidicon. The photoelectron image pattern developed at the photocathode is focused by an axial magnetic field, this produces one spiral loop onto a thin, moderately-insulating target surface. When the photoelectrons from the photocathode strike the target, secondary emission occurs, which causes the establishment of net positive charges on the target. The electron beam scans the charged target pattern, deposits some electrons on the more positively charged areas, and the modulated beam returns to an electron multiplier that surrounds the electron gun. The output signal is the amplified anode current of the electron multiplier.

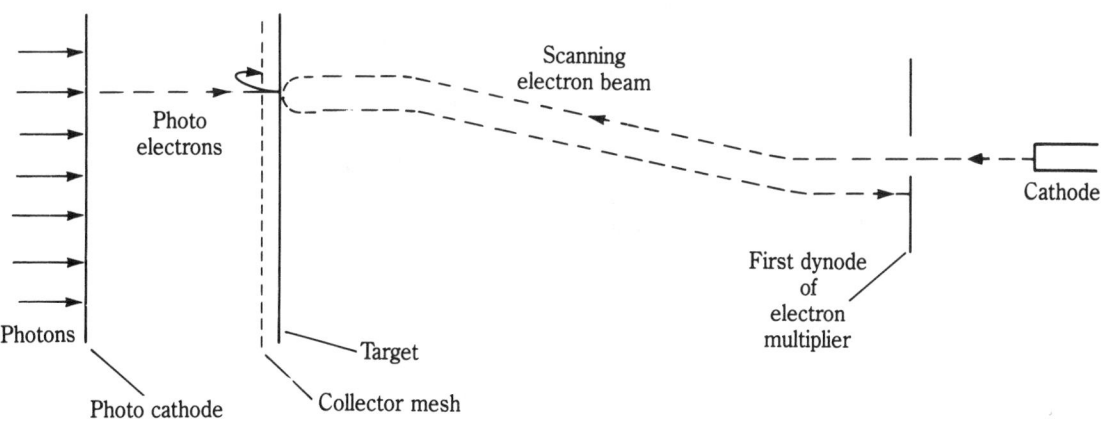

12-7   Schematic representation of Image Orthicon tube.

Monochrome picture tubes

A TV picture tube functions in essentially the same way as any other cathode ray tube (CRT). Refer to Fig. 12-8, electrons are produced by an indirectly-heated cathode, which is itself contained within a cylindrical shell that functions as a control grid. This shell has an opening for the electron beam to pass through, moving toward a similar opening in the first anode, which is also cylindrical in shape. The control grid is negative with respect to the cathode, and the first anode is positive an amount which is manually controllable. An electric field then exists in the vicinity of the two electrodes, which causes the electron beam to converge at what is called the *crossover point*.

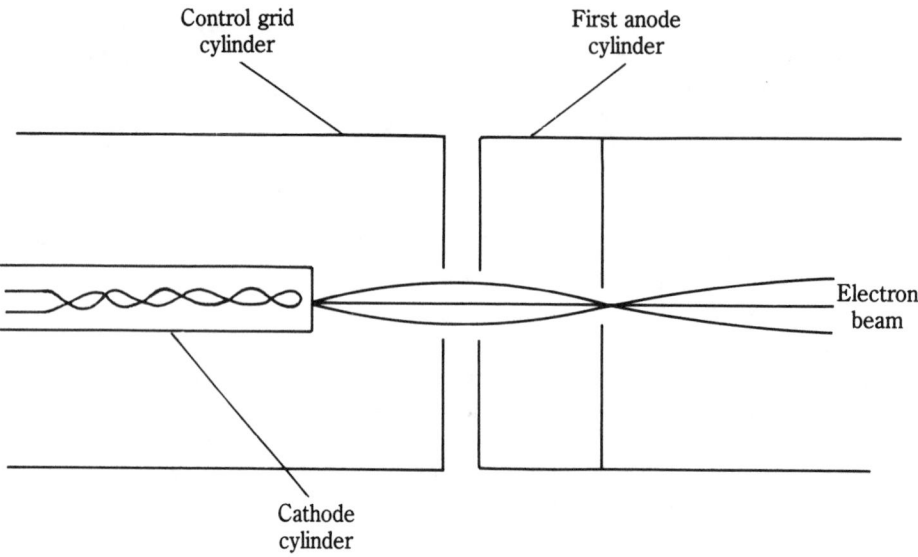

Control grid
cylinder

First anode
cylinder

Electron
beam

Cathode
cylinder

12-8   Three-element lens.

Beyond the crossover point the electron beam tends to diverge, so a second lens is constructed by adding a second anode, operating at a higher positive voltage than the first. This results in a shaped electric field in the vicinity of the two (the lens) which causes the beam to be focused and the electrons to be further accelerated.

Figure 12-9 illustrates a CRT employing magnetic focusing. This picture tube has magnetic deflection by means of coils mounted around the neck of the tube to create a vertically directed magnetic field for horizontal deflection, and a horizontally directed magnetic field for vertical deflection. Internal drive circuits in the receiver produce linearly varying currents, together with blanked retraces, at the line frequency for horizontal scanning and at the field frequency for vertical scanning.

Blanking is accomplished by driving the control grid sufficiently negative to cut off the electron beam. During the horizontal scan the control grid voltage is varied in accordance with the level of the video signal, being more negative for dark images and less negative for lighter areas. Finally, the electrons strike the fluorescent screen, converting part of their energy to light, thereby producing the original picture elements one-by-one during the 1/30 second duration of the frame.

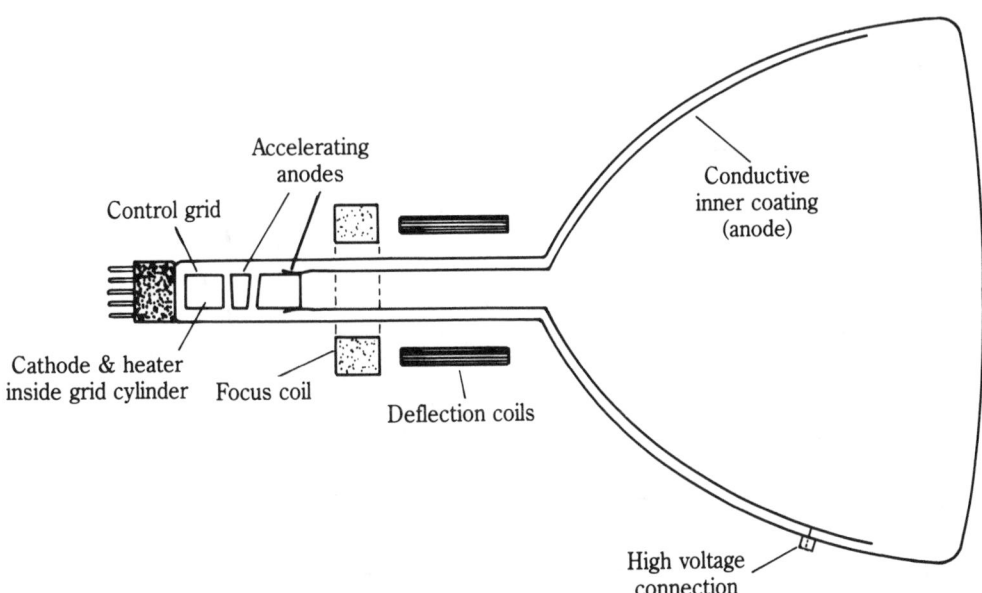

12-9 CRT having magnetic focusing and deflection.

Modulation and spectrum

The TV modulation system and signal are complex. Some simplification in explanation is possible if you think of two carriers f_v, for video (picture), and f_a for audio (sound). The video is amplitude-modulated with a vestigial lower sideband. Because the video bandwidth required is 4 MHz, the upper sideband extends 4 MHz above the carrier, which, in turn, is 1.25 MHz above the lower edge of the partially suppressed, vestigial lower sideband (Fig. 12-10). In this figure all the frequencies of the spectrum are referenced to the assigned broadcast frequency, here 60 MHz in the lower vhf band for channel 3. Television channel frequency allocations are listed in Table 12-1.

A second carrier, f_a, is located 4.5 MHz above f_v. It is frequency modulated by the audio signal with a maximum frequency deviation of 25 kHz. In speaking of TV channels, a *center frequency* or channel carrier frequency has no practical technical meaning, as contrasted with either AM or FM broadcast radio.

In anticipation of the requirements for color TV, a closer look at the upper sideband spectrum for video is in order. You will first consider a situation in which the picture consists of vertical bars only. The spacing of these alternating black and white stripes should approach the resolution of the system. In this case the spectrum will consist of the line scan fundamental frequency, 15.75 kHz, and a large number of harmonics up to 4 MHz.

You should recognize that the vertical bar pattern must be modified to more closely resemble a real-life scene in which the light intensity also changes in the vertical direction. This requires that the 15.75 kHz fundamental frequency and its many harmonics broaden into spectra clusters as shown in Fig. 12-11.

In other words, for a realistic picture the elements do not conform exactly to the

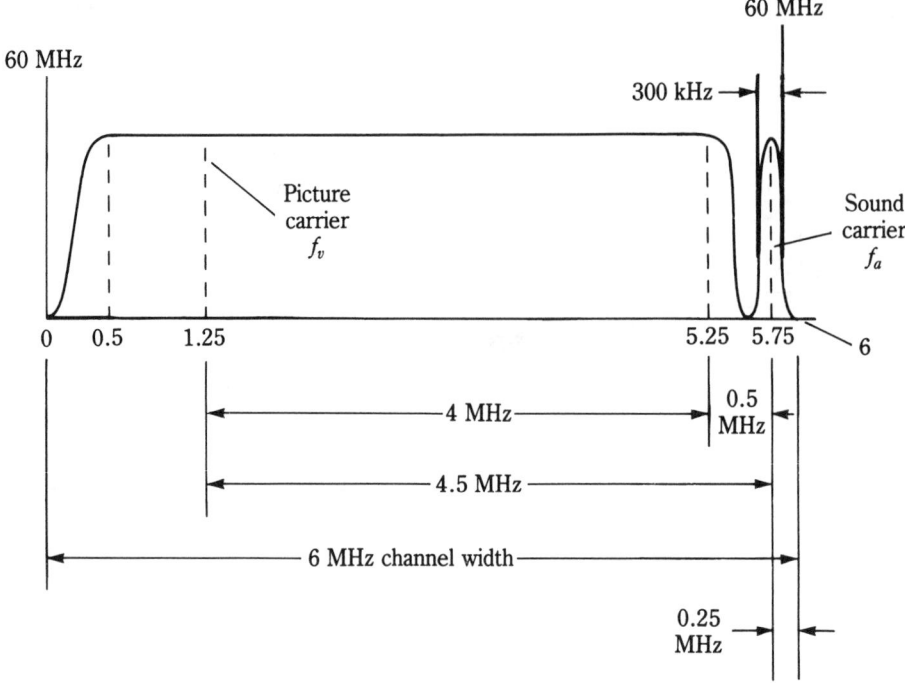

12-10   Transmitted TV signal (Channel 2) spectrum.

Table 12-1. TV Channel Frequency Allocations.

Lower VHF band		Upper VHF band		UHF band	
Channel	Lowest freq (MHz)	Channel	Lowest freq (MHz)	Channel	Lowest freq (MHz)
2	54	7	174	14	470
3	60	8	180	24	530
4	66	9	186	34	590
	(4 MHz	10	192	44	650
	skipped)	11	198	54	710
5	76	12	204	64	770
6	82	13	210	74	830
				83	884

simplistic bar pattern. The spectrum will not consist of sharply defined lines, but rather of blurred lines, that is distributions centered at 15.75 kHz and its harmonics as shown in the figure. Two particular points should be emphasized here:

1. Most of the transmitted power in the video portion of the spectrum is concentrated near 15.75 kHz and its harmonics.

2. The amount of power in each cluster decreases with the order of the harmonic.

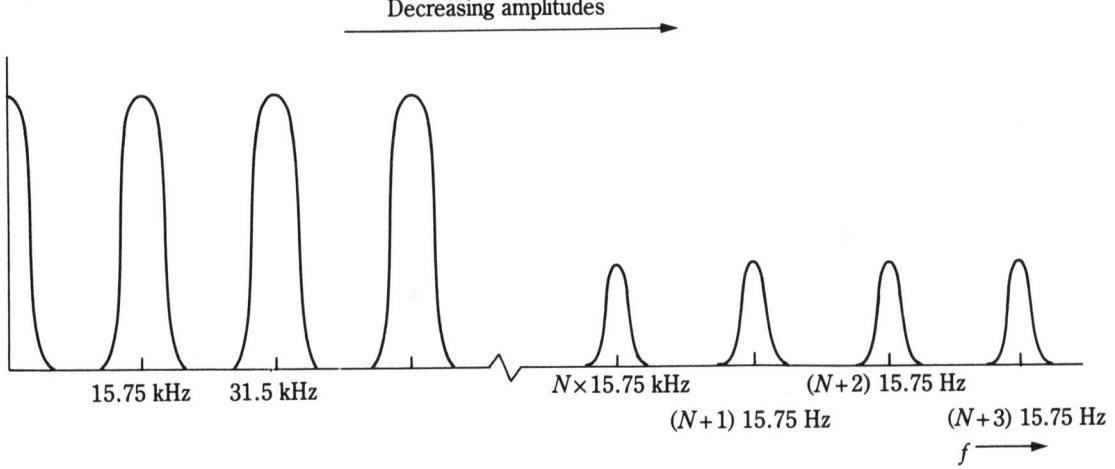

12-11   Television scanning spectra.

Color television

The present NTSC color TV system was designed long after the monochrome standards were adopted. To ensure that color broadcasts could be received and viewed in black and white (monochrome), it was necessary that the same standards be in effect with only a few cleverly designed changes.

Luminance and color

Luminance refers to the brightness as perceived by a human viewer. After careful study it was decided that picture luminance could be transmitted separately using existing monochrome standards, and that the color information could be added for color TV receivers without substantially changing the luminance quality with existing channel assignments. To do so was a large order.

The NTSC color television system derives all colors from three primary saturated colors: red (R), green (G), and blue (B). Refer to Fig. 12-12, the primary colors appear at the vertices of the triangle. Other saturated colors, consisting of primary color mixtures, appear on the perimeter with still other mixtures making up the desaturated (washed-out) colors situated within the boundaries of the triangle. White is near the center.

The luminance signal, *Y*, and two color signals, *I* and *Q*, are obtained from a matrix fed by three pickup tubes according to the formulas:

$$Y = 0.299R + 0.587G + 0.114B$$

$$I = 0.74(R-Y) - 0.27(B-Y)$$

$$Q = 0.48(R-Y) + 0.41(B-Y)$$

where R, G, and B are the signal levels from the three color cameras.

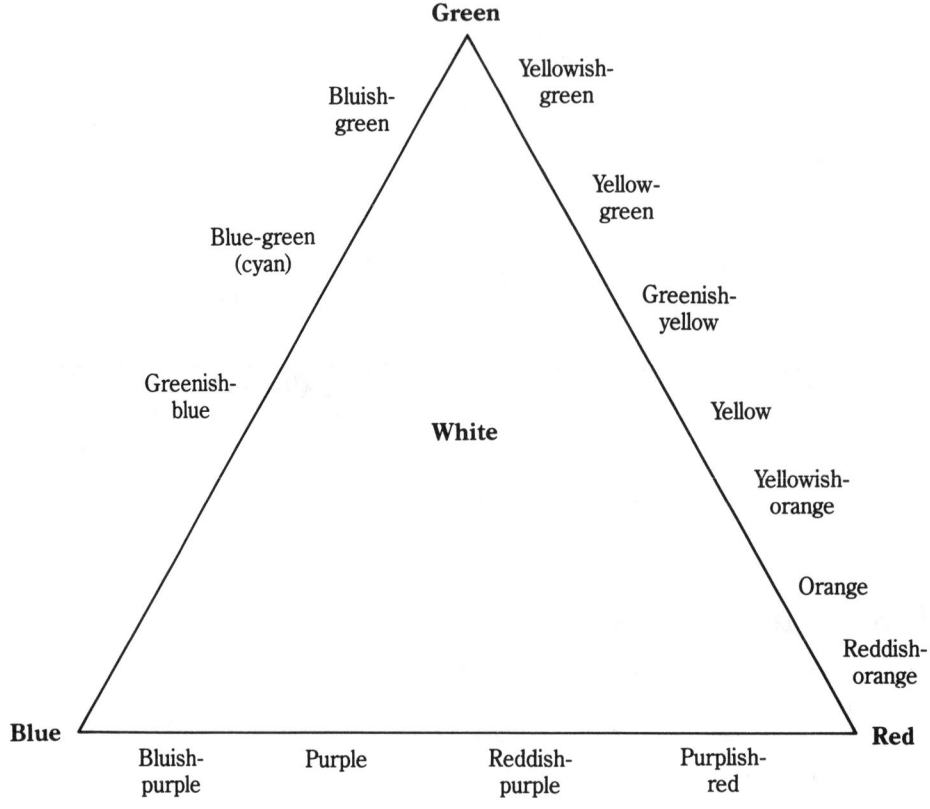

12-12 Simplified chromaticity diagram.

Neutral colors (black, gray or white) will result in *I* and *Q* being zero. A typical color camera consists of an optical system with lenses, prisms, and filters, together with three monochrome tubes such as Vidicons or Plumbicons. Each tube produces one of the three color signals.

Chrominance modulation

The problem of compatibly superimposing the color information was solved by establishing a color subcarrier frequency of 3.579545 MHz ± 10 Hz. This frequency selection, combined with a frame scanning rate of 15,734.3 Hz successfully interleaves the chrominance sidebands between the spectra clusters of the luminance sidebands in Fig. 12-11.

Double sideband, suppressed subcarrier phase modulation is used. The *I* signal has a lower sideband width of 1.2 MHz and an upper sideband width of 0.6 MHz with its suppressed subcarrier in quadrature with that of signal *I*. The overall spectra for color TV is shown in Fig. 12-13.

Color CRT

In order to reproduce the original picture in color at the receiver, phosphors producing red, green, and blue are required in the CRT. These are arranged as dots, alternating

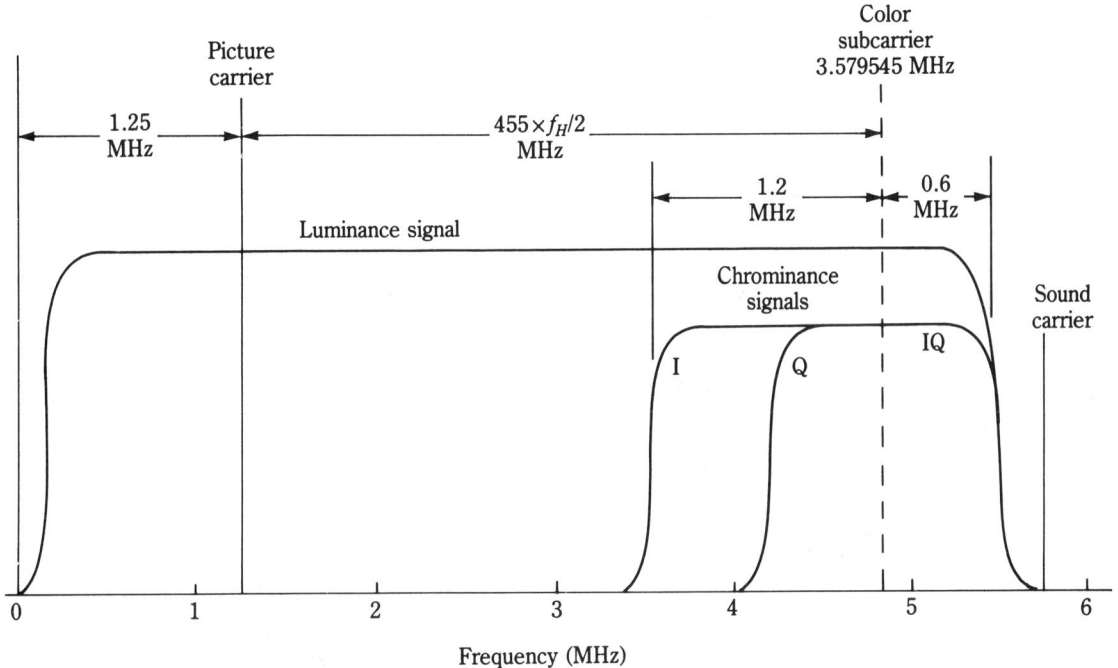

12-13 Overall frequency spectra for color television.

the three colors, row after row, so that three adjacent dots, called a *triad*, has all three colors (Fig. 12-14). One triad at a time is bombarded by three separate electron beams having the proper ratio of intensities so that light emitted by the triad phosphors form one color picture element. A 21-inch color television tube will have about 350,000 triads and three times that number of dots.

One hole corresponds to each triad in the shadow mask, which is positioned near the phosphor layer in the tube. In actual operation the three electron beams are scanned in unison at the normal horizontal and vertical rates to pass simultaneously through one hole in the shadow mask at a time; until all the picture elements of the frame have been serviced, after which the scanning is repeated. During the scanning, each beam is modulated to achieve the total color picture.

The picture tube has three separate electron guns, control grids, and focusing lenses, but has a common deflection system. It is apparent from Fig. 12-14 that the three streams of electrons follow separate paths, and merge only at the hole in the shadow mask. Another form of picture tube, the *Trinotron*, has three cathodes, and a single focusing lens with a phosphor pattern arranged in vertical bars. It will not be given further consideration here.

Transmitting systems

This section's purpose is to present a representative uhf transmitter so that its basic design and operation can be understood. Figure 12-15 has the complete block diagram of a 25 – 55 kW uhf TV transmitter. The description is broken into several separated parts.

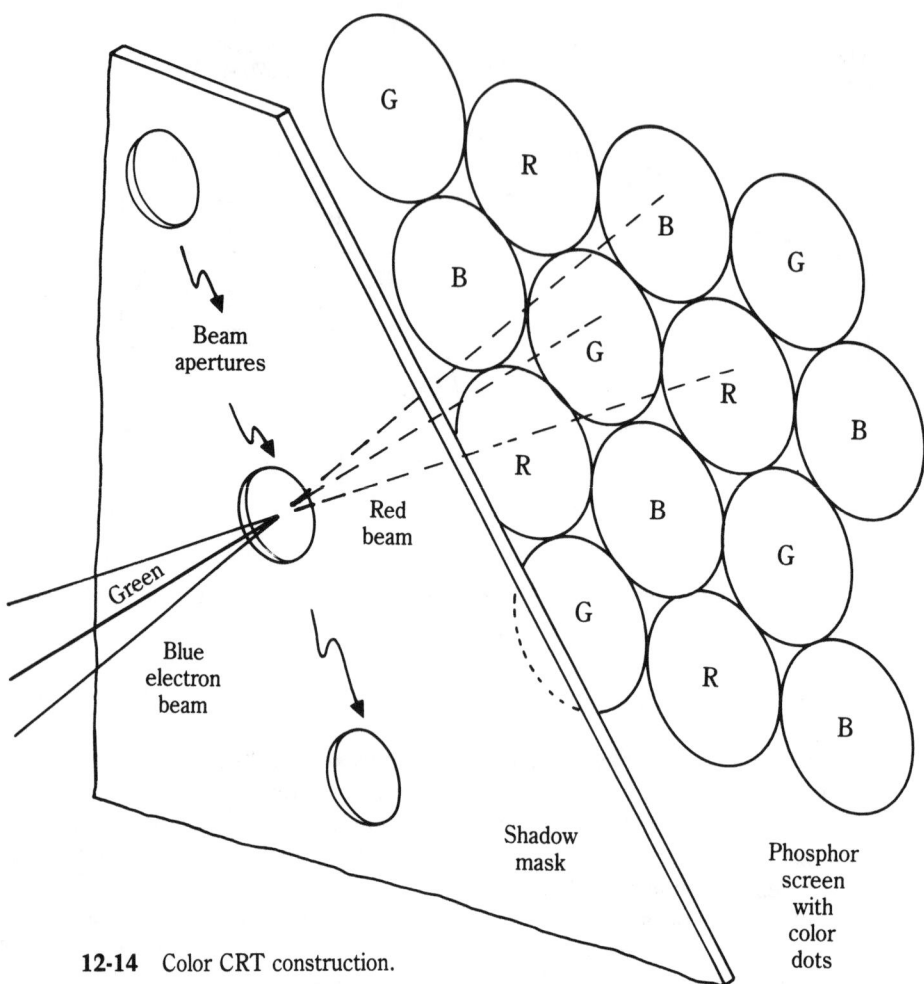

12-14   Color CRT construction.

The driver stages

The transmitter incorporates an i-f modulated solid-state drive system consisting of a modulator, a mixer, two linear amplifiers, and associated power supplies as shown in the block diagram.

The modulator accepts video and audio inputs. It produces a visual output at either of two frequencies, 38.9 MHz or 45.75 MHz. The visual system is modulated, filtered, and corrected as required by the amplifiers used. The mixer accepts the outputs of the modulator and converts them to the required carrier frequencies. Solid-state linear amplifiers then raise the power to a level suitable for driving the klystron output stages.

Video processing

The video input signal enters the video correction module where the back porch is clamped, and the sync pulses are regenerated. The sync amplitude is stabilized by circuits that do not affect the color burst. No further clamping is necessary. The video signal can be precorrected for receiver group delay distortions.

Video modulation

The processed video next modulates a carrier at the intermediate frequency. A balanced modulator is used that has good linearity, and permits the carrier to be suppressed without difficult adjustments. The crystal oscillator contains a temperature-stabilized crystal. After modulation, the signal is shaped by the vestigial sideband (VSB) filter.

Correction circuits

Phase nonlinearities produced by the VSB are compensated for in one or two group (envelope) delay correction modules. Group delay and amplitude are independently adjustable in each circuit.

Differential phase at color subcarrier frequency, and incidental phase modulation at the video carrier frequency are minimized in the next module, which also reduces intercarrier noise. Finally there is correction for amplitude nonlinearities introduced throughout the transmitter.

Power output regulation

The video signal next passes through a power-controlling amplifier intended to hold the video blanking power level of the transmitter substantially constant. A sample of the video signal is obtained from a directional coupler in the output feeder, and is passed to a detector as part of the controlling feedback circuit.

Sound modulator and frequency control

The audio input signal is fed via a variable attenuator to the sound modulator, where it frequency modulates an intercarrier frequency. The center frequency of this oscillator is derived from a crystal by means of a phase-locked loop. Preemphasis is included. The sound i-f is obtained by subtracting the intercarrier frequency from the video i-f. The sound modulator can be adapted for a stereo encoded signal.

Mixers and linear amplifiers

The video and sound i-f signals are now fed at a level of approximately 1 mW from the modulator to the mixer where they are up-converted in double-balanced mixers to the final carrier frequencies and amplified to a level of approximately 50 mW.

Next the modulated signals pass through filters that attenuate undesired products of the modulation process. The carriers are then amplified to a level of 1 W and leave through ferrite isolators.

Crystal oscillator and multiplier

The master oscillator frequency for the mixers is generated by means of a temperature-stabilized crystal oscillator. Its output is applied to a harmonic generator and filter that multiplies the crystal frequency by 12 or 16 (depending on whether the transmitter operates in the lower or upper-half of the uhf band) thus producing a mixing frequency which is higher than the carrier frequency so that harmonics can be suppressed effectively.

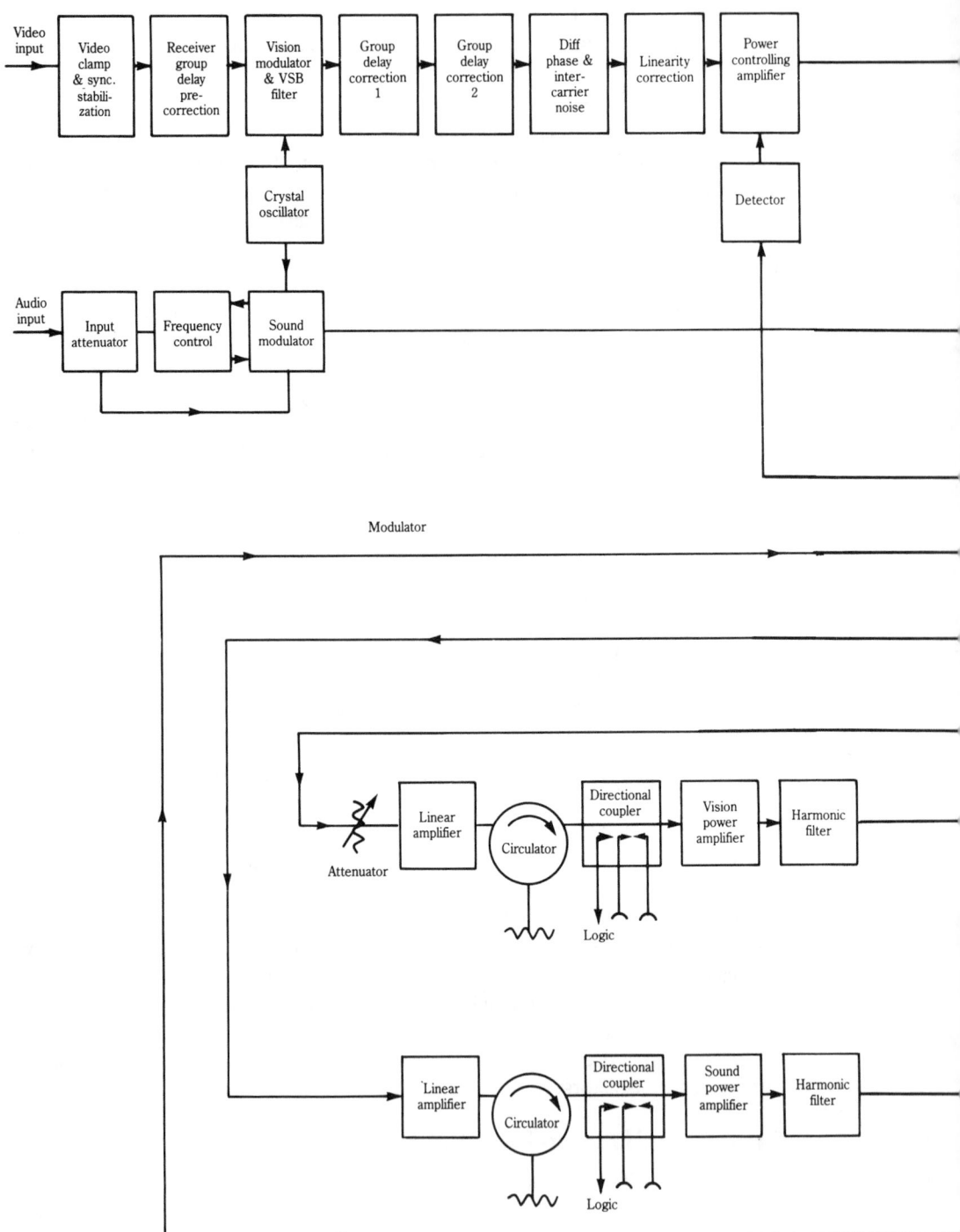

12-15 Block diagram of uhf-TV transmitter.

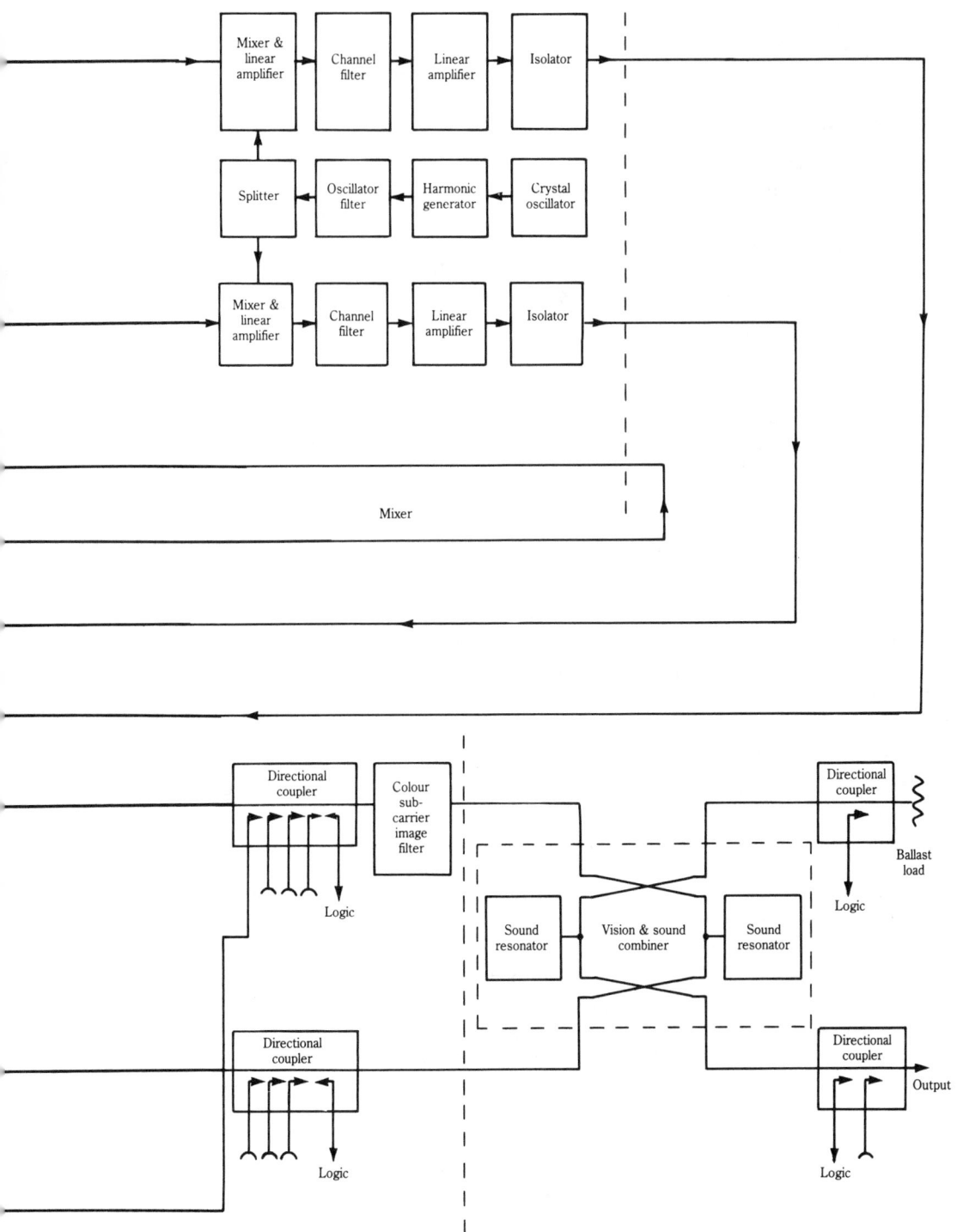

The output is split to feed the video and sound mixers.

Provision has been made to lock the frequency generating circuits to an external precision source such as a Rubidium standard by means of a synthesizer.

Solid-state linear amplifiers

Solid-state linear amplifier assemblies separately amplify the video and sound carriers obtained from the mixers. The video assembly consists of two broad-band modules in cascade. This provides adequate drive for various makes of klystron tubes. The video amplifier is preceded by a variable attenuator, which permits the adjustment of the drive level to fit the klystron requirements which vary with the channel assignments.

The sound amplifier consists of one, or two modules in cascade, depending on the drive level required.

Output stages

Versions of all three types of transmitters can be supplied that employ either external or integral cavity vapor cooled klystron tubes. In all cases the picture and the sound output stages employ identical tubes mounted together.

Output filters

The outputs of both power amplifiers pass first through harmonic filters. These are coaxial versions of low-pass M-derived filters with pi-sections and end sections.

The color subcarrier image filter consists of two loosely-coupled, short-circuited stubs in the video output feeder to attenuate output at the video frequency minus color-subcarrier frequency. These filters are provided with the means of adjusting the resonant and antiresonant frequencies.

Video and sound combiner

The video and sound combiner accepts the filtered video and sound signals and combines them without any significant interaction between the amplifiers. It consists of a constant impedance network of the Lorenz filter type, which uses 3 dB couplers as hybrids. Two couplers are joined by equal transmission lines across which two frequency-selective filters are connected in electrically-equivalent positions. The incoming video and sound signals are split equally by their respective couplers. The filters (labeled resonators in Fig. 12-15) which are resonant at the sound frequency but antiresonant at the video frequency, reflect the sound signal but permit the video frequency to pass. The split signals, both sound and video, now pass to the output coupler where their relative phases permit them to recombine into the feeder output. Unwanted signals are prevented by the filters from reaching the output, and are diverted to the ballast load. The combined signal passes via a feeder containing a directional coupler to the transmitter output.

Monochrome television receiving systems

Consideration of TV receivers will be in three parts. You will first examine the conventional receiving set as it exists today in millions of homes. You will concentrate initially on the monochrome functions and then move on to the more complex color receiver; you will complete this study with the technical innovations of the past few years which have greatly improved receiver performance.

Finally, the subject of digital signal processing will be introduced, with emphasis placed on a complete family of ICs for this purpose that are now entering the market. No attempt will be made to predict the nature of the receivers that will be necessary for the future high-resolution television systems currently under study. Figure 12-16 provides an overview of the conventional monochrome receiver. The major functional blocks will be explained briefly.

Front end

The TV tuner might more properly be called the *front end*, since it includes all of the circuits between the antenna and the i-f amplifier. Figure 12-17 illustrates the features of the front end. The uhf signal is received and converted to the 45 MHz intermediate frequency after which it is amplified by the vhf rf amplifier and fed to the i-f chain.

Reception of a vhf signal requires that the same rf amplifier be tuned to the vhf receiver. The tuning in older receivers is accomplished by means of mechanically-variable capacitors or inductors and, alternatively, by electrical control of varactor diodes. In order to compensate for the inherent local oscillator drift, *automatic fine tuning* (AFT) circuitry is added but is not shown in Fig. 12-17. AFT is similar to *automatic frequency control* (AFC) which is sometimes found in FM receivers.

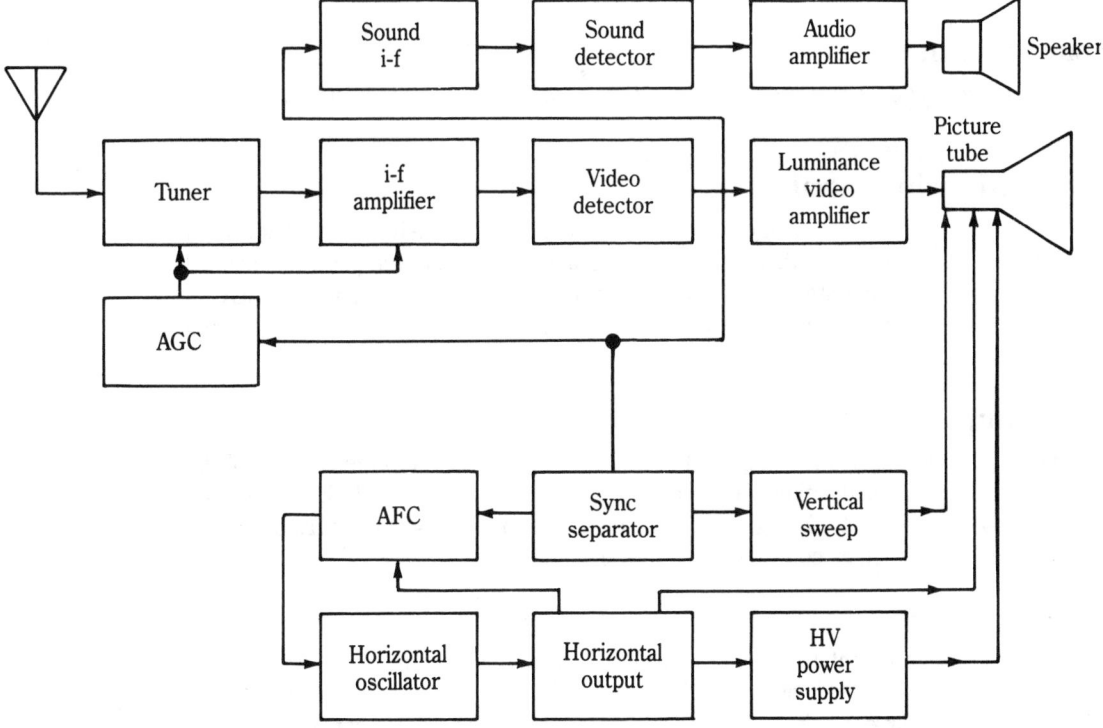

12-16 Block diagram of monochrome TV receiver.

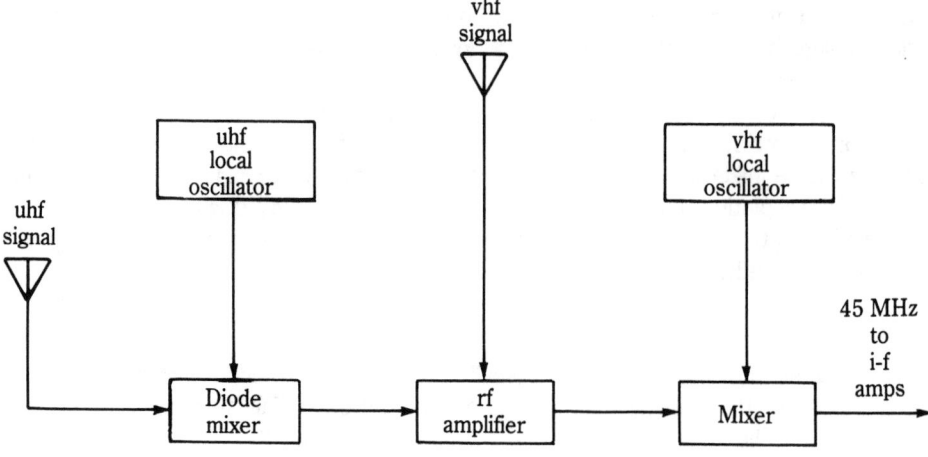

12-17   A uhf tuner block diagram.

Video detector

The *video detector* functions in the monochrome TV receiver in much the same way as the envelope detector of a broadcast AM receiver. At the output of the detector the video carrier frequency component is heterodyned with the sound carrier and sidebands to produce a 4.5 MHz sound i-f frequency which is fed to the sound i-f amplifier.

Sound system

The 4.5 MHz sound i-f amplifier is followed by a limiter, FM demodulator, and audio amplifier.

Automatic gain control

The *a*utomatic *g*ain *c*ontrol (AGC) function is to maintain the received carrier strength constant, as amplified, rather than holding the average light level of the picture constant. Therefore, the AGC operates on the synchronizing pulse level. Ordinarily the AGC is keyed so that the level is sampled at particular times, such as the time interval during horizontal flyback.

Video amplifier

The video amplifier causes the amplified video signal to be applied to the cathode of the CRT. Its effect is the modulation of the electron beam to yield light and dark areas on the tube face corresponding to the light and dark elements in the original picture or scene. Ancillary functions such as contrast control are included.

Sync separator

The separation of horizontal and vertical sync pulses can be accomplished by means of many different circuits. Their operation depends on the establishment of a bias level that allows the flow of current only during the synchronizing pulses. Note that, in Fig. 12-3, the sync pulses rise above the black level, which is the highest possible level of any video signal exclusive of the sync pulses.

Filters then separate the lower-frequency vertical sync pulses (60 Hz) from the higher-frequency horizontal sync pulses (15.75 kHz). The separated sync pulses are then passed on to the respective sweep generators.

Vertical sweep

Vertical deflection is achieved by a sawtooth waveform from an integrator driven by a multivibrator. The multivibrator, in turn, is synchronized by the selected vertical sync pulses.

Horizontal sweep

Horizontal sync pulses are used for AFC to synchronize the horizontal oscillator, whose output is amplified (horizontal output) and then applied to a high voltage transformer, called the *flyback transformer*. Its output drives the horizontal yoke windings. Moreover, the output is rectified in the HV power supply which supplies some 10 kV, or more, of acceleration voltage to the acceleration electrode of the CRT. A number of other functions are involved with the horizontal sweep and the high voltage circuits.

Color television receiving systems

As previously explained, color TV is superimposed on and compatible with the monochrome TV system. The present purpose is to explain the differences in receiver construction and operation for color as compared with monochrome.

Figure 12-18 presents the block diagram for the color television receiver. Careful comparison of Fig. 12-18 with Fig. 12-16 reveals a relatively small number of differences other than the addition of the chrominance circuitry proper.

Color oscillator

The color oscillator is crystal-controlled at the color burst frequency of 3.579545 MHz. Moreover, it must be locked in phase with the color burst, requiring that the color burst be selected from the back porch of the horizontal sync pulse, amplified, and applied to a phase detector, together with the output of the color oscillator. An error voltage then causes a fine adjustment of the oscillator frequency by means of a varactor diode placed in the oscillator circuit with the piezoelectric crystal.

Color killer

Absence of the color burst in the horizontal sync of the video signal is an indication of monochrome TV reception. The phase detector of the color oscillator then presents a large error voltage which cannot be corrected. This error voltage causes the color i-f to be turned off so that chrominance circuits are no longer effective in causing color tones to be seen on the color CRT.

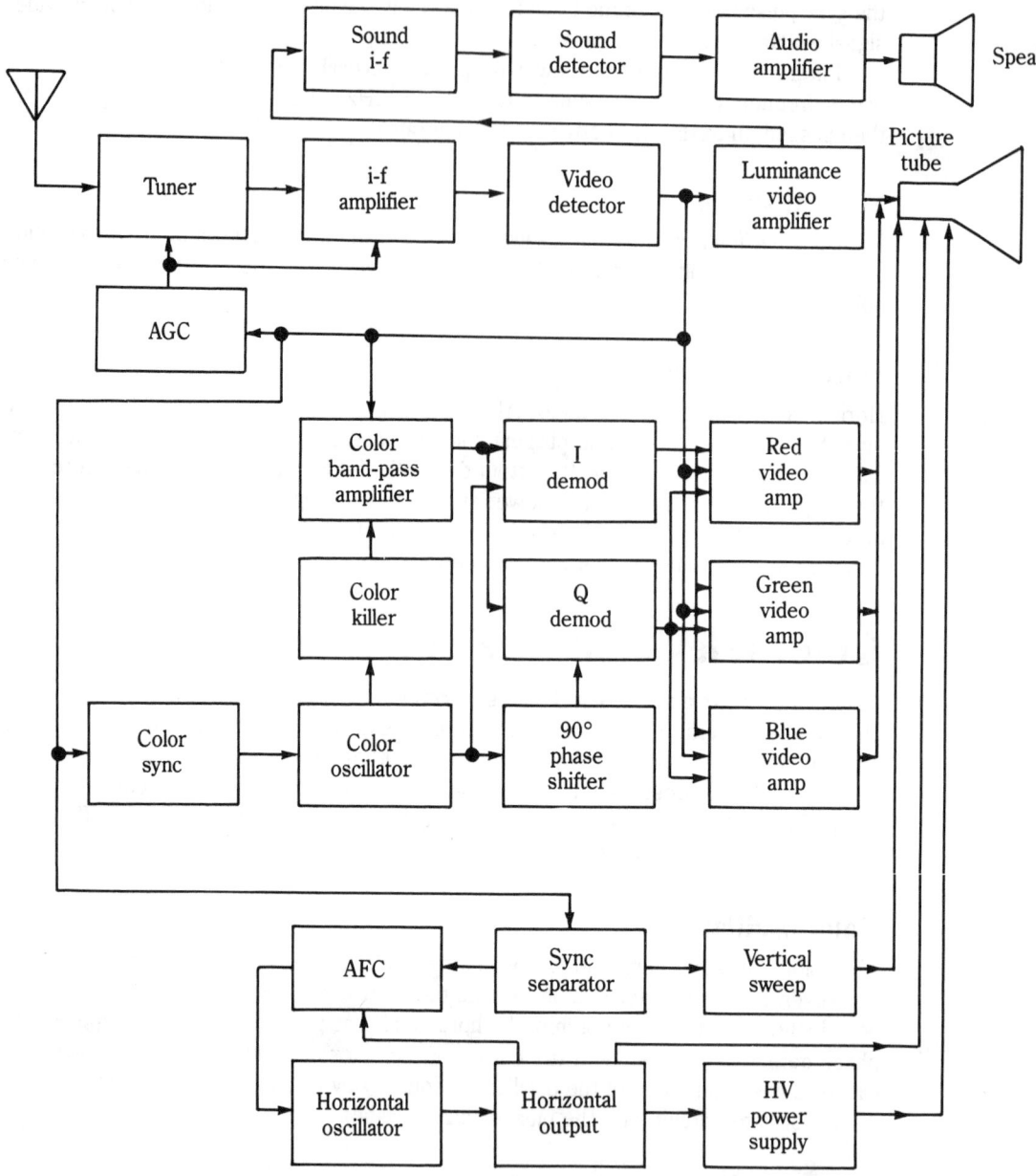

12-18 A block diagram of color TV receiver.

Color band-pass amplifier

The received video signal in the approximately 2.0 – 4.2 MHz frequency range, containing the chrominance sidebands and subcarrier, are amplified in the band-pass amplifier in preparation for demodulation of the chrominance signals.

Chrominance processing

The coherent color subcarrier from the color oscillator is used to demodulate the *I* signal and, after a 90° phase shift, to demodulate the *Q* signal. Recall that these two signals convey the chrominance information derived, such that *I* and *Q* are formed from particular combinations of the luminance signal, *Y*, and the red (R) and blue (B) signals originating in the color camera. (Green (G) has not been forgotten as it is incorporated in *Y*.)

It follows then that R, G, and B can all be obtained by summing *Y*, *I*, and *Q* in the proper proportions with 180° phase shifts injected as required to obtain negative signs. Circuitry that causes the linear mixing of these inputs is located in the red video amplifier, the green video amplifier, and the blue video amplifier of Fig. 12-18. These video amplifiers then cause the electron beams from the three electron guns in the CRT to be properly modulated in intensity to reproduce the color tones.

Improved color receivers

Substantial improvements in color TV performance have been achieved in recent years. Some of the design modifications appear in Fig. 12-19.

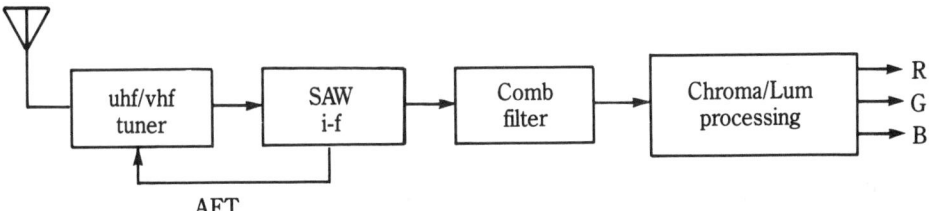

12-19   Improved color receiver.

Many new receivers now come equipped with surface acoustic wave (SAW) filters. The SAW results in improved amplitude/phase response in the i-f amplifier.

Appreciably better horizontal resolution is achieved through the use of the comb filter in separating the luminance and chrominance signals. Since the chrominance subcarrier frequency was deliberately placed midway between two horizontal sweep frequency, f_H, harmonics, the chrominance sidebands are interleaved with these harmonics. The *charge coupled device* (CCD) time delay element of Fig. 12-20 is an analog shift register with 910 stages. It is driven by a clock that has a frequency equal to four times the color burst frequency, or approximately 14.3 MHz. A time delay of one horizontal line scan period is thereby introduced in its output. When combined, by summation, with the composite undelayed video, the luminance video components are reinforced. This reinforcement occurs because the time delay of the CCD is equal to the period of the horizontal sync pulses. Every harmonic of f_H is delayed by a multiple of its period.

A signal component at the color burst frequency or the video chrominance sideband frequencies will have a time delay component of one-half the period. Recall that the chrominance video sidebands all differ in frequency from any adjacent harmonic of f_H and $f_H/2$. Therefore, the chrominance sidebands will undergo cancellation (ideally) at the output of the summing circuit of Fig. 12-20.

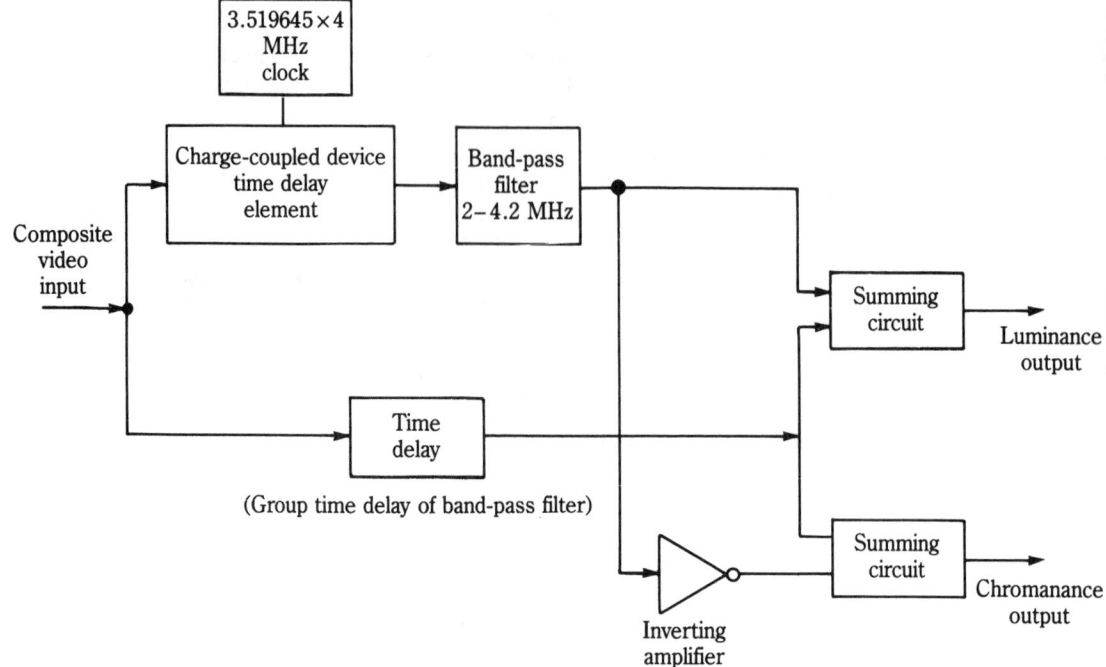

12-20   Comb filter for separating the luminance and chrominance video.

If you turn now to the lower summing circuit, the situation is quite different. The delayed composite video is inverted before being introduced into the summing circuit. For this reason the chrominance components in the summing circuit will be reinforced whereas, the luminance components will tend to cancel.

Two additional elements appear in Fig. 12-20. One is a band-pass filter which covers the range of the video chrominance sidebands so that the comb filter action only occurs in this band. The other is a time delay introduced to compensate for the band-pass filter delay of the upper path.

Digital signal processing for TV

This section deals with a new television receiver (Digivision) system created by International Telephone and Telegraph Corporation (ITT) in which the major receiver functions reside in Very Large Scale Integration (VLSI) circuits. Several large TV receiver manufacturers are involved in the development project.

The present description is included in this book with the belief that this system represents the near-future state-of-the-art in TV technology. You are also reminded of the continuing trend toward the marriage of current analog and digital technologies throughout the entire communications field.

An overview of the system is provided by the block diagram of Fig. 12-21. The five units in the center of this figure are the heart of the new system and are implemented by ICs. For the most part, the other blocks are conventional, closely resembling those for the TV receivers already described.

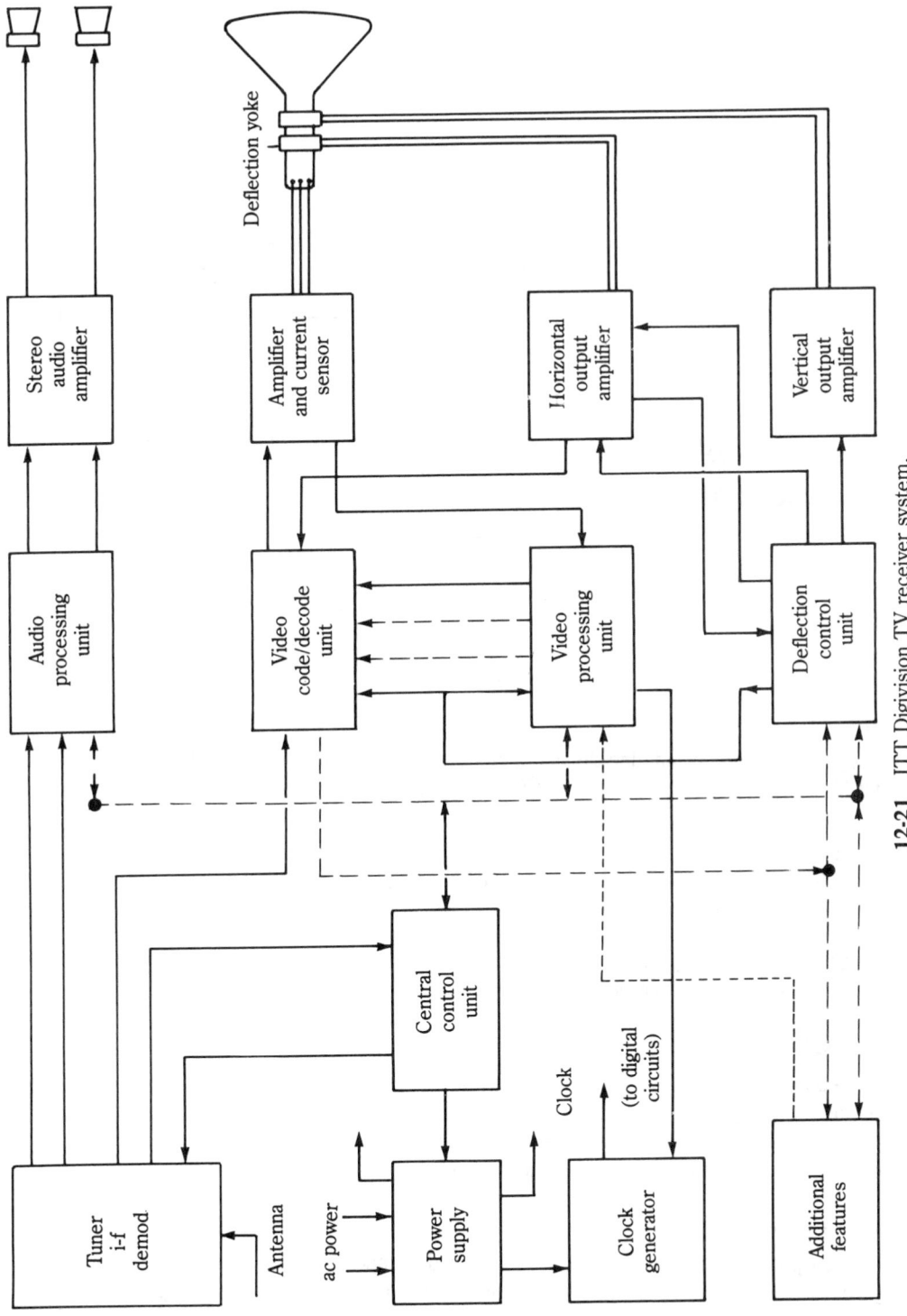

12-21 ITT Digivision TV receiver system.

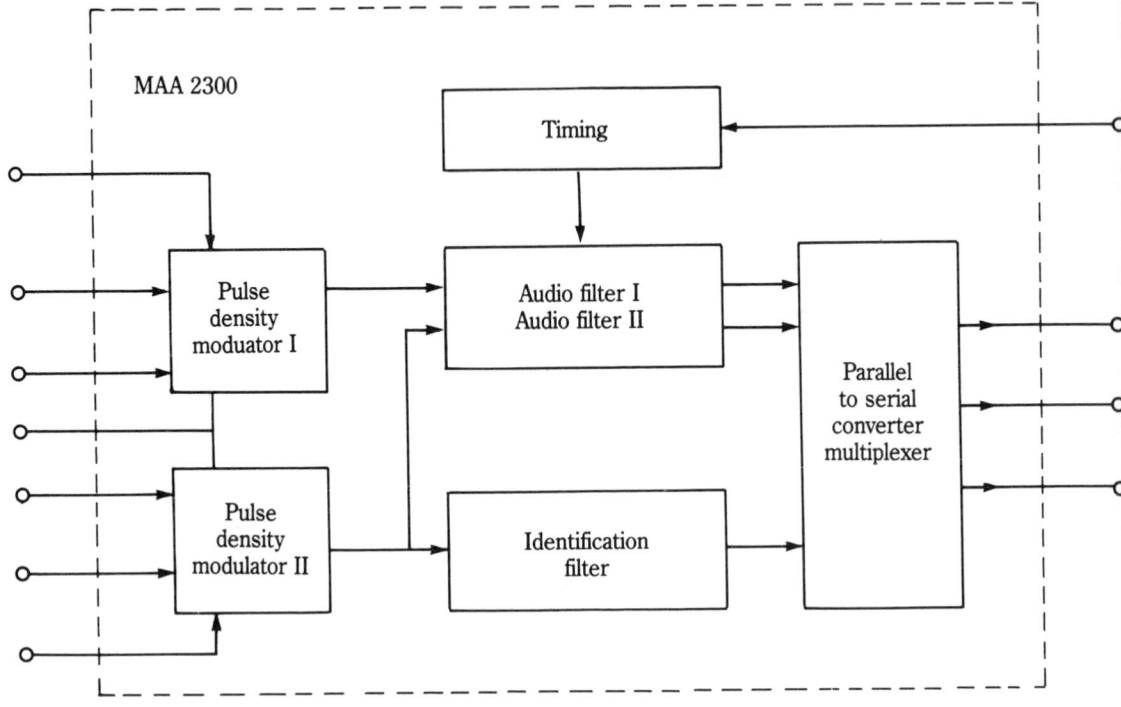

12-22 Simplified block diagram of audio A/D converter.

Audio processing unit

Two separate ICs are included in the audio processing unit. The first is the audio analog/ digital (A/D) converter which digitizes the demodulated audio signal from the tuner i-f amp-demod unit of Fig. 12-21. The audio unit block diagram appears in Fig. 12-22. The digitized audio signal has 14-bit resolution at 35 kHz. Parallel-to-serial conversion takes place at the output to reduce the number of pins required on the chip.

The audio processor (Fig. 12-23) takes the digitized audio signal from the A/D converter and divides it into the left (L) and right (R) stereo channels. Subsequent digital filtering to achieve deemphasis, tone control, loudness, and balance, for example, must be in response to user adjustments.

Video CODEC unit

The video CODEC unit (Fig. 12-24) employs parallel comparators to provide 8-bit resolution for the composite video signal, which is then passed on to the video processing unit. The video CODEC unit also converts the processed digital video output back to analog form to control the brightness and color of the CRT.

Video processing unit

The video processing unit separates the digitized composite video signal into the luminance and chrominance channels. Digital filtering is applied to these video signal

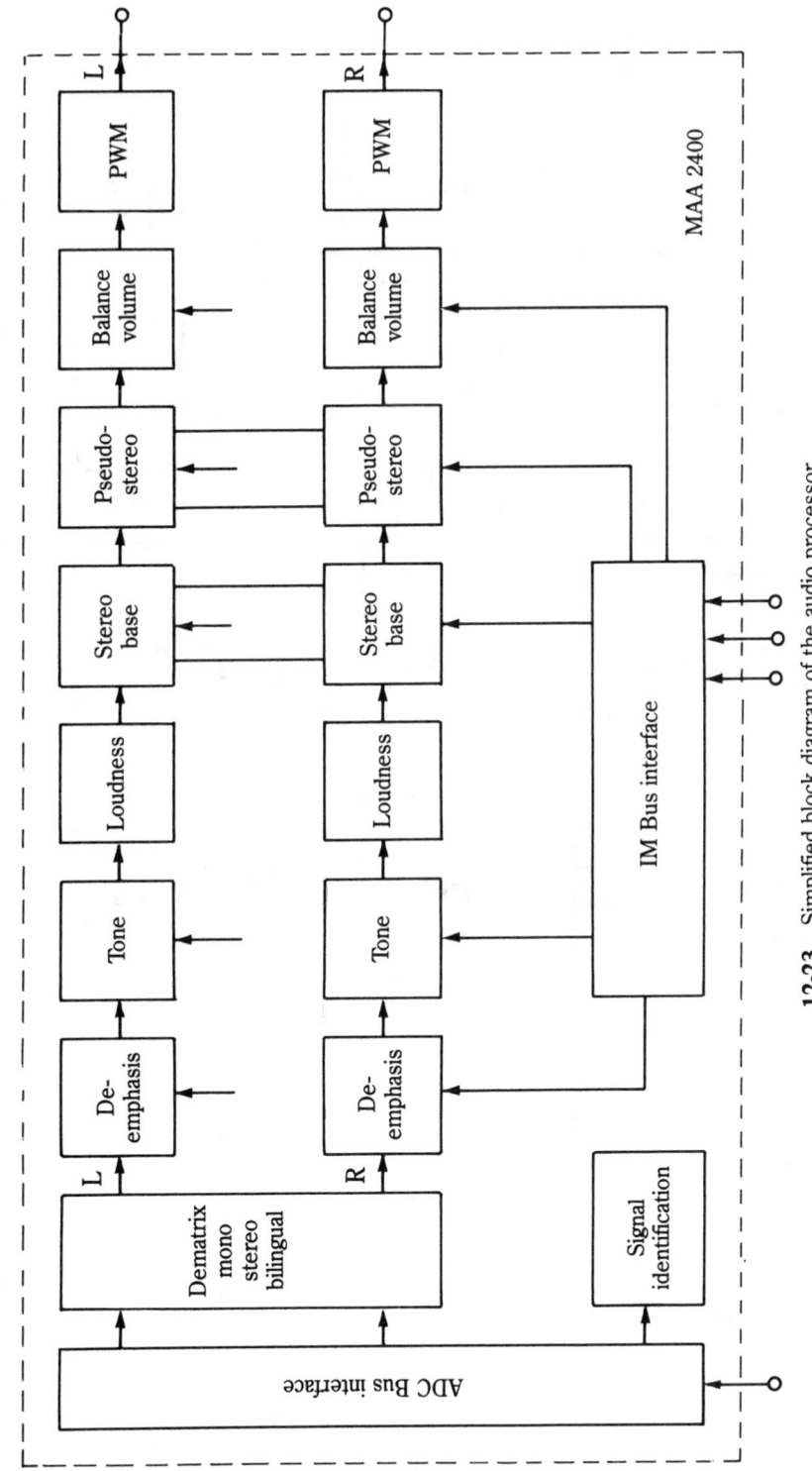

12-23 Simplified block diagram of the audio processor.

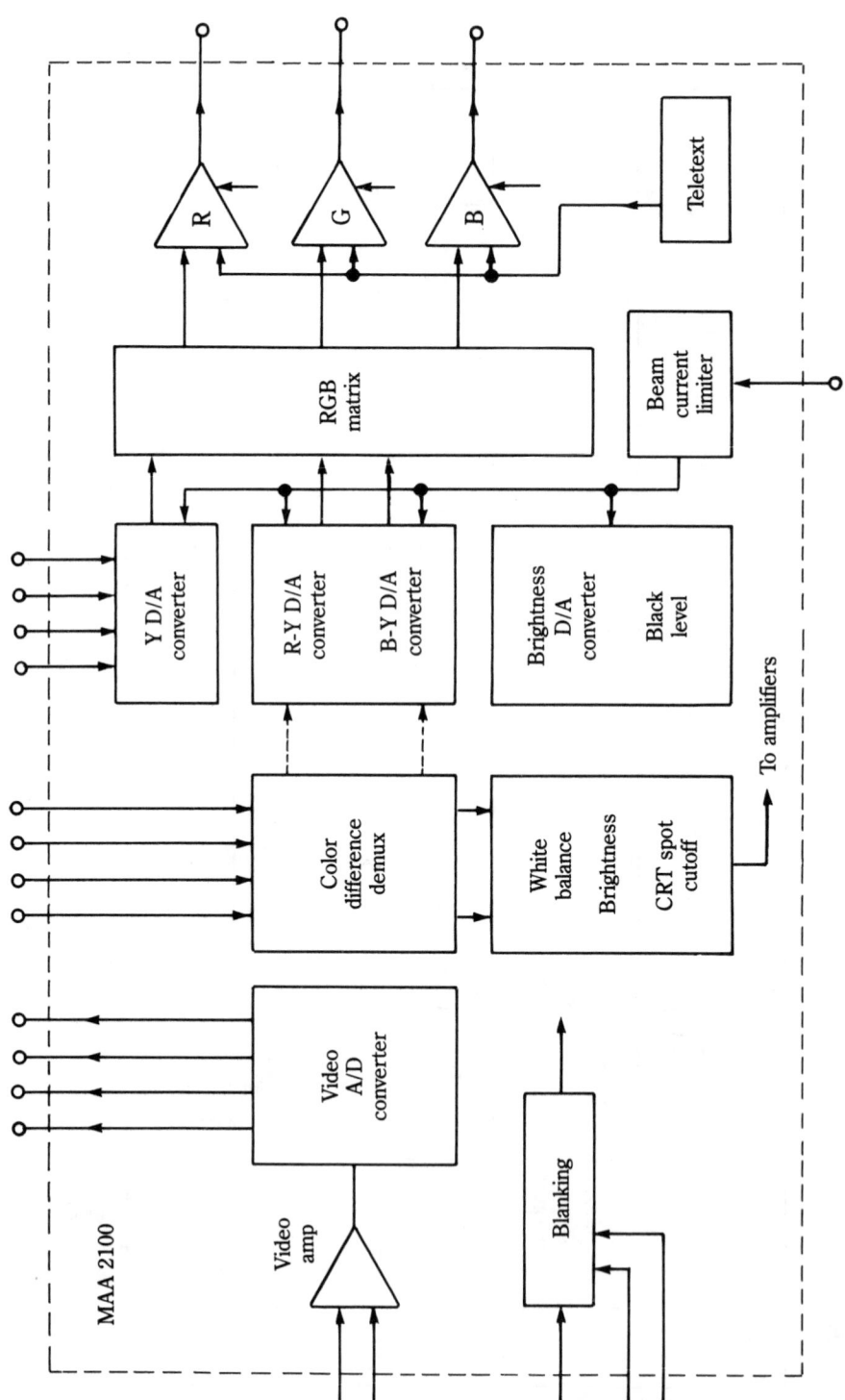

12-24 Simplified block diagram of the video CODEC unit.

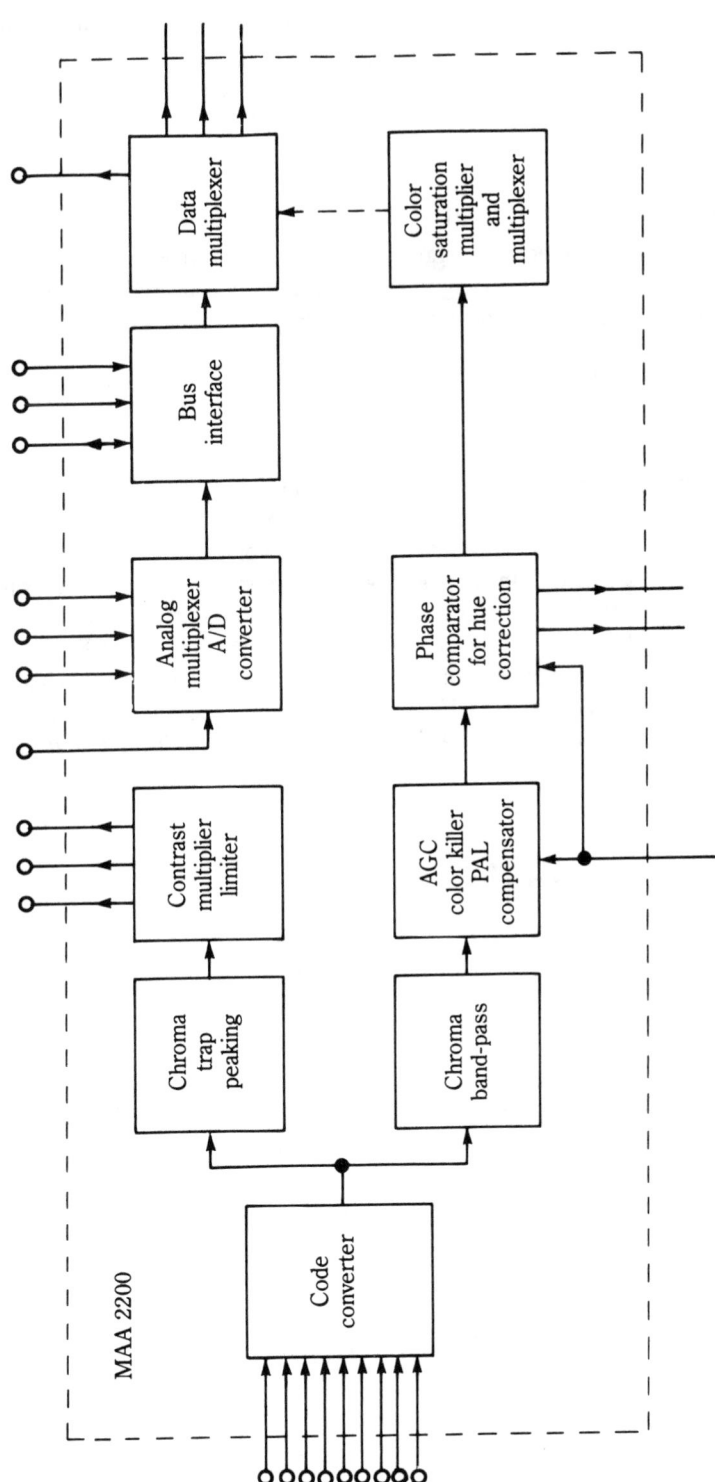

12-25 Simplified block diagram of the video processing unit.

components by delay, adding, and multiplying devices. Response to user settings for brightness and contrast is accomplished here. The block diagram is shown in Fig. 12-25.

A phase-locked loop in the system makes phase comparisons of the color signals and the color burst. User color adjustments modify the automatic corrections made to the clock.

Deflection control unit

Deflection is not directly synchronized with the sync pulses but relies on the fixed relationship between the scanning frequencies and the 3.579545 MHz color burst. Counter circuits subdivide the frequency to achieve perfect synchronism, which cannot be disturbed by temporary fading or interference.

Central control unit

The extreme flexibility of the ITT system resides in the central control unit. This unit is a microprocessor whose program is stored in an EEPROM.

As an example of this flexibility some 20 trim potentiometers can be eliminated, as well as the need to manually adjust them on the production line and readjust them as the set ages or is repaired. Instead of these adjustments to compensate for normal variations in the CRT and components, the production line or service computer will automatically load the digital equivalent of the appropriate settings into the EEPROM.

A large number of additional potential TV improvements associated with the digital control flexibility have been identified. These include digital frame storage to achieve pseudo-high resolution, adaptability to all three world color TV standards (NTSC, PAL, and SECAM), adaptability to different world stereo standards, zoom enlargement capability, and so on.

Multiple-choice questions

1. Facsimile allows remote duplication of which items?

    A. Printed page.
    B. Weather map.
    C. Photographs.
    D. All of the preceding.
    E. None of the preceding.

2. Television has a number of features in common with:

    A. FM stereo.
    B. Motion pictures.
    C. Slick magazines.
    D. Postal service.
    E. Intercontinental telephone service.

3. One major component of the television signal waveform is commonly referred to as:

    A. Video.

B. Radar.
C. Synthesized.
D. Stereo.
E. Multiplexed.

4. With facsimile the rotation of a cylinder at the sending end and that of a second cylinder at the receiving end must be:

A. Multiplexed.
B. Synchronized.
C. Geared to a common shaft.
D. Manually controlled.
E. Oriented with axis vertical.

5. In the NTSC television system the image is scanned from:

A. Right to left.
B. Left to right.
C. Top to bottom.
D. Both B and C.
E. None of the choices.

6. A complete NTSC scanning cycle, referred to as a frame, consists of:

A. 60 fields.
B. 525 lines.
C. 30 Hz.
D. Odd-numbered lines.
E. Even-numbered lines.

7. In the NTSC television system the field rate is:

A. 30 Hz.
B. 60 Hz.
C. Approximately 15.75 kHz.
D. 31.5 kHz.
E. No correct answer is given.

8. The horizontal synchronizing/blanking pulse rate is:

A. 30 Hz.
B. 60 Hz.
C. 15.75 kHz.
D. 31.5 kHz.
E. No correct answer is given.

9. The vertical blanking pulse rate is:

A. 30 Hz.

B. 60 Hz.
C. 15.75 kHz.
D. 31.5 kHz.
E. No correct answer is given.

10. Name the pulses riding on the vertical blanking pulse:

   A. Equalizing (sync) pulses.
   B. Serrated vertical sync pulses.
   C. Black level pulses.
   D. All of the preceding.
   E. Both A and B.

11. The name of a TV camera pickup tube is:

   A. Vidicon.
   B. Image orthicon.
   C. Plumbicon.
   D. Silicon-Intensifier-Target.
   E. All of the preceding.

12. In television picture tubes magnetic fields can be used for:

   A. Beam focusing.
   B. Beam deflection.
   C. Beam modulation.
   D. Field emission.
   E. Both A and B.

13. The purpose of an indirectly-heated cathode in a CRT is:

   A. Beam focusing.
   B. Beam deflection.
   C. Beam production.
   D. Both A and B.
   E. All of the choices.

14. Video modulation for TV is best described as:

   A. Amplitude modulation.
   B. Frequency modulation.
   C. Vestigial sideband.
   D. Both A and C.
   E. Both B and C.

15. Audio modulation for TV is:

   A. Amplitude modulation.
   B. Frequency modulation.

C. Vestigial sideband.
D. Both A and C.
E. Both B and C.

16. The maximum frequency deviation for audio modulation in TV transmission is:

A. 3 kHz.
B. 5 kHz.
C. 25 kHz.
D. 75 kHz.
E. There is no frequency deviation since it is amplitude modulation.

17. The separation between the video carrier and the audio carrier is:

A. 1.25 MHz.
B. 4 MHz.
C. 4.5 MHz.
D. 6 MHz.
E. None of the preceding.

18. In a practical image scanned at the NTSC rate, the spectrum will include components clustered at the fundamental and harmonics of:

A. 15.75 kHz.
B. 25 kHz.
C. 31.5 kHz.
D. 4.5 MHz.
E. 6 MHz.

19. The word which describes the brightness of an image is:

A. Radiance.
B. Chrominance.
C. Reflectance.
D. Reflectivity.
E. Luminance.

20. In the color triangle the colors at the vertices (corners) are referred to as:

A. Desaturated.
B. Primary.
C. White.
D. Washed-out.
E. Mixtures.

21. Identify one or more of the color signals which originate in the camera itself with color TV pickup:

A. R for red.

B. G for green.
C. B for blue.
D. Y for yellow.
E. Choices A, B, and C.

22. Three separate signals are derived from a matrix in a color TV transmitter. These are:

A. Y, I and Q.
B. R, S, and T.
C. M, N, and O.
D. P, D, and Q.
E. None of the choices is true.

23. A vhf television transmitter must cover a range of:

A. 54 – 88 MHz.
B. 174 – 216 MHz.
C. 54 – 216 MHz.
D. 470 – 890 MHz.
E. 54 – 90 MHz.

24. A uhf television transmitter must cover a range of:

A. 54 – 88 MHz.
B. 174 – 216 MHz.
C. 54 – 216 MHz.
D. 470 – 890 MHz.
E. 54 – 90 MHz.

25. Approximately what is the bandwidth occupied by the chrominance video signal for color TV?

A. 0.6 MHz.
B. 1.2 MHz.
C. 1.8 MHz.
D. 4 MHz.
E. 6 MHz.

26. The colors found around the perimeter of the color triangle are said to be:

A. Whitewashed.
B. Dark.
C. Monochrome.
D. Saturated.
E. Contrasted.

27. At what position on the color triangle will saturated yellow be found?

A. Between Red and Blue.
B. Between Blue and Green.
C. Between Green and Red.
D. Near the center.
E. No correct choice is listed.

28. The phase difference between the *I* and *Q* color signal carriers is:

   A. 0.
   B. 45°.
   C. 60°.
   D. 90°.
   E. 180°.

29. The most effective filter for separating luminance and chrominance frequency components is:

   A. Low-pass.
   B. Band-pass.
   C. Notch.
   D. Comb.
   E. Brush.

30. The chrominance processing circuits are deactivated when monochrome broadcasts are received by the:

   A. NOT gate.
   B. Color killer.
   C. One-shot MV.
   D. SAW filter.
   E. Both A and C.

Basic problems

1. State the relationships between the following: frame, field, and line.

2. Describe the audio carrier and modulation for television.

3. Calculate f_v and f_a for the upper vhf channel 13.

4. In analyzing the video spectrum, demonstrate the origin of the particular frequency 15.75 kHz.

5. A uhf TV transmitter delivers 55 kW visual and 11 kW aural and requires 135 kW of 3-phase power at 480 V ac. Calculate the overall power efficiency.

6. Explain what frequency components due to a vertical bar pattern might be found for monochrome TV.

7. Explain the significance of the equation:

$$Y = 0.299R + 0.587G + 0.114B$$

8. Briefly explain the method of chrominance modulation.

9. Describe how the shadow mask and color CRT triads are related.

10. In a TV transmitter, at what point are the aural and video parts combined?

Advanced problems

1. Briefly describe the nature of the horizontal blanking and synchronizing pulse.

2. Briefly describe the vertical blanking and associated pulses.

3. Briefly describe the vidicon tube.

4. The channel width for NTSC television is 6 MHz. What is the highest video frequency component that can be accommodated?

5. Describe the conditions for which the luminance signal from the matrix theoretically designates black.

6. Briefly describe the positions in the spectrum of any carrier and subcarrier employed in NTSC color TV.

7. Explain the numerical relationship between the horizontal scan rate and the color burst frequency.

8. Starting with the equations

$$
\begin{aligned}
Y &= 0.299R + 0.587B + 0.114B \\
I &= 0.74\,(R\text{-}Y) - 0.27\,(B\text{-}Y) \\
Q &= 0.48\,(R\text{-}Y) + 0.41\,(B\text{-}Y)
\end{aligned}
$$

determine the values of I and Q under the condition of no red light, i.e., $R = 0$.

9. In the comb filter described in this lesson, how is the time delay of one horizontal line period $(1/f_H)$ achieved?

10. In Digital Signal Processing for TV the received signal is digitized at some point. Where does this occur?

13

Transmission lines
and antennas

PRACTICAL CONSIDERATIONS REQUIRE THAT AN ANTENNA BE LOCATED SOME DISTANCE from its associated transmitter or receiver. A means must, therefore, be provided to transfer the rf power between the equipment and the antenna. Basically, an rf transmission line electrically connects an antenna and a transmitter or a receiver. It must do so efficiently with a minimum loss of power or signal strength.

There are four general types of transmission line—the parallel two-wire line, the twisted pair, the shielded pair, and the concentric (coaxial) line. The use of a particular type of line depends, among other things, on the frequency, the power to be transmitted, and the type of insulation. However at microwave frequencies normally specialized transmission lines are used, such as waveguides.

The two-wire line (Fig. 13-1A) consists of two parallel conductors that are maintained at a fixed distance apart by means of insulating spacers or spreaders that are placed at suitable intervals. The line has the assets of ease of construction, economy, and efficiency. In practice, such lines used in radio work are generally spaced from 5 to 15 cm apart at frequencies of 14 MHz and below. The maximum spacing for frequencies of 18 MHz and above is 10 cm. In order to reduce the radiation from the line to a minimum, it is necessary that the wires be separated by only a small fraction of a wavelength. For best results, the separation should be less than one hundredth of a wavelength.

At very high frequencies this criterion will limit the amount of the rf power that can be conveyed by the line. Consequently the principal disadvantage of the parallel two-wire line is its relatively high radiation loss.

Uniform spacing of a two-wire transmission line can be assured if the wires are embedded in a solid low-loss dielectric such as polyethylene (Fig. 13-1B). This so-called ribbon type of line is widely used to connect television receivers to their antennas.

The twisted pair transmission line is shown in Fig. 13-1C. As the name implies, it consists of two insulated wires twisted to form a flexible line without the use of spacers. It is typically used for low-frequency transmission, since at the higher frequencies there is excessive loss occurring in the insulation.

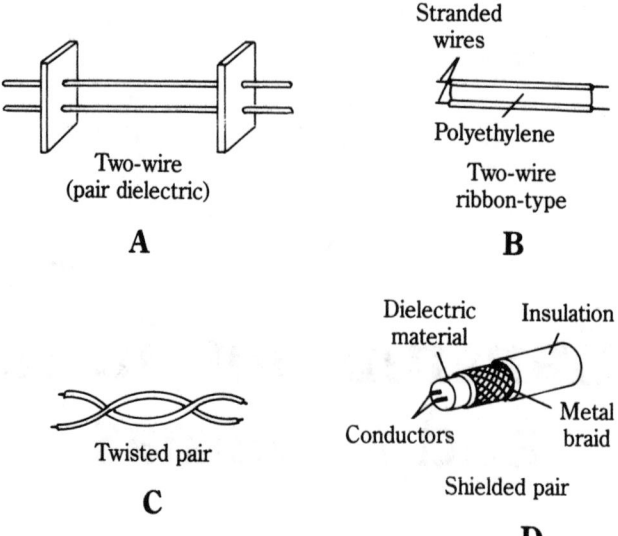

13-1   Four examples of transmission lines.

The shielded pair (Fig. 13-1D) consists of two parallel conductors separated from each other and surrounded by a solid dielectric. The conductors are contained within a copper-braid tubing that acts as a shield. This assembly is covered with a rubber or flexible composition coating to protect the line against moisture.

The principal advantage of the shielded pair is that the two conductors are balanced with respect to ground, that is, the capacitance between each conductor and the ground is uniform along the entire length of the line, and the wires are shielded against any pickup from stray fields. This balance is achieved by the grounded shield that surrounds the conductors with a uniform spacing throughout their length.

Coaxial lines, or *coaxial cables* as they are called, are the most widely used type of rf transmission line. They consist of an outer conductor and an inner conductor held in place exactly at the center of the outer conductor.

Several types of coaxial cable have come into wide use for feeding rf power to an antenna system. Figure 13-2A and B, illustrates the construction of flexible and rigid coaxial cables. In both cases one of the conductors is placed inside the other. Since the outer conductor completely shields the inner one, no radiation loss takes place. This is the main advantage of the coaxial line.

The subjects of transmission lines, antennas, and propagation (the various ways in which radio waves travel from transmitters to receivers) are explored in the following sections:

- Distributed constants of transmission lines
- The matched line
- The unmatched line
- The Hertz and the Marconi antennas
- Radiation patterns

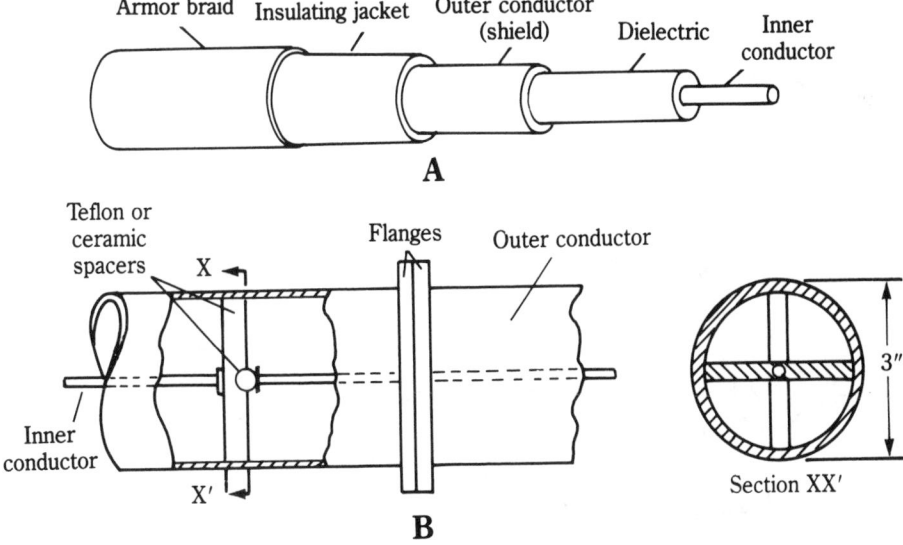

13-2   Flexible and rigid coaxial cables.

- Parasitic elements
- The loop antenna
- Driven arrays
- The nonresonant antenna
- Propagation
- The ionosphere
- Propagation under the various frequency bands

Distributed constants of transmission lines

In chapters 1 through 12 you were concerned with so-called *lumped* circuitry where the three electrical properties (resistance, inductance, and capacitance) were related to specific components. For example, you regarded a resistor as being a "lump" of resistance, and considered all connecting wires to be perfect conductors. By contrast, the resistance, R, associated with a two-wire transmission line is not concentrated into a lump but is distributed along the entire length of the line. The distributed constant of resistance is therefore measured in the basic unit of *ohms per meter*, rather than ohms.

Because the straight wires of the parallel line are conducting surfaces separated by an insulator or dielectric, the line will possess the distributed constants of self-inductance, L (henrys per meter), and capacitance, C (farads per meter). In addition no insulator is perfect and consequently there is another distributed constant, G (siemens per meter), which is the leakage conductance between the wires. These four distributed constants are illustrated in Fig. 13-3; their order of values in a practical line are R, mΩ/m; L, μH/m, C, pF/m, and G, nS/m.

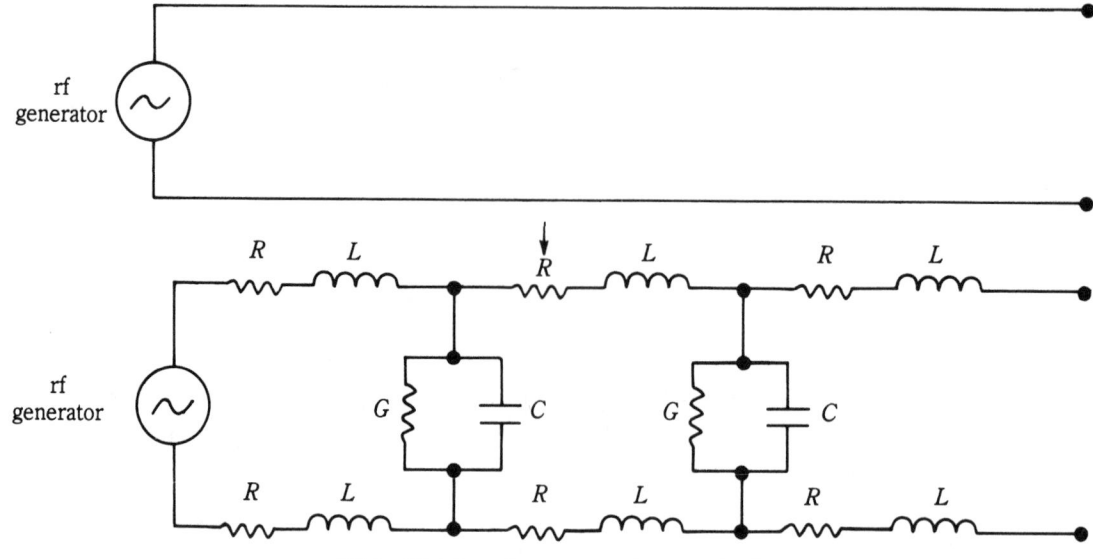

13-3   Distributed constants of a two-wire line.

In physical terms the properties of resistance and leakage conductance will relate to the line's power loss in the form of the heat dissipated and will therefore govern the degree of attenuation, measured in decibels per meter (dB/m). The self-inductance results in a magnetic field surrounding the wires, while the capacitance means that an electric field exists between the wires (Fig. 13-4); these L and C properties determine the line's behavior in relation to the frequency.

The two-wire ribbon line that connects an antenna to its TV receiver is often referred to as a 300 Ω line. But what is the meaning of the 300 Ω? Certainly you cannot find this value with an ohmmeter so you are led to the conclusion that it only appears under working conditions when the line is being used to convey rf power. In fact the 300 Ω value refers to the line's *surge* or characteristic impedance whose letter symbol is Z_o. The surge impedance is theoretically defined as the input impedance at the rf generator to an infinite length of the line (Fig. 13-5A, and B). In the equivalent circuit of Fig. 13-5B, the C and G line constants will complete a path for current to flow so that an effective current, I, will be drawn from the rf generator whose effective output voltage is E. Then the input impedance is:

$$Z_{in} = Z_o = \frac{E}{I} \; \Omega$$

Mathematically it can be shown that Z_o is a complex quantity, given by:

$$Z_o = \sqrt{\frac{R + j\omega L}{G + j\omega C}}$$

$$= \sqrt[4]{\frac{R^2 + \omega^2 L^2}{G^2 + \omega^2 C^2}} \bigg/ \frac{1}{2}\left(\text{inv tan} \frac{\omega L}{R} - \text{inv tan} \frac{\omega C}{G} \right) \Omega \qquad (13\text{-}1)$$

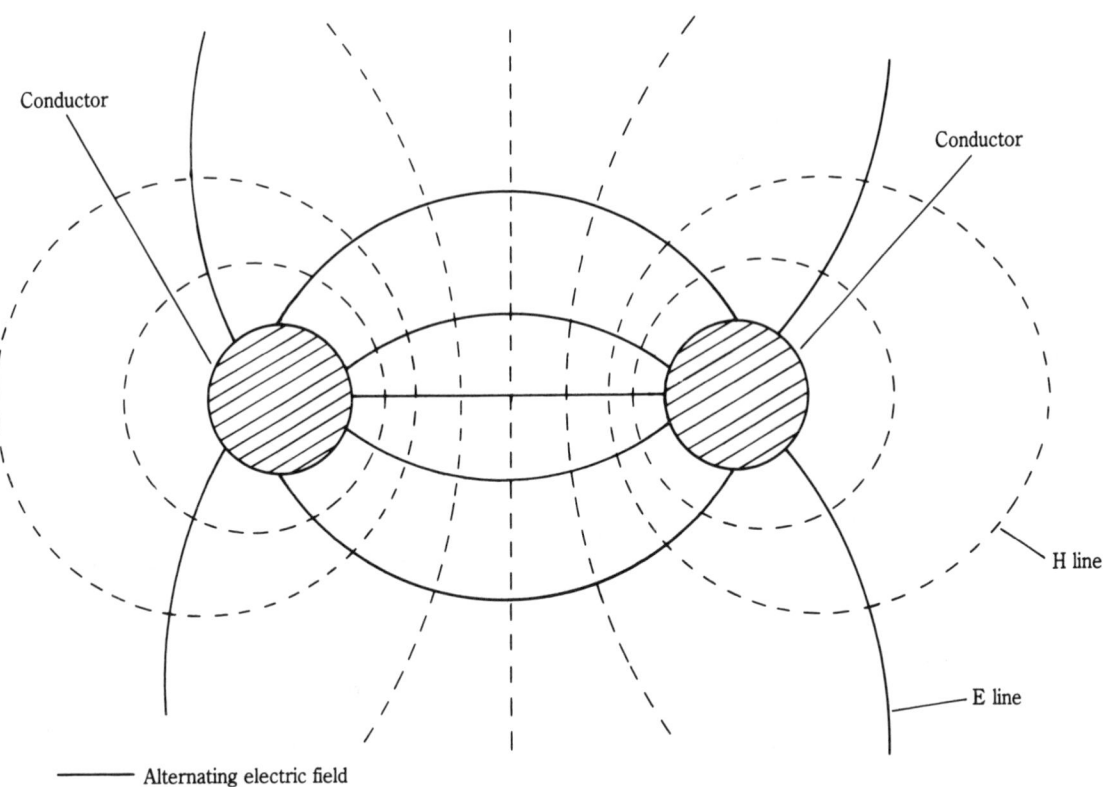

Conductor

Conductor

H line

E line

——— Alternating electric field

– – – – Alternating magnetic field

13-4 Electric and magnetic fields surrounding a two-wire line.

where,

ω, the angular frequency = $2\pi f$ (radians/sec)
f = frequency of the rf generator (Hz)

For a low loss line which is operating at radio frequencies, $\omega L \gg R$ and $\omega C \gg G$. Then:

$$Z_o \rightarrow \sqrt{\frac{L}{C}} \underline{/0°}\ \Omega \qquad (13\text{-}2)$$

An rf line therefore behaves resistively and has a surge impedance of $\sqrt{L/C}\ \Omega$. However, the value of Z_o must also depend on the lines' physical construction; the formulas are:

Two-Wire Line (Fig. 13-6)

$$Z_o = \frac{276}{\sqrt{\epsilon_r}} \log_{10} \frac{2S}{d}\ \Omega \qquad (13\text{-}3)$$

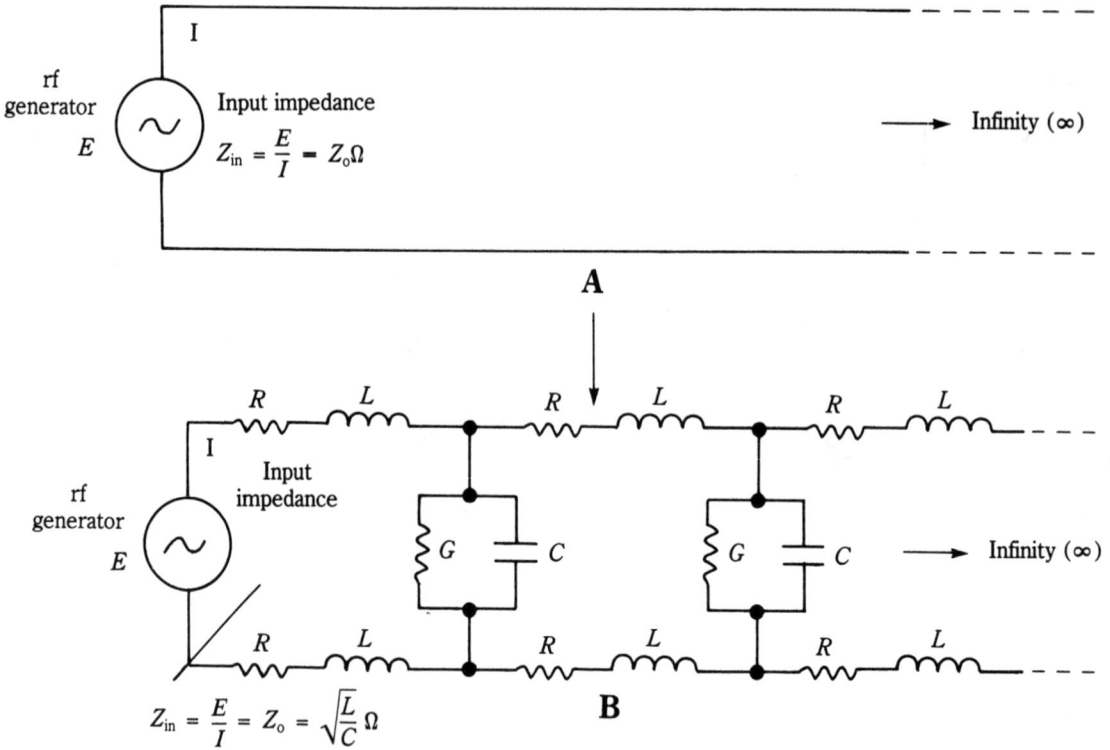

13-5   The surge or characteristic impedance, Z_o.

where

S = spacing between the conductors (m)
d = diameter of each conductor (m)
ϵ_r = relative permittivity of the insulation.

Coaxial Cable (Fig. 13-7)

$$Z_o = \frac{138}{\sqrt{\epsilon_r}} \log_{10} \frac{D}{d} \ \Omega \tag{13-4}$$

where

D = inner diameter of outer conductor (m)
d = outer diameter of inner conductor (m)

Summarizing, a transmission line is used to convey rf power from one position to another. It is an example of distributed circuitry with the four primary constants R, L, G, and C. The surge impedance, Z_o, is a secondary constant which is defined as the input impedance to an infinite length of the line; the value of Z_o depends on the line's physical construction. At radio frequencies a line behaves resistively and has a surge impedance equal to $\sqrt{L/C}\ \Omega$.

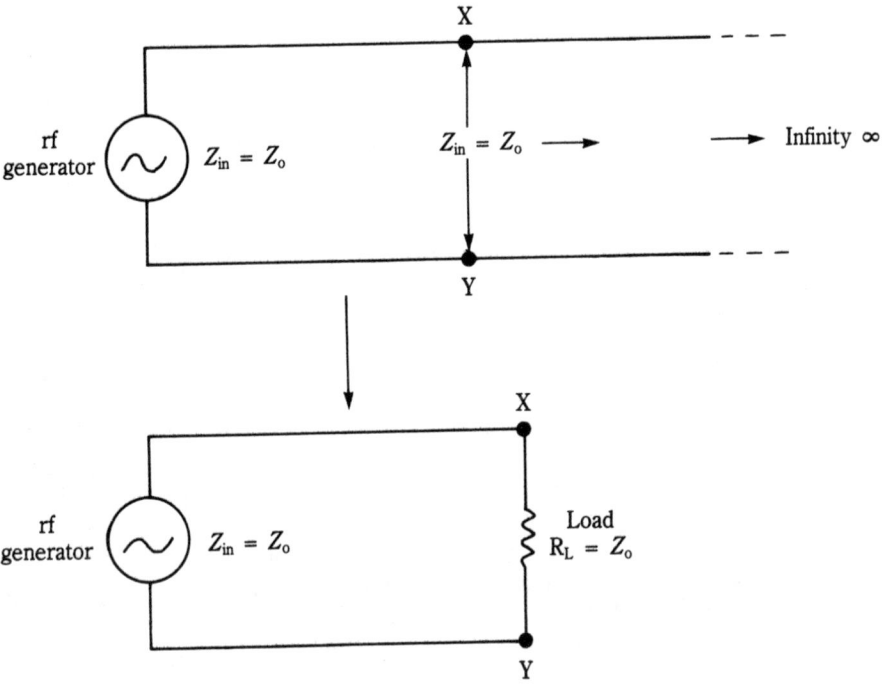

13-6 The matched line.

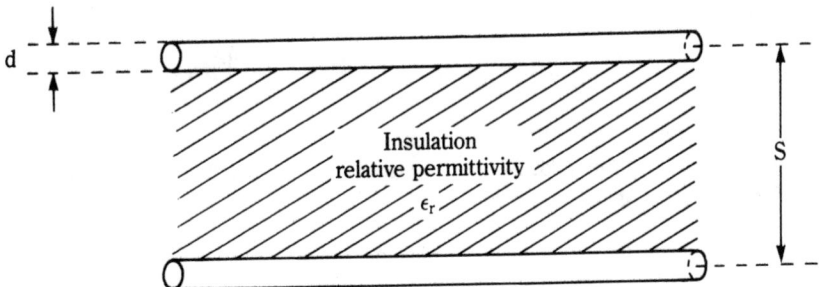

13-7 Physical factors affecting the value of the surge impedance of a two-wire line.

The matched line

Refer to Fig. 13-8, consider the conditions that exist at the position, X Y, on the infinite line. Since there is still an infinite length to the right of X Y, the input impedance at this position looking down the line will be equal to the value of Z_o. Consequently if the section of the line to the right of X Y, is removed and replaced by a resistive load whose value in ohms is the same as that of the surge impedance, it will still appear to the generator as if it is connected to an infinite line and the input impedance at the generator will remain equal to Z_o.

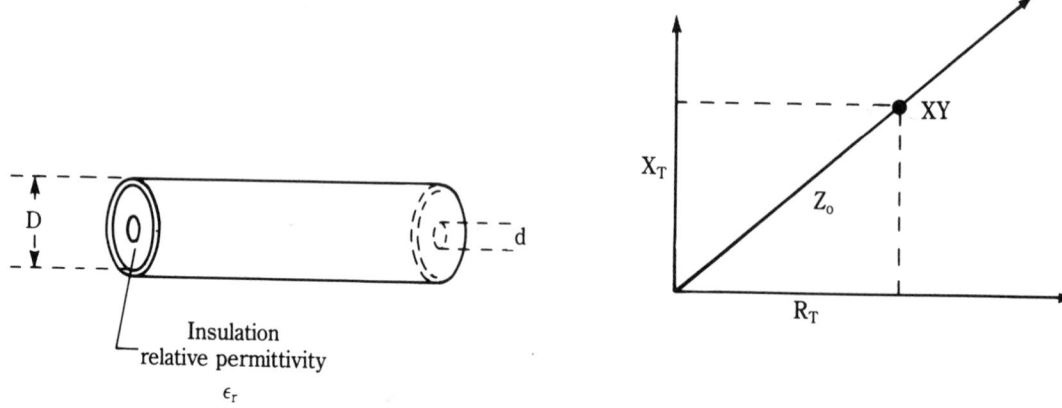

13-8 Physical factors affecting the value of the surge impedance of a coaxial cable.

When a line is terminated by a resistive load of value Z_o, the line is said to be *matched* to the load. Under matched conditions the line is most efficient in conveying rf power from the generator (for example, a transmitter) to the load, such as an antenna.

Examine in detail what happens on a matched line. Travelling sine waves of voltage and current start out from the rf generator and move down the line in phase. Due to the small amount of attenuation present on the line, the effective (rms) values of the voltage and current decay slightly but at all positions $V_{RMS}/I_{RMS} = Z_o$ (Fig. 13-9). The degree of attenuation is measured by the attenuation constant, α, given by:

$$\text{Attenuation constant, } \alpha = 8.7 \left(\frac{R}{2Z_o} + \frac{GZ_o}{2} \right) dB < m \qquad (13\text{-}5)$$

On arrival at the termination the power contained in the voltage and current waves is completely absorbed by the load. Neglecting the losses on the line, the rf power, P, conveyed down the line is given by:

$$P = V_{rms} \times I_{rms}$$

$$= I^2_{rms} \times Z_o$$

$$= \frac{V^2_{rms}}{Z_o} \text{ W} \qquad (13\text{-}6)$$

These relationships are comparable with those for a resistor; however it is important to realize that power is being conveyed down the line and is not being dissipated and lost as heat in the surge impedance.

In earlier chapters it was assumed that in a series circuit consisting of a source, a two-wire line, and a load, the current was instantaneously the same throughout the circuit. This can only be regarded as true provided the distances involved in the circuit are small compared with the wavelength of the output from the source. The wavelength whose letter symbol is the Greek lambda, λ, is defined as the distance between two con-

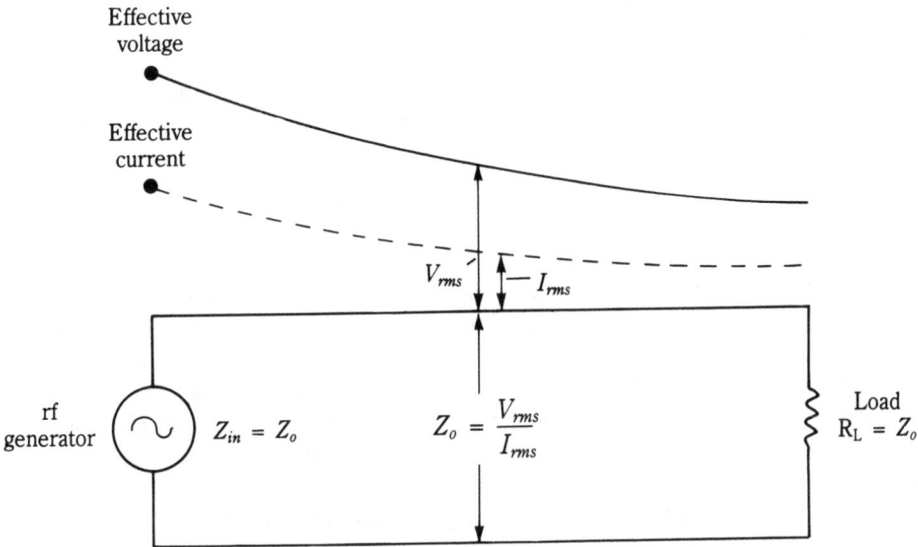

13-9 Effective voltage and current distribution on a matched transmission line.

secutive identical states in the path of the wave; for example, the wavelength of a wave in water is the distance between two neighboring crests (or troughs). On a matched transmission line the wavelength is the distance between two adjacent positions where identical voltage (and current) conditions occur instantaneously. Traveling or progressing waves exist on the line because time is involved in propagating rf energy from the source to the load. The waves travel through a distance of one wavelength in a time equal to one period. Since velocity = distance/time,

$$v = \frac{\lambda}{T}$$

$$= \lambda \times \frac{1}{T}$$

$$= f \times \lambda \tag{13-7}$$

This yields,

$$\lambda = \frac{v}{f}$$

and

$$f = \frac{v}{\lambda} \tag{13-8}$$

where

v = the velocity in meters per second (m/s)
f = the frequency in hertz (Hz)

λ = the wavelength in meters (m)
T = the period (s)

In free space the velocity of an electromagnetic wave (radio wave) is a constant approximately equal to 3×10^8 m/s (the speed of light). For example, a 100 MHz radio wave within the FM broadcast band has a wavelength of:

$$\lambda = \frac{3 \times 10^8 \text{ m/s}}{100 \times 10^6 \text{ Hz}} = 3 \text{ m}$$

On a transmission line consider the generator's output as the reference voltage. At a position which is a distance of $\lambda/4$ from the generator, the instantaneous voltage will lag by 90° on the reference level, while at distances of $\lambda/2$ and $3\lambda/4$ (positions B, and C), the voltages will be respectively 180° out of phase and 270° lagging (90° leading). A particular phase condition will then travel down at a speed which is called the *phase velocity*, v_{ϕ}; this movement of a phase condition is illustrated in Fig. 13-10. Consequently over a distance of 1 meter there will be an angular difference which is measured by the phase shift constant, β. The unit of β is the radian per meter and since there must be a difference of 360° (2 π radians) over a distance of λ meters:

$$\text{Phase shift constant, } \beta = \frac{2\pi}{\lambda} \text{ radians per meter} \tag{13-9}$$

The equations for the travelling waves of voltage and current on a loss-free line are:

$$v = E \sin\left(\omega t - \frac{2\pi d}{\lambda}\right)$$

$$= E \sin(2\pi ft - \beta d) \tag{13-10}$$

and

$$i = I \sin\left(\omega t - \frac{2\pi d}{\lambda}\right)$$

$$= I \sin(2\pi ft \beta d)$$

$$= \frac{E}{Z_o} \sin(2\pi ft - \beta d) \tag{13-11}$$

where
v = the instantaneous value of the voltage wave (V)
i = the instantaneous value of the current wave (A)
E = the peak value of the voltage wave (V)
I = the peak value of the current wave (A)
ω = angular frequency (rad/s)
f = frequency (Hz)
β = the phase shift constant (rad/m)
d = the distance from the generator (m)
t = time (s)
Z_o = surge impedance (Ω).

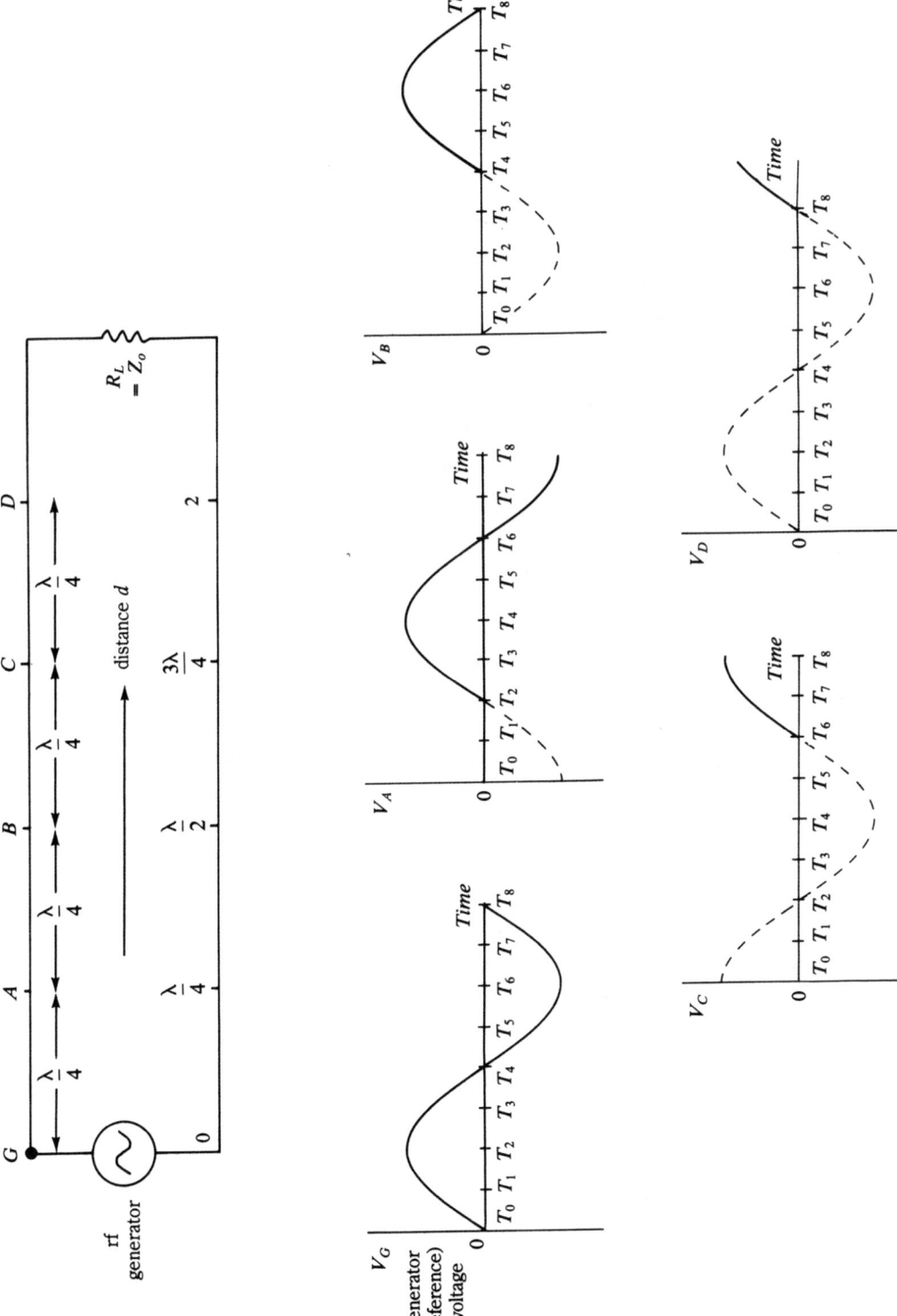

13-10 Progressive voltage wave traveling down a matched transmission line.

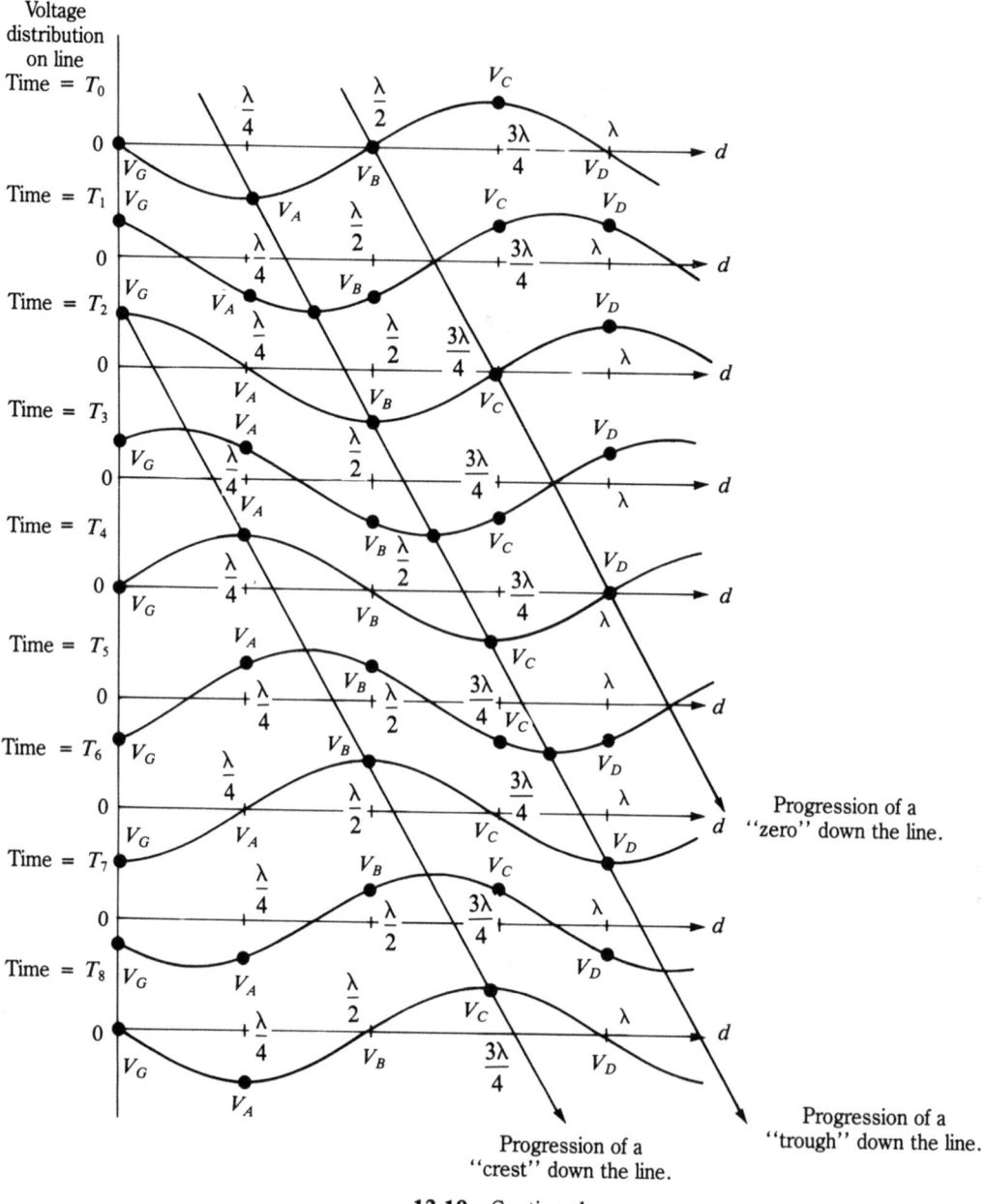

Voltage distribution on line

13-10 Continued.

Mathematically it can be shown that in a low-loss line,

$$\beta = \omega \sqrt{LC} = 2\pi f \sqrt{LC}$$

(13-12)

therefore,

$$\text{Phase velocity, } v_{\phi} = f \times \lambda$$

$$= \frac{2\pi f}{\beta}$$

$$= \frac{\omega}{\beta}$$

$$= \frac{1}{\sqrt{LC}} \text{ meters per second}$$

(13-13)

The phase velocity is the speed at which the voltage and current waves, as well as the electric and magnetic fields apparently move down the line.

The value of v_{ϕ} is always less than the velocity of light, c. The ratio of v_{ϕ}/c is called the velocity factor, δ. The value of δ varies from 0.66 for certain types of coaxial cable to 0.975 for an air-insulated two-wire line. It follows that since $v_{\phi} = f \times \lambda$, the wavelength on the line is shorter than the wavelength in free space.

The features of a matched line can be summarized as follows:

1. Travelling waves of voltage and current move down the line in phase and their power is completely absorbed by the load.
2. The ratio of the effective voltage to the effective current is constant over the entire line and is equal to the surge impedance, Z_o.
3. The input impedance at the generator is equal to the surge impedance and is independent of the line's length.
4. The power losses on the line are subdivided into:
   a. radiation and induction losses which are a problem with parallel-wire lines.
   b. the dielectric hysteresis loss which increases with frequency and depends on the type of insulator used. At microwave frequencies of a few GHz the dielectric loss is the ultimate reason for abandoning coaxial lines and using waveguides instead.
   c. the copper loss which is associated with the conductor's resistance. At high frequencies this loss is increased by the skin effect, which confines most of the electron flow to the surface (skin) of a conductor and therefore reduces the available cross-sectional area. The larger is the surface area of the conductors, the less is this type of loss.

The unmatched line

If an rf line is terminated by a resistive load which is not equal to the surge impedance, the generator will still send voltage and current waves down the line in phase, but their power will only be partially absorbed by the load. A certain fraction of the voltage and current waves which arrive (are incident) at the load, will be reflected back toward the generator. At any position on the line the instantaneous voltage (or current) will be the resultant of the incident and reflected voltage (or current) waves; these combine to produce so-called standing waves. In the extreme cases of open- and short-circuited lines, no power can be absorbed by the load and total reflection will occur.

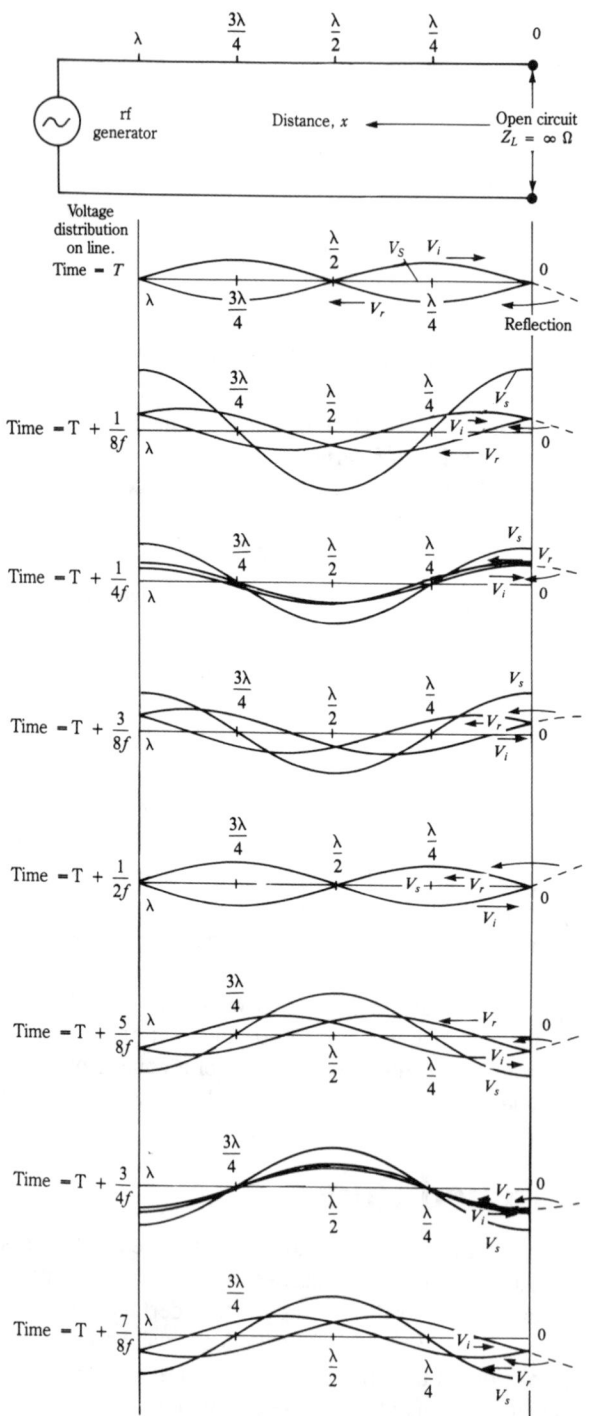

13-11A Instantaneous incident, reflected, and standing voltage waves on the last wavelength of an open-circuited transmission line.

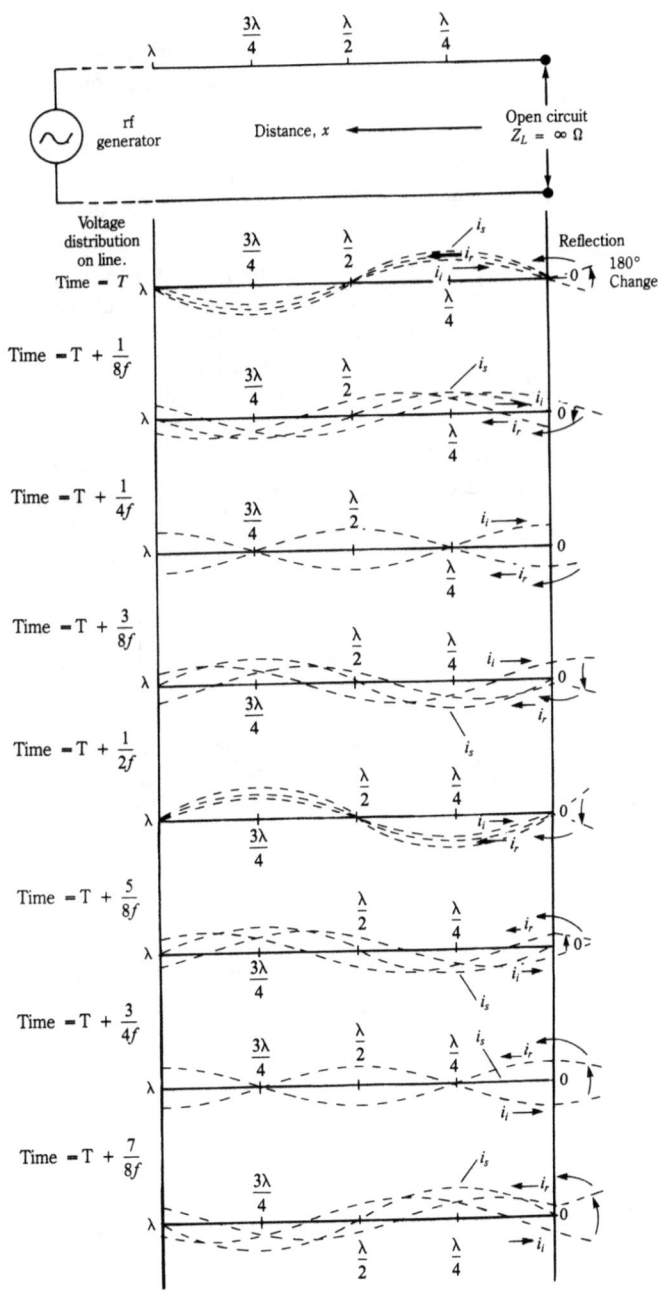

13-11B Instantaneous incident, reflected, and standing current waves on the last wavelength of an open-circuit transmission line.

Consider the production of standing waves on the last wavelength of an open-circuited line (Fig. 13-11). The arrow " → " is used to refer to an incident wave while " ← " indicates a reflected wave. All distances are measured from the line's open-circuited end, where a voltage can exist but where the current at all times must be zero. The voltage which is incident at the open circuit, cannot abruptly change its polarity as it turns back toward the generator; consequently there is an *in phase* reflection in which the incident traveling wave is regarded as extended beyond the open-circuited end and then "folded back." The incident and reflected waves are added together to produce the voltage standing wave at an arbitrarily chosen time, *T*. As the incident voltage wave moves down the line toward the open circuit, you can derive the changes in the instantaneous standing line for each subsequent interval of one-eighth of the period. The instantaneous standing wave is therefore illustrated in Fig. 13-11 for the times, T, $T + 1/(8f)$, $T + 1/(4f)$, . . . , $T + 7/(8f)$. Finally all instantaneous voltage standing waves are collected together in one representation (Fig. 13-12A). At the distances of 0, $\lambda/2$, and λ from the open circuit termination, there are *fixed* positions of maximum voltage variation with time; such variations are called *voltage antinodes* or loops. However at the distances of

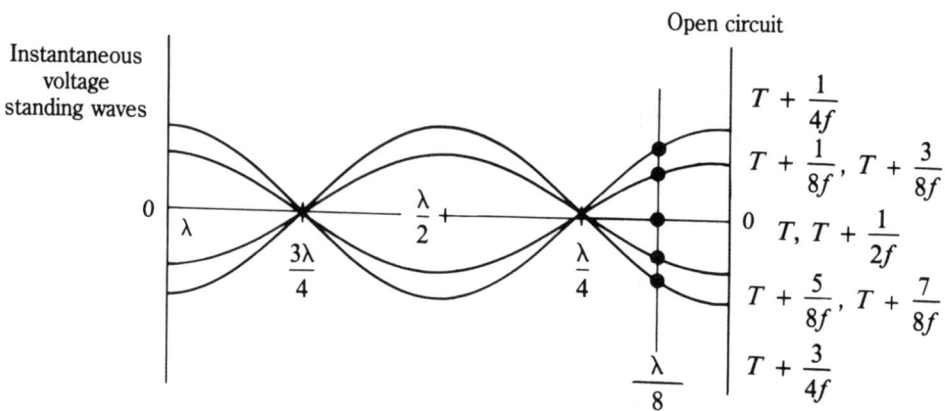

13-12A Instantaneous voltage standing waves on an open-circuited transmission line.

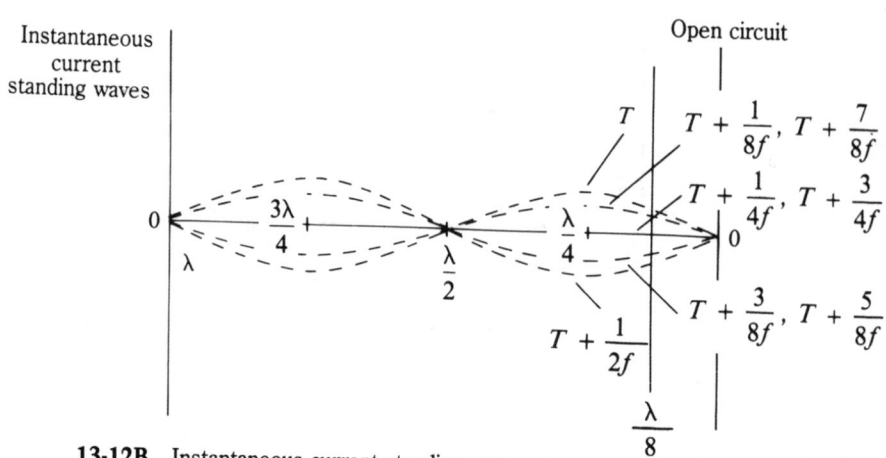

13-12B Instantaneous current standing waves on an open-circuited transmission line.

λ/4, and 3λ/4, there are *stationary* positions where the voltage is zero at all times; these are called the *voltage nodes* or *nulls*. The stationary points that occur with the standing wave, are in contrast with the traveling wave where, for example, a zero voltage condition would move down the line at a speed equal to the phase velocity.

It is important to realize that a standing wave represents a sinusoidal distribution with both time and distance. The incident traveling wave can be represented by:

$$v_i = E \sin\left(2\pi ft + \frac{2\pi x}{\lambda}\right) \tag{13-14}$$

where,

v_i = the instantaneous value of the incident voltage wave (V)
E = the peak value of the incident voltage wave (V)
f = frequency (Hz)
x = the distance measured from the open-circuit termination (m)
λ = the wavelength existing on the line (m)

The equation of the reflected wave is:

$$v_r = E \sin\left(2\pi ft - \frac{2\pi x}{\lambda}\right) \tag{13-15}$$

where,

v_r = the instantaneous value of the reflected voltage wave (V)

The instantaneous standing wave voltage, v_s, is:

$$v_s = v_i + v_r$$

$$= E\left[\sin\left(2\pi ft + \frac{2\pi x}{\lambda}\right) + \sin\left(2\pi ft \frac{2\pi x}{\lambda}\right)\right]$$

$$= 2E \sin 2\pi ft \cos \frac{2\pi x}{\lambda} \tag{13-16}$$

Sinusoidal distribution with time

Sinusoidal distribution with distance

It is customary to represent the standing wave in terms of its effective or rms value, V_s, which is expressed by $2E \cos 2\pi x/\lambda$ and is illustrated in Fig. 13-13. Notice that you only consider the magnitude of this expression and that no positive or negative sign is involved.

Now examine the current (broken line) standing wave distribution on an open-circuited line. At the open-circuit termination the circuit must reverse its direction, and

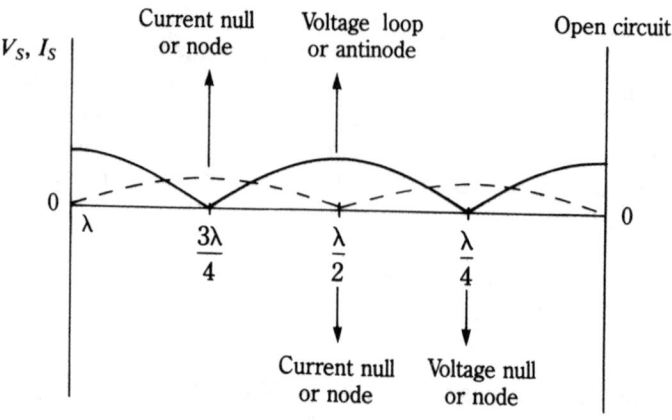

13-13 Effective voltage and current standing wave distribution on the last wavelength of an open-circuited transmission line.

therefore you have an *out of phase* reflection in which the wave is extended beyond the open circuit, then phase shifted by 180°, and afterwards folded back to provide the reflected wave. Incident and reflected waves are then combined to produce the current standing waves at times, T, $T + 1/(8f)$, $T + 1/(4f)$, ..., $T + 7/(8f)$, (Fig. 13-11B). Finally all eight instantaneous current standing waves are collected together in one presentation (Fig. 13-12B). The incident current wave, which is in phase with the incident voltage, is represented by:

$$i_i = I \sin\left(2\pi ft + \frac{2\pi x}{\lambda}\right)$$

$$= \frac{E}{Z_o} \sin\left(2\pi ft + \frac{2\pi x}{\lambda}\right) \tag{13-17}$$

where,

i_i = the instantaneous value of the incident current wave (A)

I = the peak value of the incident current wave (A)

The equation of the reflected current wave is:

$$i_r = -I \sin\left(2\pi ft - \frac{2\pi x}{\lambda}\right)$$

$$= -\frac{E}{Z_o} \sin\left(2\pi ft - \frac{2\pi x}{\lambda}\right) \tag{13-18}$$

where the minus sign preceding I indicates the "out of phase" reflection, and

i_r = instantaneous value of the reflected current wave (A)

The instantaneous standing wave current, i_s, is:

$$i_s = i_i + i_r$$

$$= I \left[\sin \left(2\pi ft + \frac{2\pi x}{\lambda}\right) - \sin \left(2\pi ft - \frac{2\pi x}{\lambda}\right) \right]$$

$$= \frac{2E}{Z_o} \cos 2\pi ft \sin \frac{2\pi x}{\lambda} \tag{13-19}$$

Notice that, because of the "sin $2\pi ft$" and "cos $2\pi ft$" terms, v_s and i_s are 90° out of phase in terms of time. The rms value of the current standing wave, I_s, has a magnitude of

$$I_s = \frac{2E}{Z_o} \sin \frac{2\pi x}{\lambda} \tag{13-20}$$

which is illustrated in Fig. 13-13. Current nulls exist at distances of zero, $\lambda/2$ and λ from the open-circuit termination, while current antinodes occur at distances of $\lambda/4$, and $3\lambda/4$.

The magnitude of the impedance, Z, at any position is given by:

$$Z = \frac{v_s}{i_s} = \frac{2E \cos \frac{2\pi x}{\lambda}}{\frac{2E}{Z_o} \sin \frac{2\pi x}{\lambda}}$$

$$= Z_o \cot \frac{2\pi x}{\lambda} \tag{13-21}$$

Because v_s and i_s are 90° out of phase, the nature of this impedance must be reactive. But at what positions on the line is the impedance capacitive, and at what positions is it inductive? Start by considering the impedance at the position which is at a distance of $\lambda/8$ from the open circuit termination (Fig. 13-12A, and B). From this position the values of the instantaneous voltage and current standing waves are plotted versus a time scale which contains T, $T + 1/(4f)$, $T + 3/(4f)$, and $T+1/F$ (Fig. 13-14). It is clear that i_s leads v_s by 90°, so the impedance is capacitive. Physically over the last eighth of a wavelength the voltage distribution, starting from a maximum at the open-circuited end, is greater than the current distribution. The electric field associated with the distributed capacitance will dominate the magnetic field produced by the distributed inductance. If $x = \lambda/8$, $\cot 2\pi x/\lambda = \cot \pi/4 = \cot 45° = 1$ and therefore the input impedance to a $\lambda/8$ line is a capacitive reactance whose value is equal to Z_o. A similar analysis for the open circuited $3\lambda/8$ line will show that i_s lags v_s and that the input impedance is an inductive reactance of value Z_o. At the $\lambda/4$ position there is a voltage null and a current antinode; theoretically the impedance is zero, but on a practical line the impedance would be

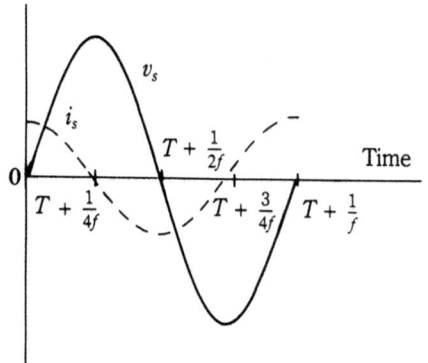

A. At a distance of $\frac{\lambda}{16}$ from the end, the input impedance is a high value of capacitive reactance.

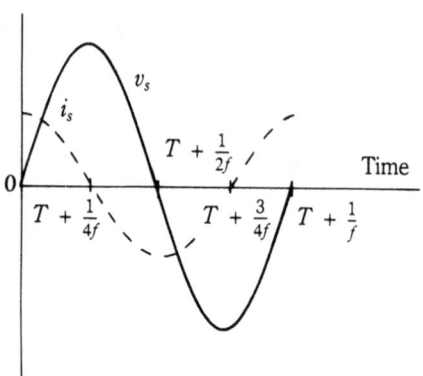

B. At a distance of $\frac{\lambda}{8}$ from the end, the input impedance is a capacitive reactance whose value is equal to the line's surge impedance, Z_0.

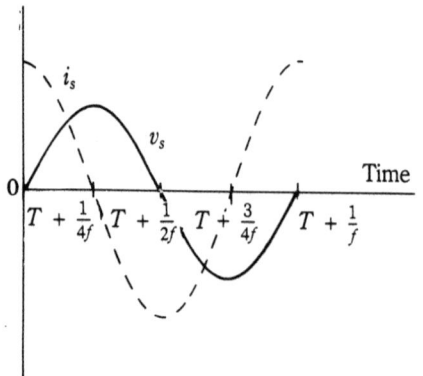

C. At a distance of $\frac{3\lambda}{16}$ from the end, the input impedance is a low value of capacitive reactance.

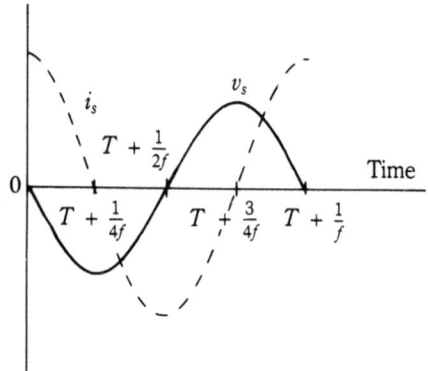

D. At a distance of $\frac{5\lambda}{16}$ from the end, the input impedance is a low value of inductive reactance.

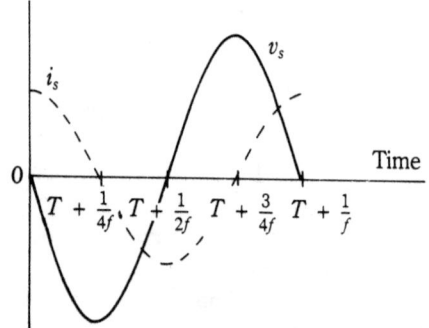

E. At a distance of $\frac{3\lambda}{8}$ from the end, the input impedance is an inductive reactance whose value is equal to the line's surge impedance, Z_0.

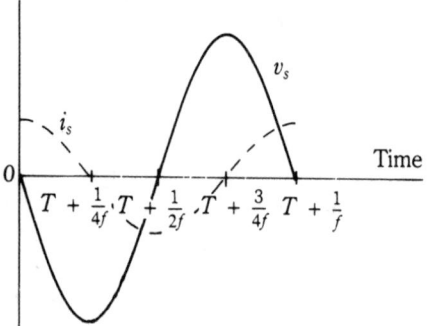

F. At a distance of $\frac{7\lambda}{16}$ from the end, the input impedance is a high value of inductive reactance.

13-14 Variation of input reactance at various positions on the last wavelength of an open-circuited lossfree transmission line.

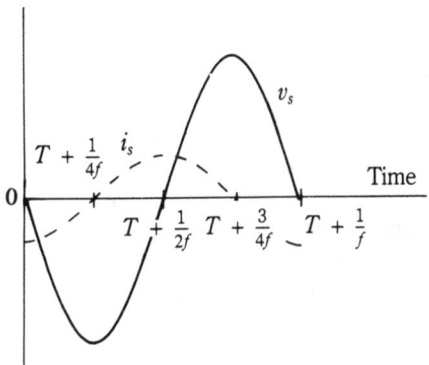

G. At a distance of $\frac{8\lambda}{16}$ from the end, the input impedance is a high value of capacitive reactance.

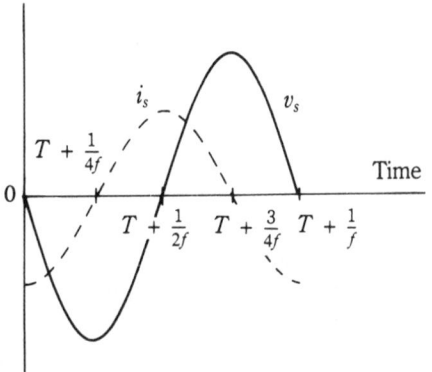

H. At a distance of $\frac{5\lambda}{8}$ from the end, the input impedance is a capacitive reactance whose value is equal to the line's surge impedance, Z_0.

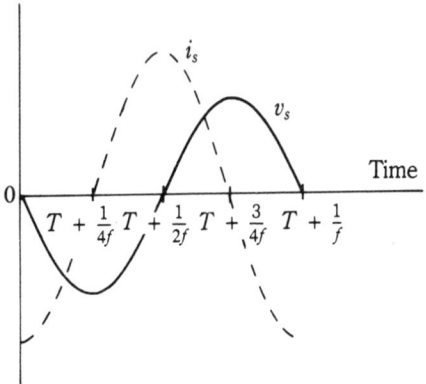

I. At a distance of $\frac{11\lambda}{16}$ from the end, the input impedance is a low value of inductive reactance.

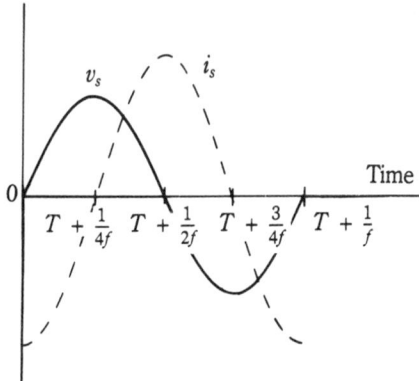

J. At a distance of $\frac{13\lambda}{16}$ from the end, the input impedance is a low value of inductive reactance.

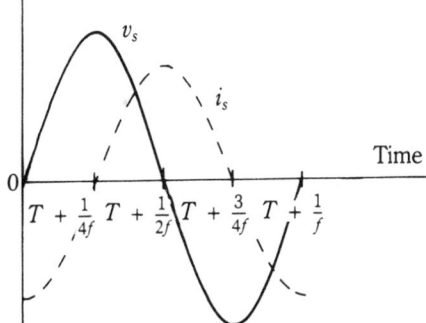

K. At a distance of $\frac{7\lambda}{8}$ from the end, the input impedance is an inductive reactance equal in value to the line's surge impedance, Z_0.

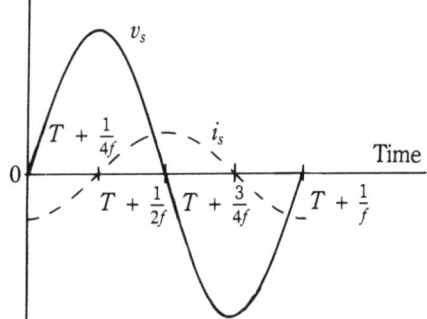

L. At a distance of $\frac{15\lambda}{16}$ from the open-circuited end, the input impedance is is a high value of inductive reactance.

equivalent to a low value of resistance. The conditions are reversed for the λ/2 position where there is a current null and a voltage antinode; the impedance is theoretically infinite but in practice is equivalent to a high value of resistance.

The variation of impedance along the open-circuited line is shown in Fig. 13-15. It is striking to find the similarity to the behavior of series and parallel LC circuits. Remember that the impedance response of the series circuit changes from capacitive through resistive (low value) to inductive; by contrast the parallel circuit has an impedance response which varies from inductive through resistive (high value) to capacitive. These equivalent LC circuits have been included in Fig. 13-15. Since the line exhibits resonant properties it might be referred to as *tuned*; by contrast the matched line is called *flat*, untuned, or non-resonant.

A short-circuited line (Fig. 13-16) also represents a complete mismatch, so total reflection will occur at the termination. However the nature of the reflection is different; the incident voltage wave will undergo an "out of phase" reflection, while the current wave will be reflected "in phase." Since the voltage and current conditions are interchanged when compared with the open-circuited line, the impedance at the λ/8 position is an inductive reactance of value Z_o. At the λ/4 position there is a voltage antinode and a current node, so the impedance is equivalent to a parallel LC circuit. (For frequencies in

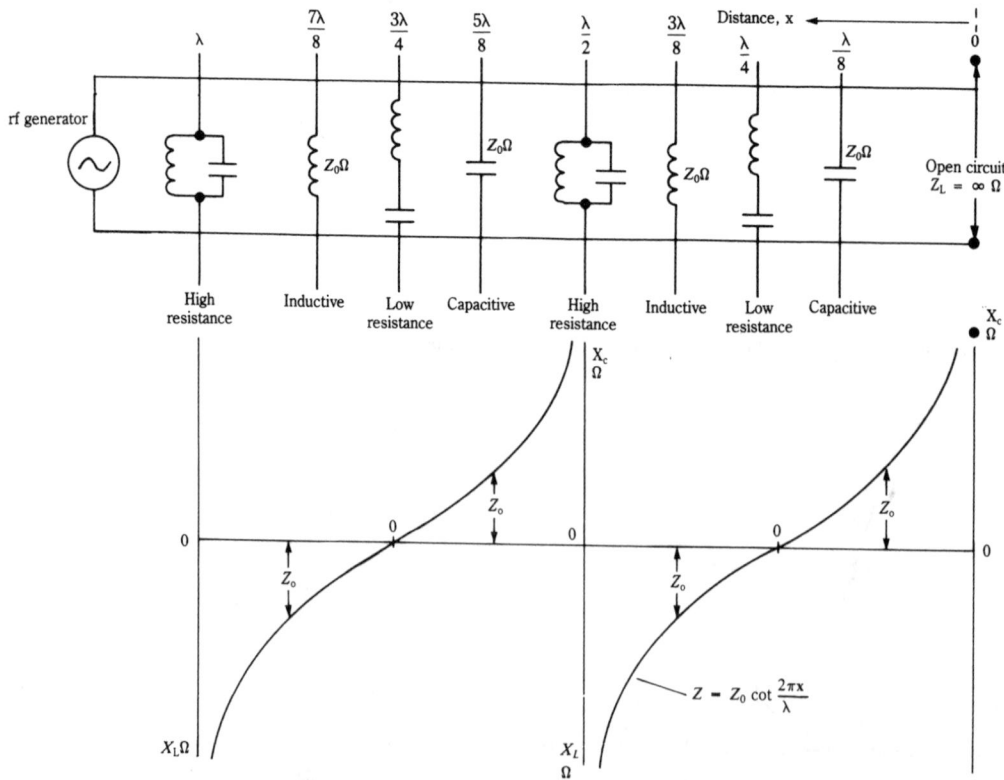

13-15 Variation of impedance along an open-circuited transmission line.

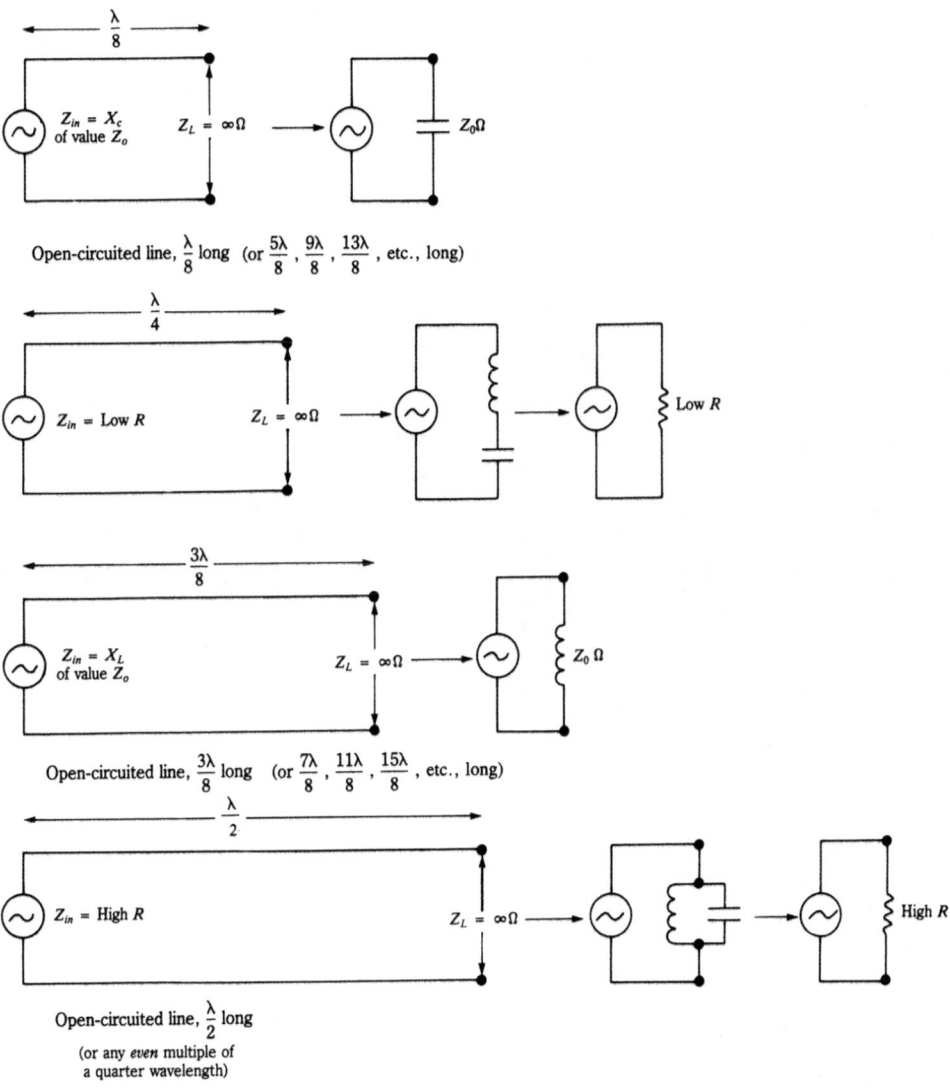

Open-circuited line, $\frac{\lambda}{8}$ long (or $\frac{5\lambda}{8}$, $\frac{9\lambda}{8}$, $\frac{13\lambda}{8}$, etc., long)

Open-circuited line, $\frac{3\lambda}{8}$ long (or $\frac{7\lambda}{8}$, $\frac{11\lambda}{8}$, $\frac{15\lambda}{8}$, etc., long)

Open-circuited line, $\frac{\lambda}{2}$ long
(or any *even* multiple of
a quarter wavelength)

Note: There is no impedance transformation over a distance
of one-half wavelength or any multiple of that distance.

13-15 continued.

the ultra-high frequency (uhf) band, lumped circuitry might be impossible and a quarter-wave section of line shorted at one end, is then used as a tank circuit.) The impedance will be a capacitive reactance equal in value to Z_o at the $3\lambda/8$ position, while at the $\lambda/2$ position the impedance is the same as that of a series LC circuit. Summarizing, there is a $\lambda/4$ shift between the impedance conditions on the open- and short-circuited lines. But which one is generally preferred? Because of ''end'' capacitance effects, it is impossible

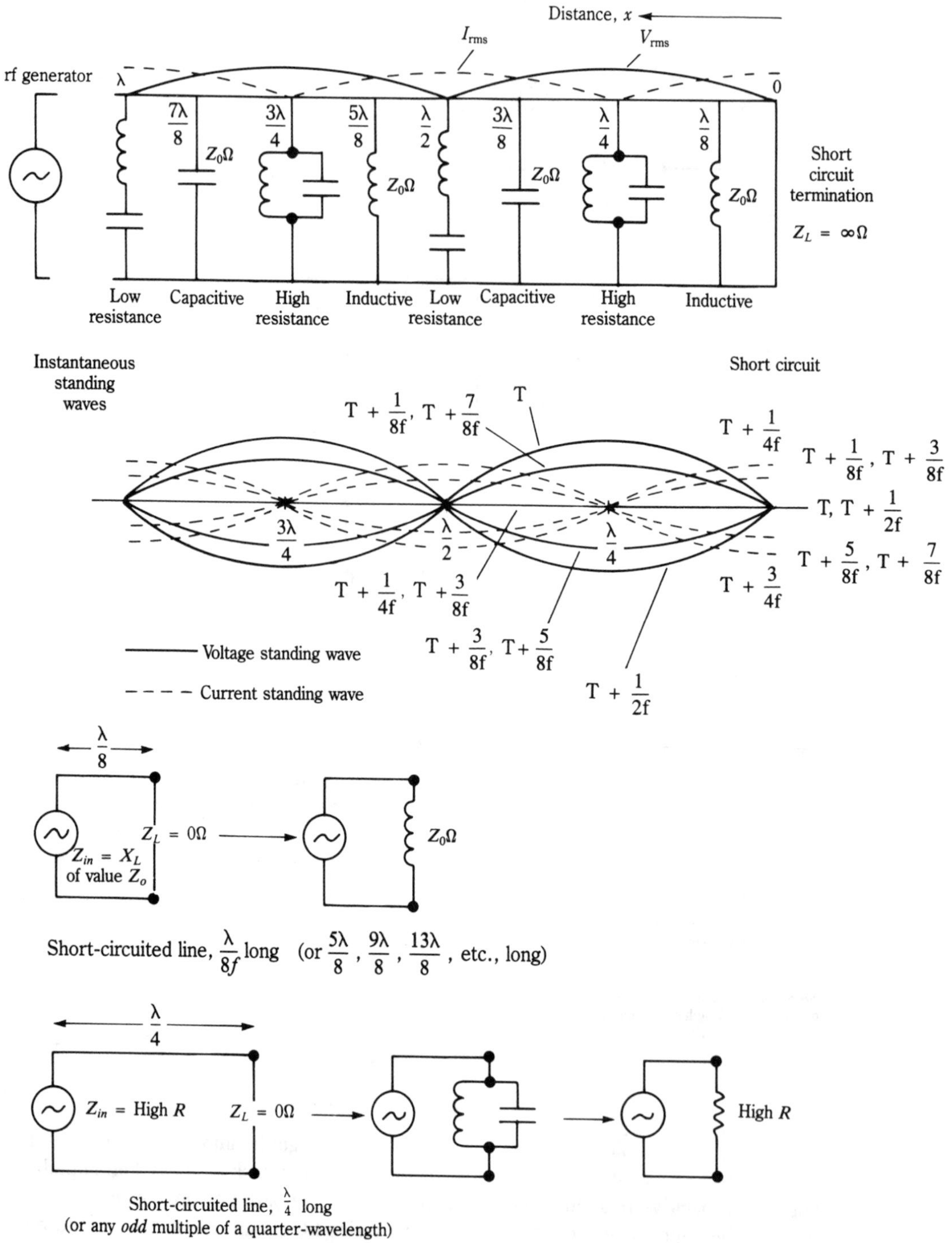

Distance, x ◄──────────

I_{rms} V_{rms}

rf generator λ 0

$\frac{7\lambda}{8}$ $Z_0\Omega$ $\frac{3\lambda}{4}$ $\frac{5\lambda}{8}$ $\frac{\lambda}{2}$ $\frac{3\lambda}{8}$ $\frac{\lambda}{4}$ $\frac{\lambda}{8}$

$Z_0\Omega$ $Z_0\Omega$ $Z_0\Omega$ $Z_0\Omega$

Short circuit termination

$Z_L = \infty\,\Omega$

Low resistance   Capacitive   High resistance   Inductive   Low resistance   Capacitive   High resistance   Inductive

Instantaneous standing waves   Short circuit

$T + \frac{1}{8f}, T + \frac{7}{8f}$ T

$T + \frac{1}{4f}$

$T + \frac{1}{8f}, T + \frac{3}{8f}$

$\frac{3\lambda}{4}$ $\frac{\lambda}{2}$ $\frac{\lambda}{4}$

$T, T + \frac{1}{2f}$

$T + \frac{1}{4f}, T + \frac{3}{8f}$

$T + \frac{5}{8f}, T + \frac{7}{8f}$

$T + \frac{3}{4f}$

$T + \frac{3}{8f}, T + \frac{5}{8f}$

─────── Voltage standing wave

─ ─ ─ ─ Current standing wave

$T + \frac{1}{2f}$

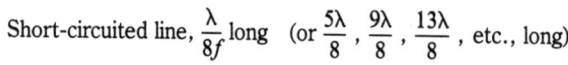

$\frac{\lambda}{8}$

$Z_L = 0\,\Omega$ ──► $Z_0\Omega$

$Z_{in} = X_L$ of value Z_o

Short-circuited line, $\frac{\lambda}{8f}$ long   (or $\frac{5\lambda}{8}$, $\frac{9\lambda}{8}$, $\frac{13\lambda}{8}$, etc., long)

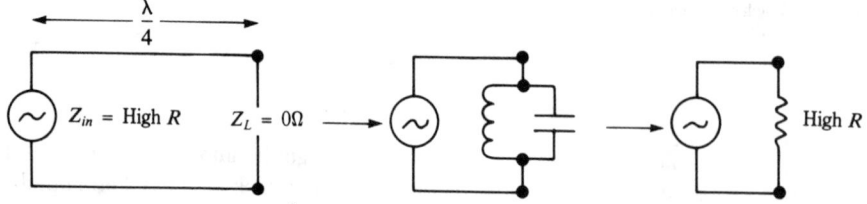

$\frac{\lambda}{4}$

Z_{in} = High R $Z_L = 0\,\Omega$ ──►   ──►   High R

Short-circuited line, $\frac{\lambda}{4}$ long
(or any *odd* multiple of a quarter-wavelength)

13-16 Standing wave distribution and impedance variation along the last wavelength of a short-circuited transmission line.

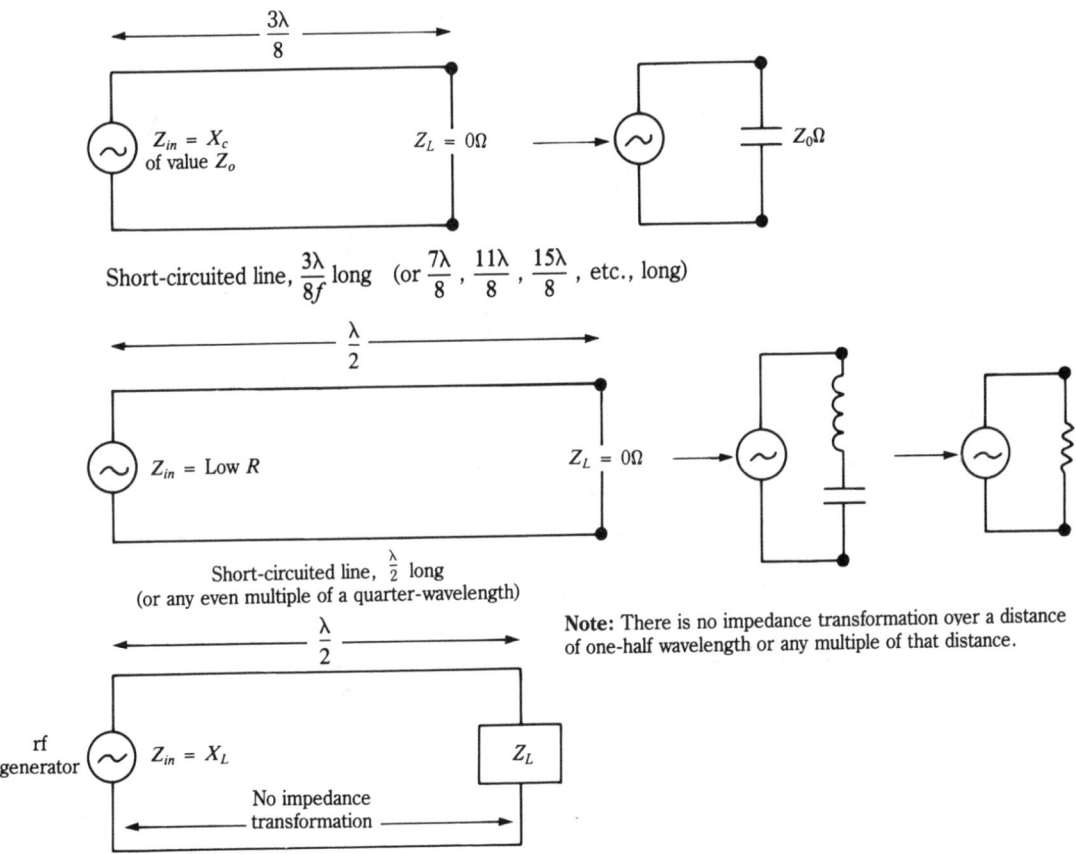

Short-circuited line, $\dfrac{3\lambda}{8f}$ long   (or $\dfrac{7\lambda}{8}$, $\dfrac{11\lambda}{8}$, $\dfrac{15\lambda}{8}$, etc., long)

Short-circuited line, $\frac{\lambda}{2}$ long
(or any even multiple of a quarter-wavelength)

Note: There is no impedance transformation over a distance of one-half wavelength or any multiple of that distance.

No impedance transformation

13-16   continued.

to achieve infinite resistance at the termination. However if a bar of low resistance is placed across a line's conductors, the result is a good approximation to a short circuit.

You will now examine the effect of terminating a lossless line with a resistive load that is not equal to the surge impedance. The amount of reflection will be reduced so that the nodes and antinodes will be less pronounced. Neglecting any losses on the line, the effective voltage and current distribution for the three possible cases of a resistive load, namely $R_L > Z_o$, $R_L \cong Z_o$, and $R_L < Z_o$, are shown in Fig. 13-17.

The fraction of the incident voltage and current reflected at the load is called the *reflection coefficient* whose letter symbol is the Greek, rho, ϱ. If $R_L > Z_o$, the effective standing wave voltage at the load is:

$$V_L = V_i + V_r = V_i + \varrho V_i = V_i (1 + \varrho) \tag{13-22}$$

where,

V_i = the effective value of the incident voltage wave
V_r = the effective value of the reflected voltage wave.

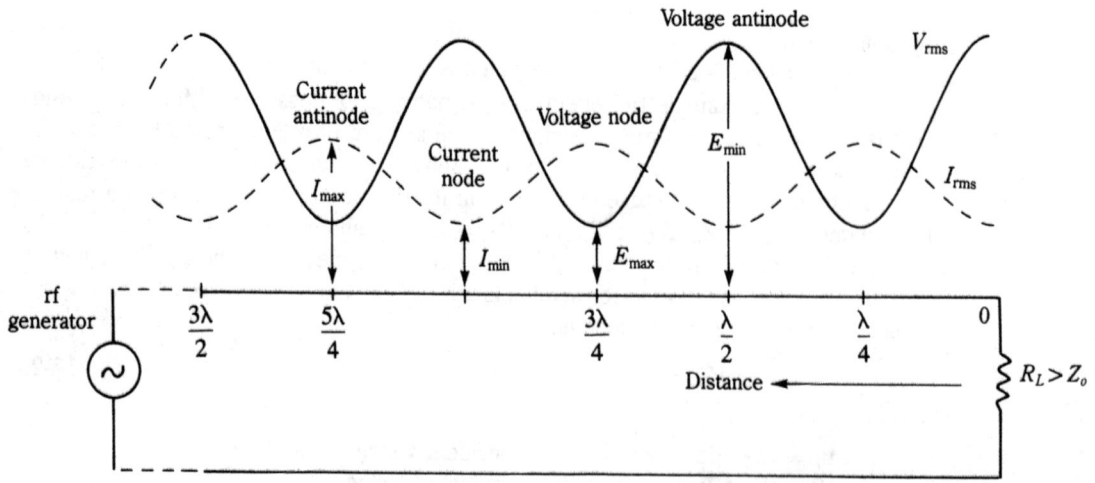

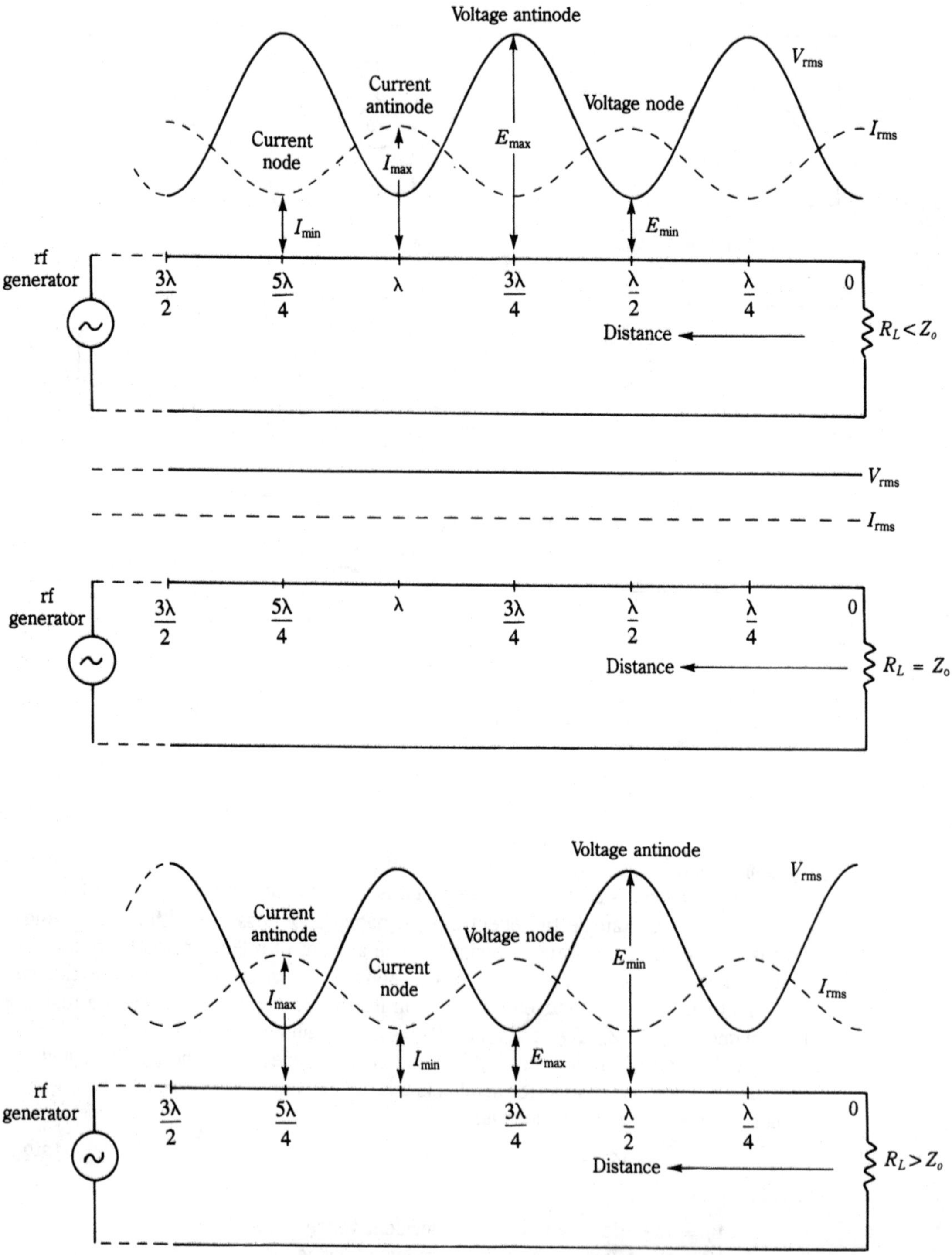

13-17 Effective voltage and current distribution on matched and unmatched lines.

The effective standing wave current through the load is:

$$I_L = I_i + I_r = I_i - \varrho\, I_i = I_i\,(1 > \varrho) \qquad (13\text{-}23)$$

The negative sign indicates the "out of phase" reflection of the current wave compared with the "in phase" reflection of the voltage wave.

Then,

$$R_L = \frac{V_L}{I_L}$$

$$= \frac{V_i\,(1+\varrho)}{I_i\,(1-\varrho)} ,$$

$$= \frac{Z_o\,(1+\varrho)}{1-\varrho)} \qquad (13\text{-}24)$$

This yields,

$$\varrho = \frac{R_L - Z_o}{R_L + Z_o} \qquad (13\text{-}25)$$

If

$R_L = \infty\ \Omega$	(open-circuited line), ϱ	$= +1$
$R_L = Z_o\ \Omega$	(matched line), ϱ	$= 0$
$R_L = 0\ \Omega$	(short-circuited line), ϱ	$= -1$
		$= 1\ \underline{/180°}$

The negative sign indicates that when $R_L < Z_o$, the current wave is reflected "in phase" but the voltage reflection is "out of phase." These results are only true for purely resistive loads.

If the load is a general impedance possessing both resistance and reactance, ϱ is not just a number but is a complex quantity with a magnitude P (capital rho), which ranges in value between 0 and 1, and a phase angle, θ, whose value lies between $+180°$ and $-180°$.

Because the reflected voltage is P times the incident voltage, and the reflected current is P times the incident current, it follows that the reflected power, P_r, is P^2 times the incident power, P_i. Then the power absorbed by the load, P_L is:

$$P_L = P_i - P_r$$

$$= (1 - P^2)\, P_i$$

$$= \left(\frac{1 - P^2}{P^2} \right) P_r$$

$$= \left(\frac{4S}{(1 + S)^2} \right) P_i \qquad (13\text{-}26)$$

where S is the VSWR.

Voltage standing wave ratio

Refer to Fig. 13-17, the effective value of a voltage antinode is E_{max} while the effective value of an adjacent voltage mode is E_{min}. The degree of standing waves is measured by the voltage standing wave ratio (VSWR) whose letter symbol is S. The VSWR is defined as the ratio E_{max}: E_{min} which is also equal in magnitude to I_{max}: I_{min}. E_{max} is the result of an in-phase condition between the incident and reflected voltages, while E_{min} is the result of a 180° out-of-phase situation.

Then,

$$E_{max} = V_i + V_r$$
$$= V_i (1 + P)$$
$$E_{min} = V_i - V_r$$
$$= V_r (1 - P)$$

This yields,

$$\text{VSWR, } S = \frac{E_{max}}{E_{min}} = \frac{1 + P}{1 - P} \qquad (13\text{-}27)$$

If

$P = 0$ (matched line), $S = 1$

$P = 1$ (open or short-circuited line), $S = \infty$

The VSWR is just a number whose value will range from 1 to ∞. On a practical matched system a value of S which is less than 1.2, is normally regarded as acceptable. It is sometimes preferable to measure the VSWR in decibels, in which case VSWR = 20 $\log_{10}S$ dB. Equation 13-27 yields:

$$P = \frac{S - 1}{S + 1} \qquad (13\text{-}28)$$

If a resistive load, R_L, is greater than Z_o, the magnitude of the reflection coefficient is given by:

$$P = \frac{R_L - Z_o}{R_L + Z_o} \qquad (13\text{-}29)$$

Combining equations (13-27) and (13-29),

$$S = \frac{1 + P}{1 - P}$$
$$= \frac{R_L}{Z_o} \qquad (13\text{-}30)$$

If R_L is less than Z_o,

$$P = \frac{Z_o - R_L}{Z_o + R_L}$$

and

$$S = \frac{Z_o}{R_L} \qquad (13\text{-}31)$$

At the position where the voltage antinodes and the current nodes coincide, the impedance of the line is a maximum and is resistive.

$$\text{Maximum impedance, } Z_{max} = \frac{E_{max}}{I_{min}}$$

$$= \frac{E_i (1 + P)}{I_i (1 - P)}$$

$$= Z_o \frac{(1 + P)}{(1 - P)} = SZ_o \qquad (13\text{-}32)$$

At the voltage node position, there is a minimum resistive impedance which is given by:

$$Z_{min} = \frac{E_{min}}{I_{max}}$$

$$= \frac{E_i (1 - P)}{I_i (1 + P)}$$

$$= Z_o \frac{(1 - P)}{(1 + P)}$$

$$= \frac{Z_o}{S} \qquad (13\text{-}33)$$

As an example, an rf power of 100 W is being conveyed down a 50 Ω loss-free line which is terminated by a 40 Ω resistive load. The following can be calculated:

$$\text{VSWR, } S = \frac{Z_o}{R_L}$$

$$= \frac{50}{40}$$

$$= 1.25$$

$$\text{Reflection coefficient, } P = \frac{S - 1}{S + 1}$$

$$= \frac{0.25}{2.25}$$

$$= 0.11$$

$$\text{Reflected power, } P_r = P^2 \times P_i$$
$$= (0.11)^2 \times 100$$
$$= 1.2 \text{ W.}$$

$$\text{Load power, } P_L = P_i - P_r$$
$$= 100 - 1.2$$
$$= 98.8 \text{ W.}$$

The presence of standing waves on a practical transmission line has the following disadvantages:

1. The incident power reaching the termination is not fully absorbed by the load. The difference between the incident power and the load power is the power reflected back toward the generator.

2. The voltage antinodes might break down the insulation (dielectric) between the conductors and cause arc-over.

3. Since power losses are proportional to the square of the voltage and to the square of the current, the attenuation on the practical line increases due to the presence of standing waves.

4. The input impedance at the generator is a totally unknown quantity which varies with the length of the line and the frequency of the generator.

By contrast, with these disadvantages, resonant lines have a number of useful applications. To review, when a quarter-wave line is shorted at one end and is excited to resonance by the circuit frequency applied at the other end, standing waves of voltage and current appear on the line. At the short circuit the voltage is zero and the current is a maximum. At the input end the current is nearly zero and the voltage is at its peak. The input impedance to the resonant quarter-wave line is extremely high, and therefore this section behaves as an insulator. Figure 13-18 shows a quarter-wave section of line acting as a standoff insulator (support stub) for a two-wire transmission line. At the particular frequency that makes the section a quarter-wavelength line, the stub acts as a highly efficient insulator. However, at other frequencies the stub is no longer resonant and will behave as an inductor or a capacitor. Such behavior causes a mismatch, and standing waves will appear on the two-wire line. At the second harmonic frequency the stub is $\lambda/2$ long and since there is no impedance transformation over the distance of one-half wavelength, the stub will place a short circuit across the main line. No power at the second harmonic frequency can then be transferred down the line. In this way a section of line which is $\lambda/4$ long at a fundamental frequency and is shorted at one end, will have the effect of filtering out all even harmonics.

The quarter-wave section can also be used to match two nonresonant lines with different surge impedances (Fig. 13-19). When a $\lambda/4$ line of surge impedance, Z_o, is terminated by a load, Z_L, the input impedance, Z_{in}, is given by:

$$Z_{in} = \frac{Z_o^2}{Z_L} \tag{13-34}$$

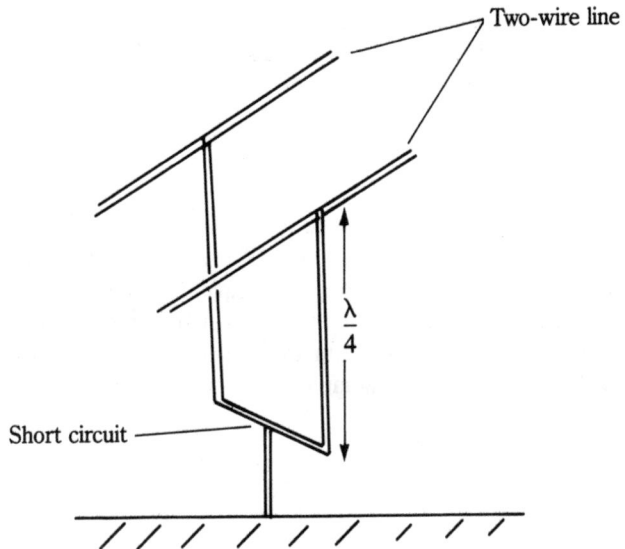

13-18 Quarter-wave support stub (metallic insulator).

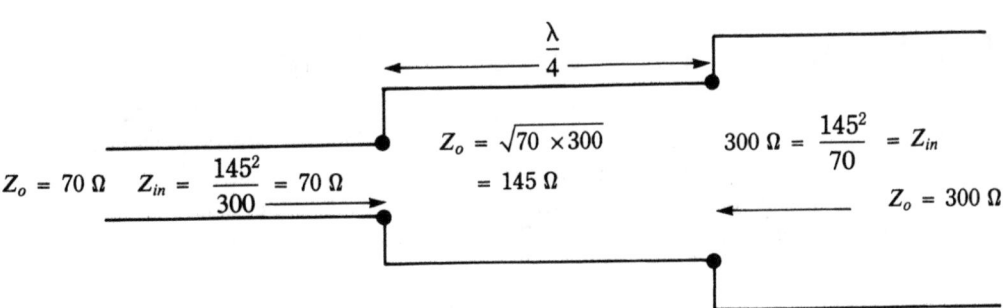

13-19 The principle of the quarter-wave impedance transformer.

This yields,

$$Z_o = \sqrt{Z_{in} \times Z_L} \qquad (13\text{-}35)$$

This formula can be used to derive some results that you have already obtained. For example,

1. If $Z_L = \infty\ \Omega$ (an open circuit), $Z_{in} = 0\ \Omega$ (a short circuit).
2. If $Z_L = Z_o\ \Omega$ (matched conditions), $Z_{in} = Z_o\ \Omega$.
3. If $Z_L = 0\ \Omega$ (a short circuit), $Z_L = \infty\ \Omega$ (an open circuit).

In the example you need to match 70 Ω and 300 Ω nonresonant lines. The $\lambda/4$ section would need to have a surge impedance of $\sqrt{70 \times 300} = 145\ \Omega$.

The effective load on the 300 Ω line as seen from the generator is $145^2/70 = 300$ Ω, while the 70 Ω line is effectively connected to a $145^2/300 = 70$ Ω termination. Both non-resonant lines are therefore matched and the standing waves only exist on the λ/4 section.

The reflectometer

The *reflectometer* is the test equipment used to obtain a direct reading of the reflection coefficient's magnitude on a transmission line. The arrangement of the equipment is shown in Fig. 13-20. One directional coupler samples the incident or forward power which is amplitude modulated while the other coupler only responds to the modulated reflected power. The voltage outputs from the deflectors are fed to the ratio meter, which is calibrated to give a direct reading of the square of the reflection coefficient's magnitude, P. Using the formula

$$S = \frac{1+P}{1-P}$$

you can calculate the corresponding value of the VSWR. The values of incident power, reflected power, voltage reflection coefficient, power reflection coefficient, and standing wave ratio are therefore obtainable from the reflectometer.

The Hertz and the Marconi antennas

An *antenna* is defined as an efficient radiator of electromagnetic energy (radio waves) into free space. The same principles apply to both transmitting and receiving antennas although the rf power levels for the two antennas are completely different. The purpose of a transmitting antenna is to radiate as much rf power as possible either in all directions (*omnidirectional* antenna) or in a specified direction (*directional* antenna). By contrast

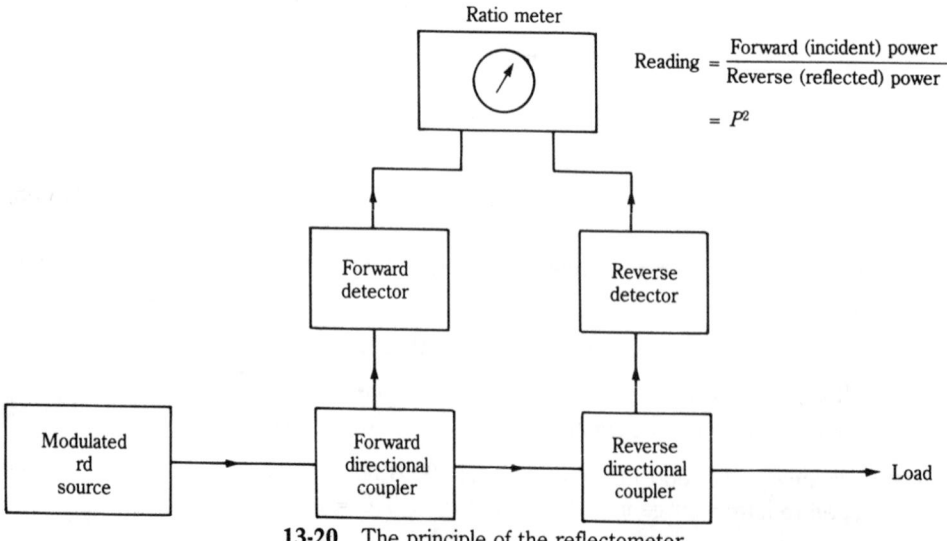

13-20 The principle of the reflectometer.

the receiving antenna is used to intercept an rf signal voltage, which is sufficiently large compared to the noise existing within the receiver's bandwidth.

The Hertz antenna

You already know that the main disadvantage of a twin-wire transmission line is its radiation loss. However, provided the separation between the leads is short compared with the wavelength, the radiation loss is small since the fields associated with the two conductors will tend to cancel out. Refer to Fig. 13-21A, the equal currents, i, which exist over the last resonant quarter-wavelength of an open-circuited transmission line, will instantaneously flow in opposite directions so that their resultant magnetic field is weak. However, if each $\lambda/4$ conductor is twisted back through 90° (Fig. 13-21B), the currents, i, are now instantaneously in the same direction, so the surrounding magnetic field is strong. In addition the conductors carry a standing wave voltage distribution with its associated electric field. The standing wave voltage and current distribution over the complete half-wavelength are shown in Fig. 13-21C; at the center feed point, the effective voltage is at its minimum level, while the effective current value has its maximum level. The effective voltage distribution is drawn on opposite sides of the two sections to indicate that these sections instantaneously carry opposite polarities. In other words, when a particular point in the top section carries a positive voltage with respect to ground, the corresponding point in the bottom section has a negative voltage. The distributed inductance and capacitance (Fig. 13-21D) together form the equivalent of a series resonant LC circuit. Notice however that the distributed capacitance exists between the two quarter-wave sections and that the ground is not involved in this distribution.

By bending the two $\lambda/4$ sections outward by 90° you form the half-wave ($\lambda/2$) dipole or Hertz (Heinrich Hertz, 1857 – 94) antenna. This antenna will be resonant at the frequency to which it is cut. You already know that the electrical wavelength in free space is given by:

$$\text{Electrical wavelength} = \frac{300}{f} \text{ meters}$$

$$= \frac{984}{f} \text{ feet} \tag{13-36}$$

This leads to,

$$\text{Electrical half wavelength} = \frac{492}{f} \text{ feet} \tag{13-37}$$

where,
  f = frequency in megahertz (MHz).

The voltage and current waves on an antenna travel at a speed which is typically 5% slower than the velocity of light. Therefore the physical half-wavelength to which the Hertz antenna should be cut, is shorter than the electrical half-wavelength.

$$\text{Physical half wavelength} = \frac{468}{f \text{ (MHz)}} \text{ feet} \tag{13-38}$$

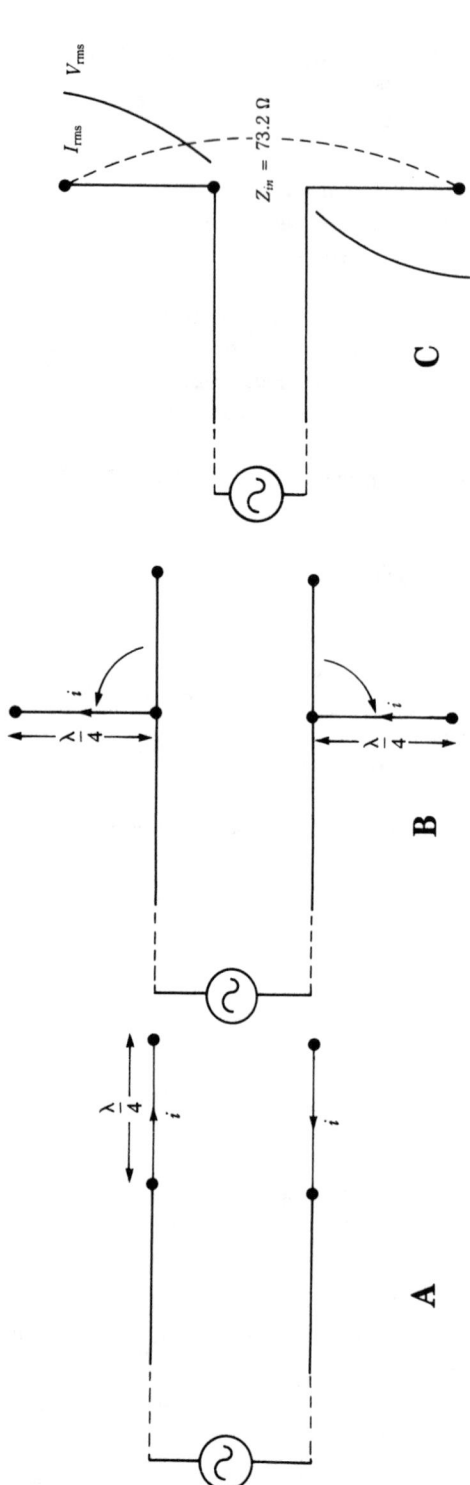

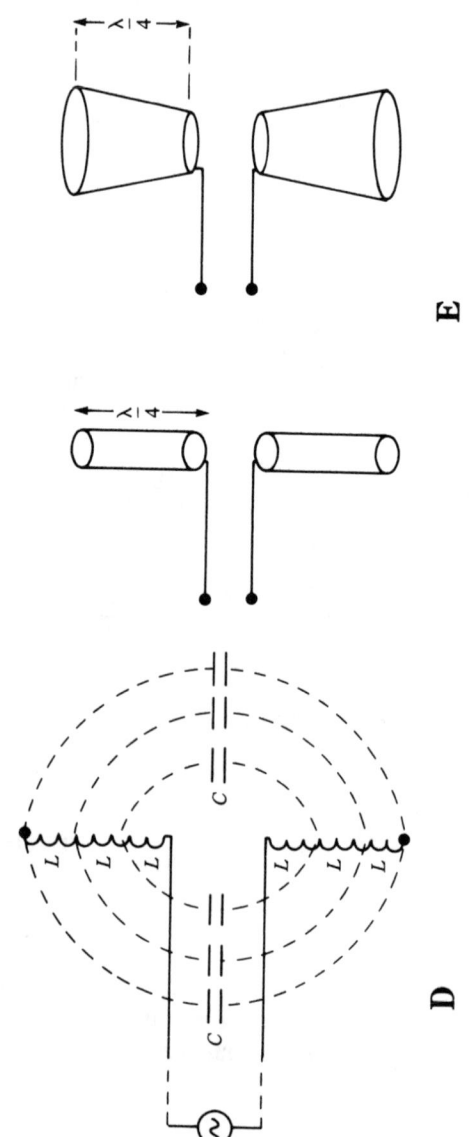

13-21   The principle of the Hertz dipole antenna.

For example, at a frequency of 100 MHz which lies within the FM commercial broadcast band of 88 to 108 MHz, the length required for the Hertz antenna is 468/100 = 4.68 feet. Such an antenna would be made from two thin conducting rods, each 2.34 ft. long, and positioned remote from ground. At the frequency of 100 MHz, the antenna behaves as a series resonant circuit with a Q of approximately 10. Therefore the thin dipole is capable of operating effectively within a narrow range that is centered on the resonant frequency. However you must remember that if the operating frequency is below the resonant frequency, the antenna will appear to be too short and will behave capacitively. Likewise at frequencies above resonance, the antenna will be too long and will be inductive.

If it is required to operate a dipole over a wide range of frequencies, it is necessary to *broad band* the antenna by lowering its Q without changing its resonant frequency. As an example, if you want to operate a dipole satisfactorily over the range of 125 to 175 MHz, the antenna should be cut to the midfrequency of 150 MHz and should possess a Q of 150/(175 − 125) = 3. Since the resonant frequency of a series LC circuit is given by $1/(2\pi\sqrt{LC})$ while $Q = 1/R \sqrt{L/C}$, you must lower the distributed inductance, L, and increase the capacitance, C. The solution is to shorten the antenna while at the same time increasing the surface area. This gives rise to such broad band shapes as the cylindrical and biconical dipoles (Fig. 13-21E).

When the Hertz dipole is resonant and the rf power is applied at the center of the antenna, the input impedance at the feedpoint is a low resistance, which mathematically can be shown to have a value of 73.2 Ω (for this reason the antenna is often spoken of as a "70 Ω dipole"). This is referred to as the *radiation resistance* of the dipole and is the ohmic load that the half-wave antenna represents at resonance; to achieve a matched condition the dipole should be fed with a 70 Ω line. The rf power, P, at the feedpoint is then given by:

$$P = I^2_A \times R_A \tag{13-39}$$

where,

I_A = effective rf current at the center of the antenna (A)
R_A = radiation resistance (Ω)

As you move from the center of the antenna toward the ends (Fig. 13-22A), the impedance increases from approximately 35 Ω (balanced either side with respect to ground) to about 2500 Ω (not infinity because of the "end" capacitance effects). It is therefore possible to select points on the antenna where the impedance can be matched by a gradual taper to a line whose Z_o is not 70 Ω; such an arrangement is known as a *delta feed* (Fig. 13-22B).

It is established that there are electric and magnetic fields in the vicinity of the antenna (Fig. 13-23A). Since the instantaneous voltage and current on the antenna are 90° out of phase, the same phase relationship applies to the E and H fields (Fig. 13-23B). These fields are continuously expanding out from the antenna and collapsing back with the velocity of light. However, because the action is not instantaneous, the collapse will only be partial, so closed electric and magnetic loops will be left in space (Fig. 13-23C). These loops represent the *radiated* electromagnetic energy that is propagated into free space and travels with the velocity of light. Those flux lines that collapse

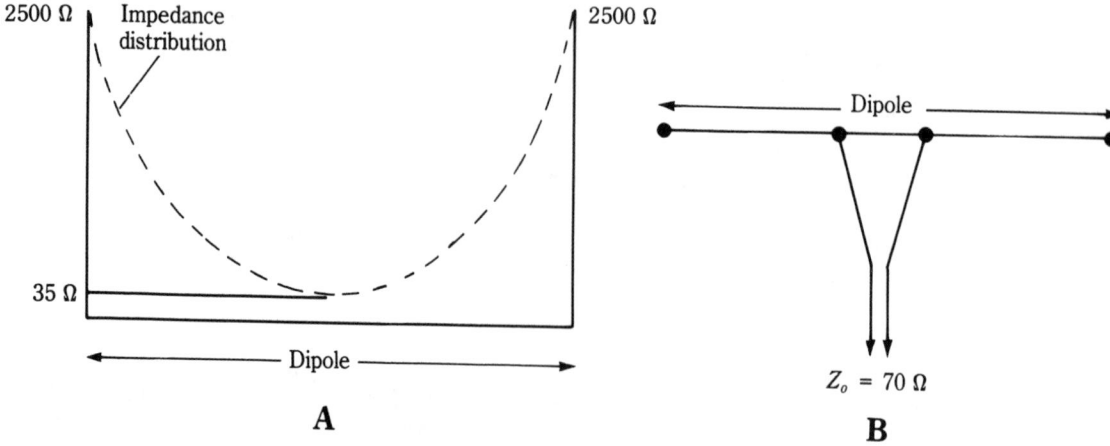

13-22 Impedance distribution along a half-wave dipole.

back into the antenna represent the *induction* field, which is only strong in the immediate vicinity of the antenna.

The fields in space around a half-wave antenna are shown in Fig. 13-24. The radiated **E** and **H** fields are in time phase but the two sets of flux lines are 90° apart in space. A vertical Hertz antenna radiates an electric field with vertical flux lines and a magnetic field with horizontal flux lines. In addition the two sets of flux lines are each at right angles to the direction in which the electromagnetic energy is being propagated. Mathematically the electric field, **E**, (measured in volts per meter), the magnetic field, **H**, (measured in amperes per meter) and the transfer of EM energy in a particular direction, represents a right-handed system of vectors. The transfer of EM energy is measured by the Poynting vector, **S**, which represents the amount of radiated energy passing through unit area (1 square meter) in unit time (1 second). **E, H,** and **S** (in that order) form a right-handed system with **S** as the vector product of **E** and **H**; in terms of units, volts/meter (**E**) × amperes/meter (**H**) = volts × amperes/(meter)^2 = watts/(meter)^2 = joules per square meter per second (**S**). If the instantaneous **E** direction is rotated through 90° to lie along the instantaneous **H** direction, the direction of **S** is that of a right-handed screw, which is subjected to the same rotation (Fig. 13-25).

The plane containing **E** and **S** is referred to as the *plane of polarization*. A vertical Hertz antenna will radiate a vertically polarized wave with a vertical **E** field and a horizontal **H** field. Similarly a horizontally polarized wave has a horizontal **E** field, a vertical **H** field and is associated with a horizontal antenna. As practical examples, AM broadcast systems have antenna systems which radiate vertically polarized waves, while TV broadcast stations use horizontal polarization. Vertical and horizontal polarization are compared in the discussion on propagation.

The Marconi antenna

At an operating frequency of 2 MHz the required physical length for a resonant Hertz dipole is 468/2 (MHz) = 234 feet. It is difficult to position a vertical antenna of this size

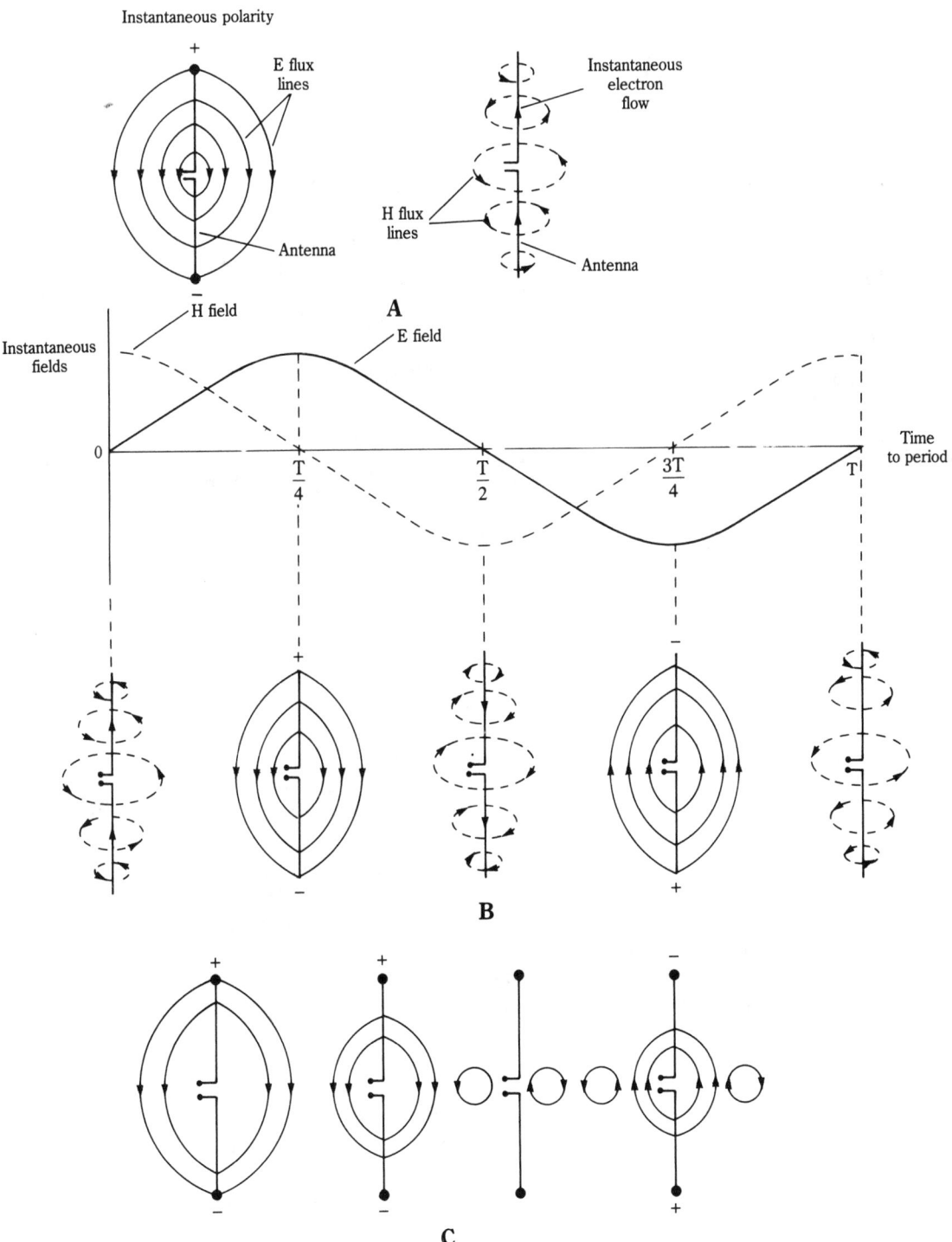

13-23 Electric and magnetic fields surrounding the half-wave antenna.

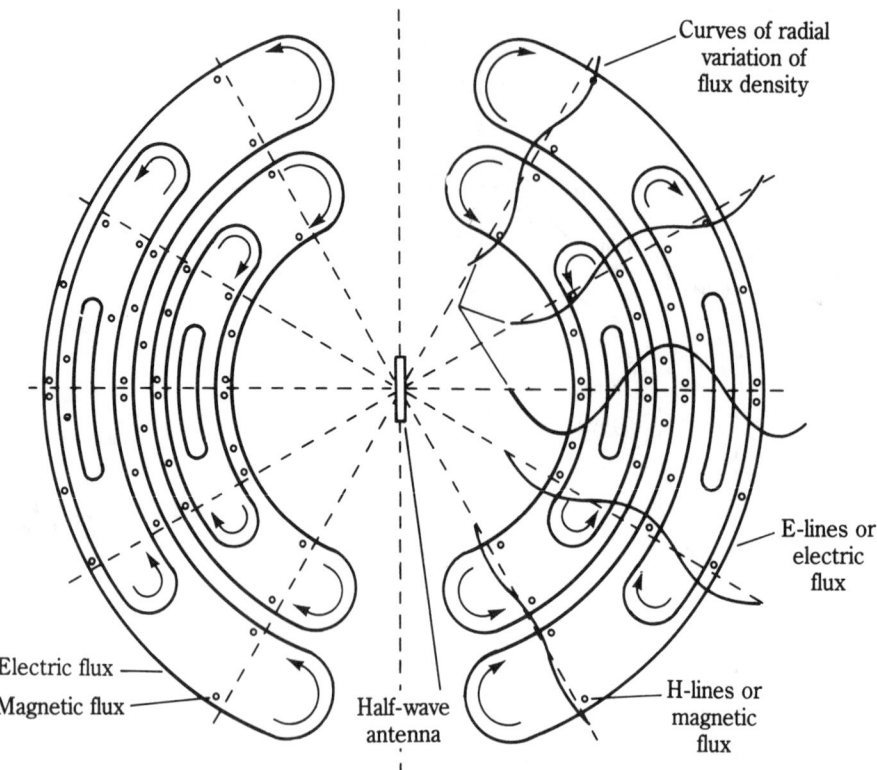

13-24 Radiated fields surrounding the half-wave antenna.

so that it is remote from ground. However the dipole could be in the form of a long wire antenna, slung horizontally between two towers from which the antenna is insulated (Fig. 13-26). Clearly the vertical Hertz antenna is not a practical proposition at low frequencies and is replaced by the $\lambda/4$ unipole or *Marconi antenna*. The Marconi antenna can be regarded as a Hertz dipole in which the lower half is replaced by a nonradiating ground image antenna (Fig. 13-27). Unlike the Hertz antenna, ground is an integral part of the Marconi antenna system; this also means that the Hertz dipole is balanced with both $\lambda/4$ sections mounted remote from ground, while the Marconi unipole is unbalanced, since part of the antenna is ground itself. In the same way the twin line with neither conductor grounded is a balanced transmission line, while a coaxial cable is unbalanced with its outer conductor grounded at intervals along its length.

The distributed inductance of the Marconi antenna is associated with the vertical $\lambda/4$ rod while the distributed capacitance exists between the rod and ground (Fig. 13-28). It is therefore impossible to operate with a horizontal Marconi antenna; this compares with the Hertz dipole which can be mounted either vertically or horizontally. Consequently all Marconi antennas radiate only vertically polarized waves.

The effective voltage and current distribution on the resonant Marconi antenna is comparable with the distribution on the upper half of the Hertz dipole (Fig. 13-29). If the $\lambda/4$ antenna is end-fed by an unbalanced coaxial cable, the radiation resistance is 73.2/2

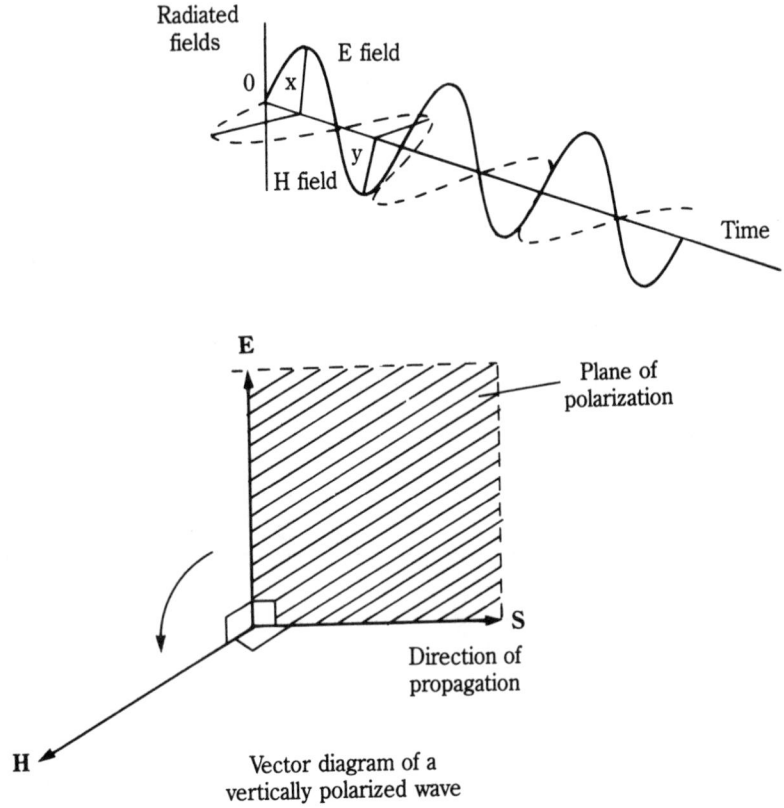

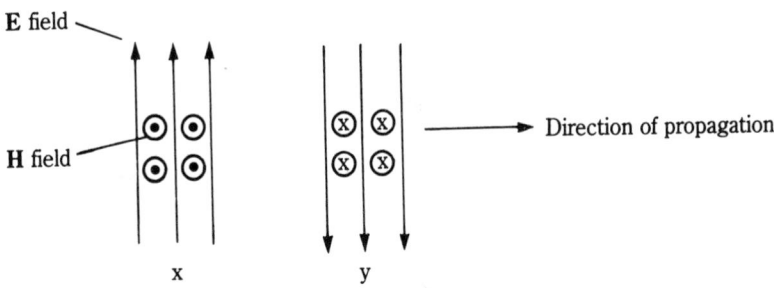

13-25   Electric and magnetic fields of a vertically polarized EM wave.

= 36.6 Ω, which is the effective resistive load of a resonant Marconi antenna. The electrical quarter-wavelength is given by the expression 246/f feet while the Marconi antenna's physical length is 234/f feet where f is the frequency in MHz. If the operating frequency is below the resonant value, the unipole is too short and behaves capacitively; however, the antenna can be tuned to resonance by adding an inductor in series (Fig. 13-30A). When operating above the resonant frequency the antenna is too long, behaves

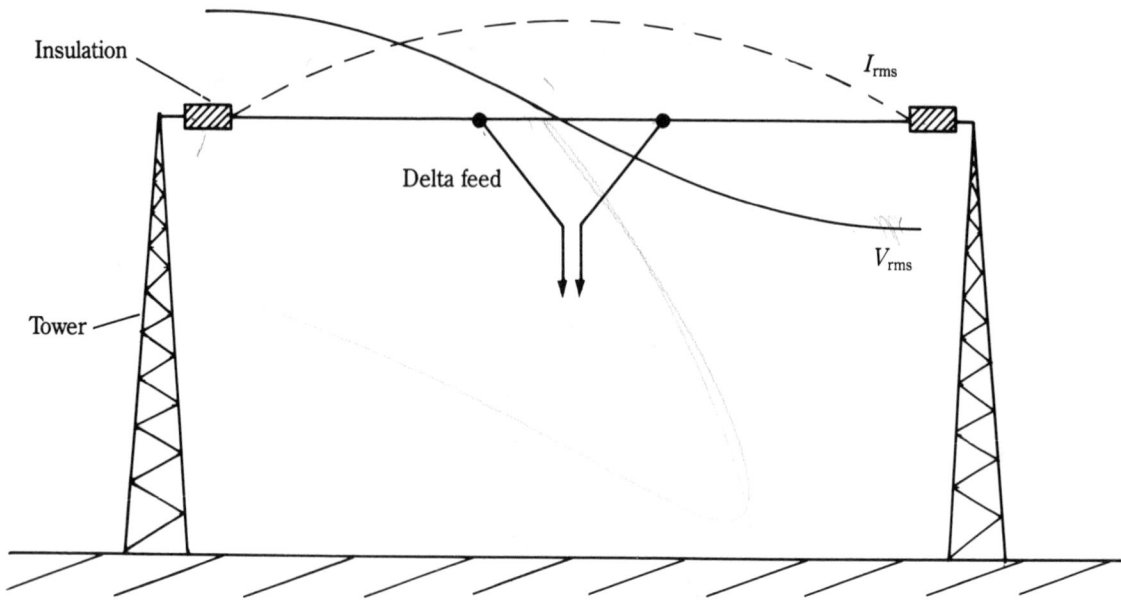

13-26   Horizontal half-wave wire antenna.

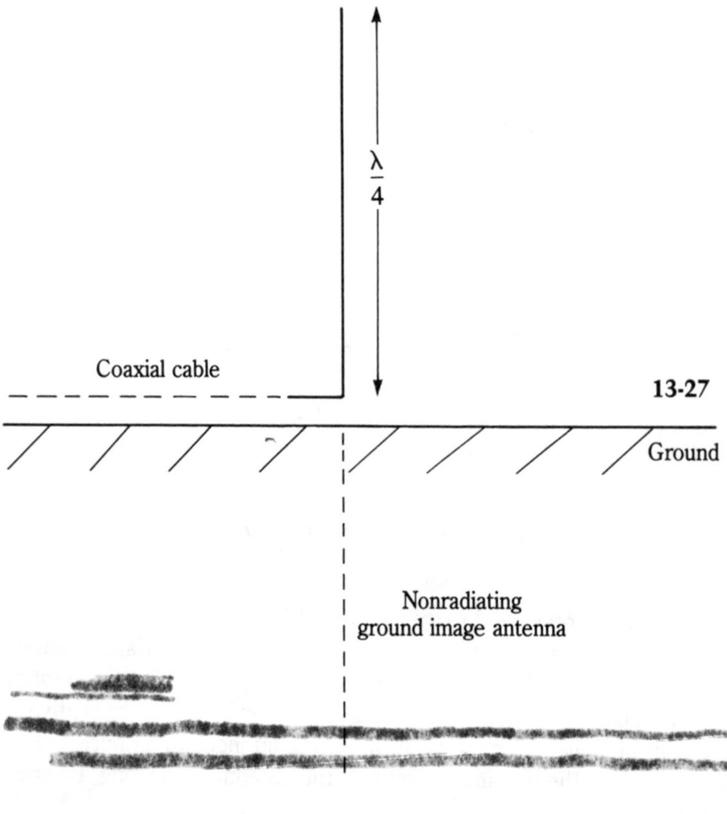

13-27   Quarter-wave Marconi unipole antenna.

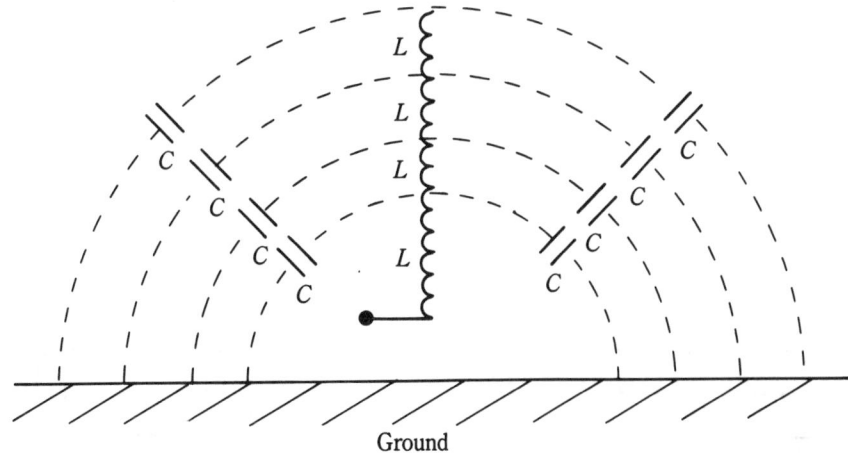

13-28   Distributed inductance and capacitance of a Marconi antenna.

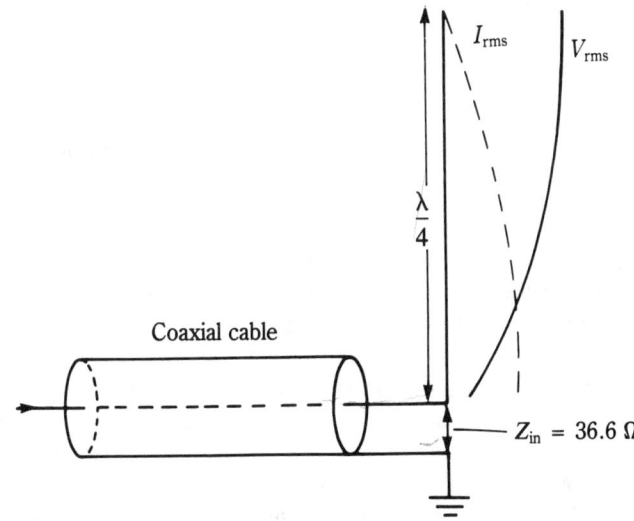

13-29   Effective voltage and current distribution on a resonant Marconi antenna.

inductively, and can be tuned to resonance with the aid of a series capacitor (Fig. 13-30B).

One type of practical Marconi antenna is the vertical whip (a rod which is flexible to a limited extent); as an example, an end-fed whip antenna is commonly mounted on the top of an automobile so that the roof can act as the required ground plane. Should the antenna be positioned in the vicinity of the rear bumper, most of the ground plane will be provided by the road surface. If it is required to operate a whip antenna at the top of a building, it is necessary to provide the antenna with an apparent ground or *counterpoise*. This normally consists of a wire structure mounted just beneath the feedpoint of the antenna and is connected to the outer conductor of the coaxial cable. The main distrib-

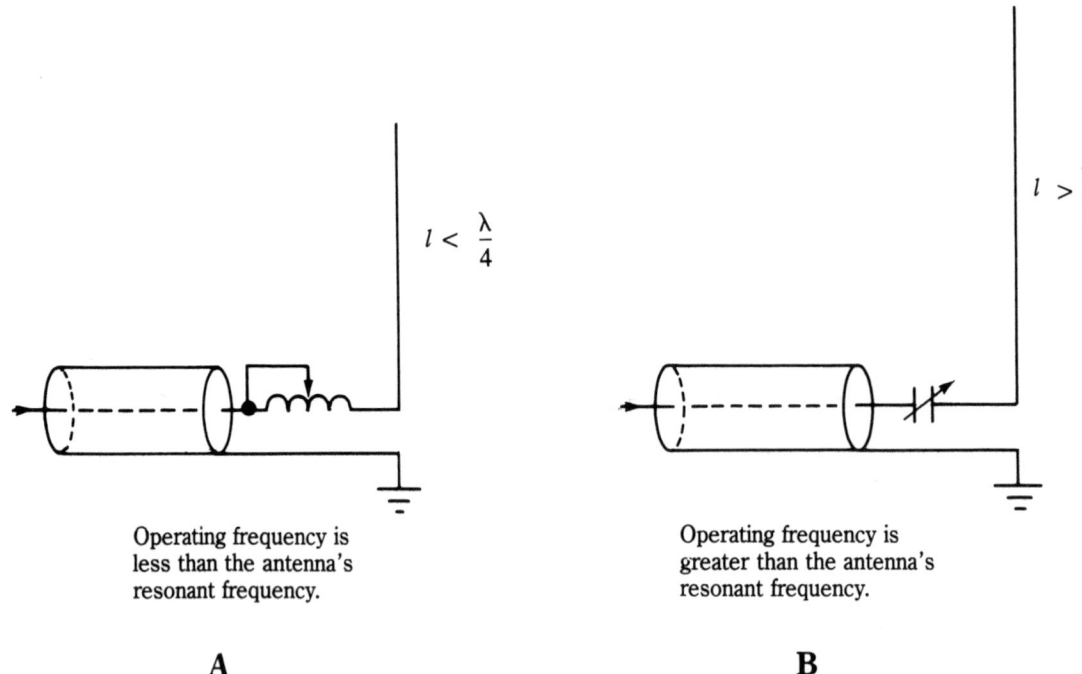

$$l < \frac{\lambda}{4}$$

$$l > \frac{\lambda}{4}$$

Operating frequency is
less than the antenna's
resonant frequency.

Operating frequency is
greater than the antenna's
resonant frequency.

A   **B**

13-30 Tuning a Marconi antenna to resonance.

uted capacitance then exists between the whip and the counterpoise, which itself acts as
a large capacitance in relation to ground. For this reason the counterpoise should be normally larger in size than the antenna, and must be well insulated from ground.

Another type of vertical radiator is a *steel tower* which can be tapered to optimize the
current distribution. Such towers can either be end-fed or shunt-fed as shown in Fig.
13-31. With shunt feeding the bottom of the tower is connected directly to ground while
the center conductor of the coaxial cable is joined to a point, *P*, on the tower, where the
resistive component of the antenna's impedance can be matched to the surge impedance
of the coaxial cable. With the grounded end behaving as a short circuit, the impedance at
the point, *P*, is inductive and this reactance is cancelled by the capacitor, *C*. It is worth
mentioning that the dc resistance between *P* and ground is virtually zero.

To improve the quality of the ground and therefore reduce ground losses, it is common practice to add a ground system, which consists of a number of bare copper conductors, arranged radially and connected to a center point beneath the antenna; the center
point is then joined to the outer conductor of the coaxial cable. These ground radials are
from $\lambda/10$ to $\lambda/2$ in length and are buried a short distance down; in addition the ground
can be further improved by laying copper mats beneath the surface. The ultimate purpose of the radial ground system is to improve the antenna efficiency, defined as the ratio
of the rf power radiated from the antenna as useful electromagnetic energy to the rf
power applied to the antenna feedpoint.

A tower can either be self-supporting or supported by guy wires. Such wires can
pick up some of the live transmitted radiation; reradiation from the guy wires will then

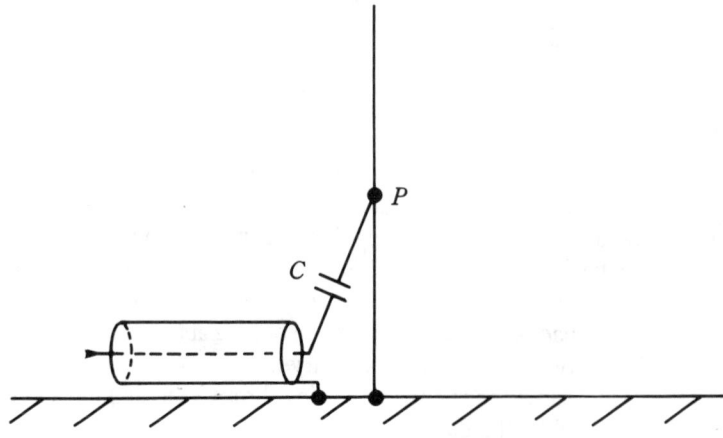

13-31 Shunt-feeding a vertical antenna.

occur and will interfere with the EM wave transmitted from the antenna. To reduce the reradiation effect to a minimum, an insulator is connected to each end of the guy wire, and intermediate insulators are attached at a spacing of about a tenth of a wavelength apart. These insulators are normally of the porcelain "egg" type so that if one of the insulators breaks, the guy wire still provides mechanical support.

If at low frequencies, the vertical antenna is too short for mechanical reasons, the amount of useful vertically polarized radiation can be increased by using *top loading*, examples are shown in Figs. 13-32A, and B. Since the current is zero at the open end of the antenna, the inverted "L" arrangement will have the result of increasing the effective current distribution over the antenna's vertical section. A more practical construc-

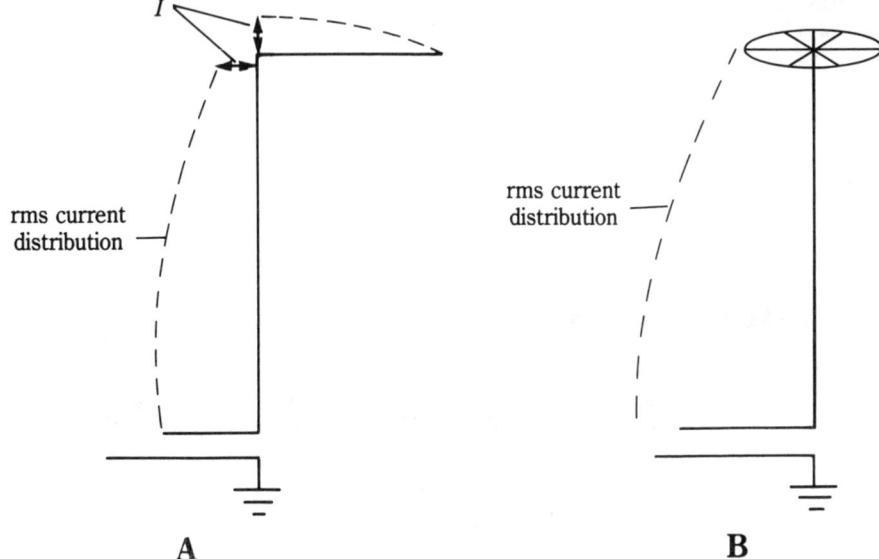

13-32 Examples of antenna "top loading."

tion is the metallic spoked wheel, which will increase the amount of the distributed capacitance to ground, and will therefore lower the antenna's resonant frequency to the required value.

Dummy antenna

The tuning and testing procedures for a transmitter must be carried out in *radio silence* and reduced power so that there is no radiation to cause interference problems. The antenna is therefore disconnected from its transmitter which is then coupled to a dummy antenna whose circuitry is shown in Fig. 13-33. The electrical properties of the dummy antenna are a noninductive resistance, *R*, (for low power transmitters a light bulb can be used) in series with a capacitance, *C*. The purpose of the capacitor is to cancel the reactance of the coupling network, *L*, to the transmitter.

The meter, M, measures the rf current flowing through the dummy antenna so that the rf power, *P*, absorbed by the dummy antenna is:

$$P = I^2 \ R \ \text{watts} \tag{13-39}$$

where

I = the rf current in the dummy antenna (A)

R = the resistance of the dummy antenna (Ω)

At full power the value of *P* should be comparable with the normal operating power delivered to the antenna.

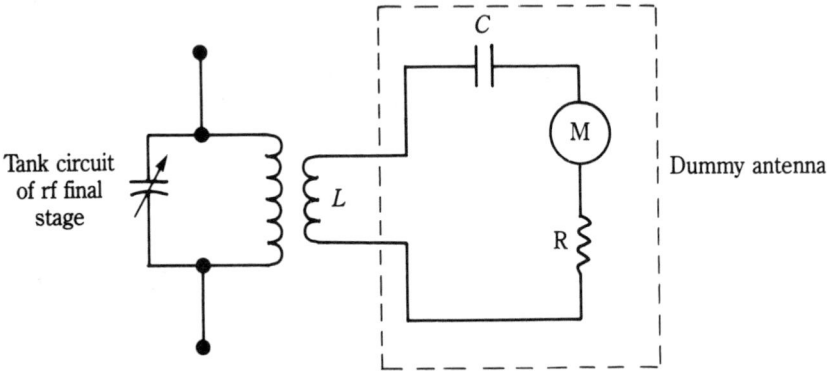

13-33 The dummy antenna.

Radiation patterns

The *radiation pattern* surrounding an antenna can either represent the field intensity, &, (volts per meter) or the power density (watts per square meter) distribution. The two patterns are similar in appearance, and you will mainly confine yourself to the field intensity, which directly measures the strength of the electric field component in the radio wave. Theoretically the value of & is equal to the voltage induced in a conductor 1 meter long and positioned parallel to the electric flux lines. Although the basic unit of the electric field intensity is the volt per meter, more practical units are millivolts per meter and microvolts per meter.

The radiation pattern is a plot of the electric field intensity at a fixed distance from the (transmitting) antenna versus an angle measured in the particular plane for which the pattern applies. In the case of the half-wave dipole, it is customary to consider two planes. The & plane contains the antenna itself and the electric flux lines, while the magnetic flux lines lie in the *H* plane which is at right angles to the antenna at its center point. Combining the results for the two perpendicular planes produces the antenna's complete three-dimensional radiation pattern.

The radiation patterns for a *vertical* dipole are shown in Fig. 13-34A, and B. The maximum field strength will be associated with the center of the antenna, where the

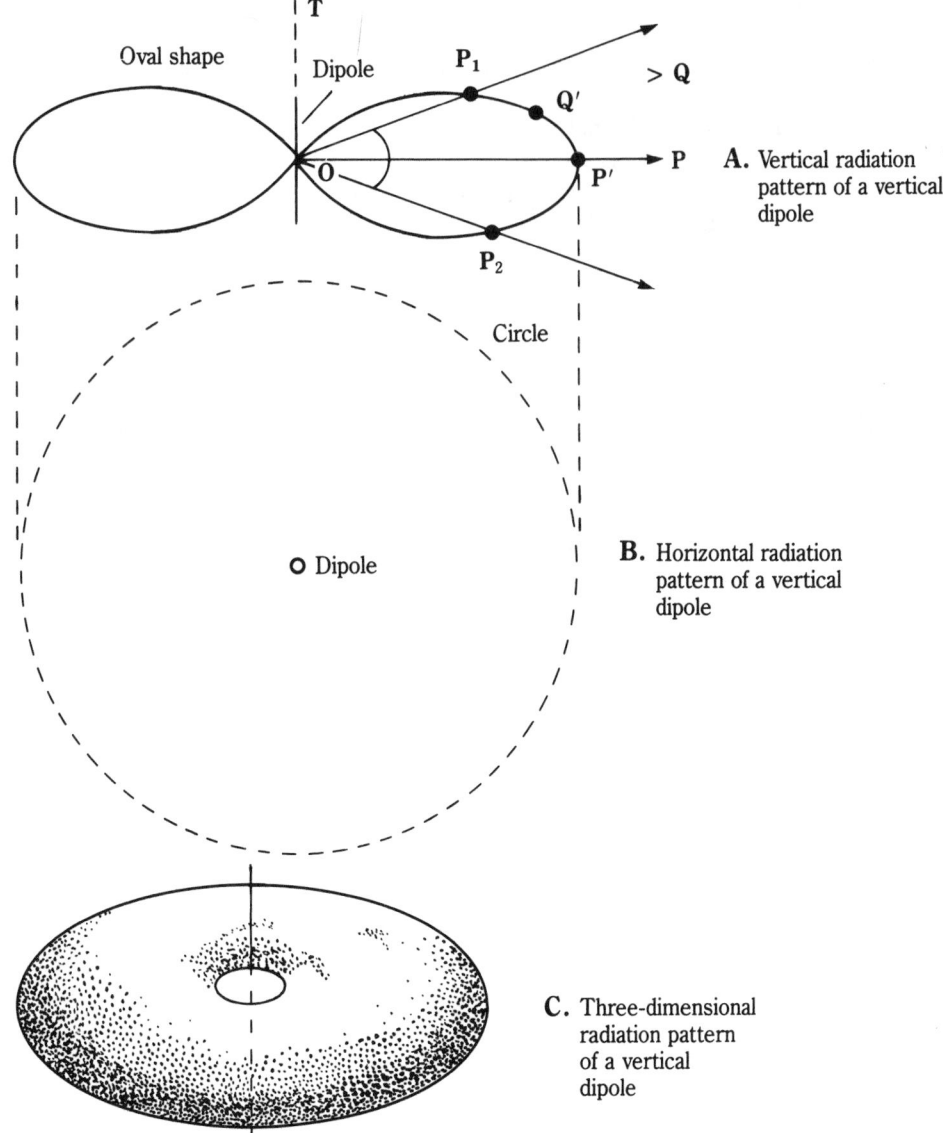

A. Vertical radiation pattern of a vertical dipole

B. Horizontal radiation pattern of a vertical dipole

C. Three-dimensional radiation pattern of a vertical dipole

13-34   Radiation patterns surrounding a vertical dipole.

highest current distribution exists; practically no radiation will occur from the ends, where the current is minimal.

The vertical radiation pattern associated with the $\mathcal{E}$ plane consists of two "oval" shapes (not circles) and is shown in Fig. 13-34A. In the direction **OP** the strength of the electric field intensity is at its maximum and is represented by the length **OP'**, while **OQ'** (which is shorter than **OP'**) is a measure of the field intensity in the direction **OQ**. For the direction **OT**, the length intercepted by the radiation plot and consequently the $\mathcal{E}$ value are both zero. The points P_1 and P_2 are those where the electric field intensity is 0.707 of its maximum value (**OP'**), the beam width is then the angle P_1 **OP**$_2$ between the half-power points and is about 50° for the half-wave dipole.

The vertical dipole radiates equally well in all horizontal directions so that its pattern in the magnetic plane is a circle (Fig. 13-34B). The complete three-dimensional pattern will then resemble a *toroidal* shape or doughnut surface with no hole in the center (Fig. 13-34C). The same pattern will apply to a vertical dipole used for reception purposes; in any particular direction the length intercepted by the pattern will be a measure of the received signal strength. For the quarter-wave Marconi antenna the vertical radiation pattern will be the same as the pattern for the top half of the vertical dipole; the horizontal pattern will again be a circle (Fig. 13-35A and B).

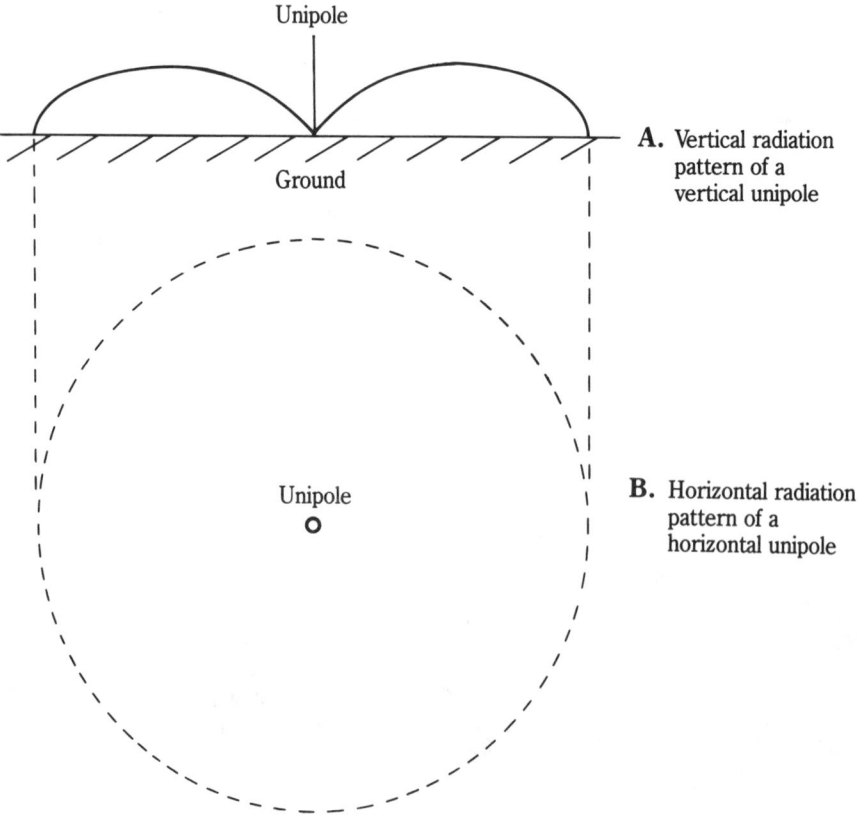

A. Vertical radiation pattern of a vertical unipole

B. Horizontal radiation pattern of a horizontal unipole

13-35   Radiation patterns surrounding a vertical unipole.

In the vicinity of a single vertical antenna the horizontal radiation can be found by joining together all positions of equal field strength; with a flat, perfectly conducting ground, the result is a circular field strength contour but in practice such a contour is distorted, owing to irregularities in the terrain.

The value of the vertical field strength, $\mathcal{E}$, in the horizontal plane surrounding a thin centerfed vertical dipole, is:

$$\mathcal{E} = \frac{60\ I_A}{d} \quad \text{volts per meter} \tag{13-41}$$

where,

$\qquad I_A$ = antenna current at the feedpoint (A)
$\qquad d$ = distance from the transmitting dipole (m)

The field strength is therefore directly proportional to the antenna current, and is inversely proportional to the distance from the transmitting antenna. This relationship is only true for the radiated field; for the induction field the field strength is inversely proportional to the *square* of the distance and therefore the induction field is only important in the immediate vicinity of the antenna.

Because the transmitter's operating power is directly proportional to the square of the antenna current, the radiated field strength is directly proportional to the square root of the transmitter power. For example, a thin centerfed dipole has a radiation resistance of 73.2 Ω, so if the rf power delivered to the antenna is 1 kW, the antenna current is:

$$I_A = \sqrt{\frac{P}{R_A}}$$

$$= \frac{1000\ W}{73.2\ \Omega}$$

$$= 3.696\ A$$

Therefore,

$$P = I_A^2\ R_A$$

and

$$R_A = \frac{P}{I_A^2} \tag{13-42}$$

where,

$\qquad P$ = the operating power by the direct method (W)
$\qquad I_A$ = the antenna current at the antenna feedpoint (A)
$\qquad R_A$ = the antenna resistance at the antenna feedpoint (Ω)

The field strength at a position one mile (= 1609.3 m) from the antenna is:

$$\mathcal{E} = \frac{60\ I_A}{d}$$

$$= \frac{60 \times 3.696}{1609.3}$$

$$= 0.1378 \text{ volts per meter}$$

$$= 137.8 \text{ millivolts per meter}$$

This value is used as the FCC standard to determine the gain of more sophisticated antennas. For example, when parasitic elements such as reflectors and directors are added to a dipole the radiated rf power is concentrated in particular directions, so the antenna gain is increased. This gain can either be considered in terms of field strength or power.

From measurements taken at a distance of 1 mile from the antenna, the field gain is the ratio of the field strength produced by the complex antenna system to the field strength (137.8 mV/m) created by a simple ideal dipole (assuming that this dipole is capable of directly replacing the complex antenna); in both cases the antenna power is 1 kW. Since the square of the field intensity is directly proportional to the radiated power, the power gain of the antenna is equal to (field gain)^2 and is either expressed as a ratio or in dB. Sometimes the term *antenna system power gain* is used and this generally takes into account the transmission line loss.

The *effective radiated power* (ERP) is the power propagated along the axis of the principal radiation lobe associated with a directional antenna.

The equations are:

ERP = (rf carrier power input to the transmission line) × (antenna system power gain ratio)

= (rf carrier power delivered to the antenna) × (antenna power gain ratio)

= (rf carrier power delivered to the antenna) × (antenna field gain ratio)^2

In addition:

Antenna power gain in dB

= 10 log$_{10}$ (antenna power gain ratio)

= 20 log$_{10}$ (antenna field gain ratio)

= power gain of antenna system in dB + transmission line loss in dB

The antenna base current ratio

Nondirectional AM stations use a single antenna tower and transmit the radio signal with equal strength in all directions from the station. Directional AM stations utilize more than one antenna tower. By establishing the position of each tower, the power radiated by each tower, and the phase of the signal in each tower, different signal strengths can be radiated in various directions. Directional antenna systems are used to improve the signal over desired areas while reducing the signal in the direction of other stations to prevent interference.

To determine if a directional antenna system is radiating the signal according to a specified radiation pattern, an instrument called an *antenna monitor* is installed at the station. The antenna monitor enables the operator to determine if the rf current in each tower is of the correct value, and if the phase of the signal radiated by each tower is also correct. Some antenna monitors indicate the ratio of current in each tower to the current in one tower, called the *reference tower*.

If the signal arrives at each tower at the same time, the current in each tower is said to be in phase. In most directional antenna systems, the time the radio frequency signal reaches each tower from the transmitter is not the same. The time difference, or phase, is measured in degrees. For each directional antenna tower, the station license contains a list of the required signal phases and antenna base current ratios. The antenna base-current ratio for a tower is calculated by dividing the current meter reading for that tower by the current meter reading of the designated reference tower. As an example,

$$\text{Antenna base current ratio for tower 2} = \frac{\text{antenna current meter reading of tower 2}}{\text{antenna current meter reading of the reference tower}}$$

This ratio, either calculated or read on the antenna monitor, normally must not deviate by more than 5% from the value on the station license. Since directional AM broadcast stations employ multiple radiating elements, the operating power of these stations, as determined by the direct method, is equal to the product of the resistance common to all the antenna towers (common point resistance R_C) and the square of the current common to all the antenna towers (common point current, I_c) Thus, the operating power of a directional AM station is:

$$P_o = I_c^2 R_c \text{ watts} \tag{13-43}$$

Part 17 of the FCC rules and regulations contains the requirements for the maintaining and lighting of antenna towers. Because the tower must contain an rf voltage distribution, it is necessary to have some means of isolating the rf potential from the lighting power line. This is achieved by an Austin-ring transformer having coils whose turns are widely spaced; this allows efficient passage of the low-frequency lighting power, but the rf coupling is negligible.

Field strength meters

At frequencies in the AM broadcast band the absolute strength can be measured by a combination of a rotatable loop antenna, a signal generator, and a sensitive receiver. Initially the loop is rotated for maximum reception, and a measure of the loop's output is recorded on a microammeter in the second detector stage of the receiver. The loop is next rotated for zero reception, and is then connected to the output of the signal generator that has been tuned to the same frequency as the incoming signal. The output of the signal generator is adjusted to give the same reading on the microammeter. Finally, after measuring the signal generator's output with an electronic voltmeter, the incoming signal strength can be obtained.

At very high frequencies a loop antenna is unsuitable and is replaced by a nondirectional vertical antenna. The receiver calibration is then achieved by using a standard field generator. This generator contains a low-power compact oscillator that is connected to an antenna whose surrounding field can be determined for a particular value of the measured antenna current.

To derive the shape of a radiation pattern surrounding an antenna, it is only necessary to measure relative rather than absolute field strength. The meter is then in the form of a receiver with a microammeter connected in the output of the second detector stage. The contour can then be formed by joining together all positions where the micro-ammeter reading is the same.

Parasitic elements

A vertical Hertz dipole has a circular (omni-directional) radiation pattern in the horizontal plane. The directional properties can be modified by adding parasitic elements to the antenna system. These parasitic elements are metal structures (for example, rods) that are not electrically connected (nondriven) and are placed in the vicinity of the driven dipole. The radiated field from the dipole induces a voltage and current distribution on the parasitic element which then reradiates the signal. At any position in space the total field strength will be the phasor resultant of the field radiated from the dipole and the re-radiated field from the parasitic element.

As a simple example, consider a vertical $\lambda/2$ dipole which is separated by a quarter-wavelength from a parasitic element (rod) which is also $\lambda/2$ long (Fig. 13-36A). Due to the separation, the field ($\mathcal{E}_{DP}$) reaching the parasite from the dipole will lag the dipole's field ($\mathcal{E}_D$) by 90°. Because the parasitic element is resonant, it behaves resistively and its current, I_P, is in phase with $\mathcal{E}_{DP}$ (Fig. 13-36B). The reradiated field, $\mathcal{E}_P$, lags I_P by 90° and is therefore 180° out of phase with $\mathcal{E}_D$. Due again to the $\lambda/4$ separation, the reradiated field reaching the dipole, $\mathcal{E}_{PD}$, lags $\mathcal{E}_P$ by 90°. In the direction X the resultant field, $\mathcal{E}_X$, is the phasor combination of $\mathcal{E}_D$ and $\mathcal{E}_{DP}$; the magnitude of $\mathcal{E}_X$ is the same as that of $\mathcal{E}_Y$ which is the resultant of $\mathcal{E}_{DP}$ and $\mathcal{E}_P$ in the direction of Y. However, in the two directions that are perpendicular to the plane of the paper, and passing through the point, 0, $\mathcal{E}_D$ and $\mathcal{E}_P$ are 180° out of phase and the two fields will tend to cancel. The result is to change the shape of the horizontal radiation pattern from a circle (omnidirectional) into two ovals or lobes (bi-directional), such that the area of the two ovals is equal to the area of the original circle (Fig. 13-36C). In fact this treatment is oversimplified, since it ignores the reduction in the field strength due to the separation between the dipole and the parasitic element.

Let the parasitic element now be increased in length so that it is more than $\lambda/2$ long and therefore behaves inductively (Fig. 13-37A). The parasitic current, I_P, lags $\mathcal{E}_{DP}$, so the various phase relationships are modified as in Fig. 13-37B. The magnitude of $\mathcal{E}_X$ is now greater than $\mathcal{E}_Y$, so the major radiation lobe is pointed toward the X-direction while the minor lobe is in the Y-direction. The parasitic element is therefore behaving as a reflector, since its effect is to increase the radiation in the forward direction (from the parasitic rod toward the dipole) at the expense of the radiation in the backward direction. By comparing the field strengths along the axes of the two lobes you can calculate the *front-to-back* ratio as defined by (forward field strength/backward field strength). For a single reflector rod the front-to-back ratio can be adjusted up to a value of 3.1; to take into account the reduction in field strength between the dipole and the parasitic elements, the separation and the length of the reflector rod are typically optimized to 0.1 λ and 0.55 λ respectively.

If the length of the parasitic rod is reduced to less than half of the physical wavelength (Fig. 13-38A), the rod behaves capacitively, I_P leads $\mathcal{E}_{DP}$ and $\mathcal{E}_X$ is again greater than $\mathcal{E}_Y$ (Fig. 13-38B). The parasitic element now behaves as a director, since it directs

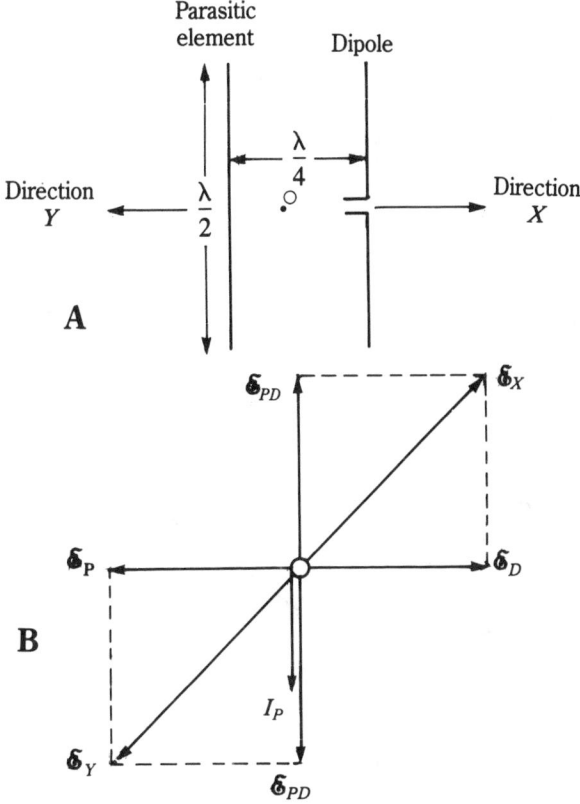

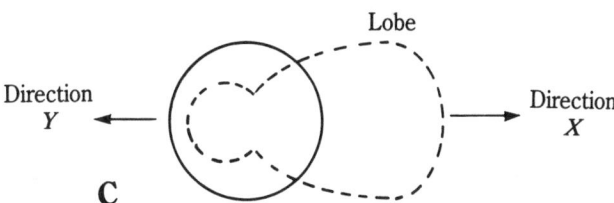

13-36 Principle of the parasitic element.

the radiation toward the major lobe. Typically the length of the director rod is 0.45 of the physical wavelength and its separation from the driven dipole is 0.1 λ.

A combination of a driven dipole, a reflector rod, and a director rod forms a Yagi array, (Fig. 13-39A) which has a power gain of 5 to 7 dB. The power gain can be further increased by adding additional director rods. With the major lobe increased at the expense of the minor lobe the Yagi system is a unidirectional antenna.

It is common practice to use a folded dipole (Fig. 13-39B) which has a lower Q than a thin rod and can therefore be operated effectively over a wider range of frequencies.

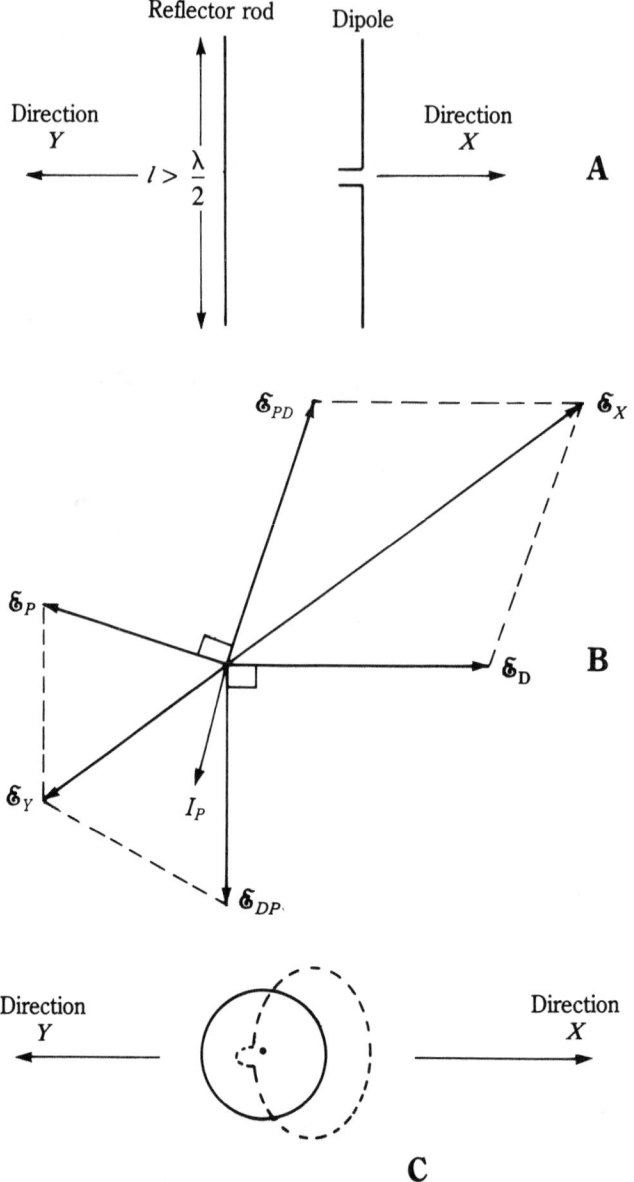

13-37   Principle of the reflector element.

The input resistance of a resonant folded dipole is approximately 300 Ω, and it therefore can conveniently be attached to a 300 Ω twin line. However if the balanced dipole is fed by a coaxial cable with its outer conductor grounded, it will be necessary to change from the balanced (dipole) to the unbalanced (coaxial cable) condition. This is achieved by means of a *balun* (*ba*lanced, *un*balanced), one example of which is the phase matching

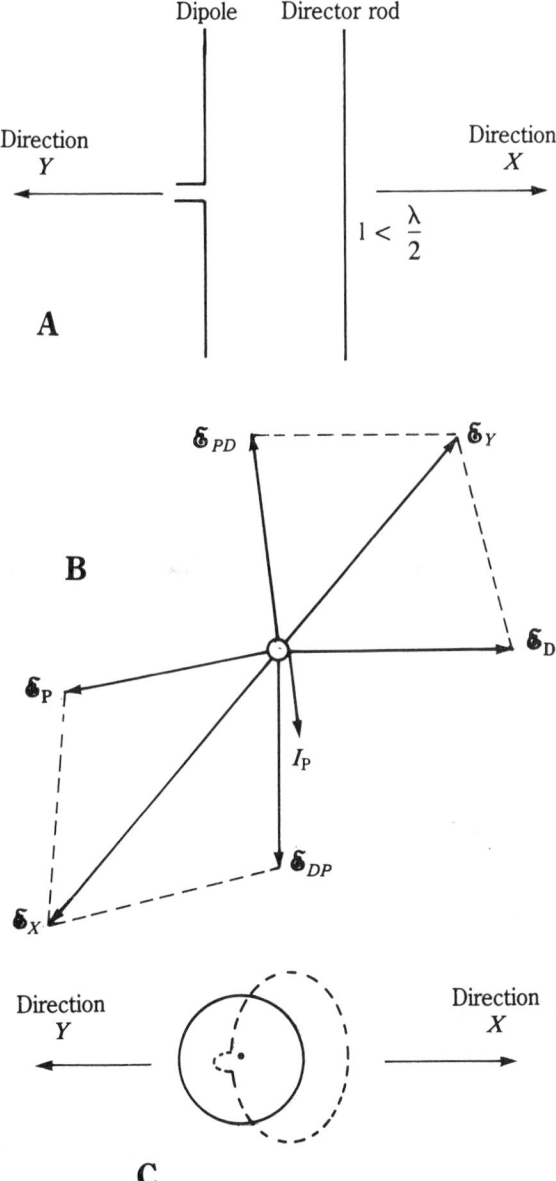

13-38 Principle of the director element.

transformer (Fig. 13-40). The λ/2 and λ sections provide the necessary 180° phase difference, so the two lines feeding the dipole are balanced with respect to ground. The equivalent circuit shows that for an impedance match, one-quarter of the dipole load must equal the surge impedance of the coaxial cable. In the case of a folded dipole the coaxial cable's surge impedance must equal 300/4 = 75 Ω.

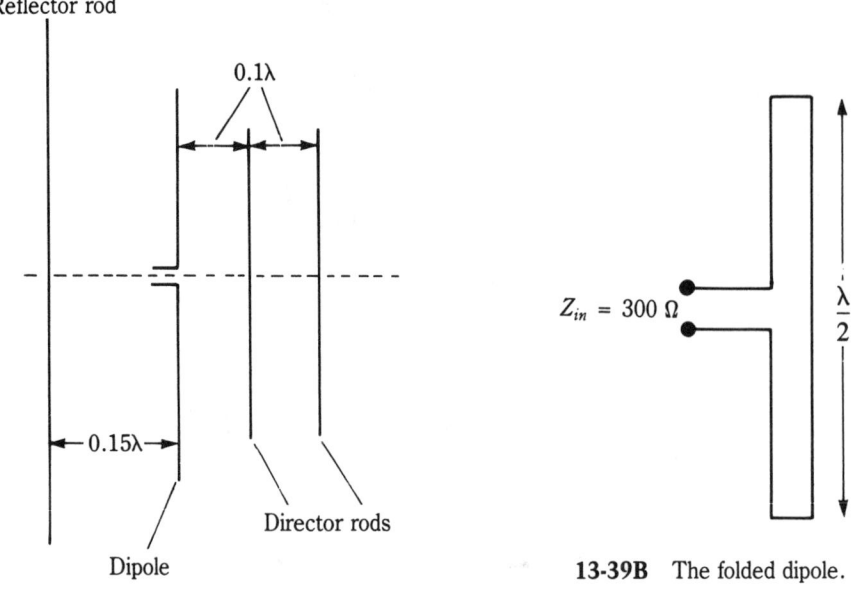

13-39A   The Yagi antenna.

13-39B   The folded dipole.

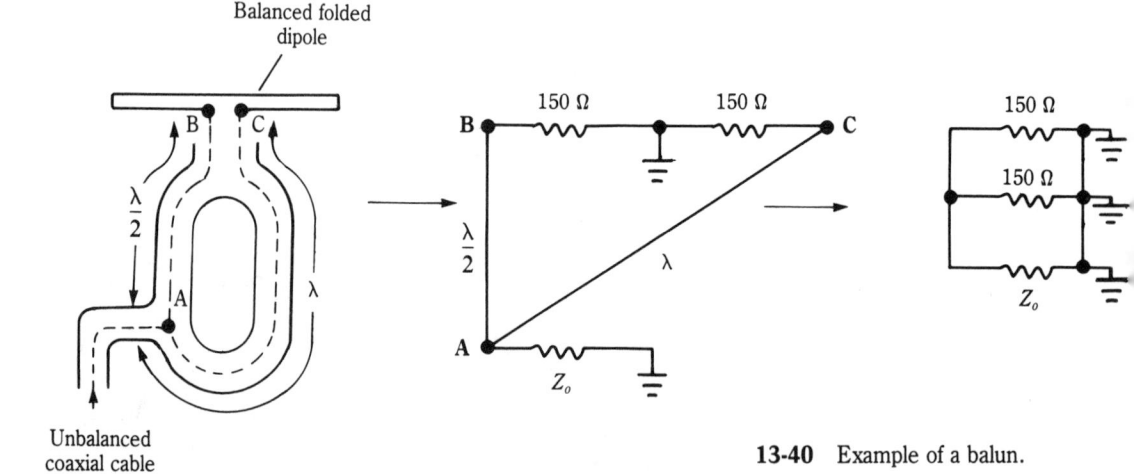

13-40   Example of a balun.

The loop antenna

A vertical loop antenna (Fig. 13-41A) has a radiation pattern which is bidirectional in the horizontal plane. Such an antenna is used to receive vertically polarized waves, and its main application is in direction finding (DF) to locate the position of a transmitter employing ground or surface-wave propagation.

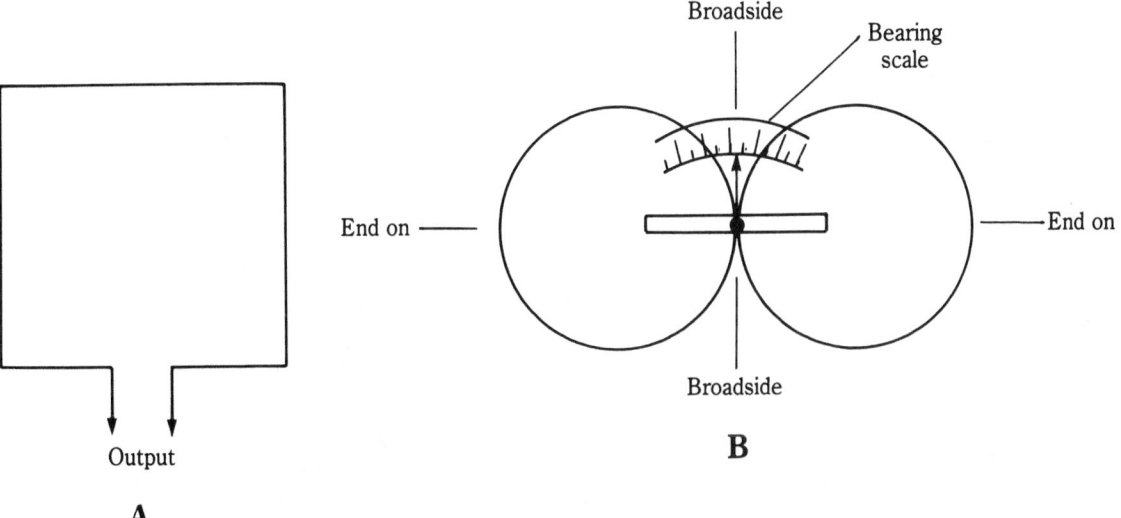

13-41 The loop antenna.

If the vertically polarized radio wave is coming from a horizontal direction at right angles to the plane of the loop (broadside on to the loop), the signals in the two vertical sections will be in phase and equal in magnitude. At the same time there will be zero signal in the horizontal sections. Under these conditions the loop's output will be zero because the two signals in the vertical sections will cancel when summed around the loop. However, if the wave arrives end on to the plane of the loop, there will be a maximum phase difference between the two signals in the vertical sections; this results in maximum output voltage from the loop. Because the loop is used at relatively low frequencies, the separation between the vertical sections is typically less than one-tenth of the physical wavelength. With two nulls and two maxima, the horizontal radiation pattern (Fig. 13-41B) of the vertical loop consists of two equal circles and is therefore bidirectional in nature. A similar bidirectional pattern occurs in the vertical plane with a horizontal loop antenna.

At the null there is maximum variation of the loop's output with the change in the direction of the transmitter. Consequently the loop is rotated to a null position so that the pointer which rotates with the loop, can provide an indication on a bearing scale. However, since there are two nulls, the pointer can indicate either the true or the reciprocal bearing of the transmitter (the true and reciprocal bearings are 180° apart). By using two direction finding receivers, each in a different location, the two bearing lines can be transferred to a map. Provided the locations are sufficiently far apart, the intersection of the bearing lines will accurately indicate the transmitter's position. It is also possible to distinguish between the true and reciprocal bearings by combining the loop output with the output of the *sensing* unipole, which is placed at or near the center of the loop. The resultant radiation pattern has a *cardioid* or heart shape which can be used in some procedure to separate the true bearing from its reciprocal.

The loop antenna can occur in a variety of shapes (rectangular, triangular, hexagonal, diamond, and circular) and can consist of one or more turns. The purpose in using a greater number of turns is to increase the loop's output. A loop of many turns was commonly used in the past as the internal antenna of AM broadcast receivers. The corresponding modern antenna has the turns wound on a ferrite core which is strongly magnetic; the result is to reduce the physical size required by the antenna.

Driven arrays

You have seen that a unidirectional antenna system can be formed by using a driven dipole together with one or more parasitic elements. Even greater directivity and power gain can be obtained from an array of driven antennas whose fields tend to reinforce in the required direction and to cancel in other directions. This effect is achieved by optimizing the spacing of the antennas and the phase relationships between their currents.

The two basic arrangements of driven antennas are called *broadside* and *collinear*. In the simplest broadside example two vertical dipoles are spaced λ/2 apart horizontally and are then fed in phase. Maximum radiation will then occur in the two horizontal directions, **X**, and **Y**, (Fig. 13-42A) where the distances to the centers of the dipoles are equal. By contrast there would ideally be no resultant radiation in the horizontal directions **P, and Q,** since the two distances would differ by a half-wavelength. This would introduce 180° phase difference between the two dipole fields at any position along the **P**, and **Q** directions so that the fields would tend to cancel; this is illustrated in the horizontal radiation pattern of Fig. 13-42B. Relative to the antennas the horizontal radiation pattern is rotated through a right-angle if the dipoles are fed 180° out of phase.

Increased directivity is obtained by using a greater number of broadside antennas. For example if five vertical dipoles are arranged along a line with λ/2 spacing between adjacent dipoles (Fig. 13-43A), the total length of the array is 2λ and its radiation pattern

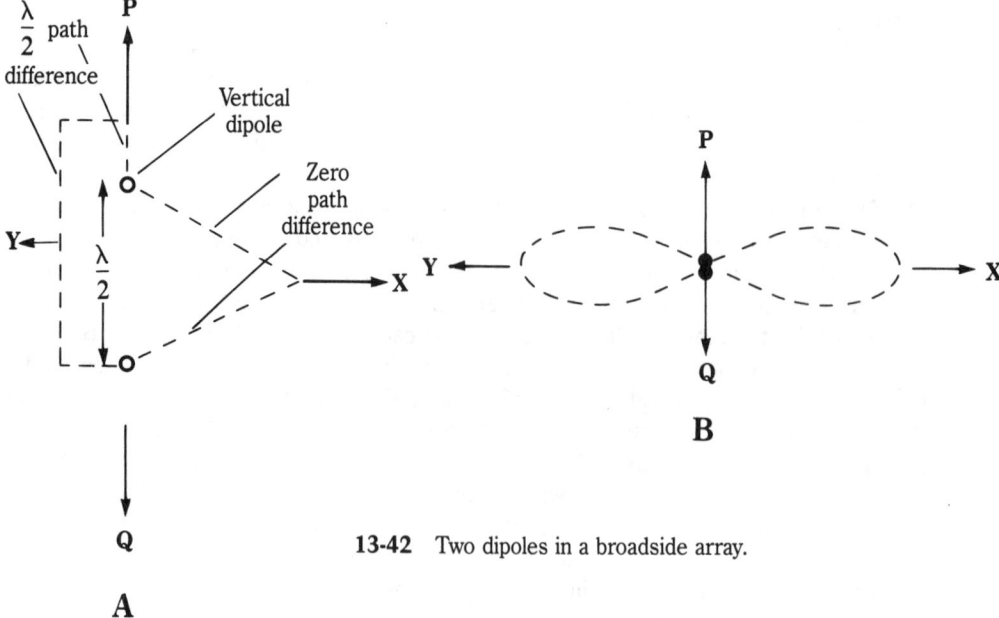

13-42 Two dipoles in a broadside array.

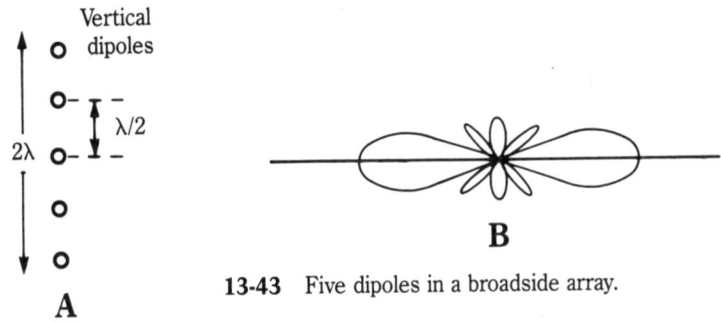

13-43 Five dipoles in a broadside array.

is illustrated in Fig. 13-43B. Although the major lobes are more pronounced, there are also a number of parasitic or minor lobes.

In a simple collinear arrangement two horizontal dipoles are placed side by side with their centers separated by about a wavelength. Along certain directions in the horizontal plane the distances to the center of the dipole will differ by half a wavelength so that cancellation will occur. If the collinear array is increased to four dipoles which are spaced one-half wavelength apart, the horizontal pattern consists of two major and four parasitic lobes (Fig. 13-44).

A combination of broadside and collinear arrays is shown in Fig. 13-45. This horizontal array consists of four driven dipoles that are delta fed in phase. These dipoles have a vertical spacing of half a wavelength and are mounted about a wavelength apart horizontally. Parasitic reflectors reduce the back radiation, so the single major lobe is concentrated along the forward, **X**, direction.

Turnstile antenna

This antenna name is derived from its appearance. In its basic form it consists of two horizontal dipoles which cross at right angles and are fed 90° out of phase (Fig. 13-46A).

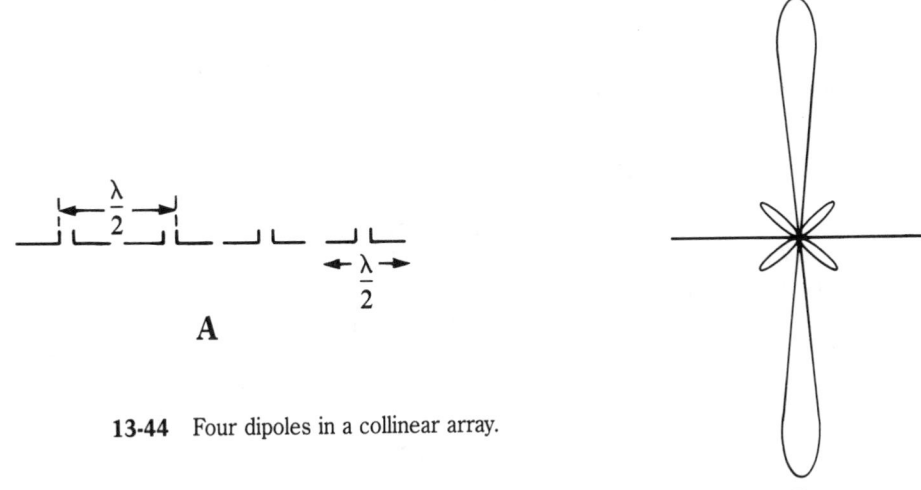

13-44 Four dipoles in a collinear array.

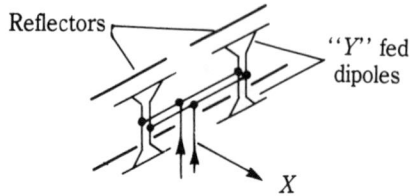

"Y" fed dipoles

13-45 Combination of broadside and collinear arrays.

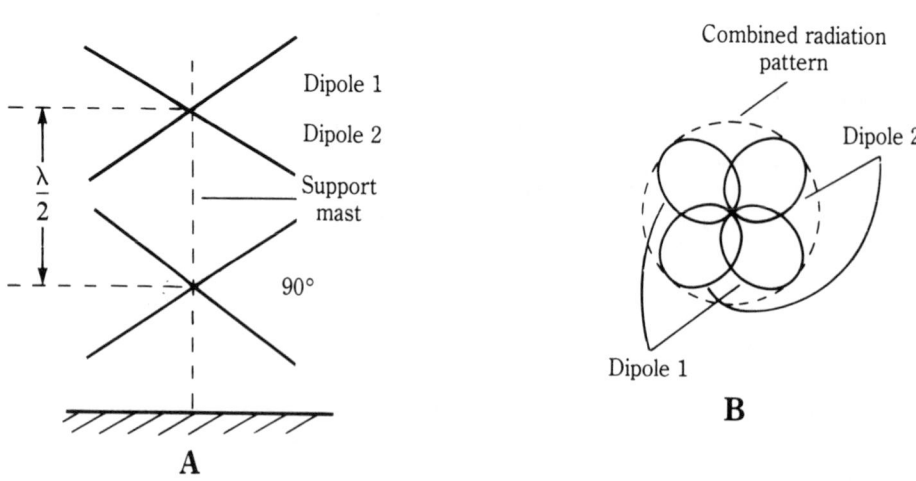

13-46 The turnstile antenna.

The horizontal radiation pattern for each dipole consists of two ovals but when these are combined, the result is roughly circular (Fig. 13-46B). This is due to feeding the antennas 90° out of phase so that the resultant field strength must be found by using the Pythagorean equation. To increase the power gain a number of turnstile antennas may be mounted on a mast with a vertical spacing of half a wavelength between adjacent antennas; this will provide more directivity in the vertical plane. Modified turnstile antennas are commonly used for TV transmission.

The nonresonant antenna

In Fig. 13-47A a single wire in space (remote from earth) is terminated by a resistive load of value Z_o; the return line consists of a nonradiating ground. The three-dimensional pattern surrounding the wire has a major lobe and a number of minor lobes. Assuming negligible attenuation along the wire, a cross-section of these lobes is shown in Fig. 13-47B, the pattern is the same for any plane containing the wire. The angle between the axis of the major lobe and the wire depends on the number of wavelengths that exist between the generator and the matched termination. If $L = 2\lambda$ (Fig. 13-47B) this angle is about 35° and the radiation resistance is approximately 170 Ω. However if L is raised to 4λ (or the generator frequency is doubled) the angle is reduced to 25°, the radiation resistance increases to 210 Ω, and there is a greater number of minor lobes

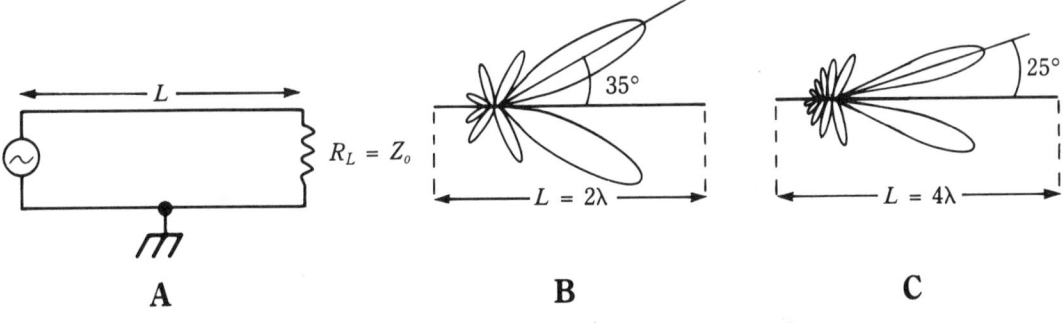

13-47 Radiation from a nonresonant line.

(Fig. 13-47C). These results assume that the radiation resistance is small compared with the surge impedance of the single-wire line.

The principle of the nonresonant line is used in the rhombic antenna which is primarily employed for long-distance fixed-service communication. With this type of traveling-wave antenna the radiated sky wave is directed upward at an angle to the earth and is refracted back by the ionosphere. Figure 13-48A shows a rhombic antenna which consists of four nonresonant wires or legs, each four wavelengths long and arranged in the form of a diamond or rhombus.

The corners of the antenna can be supported by wooden poles so that the plane of the diamond is normally parallel to the ground and the radiation is in the form of a horizontally polarized wave. If the tilt angle, ϕ, is optimized, the four legs all have major lobes which are directed parallel to the line joining the termination to the feedpoint. When the radiations from the four wires are combined, there is a resultant major lobe whose axis is at a vertical angle, β, relative to the plane of the diamond (Fig. 13-48B).

In practice the value of the tilt angle is not particularly critical, especially if the length of each leg is greater than two wavelengths. For example, if the design of the rhombic antenna is optimized for communication at 10 MHz, it can be operated without adjustment between 7.5 MHz and 15 MHz. At the higher frequencies there will be a sharper directional pattern and the main lobe's vertical angle will be less.

The surge impedance of the rhombic antenna is typically of the order of 600 Ω, and the terminating load must be capable of dissipating between one-quarter and one-half of the rf power supplied to the antenna. The necessary resistance can be provided by a high-loss twin line that is sufficiently long to dissipate the necessary amount of power and is then terminated by a low-wattage resistor.

Propagation

The word *propagation* is derived from the Latin verb "propagare"—to travel, and describes the various ways by which a radio wave travels from the transmitting antenna to the receiving antenna. Start by considering a long half-wave antenna which is fed at one end by a coaxial cable with its outer conductor grounded. The three-dimensional radiation pattern of the wave resembles a large "doughnut" which is laying on the ground with the antenna as its center (Fig. 13-49A). Part of the radiated wave moves downward and outward in contact with the ground and is affected by the conditions at and

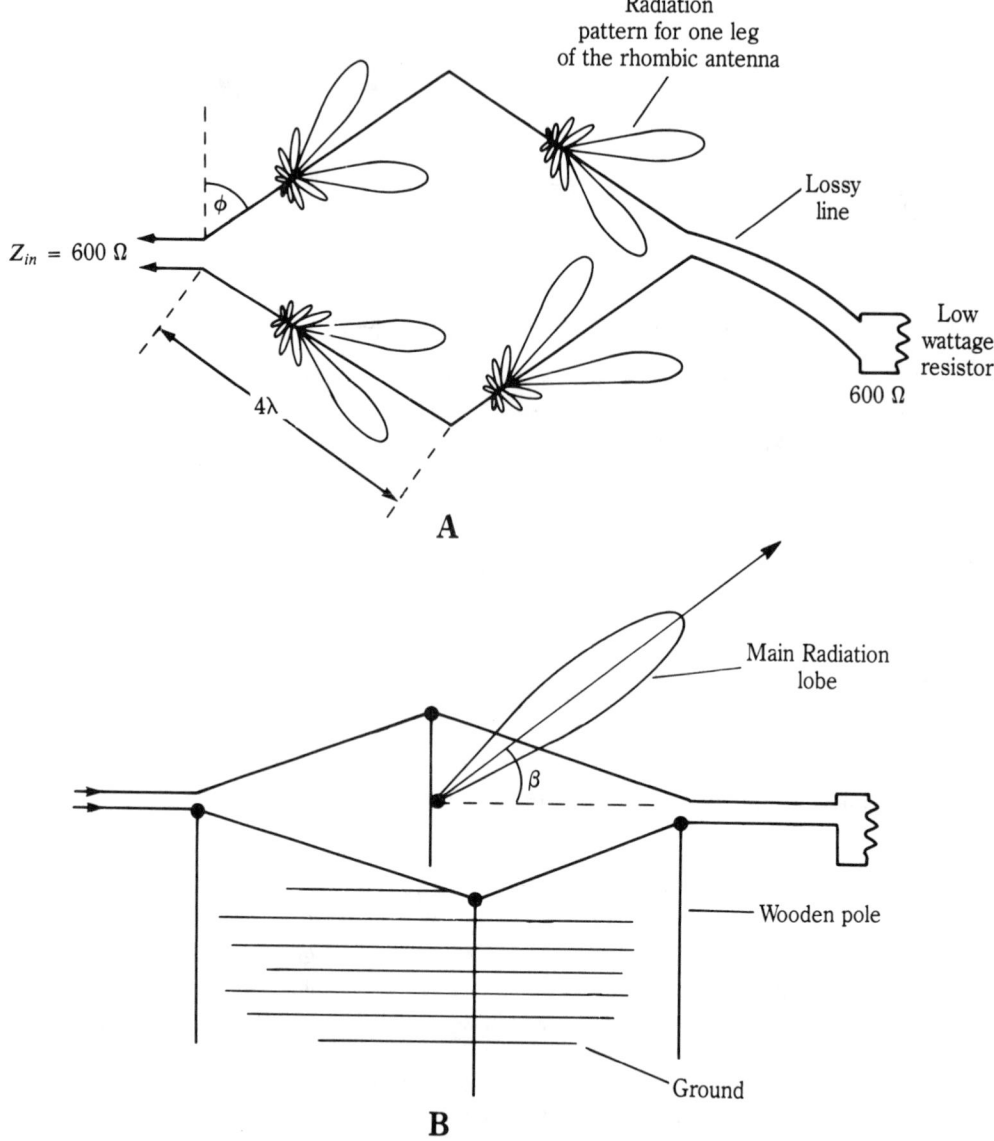

Radiation
pattern for one leg
of the rhombic antenna

Lossy
line

$Z_{in} = 600\ \Omega$

Low
wattage
resistor

$600\ \Omega$

ϕ

4λ

A

Main Radiation
lobe

β

Wooden pole

Ground

B

13-48   The rhombic antenna.

below the surface of the earth. This component of the radiation is referred to as the *ground* or *surface wave*.

Higher up the antenna, the radiation is little affected by the ground conditions. This portion is called the *space* or *direct wave*, which travels in a practically straight line from the transmitting antenna to the receiving antenna (direct line propagation). Operating in conjunction with the space wave is the ground reflected wave (Fig. 13-49B) so that the signal arrives at the receiving antenna through two different paths.

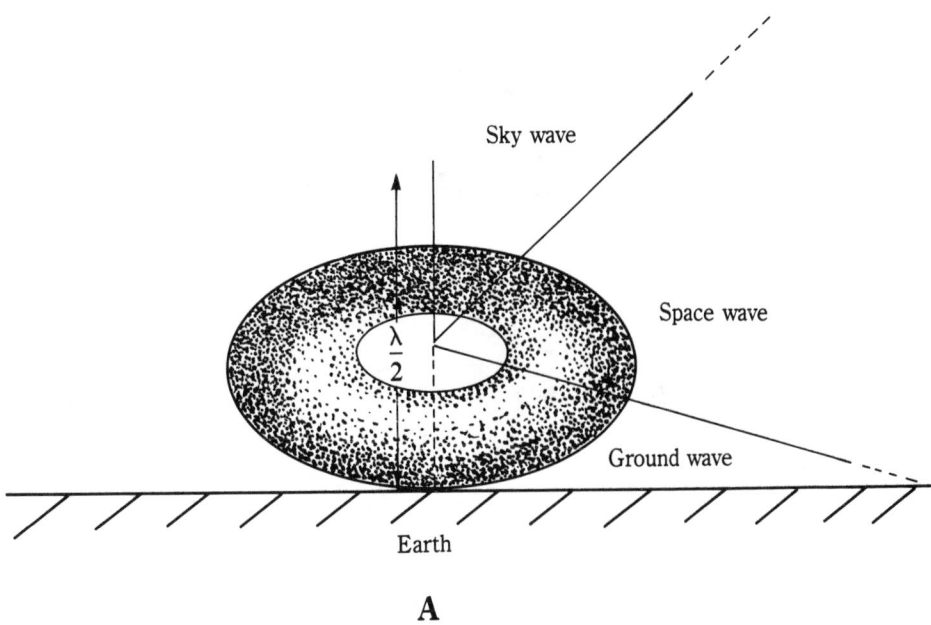

A

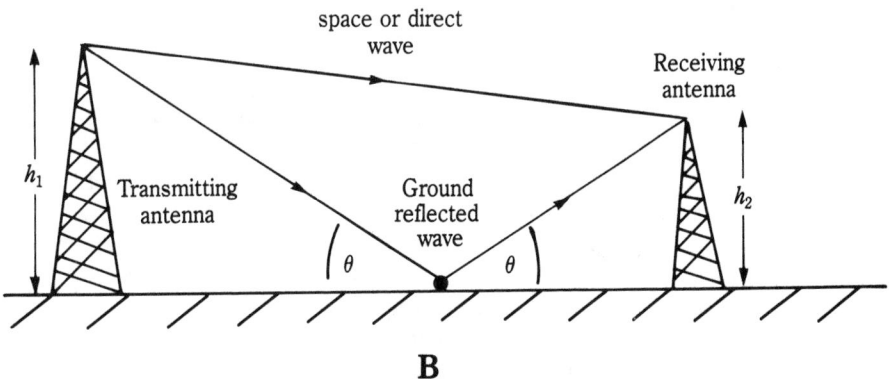

B

13-49 Component of radiation from a half-wave antenna.

Toward the top of the antenna the radiation moves outward and upward to the iono-sphere, which is an ionized layer of gas (primarily hydrogen) which extends from about 35 miles to 250 miles above the earth's surface. This portion of the radiation is referred to as the *sky* or *indirect wave* which is refracted by the ionosphere back to earth.

The ground, space, and sky waves are illustrated in Fig. 13-50. Primarily the ground wave is used for long-distance communication by high-power transmitters at relatively low frequencies; as an example, the signals from commercial AM stations are carried by the ground wave. The space wave mainly operates at very high frequencies for both

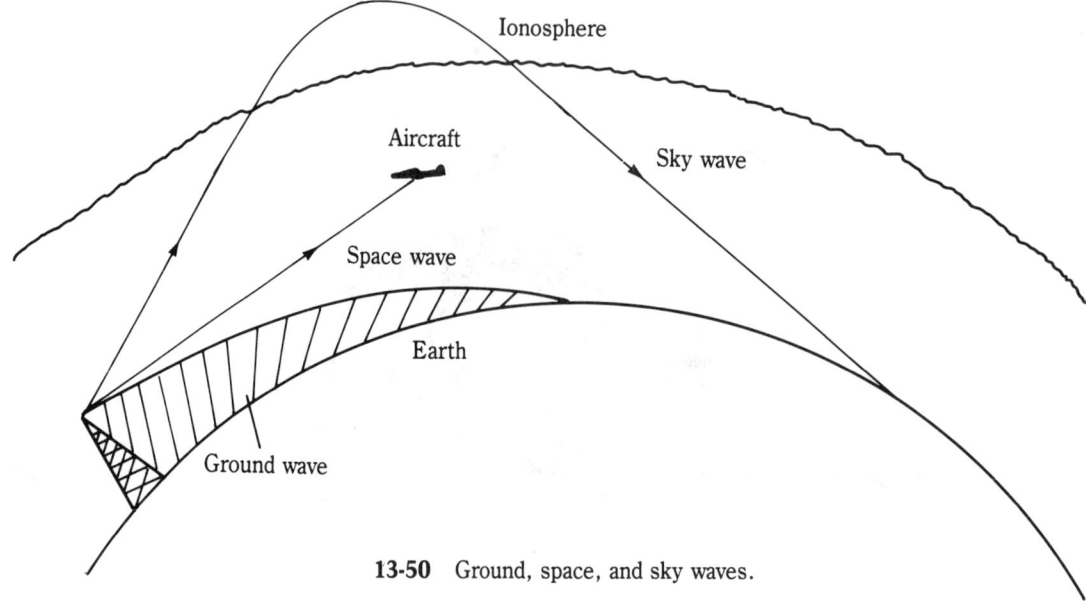

13-50 Ground, space, and sky waves.

short-distance and long-distance communications; practical examples are TV and FM broadcasts, as well as communication systems using satellites. The sky wave which travels through the ionosphere is used for long-distance fixed service communications and operates in the short-wave band.

Ground wave

The ground wave can only be used with vertically polarized signals which are radiated from a vertical antenna system. As the wave passes over and through the ground, there is a velocity difference between the portions that travel through the air and through the earth. As a result the wave front leans forward and there is an electric field component which is parallel to the earth's surface. This creates a voltage stress so that ground currents flow and energy is absorbed from the wave. At the same time the magnetic field component of the radio wave cuts the earth and induces eddy currents, so further energy is absorbed. With a horizontally-polarized wave the losses would be enormous and the wave could be completely wiped out at a short distance from the transmitting antenna.

For a vertically-polarized wave the attenuation due to ground losses depends on the conductivity of the ground and the frequency used. The attenuation rises rapidly as the frequency is increased so that the ground wave transmission is limited to frequencies between 12 kHz and 2 MHz. The lower end of this spectrum is used for long-distance submarine broadcast with extremely high power transmitters. The wavelengths of these broadcasts are of the order of miles so that the ground wave can be *diffracted* or bent around the earth's surface, and the signal can be received well beyond the horizon by a submarine operating at periscope depth. In fact providing the transmitter power is sufficient (up to 1 MW), the ground wave coverage might be worldwide. The disadvantage of using ground wave under these conditions is the need for enormous antenna systems that are relatively inefficient; as an example, one such system is slung between two mountains which are 20 miles apart and has an efficiency of about 15%.

Since the electrical properties of the earth along which the ground wave travels, are relatively constant, the signal strength from a particular station does not vary greatly at any given point. This is essentially true in most localities, and is the main advantage of using ground wave as opposed to sky wave for long-distance communications. Exceptions would be those areas with distinct rainy and dry seasons; in these cases the difference in the amount of moisture causes the soil's conductivity to change. The conductivity of salt water is approximately 5000 times greater than that of dry soil. Transmitters for submarine communications are therefore built as close to the ocean's edge as possible to take advantage of the superior groundwave propagation over salt water.

Space wave

The space wave is the radiation component which travels in a practically straight line from the transmitting antenna to the receiving antenna. This method of propagation is the one most commonly used in modern communications; everyday examples are TV and FM broadcasts, as well as radar and microwave links. The frequencies employed in these systems extend from 30 MHz upwards, and at these high frequencies the ground-wave range is negligible. Radio waves in this part of the frequency spectrum cannot *normally* be refracted back to earth by the ionosphere, so sky wave propagation is impossible.

If the transmitting and receiving antennas are both located on the ground, the range of the space wave is limited by the curvature of the earth (Fig. 13-51A). For example, the "hump" of the Atlantic ocean between England and America is approximately 200 miles high and therefore communication by space wave between these two countries can only occur via satellite. Figure 13-51B shows the limiting line-of-sight condition, in which the line joining the tops of the antennas just grazes the earth's surface. The optical range is then given by:

$$D = 1.23 \left(\sqrt{h_1} + \sqrt{h_2} \right) \text{ land miles} \tag{13-41}$$

where,

h_1, h_2 = antenna heights (feet)

In practice the space wave is refracted by the earth's atmosphere so that 10% to 15% must be added to the optical range. As an example, calculate the maximum range when the transmitting and receiving dipoles are mounted on board two ships and each dipole is 64 feet above the surface of the sea. The optical range is $1.23 (\sqrt{64} + \sqrt{64}) = 1.23 \times 16 = 19.7$ land miles and, after allowing an additional 10% for atmospheric refraction, the maximum radio range is approximately 22 land miles or 20 sea miles. However, if one of these ships was communicating with an aircraft flying at 10,000 feet, the optical range increases to $1.23 (\sqrt{64} + \sqrt{10000}) = 1.23 \times 108 = 133$ land miles and the maximum radio range is about 145 miles. For the same reason it is customary to increase the possible range and service area by locating TV and FM antenna systems on mountain tops. In addition to the line-of-sight transmission and the atmospheric refrac-

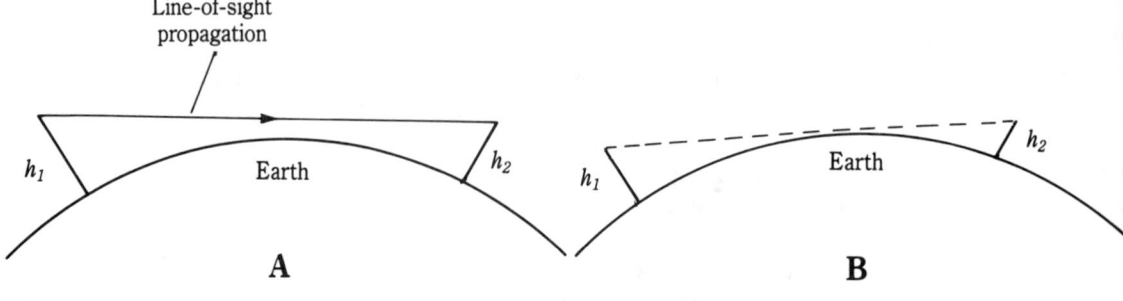

13-51   Space-wave propagation.

tion, there is a further small increase in the range due to the diffraction around the earth's surface.

Space waves are readily reflected from large metal obstacles so that the received signal will be the result of the space wave traveling directly from the transmitting antenna and all those radio waves that have been reflected from the various obstacles. Since all these waves have taken different paths to the receiving antenna, their signals are not in phase and the result is various forms of interference. One example is TV *ghosting* which appears as a double-image pattern on the receiver screen.

Tropospheric ducting

Unusual ranges, well beyond the predictions of line-of-sight transmission, can be caused by abnormal atmospheric conditions a few miles above the earth. These conditions occur in the troposphere, which extends from the earth's surface to a height of approximately 35,000 feet. Under normal conditions the warmest air is found near the surface of the ocean, and the temperature subsequently decreases with height. However, in the tropics a situation can occur where pockets of warm air are trapped between layers of cooler air. The boundaries between the warm air and the cooler air are called *temperature inversions*; these are capable of refracting space waves which would otherwise continue their line-of-sight propagation. The result is a high tropospheric duct in which the space wave is trapped and transmitted for hundreds of miles. These high-level ducts typically start at heights of 500 to 1,000 feet and extend an additional 500 to 1,000 feet into the atmosphere. The signal might then be received by an antenna located in the duct or an antenna on the earth's surface provided the lower temperature inversion has disappeared (Fig. 13-52A). Under other tropical conditions, the temperature of the air above the surface of the sea will initially increase with the elevation, but at a certain height there will be an inversion and the temperature will subsequently decrease. The result is a surface duct which might extend to an elevation of a few hundred feet. A space wave can be trapped in the surface duct and can be received well beyond the horizon (Fig. 13-52B).

Little use can be made of ducts for reliable communications because their occurrence and duration cannot be predicted. Both surface and high-level ducts represent *anomalous propagation* (sometimes referred to as *anaprop*) in which ranges far in excess of normal are experienced.

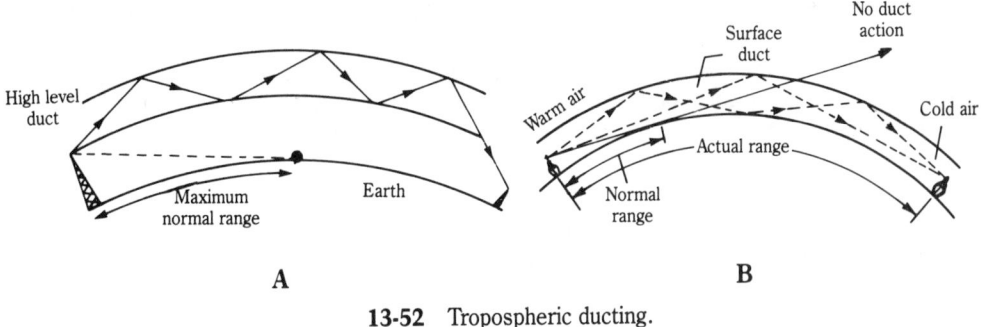

13-52 Tropospheric ducting.

Tropospheric scatter

Tropospheric scatter is a method of propagation which finds an increasing use in modern communications, and relies on conditions in the troposphere at an elevation of a few thousand feet. Owing to discontinuities of temperature and humidity in the troposphere, a scattering region is formed, as shown in Fig. 13-53. Unlike the creation of ducts, the scattering region always exists and is capable of returning a small fraction of the transmitted power back to earth. An analogy of the process is to consider a stream of water that is directed toward a ceiling with a rough surface. However, most of the transmitted energy travels straight out into space and only a small amount is scattered in the required forward direction; there is also some scattering in other undesired directions so that this energy will also not be intercepted by the receiving system. Typically the received power is only 10^{-6} and 10^{-9} times the transmitted power. Consequently high-power transmitters and extremely sensitive receivers are essential to operate a reliable tropospheric scatter communications system.

The possible frequency range for this propagation method extends from 300 MHz to 10 GHz, although the most commonly used frequencies are 900 MHz, 2 GHz, and 5 GHz. This wide range is possible because the attenuation only rises slowly as the frequency is increased. The minimum range is about 50 miles and is determined by the size of the scattering volume. The maximum range is 400 to 500 miles, and is a geometrical result of the average height of the region from which the energy is scattered.

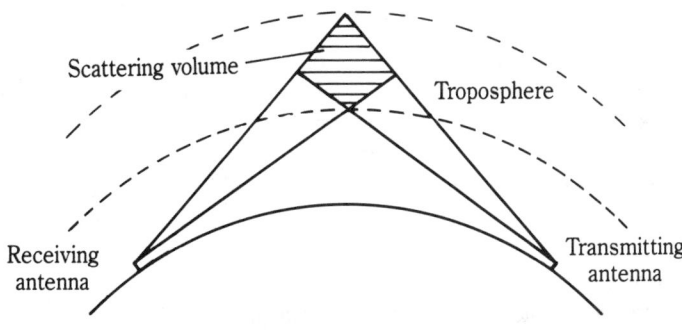

13-53 Tropospheric scatter.

Apart from the high level of transmitted power the main disadvantage of tropospheric scatter is the severe fading due to atmospheric changes and the many possible paths by which the energy travels through the scattering volume. The problem of fading can be overcome by some form of diversity reception in which a number of intercepted signals are combined at the receiving system, or the strongest of these signals at any instant is automatically selected.

Frequency diversity   The same information is simultaneously transmitted on a number of separate frequencies which suffer different and varying degrees of fading.

Space diversity   A number of receiving antennas are positioned several wavelengths apart. The strongest signal is automatically selected and fed to the receiver.

The ionosphere

The upper parts of the earth's atmosphere absorb large quantities of radiant energy from the sun, which not only heats the atmosphere but also produces some ionization in the form of free electrons and positive ions.

When an electromagnetic wave strikes an atom, it is capable of moving an electron from an inner to an outer orbit. When this occurs, the electron has absorbed energy from the wave. If the frequency of this incident wave is sufficiently high, such as in the ultraviolet region, an electron might be knocked completely out of an atom. When this occurs, a positively charged atom or ion remains in space together with a free electron. The rate of ion and free electron formation depends on the density of the atmosphere and the intensity of the ultraviolet wave. As the ultraviolet wave produces positive ions and free electrons, its intensity diminishes. Therefore, the ionized region will tend to form in a layer due to:

1. forming few positive ions and free electrons due to the less dense atmosphere when the ultraviolet is most intense

2. forming more positive ions and free electrons due to the more dense atmosphere when the ultraviolet wave is of moderate strength

3. again forming few positive ions and free electrons due to the low intensity of the ultraviolet wave when in the most dense atmosphere.

This relationship between ultraviolet intensity, rate of ionization, and atmospheric density is shown in Fig. 13-54.

The formation of ions and free electrons is not, in itself, sufficient reason to account for the existence of an ionized layer, because the positive ions and free electrons tend to recombine due to the inherent attraction of unlike charges. The recombination rate is directly related to the molecular density of the atmosphere, since the more dense the atmosphere, the smaller is the mean free path of the electrons.

The recombination rate is also directly proportional to the density of the positive ions and the free electrons. Therefore, as the ultraviolet waves continue to produce ions and free electrons, a free electron density will be reached where the recombination rate just equals the rate of formation. In this state of equilibrium, a free electron density exists for every set of given conditions.

The formation of more than one layer is explained by the existence of different ultraviolet wave frequencies. The lower-frequency ultraviolet waves tend to produce a higher

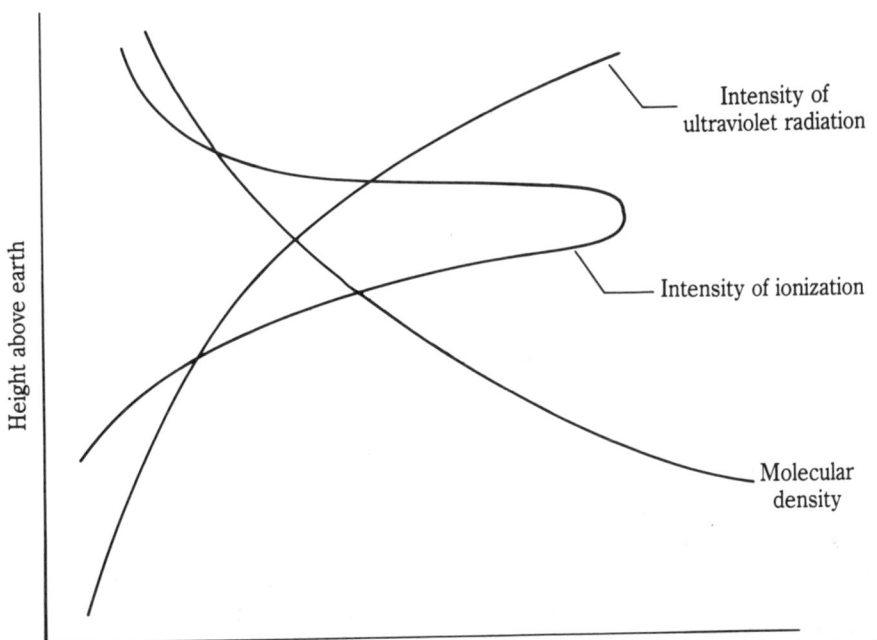

Intensity of
ultraviolet radiation

Intensity of ionization

Molecular
density

Height above earth

13-54 Relative magnitudes of UV radiation intensity, ionization intensity, and molecular density.

altitude layer, expending all their energy at this height. The higher-frequency ultraviolet waves tend to penetrate deeper into the atmosphere before producing appreciable ionization. In addition to the ultraviolet waves from the sun, particle radiation, cosmic radiation, and meteors all produce ionization in the earth's atmosphere, particularly at a higher-altitude layer. In addition, there is no mixing process in the ionosphere and consequently gases are arranged in layers according to their different densities; each gas requires a particular ultraviolet intensity to produce a strong ionization density.

The three principal layers formed in the daytime are called the E-, F_1-, and F_2-layers (Fig. 13-55A). In addition to these regular layers, there is a region below the E-layer that is responsible for much of the daytime attenuation of high-frequency radio waves. Called the D-region or layer, it lies between heights of 50 and 90 km. The heights of the maximum density of the regular E- and F_1-layers are relatively constant, with only small diurnal and seasonal changes. The F_2-layer is more variable with typical heights within the range of 200 to 400 km. The F-layers are composed of the lightest gas (hydrogen) in an ionized state.

At night (Fig. 13-55B) the F_1- and F_2-layers join to form a single F-layer. The regular E-layer is governed closely by the amount of ultraviolet radiation from the sun, and at night tends to decay uniformly with time. The D-layer almost entirely disappears at night due to recombination.

An anomalous ionization, termed *sporadic-E*, is often present in the E-region in addition to the regular ionization. Sporadic-E ionization usually exhibits the characteristics of patches of intense ionization, which might appear anywhere in the height range of 90 to 130 km. These patches can be from 1 km to several hundred km across. The occurrence of sporadic-E is quite unpredictable and although very high frequencies are

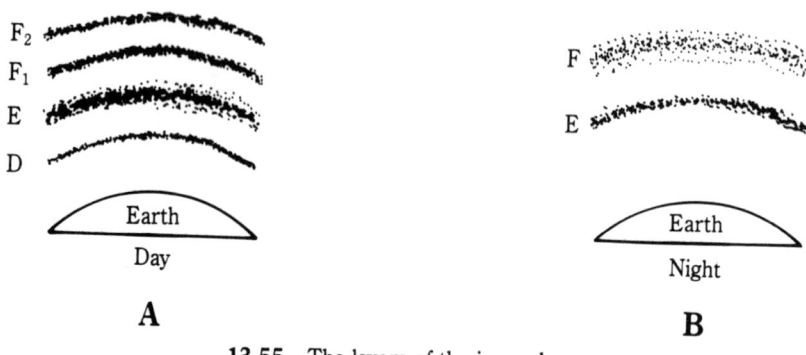

13-55   The layers of the ionosphere.

regularly returned, there is no possibility of predicting the conditions, so they cannot be used for reliable communications.

In contrast with the unpredictable sporadic conditions the E-layer consistently represents a scattering region comparable with the troposphere conditions already described in this section. However, with ionospheric scatter the attenuation increases rapidly with frequency so that the usable range only extends from approximately 30 MHz to 100 MHz. The minimum range is determined by the size of the scattering volume, and is typically of the order of 250 miles. The maximum range is about 1500 miles and is the geometrical result of the E-layer's average height.

The changes that can occur in the ionosphere can be loosely tabulated:

1. Diurnal changes. Day and night changes in the height and the density of the layers.

2. Seasonal changes. These are obviously very much tied in with the geographical position. For example, the ionosphere is very weak and irregular near the earth's poles.

3. Sporadic-E. Patches of intense ionization in the E-layer.

4. Sudden *i*onospheric *d*isturbances (SIDs) or Dellinger fade-outs. These are caused by intense ultraviolet radiation given out by the sun during a solar flare. The result of a SID is a sudden large increase in the ionic density of the highly absorptive D-region and an increase in the ionic density of the moderately absorptive E-layer.

5. Magnetic storms should not be confused with SIDs, although they both have the same effect of reducing the probability of any communication by means of the sky wave. Magnetic storms are apparently caused by the emission of particles by the sun. These particles are emitted at the same time as the flare, but being of finite mass, they take up to about 36 hours to arrive after a SID. A magnetic storm might last for several days, with its appearance being very sudden and the recovery of the layers to normal being very slow. Because the emitted particles are mostly magnetic, the effects of a magnetic storm are most severe in the geomagnetic polar regions. Magnetic storms are more likely to occur during periods of maximum sunspot activity, and this occurs in a regular 11 year pat-

tern. At the height of the 11 year sun spot cycle, the emission of ultraviolet waves from the sun is considerably greater than during years of the "quiet sun." During periods of intense solar activity, ionization is much greater and much higher frequencies are regularly returned in the high-frequency spectrum. However, communications are more likely to be interrupted by SIDs and magnetic storms.

Refraction of a radio wave by the ionosphere

When sky wave propagation is used for communication the electromagnetic wave from the antenna is transmitted toward the ionized layer at an oblique angle. The incident wave is then apparently reflected back from a certain angle toward the receiving antenna. Actually the wave is not reflected, though this term is commonly used for convenience; the wave is in fact bent back toward the earth by refraction, just as a prism refracts light. As far as propagation is concerned, the all-important effect of ionization is to reduce the refraction index (μ) of the ionosphere; this causes the wave (travelling from a medium of high μ to a medium of low μ) to be refracted in accordance with Snell's optical law. Refer to Fig. 13-56:

$$\mu = \frac{\sin i}{\sin r} \tag{13-44}$$

where
  i = angle of incidence measured relative to the normal.
  r = angle of refraction measured relative to the normal.

When the radio wave approaches the layer at an oblique angle, the upper part of the wave front will pass through a region of stronger ionization, compared with the lower part. Since the wave's phase velocity increases with a greater ionization density, the top of the wave front will move faster than the bottom and refraction will occur.

The refractive index of the ionosphere is a function of the frequency, f, of the wave:

$$\mu = \sqrt{1 - \frac{81N}{f^2}} \tag{13-45}$$

where
  N = the number of free electrons per cubic centimeter of the ionized region.

For a given frequency μ decreases as the density of ionization increases toward the center of the ionized layer.

At a certain value of μ, depending on the angle of incidence, $\sin r = 1$ and $r = 90°$ (Fig. 13-57). This makes the path of the wave more normal to the earth's radius and further refraction will cause the wave eventually to leave the ionized layer at the same angle as it entered. The further refraction occurs provided the top of the wave is still traveling in a region of greater ionospheric density.

In the special case of a vertically incident wave, μ must reach a value of zero for the wave to be returned to earth. The highest value at which this occurs in a given layer is called the *critical frequency*, f_o, of the layer.

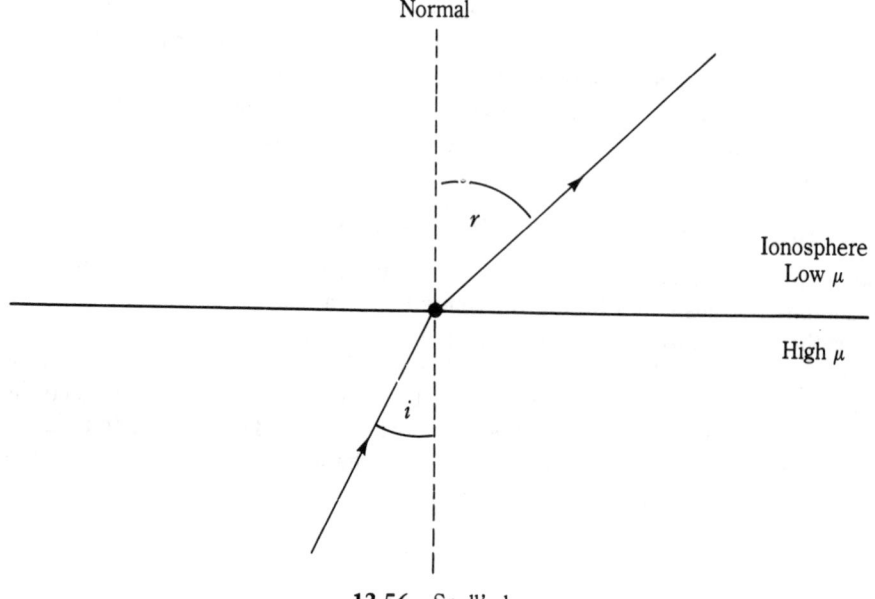

13-56   Snell's law.

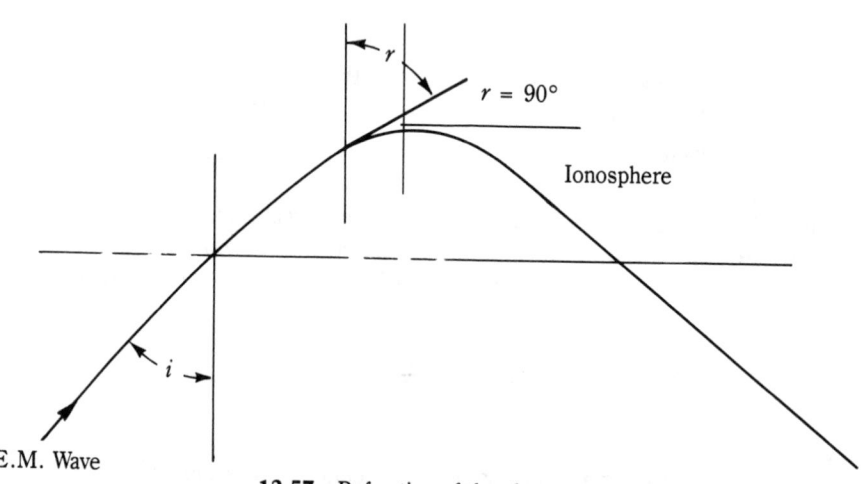

13-57   Refraction of the sky wave.

From equation 13-46:

$$0 = \sqrt{1 - \frac{81N_{max}}{f_o{}^2}}$$

This yields,

$$f_o = \sqrt{81N_{max}}$$

(13-47)

where
N_{max} = maximum free electron density in the layer.

From a knowledge of the critical frequency of a layer, the highest frequency at an oblique angle of incidence, i, which will be returned to earth, can be calculated. Since $r = 90°$, $\sin r = 1$, and,

$$\sin i = \mu$$

$$= \sqrt{1 - \frac{f_o{}^2}{f^2}}$$

$$\sin^2 i = \mu^2$$

$$= 1 - \frac{f_o{}^2}{f^2}$$

This yields,

$$\cos^2 i = \frac{f_o{}^2}{f^2}$$

$$f = f_o \sec i \tag{13-48}$$

This is known as the secant law, which is the basic equation used in choosing operating frequencies for a given transmission path. It is only true for a flat earth and a flat layer and must be modified for accurate calculations.

The secant law shows that for a particular frequency, there is a certain maximum angle of incidence for which the first sky wave is returned to the earth. As the angle is increased, the sky wave will penetrate the ionosphere layer to a lesser depth. However, the length of the path through the ionosphere will increase and the absorption will be greater. Consequently there is a limit to the reception zone which is covered by the sky wave.

The *skip distance* (Fig. 13-58A) is defined as the separation between the transmitting antenna and the point where the first sky wave is returned to the earth's surface. If the frequency is increased, the refraction in the ionosphere is less, and the skip distance increases. Between all the positions where the intensity of the ground wave has fallen below the noise level and the places where the sky waves are first returned, there is a skip or silent zone in which no reception is possible (Fig. 13-58B). The amount of the silent zone can be reduced by lowering the frequency. However, if the ground and sky wave coverages overlap, the two received signals will have taken different paths and the result will be a form of interference fading.

If the receiver is located at the position where the first sky wave is returned, it creates the condition of the *maximum usable frequency* (MUF), since if the transmitter frequency is increased, the skip distance will increase, and the receiver will be placed in the silent zone. Because the intensity of the ionosphere decreases at night, it is common practice to lower the frequency used for night time communications.

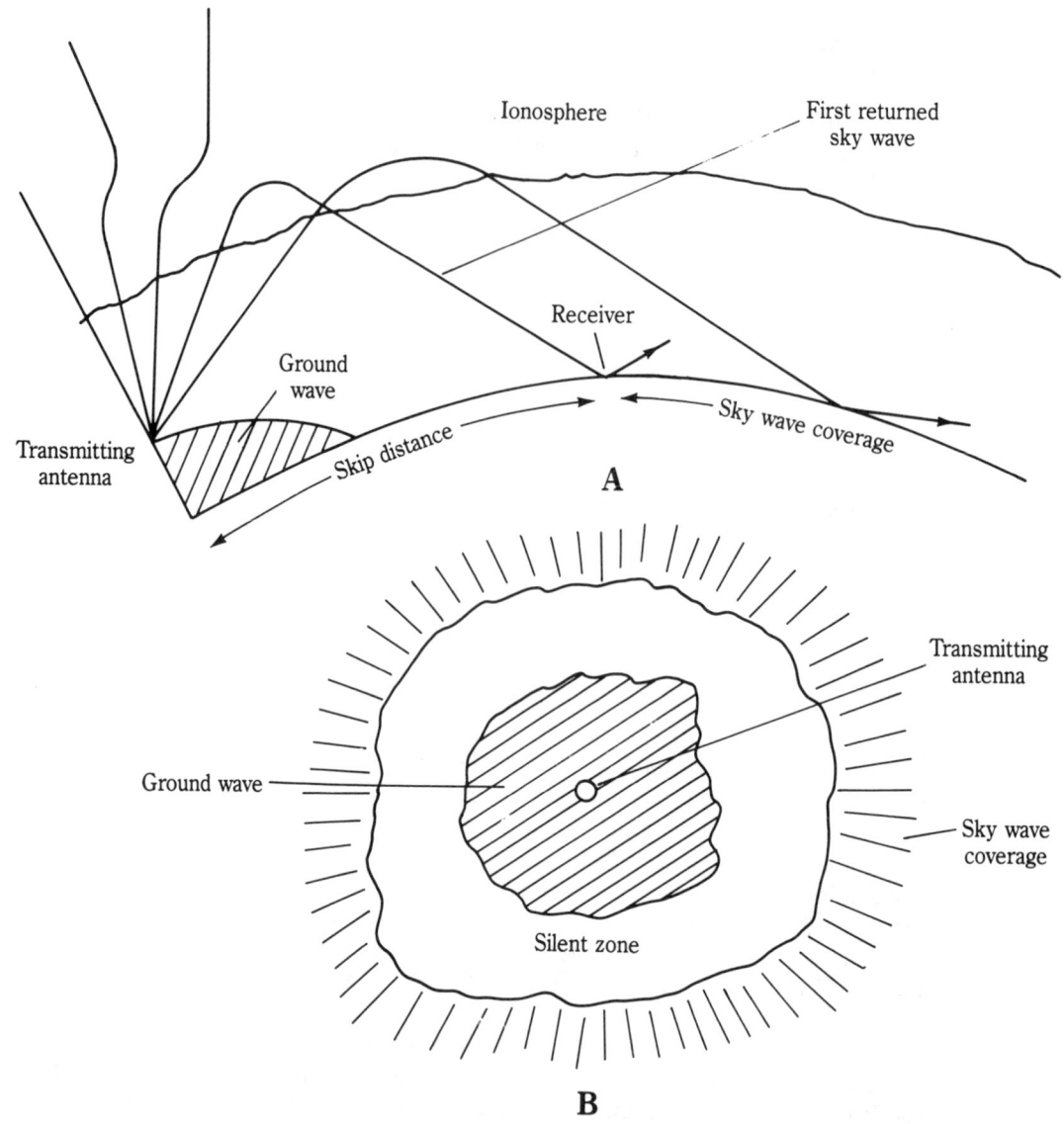

13-58 Skip distance and silent zone.

Multihop transmission

The sky waves described so far have been single-hop transmissions where the wave has been refracted only once in the ionosphere before being picked up by the receiver. However, it is possible for the wave to be reflected by the earth back into the ionosphere so that refraction occurs for a second time (Fig. 13-59). This process might be repeated for a number of times, and is referred to as *multihop transmission* which can be used to transmit over long distances within the frequency range of 9 to 30 MHz. However, in general, single-hop transmission results in greater field intensities than the multihop type.

As illustrated in Fig. 13-59, it is possible for a receiver to pick up the signal by both single-hop and multihop transmission. Because the two paths are different, there will be interference fading which might be overcome by using diversity reception. Over extremely long distances there can be interference between two multihop paths, one of which has travelled the "short way" around the earth's surface, while the other has taken the "long way." This form of interference can be avoided by employing a rhombic antenna whose one major lobe is directed at a specific angle toward the ionosphere.

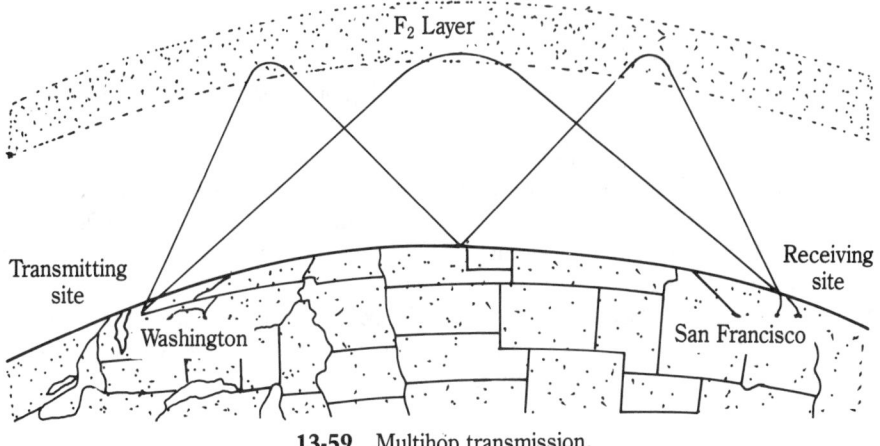

13-59 Multihop transmission.

Absorption

A radio wave entering an ionized layer interchanges energy with the free electrons and the ions. If the ions do not collide with gas molecules or other ions, all the energy transferred to the ionosphere is reconverted back to electromagnetic energy, and the wave continues to be propagated with undiminished intensity. On the other hand, where ions and electrons engage in collisions, they dissipate some of the energy they have acquired from the wave, so the result is a certain degree of attenuation. This attenuation, or absorption, is proportional to the product of the number of ions and the collision frequency. Therefore the attenuation is ordinarily greatest in the region where the product of ion intensity and the collision frequency is greatest. In fact most absorption occurs in the D-layer, which lies closest to the earth. Considerably less absorption occurs in the E-layer, and very little in the layers above. It follows that absorption is restricted mainly to the hours of daylight and falls to a low value at night. In general, absorption, during daylight hours, increases to a maximum value at around 1.5 MHz and then decreases rapidly with an increase in frequency.

Maximum usable frequency

The basis for all frequency planning is a layer's critical frequency, f_o, and measurements are collected systematically by ionospheric sounding centers through the world. The practical parameter, however, is the *maximum usable frequency* of which there are several definitions; in each case MUF designates the highest frequency which, in a particular set of conditions, can be used to propagate radio waves over a given route.

Predictions for f_o and MUF for the lower layers can be obtained from charts for any given sun spot condition and a knowledge of the sun's elevation. The MUF that applies to the F_2-layer is not so easily determined, because it obeys complex laws and its correlation with solar activity and geographical factors is not so close.

However, figures are published based on a maximum usable frequency that will give a reasonable prospect of good communication. The published figures are the *optimum working frequencies* (OWF) and in general this figure is about 85 to 90% of the maximum usable frequency; this allows for slight changes in the ionosphere.

Lowest usable frequency

The *lowest usable frequency* (LUF) is the lower limiting frequency which will provide satisfactory communication for a given link. The LUF is the frequency at which the received field intensity just equals the required field intensity for reception. The received field intensity depends on the receiver antenna system, the path length, and the absorption, which generally increases as the frequency is decreased. The required field intensity for reception depends on noise limitations at the site and these generally decrease as the frequency is raised.

From previous discussion it is clear that the MUF is a natural limitation while the LUF is a man-made limitation. It is possible that conditions can arise where the LUF is higher than the MUF; in this situation, propagation is impossible.

Propagation under the various frequency bands

The following frequency bands were named as the result of the Atlantic City Conference (1947).

very low frequency (vlf) band	3 – 30 kHz
low frequency (lf) band	30 – 300 kHz
medium frequency (mf) band	300 kHz – 3 MHz
high frequency (hf) band	3 – 30 MHz
very high frequency (vhf) band	30 – 300 MHz
ultra-high frequency (uhf) band	300 MHz – 3 GHz
super high frequency (shf) band	3 – 30 GHz
extremely high frequency (ehf) band	30 – 300 GHz

vlf and lf bands

Propagation in the vlf and lf bands mainly depends on the use of the ground wave. Vertically polarized radiation is invariably used, and the coverage might be worldwide provided the transmitter's power is sufficient. Sky waves are refracted back by the ionosphere's D-layer and, on their return to earth, will establish further ground waves.

mf band

The range of the ground wave can extend to several hundred miles. For frequencies up to 2 MHz the sky waves are returned by the E-layer. During day conditions the absorption is high but at night the sky wave can be received well beyond the ground wave range. However, if the sky waves are returned within the ground wave coverage, severe interference fading can result.

hf band

The ground wave range is of the order of 100 miles or less. Sky waves normally pass through the D- and E-layers to be refracted back by the F-layer(s). Single-hop transmissions can cover distances of 2,000 to 2,500 miles with frequencies up to 20 MHz. Multihop transmission can cover any distance in the frequency range of 10 to 30 MHz.

Sky wave communication can be blacked out by sudden ionospheric disturbances which are caused by solar flares and the arrival of charged particles emitted by the sun.

vhf and higher frequency bands

The ground wave coverage can virtually be ignored. Communication is by space wave with its range primarily limited by "line-of-sight" conditions. Above 30 MHz atmospheric noise is less than in the lower frequency bands.

Within this frequency range the radio waves normally pass straight through the ionosphere. However, at the peak of the sunspot cycle, vhf waves can be refracted back to earth so that ranges far in excess of normal are experienced; this can also occur as the result of sporadic-E conditions.

Anomalous and unpredictable propagation conditions frequently occur as the result of high-level or surface ducting in the troposphere. However reliable communications can be established through the scattering process in either the troposphere or the ionosphere.

Multiple-choice questions

1. How is it possible to lower the resonant frequency of a Hertz dipole antenna?

A. Reduce the frequency at the transmitter.
B. Connect a capacitor in series with the antenna.
C. Connect a resistor in series with the antenna.
D. Connect an inductor in series with the antenna.
E. Reduce the length of the antenna.

2. A quarter-wave line is connected to an rf generator and is shorted out at the far end. What is the input impedance to the line at the generator?

A. A low value of resistance.
B. A high value of resistance.
C. A capacitive resistance which is equal in value to the line's surge impedance.
D. An inductive reactance which is equal in value to the line's surge impedance.
E. A value of resistance equal to the Z_o of the line.

3. If the SWR on a transmission line has a high value, the reason could be:

    A. An impedance mismatch between the line and the load.
    B. That the line is nonresonant.
    C. A reflection coefficient of zero at the load.
    D. That the load is matched to the line.
    E. A high degree of attenuation between the load and the position where the SWR is measured.

4. In Fig. 13-60, what does the broken line represent?

    A. The current distribution on a half-wave Hertz antenna.
    B. The current distribution on a quarter-wave Marconi antenna.
    C. The voltage distribution on a half-wave Hertz antenna.
    D. The voltage on a quarter-wave Marconi antenna.
    E. The impedance distribution on a half-wave Hertz antenna.

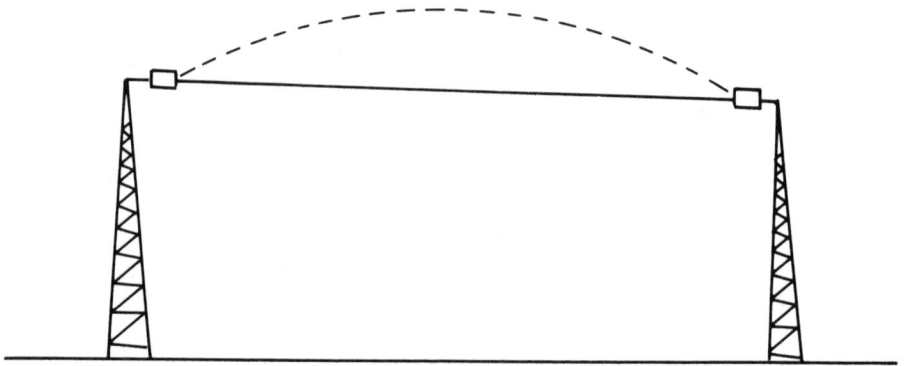

13-60 Diagram for Multiple-Choice question 4.

5. If a quarter-wave transmission line is shorted at one end:

    A. There is minimum current at the shorted end.
    B. The line behaves as a parallel-tuned circuit in relation to the generator.
    C. The line behaves as a series-tuned circuit in relation to the generator.
    D. There is a maximum voltage at the shorted end.
    E. There is a minimum voltage at the open end.

6. Which of the following antennas receives signals in the horizontal plane equally well from all directions?

    A. Horizontal Hertz antenna.
    B. Vertical loop antenna.
    C. Vertical Yagi antenna.
    D. A vertical antenna which is a quarter-wavelength long.
    E. Rhombic antenna.

7. If the length of a Hertz dipole is decreased:

  A. Its resonant frequency will be lowered.
  B. Its distributed inductance will be increased.
  C. Its resonant frequency will be increased.
  D. Its distributed capacitance between the antenna and ground will be increased.
  E. The antenna will present an inductive load, if the generator frequency is unchanged.

8. A 50-ohm transmission line is feeding an antenna which represents a 50 Ω resistive load. To shorten the line, the length must be:

  A. Any convenient value.
  B. An odd multiple of three-quarters of a wavelength.
  C. An odd multiple of half a wavelength.
  D. An even multiple of a quarter of a wavelength.
  E. An odd multiple of an eighth of a wavelength.

9. A final amplifier of a radio transmitter draws 250 mA of plate current when the plate supply voltage is 1400 volts. If the plate circuit efficiency is 80% and the transmitter is connected to an antenna having a feed-point impedance of 70 ohms, what is the antenna current at the feed-point?

  A. 4.0 A.
  B. 2.2 A.
  C. 1.25 A.
  D. 2.0 A.
  E. 5.0 A.

10. The outer conductor of the coaxial cable is usually grounded:

  A. At the beginning and at the end of the cable.
  B. Only at the beginning of the cable.
  C. Only at the end of the cable.
  D. At the middle of the cable.
  E. The outer conductor must never be grounded.

11. If the transmitter power remains constant, an increase in the frequency of the sky wave will:

  A. Lengthen the skip distance.
  B. Increase the range of the ground wave.
  C. Reduce the length of the skip distance.
  D. Have no effect on the ground wave range.
  E. Increase the absorption in the ionosphere.

12. A one-quarter wavelength shunt-fed vertical Marconi antenna:

  A. Has maximum radiation in a vertical direction.

B. Must have a horizontal receiving antenna for the best reception.
C. Must use a receiving antenna which has an electric field in a horizontal direction.
D. Must have a vertical receiving antenna for the best reception.
E. Represents a 70 Ω load in its resonant condition.

13. A shunt-fed quarter-wavelength Marconi antenna:

A. Has maximum rf impedance to ground at its feedpoint.
B. Has a current null at its feedpoint.
C. Has zero dc resistance to ground.
D. Has zero rf resistance to ground.
E. Uses a balanced twin line as its feeder cable.

14. A feature of an infinite transmission line is that:

A. Its input impedance at the generator is equal to the line's surge impedance.
B. Its phase velocity is greater than the velocity of light.
C. No rf current will be drawn from the generator.
D. The impedance varies at different positions on the line.
E. Its input impedance is equivalent to a short circuit.

15. The parasitic element of an antenna system will:

A. Decrease its directivity.
B. Increase its directivity.
C. Give the antenna unidirectional properties.
D. Make the antenna more omnidirectional.
E. Both B and C.

16. What does Fig. 13-61 show?

A. The current distribution on a half-wavelength Hertz antenna.
B. The voltage distribution on a half-wavelength Hertz antenna.
C. The voltage distribution on a quarter-wavelength Marconi antenna.
D. The current distribution on a quarter-wavelength Marconi antenna.
E. The impedance distribution on a quarter-wavelength Marconi antenna.

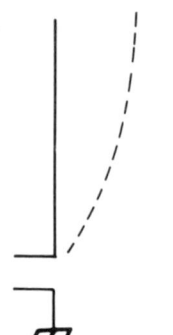

13-61 Diagram for Multiple-Choice question 16.

17. If the length of an antenna is changed from 2.5 meters to 2.8 meters, its resonant frequency will:

    A. Increase.

    B. Depend on the velocity factor so the resonant frequency can either be increased or decreased.

    C. Will be unchanged.

    D. Decrease.

    E. Lie in the hf band.

18. What is the effect of adding a capacitor in series with an antenna?

    A. The antenna's resonant frequency will increase.

    B. The antenna's resonant frequency will decrease.

    C. The antenna will be physically 5% longer than its electrical length.

    D. A capacitor is never added in series with an antenna.

    E. The purpose is to block dc from being applied to the antenna.

19. How does the electric field strength of a standard broadcast station vary with the distance from the antenna?

    A. The field strength varies inversely as the square of the distance from the antenna.

    B. The field strength is directly proportional to the distance from the antenna.

    C. The field strength remains constant regardless of the distance from the antenna.

    D. The field strength varies inversely as the distance from the antenna.

    E. The field strength is directly proportional to the square of the distance from the antenna.

20. Stacking elements in a transmitting antenna system:

    A. Increases the field strength at the receiving antenna.

    B. Increases the directivity of the transmitter antenna.

    C. Decreases the size of the major lobe in the radiation pattern.

    D. Produces an omnidirectional radiation pattern.

    E. Both A and B.

21. When the surge impedance of a line is matched to a load, the line will:

    A. Transfer maximum current to the load.

    B. Transfer maximum voltage to the load.

    C. Transfer maximum power to the load.

    D. Have a VSWR equal to zero.

    E. Carry standing waves.

22. A vertical loop antenna has a:

    A. Unidirectional radiation pattern in the horizontal plane.

    B. Unidirectional radiation pattern in the vertical plane.

C. Omnidirectional radiation pattern in the horizontal plane.
D. Omnidirectional radiation pattern in the vertical plane.
E. A bidirectional radiation pattern in the horizontal plane.

23. In order to get maximum radiation to all surrounding points in the horizontal plane, the antenna used is a (an):

   A. Vertical loop.
   B. Vertical quarter-wavelength rod.
   C. Array which includes parasitic elements.
   D. A horizontal Hertz dipole.
   E. System containing a number of vertical towers.

24. What is the electrical wavelength of a 500 MHz signal?

   A. 60 centimeters.
   B. 6 meters.
   C. 0.06 meter.
   D. 60 meters.
   E. 6 centimeters.

25. If the antenna current is doubled, the field strength at a particular position is:

   A. Doubled.
   B. Halved.
   C. Multiplied by a factor of four.
   D. Divided by a factor of four.
   E. Multiplied by a factor of 1.414.

26. The physical length of a Hertz dipole resonant at 100 MHz, is:

   A. 9.36 feet.
   B. 9.84 feet.
   C. 4.92 feet.
   D. 4.68 feet.
   E. 2.46 feet.

27. The rhombic antenna is primarily used for:

   A. Ground wave propagation.
   B. Space wave propagation.
   C. Ionospheric scatter propagation.
   D. Tropospheric scatter propagation.
   E. Sky wave propagation.

28. A lossless line is terminated by a resistive load which is not equal to the surge impedance. If the value of the reflection coefficient is 0.5, the VSWR is:

   A. 2.

B. 3.
C. 4.
D. 5.
E. 1.5.

29. Ground wave propagation:

A. Uses horizontal antenna systems.
B. Is limited to "line-of-sight" propagation.
C. Has a range which increases as the frequency is raised.
D. Has a range which increases as the power is raised.
E. Both C and D.

30. A lossless line has a surge impedance of 500 Ω and is terminated by a resistive load of 250 Ω. The VSWR is equal to:

A. 0.33.
B. 0.5.
C. 1.0.
D. 2.0.
E. 2.5.

Basic problems

1. On a lossless rf line the distributed primary constants are $L = 6.5\ \mu H/m$ and $C = 8.7\ pF/m$. What is the value of the line's surge impedance? If the line is matched and the line current is 600 mA (rms), what is the amount of power conveyed down the line?

2. A matched rf line (assumed to be lossless) is terminated by a 300 Ω resistive load. If the generator voltage is 150 V (rms), what is the input impedance to the generator, and what is the amount of power delivered to the load?

3. A matched rf line has a velocity factor of 0.85 and the generator feeding the line has a frequency of 100 MHz. What is the value of the wavelength on the line?

4. A lossless 100 Ω rf line with an air dielectric is terminated by an open circuit. If the frequency of the rf generator feeding the line is 150 MHz, how far from the open-circuited end is the first voltage null? If the line is 2.25 m in length, what is the input impedance at the generator?

5. A lossless 50 Ω rf line with an air dielectric is terminated by a short circuit. If the frequency of the rf generator is 75 MHz, what is the distance between two adjacent current nulls? If the line is 4.5 m long, what is the input impedance at the generator?

6. A lossless 600 Ω line is terminated by a resistive load of 200 Ω. What are the values of the VSWR on the line and the reflection coefficient at the load?

7. In Basic Problem 6 the rf power incident at the load is 50 W. What are the amounts of the reflected power and the power absorbed by the load?

8. What is the electrical length in feet of a Hertz antenna designed to resonate at 60 MHz?

9. What is the physical length in feet of a resonant Marconi antenna operating on 9 MHz?

10. A λ/4 transformer is used to match a 70 Ω dipole load to a balanced 100 Ω line. What is the value of the transformer's surge impedance?

Advanced problems

1. Each conductor of a twin-lead transmission line has a radius of 1.6 mm and the spacing between the centers of the conductors is 0.85 cm. If an air dielectric is used, what is the value of the line's surge impedance? If the line (assumed to be lossless) is matched and the generator voltage is 125 V (rms), what are the values drawn from the generator and the power delivered to the load?

2. The outer conductor of a coaxial cable has an inner diameter of 1.8 cm while the inner conductor has an outer diameter of 2.4 mm. What is the value of the cable's surge impedance if its dielectric has a relative permittivity of 2.2? If the line (assumed to be lossless) is matched and 50 W of rf power is delivered to the load, what are the effective values of the line voltage and the line current?

3. The primary constants of an rf transmission line are $L = 1.65 \ \mu H/m$, $C = 7.75 \ pF/m$, $R = 55 \ m\Omega/m$, and $G = 12 \ nS/m$. Calculate the values of the surge impedance and the attenuation constant.

4. In Advanced Problem 3 the frequency of the generator feeding the line is 300 MHz. Find the value of the phase shift constant, the wavelength on the line, and the phase velocity.

5. A transmission line with an air dielectric and a surge impedance of 100 Ω is terminated by a load consisting of 80 Ω resistance in series with 130 Ω capacitive reactance. Calculate the phasor value of the reflection coefficient. What is the magnitude of the VSWR?

If a 100 MHz generator delivers 90 W to the line which is assumed to be lossless, what are the amounts of the reflected power and the power absorbed by the load? What is the distance between two adjacent voltage nulls on the line?

6. At a position which is 2.5 miles from a transmitting antenna, the field strength is 550 μV/m. If the antenna current is halved, what are the field strengths at the same position and at a position 1.5 miles from the transmitting antenna?

7. At a position 4 miles from the antenna of a 5 kW transmitter, the field strength is 200 mV/m. If the power is now reduced to 3.0 kW, what is the new distance of the 500 mV/m contour from the antenna?

8. 15 kW of rf power are delivered to an antenna whose field gain is 3.5. What is the value of the effective radiated power?

9. The rf power delivered to an antenna system is 8.5 kW. If the antenna power gain is 7.5 dB, calculate the value of the effective radiated power.

10. A directional antenna system consists of four towers whose common point resistance is 75 Ω. If the associated common point current is 8.5 A, calculate the value of the operating power.

    Tower one is the reference tower and its antenna current is 2.5 A. The antenna currents of towers two, three, and four are respectively 3.0 A, 4.5 A, and 1.9 A. What is the antenna base-current ratio for tower three?

14

Microwave techniques

THE PRIMARY PURPOSE OF THIS CHAPTER IS TO PROVIDE YOU WITH AN INTRODUCTION to various aspects of microwave techniques. The frequency range of the microwave region is open to debate. Some textbooks state that the range commences where the uhf-band TV band finishes, at 890 MHz, while others contend that the start of the microwave band coincides with the beginning (300 MHz) of the uhf band. The most popular view is that microwave extends from 1 GHz (10^9 Hz) to 1 THz (10^{12} Hz); this corresponds to a range of wavelengths from 30 cm to 0.3 mm. However the majority of microwave equipment operates between 1 and 100 GHz, which includes the *super high frequency* (SHF) band of 3 to 30 GHz. As already discussed, all microwave frequencies will be propagated by the space or direct wave, which travels in a practically straight line from the transmitter.

Within the microwave region there are a number of designated bands; of these the most common are shown in Table 14-1.

In Table 14-1 the top limit of the microwave frequencies is the beginning of the infrared or heat region, which ranges from 1 to 375 THz. Next comes the narrow visible light spectrum which extends from red, 375 THz (wavelength: 0.8 micron = 8×10^{-7}m), to violet, 750 THz (wavelength: 0.4 micron = 4×10^{-7}m). Rounding off the electromagnetic (EM) waves there is ultraviolet (UV), X-rays, gamma (γ) rays, and cosmic rays.

Before covering microwave techniques in depth, here are some of the features which distinguish microwave from the lower rf bands. These are:

1. The tuned circuits. Knowing that the resonant frequency is,

$$f_r = \frac{1}{2\pi\sqrt{LC}}$$

you will find that the values of L and C required for a resonant frequency of 10 GHz (X-band) are respectively much less than 1 μH and 1 pF. With these small values it is impossible to use "lumped" circuits, since the leads connecting the inductor and the capacitor would probably have much more inductance and capac-

**Table 14-1. Band
Frequency Ranges.**

Band	Frequency Range (GHz)
L	1.2 - 2.7
S	2.6 - 3.95
C	3.95 - 5.85
X	8.2 - 12.4
K	18.0 - 26.5

itance than the total values required. You therefore use distributed circuits which are in the form of resonant cavities.

2. The insulators. All insulators suffer to some degree from dielectric hysteresis loss which increases as the frequency is raised. In the microwave region this type of loss is too severe to permit you to use such insulators as porcelain, lucite, bakelite, waxed paper, etc. You are therefore required to introduce special insulators, examples of which are polyethylene, polystyrene, and teflon.

3. The transmission lines. At microwave frequencies it is impossible to use a conventional coaxial cable, except for very short lengths. The principal reason is the severe dielectric loss which results in a high degree of attenuation. The main type of practical microwave line is the waveguide, which is basically a hollow metal pipe with a rectangular or circular cross-section.

4. The active devices. In the discussion on the lighthouse triode in chapter 3 you learned that the conventional tubes were limited to below 2 GHz as a result of the transit time effect. At higher frequencies you must employ active devices which actually use transit time to achieve amplification or oscillation. Such tubes include magnetrons, klystrons, *traveling wave tubes* (TWTs), and *backward wave oscillators* (BWOs). There are also various solid-state devices which can achieve amplification, frequency multiplication, and oscillation.

The features of microwave techniques will be covered in the following topics:

- Resonant cavities
- A comparison between transmission lines and waveguides
- Field pattern development in a rectangular waveguide
- Propagation in a rectangular waveguide
- The circular waveguide
- Joints, bends, twists, irises, posts, and screws
- Waveguide impedance
- Matching stubs
- Waveguide and hybrid couplers
- The ferrite isolator

- Microwave measurements
- Microwave tubes
- The Gunn solid-state diode oscillator

Resonant cavities

In chapter 13 the equivalence between the quarter-wave line and the parallel LC tank circuit was discussed (Fig. 14-1A). At the X-band frequency of 10 GHz, the wavelength is 3 cm, and therefore the quarter-wave line would appear to be 0.75 cm long before being terminated by a short circuit; in practice the phase velocity on the line is lower than the free space velocity so that the actual length is somewhat less. Irrespective of the length considerations, the skin effect loss on such a line is severe and its Q is correspondingly low. However, if a second identical quarter-wave line were paralleled with the first at the open end (Fig. 14-1B), the total inductance is theoretically halved, while the total capacitance, C_T, is doubled. This means that the resonant frequency of the parallel combination

$$f_r = \frac{1}{2\pi\sqrt{L_T C_T}}$$

is unchanged, while

$$Q = R \times \sqrt{\frac{C_T}{L_T}}$$

is increased.

It follows that more and more parallel quarter-wave lines can be added until you finally create a solid cylindrical "can" (Fig. 14-1C). Such a structure has no radiation loss and its rf resistance is extremely low; the result is a high Q value of several thousand. The larger the diameter of the can, the lower is its resonant frequency.

If the cylindrical cavity is excited, the simplest pattern of the magnetic flux is a series of concentric circles with zero intensity at the center and the highest intensities at the curved wall and the ends of the structure. By contrast the lines of the electric flux are at all positions transverse and parallel to the Z-axis of the cylinder, with the maximum intensity at the center and zero at the curved wall. Such a distribution of flux lines is shown in Fig. 14-2A and is referred to as the TM_{010} mode. The designation "TM" means that all the magnetic flux lines are entirely transverse (at right angles) to the direction of the cylinder's Z-axis; likewise, if all the electric flux lines were transverse, the designation would be "TE."

The first subscript (zero) in TE_{010} denotes the number of full-wave changes for the electric field around the circumference. The second subscript (one) is the number of half-wave changes for the electric field across the diameter, while the third subscript (zero) is the number of half-wave changes along the Z-axis. There are four other possible field distributions (modes) in Fig. 14-2, and you should verify their designations.

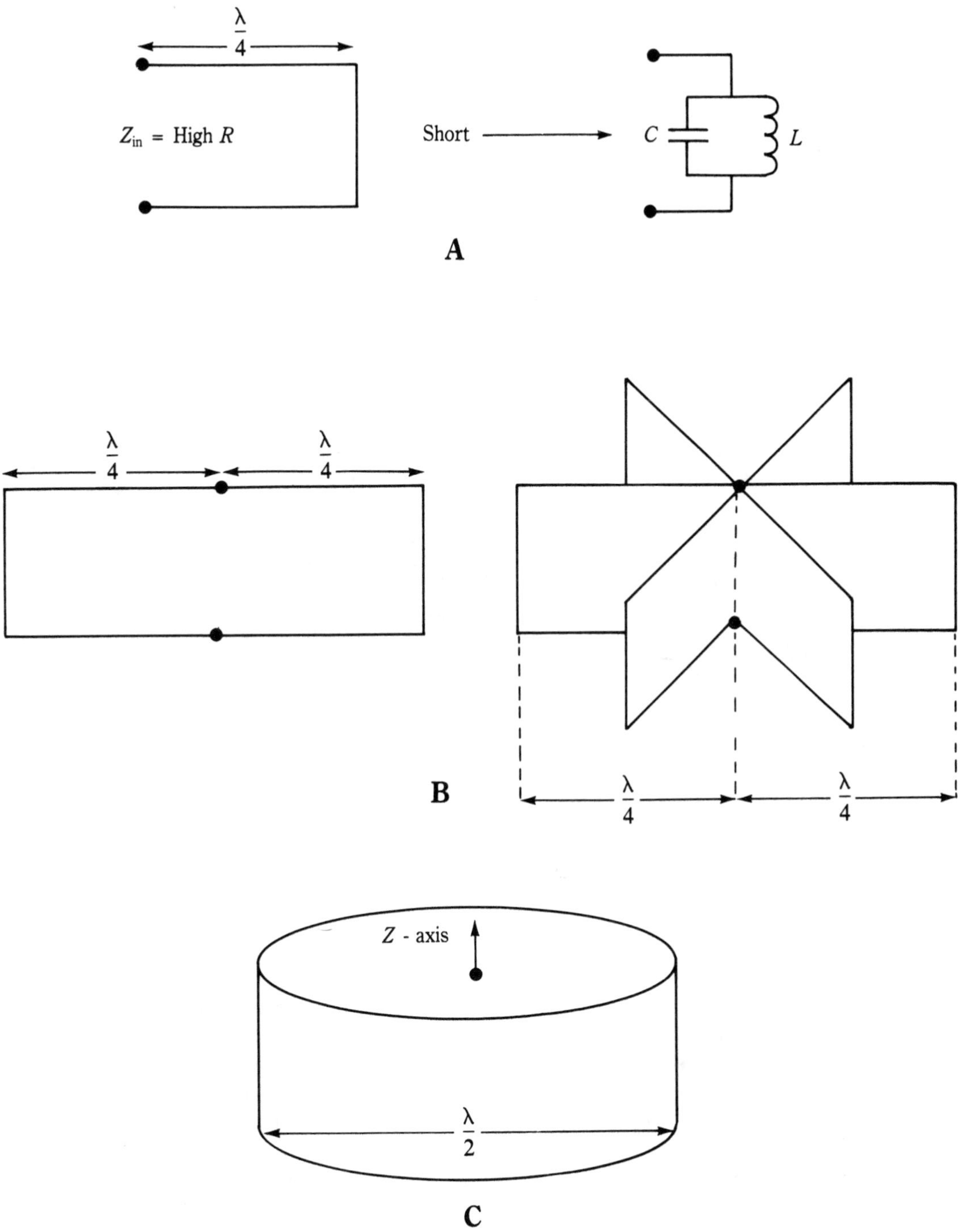

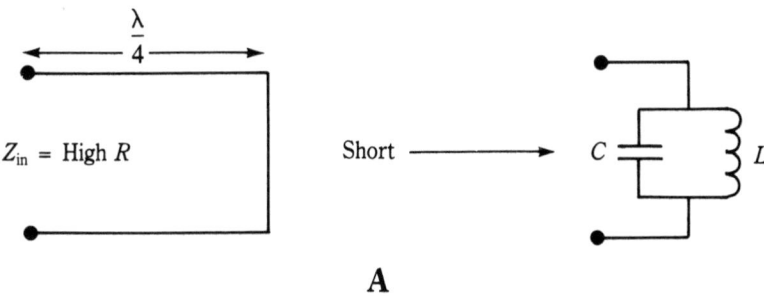

A

B

C

14-1 The cylindrical "can" as a resonant cavity.

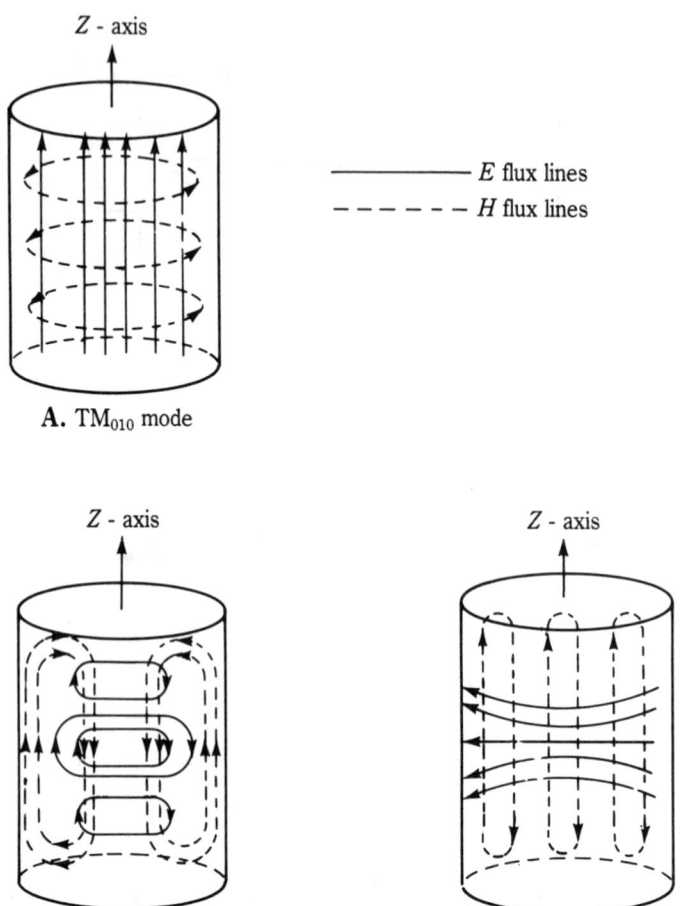

Z - axis

— — — — — *E* flux lines

– – – – – – *H* flux lines

A. TM$_{010}$ mode

Z - axis

Z - axis

B. TE$_{011}$ mode

C. TE$_{111}$ mode

14-2 Examples of cylindrical cavity modes.

The resonant "can" or cavity can be mechanically tuned to resonance by a number of methods:

1. The top of the can can be made in the form of an adjustable plunger, which can then alter the operating dimensions of the cavity (Fig. 14-3A) and therefore change its resonant frequency; for example, if the distance, D, is reduced, the cavity is smaller and its resonant frequency is higher. This is the principle behind a common type of absorption microwave frequency meter that is directly attached to the waveguide. A radiating slot or hole is then cut so that, providing the cavity is at or near resonance, a small amount of rf energy can be absorbed from the waveguide into the cavity. This absorption can be monitored by coupling a detector to the cavity.

   One method of coupling is to insert a loop which links with the cavity's magnetic field; the pickup from the loop can then be detected, amplified, and fed to a

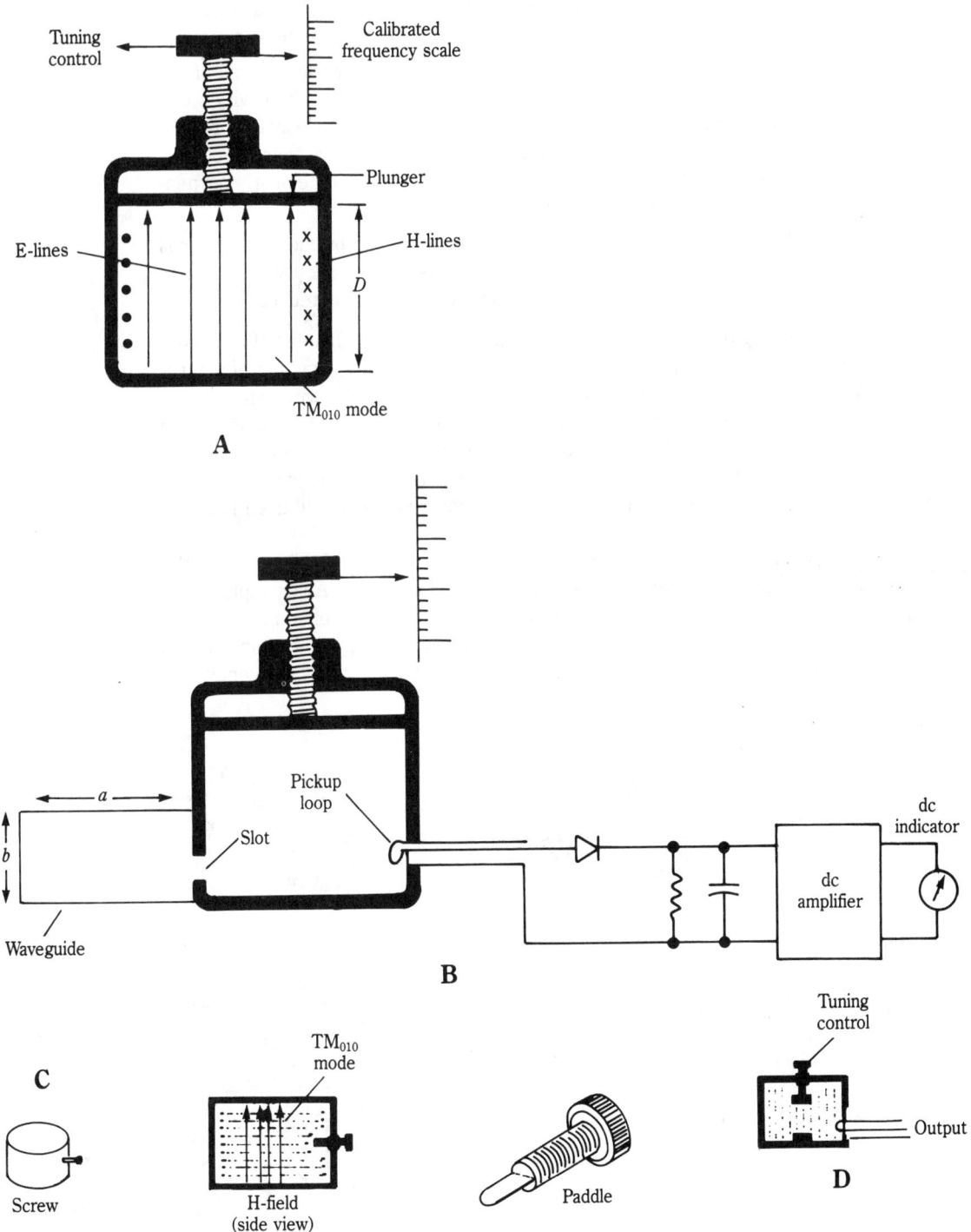

14-3 Methods of tuning the cylindrical cavity.

dc indicator (Fig. 14-3B). The plunger is adjusted for a maximum reading in the dc indicator and the frequency can then be read directly off a calibrated scale.

2. To tune some klystrons a metallic screw is inserted into the curved wall of the cavity (Fig. 14-3C). Because the screw is not ferro-magnetic, the cavity's distributed inductance is lowered and the resonant frequency of the cavity is raised. Alternatively, a paddle can be used in place of the screw. The resonant frequency can be increased by turning the paddle to a position which is more nearly perpendicular to the direction of the H flux lines.

3. The capacitance at the center of the cavity can be varied by means of a small plate whose position is controlled externally (Fig. 14-3D). If the plate is screwed in, the capacitance is increased and the resonant frequency is reduced.

The cylinder is only one of a variety of possible cavity shapes. Instead of a shorted quarter-wave twin lead, you can start with a curved piece of wire (fraction of a turn) which will mainly represent the required inductance. The ends of the wire can then run geometrically in parallel to provide most of the capacitance (Fig. 14-4A). This shape can then be revolved to form a surface of revolution; the result is the *rhumbatron* or *reentrant cavity* whose electric and magnetic fields are indicated in Fig. 14-4B. Figure 14-4C, and D, illustrates other versions of the reentrant cavity which is commonly used in reflex klystrons.

To tune a reentrant cavity, one or more of the surfaces is made flexible so that the cavity can be mechanically stressed to alter its operating dimensions. An example of this method is shown in Fig. 14-4E, and is the principle behind the tuning of some klystrons. As the tuner screw is rotated the distance between the cavity's top and bottom walls is changed; this alters the distributed capacitance and varies the resonant frequency. For

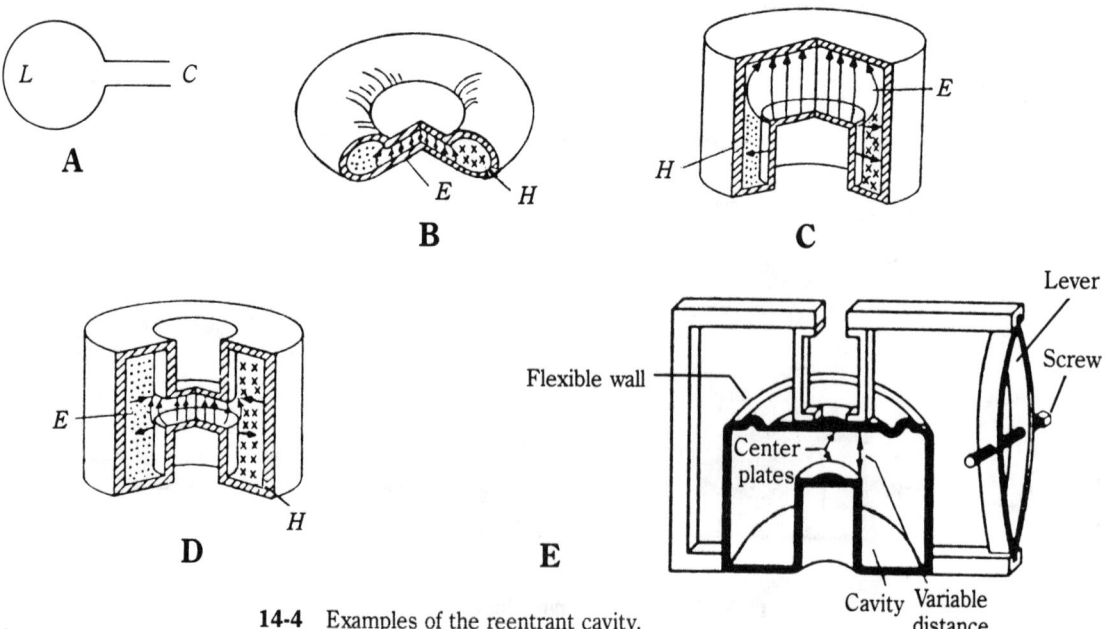

14-4   Examples of the reentrant cavity.

example, if the distance, D, is increased, the distributed capacitance is decreased and the resonant frequency is raised. By this means the frequency of an X-band cavity can be changed by up to 4 GHz. The energy from the cavity can then be passed to a waveguide by means of a loop or a radiating slot.

If the shape of Fig. 14-4A is extended into the paper, the result is a *hole and slot* cavity (Fig. 14-5) which is cut from a block of solid copper, and is commonly used in magnetrons.

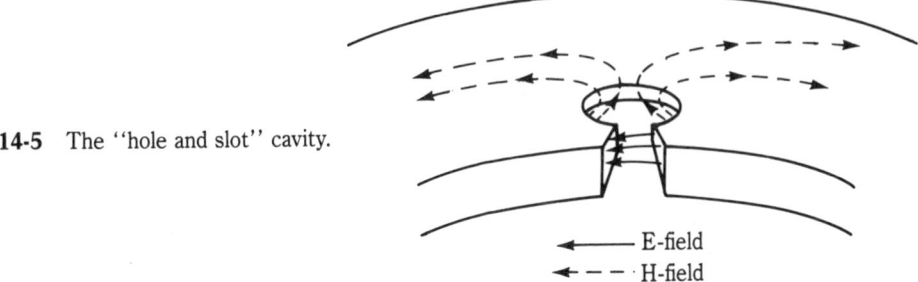

14-5 The "hole and slot" cavity.

Another possible cavity shape is the rectangular box whose cross-section consists of a narrow dimension "b" and a wide dimension "a" which is longer than half a wavelength. This cross-section is then extended a distance of half a wavelength along the Z-axis to form a closed box (Fig. 14-6A). The resulting distribution of the electric and magnetic fields are shown in Fig. 14-6B and C. The mode shown is designated as TE_{101} where the first suffix indicates the number of half-wave variations along the wide dimension a, the second suffix denotes the number of half-wave variations across the narrow dimension, b, and the third suffix is the number of half-wave variations in the direction of the Z-axis.

In the various applications of cavity resonators it is both necessary to excite a cavity by introducing energy and also to have ways of removing the energy. The most common methods of coupling are the E-probe, the H-loop, and the slot or aperture.

Figure 14-7A shows the E-probe, which represents capacitive coupling and is introduced into the cavity at the position where the electric field is the strongest; this probe can then set up E-lines and maintain a continuous oscillation. Likewise the H-loop (Fig. 14-7B) is a form of inductive coupling and is positioned where the magnetic field is the greatest. The third method is to introduce or remove energy by means of an aperture or slot which is common to the waveguide and the resonator (Fig. 14-7C).

In a reentrant cavity (Fig. 14-7D) energy can be introduced by a stream of electrons which pass through holes in the center of perforated plates or flat mesh grids. The energy can then be removed by an H-loop. This method of exciting a cavity into oscillation is used in the klystron tube.

A comparison between transmission lines and waveguides

In chapter 13 the theory of the twin lead transmission line which is operated successfully at frequencies well below the microwave region was covered. Such a line suffers from the

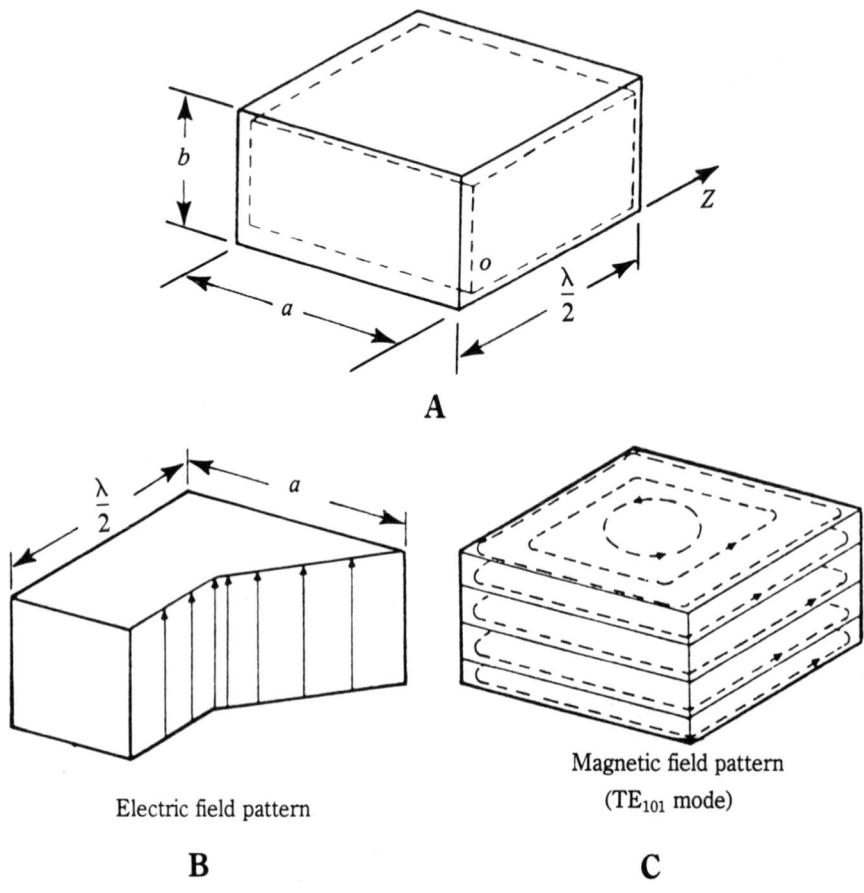

A

Electric field pattern

B

Magnetic field pattern
(TE$_{101}$ mode)

C

14-6 The rectangular box cavity.

following losses:

1. The copper loss. This loss is associated with the resistance of the conductors which will dissipate energy in the form of heat. At microwave frequencies the copper loss is primarily the result of the "skin effect" which restricts the flow of the current to a very small layer at the conductor's surface. The greater is the surface area of the conductor, the less are its skin effect and the consequent copper loss. A twin lead line uses a pair of conductors with a small surface area so that the copper loss is severe. By contrast, the waveguide is a hollow metal pipe with a cross-section which is rectangular, circular, or elliptical. In any case the surface area involved is much larger so that the copper loss is limited to a low value. In addition the loss can be further reduced by plating the inside surfaces of the guide with gold or silver.

2. The dielectric loss. This loss is related to the type of insulator between the conductors. All insulating materials suffer to some degree from dielectric hysteresis loss which causes the insulator to dissipate energy in the form of heat. This

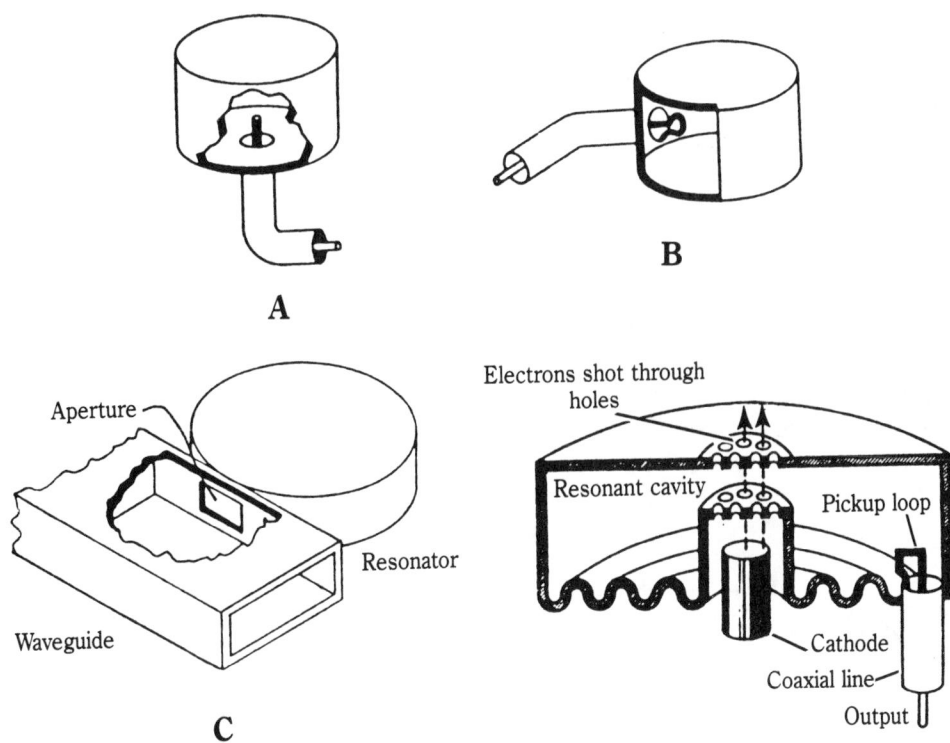

14-7 Methods of coupling to and from a resonant cavity.

effect increases with the frequency, and in the microwave region the hysteresis loss for most insulators is extremely severe. However, a waveguide uses a dielectric of dry air or an inert gas such as nitrogen; for these dielectrics the hysteresis loss is relatively low.

3. The radiation loss. The twin lead line does not confine the electric and magnetic fields in a direction perpendicular to the plane which contains the conductors (Fig. 14-4). As previously discussed in chapter 13, the electric and magnetic fields are respectively associated with the voltage and the current on the line. When the separation between the conductors becomes an appreciable fraction of a wavelength, energy will escape from the line by radiation. The higher the frequency, the shorter the wavelength and the greater the radiation loss. By contrast a waveguide represents a closed metallic surface; the electric and magnetic fields are therefore confined and the radiation loss is negligible.

The coaxial cable is an improvement over the two-wire line. Since the outer conductor is extended around the inner conductor, the electric and magnetic fields are contained, so there is virtually no radiation loss. The copper loss is reduced because of the outer conductor's larger surface area. However, there remains the dielectric loss which cannot be eliminated, because even if an air dielectric were used, the inner conductor would have to be supported at intervals by insulating spacers. Dielectric hysteresis is the ultimate reason why you must finally turn to waveguides and abandon coaxial cables. At 3 GHz the typical attenuation for a coaxial cable is 0.6 dB per meter, while with a waveguide it is only 0.02 dB per meter. To look at this situation in practical terms, a 5-meter length of coaxial cable would have an attenuation of $5 \times 0.6 = 3$ dB which is equivalent to a half-power loss. In other words, if 100 watts of rf power were introduced into a 5-meter length of coaxial cable, only 50 watts would emerge. By contrast the waveguide would have to be $3/0.02 = 150$ meters long to produce a half-power loss. At 10 GHz the figures are even more dramatic; 3 dB per meter for the coaxial cable but only 0.03 dB per meter for the waveguide.

There is a further advantage for the waveguide in terms of power capability. With a coaxial cable the power capability is primarily determined by the spacing between the conductors. If the power being conveyed is excessive, large voltages are established between the conductors and arcing might occur across the dielectric. A circular waveguide is regarded as a coaxial cable with the inner conductor removed. Consequently the spacing and the power capability of a circular waveguide are greater than those of a coaxial cable with the same external dimensions (Fig. 14-8).

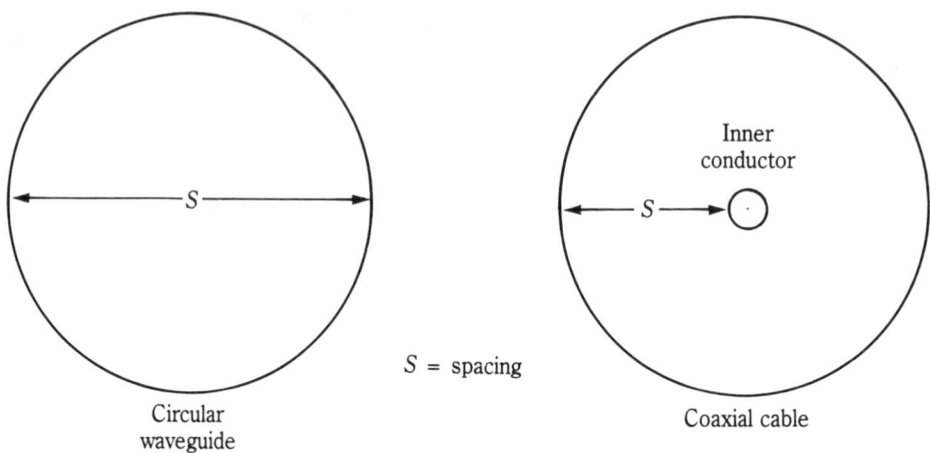

14-8 Comparison between the circular waveguide and the coaxial cable.

By now you must be thinking "waveguides have all the advantages with regards to attenuation and power capability, so why ever bother with coaxial cables?" The answer, in one word, is *size*. It will be shown in this section that in order for an electromagnetic wave to propagate successfully down a rectangular guide (Fig. 14-9), the wide dimension, *a*, must exceed one-half-wavelength corresponding to the frequency of the wave. Consequently at a frequency of 100 MHz, the wavelength is 3 meters, and therefore *a* must exceed 1.5 meters or roughly 5 feet! Such a waveguide would be extremely cumbersome and expensive.

If the wide dimension must be greater than half a wavelength long, it follows that for a particular waveguide there must be a *lower* frequency limit which is known as the cutoff value, f_c; at this frequency the wide dimension is exactly one-half wavelength long.

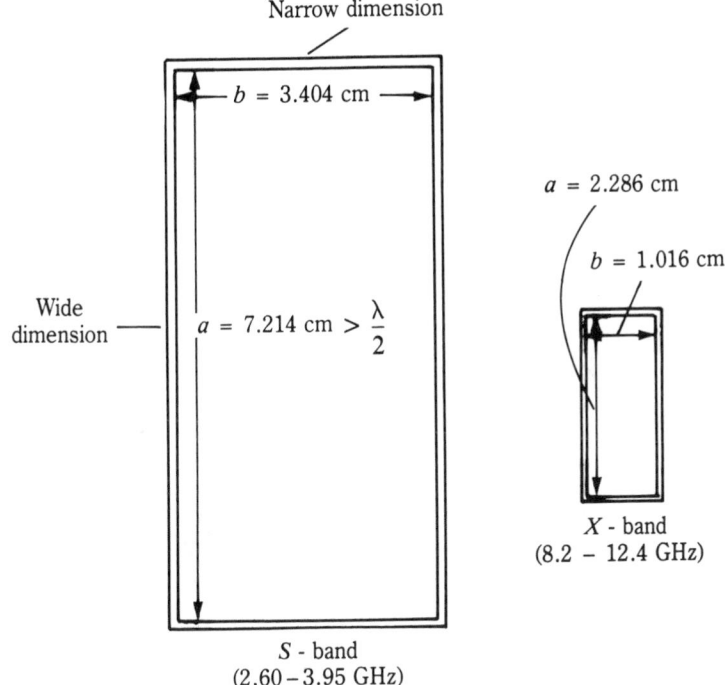

Narrow dimension

$b = 3.404$ cm

$a = 2.286$ cm

$b = 1.016$ cm

Wide dimension

$a = 7.214$ cm $> \dfrac{\lambda}{2}$

X - band
(8.2 – 12.4 GHz)

S - band
(2.60 – 3.95 GHz)

14-9 Comparison between the sizes of rectangular waveguides.

For example an S-band rectangular guide which operates over 2.6 to 3.95 GHz, has an inner wide dimension of 7.214 cm, which corresponds to a cut-off frequency of $300/(2 \times 7.214 \times 10^{-2} \times 10^3) = 2.078$ GHz. In the X-band of 8.2 to 12.4 GHz the waveguide's inner wide dimension is 2.286 cm and the cut-off frequency is $150/22.86 = 6.557$ GHz. For reasons which you will later see, the operating frequency range of a rectangular guide is limited between $1.25 f_c$ and $1.9 f_c$. By contrast a coaxial cable can be used from dc up to and including the lower part of the microwave region.

Apart from considerations of size and frequency range, a waveguide system is expensive and mechanically rigid. The ideal situation would be one continuous section of waveguide between the transmitting and receiving points. However this is normally not practical, and it is necessary to join various sections together to form a complete line. At the joins care must be taken to prevent discontinuities inside the waveguide, and to avoid any leakage of the rf energy.

Development of a rectangular waveguide from a two-wire line

A two wire line with an air dielectric must be mechanically supported at intervals by some form of insulator (Fig. 14-10A). However, at high radio frequencies insulators such as porcelain or plastic have a large dielectric loss, so they act as a low impedance across the line and represent a discontinuity. A superior high-frequency insulator is a quarter-wavelength of rf line which is shorted at one end (Fig. 14-10B). You discovered that the input impedance to such a line was theoretically infinite, although in practice the input

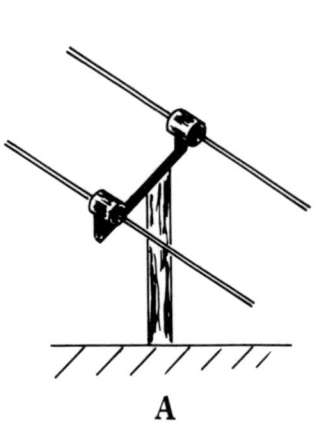

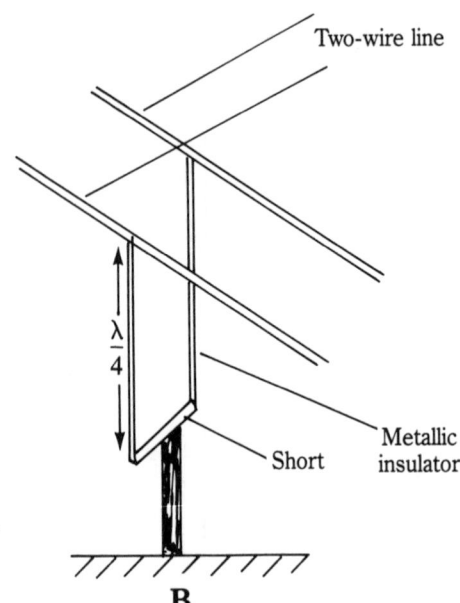

14-10   The principle of the metallic insulator.

impedance is a very high value of resistance which will have little effect on the two-wire line. This type of metallic insulator is sometime referred to as a quarter-wave (λ/4) support *stub*.

The λ/4 support stubs can be joined both above and below the twin line. To increase the rigidity of the line, you can add more and more stubs until each makes contact with its neighbor. The result is a hollow rectangular box (Fig. 14-11) with the line itself forming part of the walls. It is clear that the wide dimension, a, must be longer than a half-wavelength, because if *a* is less than λ/2, the stubs would behave inductively and appear as a severe discontinuity across the line. At the cutoff condition, *a* equals λ_c/2 or $\lambda_c = 2a$; the cutoff frequency is then given by:

$$\text{Cut-off frequency, } f_c = \frac{c}{\lambda_c}$$

$$= \frac{c}{2a} \tag{14-1}$$

where,

f_c =   cutoff frequency (Hz)
λ_c =   cutoff wavelength (m)
a =   the waveguide's wide dimension (m)
c =   300 $\times$ 10^6 which is the velocity of electromagnetic waves in free space (m/s)

For most waveguides the value of "a" is normally about 0.8λ.

The narrow b dimension primarily governs the power handling capacity of the guide. If the value of b is too low, the voltage established across the narrow dimension might exceed the breakdown potential of the air dielectric so that arcing will occur. You will also

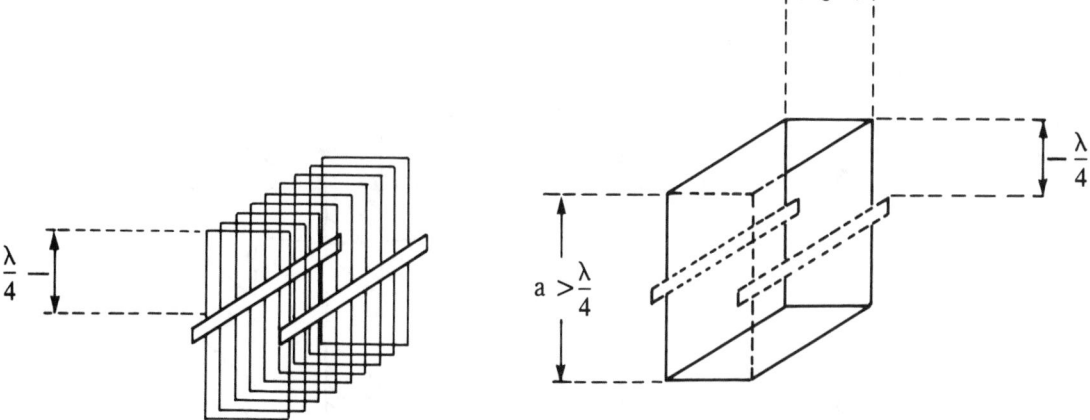

14-11   The formation of a rectangular waveguide by the addition of quarter-wave support stubs.

see that the value of *b*, typically 0.4λ (or $a/2$), determines the upper frequency at which the waveguide can operate.

Field pattern development in a rectangular waveguide

The E-field

Energy is transferred down a rectangular waveguide as an electromagnetic wave which consists of alternating electric (*E*) and magnetic (*H*) fields. Associated with the electric field is a voltage distribution across the guide's narrow dimension, while "wall" currents flow along the inner surfaces to create the magnetic field. To further your understanding of the propagation process you must first derive the patterns of the E and H flux lines inside the waveguide.

Start by considering a lossless two-wire line which is one wavelength long and is terminated by a resistive load, R_L, equal in value to the line's surge impedance, Z_o. As you learned in chapter 13, this line is matched so that the voltage and the current waves move down the line in phase and their energy is totally absorbed by the load. Figure 14-12 shows such a line to which a number of $\lambda/4$ support stubs have been added. It must be emphasized that the voltage wave shown is an *instantaneous* distribution along the line. This wave would in fact travel down the line with its phase velocity which is only slightly less than the speed of light.

At position *A* the voltage has reached its peak value so that there is a maximum electric field between the wires; this is indicated by the high density of the flux lines. However, no voltage can ever exist at the shorted ends of the support stubs, so as you move from the two-wire line to the ends of the support stubs at position *A*, the E-field must increase from its maximum value to zero.

Turn now to position *B*, the voltage is instantaneously zero so that there are no E-lines between the conductors and none exist on the stubs. Position *C* is separated in

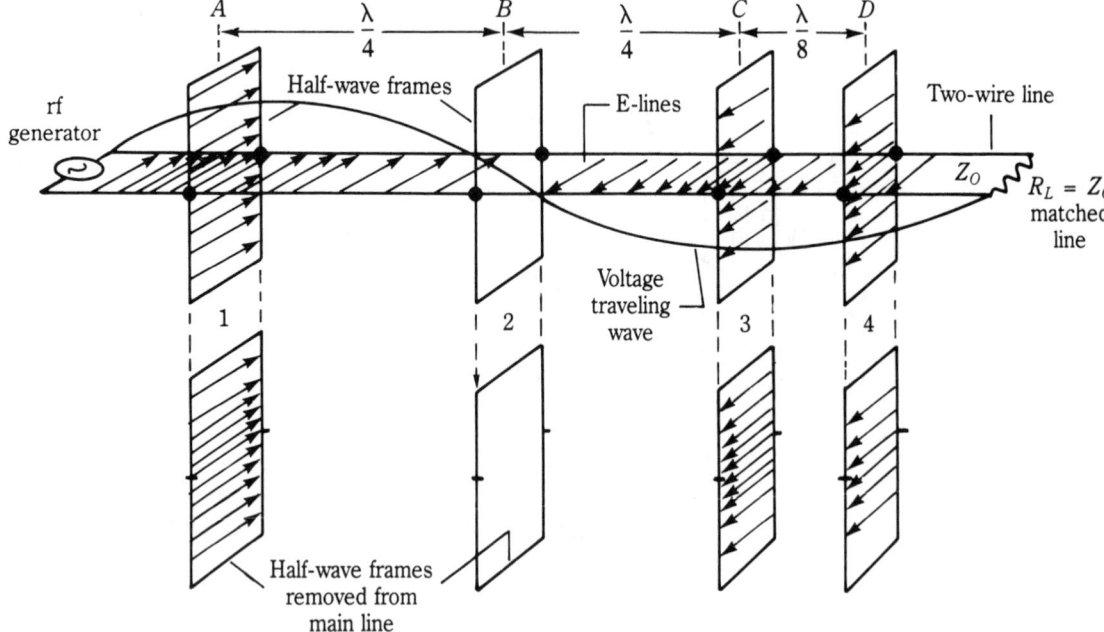

14-12 Instantaneous distribution of E-lines on a matched line with support stubs.

distance by one half-wavelength from position A and by one quarter-wavelength from position B. At C the voltage is again at its instantaneous peak value but with opposite polarity when compared with the voltage at position A. Consequently the distributions of the E-lines are the same at A and C but the directions of the two sets of lines are reversed. Position D is intermediate between position C and the load, so the number of flux lines is correspondingly less.

You can now add further support stubs to create a hollow rectangular waveguide which is one-wavelength long. The pattern of the instantaneous E-lines inside the waveguide is illustrated by the various views of Fig. 14-13.

The H-field

To derive the pattern of the H-field inside a rectangular waveguide you can start by considering a matched two-wire line (Fig. 14-14) which is one-wavelength long. The current wave is in phase with the voltage wave of Fig. 14-12 and support stubs have been added between two positions, A and E, which are a half wavelength apart. At these positions the current (electron flow) has instantaneously reached its peak value on the line. The directions of these currents and of the currents on the support stubs are indicated by arrows. Each of these currents then contributes to the pattern of the H-field inside the waveguide.

Around each conductor is a small individual loop whose arrow indicates the direction of the magnetic field surrounding that conductor (left-hand rule). When two adjacent loops have arrows in opposite directions their magnetic fields will tend to cancel, but when the arrows are in the same direction, the magnetic fields combine. Inside the

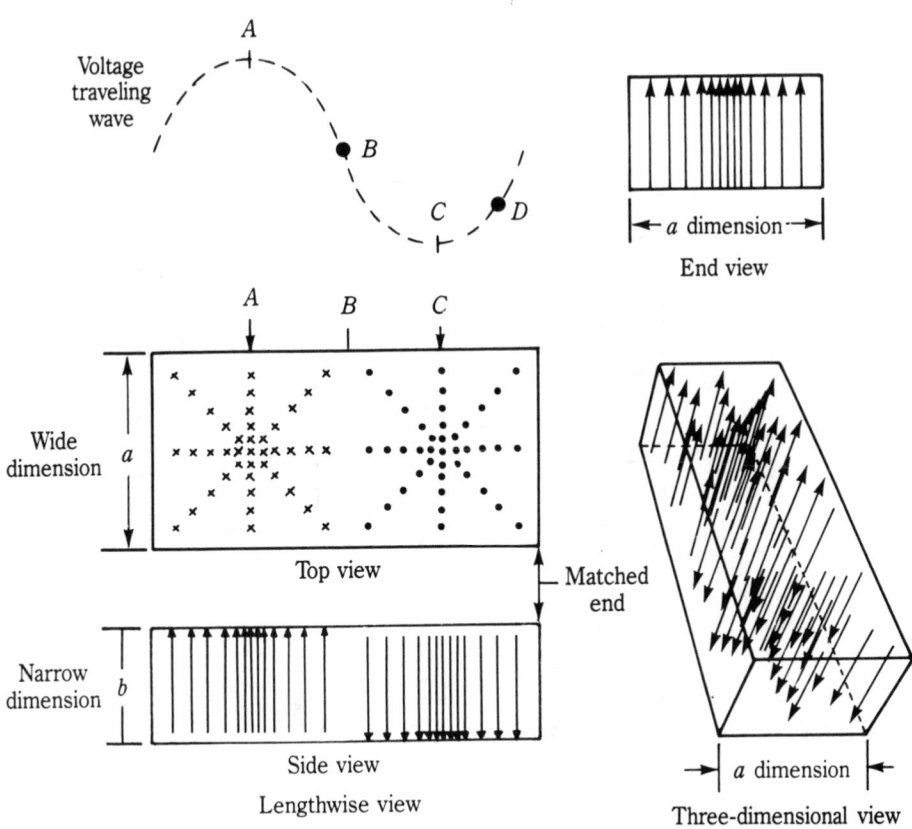

14-13 Instantaneous distribution of E-lines in a one-wavelength section of a rectangular wave-guide.

waveguide the resultant magnetic field is in the form of an H-loop which extends over a distance of half a wavelength; the various views of this H-loop are illustrated in Fig. 14-15. Outside the waveguide the individual loops cannot join to form a continuous flux path, so there is no external magnetic field.

When you combine the E- and H-fields over a distance of a wavelength, you obtain the patterns of Fig. 14-16. The E-field is entirely transverse to the direction of propagation, and its peak coincides in position with the peak of the transverse component of the H-field. However, the H-field also has a longitudinal component in the same direction as the energy is being propagated.

The field patterns which you have established obey two important boundary conditions which were originally stated by Clerk Maxwell:

1. In order to exist, all *electric* flux lines must be *perpendicular* to the walls of the guide. This also means that no electric flux line whose direction is parallel to a wall, can be positioned at the surface of that wall. Were such a line to exist, the result would be a voltage stress at the wall; a surface current would then flow and the line would be eliminated.

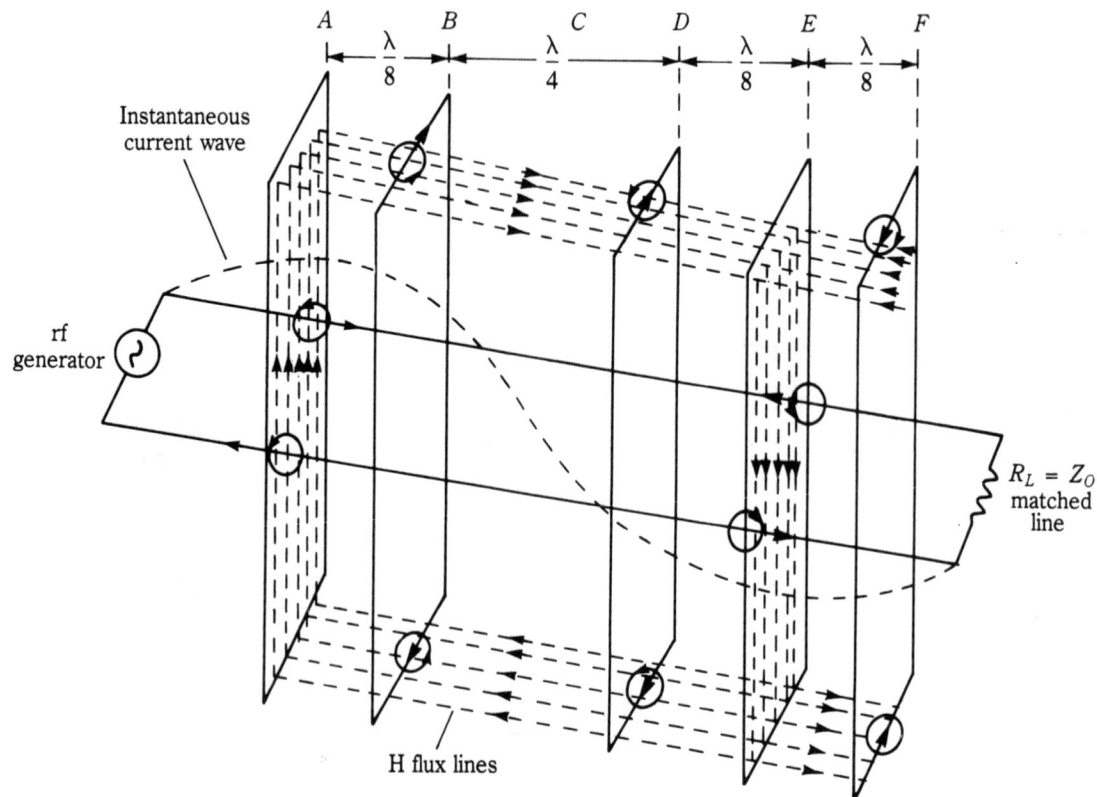

14-14 Instantaneous distribution of H-lines on a matched line with support stubs.

2. In order to exist, all *magnetic* flux lines at a wall must run *parallel* to the surface. Consequently no magnetic flux line can have a component which is perpendicular to the surface. Were such a component to exist, it would cut the wall, eddy currents would be induced, and the component would be removed.

You now know the field patterns which must exist inside a waveguide. The question is, "How does the EM wave move down the waveguide in order to create the required field patterns and obey Maxwell's boundary conditions?" The answer lies in the discussion of the next section.

Propagation in a rectangular waveguide

In chapter 13 the EM wave, which is radiated out into free space, was discussed. Such a wave consists of alternating electric and magnetic fields which are in time phase, although the two sets of flux lines are 90° apart in space. Both fields are entirely transverse in the sense that their directions are always at right-angles to the direction of propagation. This is referred to as a *transverse electric, transverse magnetic* (TEM) wave which only exists on a transmission line or in free space.

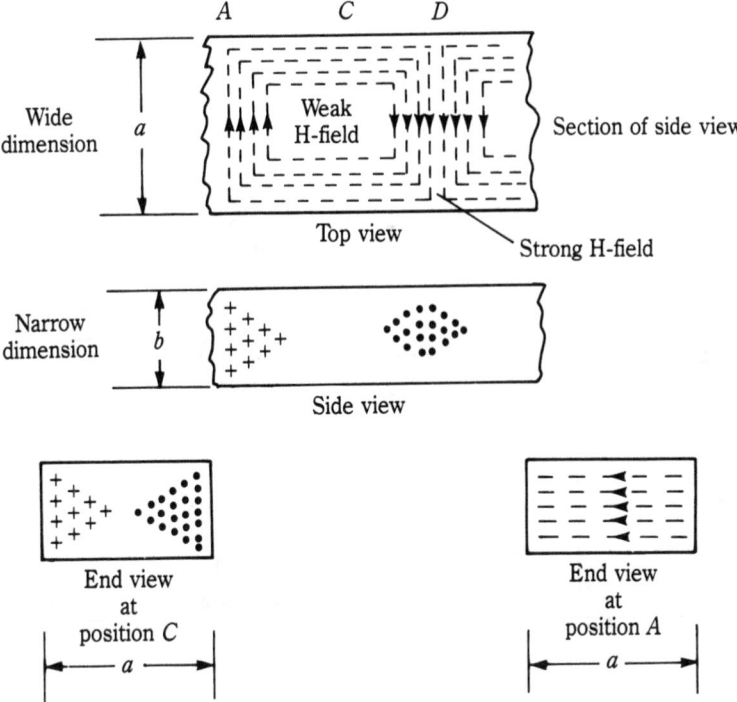

14-15   Instantaneous views of the magnetic flux lines in a rectangular waveguide.

The next question: "Does the energy move straight down the rectangular wave-guide as a TEM wave?" The answer is "no." If you enclose a rectangular waveguide around a TEM wave, you cannot obey Maxwell's boundary conditions. You also know that the H-field pattern contains a longitudinal component and no such component exists in the TEM wave. To cut down on the suspense, the answer is "The electromagnetic energy progresses down the guide by a series of reflections off the internal surface of the narrow dimension (Fig. 14-17)". At each reflection the angles (θ) of incidence and reflection are equal.

Consider a TEM wave which approaches a plane conductor (flat metal surface) at an angle (Fig. 14-18). The dark lines with their arrows represent the directions of the incident and reflected wavefronts, which are moving with the free space velocity (virtually at the speed of light). The full lines are $\lambda/2$ apart in free space and represent the incident H flux lines. The broken lines are also $\lambda/2$ apart and are used to show the reflected H-field. The symbols for the incident and reflected E-fields are respectively $\otimes$, $\odot$, $\times$, and $\bullet$. These incident and reflected waves are of the same amplitude but have a phase reversal of 180°.

The fields of the incident and reflected waves are superimposed, and must be combined to produce the resultant E- and H-field patterns. At the metal surface itself, the phase reversal causes cancellation between the incident and the reflected E-fields so that there are no resultant electric flux lines that are parallel to the surface. This obeys Maxwell's first boundary condition.

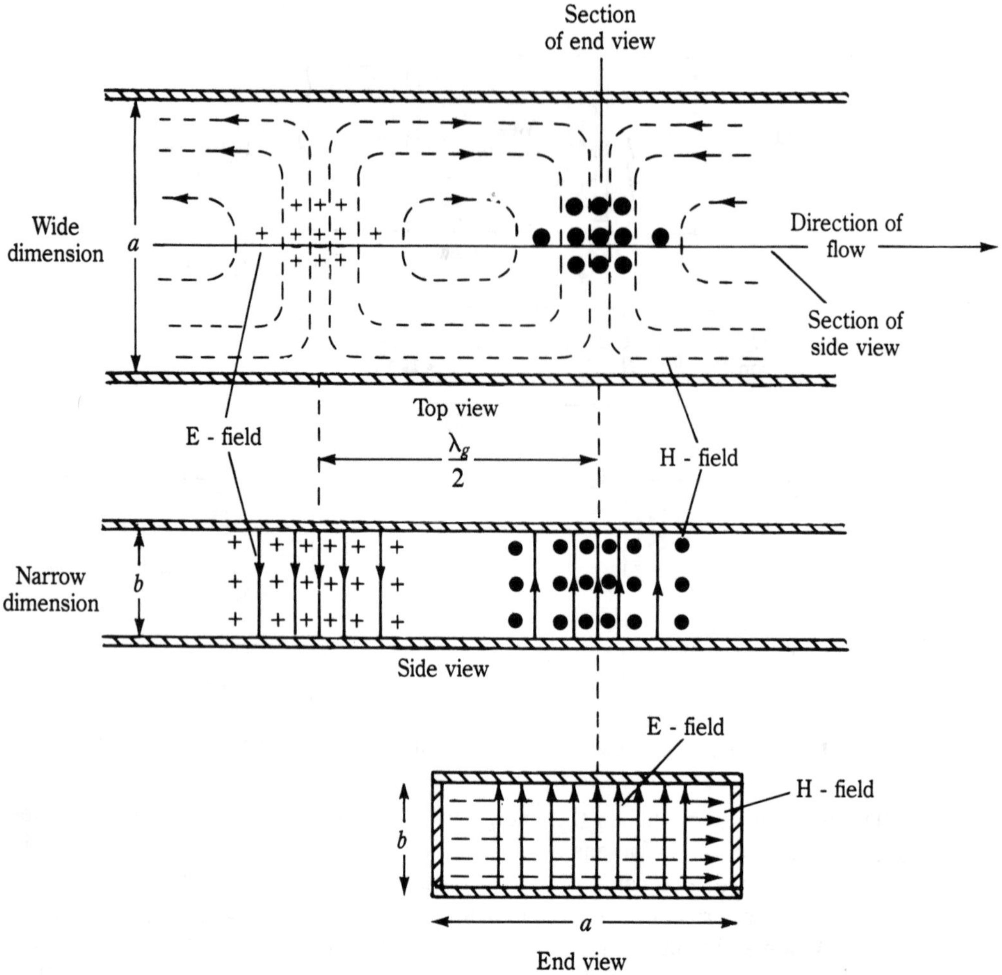

Section
of end view

Wide
dimension a

Direction of
flow

Section of
side view

Top view

E - field

$\dfrac{\lambda_g}{2}$

H - field

Narrow
dimension b

Side view

E - field

H - field

b

a

End view

14-16 Distribution of electric and magnetic fields in a rectangular waveguide (TE_{10} mode).

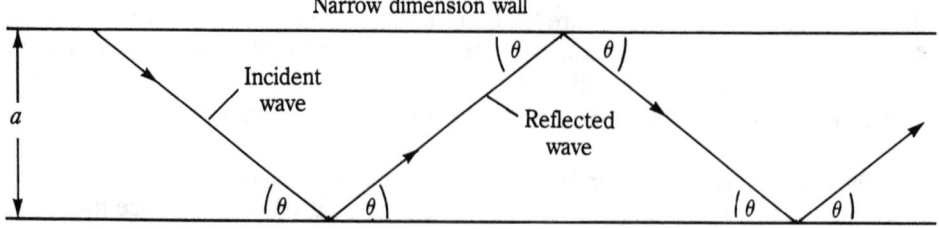

Narrow dimension wall

a

Incident
wave

θ θ

Reflected
wave

θ θ

θ θ

14-17 Path of an EM wave in a rectangular waveguide.

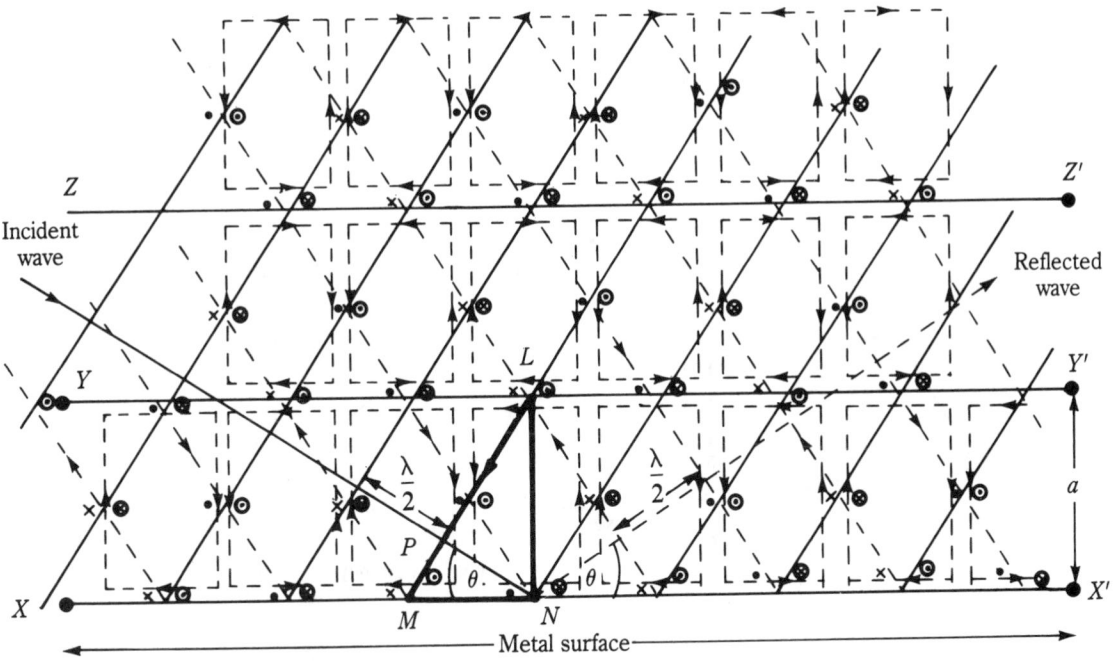

14-18 Combination of incident and reflected waves at the metal surface of a rectangular wave guide.

When the incident and the reflected waves are combined the resulting pattern of the E- and H-fields between the lines *XX'* and *YY'* is exactly the same as we previously derived from the support stub theory. You can therefore infer that the EM wave must progress down the guide by a series of reflections off the narrow dimension.

Now, you must find out the factors which determine the angle of incidence and the velocity with which the energy progresses down the guide. To do this you extract the triangle *LMN* from Fig. 14-18 and display its magnified form in Fig. 14-19. As the wave-front moves from *P* to *N* (a distance of λ/2), the field pattern progresses a greater distance from *M* to *N*. The distance *MN* is a half-wavelength of the field pattern as it exists inside the waveguide and is termed $\lambda_g/2$ where λ_g is the guide wavelength. For an analogy, think of sea waves approaching the shore-line at an angle (Fig. 14-20). As the wave-front moves from one crest to the next through the distance *CC'* (one wavelength), the pattern of the crests at the shore-line covers the greater distance *SS'*.

In the right-angled triangle *LPN*,

$$\sin \theta = \frac{PN}{LN}$$

$$= \frac{\lambda/2}{a}$$

$$= \frac{\lambda}{2a} \tag{14-2}$$

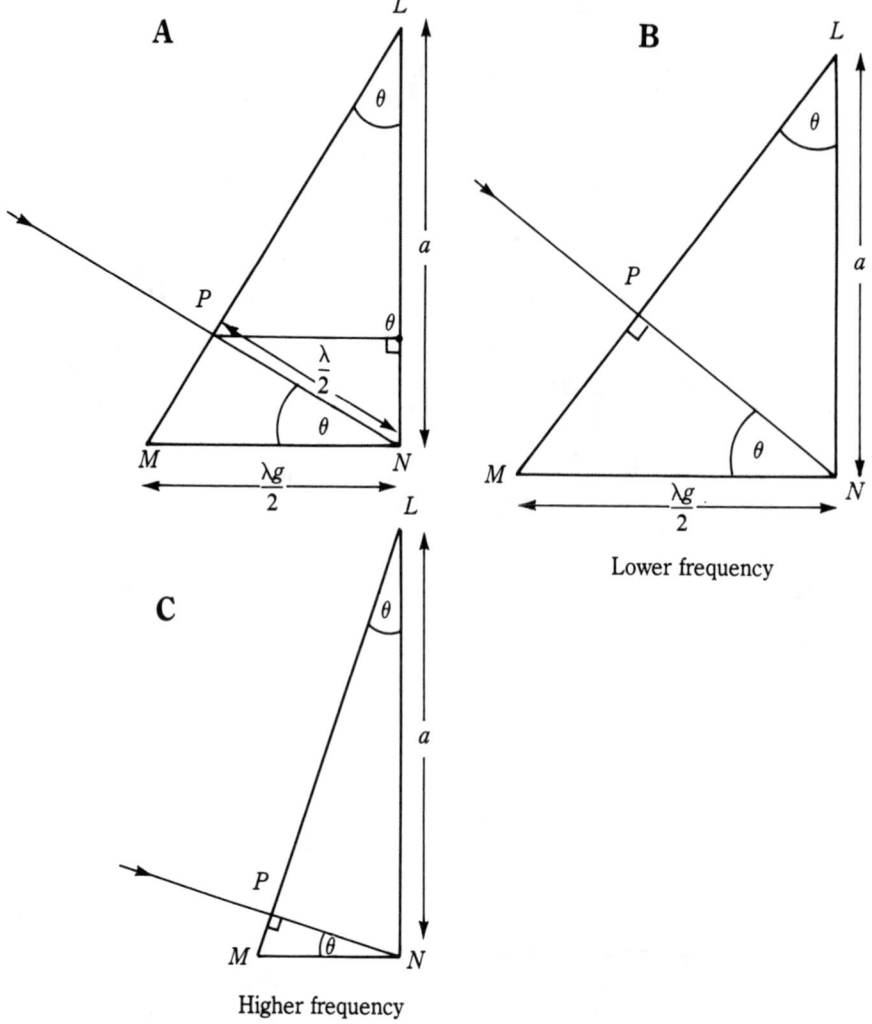

14-19   The effect of the frequency on the angle of incidence.

For a given rectangular waveguide, the "*a*" and "*b*" dimensions are fixed. If the frequency is lowered, the wavelength is longer and the value of θ is increased (Fig. 14-19B). In the limiting cutoff condition, $\lambda_c = 2a$ and $\theta = 90°$; the wave will then bounce back and forth between the narrow dimensions and will not progress down the guide. When the frequency is raised, the wavelength decreases and the value of θ is lowered (Fig. 14-19). This could be continued indefinitely; however at a frequency of approximately $1.9 f_c$, the narrow "*b*" dimension comes into play and limits the top frequency at which the waveguide can successfully operate.

In the right-angled triangle MPN,

$$\cos \theta = \frac{PN}{MN}$$

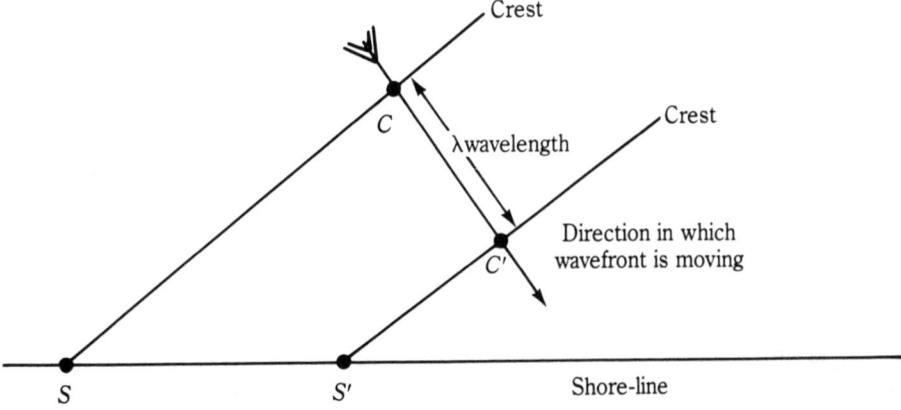

14-20   Sea-wave analogy.

$$= \frac{\lambda/2}{\lambda_g/2}$$

$$= \frac{\lambda}{\lambda_g} \tag{14-3}$$

Therefore,

$$\tan \theta = \frac{\sin \theta}{\cos \theta}$$

$$= \frac{\lambda/2a}{\lambda/\lambda_g}$$

$$= \frac{\lambda_g}{2_a} \tag{14-4}$$

Equation 14-4 shows that, if you decrease the frequency, θ is raised and the guide wavelength, λ_g, is increased. In the cutoff condition, $\theta = 90°$ and the guide wavelength, λ_g, is infinitely long.

Using the trigonometrical relationship $\sin^2\theta + \cos^2\theta = 1$,

$$\left(\frac{\lambda}{2a} \right)^2 + \left(\frac{\lambda}{\lambda_g} \right)^2 = 1$$

This yields,

$$\lambda = \frac{\lambda_g}{\sqrt{1 + \left(\dfrac{\lambda_g}{2a} \right)^2}} \tag{14-5}$$

and

$$\lambda_g = \frac{\lambda}{\sqrt{1 - \left(\dfrac{\lambda}{2a} \right)^2}} \tag{14-6}$$

Remembering that

$$\lambda = c/f$$

and

$$\lambda_c = 2a$$

$$= c/f_c$$

$$\lambda_g = \frac{\lambda}{\sqrt{1 - \left(\frac{f_c}{f}\right)^2}} \tag{14-7}$$

In the time that the wavefront moves from P to N at the speed of light c, the field pattern progresses from M to N (a distance of $\lambda_g/2$) with the *phase* velocity, v_ϕ. With the same time interval the electromagnetic energy has moved down the guide a distance PQ; this physical movement of energy takes place at the *group* velocity, v_g.

Since $MN = PN \sec \theta$, and $PQ = PN \cos \theta$,

$$\text{Phase velocity, } V_\phi = \frac{c}{\cos \theta}$$

$$= c \sec \theta \tag{14-8}$$

and

$$\text{Group velocity, } v_g = c \cos \theta \tag{14-9}$$

Therefore,

$$v_\phi \times v_g = \frac{c}{\cos \theta} \times c \cos \theta$$

$$= c^2 \tag{14-10}$$

where v, v_g, and c are all measured in meters per second.

At the cutoff condition the group velocity is zero and the phase velocity is infinite. This result does not contravene any physical laws since the phase velocity involves the *apparent* movement of a field pattern and not the movement of any physical quantity. As the frequency is raised, v_ϕ decreases and v_g increases as both these velocities approach the velocity of light, c. For an X-band waveguide the graphs of v_g and v_ϕ versus frequency are shown in Fig. 14-21.

You are left to consider the waveguide's phase shift constant, β. A phase shift of 2π radians occurs over a distance equal to the guide wavelength, λ_g. Therefore,

$$\text{Phase shift constant, } \beta = \frac{2\pi}{\lambda_g} \tag{14-11}$$

where

β = phase shift constant (rad/m)
λ_g = guide wavelength (m)

Calculate the results for a practical rectangular waveguide which is designed to operate over the X-band, 8.2 to 12.4 GHz. The inner guide dimensions are 2.286 cm (a) and 1.143 cm (b). If the transmitted frequency is 10 GHz,

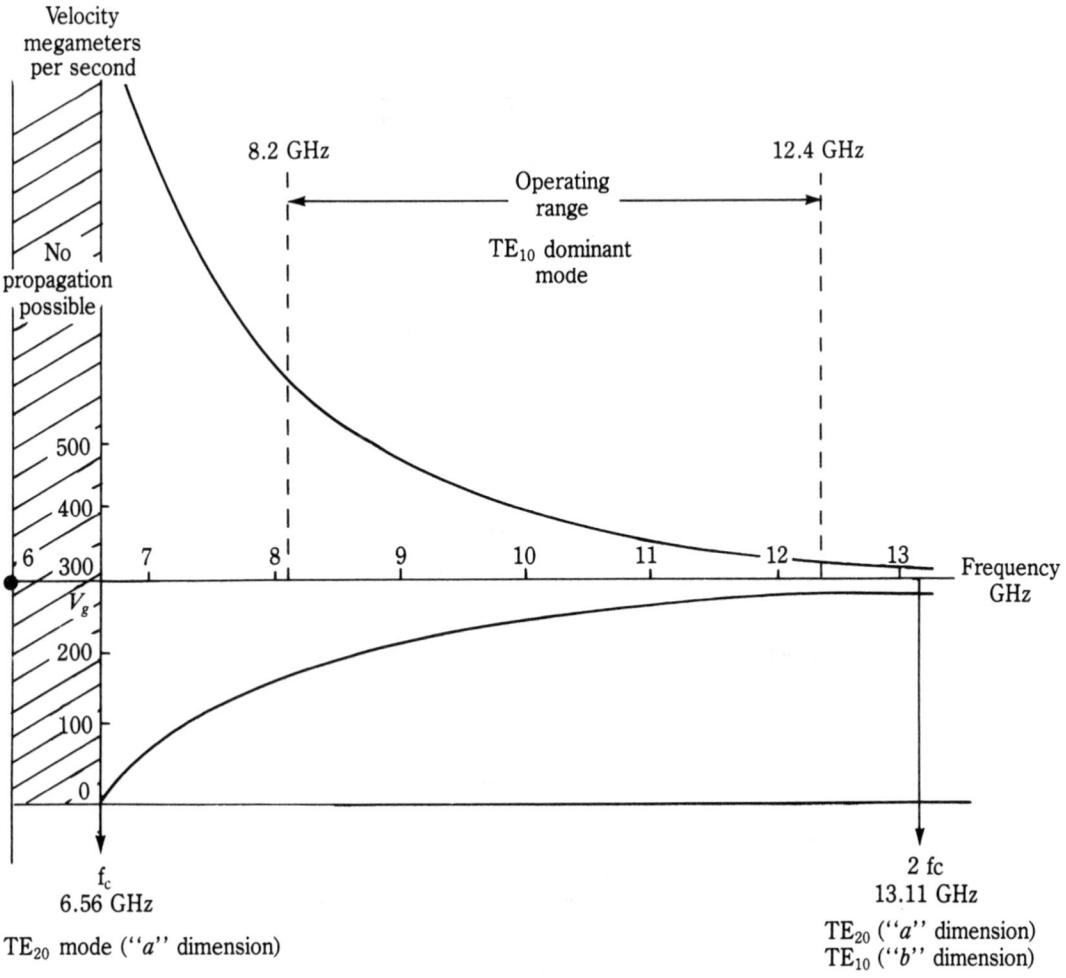

14-21 Variation of group velocity and phase velocity with frequency in an X-band rectangular waveguide.

Free space wavelength, $\lambda = \dfrac{c}{f}$

$$= \frac{3 \times 10^8 \; m/s}{10 \times 10^9 \; Hz}$$

$$= 0.03 \; m$$

$$= 3 \; cm$$

Cutoff wavelength, $\lambda_c = 2a = 2 \times 2.286$

$$= 4.572 \; cm$$

$$\text{Cutoff frequency, } f_c = \frac{c}{\lambda_c}$$

$$= \frac{3 \times 10^8 \ m/s}{4.572 \times 10^{-2} \ m}$$

$$= 6.562 \ GHz$$

$$\text{Angle of incidence, } \theta = \text{inv sin} \frac{\lambda}{2_a}$$

$$= \text{inv sin} \frac{3}{4.572}$$

$$= 41.01°$$

$$\text{Guide wavelength, } \lambda_g = \frac{\lambda}{\cos \theta}$$

$$= \frac{3}{\cos 41.01°}$$

$$= 3.98 \ cm$$

$$\text{Phase velocity, } v_\phi = \frac{c}{\cos \theta}$$

$$= \frac{300}{\cos 41.01°}$$

$$= 397.6 \ \text{megameters per second}$$

$$\text{Group velocity, } v_g = c \cos \theta$$

$$= 300 \cos 41.01°$$

$$= 226.4 \ \text{megameters per second}$$

$$\text{Phase shift constant, } \beta = \frac{2\pi}{\lambda_g}$$

$$= \frac{2}{3.98 \times 10^{-2} m}$$

$$= 159 \ \text{radians per meter}$$

The dominant mode and higher order modes

The field patterns shown in Fig. 14-18 represent only one of the infinite number of possible ways in which E- and H-fields can exist inside a rectangular waveguide. Each such field configuration is known as a *mode of operation*. For example in Fig. 14-18 you could have inserted the upper metal surface at the line ZZ' rather than the line YY'. If this had been done, there would have been two H-loops instead of one across the wide dimension. However, in order to accommodate this new mode, you would need to double the length of the wide dimension, *a*.

The field configurations described in the previous section represent the *dominant* mode. For this mode the inner dimensions of the waveguide are the least possible so that all other modes require a larger waveguide—an obvious disadvantage. Furthermore, the dominant mode is the easiest to excite in the waveguide, is most efficient, and has the lowest cutoff frequency. Waveguides are normally designed so that only the dominant mode is propagated. For example, the X-band waveguide with internal dimensions of 2.286 cm (*a*) and 1.143 cm (*b*) has a cutoff frequency of $f_c = 6.56$ GHz at the dominant mode. If the frequency is raised above $2f_c = 2 \times 6.56 = 13.1$ GHz, a higher-order mode will be excited in the "*a*" dimension and the dominant mode will at the same time appear in the "*b*" dimension (because in the example, $b = a/2$). Consequently the energy introduced into the waveguide will be split between three possible modes; this is highly undesirable because

1. the methods of joining waveguide sections together might solely respond to the dominant mode

2. the means of removing the energy might only be effective at the dominant mode

To classify the various waveguide modes a number of systems are used that are similar to that for the cavity resonator. You must first decide whether the mode is *t*ransverse *e*lectric (TE) or *t*ransverse *m*agnetic (TM). In a TE mode all electric flux lines are at right-angles to the direction of propagation so that there are no longitudinal lines which are parallel to the direction of travel. This fits the description of the E-lines in the rectangular waveguide which is operating at the dominant mode.

For a TM mode the H-lines are entirely transverse. Such is not the case with the dominant mode where the H-lines have a longitudinal component.

In addition to the designation TE or TM, subscript numbers are used to complete the description of the field patterns. In a rectangular guide, the first subscript is the number of half-wave patterns in the wide "*a*" dimension, while the second subscript is the number of half-wave patterns in the narrow "*b*" dimension. Figure 14-22 shows the cross-sectional distribution of the transverse E-lines for the dominant mode. Clearly there is one half-wave pattern along the "*a*" dimension but no change across the "*b*" dimension; the complete designation for the dominant mode is therefore TE_{10}. Two higher-order modes are shown in Fig. 14-23A, and B, and you should verify their designations. Such modes might rarely be used under very special circumstances but you should concentrate on the dominant mode.

14-22   The dominant TE_{10} mode.

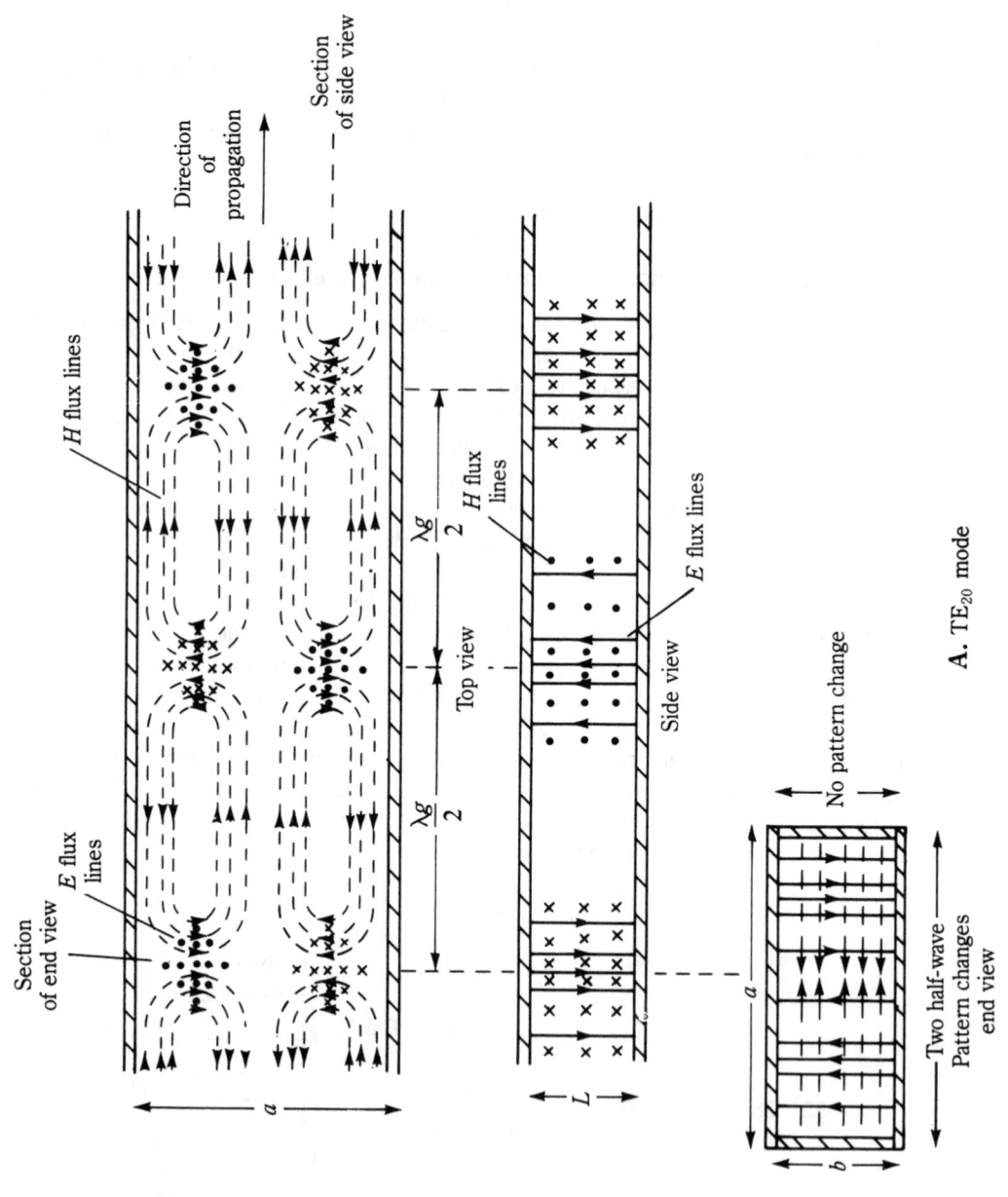

A. TE$_{20}$ mode

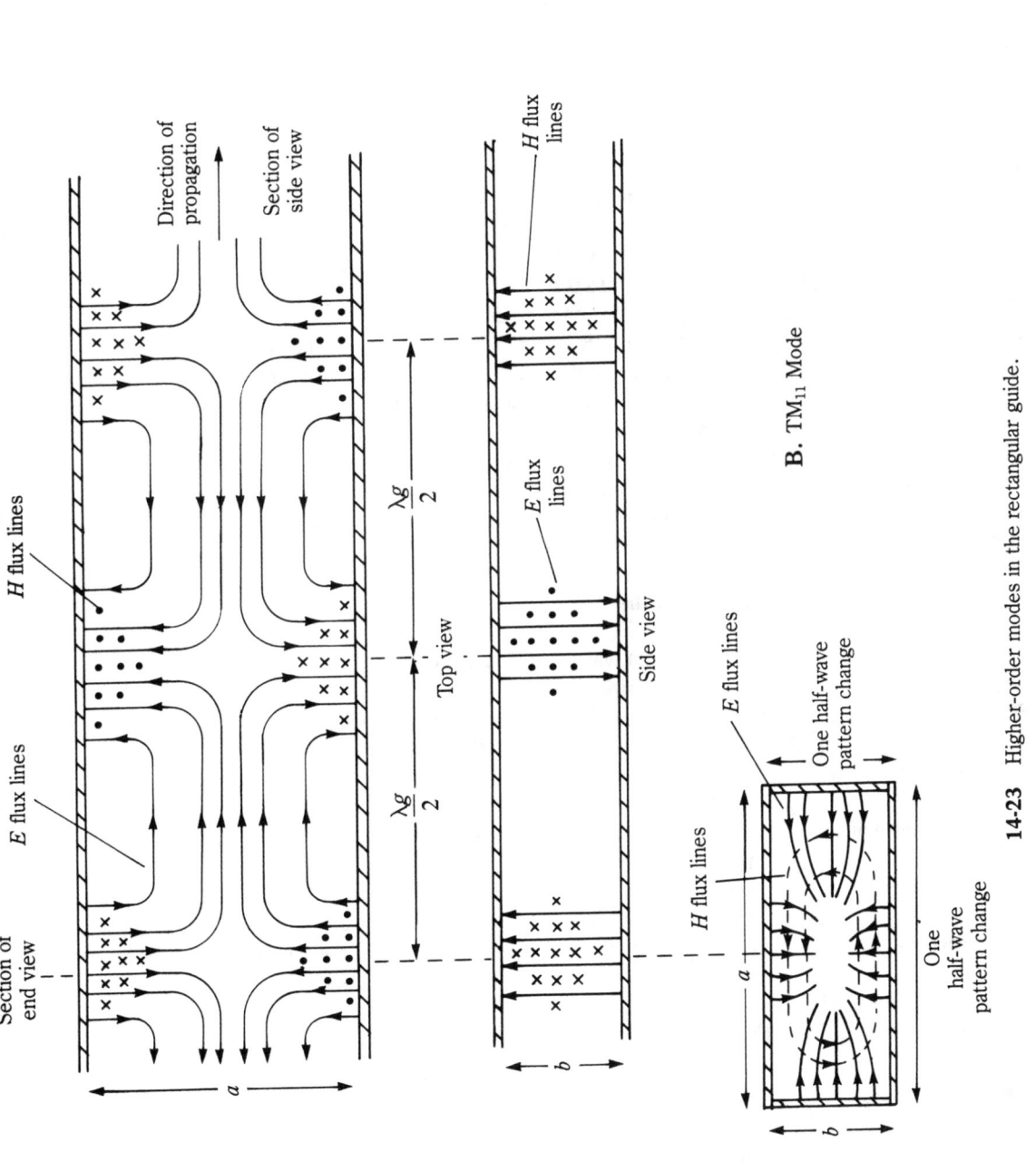

B. TM$_{11}$ Mode

14-23 Higher-order modes in the rectangular guide.

Methods of coupling

Fundamentally there are three methods of coupling rf energy into or out of a wave-guide—probe, loop, and aperture or slot; these same methods were used with cavity resonators. The E-probe represents capacitive coupling and is illustrated in Fig. 14-24A. Its action is similar to that of a λ/4 antenna so that when the probe is excited by the rf signal, an alternating field is established. The probe is normally positioned in the center of the "*a*" dimension and is a quarter-wavelength, $\lambda_g/4$ (or an odd multiple of a quarter-wavelength) from the sealed end, which can be in the form of an adjustable piston. At this position there is a maximum density of the E-lines, so there is maximum coupling between the probe and the guide. Usually the probe is fed by a short length of a coaxial cable with its outer conductor connected to the wall of the waveguide. The inner conductor is joined to the probe, which extends into the guide but is insulated from the walls. The degree of coupling can be altered, by varying the depth of the probe's insertion, by shifting it from the position of the maximum E-line density, or by partial shielding.

In a pulse-modulated radar system the bandwidth is large and is of the order of MHz. So that the probe shall not discriminate appreciably against any of the numerous side-bands, you can use a variety of wideband probes (Fig. 14-24A).

Figure 14-24B illustrates *loop* or *inductive coupling*. The loop is placed at a point of the maximum H-field, for which there is a number of possible locations; it is normally part of a coaxial cable whose outer conductor is connected to the waveguide. The inner conductor then forms the loop inside the guide, and the end of the loop is connected to the internal wall. The degree of coupling can be varied by altering the position of the loop's plane relative to the direction of the H-lines.

Aperture or *slot coupling* is shown in Fig. 14-24C. Slot *X* is in an area where the E-field is a maximum (electric field coupling) while slot *Y* is in a region of maximum H-field density (magnetic field coupling). The position of slot *Z* coincides with maximum E- and H-fields so that the coupling is electromagnetic. *X*, *Y*, and *Z* are all radiating slots which are normally cut at right angles to the direction of the wall currents flowing along the inner surfaces of the guide; this produces maximum distortion of the wall currents so that radiation occurs (Fig. 14-24D).

The degree of coupling increases with the size of the slot, especially its width, and it also depends on the angle of the slot in relation to the directions of the wall currents. Based on these principles a whole variety of radiating slots is illustrated in Fig. 14-24E. For loose coupling the slots can be replaced by small circular holes. It should be emphasized that the methods of introducing energy into a waveguide are equally effective in removing it. By contrast, nonradiating slots produce minimum disturbance of the wall currents and are used for monitoring purposes such as VSWR measurements; examples of such slots are shown in Fig. 14-24F.

The circular waveguide

Although rectangular guides are used almost exclusively in radar systems, there are special cases in which circular waveguides find their application. A good example occurs in a radar antenna system that is required to revolve relative to the stationary transmitter and receiver. When one waveguide section is rotated relative to another, it is impossible to

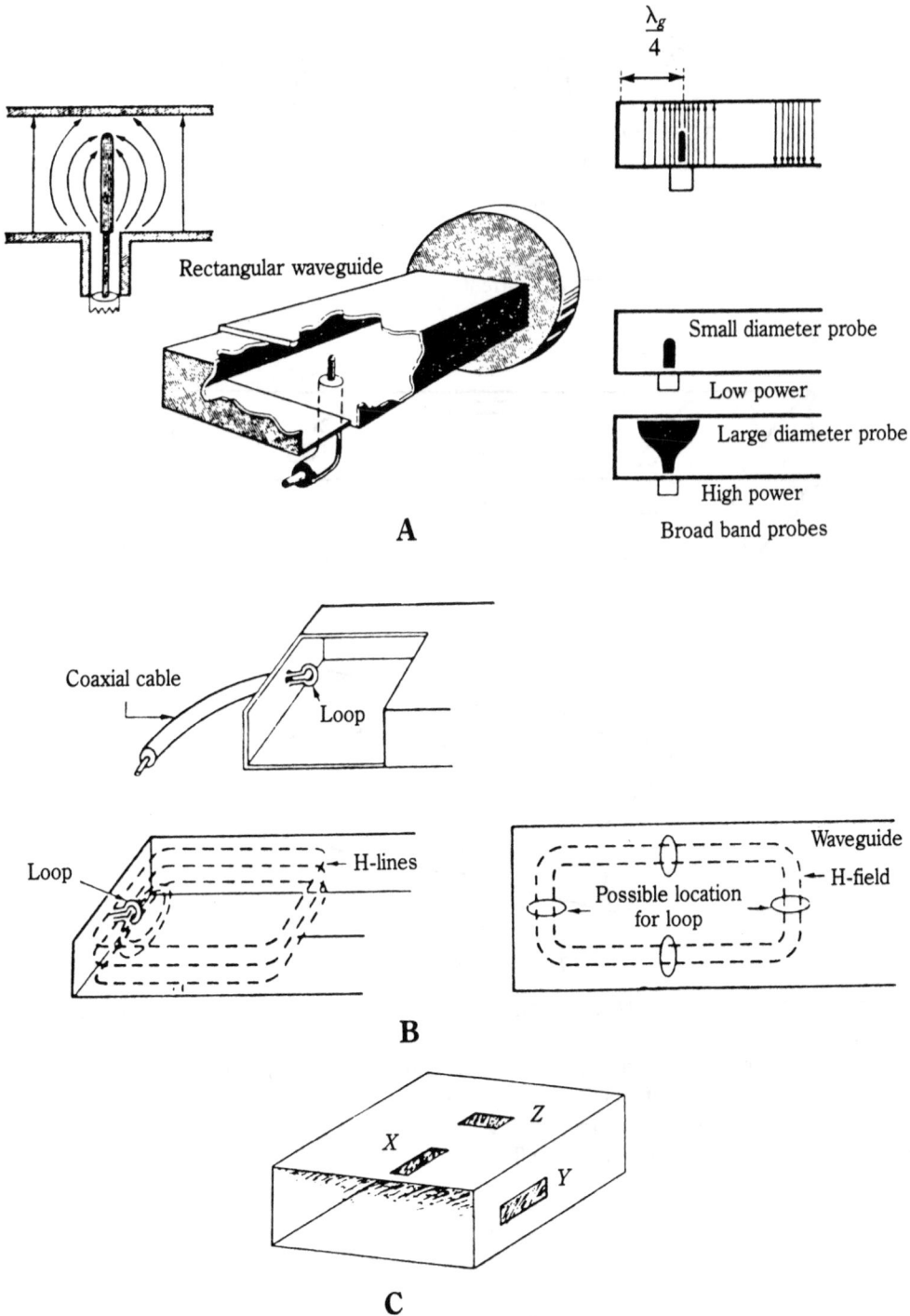

14-24 Methods of coupling to and from a rectangular waveguide.

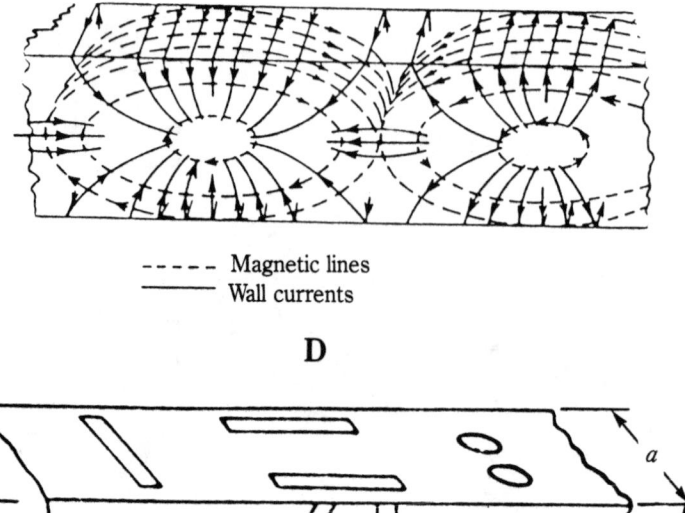

- - - - - Magnetic lines
——— Wall currents

D

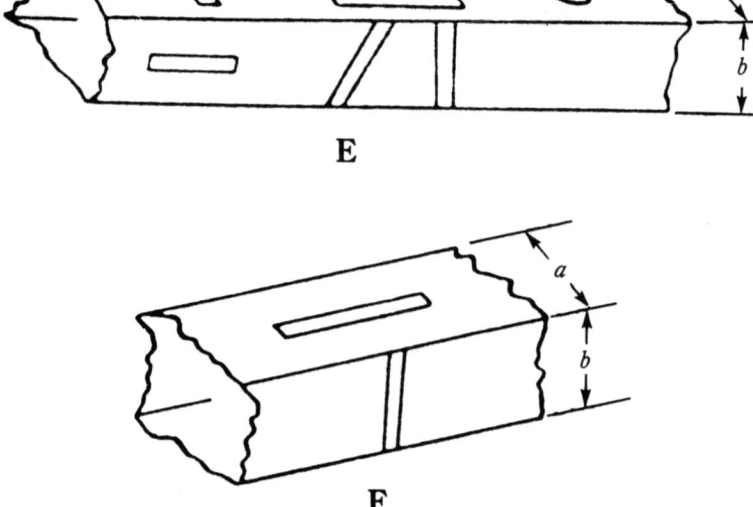

E

F

14-24   Continued

maintain continuity with rectangular guides, so the only solution is to use a circular guide.

The dominant mode of a circular waveguide is illustrated in Fig. 14-25. For this mode the cutoff wavelength is 1.71 times the inner diameter, d, of the waveguide (quoted result). Remember that the cutoff wavelength of a rectangular guide is twice the wide dimension when operating in the TE_{10} dominant mode. For the same cutoff frequency,

$$1.71\, d = 2a$$

or

$$d = \frac{2a}{1.71}$$
$$= 1.17a$$

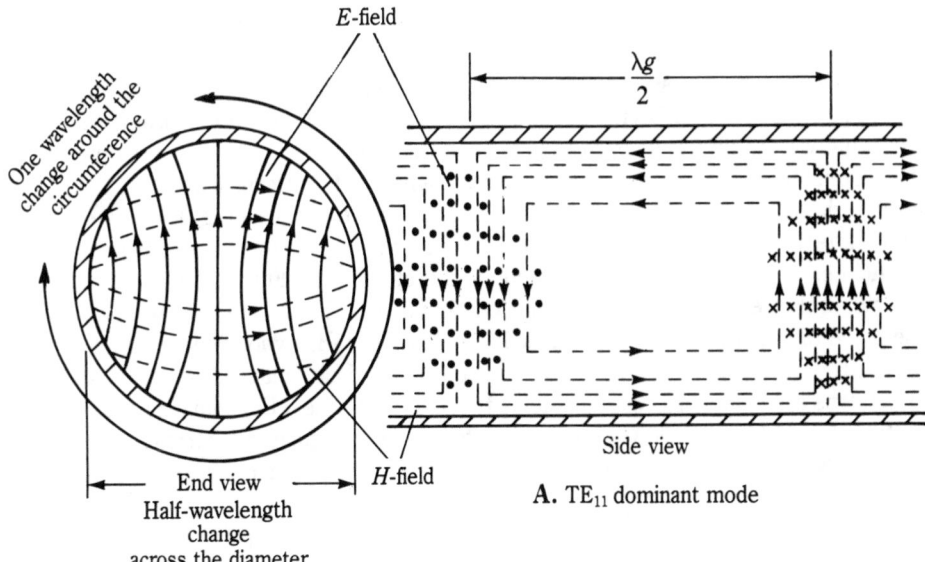

E-field

One wavelength change around the circumference

λg/2

A. TE₁₁ dominant mode

End view
Half-wavelength change across the diameter

H-field

Side view

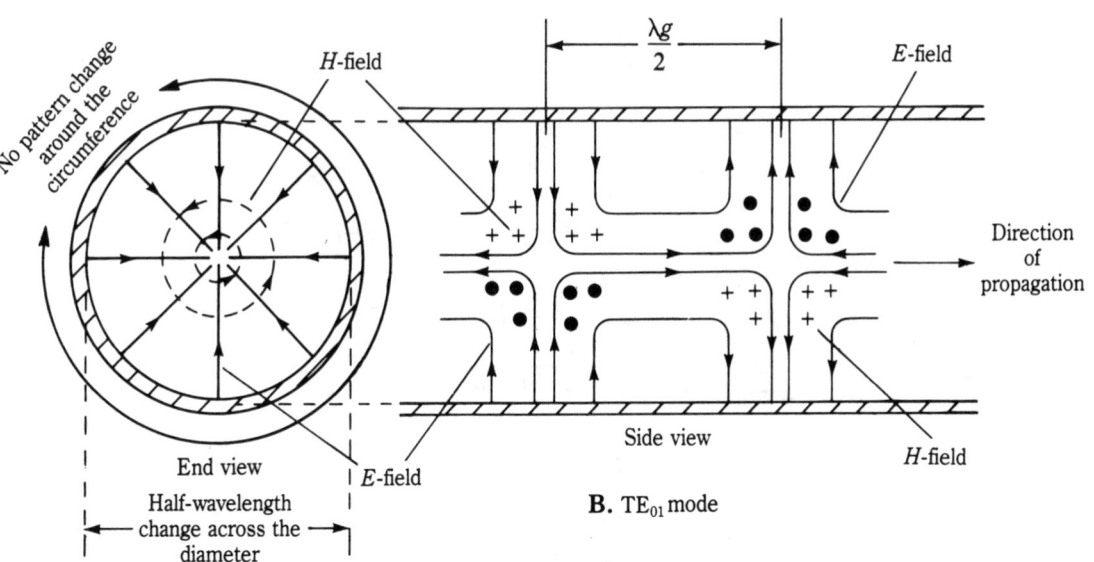

No pattern change around the circumference

H-field

λg/2

E-field

Direction of propagation

End view
Half-wavelength change across the diameter

E-field

Side view

H-field

B. TE₀₁ mode

14-25 Modes in the circular waveguide.

so that the circular guide is larger than the corresponding rectangular waveguide—an obvious disadvantage. However, the circular waveguide suffers from a far more serious problem. In a rectangular guide the directions of the E- and H-lines can be referred to as the directions of the narrow and wide dimensions. With a circular guide no such references exist and consequently the wave's plane of polarization tends to rotate as the wave moves down the guide. It can then be difficult to remove the energy from the guide since, for example, an E-probe requires that the direction of the probe is parallel to the direction of the E-lines.

To classify the various modes which can exist in a circular guide, again divide the modes into TE and TM. On the numbering system the first subscript indicates the number of full-wave E patterns which you encounter as you move around the circumference; the second subscript refers to the number of half-wave E patterns across the diameter. For the dominant mode the electric flux lines are entirely transverse, there is one full-wave pattern around the circumference and one half-wave pattern across the diameter; the full designation of the dominant mode is therefore TE_{11} (Fig. 14-25A).

When the center conductor is removed from a coaxial to create a circular waveguide the mode of operation is designated as TM_{01} (Fig. 14-25B), whose cutoff wavelength is only $1.31 \times$ the guide's diameter. This mode is of particular interest because its E- and H-patterns are compatible with the fields that exist with the TE_{10} dominant mode of the rectangular guide. This allows you to convert from a rectangular guide to a circular guide, then carry out a rotation, and afterwards convert back to a rectangular guide (Fig. 14-26).

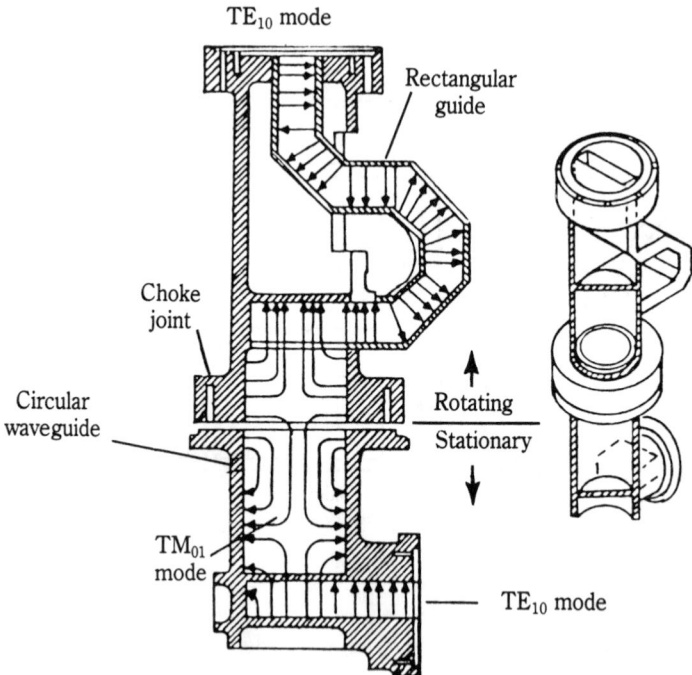

14-26   Conversion between a rectangular guide and a circular guide.

Joints, bends, twists, irises, posts, and screws

Because it is practically impossible to build a waveguide system in one piece, it is normally constructed in sections which must be connected by joints. Any irregularities in the joints cause reflection effects, create standing waves, and increase the attenuation. A proper permanent joint affords a good connection between the two sections of the wave-guide and has very little effect on the E- and H-field patterns. During manufacture the waveguide sections are machined to within a few mils (0.001″) and then welded together. The result is a hermetically sealed and mirror smooth joint.

In locations where sections of a waveguide system must be taken apart for normal maintenance and repair, you obviously cannot use a permanent joint. In order to allow portions of the waveguide to be separated, the sections are connected by semiperma-nent joints, of which the simplest is the *bolted flange* (Fig. 14-27A). Here the two sec-tions are merely bolted together with a gasket to exclude moisture. However, a superior solution is the *choke joint* (Fig. 14-27B, and C) which is the one most commonly used. This consists of flanges which are connected to the waveguide at its center. In Fig. 14-27C, the right-hand flange is flat and the one on the left is slotted one-quarter wave-length, $\lambda_g/4$, deep from the inner surface of the waveguide. This slot is positioned at a distance of one-quarter wavelength ($\lambda_g/4$) from the point where the flanges are joined. Since the quarter-wavelength slot is terminated by a short circuit, the two quarter-wave-

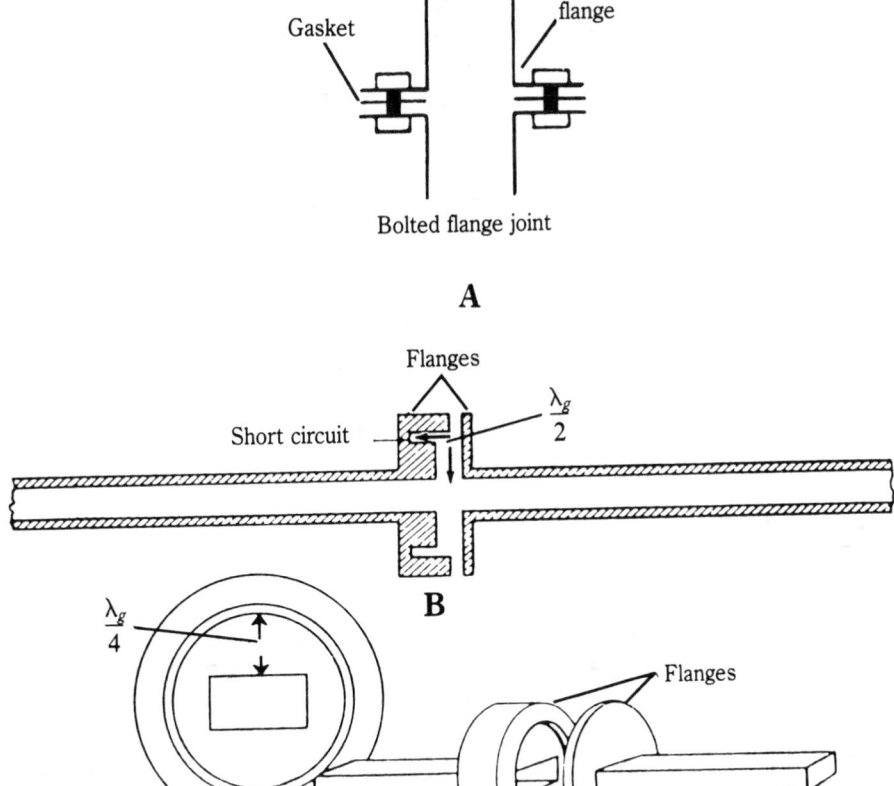

14-27 Waveguide joints.

lengths together form a half-wavelength section which presents a short circuit at the place the walls are joined together. The result is an *electrical* short circuit at the junction of the two waveguides. The two sections might actually be separated by as much as a tenth of a wavelength which is 3 mm at 10 GHz in the X-band. This separation allows us to seal the interior of the waveguide with a rubber gasket to exclude moisture. Any moisture introduces a discontinuity, creates standing waves, alters the guide's impedance, increases the attenuation and can cause corrosion. For this reason long, level runs of a waveguide should be avoided; in addition a small hole is sometimes drilled at a waveguide's lowest point so that any accumulation of moisture can be drained out. As an example, the waveguide run from a radar transmitter to its antenna should be kept to a minimum length.

The quarter-wavelength distance from the walls to the slot is modified slightly to compensate for the small reactance introduced by the short space and the open circuit from the slot to the periphery of the flange. The loss introduced by a well designed choke joint is less than 0.03 dB, while a flange joint has a loss of 0.05 dB or more.

It has already been mentioned that rotating joints are usually required in a radar set where the transmitter and the receiver are stationary but the antenna system is revolving. A simple method for rotating one waveguide section relative to another is to use a mode of operation whose field distribution is symmetrical about the axis of rotation. This requirement is met by using a circular waveguide operating in the TM_{01} mode (Fig. 14-26). A choke joint then separates the sections mechanically but joins them electrically. As one section rotates, the field distributions are fixed so that there are minimum reflections. Because radar systems mainly employ rectangular waveguides, the circular rotating joint must be inserted between two rectangular sections (Fig. 14-26). The joint then consists of two sections of circular guide; one section is rotating while the other is stationary. At the end of each of the sections there is a transition between the circular guide and the rectangular guide.

In Fig. 14-26 the rectangular sections are operating in the dominant TE_{10} mode. The E-lines of the bottom rectangular section penetrate into the circular section and excite the TM_{01} mode, which provides the required axial symmetry for rotating joints. At the top of the revolving circular guide the E-lines couple the energy in the rectangular section which leads to the antenna; you are now operating once more in the dominant TE_{10} mode.

Bends and twists

Any sudden change in the size, shape, or direction of the waveguide system will result in the introduction of reflection effects and an increase in the value of the SWR. However, most waveguide systems normally require bends and these might be in the direction of the wide dimension (Fig. 14-28A) so that the H-loops are primarily affected (H-bend). Alternatively the bend might be in the narrow dimension (Fig. 14-28B) so that the E-lines are mainly distorted (E-bend). In order that either of these bends shall not produce an individual increase in the SWR of more than 1.05:1, the bend must have a minimum radius of at least $2\lambda_g$.

A twist (Fig. 14-29) is used to rotate the wave's plane of polarization through 90°; so that the twist cannot introduce an SWR of more than 1.05:1, the actual distance over which the twist occurs must exceed $2\lambda_g$.

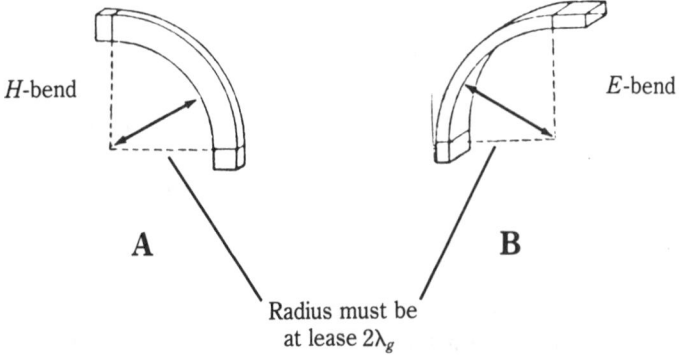

H-bend

E-bend

A

B

Radius must be
at lease $2\lambda_g$

14-28 Waveguide gradual bends.

At least $2\lambda_g$

14-29 A 90° twist in a rectangular waveguide.

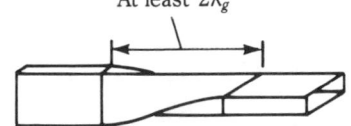

A sharp 90° bend (Fig. 14-30A) is not permissible since total reflection would occur at such a bend and the result would theoretically be an infinite value of SWR. In one solution (Fig. 14-30B) the guide is bent through 45° twice with the two bends one-quarter of the guide wavelength apart. The combination of the direct reflection at one bend and the inverted reflection from the other bend tend to cancel. It will then appear as if no reflection had occurred although in practice the individual SWR introduced by the double bend is approximately 1.1:1.

A superior solution to the two 45° bends is the mitered 90° bend of Fig. 14-30C. Provided $d = 0.65\,D$ the individual SWR introduced by the mitered bend is only 1.05:1. These mitered bends can either be of the E or H variety without changing the mode of operation.

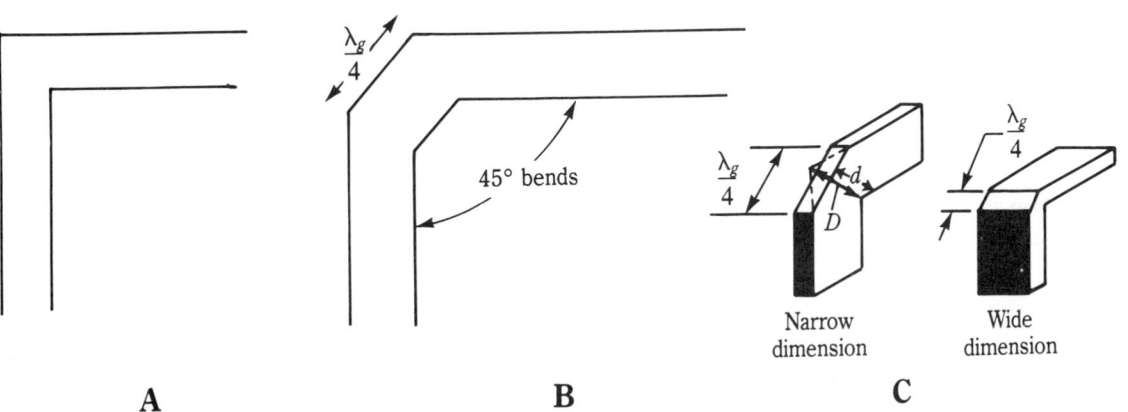

$\dfrac{\lambda_g}{4}$

$\dfrac{\lambda_g}{4}$

45° bends

$\dfrac{\lambda_g}{4}$

d

D

$\dfrac{\lambda_g}{4}$

Narrow
dimension

Wide
dimension

A

B

C

14-30 Waveguide bends.

Irises, posts and screws

No bend, joint, or twist is ever perfect, so there is a certain amount of reflection which causes a reactance to appear in the guide and present some degree of mismatch.

Irises are metal diaphrams which are placed at permanent positions within a waveguide. Their purpose is to eliminate the reactances caused by reflection effects which appear at bends, joints, etc. Each iris has an opening or window through which the EM wave passes.

As an example a reflection effect can produce a certain capacitive reactance at the entrance to a bend. This can be eliminated by an inductive iris, which tends to concentrate the H flux lines (Fig. 14-31A). By contrast the capacitive iris of Fig. 14-31B concentrates the E-lines, while the resonant irises of Fig. 14-31C can be used to eliminate unwanted modes.

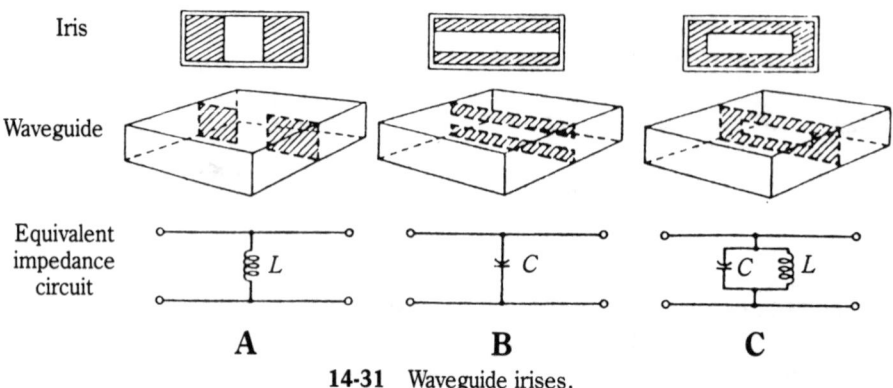

14-31 Waveguide irises.

A similar effect is provided by conducting posts or screws. These posts which only penetrate part of the way into the guide (Fig. 14-32A), form shunt capacitive reactances that can tune out any inductive reactance effects produced by reflections. If the post or screw makes contact with both the top and the bottom walls, it behaves as an inductive reactance at high microwave frequencies (Fig. 14-32B). The only difference between the fixed posts and the variable screw is that the screw can be adjusted to change its reactance.

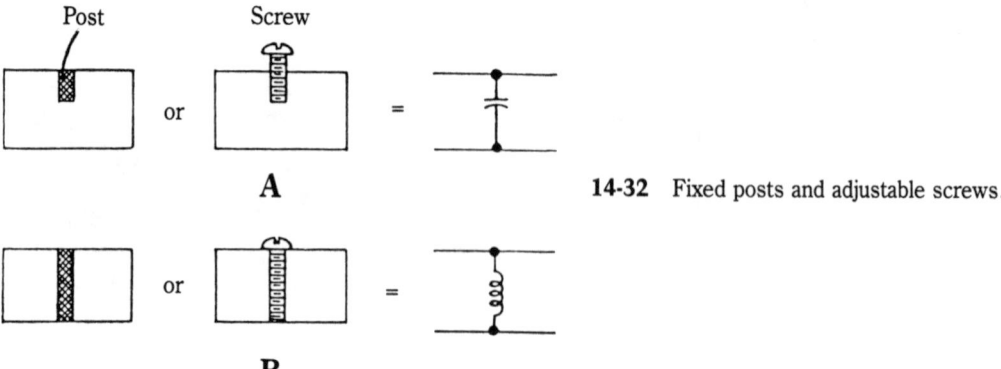

14-32 Fixed posts and adjustable screws.

Waveguide impedance

For a TEM wave traveling in a vacuum the intrinsic impedance of free space is defined by:

$$\text{intrinsic impedance of free space, } \eta_o = \frac{\mathcal{E}}{H} \tag{14-12}$$

where

η_o = intrinsic impedance of free space (Ω)
$\mathcal{E}$ = electric field intensity of the EM wave (V/m)
H = magnetic field intensity (A/m)

It can be shown that the intrinsic impedance,

$$\eta_o = \sqrt{\frac{\mu_o}{\mathcal{E}_o}} \tag{14-13}$$

where

$$\mu_o = 4\pi \times 10^{-7} \, H < m$$

and

$$\frac{1}{\sqrt{\mu_o \times \mathcal{E}_o}} = c \text{ (velocity of light, m/s)}$$

Then

$$\frac{1}{\sqrt{\mathcal{E}_o}} = c \times \sqrt{\mu_o}$$

and

$$\text{Intrinsic impedance, } \eta_o = \frac{\mathcal{E}}{H}$$

$$= \sqrt{\mu_o} \times \sqrt{\mu_o} \times c$$
$$= 4\pi \times 10^{-7} \times 300 \times 10^6$$
$$= 120\pi \approx 377 \, \Omega$$

For waveguides the specific wave impedance is comparable with the surge impedance of transmission lines. However this wave impedance, η, depends on the particular mode of operation. For example,

TE Modes. Specific wave impedance,

$$\eta = \eta_o \times \frac{\lambda_g}{\lambda} \tag{14-14}$$

TM Modes. Specific wave impedance,

$$\eta = \eta_o \times \frac{\lambda}{\lambda_g} \qquad (14\text{-}15)$$

Notice that the value of the wave impedance depends on the values of η_o, the frequency and the wide dimension. This means that the concept of the wave impedance could not be used to match two waveguide sections with different narrow dimensions.

The main use of the wave impedance is in the correct design of waveguide terminations. Of course, on a waveguide there is no way of attaching a fixed resistive load as the termination. However, the end of the waveguide can be filled with graphited sand (Fig. 14-33A) which will then dissipate the required energy. Virtually no energy is reflected back into the waveguide and the SWR is less than 1.01:1.

Another method is to include a resistive rod (Fig. 14-33B) which is positioned at the point where the density of the E-lines is the greatest. Yet a third method is to terminate the waveguide with a taper (Fig. 14-33C) which is aligned with either the E- or H-lines. Such a taper is made from either powdered iron or carbon mixed with a binder which is deposited on a dielectric strip. When the flux lines cut the wedge, the induced currents create the required energy loss. For all these matched terminations there is virtually total absorption of the energy and little reflection occurs.

For test and monitoring purposes it is often desirable for all the energy to be reflected back from the end of the waveguide. This can be accomplished by permanently welding a metal plate (short circuit) at the end of the waveguide (Fig. 14-33D). If it is necessary that the end plate is moveable, the contact between the guide and the plate must be exceptionally good so that the H-field will not be attenuated (Fig. 14-33E).

If you wish to use a moveable short the required arrangement is similar to that of the choke joint previously discussed. Basically it consists of an adjustable plunger that fits into the guide as shown in Fig. 14-33F. The walls of the waveguide and the plunger form a half-wave channel. Since the half-wave channel is closed at one end, the other end also behaves as a short circuit. The result is a perfect connection between the wall and the plunger. The actual physical connection is made at a quarter-wavelength from the short circuit where the standing wave current is at its minimum level. This makes it possible for the plunger to slide loosely in the guide at the point where the contact resistance is very low.

With certain matching and coupling requirements it might be necessary to change the wave impedance of the guide before attaching the matching termination. This is commonly done by including a ridged waveguide section (Fig. 14-34) which is capable of varying the waveguide impedance by a factor of 20 or more and by multiplying the attenuation by a few hundred. These changes are brought about by gradually tapering the ridge.

Other examples of terminating a waveguide are the crystal detector for signal demodulation, and the thermistor or barretter mount for the measurement of microwave power. Such terminations are designed so that they create an SWR of less than 1.1:1.

Consider two of the two waveguide sections which have the same wide dimensions, frequency, and the same mode of operation. However, if the narrow dimensions are different, the two sections can only be matched if their current impedances (characteristic

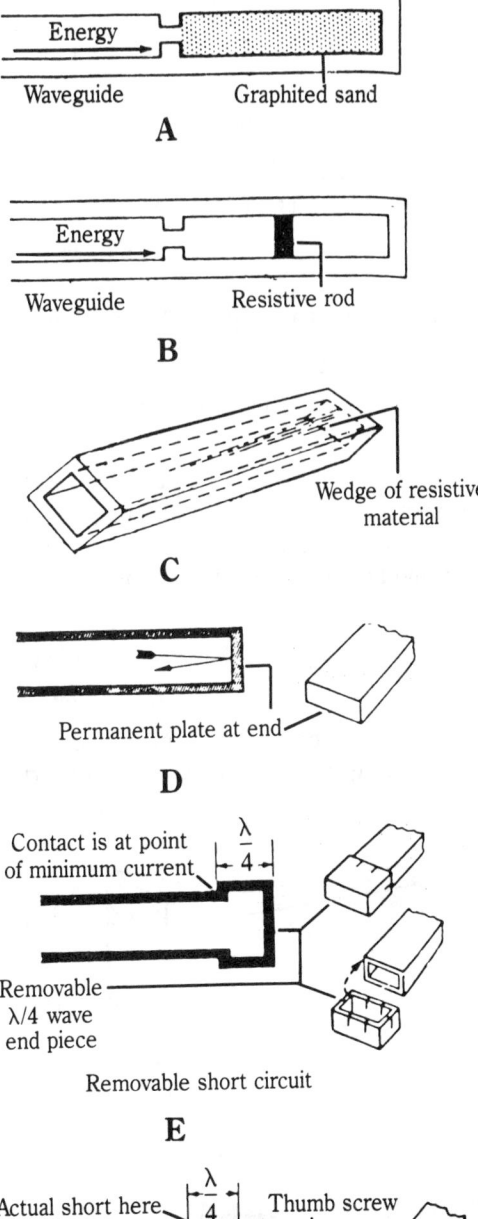

A

B

C

14-33 Examples of waveguide terminations.

D

E

F

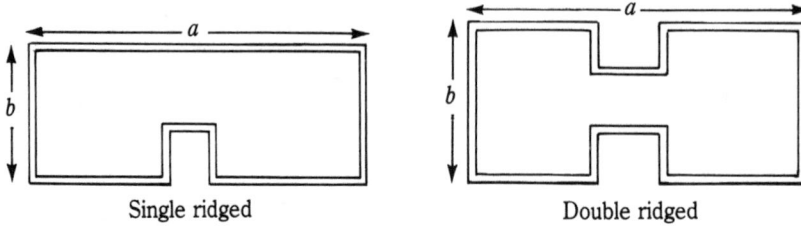

| Single ridged | Double ridged |

14-34 Examples of ridged waveguides.

impedances) are made equal. The characteristic impedance is defined as:

$$\text{Characteristic impedance} = v_o \times \frac{\pi b \lambda_g}{2a\lambda} \tag{14-16}$$

The match can be achieved by adjusting the values of "b" and "a" until the characteristic impedances of the two sections are the same.

In another application it is possible to flare out the guide until the factor

$$\frac{\pi b \lambda_g}{2a\lambda}$$

is equal to 1. The characteristic impedance is then matched to the intrinsic impedance of free space (Fig. 14-35). The match can be further improved by including a dielectric baffle plate across the end face of the guide.

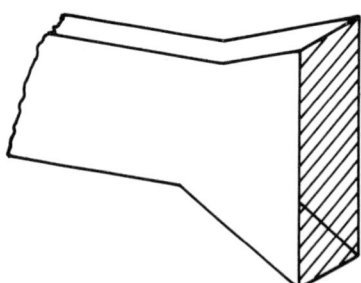

14-35 Waveguide flare and dielectric baffle.

Matching stubs

The purpose of a metallic insulator in supporting a two wire line has already been discussed. However, the main use of a stub is to match a general impedance, such as an antenna load, to the surge impedance of the cable which feeds the load. The stub itself can be regarded as a short section of line, about one quarter-wavelength long, which is terminated by a moveable short (Fig. 14-36). The single moveable stub is placed across (in parallel with) the twin line, and by sliding the stub along the line a position is found

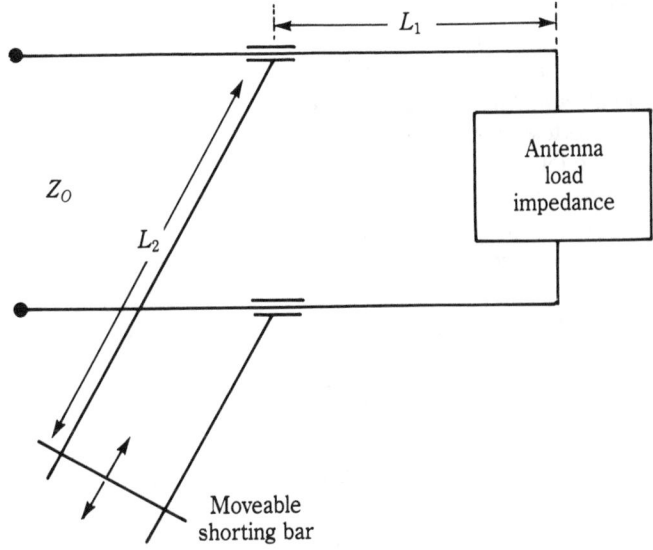

14-36   The principle of the single moveable stub.

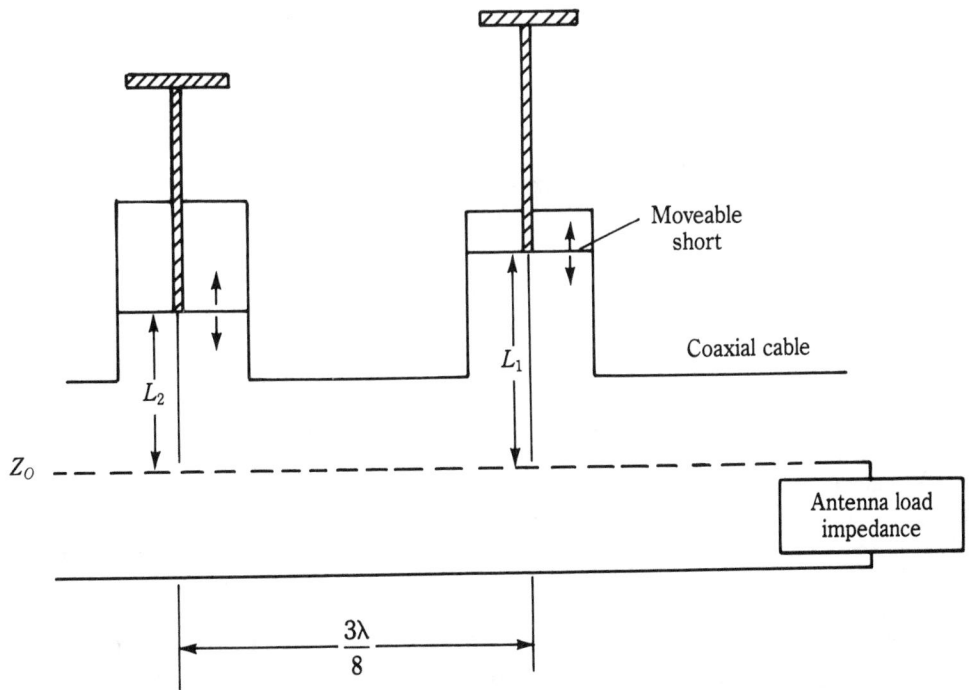

14-37   Two fixed stubs which are $3\lambda/8$ apart.

such that the combination of the stub and the load presents an entirely resistive impedance whose value is equal to the feeder's Z_o. The requirement is therefore to determine the values of L_1 and L_2; this can be done by trial and error.

The application of a single moveable stub is limited to the two wire line, and it is preferable to use two fixed stubs which are typically $3\lambda/8$ apart (Fig. 14-37). It is then necessary to find the lengths of the stubs to achieve the required impedance match. At microwave frequencies the stubs have pistons which use the principle of the choke joint's half-wave channel to achieve the necessary short circuit (Fig. 14-38). When the two stubs are adjusted for their correct lengths, the combination of the stubs and the antenna load produces an overall impedance that is entirely resistive and is equal to the value of the waveguide's impedance.

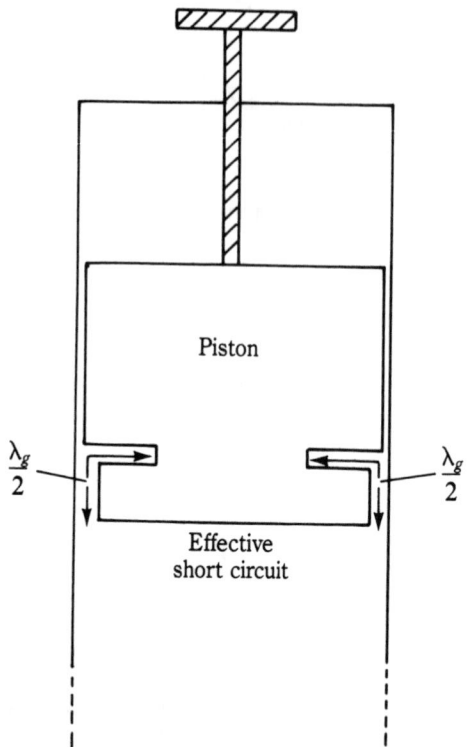

14-38 Cross-section of a microwave matching stub.

Waveguide stubs can either be of the E- or H-variety. For an E-stub the electric flux lines penetrate from the waveguide to set up standing waves on the stubs. Because the stub is across the wide dimension, its reactance is inserted in series with the equivalent line and is not connected in parallel (Fig. 14-39).

The H-stubs are joined to the waveguide's narrow dimension so that the H-loops can penetrate from the waveguide into the stub. By contrast with the E-stub, the H-stub is across the equivalent main line and represents a parallel susceptance (Fig. 14-40).

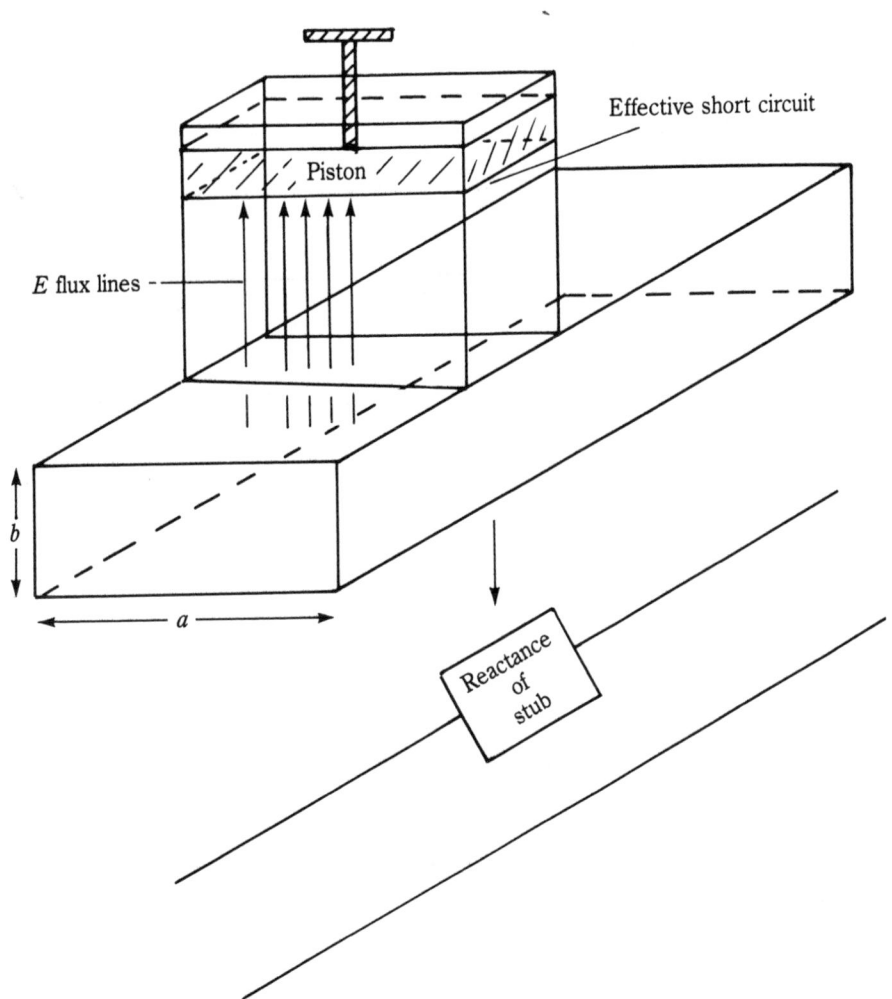

14-39   The E-plane stub and its equivalent circuit.

The stub positions that have been described are compatible with the main wave-guide in the sense that their T-junctions do not create any field discontinuities. In Fig. 14-41 you have compatible E- and H-junctions but there is also an incompatible junction, since any input signal at A will produce equal outputs at B and D but zero output from C. This is the principle behind some of the directional couplers used with waveguides.

Waveguide and hybrid couplers

A hybrid combination of E- and H-type junctions is the *magic-tee coupler,* as shown in Fig. 14-41. Energy passes freely between all four guides except between A and C, since the narrow section of guide A acts as an attenuator for the TE_{10} mode in the guide C.

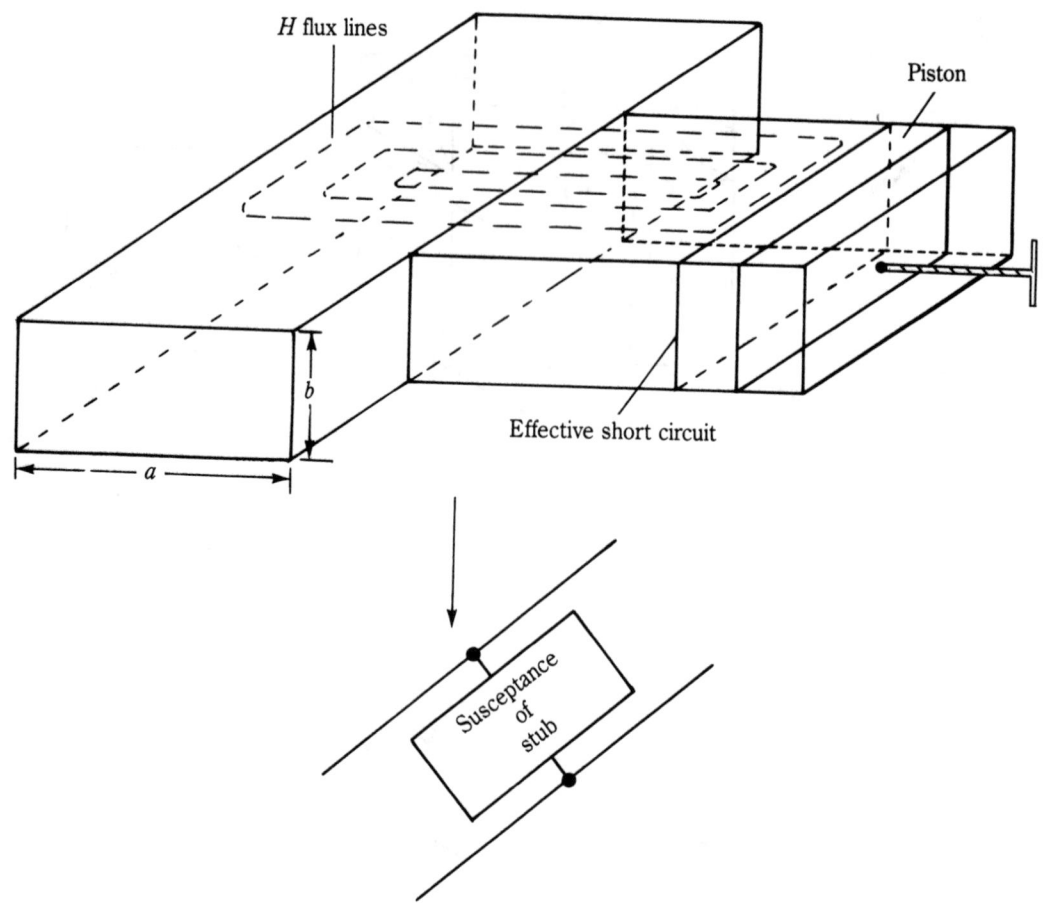

14-40 The H-plane stub and its equivalent circuit.

The directional properties of the magic-tee coupler are summarized as follows:

Input Signal	Output Signals
A	B and D output signals are in phase.
C	B and D outputs are 180° out of phase; no output from A.

Notice that these directional properties are independent of the frequency.

One purpose of the magic-tee coupler is to match the B and D terminations to their guides. If the original signal is introduced at A, there will be equal outputs from B and D but zero output from C. If the distances from C to the B and D terminals are equal, the reflected signals will emerge from C. For a perfect match the terminations must be adjusted until the total output from C is zero.

If the distances from C to the two B and D terminations differ by a quarter-wavelength you can use the magic-tee to match these terminations. If the match is correct, the two signals arriving at C from the terminations will be equal in magnitude but 180°

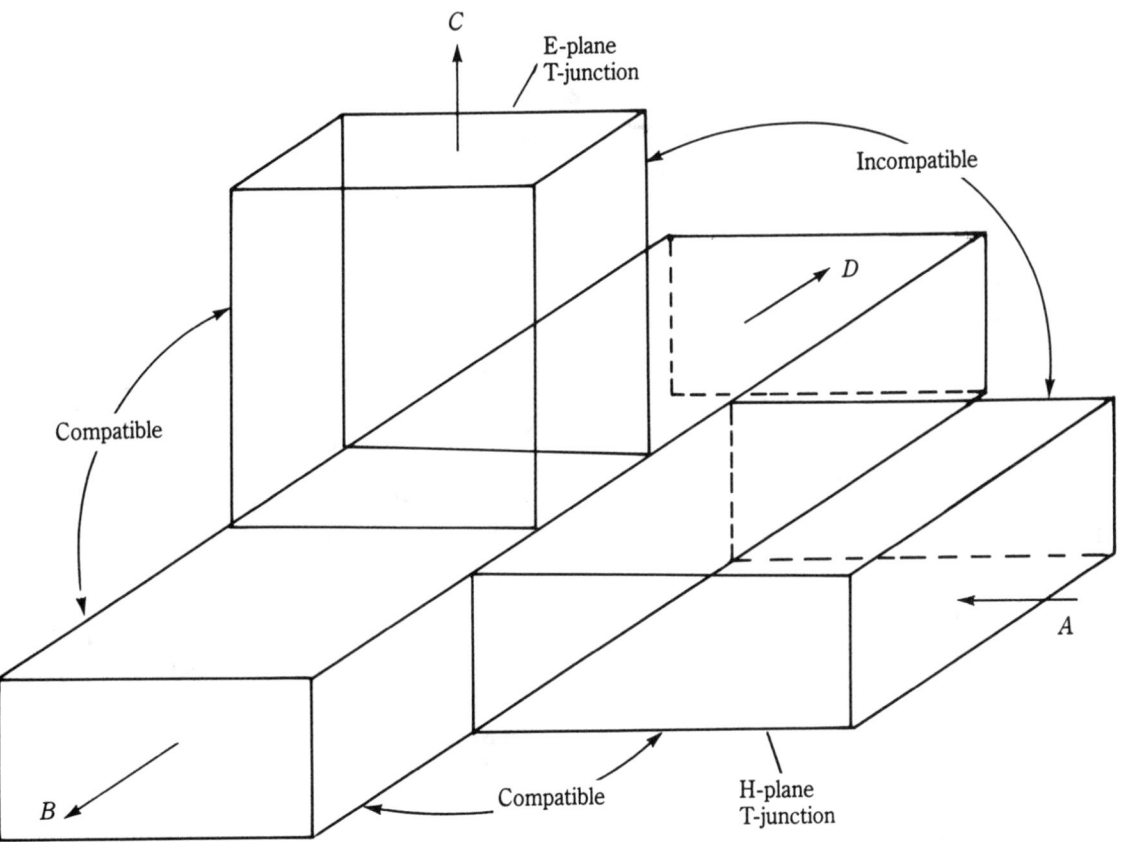

14-41   The magic-tee directional coupler.

out of phase (due to the half-wavelength difference between the two paths). The terminations can then be adjusted for zero output from *C*.

Directional slot couplers

Directional slot couplers are primarily used for monitoring purposes. In the device of
Fig. 14-42A, a small part of the energy passing along the lower waveguide is coupled to
the other guide through small slots or holes as shown. By virtue of the difference in the
path lengths, this energy will combine in the direction of *C* but will tend to cancel out in
the direction of *D*; hence *A* is loosely coupled to *C* but not to *D*. For the energy
reflected back, *D* is loosely coupled to *B* but not to *C*. Since *C* can sample the incident
energy while *D* samples the reflected energy, you can use this type of directional coupler
to measure the incident power, the reflected power, the power reflection coefficient, the
voltage reflection coefficient, and the VSWR.

In the alternative arrangement of Fig. 14-42B, and C, the two amounts of energy
passing through the small circular holes are 180° out of phase by virtue of the fact the
two holes are set on either side of the center-line of the guide. Allowing for these differences in the path lengths, some of the energy entering at A passes through to C but not
through to *D*.

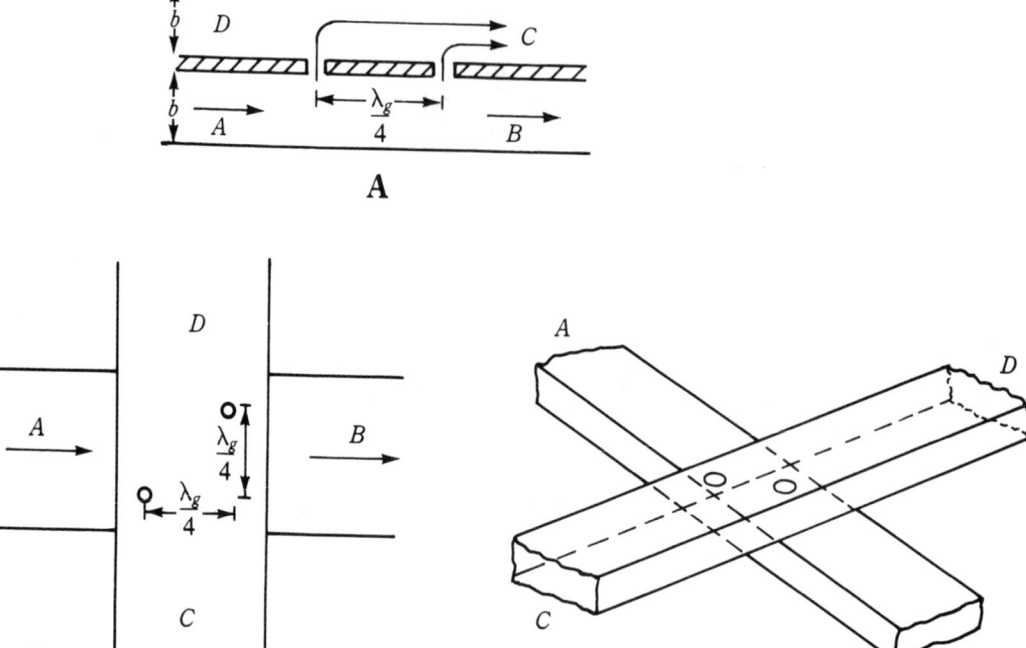

14-42   Directional slot coupler.

The hybrid ring or rat race coupler

The *Hybrid Ring* or *rat race Coupler* arrangement is illustrated in Fig. 14-43A and is a combination of E- and H-type junctions. Consider the path lengths and their consequent phase differences, the direction properties are the same as those listed for the magic-tee. However, the rat race, unlike the magic-tee, is sensitive to changes in the wavelength.

A *coaxial hybrid ring* (Fig. 14-43B) is a simple coaxial equivalent to the rat race circuit. In Fig. 14-43C a phase inversion is introduced by inductive coupling, which produces an effective path difference of $3\lambda/4$ between points C and D. Consideration of the various path lengths show the same directional properties as those listed for the magic-tee.

The ferrite isolator

The *isolator* is a device which has a low forward loss but a high reverse loss. This effect is the result of the nonreciprocal phase shift created by a ferrite material that is the heart of the isolator circuit. In 1845 Michael Faraday demonstrated that the plane of polarization of a linearly polarization light wave rotated when the light is passed through certain materials in a direction parallel to the flux lines of an external magnetic field. The same effect is produced in the microwave region when operating with ferrite materials. Such

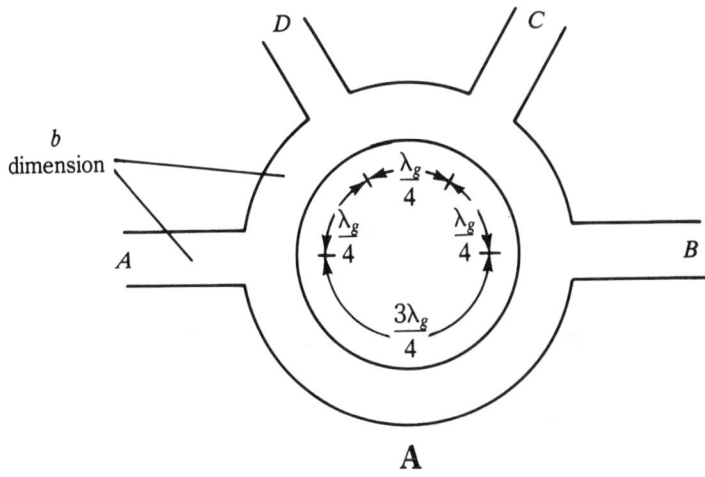

A

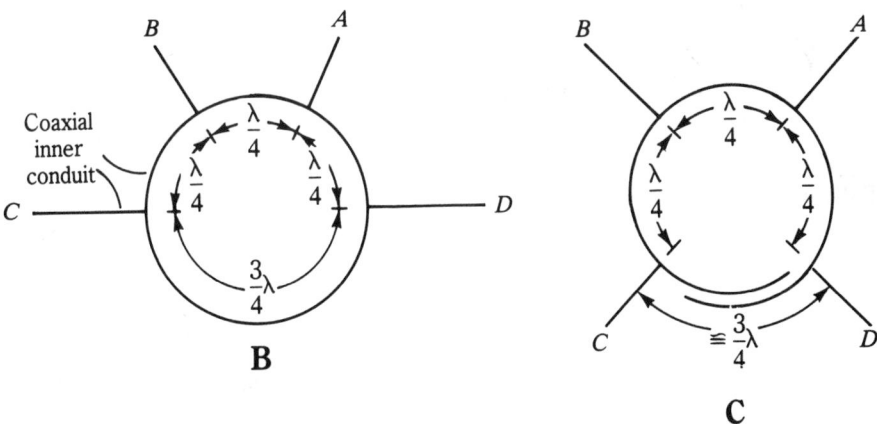

B

C

14-43 Hybrid rings.

materials are transparent to electromagnetic waves, have excellent magnetic properties, and have very high values of specific resistance.

The basic principle of a microwave isolator is illustrated in Fig. 14-44 A. At the input end there is a rectangular waveguide which (relative to the plane of the paper) is fed by an EM-wave with vertical electric flux lines. A transition is now made to a circular waveguide operating in the TE_{11} mode, and a resistive attenuator vane is inserted parallel to the original rectangular guide's wide dimension. Since the E-lines are perpendicular to the plane of the vane, no attenuation occurs.

The next step is to pass the wave through the ferrite specimen that is situated in the presence of the external magnetic field. The result is to twist the plane of polarization through 45° in the clockwise direction. The emerging electric field is also at right-angles to the second resistive vane, so the attenuation of the isolator in the *forward* direction has been kept to a minimum. Finally there is another conversion from a circular to a

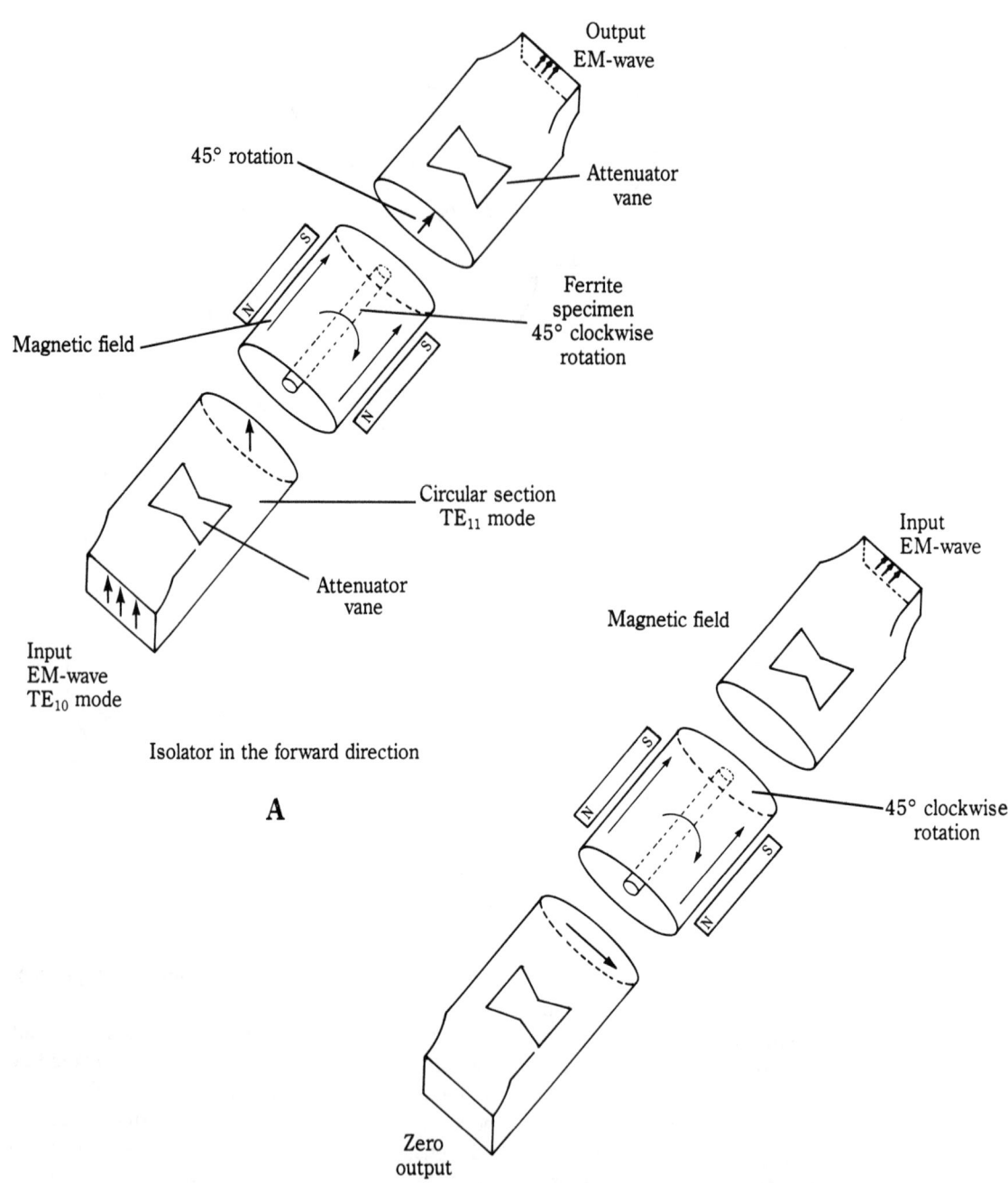

Output
EM-wave

45° rotation

Attenuator
vane

Magnetic field

Ferrite
specimen
45° clockwise
rotation

Circular section
TE$_{11}$ mode

Attenuator
vane

Input
EM-wave
TE$_{10}$ mode

Isolator in the forward direction

A

Input
EM-wave

Magnetic field

45° clockwise
rotation

Zero
output

Isolator in the reverse direction

B

14-44   The rotation isolator.

rectangular waveguide whose orientation corresponds to the emerging wave's plane of polarization.

If you now attempt to send that same signal as a reverse wave back through the isolator, the plane of polarization will be twisted another 45° in the *same* direction as the shift of the forward wave (Fig. 14-44B). Consequently, when the reverse wave emerges from the ferrite specimen the E-lines will be parallel to the first attenuator vane, and severe attenuation is the result. Typically the low forward loss is about 1 dB or less, whereas the high reverse loss is 20 dB or more. If the external magnetic field is supplied by an electromagnet whose current can be varied, the characteristics of the isolator can be changed.

The circulator

The properties of the ferrite specimen used in the isolator can be adapted to the *circulator*. There are many applications of circulators but the principle is to establish various entry/exit points or *ports* where the rf power can either be fed or extracted.

Figure 14-45 shows a circulator that employs the principle of the 45° Faraday rotation. The results of the circulator are:

1. When a vertically polarized wave enters port 1, its plane of polarization is rotated through 45° by the ferrite specimen and the wave then leaves through port 2. Port 3 and port 4 represent incompatible junctions.

2. A 45° polarized wave entering port 2 is horizontally polarized after passing through the ferrite specimen and will therefore exit from port 3. Likewise a wave entering port 3 has its plane of polarization rotated through 45° and only leaves through port 4.

3. A wave entering port 4 will be vertically polarized after leaving the ferrite specimen and will emerge from port 1.

These results are summarized in the circular network diagram of Fig. 14-46. As a simple rule, a 90° rotation from the entrance port in the clockwise direction of the circular arrow will automatically locate the single exit port. Such an arrangement could be used as the duplexer of a radar system which uses the same antenna for both the transmitter and the receiver.

Microwave measurements

A number of devices are used to measure the E- and H-field intensities in a waveguide with a view to obtaining such values as the incident power, the reflected power, and the VSWR.

The most common devices include:

1. The crystal detector shown in Fig. 14-47. The detector current is proportional to the rf voltage induced in the E- or H-probe, which in turn is proportional to the strength of the electric (or magnetic) field intensity at that point in the waveguide. The detector characteristic is of the *square law* variety, so the reading of the ammeter, A, is a measurement of the rf power.

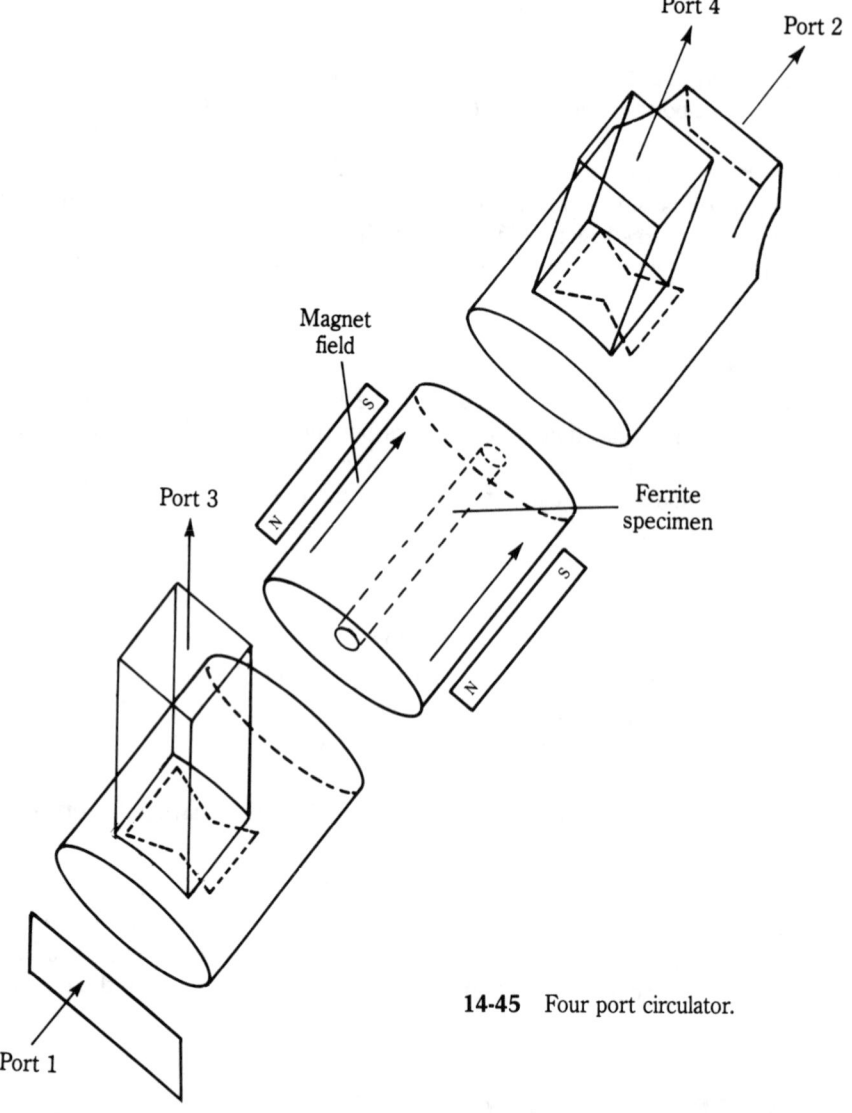

Port 4

Port 2

Magnet
field

Port 3

Ferrite
specimen

Port 1

14-45 Four port circulator.

2. The neon indicator. This type of indicator (Fig. 14-48A) takes the form of a pen-cil-type neon which is inserted through a nonradiating slot into the guide's E-field. The height to which the ionization extends, is proportional to the electric field intensity at that point. By moving the neon along the slot it is possible to obtain the value of the SWR. The multineon indicator of Fig. 14-48B uses several such neons so that the SWR is readily measured without requiring any move-ment.

3. The thermocouple power meter. The dc output of the thermocouple is propor-tional to the temperature of the heater wire, and therefore to the square of the

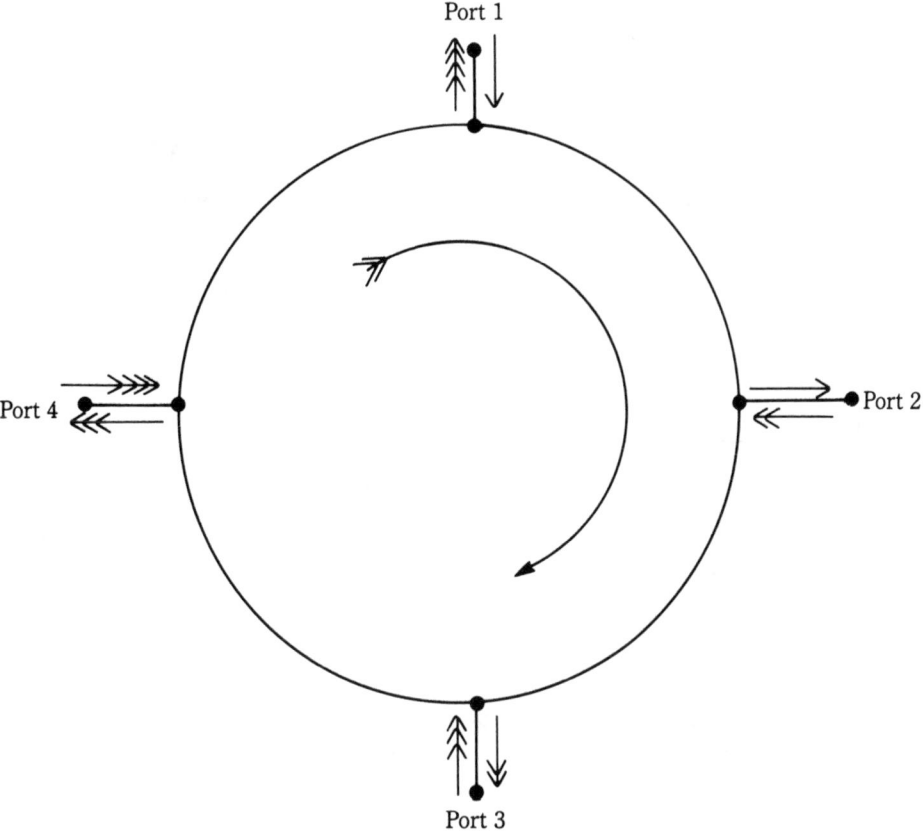

14-46 The principle of the circulator.

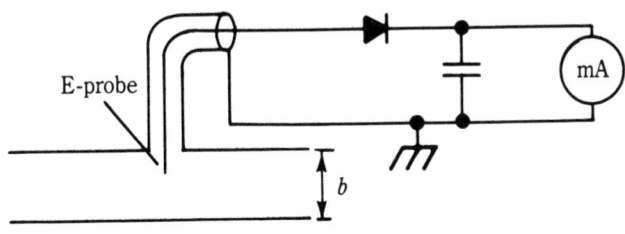

14-47 The crystal detector.

current induced in the H-probe of the waveguide in Fig. 14-49A. A reading of the meter, M, is therefore proportional to the rf power associated with the H-field at that particular point in the waveguide.

A measurement of the total apparent rf power (the sum of the incident and reflected powers) can be obtained by adding the outputs of the two thermocouples which are spaced $\lambda_g/4$ apart as in Fig. 14-49B.

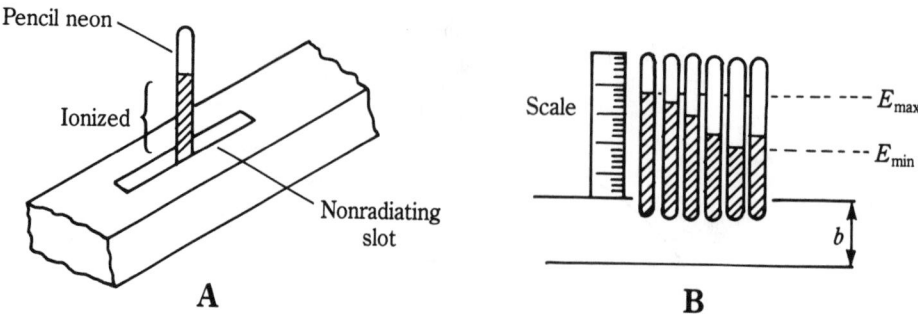

14-48   SWR neon indicators.

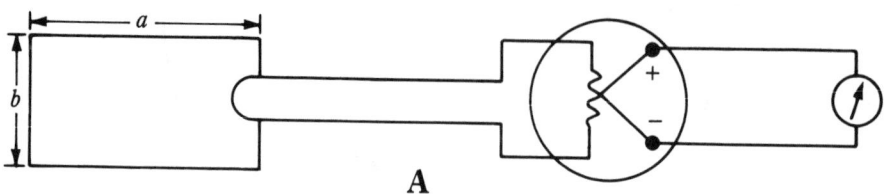

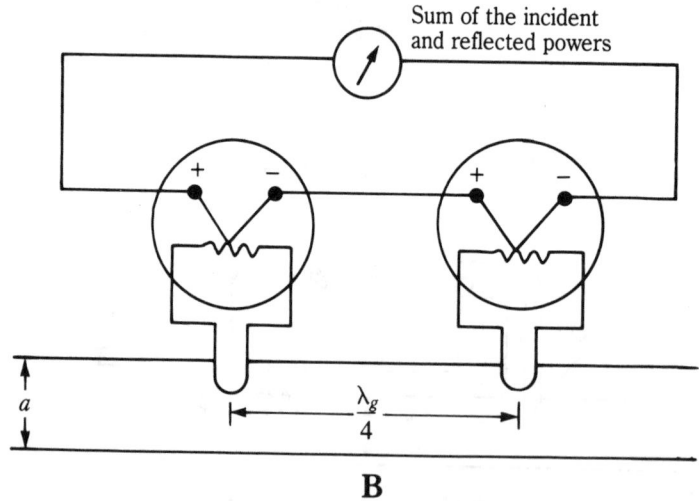

14-49   Thermocouple power meters.

The standing wave ratio, S, can then be used to obtain the true (load) rf power output since:

$$\text{True Power Output} = \text{Total Apparent Power Output} \times \frac{2S}{1 + S^2} \qquad (14\text{-}17)$$

where,

true (load) power = incident power – reflected power

and S is the value of the SWR.

4. The slot hybrid meter. The two slot hybrid arrangement can provide a more accurate method of measuring the rf power output. In Fig. 14-50, a small fraction of the incident rf power from the source is fed via a slot hybrid to a thermistor probe which has a large negative coefficient of resistance. The thermistor resistance is therefore inversely proportional to the rf power and forms part of the Wheatstone bridge.

When R is adjusted for zero balance, the value of R is calibrated to read the incident rf power. Any reflected power passing through the slot is absorbed by the wedge, W.

The hybrid can be reversed to measure the reflected power. Then the load power output = incident power – reflected power, the voltage reflection coefficient,

$$P = \sqrt{\frac{\text{reflected power}}{\text{incident power}}}$$

and the voltage standing wave ratio, $S = (1 + P)/(1 - P)$

Microwave tubes

You have already learned that the ultimate limitation on conventional tubes is the transit time taken by an electron to cross from the cathode to the plate. For example, if an electron is accelerated from the rest through a distance, d (in meters) by a potential difference of V (volts), the electric field intensity is:

$$\mathcal{E} = \frac{V}{d} \, \text{V/m}$$

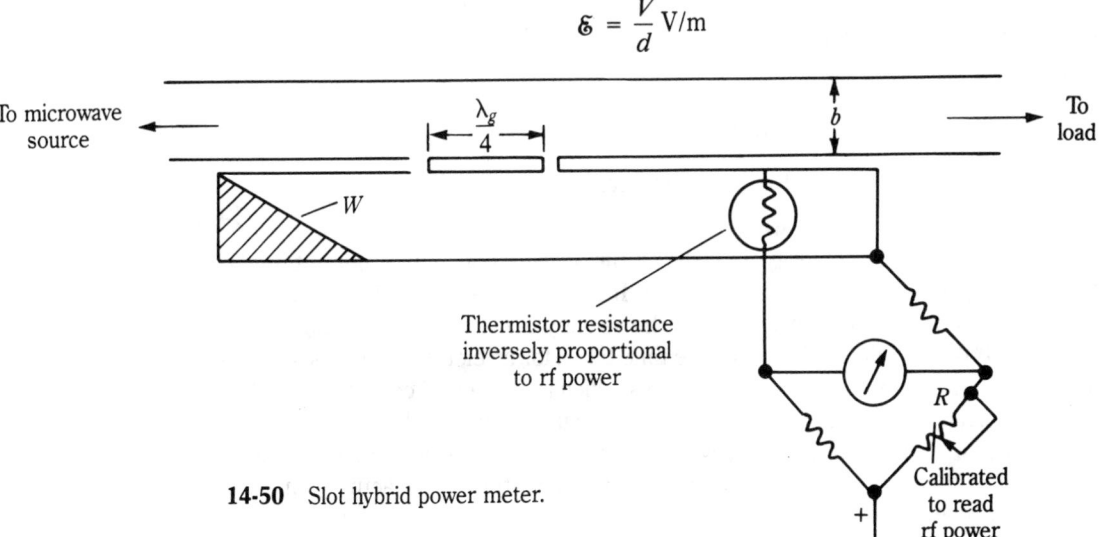

14-50   Slot hybrid power meter.

and the force exerted on the electron,

$$F = Q\mathscr{E}$$
$$= ma \tag{14-18}$$

This yields,

$$a = \frac{Q\,V}{d} \tag{14-19}$$

where,

F = accelerating force (N)
a = electron acceleration (m/s²)
V = accelerating voltage (V)
Q = electron charge (1.602×10^{-10} C)
m = electron mass (9.1096×10^{-31} kg)
d = distance through which the electron is accelerated (m)

The electron's terminal velocity, v, at the end of the distance, d, is given by:

$$v^2 = 2ad$$
$$= 2\frac{QV}{md} \times d$$
$$= \frac{2Q}{m} \times V \tag{14-20}$$

This leads to,

$$v = \sqrt{\frac{2Q}{m}} \times \sqrt{V} \tag{14-21}$$

substituting, $Q = 1.602 \times 10^{-19}$ C

and,

$$m = 9.1096 \times 10^{-31} \text{ kg}$$
$$v = 5.933 \times 10^5 \times \sqrt{V} \text{ meters/sec}$$
$$\tag{14-22}$$

Notice that the terminal velocity depends on the accelerating voltage but not on the distance through which the acceleration occurs. If the voltage is 100 V, the terminal velocity is 5.933×10^6 m/s (approximately 2% of the velocity of light) and the average velocity is $5.933 \times 10^6/2 = 2.9665 \times 10^6$ m/s. The time taken for an electron to cross a distance of 1 mm is $10^{-3}/(2.9665 \times 10^6) = 10^{-9}/2.9665$ s which is the period corresponding to a frequency of nearly 3 GHz. Consequently conventional uhf tubes such as lighthouse triodes are limited to about 2 GHz and below. Since the transit time is the limiting factor for conventional tubes, it follows that an active microwave tube must actually make use of the transit time in order to achieve amplification and/or oscillation.

Examples of such devices are:

1. The *magnetron tube* is normally the heart of a pulsed radar system. The magnetron is an oscillator which is capable of generating a short duration rf pulse with a peak power output of the order of megawatts. However, there is a long time interval between pulses, so the duty cycle is low. For example, a magnetron generates 500 pulses per second but the time interval for each pulse is only $1\mu s$. The period between pulses is $1/500$ s $= 2000$ μs and therefore the duty cycle is only:

$$\text{Duty cycle} = \frac{\text{Active interval}}{\text{Total period}}$$

$$= \frac{\text{Active interval}}{\text{Sum of active and inactive intervals}}$$

$$= \frac{1\mu s}{2000\ \mu s} = 0.005$$

If the peak power is 1 MW,

$$\text{Average power} = \text{Peak power} \times \text{Duty cycle}$$
$$= 1000000 \times 0.0005 = 500 \text{ W}.$$

The magnetron is also the active device in the microwave oven, which operates at a frequency of 2.45 MHz. The space where the food is placed is coupled to the magnetron's resonant cavity.

The frequency range of magnetrons as a whole covers 0.6 to 30 GHz. For a particular magnetron the frequency is normally fixed, but there are methods of obtaining a limited tuning range as high as 10% of the center frequency.

2. The multicavity klystron tube is a stable microwave power amplifier that provides high gain (3 to 90 dB) at medium efficiency (30 to 50%). The frequency range of multicavity klystrons extends from 3 to 30 GHz.

3. The reflex klystron tube is a microwave oscillator with a power output of only a few milliwatts, and a very low efficiency of 5% or less. Its frequency can be tuned mechanically over a 30% range, and electronically over a range of 2% or less. Its main use is in test equipment, and as the local oscillator in a radar receiver.

4. The *traveling wave tube* (TWT) is a high-gain, low-noise, wide-band microwave amplifier. Since no resonant cavities are employed, the upper frequency limit of a TWT can be twice its lower limit.

The primary use of the TWT is voltage amplification (although power TWTs with characteristics similar to those of a power klystron have been developed). The wide bandwidth and low-noise characteristics have made the TWT ideal for use as rf amplifiers in microwave and electronic countermeasures equipment. For these purposes TWTs have been designed for frequencies as low as 300 MHz but as high as 50 GHz.

5. The *backward wave oscillator* (BWO) consists of two basic types:

   (a) The "O-" type which is frequently used as a low-power oscillator with an output of a few hundred milliwatts over a normal frequency range from 1 to 15 GHz. However such oscillators have been made to operate as high as 200 GHz.

   (b) The "M-" or *crossed field type*. This is primarily used as a transmitting tube with a power output of a few hundred watts and an efficiency of about 30%. The output frequency is of the order of a few GHz.

The discussion of microwave tubes begins by concentrating on the magnetron.

The magnetron

Basically, the magnetron is a diode and has no grid. A magnetic field in the space between the plate (anode) and the cathode serves as the controlling mechanism. The plate of a magnetron does not have the same physical appearance as the plate of an ordinary electron tube. Since conventional LC networks become impractical at microwave frequencies, the plate is fabricated into a cylindrical copper block containing resonant cavities that serve as tuned circuits. The magnetron's base differs greatly from the conventional base. It has short, large-diameter leads that are carefully sealed into the tube and shielded, as shown in Fig. 14-51.

The cathode and the filament are at the center of the tube and are supported by the filament leads, which are large and rigid enough to keep the cathode and the filament fixed in position. The output lead is usually a probe or loop extending into one of the tuned cavities and coupled into the waveguide. The plate structure, as shown in Fig. 14-52, is a solid block of copper. The cylindrical holes around its circumference are the resonant cavities. A narrow slot runs from each cavity into the central portion of the tube, and divides the inner structure into as many segments as there are cavities. Alternate segments are strapped together to put the cavities in parallel with regard to the output. These cavities control the output frequency. The straps are circular metal bands that are placed across the top of the block at the entrance slots to the cavities. Because

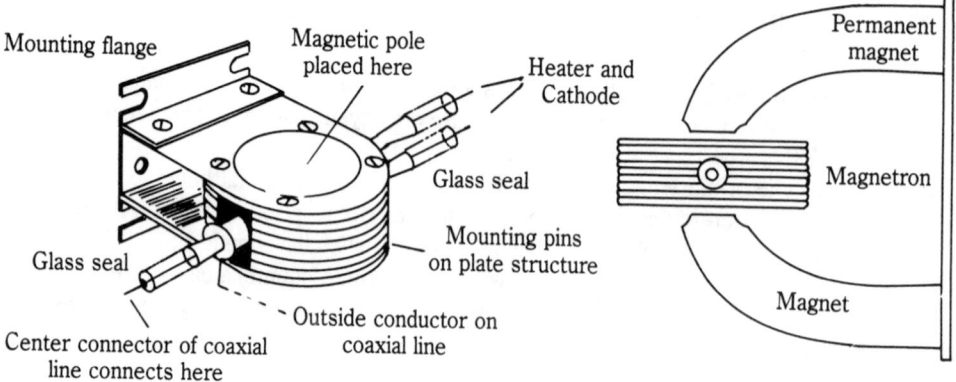

14-51   Construction of the magnetron.

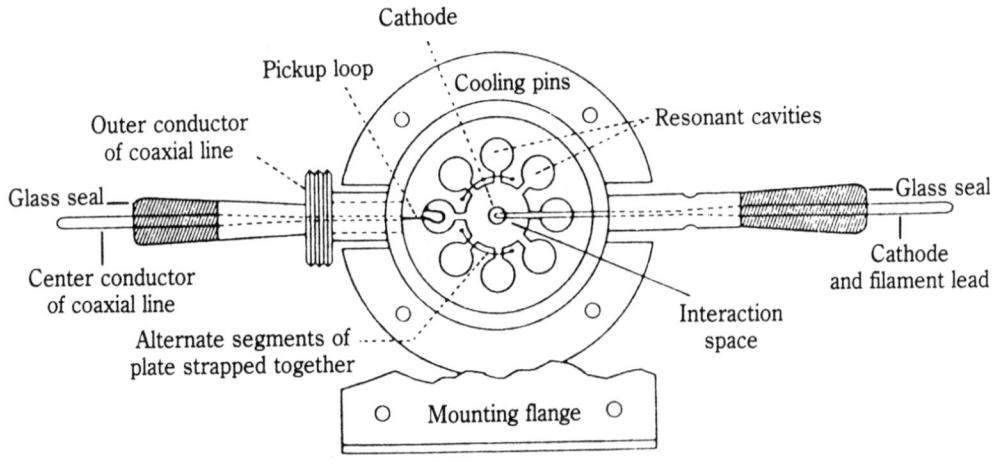

14-52 Cutaway view of the magnetron.

the cathode current is of the order of several amperes, the cathode must be large and must be able to withstand high operating temperatures. Such a cathode must also have good emission characteristics, particularly under back bombardment, because much of the output power is derived from the large number of electrons emitted when high-velocity electrons return to strike the cathode.

The cathode is indirectly heated, and is constructed of a high-emitting material. The open space between the anode and the cathode is called the *interaction space* because it is in this space that the electric and magnetic fields interact to exert a force on the electrons. The magnetic field is normally provided by a strong permanent magnet, mounted around the magnetron so that the direction of the magnetic field is parallel with the axis of the cathode. The cathode is mounted in the center of the interaction space.

Since the anode is exposed while the cathode leads are carefully sealed, the anode is grounded and a large negative voltage of several kV is applied to the cathode during the time of the magnetron's oscillation.

Basic magnetron operation

The theory of the operation of the magnetron is based on the motion of electrons under the influence of combined electric and magnetic fields.

The following laws govern this motion:

1. The direction of an electric field is from the positive electrode to the negative electrode. The law governing the motion of an electron due to an electric, or E-field, states that the force exerted by the electric field on an electron is proportional to the electric field intensity. Electrons tend to move from a point of negative potential toward a positive potential as shown in Fig. 14-53. In other words, the electrons tend to move against the direction of the E-field.

2. When an electron is being accelerated by an E-field, as shown in Fig. 14-53, energy is taken from the field by the electrons.

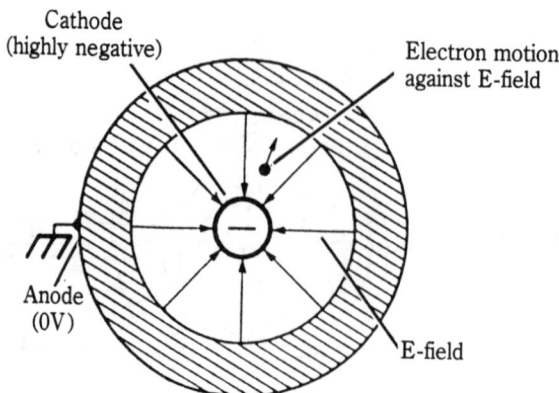

14-53 The motion of the electron in the magnetron's E-field.

3. The law of motion of an electron in a magnetic or H-field, states that the force exerted on the electron in a magnetic field is at right-angles to both the directions of the field and the path of the electron. The direction of the force is such that the electron trajectories are clockwise when viewed in the direction of the magnetic field as shown in Fig. 14-54; this is an example of the right hand motor rule. If the permanent magnetic field strength is increased, the electron path will take a sharper bend. Likewise, if the velocity of the electron increases, the field around it increases, and its path will again bend more sharply.

Examples of anode blocks

The first type shown in Fig. 14-55A has cylindrical cavities, and therefore possesses a hole-and-slot anode. The second type uses the *vane anode* (Fig. 14-55B) which has trapezoidal cavities. These first two anode blocks operate in such a way that alternate segments must be connected, or strapped, to ensure that each segment is opposite in polarity to its neighboring segment on either side (as shown in Fig. 14-56). This also requires an even number of cavities.

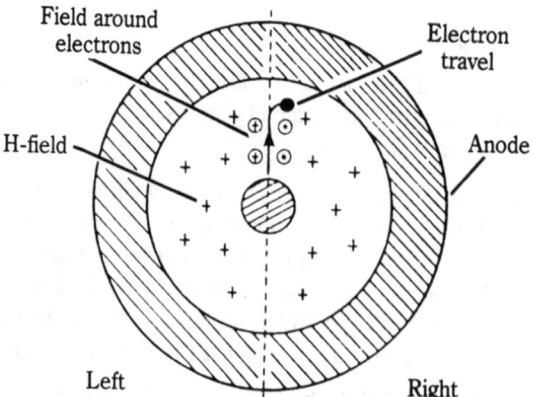

14-54 The motion of the electron in the magnetron's H-field.

Straps

Straps

Unstrapped

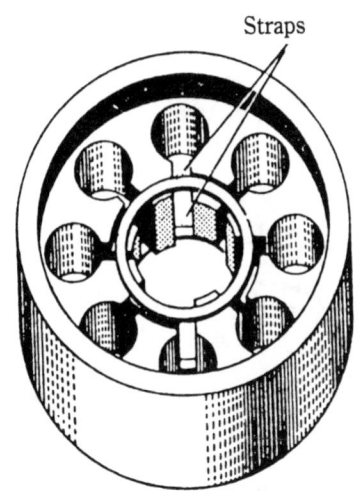

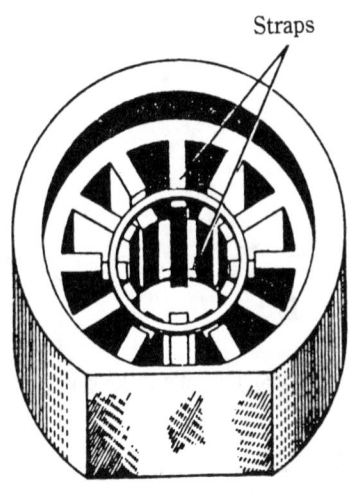

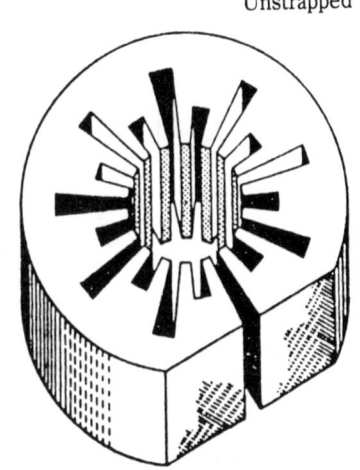

A. Hole and slot
block

B. Regular trapezoidal
block

C. Rising sun block

14-55   Examples of anode blocks.

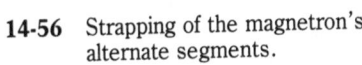

14-56   Strapping of the magnetron's
alternate segments.

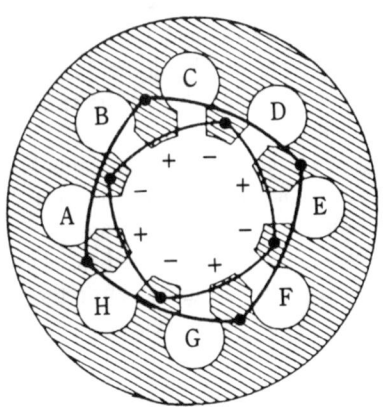

The third type, illustrated in Fig. 14-55C is called a *rising sun block* because of its appearance. The alternative large and small trapezoidal cavities in this block result in a stable frequency which lies between the resonant frequencies of the large and the small cavities.

Figure 14-57A shows the physical appearance of the resonant cavities contained in the hole-and-slot anode that you will use when analyzing the operation of the magnetron. Notice that the cavity consists of a cylindrical hole in the copper anode, and a slot which connects the cavity to the interaction space.

The electrical equivalent circuit of the cavity and slot is shown in Fig. 14-57B. The parallel sides of the slot form the plates of a capacitor, while the walls of the hole act as an inductor. The hole and the slot then form a high-Q resonant LC circuit. As shown in Fig. 14-55A, the anode of the magnetron contains a number of these cavities.

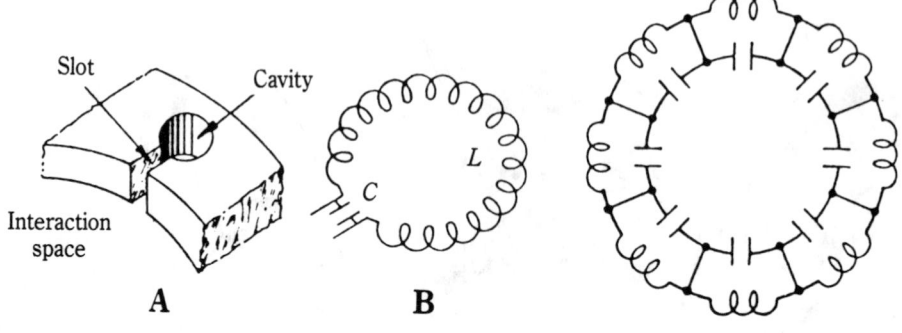

14-57   Equivalent electrical circuit of the "hole and slot" cavity.   **C**

An analysis of the anode in Fig. 14-55A shows that the LC tank circuits of the cavities are in series, as shown in Fig. 14-57C. This is assuming that the straps have been removed. However, an analysis of the anode block after alternate segments have been strapped (Fig. 14-56), will reveal that the cavities are now connected in parallel. The result of the strapping is shown in Fig. 14-58.

Operation of the magnetron

The resultant electric field in the electron resonant magnetron is a combination of the ac fields and a dc field. The dc field extends radially between the anode and the cathode while the ac fields are due to the rf oscillations induced in the resonant cavities of the anode block.

Figure 14-59 shows the ac fields between adjacent segments at the instant of the peak value of one alternation in the rf oscillations occurring in the cavities.

A strong dc field extends from the anode to the cathode, and is due to a large negative dc voltage pulse applied to the cathode. This strong dc field causes electrons to accelerate toward the anode after they have been emitted from the cathode. These accelerated electrons take energy from the dc electric field. Oscillations are sustained in

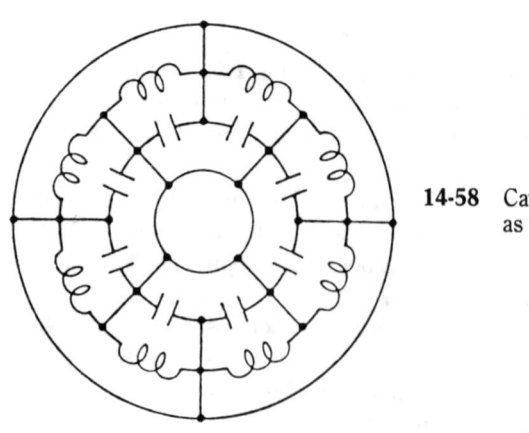

14-58   Cavities connected in parallel as the result of strapping.

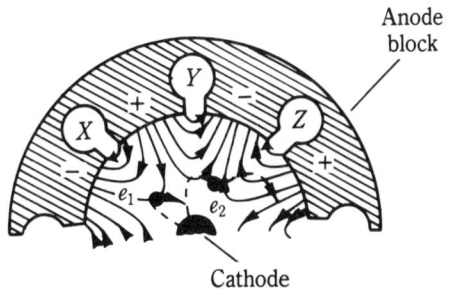

14-59 Motion of the electrons in the presence of the magnetron's E- and H-fields.

a magnetron because the electrons gain energy from the dc field, and give up this energy to the ac fields as they pass through these fields.

In Fig. 14-59 consider an electron e_1, which is shown entering the field around the slot entrance to the cavity X. The clockwise rotation of the electron path is due to the interaction of the magnetic field around the moving electron with the permanent magnetic field, which is assumed to be entering the paper in Fig. 14-59. Notice that the electron, e_1, which has entered the ac field around the cavity X, is going against the direction of the ac field. This electron will therefore take energy from the ac field and will be accelerated, so that it turns more sharply as its velocity increases. The electron e_1, will therefore turn away from the anode, and when it strikes the cathode, it will give up the energy it received from the ac field in the form of heat. This will force more electrons to leave the cathode and accelerate toward the anode. By contrast the electron e_2, is slowed down by the ac field and therefore gives up some of its energy to that field. Because the electron e_2, loses some of its velocity, the deflection force exerted by the H-field is reduced and the electron path deviates to the left in the direction of the anode, and not in the direction of the cathode as was the case with the electron, e_1.

The cathode to anode potential and the magnetic field strength (E-field to H-field relationship) determines the time taken by the electron, e_2, to travel from a position in front of the cavity, Y, to a position in front of the cavity, Z. This time is equal to approximately one-half period of the rf oscillation of the cavities. When the electron e_2, reaches a position in front of the cavity Z, the ac field of that cavity will be reversed from that shown in Fig. 14-59. As a result the electron e_2 will give up energy to the ac field of cavity Z and will slow down still further. The electron, e_2, will actually give up energy to each cavity as it passes, and will eventually reach the anode when its energy is expended. Therefore, the electron, e_2, will have helped to sustain the oscillation because it has taken energy from the dc field and given it to the ac fields. The electron, e_1, which took energy from the ac field around the cavity X did little harm because it immediately returned to the cathode.

Electrons such as e_2 that give energy to the ac field as they rotate clockwise from one ac field to the next, stay in the interaction space for a considerable time before striking the anode.

The cumulative action of so many electrons with some being returned to the cathode while others are directed toward the anode, forms a pattern resembling the spokes of a wheel, as shown in Fig. 14-60.

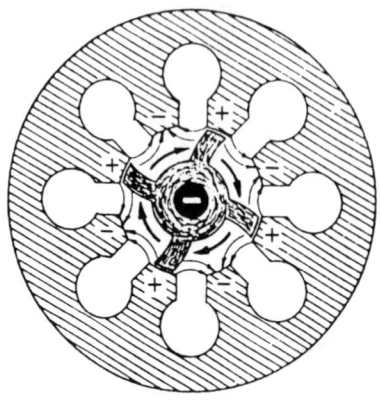

14-60 Concentrations of the electrons in the interaction space.

This overall space charge "wheel" rotates about the cathode at an angular velocity of two anode segments for each cycle of the ac field, and it also has a phase relationship that enables the electron concentration to deliver energy continuously and therefore sustain the rf oscillations. Electrons emitted from the area of the cathode between the spokes are, as previously discussed, quickly returned to the cathode.

In Fig. 14-60 it is assumed that alternative segments between cavities are at the same potential at the same instant, and that there is an ac field existing across each individual cavity. This type of mode operation is called the *pi* (π) *mode*, since adjacent segments of the mode have a phase difference of 180° or π radians. There are, in fact, several other possible modes of oscillation, but the π mode has the greatest power output and is the one which is most commonly used.

In order to ensure that alternate segments have identical polarities, an even number of cavities, usually six or eight, are used, and alternate segments are strapped. The frequency of the π mode is separated from the frequency of the other modes by the strapping.

For the π mode, all parts of each strapping ring are at the same potential, but the two rings have alternately opposing potentials, as shown in Fig. 14-61. The stray capacitance between the rings then adds capacitive loading to the resonant mode; however, if

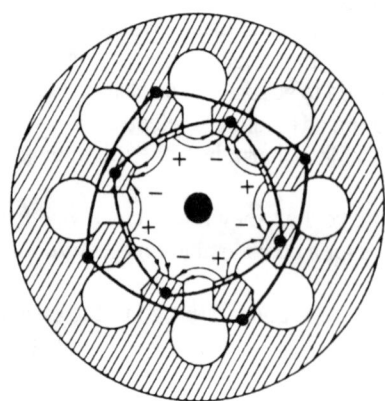

14-61 Separation of modes due to strapping.

there is a phase difference between the successive segments connected to a given strapping ring, current is caused to flow in that strap, which then has an inductance effect. As a result an inductive shunt is placed in parallel with the cavity's equivalent circuit, thereby lowering the total inductance and increasing the frequency for all modes other than the π mode.

Coupling methods

A magnetron's rf energy can be removed by means of a coupling loop. At frequencies lower than 10,000 MHz, the coupling loop is made by bending the inner conductor of a coaxial cable into a loop and soldering the end to the outer conductor, so that the loop projects into the cavity (Fig. 14-62A). To obtain sufficient pickup at higher frequencies the loop is located at the end of the cavity as shown in Fig. 14-62B.

The *segment-fed loop method* is shown in Fig. 14-62C. Here the loop intercepts the flux passing between the cavities. By contrast the *strap-fed loop* method (Fig. 14-62D) intercepts the energy between the strap and the segment. On the output side, the coaxial cable feeds directly into a waveguide with the vacuum seal at the inner conductor helping to support the line.

Aperture, or *slot coupling* is illustrated in Fig. 14-62E. This method allows the rf energy to be coupled directly to a waveguide with an iris feeding into the waveguide connector through the slot.

Tuning

A tunable magnetron permits the system to be operated at a precise frequency anywhere within a band of frequencies, as determined by the magnetron's characteristics.

The resonant frequency of a magnetron can be varied by changing the inductance or capacitance of the resonant cavities. In Fig. 14-63 an inductive tuning element is inserted into the hole portion of the hole-and-slot cavities. It changes the inductance of the resonant circuits by altering the surface to volume ratio in a region of high current. The type of tuner in Fig. 14-63 is called a *sprocket tuner* or *crown of thorns* tuner. All of its tuning elements are attached to a frame, which is positioned by means of a flexible bellows arrangement. The insertion of the tuning elements into each anode hole decreases the inductance of the cavity and therefore increases the resonant frequency. One of the limitations of inductive tuning is that it lowers the unloaded Q of the cavities and therefore reduces the efficiency of the tube.

The insertion of an element (ring) into the cavity slot as shown in Fig. 14-64 increases the slot capacitance and decreases the resonant frequency. Because the gap is narrowed in width, the breakdown voltage will be lowered, and capacity magnetrons must be operated with low voltages, hence the low power outputs. The type of capacity tuner illustrated in Fig. 14-64 is called a *cookie cutter* tuner. It consists of a metal ring inserted between two rings of a double-strapped magnetron, thereby increasing the strap's capacitance. Because of the mechanical and voltage breakdown problems associated with the cookie cutter tuner, it is more suited for use at the longer wavelengths.

Both the capacitance and inductance tuners described are symmetrical. Each cavity is affected in the same manner, and the angular symmetry of the π mode is preserved.

A 10% frequency range can be obtained with either of the two tuning methods described. There is some indication that the ''cookie cutter'' tuner is more restricted

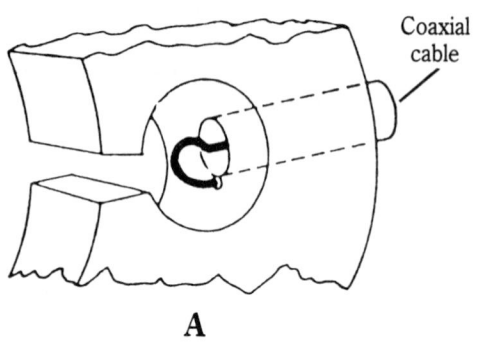

Coaxial
cable

A

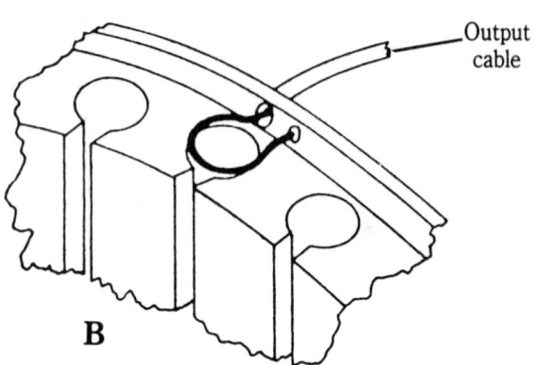

cable

B

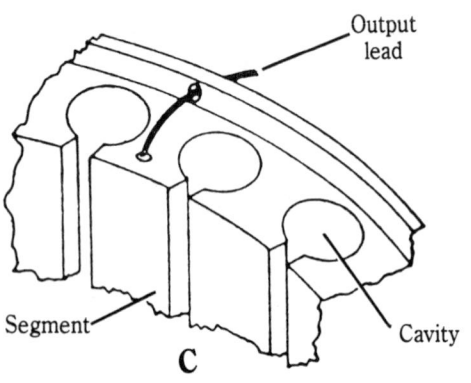

lead

Cavity

C

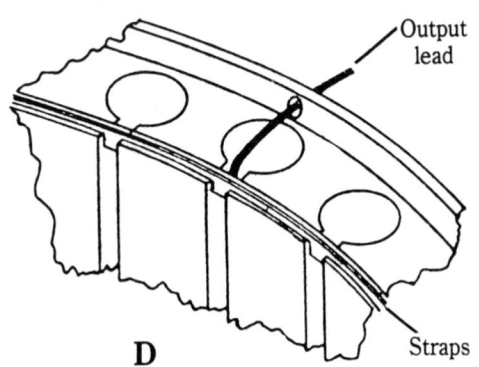

lead

Straps

D

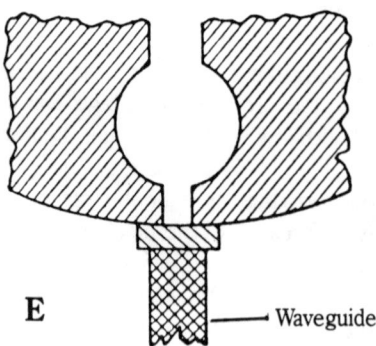

E

Waveguide

14-62 Methods of coupling from the "hole and slot" magnetron.

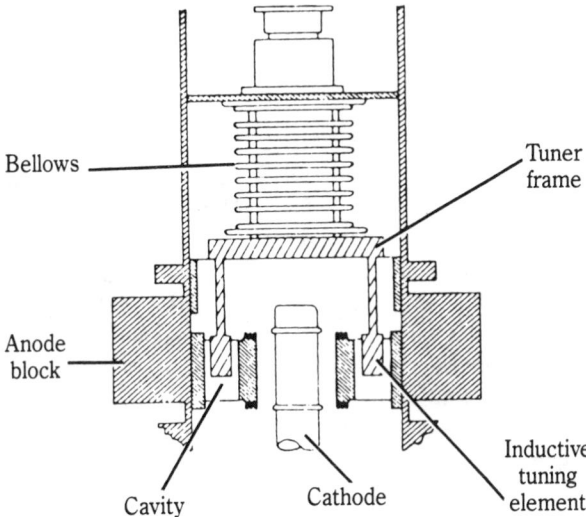

Bellows

Tuner frame

Anode block

Inductive tuning element

Cavity

Cathode

14-63 Inductive magnetron tuning.

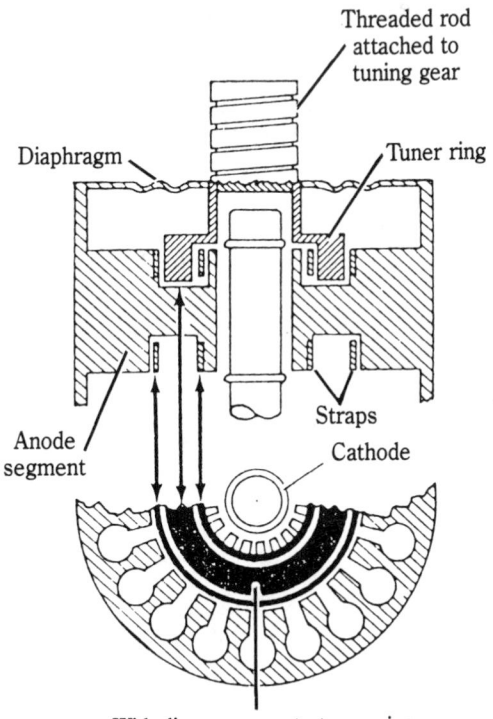

Threaded rod attached to tuning gear

Diaphragm

Tuner ring

Anode segment

Straps

Cathode

Wide line represents tuner-ring
position between the magnetron straps

14-64 Capacitive magnetron tuning.

than the "crown of thorns" tuner. The two tuning methods can be used in combination to cover a larger tuning range than is possible with either one alone.

Defects in the magnetron

The main indication of a defective magnetron is a low reading of the magnetron current so that the undercurrent relay drops out. The CRT will then show noise and the sweep, but the radar echoes will be unusually weak and fuzzy or will not be seen at all. In addition, the automatic frequency control (AFC) system will be ineffective.

If the magnetron itself is within its operating limits but the external magnet is too weak, the magnetron current will increase but the frequency of the oscillation will drift, and the AFC system will not be effective. Under extreme conditions the oscillation might entirely cease. To avoid such weakening, the technician should not subject the magnet to extreme heat or any physical shocks. Furthermore, all metal tools should be kept well away from the magnet's presence.

In terms of general precautions, the technician when servicing or maintaining a radar set on board a ship, should make certain that all power is shut off, and that all capacitors are fully discharged. In particular, CRTs must be handled with great care. One further point—arcing might occur in the modulator unit, the magnetron, the waveguide assembly, and other parts of the radar system. Such arcing can present a hazard if the ship is handling explosive or inflammable material.

The principles of the magnetron can be summarized by considering a typical set of operating conditions. Let a magnetron be pulsed 500 times per second by an accompanying discharge line, which consists of three LC sections each having an inductance of $L = 6.5\ \mu H$, and a capacitance of $C = 4000\ pF$ (Fig. 14-65).

The pulse length, $2N\sqrt{LC}$, is about 1.0 μs, and the characteristic impedance

$$Z_o = \sqrt{\frac{L}{C}} \approx 40\ \Omega$$

An average magnetron requires an anode voltage of 26 kV and an anode current of 40 A, so the magnetron impedance,

$$Z_M = \frac{26000\ V}{40\ A}$$
$$= 640\ \Omega$$

Hence the required turns ratio for the pulse transformer,

$$T = \sqrt{\frac{Z_M}{Z_o}}$$
$$= 4{:}1$$

The primary pulse voltage $= \dfrac{26\ kV}{T} = 6.5\ kV$

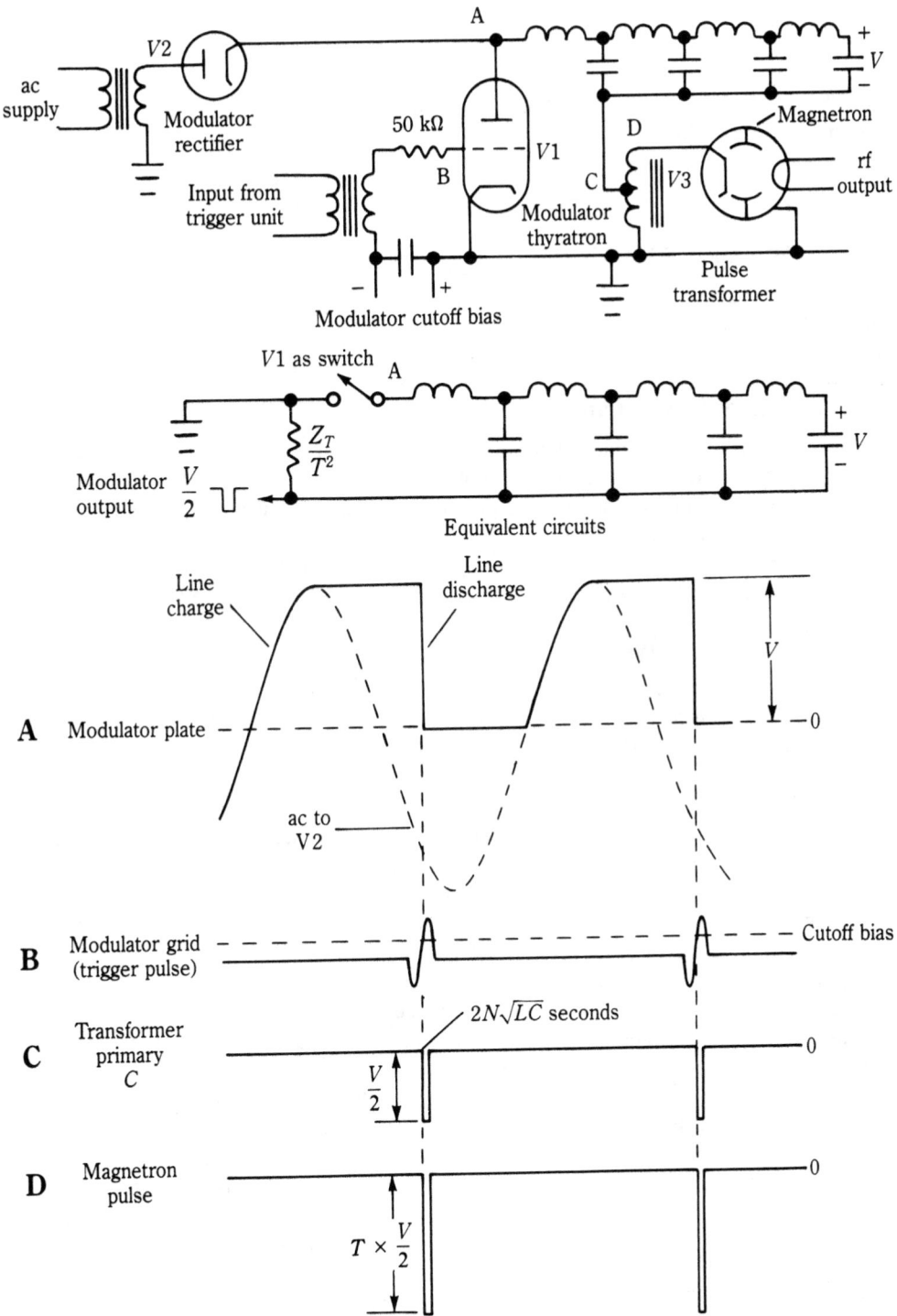

14-65 Discharge line modulator using thyratron tube.

The primary pulse current $= 40 \text{ A} \times T$

$$= 160 \text{ A}$$

The required discharge line voltage $= 2 \times 6.5 \text{ kV} = 13 \text{ kV}$
and this is achieved by the ac supply and the diode $D1$.

The peak of the pulse power input to the magnetron $= 26000 \text{ V} \times 40 \text{ A}$
$$= 1040 \text{ kW}$$

The average power input to the magnetron $= 1040 \text{ kW} \times 1 \text{ } \mu s/2000 \text{ } \mu s$
$$\approx 500 \text{ W}$$

The typical rf peak value power output $= 400 \text{ kW}$, so the magnetron efficiency is approximately 40%.

The multicavity klystron

The klystron tube is a stable microwave power amplifier that provides high gain at medium efficiency. Depending on the type of tube, klystron power outputs range from a few milliwatts to several megawatts peak power, and over 100 kilowatts average power. The power gains vary from 3 to 90 dB.

Klystron amplifiers are somewhat noisy and are therefore used mainly as power amplifiers. However, they have applications in many facets of microwave technology.

Operation of the klystron amplifier

The klystron tube makes a virtue of the very thing that defeats the triode—the transit time of the electrons as they cross from the cathode to the plate. By contrast, the klystron modulates the velocity of the electrons, so that as the electrons travel through the tube, electron bunches are formed (density modulation). These bunches deliver a positive feedback voltage to the output resonant circuit of the klystron.

Figure 14-66 shows a cutaway representation of the basic klystron amplifier. The klystron amplifier consists of three separate sections: the electron gun, the rf section, and the collector.

First consider the electron gun structure. It consists of a heater, cathode, control grid, and anode. Electrons are emitted by the cathode and drawn toward the anode, which is operated at a positive potential with respect to the cathode. The electrons are formed into a narrow beam by either electrostatic or magnetic focusing techniques; this ensures that the electron beam does not spread out. The control grid is used to govern the number of electrons that reach the anode region. It can also be used to turn the tube completely on or off in certain pulsed amplifier applications.

The electron beam is well formed by the time it reaches the anode. The beam passes through a hole in the anode and on to the rf section of the tube, and eventually strikes the collector. The electrons are then returned to the cathode through an external power supply. It is evident that the collector of a klystron acts much like the plate of a triode insofar as the collection of electrons is concerned. However, there is one important difference. The plate of the triode is normally connected, in some fashion, to the output rf circuit, while, in a klystron amplifier, the collector has no connection to the rf circuitry at all.

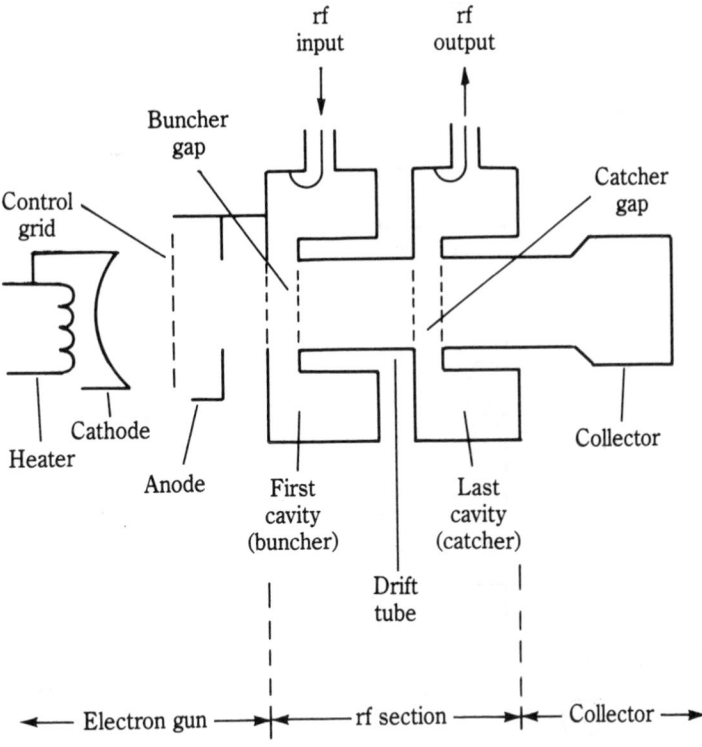

14-66   The basic klystron amplifier.

Look at the rf section of the basic klystron amplifier. This part of the tube is quite different from a conventional triode circuit, since the resonant circuits used in a klystron amplifier are reentrant cavities.

Refer to Fig. 14-66, the electrons pass through the cavity gaps in each of the resonators as well as the cylindrical metal tube between the gaps. This metal tube forms the so-called drift space. In a klystron amplifier the low level rf input signal is coupled to the first resonator, which is called the *buncher cavity*. The signal can be coupled through either a waveguide or a coaxial connection. If the cavity is tuned to the frequency of the rf input, it will be excited into oscillation. An electric field will exist across the buncher gap, alternating at the input frequency. For half a cycle, the electric field will be in a direction that will cause the field to increase the velocity of electrons flowing through the gap. On the other half-cycle, the field will be in a direction which will cause the field to produce a decrease in electron velocity. This effect is called velocity modulation, and is illustrated in Fig. 14-67A. Note that when the voltage across the cavity gap is negative, the electrons will decelerate; when the voltage is zero, the electrons will be unaffected, and when the voltage is positive, the electrons will accelerate.

After leaving the buncher gap (Fig. 14-67B) the electrons proceed through the tube's drift region, then on to the collector. In the drift region, the electrons which have been speeded up by the electric field in the buncher gap, will tend to overtake electrons which have been slowed down. Because of this action, bunches of electrons will begin to

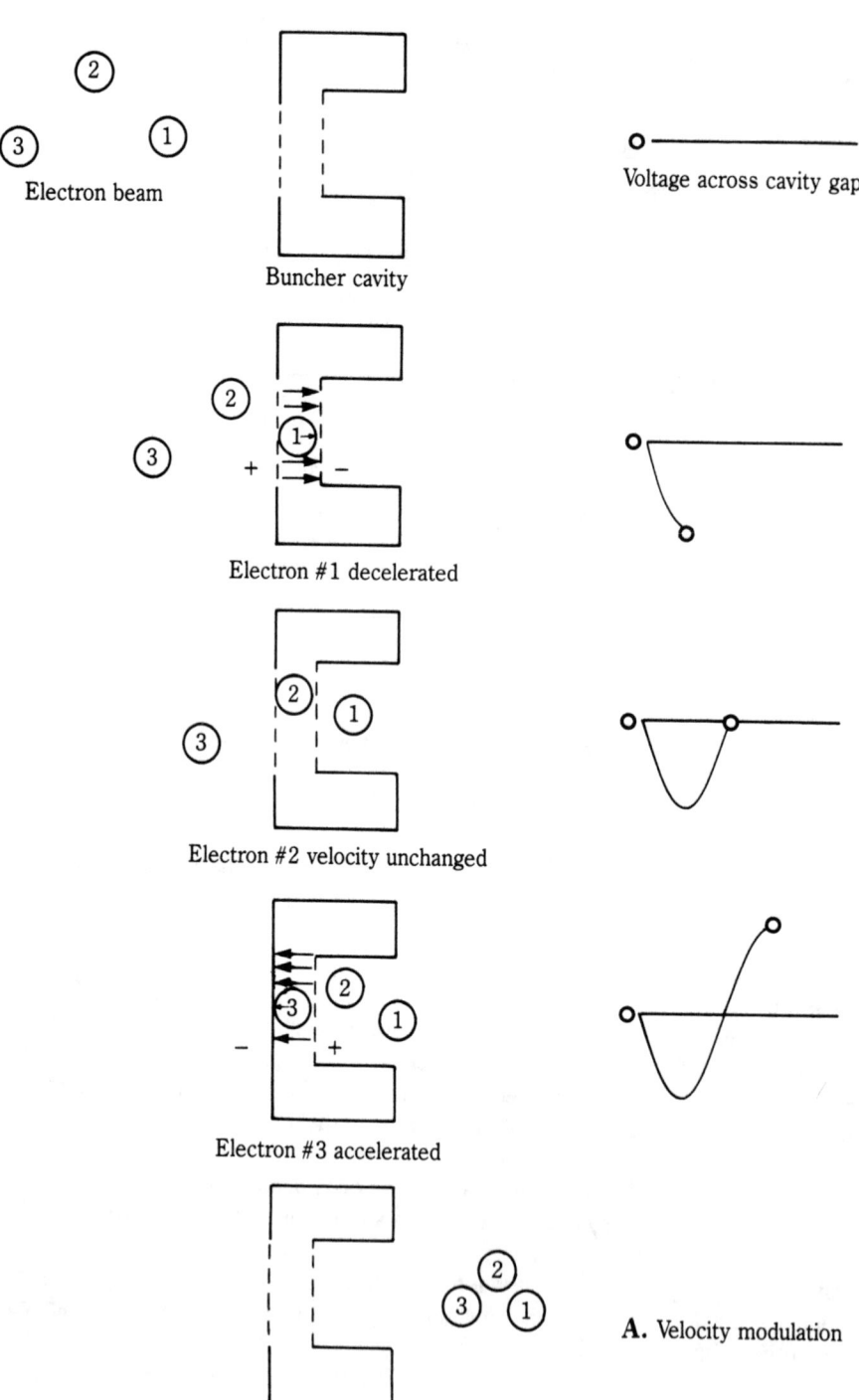

Electron beam

Voltage across cavity gap

Buncher cavity

Electron #1 decelerated

Electron #2 velocity unchanged

Electron #3 accelerated

A. Velocity modulation

Electrons beginning to bunch, due to velocity differences

14-67 The amplifying action of the klystron.

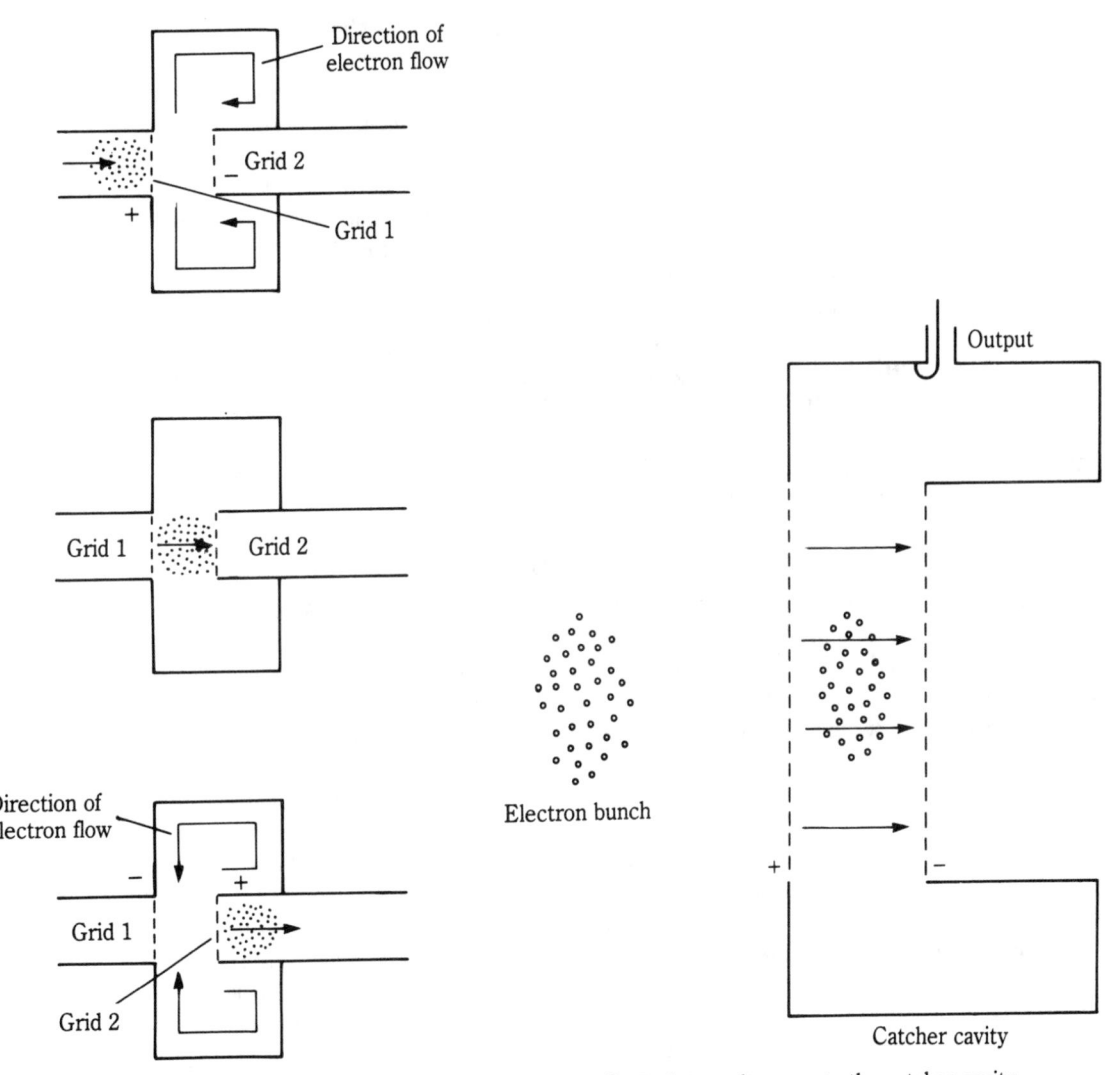

B. Density modulation

C. Delivery of energy to the catcher cavity

14-67 Continued

form in the drift region, and will be completely formed by the time that they reach the gap of the last cavity, which is called the *catcher cavity*. Bunches of electrons periodically flow through the gap of this catcher cavity, and during the time between bunches, relatively few electrons flow through the gap. The time between the arrival of the electron bunches is equal to the period of the cycle of the rf input signal.

The initial bunch of electrons flowing through the catcher cavity will cause the cavity to oscillate at its resonant frequency. This sets up an alternating electric field across the catcher cavity gap, as illustrated in Fig. 14-67C. With the right design and operating potentials, a bunch of electrons will arrive in the catcher cavity gap at the proper time to be retarded by the rf field, and therefore energy will be given up to the catcher cavity.

The rf power in the catcher cavity will be much greater than that applied in the buncher cavity. This is due to the ability of the concentrated bunches of electrons to deliver large amounts of energy to the catcher cavity. Because the electron beam delivers some of its energy to the output cavity, it arrives at the collector with less total energy than it had when it passed through the input cavity. This difference in beam energy is approximately equal to the energy delivered to the output cavity.

It is appropriate to mention here that velocity modulation does not form perfect bunches of electrons. There are some electrons that come through the input cavity with the wrong phase relationship, and show up in the output cavity gap between the bunches. The electric field across the gap at the time these ''out-of-phase'' electrons come through, is in a direction to accelerate them. This causes some energy to be taken from the cavity. However, much more energy will be contributed to the output cavity by the concentrated bunches of the electrons than will be withdrawn from it by the relatively small number of ''out-of-phase'' electrons.

The multicavity power klystron amplifier

In the foregoing discussion, only a basic two-cavity klystron has been considered. This simple type of klystron amplifier is not capable of high gain, high output power, or suitable efficiency. With the addition of intermediate cavities and other physical modifications the basic two-cavity klystron can be converted to a multicavity power klystron. This amplifier is capable of high gain, high power output, and satisfactory efficiency. Figure 14-68 illustrates a typical multicavity power klystron amplifier.

In addition to the intermediate cavities, there are several physical differences between the basic and the multicavity klystron. The cathode of the multicavity power klystron must be larger, in order to be capable of emitting large numbers of electrons. The shape of the cathode is usually concave, to aid in focusing the electron beam. The collector must also be larger to allow for greater heat dissipation. In a high-power klystron, the electron beam can strike the collector with sufficient energy to cause the emission of X-rays from the collector. Many klystrons have a lead shield around the collector as protection against these X-rays. Most high-power klystrons are liquid cooled, and must be constructed to facilitate the cooling system.

Klystron amplifiers have been built with as many as five intermediate cavities, and therefore a total of seven cavities. The effect of the intermediate cavities is to improve the bunching process. This results in increased amplifier gain, and to a lesser extent increased efficiency. Adding more intermediate cavities is roughly analogous to adding more stages to an i-f amplifier. The overall amplifier gain is increased and the overall bandwidth is reduced, if all the stages are tuned to the same frequency. The same effect occurs with klystron amplifier tuning. A given klystron amplifier tube will deliver high gain and narrow bandwidth if all the cavities are tuned to the same frequency. This is called *synchronous tuning*. If the cavities are tuned to slightly different frequencies, the gain of the klystron amplifier will be reduced and the bandwidth can be appreciably increased. This is called *stagger tuning*. Most klystron amplifiers that feature relatively wide bandwidths, are stagger tuned.

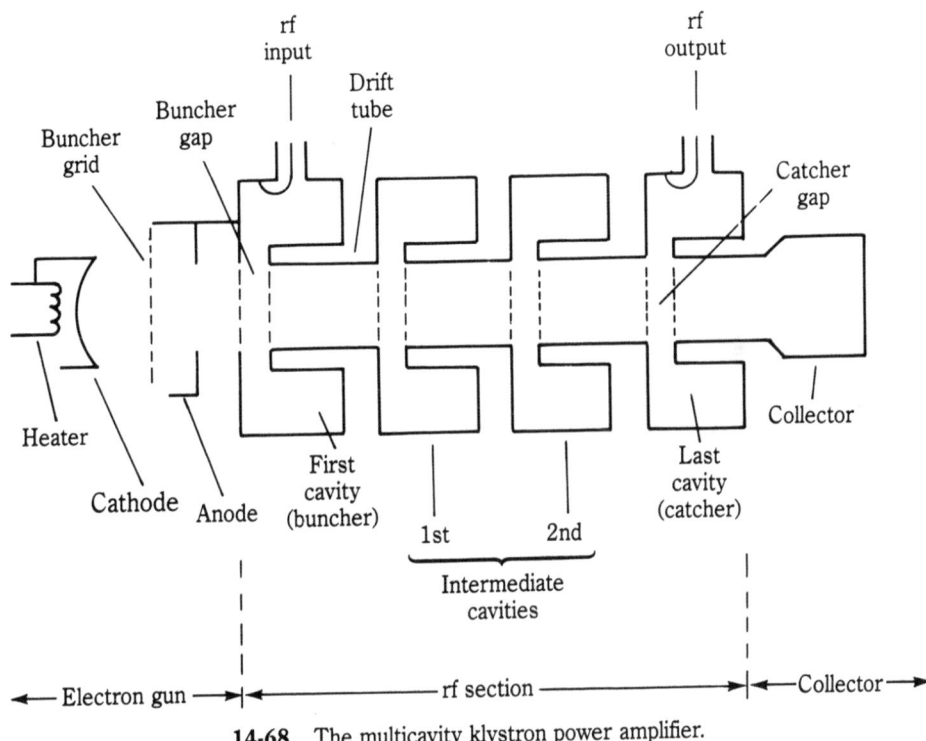

14-68 The multicavity klystron power amplifier.

The klystron is not a perfectly linear amplifier, so the rf power output is not linearly related to the rf power input at all operating levels. Another way of stating this is that the klystron amplifier will saturate, just as a triode amplifier will have a limiting action if the input signal becomes too large. In fact, if the rf input is increased to levels above saturation the rf power output will actually decrease. Figure 14-69 shows a plot of typical klystron amplifier performance for various tuning conditions. The rf output is plotted as a function of the rf input. Curve *A* of Fig. 14-69 shows typical performance for synchronous tuning. Under these conditions the tube has maximum gain. The power output is almost perfectly linear with respect to the power input, up to about 70 percent of saturation. However, as the rf input is increased beyond that point, the gain decreases and the tube saturates. As the rf input is increased beyond saturation, the rf output decreases.

To understand the reason for this decrease, you will recall that in the previous discussion, the electron bunches were formed by the action of the rf voltage across the buncher cavity gap. This rf voltage accelerated some electrons and slowed down other electrons, resulting in the formation of the bunches in the drift region. Obviously, this speeding up and slowing down effect will be increased as the rf drive's power is increased. The saturation point shown in Fig. 14-69 is reached when the bunches are most perfectly formed at the instant they reach the output cavity gap. This results in the maximum power output condition. When the rf input is increased beyond this point, the bunches are perfectly formed before they reach the output gap. Consequently, they form too soon, and by the time the bunches have reached the output gap, they tend to

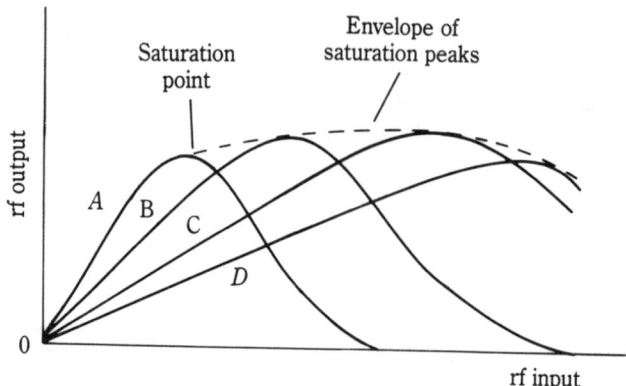

14-69   Comparison between synchronous and staggered tuning.

debunch because of the mutual repulsion of the electrons and because the faster electrons have overtaken and passed the slower electrons. This is the reason for the decrease in output power.

If a multicavity klystron power amplifier is synchronously tuned, and the next to last cavity is then tuned to a higher frequency, the gain of the amplifier is reduced, but the saturation power output level might be increased. This effect is shown by curves *B* and *C* of Fig. 14-69. Curve *B* represents a small amount of detuning of the next to last cavity, and curve *C* represents even more detuning. Note that the gain of the tube has been reduced, but that the saturation output power is higher than that obtained with synchronous tuning. Many klystron amplifiers are stagger tuned because of the resulting higher output power capability with the same beam power input. This increases the efficiency, provided that enough rf drive power is available to operate under the stagger-tuned condition. Also, stagger tuning results in a wider amplifier bandwidth. As might be expected, stagger tuning can be carried too far, at which point the saturation output power will drop. This is illustrated by curve *D* of Fig. 14-69.

In the simple klystron amplifier the electrons were formed into a narrow beam by either electrostatic or magnetic focusing techniques. In a multicavity power klystron it is even more important to focus the electron beam so that the spreading effect is kept to a minimum. This is normally achieved by an external permanent magnet or a number of electromagnets.

In a pulsed microwave system using a power klystron, one of three methods can be used to accomplish modulation. The first method is to switch the beam accelerating voltage on and off. The second is periodic interruption of the rf input signal. The last method is to turn the klystron beam current on and off.

When a klystron is pulsed by turning the accelerating voltage on and off, the entire beam current must be pulsed as well. This action is similar to modulating a magnetron, and requires a modulator capable of handling the full power of the beam.

If modulation is accomplished by switching of the rf input signal, the beam current must also be pulsed. If this is not done, some beam power will be dissipated, to no useful purpose, in the interval between the rf input pulses. This reduces the efficiency of the tube.

Of the three methods, pulsing the modulating grid or anode is the most commonly used. For communications use, the klystron is usually modulated by applying the intelligence to the modulating grid.

The chief advantage of a klystron amplifier is that it is capable of high-power output along with good stability, efficiency, and gain. Because a klystron is basically a power amplifier, it can be driven by a stable oscillator, operating at a lower frequency, followed by a frequency multiplier chain. This arrangement results in a more stable operation than is possible with a self-excited power oscillator.

Another advantage of a klystron is that its dc and rf sections are separate. This allows the cathode and collector regions to be designed for optimum performance, without concern for their effect on the rf fields. As a result, the life of a klystron is increased over other types of microwave generators.

The chief limitations of klystron amplifiers are their large size, high operating voltages, and the complexity of their associated equipment, such as is required for cooling (Fig. 14-70).

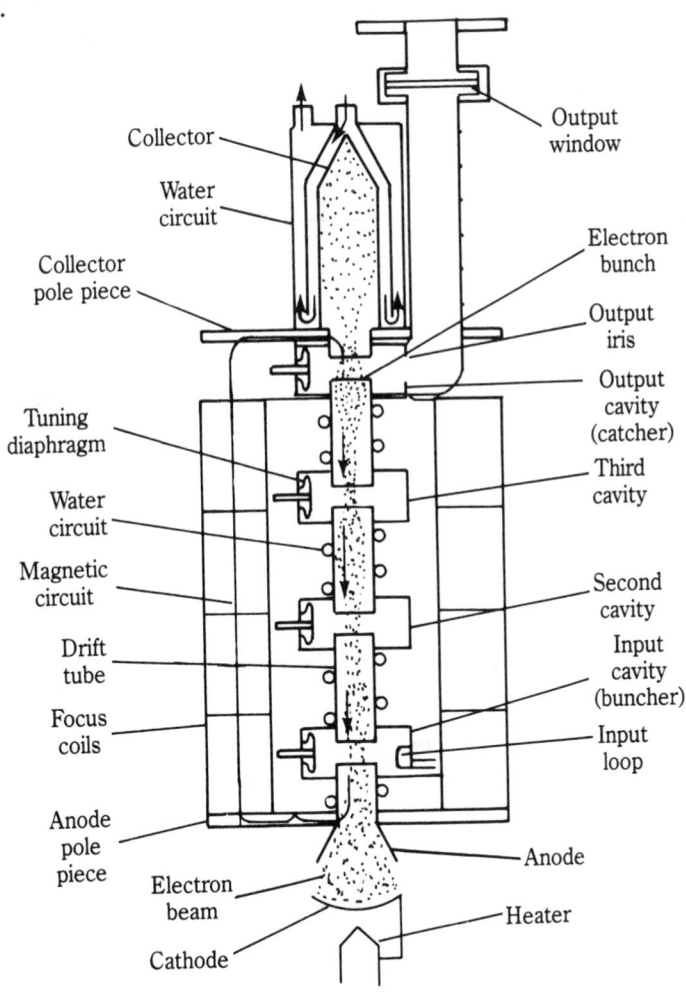

14-70 Construction of the multicavity power klystron.

The reflex klystron

In a radar system, most receivers use 30 or 60 MHz as their intermediate frequencies. A highly important factor in receiver operation is the tracking stability of the local oscillator which generates the frequency that beats with the incoming signal to produce the i-f. For example, if the local oscillator frequency is 3000 MHz, a frequency shift of as much as 0.1% would be a 3 MHz frequency shift. This is equal to the bandwidth of most receivers, and would cause a considerable loss in gain.

In receivers that use crystal mixers, the power required of the local oscillator is small, being only 20 to 50 milliwatts in the 4 GHz region. Due to the very loose coupling, only about one milliwatt reaches the crystal.

Another requirement of a local oscillator is that it must be tunable over a range of several megahertz. This is to compensate for changes in the transmitted frequency, and in its own frequency. It is desirable that the local oscillator has the capability of being electronically tuned by varying the voltage applied to one of its electrodes.

Because the reflex klystron (Fig. 14-71) meets all of these requirements, it is commonly used as the local oscillator in microwave receivers. Basically, this type of klystron consists of a source of electrons, a reentrant cavity, and a repeller or reflector plate.

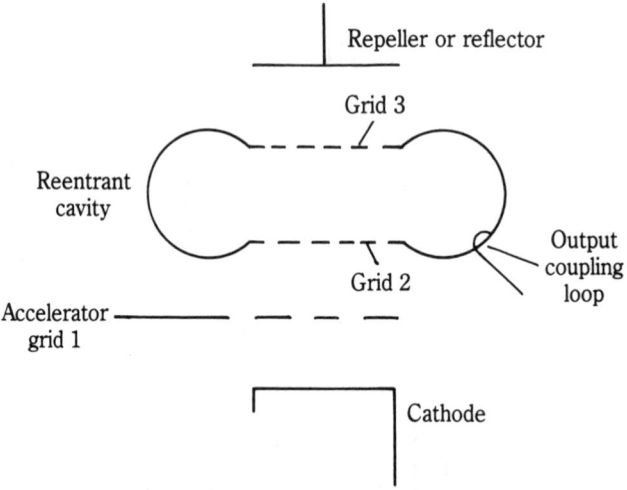

14-71   The principle of the reflex klystron.

The electrons that are accelerated by grid 1 will be velocity modulated as they pass through the cavity grids (grids 2 and 3). After moving through the cavity grids, the electrons will be traveling at different velocities. Since the reflector plate is made highly negative with respect to the cathode, these electrons will be repelled from the reflector and will reverse their direction. The high-velocity electrons will come physically closer to the reflector plate than either the medium- or low-velocity electrons. After repulsion, all electrons will be directed back toward the cavity grids. A bunching action then occurs on the return trip of the electrons. The distance that the electrons move before they are repelled by the negative reflector is a function of the voltage on the accelerating grid, the dc value of the negative cathode voltage, the negative dc voltage applied to the reflector, and the magnitude of the rf voltage existing on the cavity grids due to the oscillation in the cavity resonator. These applied voltages and the physical construction of the klystron should be such that the electrons will return to the cavity grids in bunches.

The bunching process itself is illustrated in Fig. 14-72A, which is sometimes referred to as an Applegate diagram. Figure 14-72B shows a velocity-time diagram of the electrons during their transit. The electron at time 3 (center electron) passes the cavity as the rf field (bunching voltage) is zero, and its velocity is unaffected. Therefore, a bunch will form about this electron. The electrons at times 1 and 2 pass through the cavity with higher velocities, because they move through at a time when the rf voltage across the cavity grids is producing an accelerating field. Therefore, these electrons penetrate further into the drift space and return to the cavity at essentially the same time as the center electron. Similarly, the electrons at times 4 and 5 leave with a lower velocity, penetrate a shorter distance, and return at the same instant as the previous three electrons.

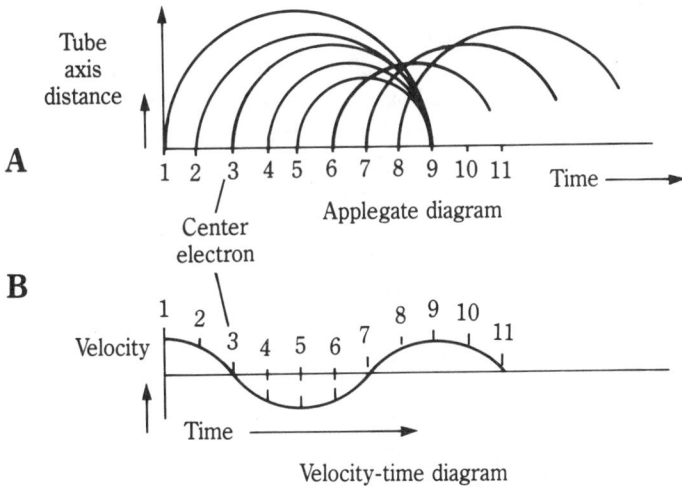

14-72   The bunching process in the reflex klystron.

When the bunch of electrons return, the potential of the cavity grids is important. At the time of return the potential applied to the cavity grids must be such that some of the energy of the bunch is absorbed. The maximum absorption of energy will occur when the bunched electrons reach the midpoint between the cavity grids as the rf voltage between these grids reaches its peak value.

As the electron bunch arrives at the midpoint, the grid nearest the repeller plate must be positive in relation to the other buncher grid. The electron bunch will then be decelerated in this field, so that some of its energy is expended in sustaining rf oscillations within the grid's cavity. For example, in Fig. 14-72B, the center electron (time 3) remains in the drift space for three-quarters of a period, and the bunch (electrons at times 1, 2, 3, 4, and 5) return at time 9, when the cavity field has a maximum value in the direction that decelerates the returning bunch.

Under these conditions, electrons leaving the cathode will receive maximum acceleration from the cavity field, while the returning electron bunches will receive maximum deceleration. If the grids are separated by approximately one half-wavelength, the electron bunch will pass through the first grid (the one nearest to the reflector) as its rf potential is zero, and changes from negative to positive. The electron bunch will pass through the second grid when its potential is zero and is changing from negative to positive. After the returning electron bunches have given their energy up to the cavity, they are absorbed by the cavity grid nearest the cathode, and are returned to the power supply.

The cavity grids perform a dual function—velocity modulation, followed by density modulation at the catcher grid. The output from the tube is extracted by means of the coupling loop shown in Fig. 14-71. By proper adjustment of the negative voltage applied to the reflector plate, the electrons that have passed through the bunching field can be made to pass through the resonator again at the proper time to deliver energy to the circuit of Fig. 14-71. The result is the positive feedback needed to sustain oscillations in the cavity. Spent electrons are removed from the tube by the accelerating grid, or by the grids of the resonator.

The operating frequency of the tube can be varied over a small range by changing the voltage on the reflector plate. This potential determines the transit time of the electrons between their first and second passages through the resonator. However, the output power of the oscillator is affected considerably more than the frequency by changes in the magnitude of the reflector voltage. This is because the output power depends on the fact that the electrons are bunched at exactly during the time of the deceleration half-cycle of the oscillating voltage. The volume of the resonant cavity is mechanically altered to change the oscillator's frequency. The reflector voltage can be varied over a narrow range to provide minor frequency adjustments.

It was mentioned that the electron bunches should arrive at the grid's midpoint when the rf swing is at its maximum positive value on the grid closest to the reflector plate.

It is, however, not necessary for the electron bunches to return on the first positive half-cycle. They might be returned on the second, third, or fourth positive half-cycle. The positive half-cycle in which the electrons are returned and bunching occurs, determines the mode of operation.

The mode of operation is governed by the transit time of the electrons. This transit time means the time between the electrons leave the bunching grids, and the time when the bunches deliver their energy to the cavity grids. Figure 14-73 shows the electrons being returned for the different operational modes. For the first mode, the bunching should occur three-quarters of a period after the average velocity electrons leave the bunching grids, the second mode of operation occurs after one-and-three-quarter periods, the third after two-and-three-quarter periods and the fourth after three-and-three-quarter periods. In practical operation, either the second, third, or fourth modes are used. Because of the time intervals, these are commonly referred to as the $3/4$, $1 3/4$, $2 3/4$, $3 3/4$, modes.

The mode of operation is determined by the electron transit time, which is the function of both the accelerating voltage and the reflector voltage. Consequently, the mode of

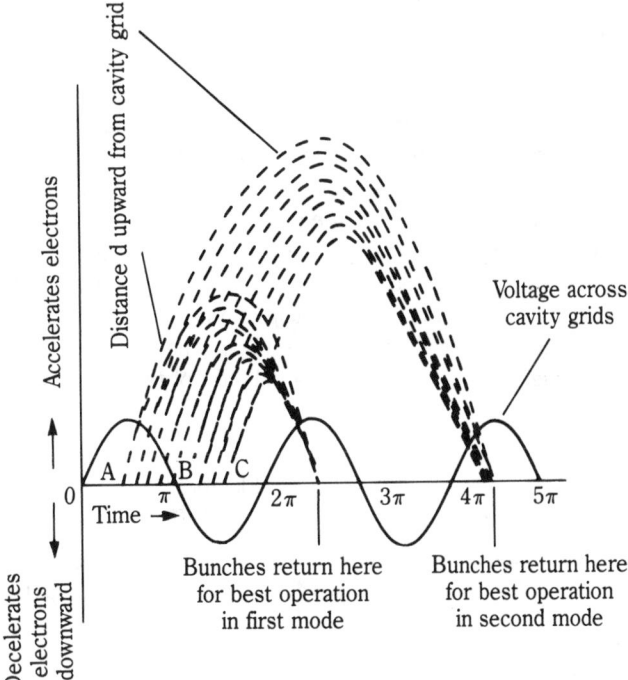

14-73 Formation of different modes in the reflex klystron.

operation is in fact controlled by the reflector voltage, because the accelerating voltage is a fixed quantity.

The variations in the power output and the frequency that occur as the reflector voltage is changed, are particularly important characteristics of the klystron. Figure 14-74 illustrates curves that can be obtained if the power output and the frequency of a reflex klystron are measured as its reflector voltage is varied from a low to a high negative value, while the accelerating voltage is kept constant. Oscillations only occur for certain values of the reflector voltage (corresponding to the various modes) for which the electron bunches return in the proper phase to deliver their energy to the cavity.

The center points of the modes, labeled A, B, and C, in Fig. 14-74, correspond to those reflector voltages for which the time spent by the electrons in the drift space is a whole number of periods together with three-quarters of a period. At these points the oscillation frequency is the resonant frequency of the cavity, and the power output is the maximum power of that mode. Note that the power outputs for the various modes at the resonant frequency are not the same, and the output is least in the highest mode. This can be explained by examining the factors that limit the amplitude of the oscillations and which, in turn, limit the output power.

The power limitation is due to overbunching as well as the usual losses in the oscillatory circuit. Overbunching occurs as oscillations build up, and the bunching voltage becomes greater, and increases the amounts of acceleration and deceleration. This causes bunching to occur in a shorter period of time (before the electrons reach the grid on the return trip) and tends to reduce the magnitude of the oscillations. In the higher modes where the bunches are formed more slowly, the electrons are more susceptible to overbunching.

As shown in Fig. 14-74, the frequency of the oscillations is variable to a limited degree in any of the modes of operation by changing the reflector voltage. When the reflector voltage is altered, a bunch is caused to return either a little sooner or a little later than normal. Away from resonance, the amplitude of the oscillation decreases by an amount depending on the Q of the cavity. The tuning range is small in comparison with the frequency of the oscillations and varies somewhat from one mode to another. It is greatest in the highest mode, because bunching and debunching take place at a slower rate, and because a greater variation from the ideal time of return is possible without appreciable debunching. This will cause the amplitude of the oscillations to drop below the usable output level.

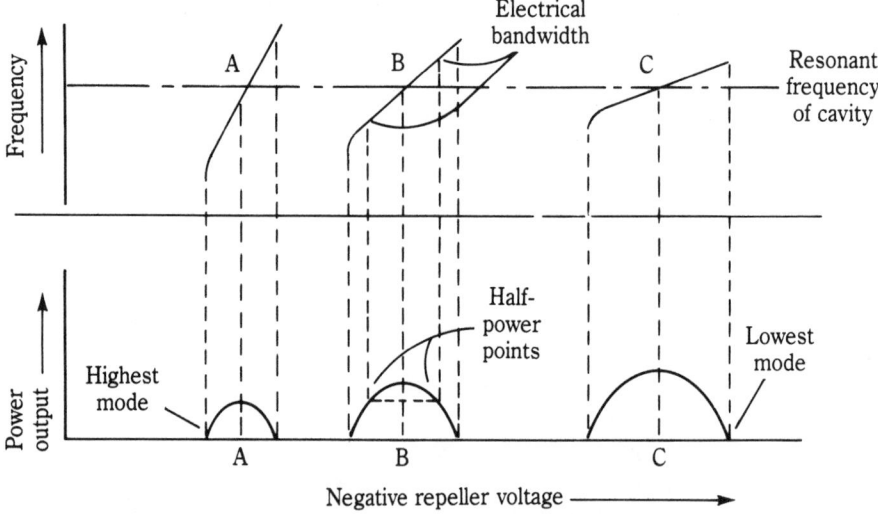

14-74   Frequency and power characteristics of the reflex klystron.

Another way to look at this is to consider that in the highest mode, the time taken by an electron in the drift space is greater, and the change in the period and its accompanying change in frequency occur in a relatively shorter interval of time. For example, in the third mode, the interval before the return must be about two-and-three-quarter periods (11 quarters of a period). A small change in the timing of the bunching voltage would therefore be only $3/11$ as great a portion of the interval as it would be if the operation were in the first mode, where the ideal time interval is only three-quarters of a period.

The band of frequencies that can be obtained by varying the reflector voltage lies between the half-power points as shown in Fig. 14-74. This range of frequencies is known as the *electronic tuning bandwidth* (of the order of 1% of the cavity frequency). The power output curve of the bandwidth is asymmetrical for the lower-order voltage modes. This results from the fact that as the negative reflector voltage is increased, not only does the bunching voltage decrease and cause the bunches to form at a later time, but the reflector voltage causes a quicker return. The effects of the two actions combine to cause poor bunching at the return of the electrons, resulting in a rapid drop in the output on the high side of the mode's peak. At lower voltages, however, even though the bunching voltage decreases and causes slower bunching, the decreased reflector voltage causes a later return to the grids. In this way, the two effects are counteracting, and a greater change in the reflector voltage is possible before the output drops below the usable level. The asymmetry is not noticeable in the higher-order modes, because the percentage change in the bunching that can occur in a higher-order mode is negligible.

As the local oscillator in a microwave receiver, a reflex klystron need not supply large amounts of power, but should oscillate at a frequency that is relatively stable and easily controlled. The efficiency of a reflex klystron is normally of the order of 2% or less.

The need for a wide electronic tuning range suggests the use of a voltage mode of a high order. However, if a mode of excessively-high order is selected, the power available is too small for local oscillator applications, and a compromise between the wide range and the power is necessary. The use of a very-high-order mode is also undesirable, because the noise output of a reflex klystron is essentially the same for all voltage modes. Therefore, the closer coupling to the mixer required with high-order, low-power modes, increases the receiver's noise figure. Usually the $1^{3}/_{4}$ or $2^{3}/_{4}$ voltage mode is found to be suitable.

Because the modes are asymmetrical, the point of operation is usually a little below the resonant frequency of the cavity. Tuning above the operating frequency is then possible to a greater degree than if the precise resonant frequency were used.

In practice, the reflex klystron is operated in conjunction with an AFC circuit. Since the reflector voltage is effective in making small changes in frequency, the AFC circuit is used to control the reflector voltage to maintain the correct intermediate frequency. It should be noted that the coarse frequency of oscillation is determined by the dimensions of the cavity and there is, on most reflex klystrons, a mechanical adjustment that varies the cavity size. For example a 10 GHz X-band klystron can be mechanically tuned between 8.5 and 11.5 GHz (Fig. 14-75).

The traveling wave tube

The *traveling wave tube* (TWT) is a high-gain, low-noise, wide-bandwidth microwave amplifier. TWT's are capable of gains of 40 dB or more, with bandwidths in which the upper frequency is twice the lower frequency. TWTs have been designed for frequencies as low as 300 MHz, and as high as 50 GHz.

The primary use for the TWT is broadband voltage amplification (although high-power TWTs, with characteristics similar to those of a power klystron, have been developed). Their wide bandwidth and low-noise characteristics make them ideal for use as rf amplifiers in microwave equipment.

Figure 14-76 is a pictorial diagram of a TWT. Notice that there are no resonant cavities; this accounts for the wide bandwidth available. The electron gun produces a stream of electrons, which are focused into a narrow beam by an axial magnetic field, much the same as in a klystron tube. This field is produced by a permanent magnet or an electromagnet that surrounds the helix portion of the tube. The narrow beam is accelerated, as it passes through the helix, by a high potential on the helix and the collector.

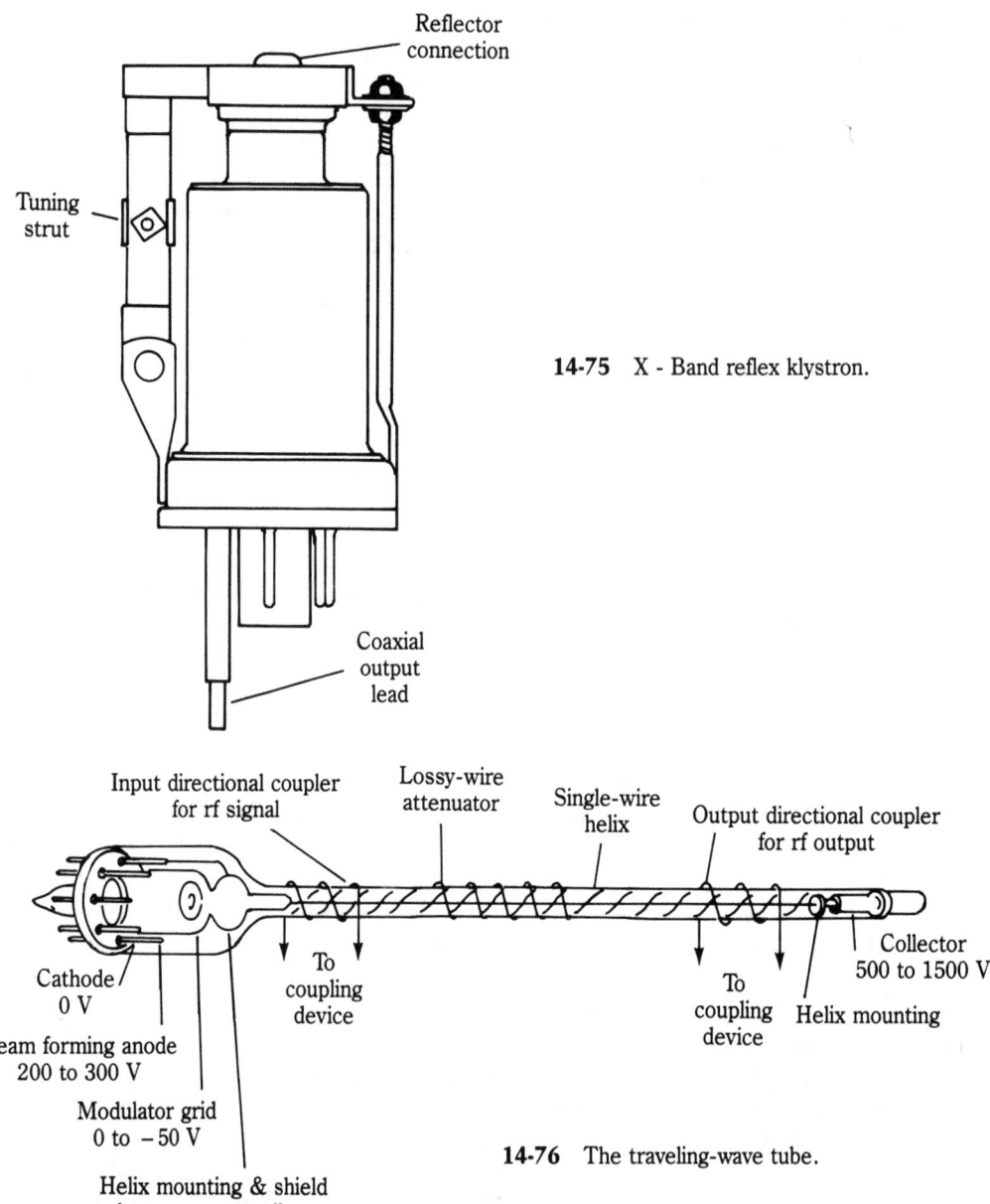

14-75 X - Band reflex klystron.

14-76 The traveling-wave tube.

TWT operation

The electron beam in a klystron travels, for the most part, in regions which are free from rf electric fields; the beam in a TWT is continually interacting with an rf electric field that propagates along an external circuit surrounding the beam.

To achieve amplification, the TWT must propagate a wave whose phase velocity is nearly synchronized with the velocity of the electron beam. It is difficult to accelerate the beam to more than about one-fifth the velocity of light. Therefore, the forward velocity of the rf field propagating along the helix must be reduced to nearly that of the beam.

The phase velocity in a waveguide which is uniform in the direction of propagation, is always greater than the velocity of light. However, this velocity can be reduced below the velocity of light by introducing a periodic variation of the circuit in the direction of propagation. The simplest form of variation is obtained by wrapping the circuit in the form of a helix which acts as a *slow wave* structure.

As previously explained, the electronic beam is focused and constrained to flow along the axis of the helix. The longitudinal components of the input signal's rf electric field, along the axis of the helix or slow wave structure, continually interact with the electronic beam to provide the amplification of the TWT. This interaction is pictured in Fig. 14-77, which illustrates the rf electric field of the input signal, as it propagates along the helix, and penetrates into the region occupied by the electron beam.

First consider the case where the electron beam velocity is exactly synchronized with the circuit's phase velocity. The electrons then experience a steady dc electric force, which tends to bunch them around position A, and debunch them around position B. This action is due to the accelerating and decelerating electric fields, and is similar to velocity and density modulation which was previously discussed. In this case, as many electrons are accelerated as are decelerated; hence, there is no net energy transfer between the beam and the rf electric field. To achieve amplification, the electron beam is adjusted to travel slightly faster than the rf electric field propagating along the helix. The bunching and debunching mechanisms just discussed are still at work, but the bunches now move slightly ahead of the fields on the helix. Under these conditions more electrons are in the decelerating field to the right of A than are in the accelerating field to the right of B. Because more electrons are decelerated than are accelerated, the energy balance is no longer maintained. Therefore, energy is transferred from the beam to the rf field, and the signal is amplified.

The fields can propagate in either direction along the helix. This leads to the possibility of oscillation due to the reflections back along the helix. This tendency is minimized by placing some resistive material near the input end of the slow wave structure. This

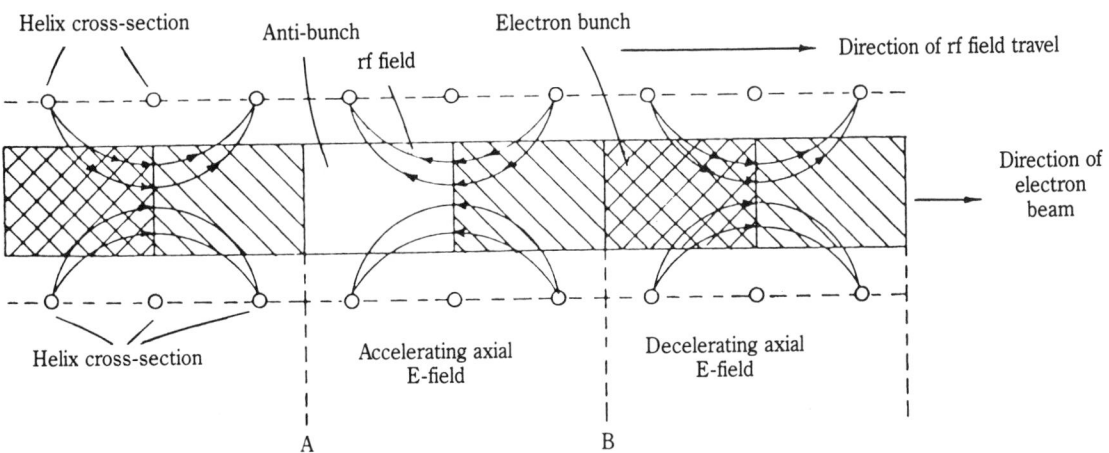

14-77 Interaction between the rf signal and the electron beam.

resistance might take the form of a lossy attenuator (Fig. 14-75) or a graphite coating placed on insulators adjacent to the helix. Such lossy sections completely absorb any backward traveling wave. The forward wave is also absorbed to a great extent, but the signal is carried past the attenuator by the bunches of electrons. Since these bunches are not affected as they pass by the attenuator, they are capable of reinstituting the signal on the helix.

Methods of coupling

Some means must be provided to apply the rf signal to one end of the helix, and to remove it from the other end. Four methods of coupling are illustrated in Fig. 14-78.

Figure 14-78A illustrates waveguide matching. The waveguide is terminated in a nonreflecting impedance, and the helix is inserted into the waveguide. The efficiency of the system is good, but the waveguide has a far higher Q than the traveling-wave tube. This means that the broadband characteristics of the TWT suffer, in that the entire bandwidth is not available for amplification, since the waveguide will not respond over such a wide spectrum.

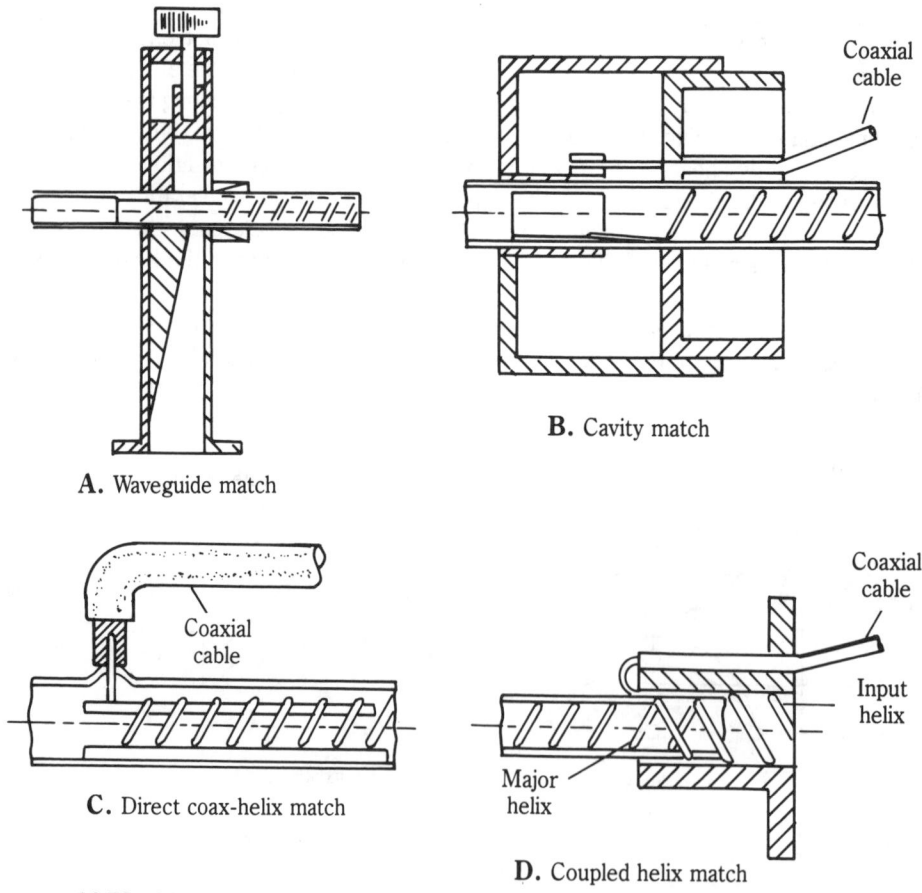

A. Waveguide match

B. Cavity match

C. Direct coax-helix match

D. Coupled helix match

14-78   Methods of coupling the rf signal to and from the traveling-wave tube.

The cavity match, illustrated in Fig. 14-78B, is very similar to the waveguide match. Cavities can be made to resonate over wider ranges than waveguides, but they still have a high Q compared with the TWT. The helix is placed at the mouth of the cavity, thereby absorbing energy so that an E-field is produced. The rf signal is fed into the cavity by a coaxial cable.

Figure 14-78C illustrates a direct coaxial cable-helix match, and is the simplest system of all. The center conductor of the input coaxial cable is connected directly to the helix. Although this method is used quite frequently, it has a disadvantage. A high VSWR is set up by this type of match, and this causes heating around the input connection. Since this connection passes through the glass envelope, the envelope is subject to heating and possible breakage at this point. However, this is a major problem only in the higher-power TWT's.

Figure 14-78D illustrates the coupled-helix match. In this system, the coaxial center conductor is attached to a small helix. The major helix is inserted within this input helix, where it acts as the secondary of a transformer. This system has a good VSWR and is broader in bandwidth than cavities or waveguides, although it is unable to handle large amounts of power. You should note that any of these methods can be used for output as well as input coupling.

The traveling-wave tube has also found application as a microwave mixer. By virtue of its wide bandwidth, the TWT can accommodate the frequencies generated by the heterodyning process (provided of course, that the frequencies have been chosen to be within the range of the tube). The desired frequency is selected by the use of a filter on the output of the helix. Such a circuit has the added advantage of providing gain, as well as simply acting as a mixer.

The TWT can be modulated by applying the modulating signal to a grid. This modulator grid can be used to turn the electron beam on and off, as in pulsed microwave applications, or to control the density of the beam and its ability to transfer energy to the traveling-wave. The grid can be used to amplitude modulate the output signal.

A forward-wave traveling-wave tube can be constructed to serve as a microwave oscillator. Physically, TWT amplifiers and oscillators differ in two major ways. The helix of the oscillator is longer than that of the amplifier, and there is no input connection to the oscillator.

The operating frequency of a TWT oscillator is determined by the pitch of the tube's helix. The oscillator can be tuned, within limits, by adjusting the operating potentials of the tube.

The electron beam, passing through the helix, induces an electromagnetic field in the helix. Although initially weak, this field will, through the action previously described, cause bunching of succeeding portions of the electron beam. With the proper potentials applied, the bunches of electrons will reinforce the signal on the helix. This, in turn, increases the bunching of succeeding portions of the electron beam. The signal on the helix is sustained and amplified by this positive feedback resulting from the exchange of energy between the electron beam and the helix.

O-type backward-wave oscillator

A typical O-type backward-wave oscillator is shown in Fig. 14-79. The slow wave circuit in the middle of the microwave region is a tape helix of such dimensions as to operate in

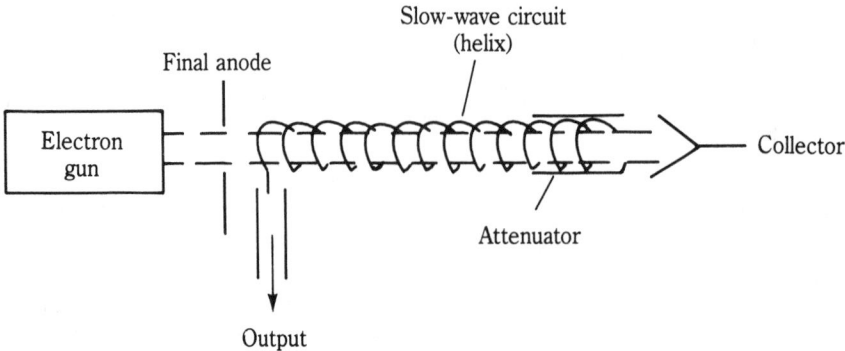

14-79 The O-type backward-wave oscillator.

the backward-wave fundamental mode, or if the impedance is high enough to make its use practical, in a backward harmonic of the forward-wave's fundamental component.

The electron beam is shot into the helix and synchronizes with the backward-wave on the helix, whose phase velocity is equal to the beam velocity.

The oscillations begin in the backward-wave oscillator in much the same way as they begin in other oscillators. Noise components are established on the helix or other slow-wave structure, as a result of the shot noise coupled from the electron beam, and from the thermal energy developed at the collector. The waves which travel backwards on the tube, velocity modulate the beam, and cause density modulation so that the electrons tend to bunch. These bunches of electrons reinforce the wave, which exists on the line. In this way, oscillations are built up at a single frequency determined solely by the electron beam velocity, which in turn is a function of the accelerating voltage.

If a wave is propagated forward at the same speed as the electrons in the beam, an individual electron will experience a force of the same magnitude and direction at certain points on the periodic structure. Provided that this synchronism is maintained, some kinetic energy will be extracted from the beam and the wave will be amplified. This is the basic principle of the traveling-wave tube.

If, however, an electron travels in the opposite direction to the wave, but this time with a reduced velocity, then there is again the possibility of an individual electron experiencing a force of the same magnitude and direction at certain points along the periodic structure.

Assume that v(m/s) is the electron velocity required to synchronize with a forward-wave, whose phase change per segment is ϕ radians. Then, to synchronize in the reverse direction, the electron velocity must be v $\phi/ (2\pi - \phi)$ m/s.

If the phase difference per segment in Fig. 14-80 is $\pi/2$, or four segments for one wavelength, then, for an electron to synchronize with the wave at the gaps, it must travel forward one segment, while the wave travels back three segments. It is then possible for the electron beam to be coupled to a component of the backward-wave.

In producing oscillations, the electron beam gives up a proportion of its kinetic energy of the rf field on the line, and this power can be taken from the line at the end nearest to the beam's point of entry. At the collector end of the tube there is a matched load to absorb any reflected power.

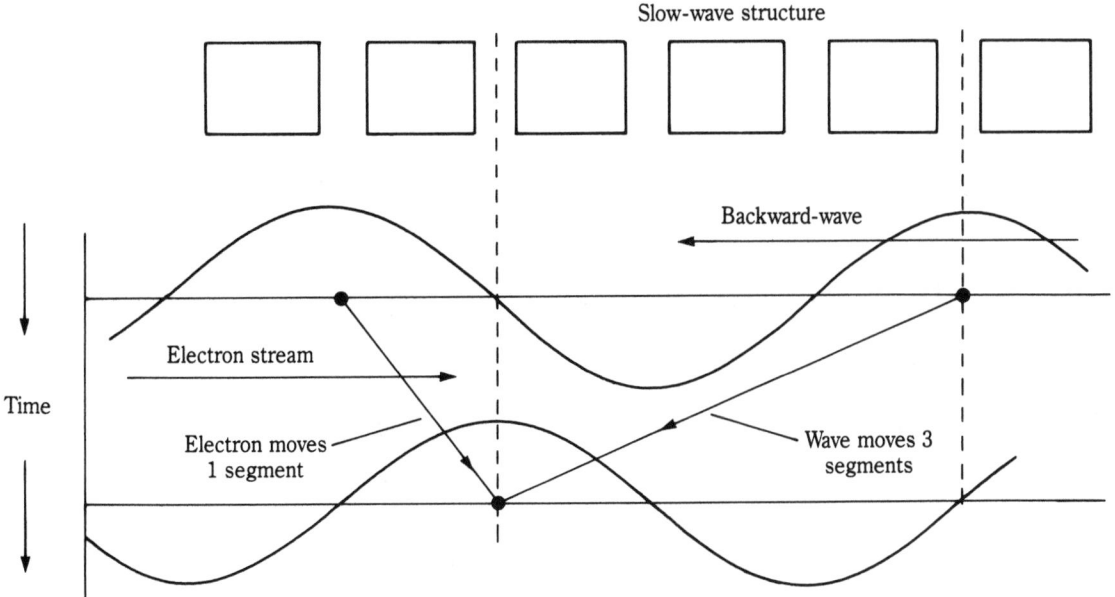

14-80 Electron beam interaction with the backward-wave.

As shown in Equation 14-22, the electron velocity is proportional to the square root of the accelerating voltage. By contrast, the phase velocity of the wave on the periodic structure is a linear function of the wave's frequency. Therefore, the variation of the frequency with the voltage will tend to be rapid at the low voltages, but will change more slowly at the higher voltages. Furthermore, unless the current is varied across the frequency band by the use of a grid, any shift in the frequency brought about by a change in the accelerating voltage, will also produce considerable difference in the output power over the frequency band.

The O-type backward-wave oscillator is very useful as a low-power active device, and is frequently used as a local oscillator or as a low-power modulator. The characteristics of this device are illustrated in Fig. 14-81A, B, and C.

M-type backward-wave oscillator

It has been shown that not only does the traveling-wave tube possess a limited power output, but the actual output efficiency is low. This is because the electrons are continually slowed down by the wave as they give up kinetic energy, and there comes a point where velocity synchronism between the beam and the wave no longer exists. Under these conditions there can be no further amplification however long the tube, and therefore the tube saturates. Because the acceptable velocity range is so small (even when the tube is saturated), the emergent electron beam still possesses considerable kinetic energy, which is dissipated as heat at the collector. The tube is thus inefficient and the only way that the efficiency can be increased is:

1. To enable the electrons to give up a higher fraction of their energy to the interaction process

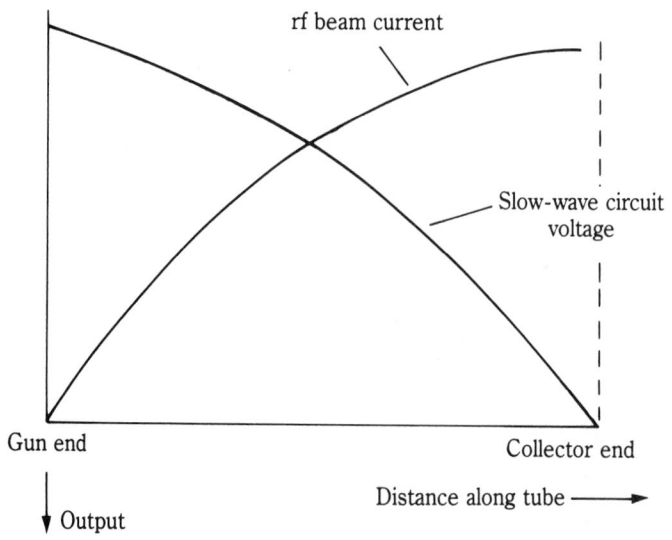

Distributions of rf beam current, slow-wave circuit
voltage along a backward-wave oscillator tube

A

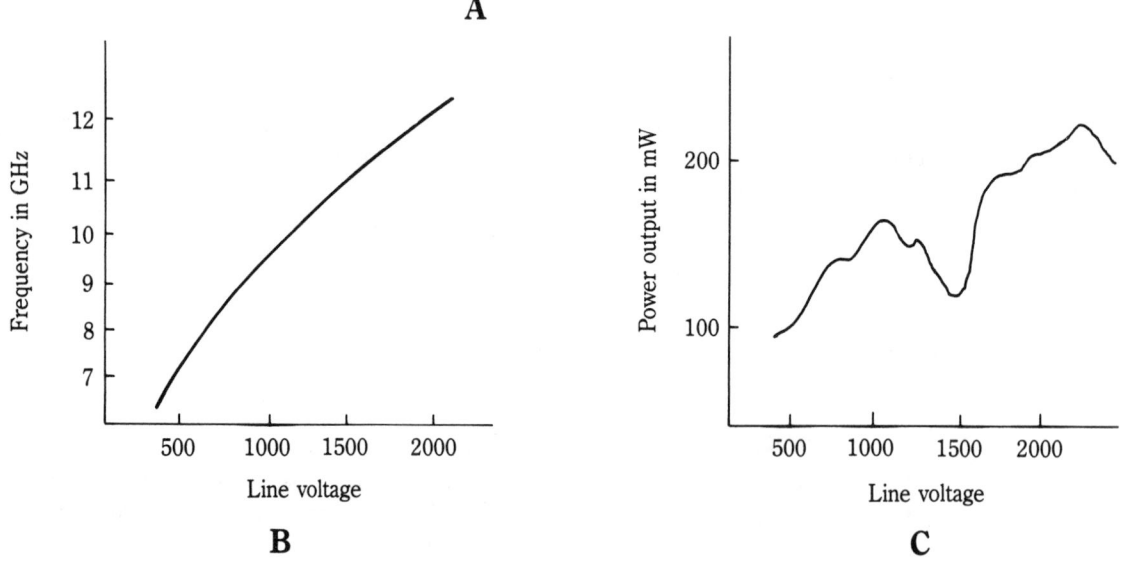

B

C

14-81   Typical characteristics for an O-type backward-wave oscillator.

2. To reduce the beam velocity after the interaction has taken place and before the
    electrons reach the collector.

At the moment the best way of improving the efficiency is to modify the process to
M-type interaction. This dates back to the early days of magnetrons, though its role in
improving efficiency has only recently been recognized.

In order to make traveling-wave tubes and klystrons function, it is generally necessary to add some kind of magnetic focusing that operates in an axial direction; this limits the radial electron motion, but does not interfere with the interaction mechanism. However, if a transverse rather than a longitudinal magnetic field is applied to the electron beam, the result is a new type of interaction.

Consider the linear crossed-field system shown in Fig. 14-82. If the action of the rf field is ignored for the moment, then an electron beam injected into the crossed-field region, will drift to the right with a mean velocity, $v = \mathcal{E}/B$. The actual path described by the electrons depends on the injection velocity but their mean drift velocity does not. Since the mean drift velocity is independent of the injection velocity, an electron beam traveling as shown will remain in synchronism with a wave propagating on the slow-wave structure over the whole of its length. It follows that no matter how much energy is extracted from the beam by the wave, the electrons will always maintain their drift velocity as a result of the transverse electric field.

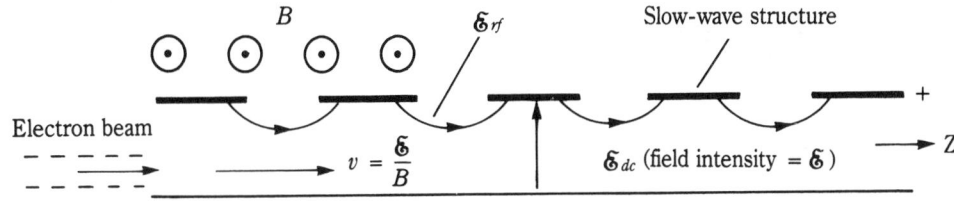

14-82 Linear crossed field system.

Under dc conditions the magnetic field, provided it is strong enough, prevents an electron from moving to the line of the slow-wave structure under the action of the transverse electric field. Therefore, an electron in the crossed-field region at any point between the electrodes possesses considerable potential energy. Without the magnetic field the electron would immediately move to the line of the slow-wave structure and arrive with a kinetic energy exactly equal to the potential energy it possessed when the magnetic field was present. Consequently, since the action of the rf wave is to draw energy from the electron beam (by endeavoring to slow it down) and since the drift velocity is compensated by the energy drawn from the transverse electric field, the electron trajectories must move out toward the line as the distance increases in the direction toward the collector. By moving toward the line the electrons can give up their potential energy and make good the losses in the longitudinal drift that the slow wave is causing.

Although the O-type device functions as a result of the electron beam giving up *kinetic* energy to the wave, the M-type device functions as a result of the electrons yielding *potential* energy to the wave.

The basic device so far described would function as an M-type amplifier. However, such an amplifier is very noisy, and this form of interaction finds its main use in the M-type backward-wave oscillator or carcinotron.

The slow-wave structure used is commonly an interdigital line (Fig. 14-83A). Although the whole system is curved into an arc of a circle for convenience (Fig. 14-83B), the electrical properties and the operation are unchanged. The first bar of the

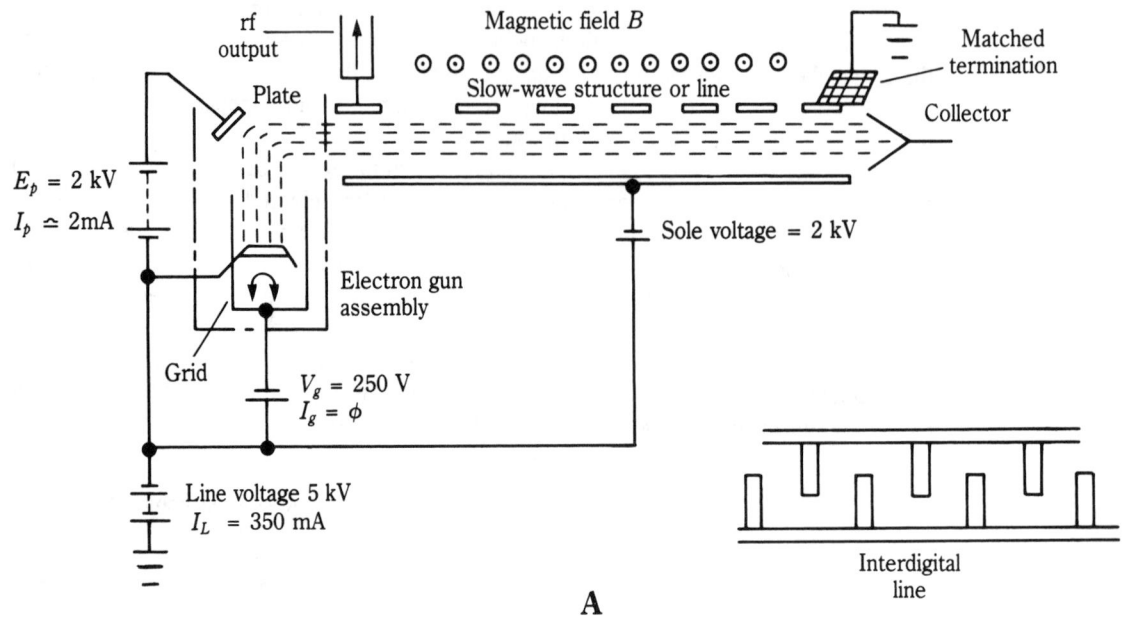

A

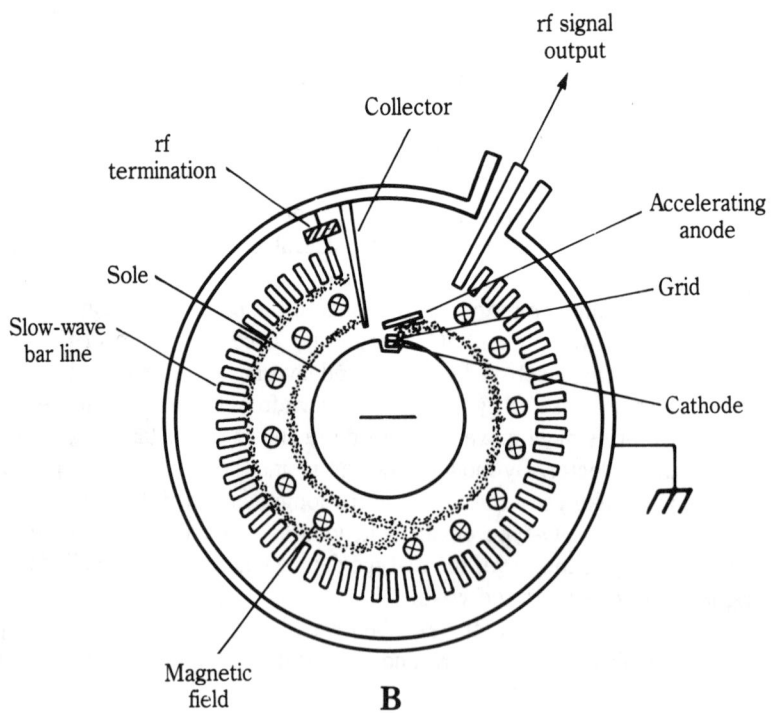

14-83 The M-type backward-wave oscillator.

line, from which the output is taken, is commonly called the *anode*, which serves as a grounded shield to the first fingers of the line, and intercepts a large number of electrons that might otherwise melt the first fingers of the line.

The gun system used in the M-type device is completely different from that used in the O-type. This is because of the effect of the magnetic field. The object is to inject electrons into the interaction space with their required drift velocity. Electrons are accelerated from the cathode by applying a voltage to the plate. The electrons start off in a direction that is normal to the cathode, but are immediately bent into their cycloidal orbits by the magnetic field. At the top of their orbits the electrons pass into the region between the line and the sole plate and (in the absence of the rf field on the line) the electrons travel along the equipotential line with a mean velocity of v = $\mathcal{E}/B$. Even if the initial velocity is not quite correct, the electrons will still have the same average velocity, so that instead of following a linear path, the electrons will oscillate around the mean path. This allows a simple gun to be used, and no compensation is required for the small variations in the line voltage.

When an rf field is excited on the interdigital line it is so designed that the dominant mode is a backward-wave. The tangential component of the rf field will alternatively force the electrons closer and farther from the line. When the electrons approach the line, they give up potential energy to the wave, as they move away, energy will be extracted.

As the favorable electrons approach the line they experience stronger fields and give up energy more quickly. By contrast the unfavorable electrons, which extract energy, tend to move farther away from the line, and therefore experience weaker fields.

As an approximation, no density modulation occurs in the M-type interaction; however the beam is drawn up into the wave (Fig. 14-84) while its space-charge density and velocity remain the same. This is in direct contrast with the O-type where density modulation is essential.

The efficiency of the M-type backward oscillator is a function of $\mathcal{E}/B$, and can be increased by raising B or alternatively by reducing $\mathcal{E}$ for a given value of B.

Theoretical efficiencies of about 80% should be possible, but in practice the efficiencies are of the order of 20 to 30%. These characteristics are illustrated in Fig. 14-85.

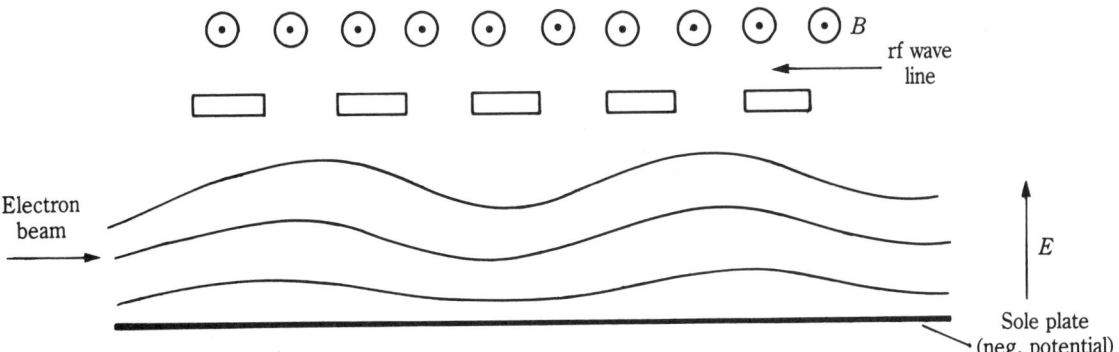

14-84   Electron beam formed into waves by the effect of the rf wave on the line.

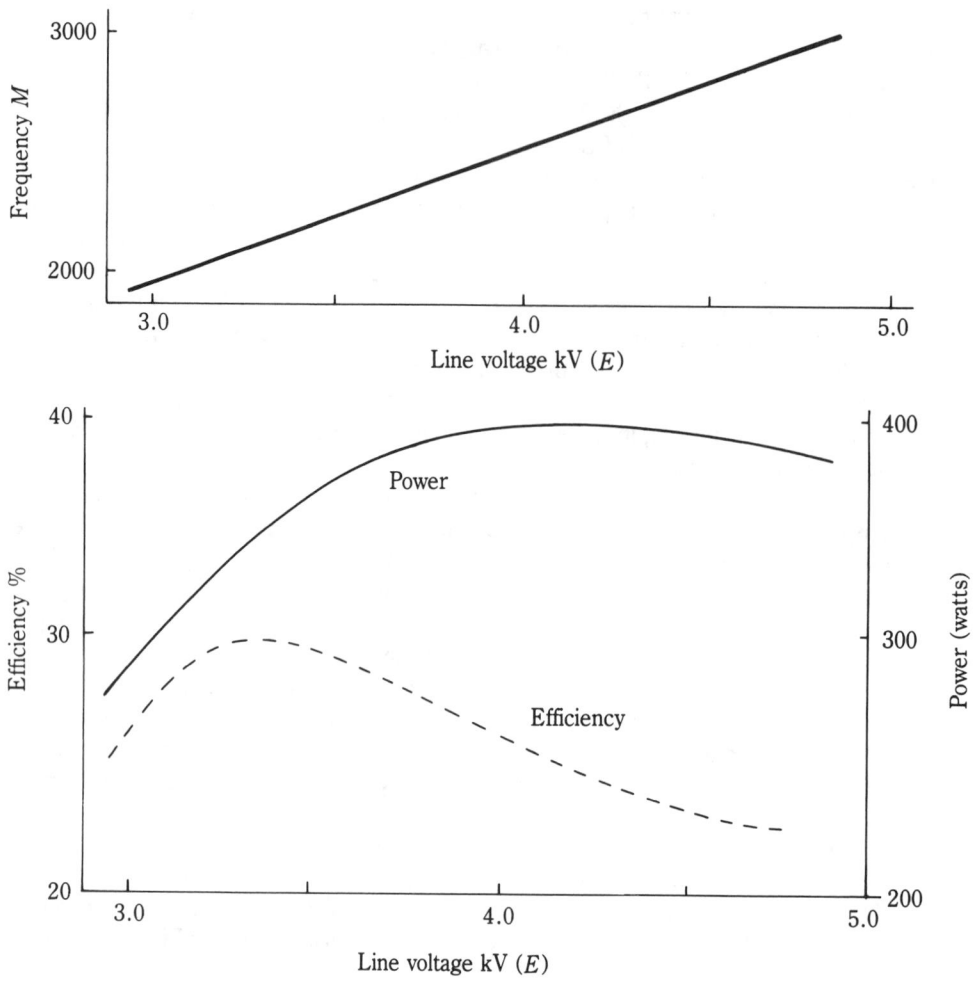

14-85   Characteristics of M-type backward-wave oscillator.

The Gunn solid-state diode oscillator

The Gunn oscillator is a solid-state (gallium-arsenide crystal) bulk-effect source of microwave energy. The discovery that microwaves could be generated by applying a steady voltage across a chip of an n-type gallium-arsenide crystal was made in 1963 by J. B. Gunn. The operation of this crystal device results from the excitation of electrons into energy states higher than those in which they normally occur.

In a gallium-arsenide semiconductor there exist empty valence bands, which are higher than those occupied by the electrons. These higher valence bands have the property that their electrons are less mobile under the influence of an electric field than when they exist normally in a lower valence band.

To simplify the explanation of this effect, assume that the electrons in the higher valence band have essentially no mobility. If an electric field is applied to the gallium-arsenide semiconductor, the current will increase with a rise in the voltage, provided that the voltage is at a low level. However, if the voltage is made high enough, it is possible to excite electrons from their initial band to the higher band where they become immobile. If the rate at which electrons are removed is high enough, the current will decrease even though the voltage is being increased. This is an equivalent negative resistance effect which can form the basis for an oscillator circuit.

If a voltage is applied across an unevenly doped n-type gallium-arsenide crystal, the crystal will break up into regions or domains with electric fields of different intensities. In particular, small domains can form within which the field will be very strong, while in the surrounding crystal material the electric field will be comparatively weak.

It is not difficult to see that such a domain is unstable. For example, assume that there is a sudden increase in the electron density at some point in the crystal which tends to reduce the electric field to the left of the disturbance, while increasing the electric field to the right. In a negative-resistance material the decreasing field to the left disturbance will cause an increase in the current flowing into the disturbed region, while the increase in the field to the right will tend to lower the current outside this region. This current pattern will have the effect of building up the charge disturbance even more; the situation will then become unstable, and will result in a redistribution of the electric field within the crystal.

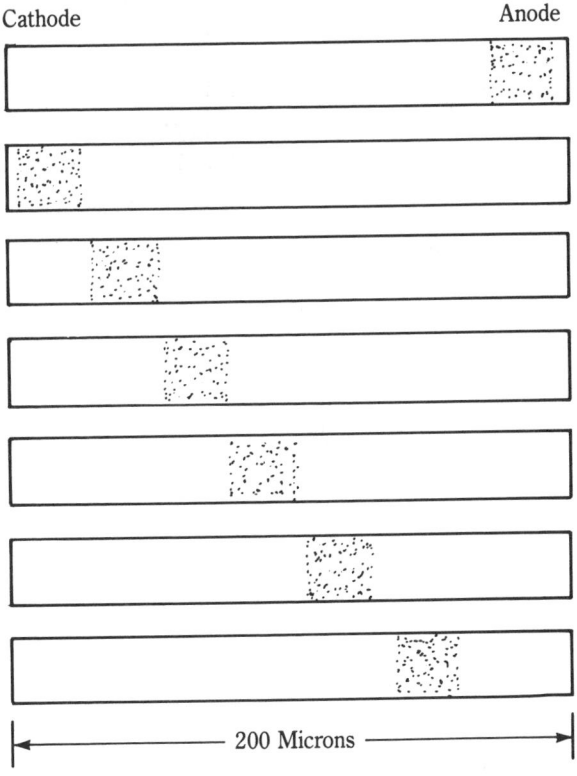

14-86 Movement of a charge domain through a gallium-arsenide chip.

The domains formed in the gallium-arsenide crystal are not stationary, since the electric field acting on the electron charge will cause the domain to move across the crystal. This is illustrated in Fig. 14-86. The domain will travel across the crystal from one electrode to the other, and as it disappears at the anode, a new domain will form near the cathode.

The Gunn oscillator will have a frequency inversely proportional to the time required for the domain to cross the crystal. This time is proportional to the length of the crystal, and to some degree to the applied voltage. Each domain results in a pulse of current at

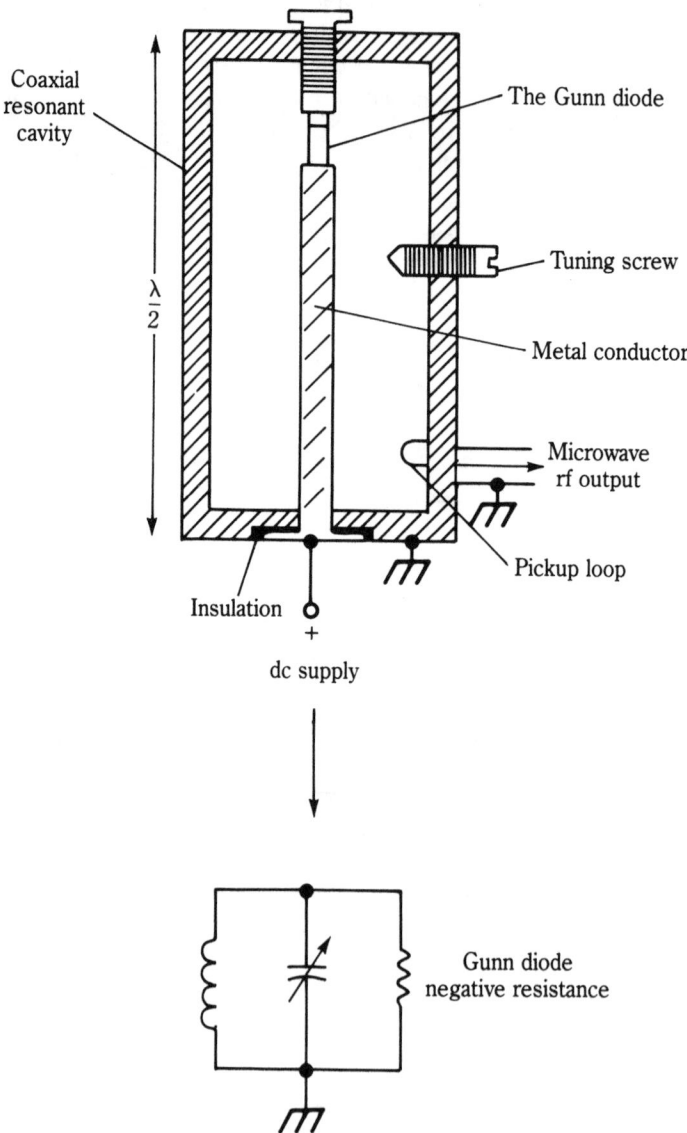

14-87   The Gunn diode oscillator with coaxial resonant cavity.

the output, so that the output of the Gunn oscillator is a microwave frequency which is determined, for the most part, by the physical length of the chip.

It is possible for the Gunn diode to oscillate by using a simple resistor as the load. However, the efficiency of such an arrangement is only a few percent. Preferably, the diode should be placed in a resonant cavity, which improves the efficiency and allows variation in the operating frequency. Such an oscillator (Fig. 14-87) is capable of delivering power outputs of a few watts at 30 to 40 GHz (continuous operation), and up to 200 watts in pulsed operation. The power output capability of this device is limited by the difficulty of removing heat from the small chip. The advantages of the Gunn oscillator are its small size, ruggedness, and low cost of manufacture.

Multiple-choice questions

1. The cavity resonator:

   A. Is equivalent to an LC resonant circuit.
   B. In a reflex klystron has its output taken from the reflector plate.
   C. Produces a frequency which is independent of the cavity size.
   D. Is confined to frequencies below 100 MHz.
   E. Has a low Q factor for narrow-band operation.

2. Why is nitrogen gas sometimes used in waveguides?

   A. To increase the distributed capacitance.
   B. To keep the waveguide dry.
   C. To reduce the skin effect at the walls of the guide.
   D. To lower the voltage rating.
   E. To raise the guide's wave impedance.

3. At which position is the input signal inserted into a traveling-wave tube?

   A. At the cathode end of the helix.
   B. At the collector.
   C. At the collector end of the helix.
   D. At the control grid of the electron gun.
   E. At the final anode of the electron gun.

4. Why is it impossible to use a waveguide at low radio frequencies?

   A. High dielectric loss.
   B. Severe attenuation.
   C. Excessive skin effect.
   D. Excessive radiation.
   E. The size of the waveguide.

5. Coupling into and out of a traveling-wave tube can be accomplished by a:

   A. Waveguide match.
   B. Cavity match.

C. Direct coax-helix match.
D. Coupled-helix match.
E. All of the above.

6. A high-power microwave pulse of the order of megawatts can be generated by a:

   A. Traveling-wave tube.
   B. Lighthouse triode.
   C. Magnetron.
   D. Reflex klystron.
   E. Gunn diode.

7. A traveling-wave tube (TWT) amplifies by virtue of:

   A. The absorption of energy by the signal from an electron stream.
   B. The effect of an external magnetic field.
   C. The energy contained in the cavity resonators.
   D. The energy liberated from the collector.
   E. The helix behaving as a fast-wave structure.

8. What is the purpose of the electromagnetic field which surrounds a traveling-wave tube?

   A. To accelerate the electron.
   B. To velocity modulate the electron beam.
   C. To keep the electrons from spreading out.
   D. To slow down the signal on the helix.
   E. To density modulate the electron beam.

9. Which of the following is used as an oscillator device in the SHF band?

   A. Thyratron tube.
   B. Unijunction transistor.
   C. Tunnel diode.
   D. Klystron tube.
   E. Both C and D.

10. How do you couple in and out of a waveguide?

    A. Wrap a coil of wire around one end of the waveguide.
    B. Insertion of an E-probe into the waveguide.
    C. Insertion of an H-loop into the waveguide.
    D. By forcing electrons through the wall of the waveguide.
    E. Both B and C.

11. Microwave frequencies are normally regarded as those in the range of:

    A. 1 to 500 MHz.
    B. 1000 to 10,000 GHz.

C. 1 to 1000 GHz.

D. 10 to 100 GHz.

E. 300 to 3000 MHz.

12. For electromagnetic waves which are traveling in free space, the:

    A. Direction of the magnetic field is parallel to the plane of polarization.

    B. Direction of propagation is at 45° to the plane of polarization.

    C. Electric field is parallel to the direction of propagation.

    D. Electric and magnetic fields are in the same direction.

    E. Electric and magnetic fields are at right-angles to each other.

13. The highest frequency which a conventional vacuum-tube oscillator can generate is *not limited by* the:

    A. Electron transit time.

    B. Distributed lead inductance.

    C. Inter-electrode capacitance.

    D. Electron velocity.

    E. Degree of emission from the cathode.

14. A rectangular waveguide is operating in the dominant TE_{10} mode. The associated flux lines are established:

    A. Transversely across the narrow dimension of the waveguide.

    B. Transversely across the wide dimension of the waveguide.

    C. In the metal walls parallel to the direction of propagation.

    D. In the metal walls perpendicular to the direction of propagation.

    E. Transversely in the wide dimension and longitudinally in the direction of propagation.

15. For the dominant mode of a rectangular waveguide, the distance between two instantaneous consecutive positions of maximum field intensity (in a direction parallel to the walls of the waveguide) is referred to as half of the:

    A. Free-space wavelength.

    B. Cutoff wavelength in the wide dimension.

    C. Guide wavelength.

    D. Group wavelength.

    E. Cutoff wavelength in the narrow dimension.

16. The guide wavelength, λ_g, in a rectangular waveguide is:

    A. Zero at the cutoff frequency.

    B. Equal to the free-space wavelength at the cutoff frequency.

    C. Equal to the free-space wavelength for the same signal frequency.

    D. Less than the free-space wavelength at the cutoff frequency.

    E. Greater than the free-space wavelength at the same signal frequency.

17. Using the TE_{10} mode, microwave power can only be transmitted along a rectangular guide provided:

    A. The wider dimension is less than one-half of the wavelength in free space.
    B. The narrow dimension is less than one-quarter of the wavelength in free space.
    C. The wide dimension is greater than one-half of the guide wavelength.
    D. The wide dimension is greater than one-half of the wavelength in free space.
    E. The group velocity is greater than the phase velocity.

18. If a signal frequency applied to a rectangular guide is increased and the dominant mode is employed:

    A. The reflection angle, θ, is increased.
    B. The free space wavelength, λ, is increased.
    C. The phase velocity, v_ϕ, is increased.
    D. The guide wavelength is increased.
    E. The group velocity, v_g, is increased.

19. If a 6 GHz signal is applied to a rectangular waveguide and the reflection angle, θ, is $20°$, what is the value of the guide wavelength, λ_g?

    A. 6.10 cm.
    B. 5.32 cm.
    C. 5.00 cm.
    D. 4.78 cm.
    E. 5.47 cm.

20. The inner dimensions of a rectangular waveguide are 1.75 cm by 3.5 cm. The cutoff wavelength for the dominant mode is:

    A. 0.875 cm.
    B. 1.75 cm.
    C. 3.5 cm.
    D. 7.0 cm.
    E. 0.4375 cm.

21. A signal whose wavelength is 3.5 cm is being propagated along a guide whose inner dimensions are 2 cm by 4 cm. What is the value of the guide wavelength, λ_g?

    A. 3.15 cm.
    B. 3.89 cm.
    C. 3.57 cm.
    D. 6.30 cm.
    E. 7.14 cm.

22. The frequency range over which a rectangular waveguide is excited in the dominant mode is limited to:

    A. The difference between the frequency for which the reflection angle, θ, is $90°$ and the frequency for which θ is zero.

B. The difference between the frequency for which the free-space wavelength is equal to the cutoff value and the frequency for which the free-space wavelength is equal to the guide wavelength.

C. The difference between the frequency for which the group velocity is zero and the frequency for which the phase velocity is infinite.

D. The difference between the frequency at which the cutoff wavelength is twice the wide dimension and the frequency for the cutoff wavelength is twice the narrow dimension.

E. The difference between the frequency at which the guide wavelength equals the wide dimension and the frequency at which the guide wavelength equals the narrow dimension.

23. If a rectangular waveguide is to be excited in the dominant mode, the E-probe should be inserted:

A. At the sealed end.

B. At a distance of one quarter-wavelength from the sealed end.

C. At a distance of one half-wavelength from the sealed end.

D. At a distance of three-quarters of a wavelength from the sealed end.

E. Both B and D.

24. Which of the following statements is true regarding waveguide bends and twists?

A. Increasing the length over which a 90° twist occurs, raises the value of the SWR.

B. An H-bend results in a much higher SWR value than the E-bend.

C. A sharp-angled turn causes a much lower SWR value than a curved bend.

D. Lowering the inner radius of a 90° bend decreases the SWR value.

E. A 90° bend can only be used if the corner is mitered.

25. The M-type backward-wave oscillator depends on:

A. Velocity modulation for its operation.

B. Density modulation for its operation.

C. The correct ratio of the electric field intensity to the magnetic flux density to create the necessary electron drift velocity.

D. Resonant cavities for its operation.

E. A fast-wave structure to increase the phase velocity.

26. As the electron beam moves through a klystron's intercavity drift space:

A. Frequency modulation at the input cavity creates velocity modulation at the output cavity.

B. Velocity modulation at the input cavity creates density modulation at the output cavity.

C. Density modulation at the input cavity creates velocity modulation at the output cavity.

D. Phase modulation at the input cavity creates velocity modulation at the output cavity.

E. Velocity modulation at the input cavity creates amplitude modulation at the output cavity.

27. The frequency of the oscillation generated by a magnetron, is mainly determined by:

A. The flux density of the external magnet.
B. The volume of the interaction space.
C. The ratio of the dc cathode voltage to the magnetic flux density.
D. The number of cavity resonators.
E. The dimensions of each cavity resonator.

28. If the instantaneous rf potentials on the two sides of a magnetron cavity are of opposite polarity, the operation is in the:

A. π mode.
B. $\pi/2$ mode.
C. 2π mode.
D. $\pi/4$ mode.
E. $3\pi/2$ mode.

29. The Gunn diode oscillator:

A. Is capable of generating continuous microwave power of the order of kilowatts.
B. Generates frequencies which are below 100 MHz.
C. Operates over a positive resistance characteristic.
D. Depends on the formation of charge domains.
E. Both A and D.

30. The traveling-wave tube amplifies over a broad band of frequencies because:

A. It contains low-Q resonant cavities.
B. Of the external electromagnetic field.
C. The pitch of the helix is varied over its length.
D. The device does not employ any resonant cavities.
E. Of the attenuation introduced along part of the helix.

Basic problems

1. A microwave signal has a frequency of 6.2 GHz. What is the wavelength of the signal in free space?

2. The internal dimensions of a rectangular waveguide are 4.755 cm and 2.215 cm. What are the values of the cutoff wavelength and the cutoff frequency for the TE_{10} dominant mode in the wide dimension? For which band would this waveguide be suitable?

3. In Basic problem 2 the frequency of the signal feeding the waveguide is 4.2 GHz. What is the angle of the incidence inside the waveguide?

4. The internal dimensions of an S-band rectangular waveguide are 7.214 cm and 3.404 cm. The frequency of the signal propagating down the waveguide is 3.3 GHz. What is the guide wavelength for the dominant TE_{10} mode in the wide dimension?

5. In Basic problem 4, calculate the values of the group velocity, the phase velocity, and the phase shift constant.

6. A rectangular K-band waveguide has internal dimensions of 1.067 cm and 0.432 cm. At a frequency of 21.7 GHz what is the value of the specific waveguide impedance in the dominant TE_{10} mode?

7. In Basic problem 6 what is the value of the waveguide's characteristic impedance at the frequency of 23.8 GHz?

8. A circular waveguide has an internal diameter of 5.276 cm. What is the value of the cutoff frequency in the TE_{11} dominant mode?

9. A circular waveguide has an internal diameter of 3.815 cm. What is the value of the cutoff frequency in the TM_{10} mode?

10. A peak voltage pulse of 25 kV is applied to a magnetron whose anode current is 50 A. If the pulse repetition rate is 750 Hz, and the pulse duration is 1.5 μs, calculate the average power input to the magnetron.

Advanced problems

1. The internal dimensions of an S-band rectangular guide are 7.214 cm and 3.404 cm. Calculate the cutoff frequencies for the TE_{10}, and TE_{20} modes in the wide dimension, and the TE_{10} mode in the narrow dimension.

2. Using a slotted line the measured value of the VSWR on a rectangular guide is 1.8. Express this value in decibels. If the incident power being propagated down the line is 15 W, calculate the amount of power absorbed by the load.

3. The internal dimensions of a rectangular X-band waveguide are 2.286 cm and 1.016 cm. If a 9.2 GHz signal is being propagated down the line in the dominant mode, what are the values of the angle of incidence, guide wavelength, group velocity, phase velocity, and phase shift constant? What is the ratio of the signal frequency to the cutoff frequency?

4. In Advanced problem 3 calculate the values of the wave impedance and the characteristic impedance for the TE_{10} mode.

5. A microwave reflectometer shows that 8% of the waveguide's incident power is reflected from the load. Neglecting any attenuation, calculate the value of the VSWR on the waveguide.

6. A rectangular waveguide is terminated by a brass plate which behaves as a short circuit. The internal dimensions of this C-band waveguide are 4.755 cm and 2.215 cm. If the signal frequency is 4.8 GHz, what is the separation in centimeters between two adjacent E-field nulls?

7. The angle of incidence is 25° for an EM wave propagating down on a rectangular guide in the TE_{10} mode. If the wide dimension of the guide is 2.8 cm, calculate the frequency of the EM wave.

8. A waveguide is sealed at one end and excited by an E-probe. If the angle of incidence is 20°, and the signal has a free space wavelength of 12 cm, how far from the sealed end should the probe be positioned?

9. A rectangular waveguide operating in the TE_{10} mode is terminated by a short circuit. Using a slotted line the distance between two adjacent E-field nulls is 2.5 cm. If the frequency of the microwave signal is 9.8 GHz, what is the length of the guide's inner wide dimension?

10. A microwave source feeds a section of lossy waveguide which introduces a one-way attenuation of 2.2 dB. If the waveguide section at one end is terminated by a short circuit, calculate the value of the VSWR at the other end.

Appendices

Appendix A
Elements of a
remote-control AM station

Elements of a Remotely Controlled AM Station

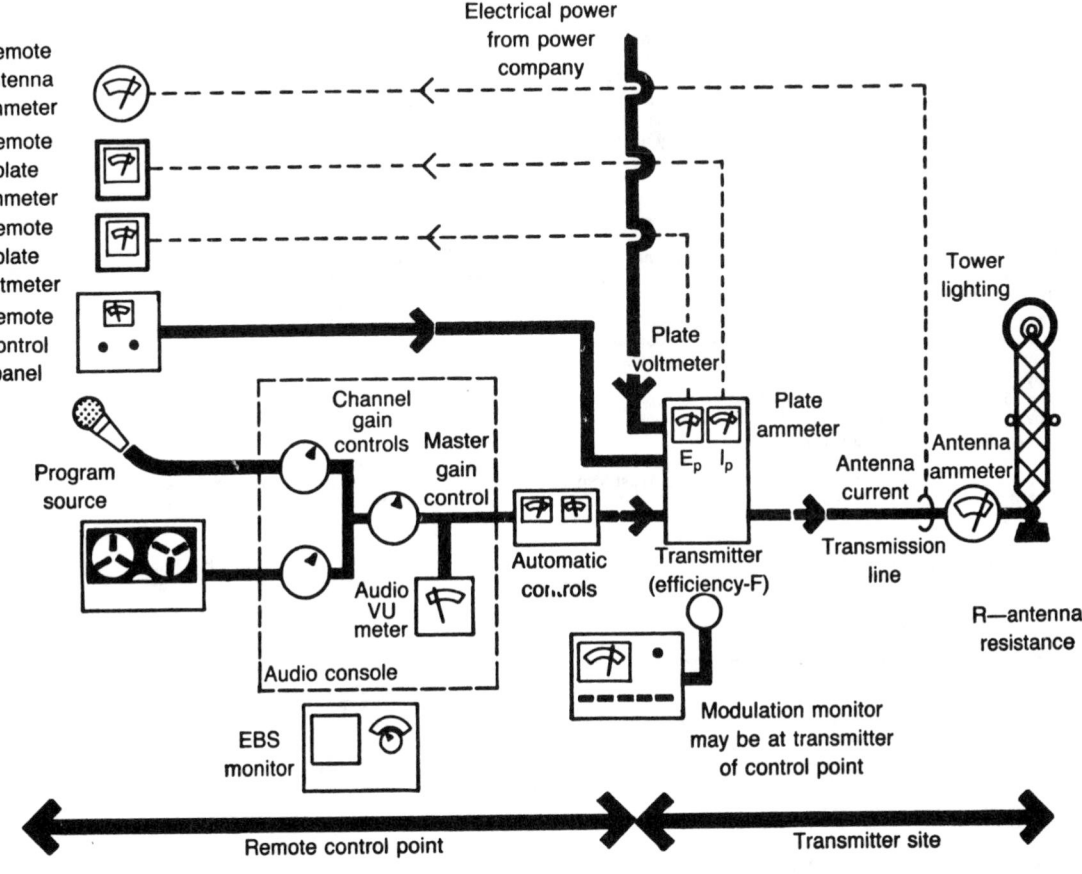

Appendix B
AM station license

FCC Form 352

UNITED STATES OF AMERICA
FEDERAL COMMUNICATIONS COMMISSION

File No.: BL-13,040

Call Sign: W S L W

STANDARD BROADCAST STATION LICENSE

Subject to the provisions of the Communications Act of 1934, subsequent Acts, and Treaties, and Commission Rules made thereunder, and further subject to conditions set forth in this license, [1]the LICENSEE
REGIONAL RADIO, INC.
is hereby authorized to use and operate the radio transmitting apparatus hereinafter described for the purpose of broadcasting for the term ending 3 a.m. Local Time October 1, 1972
The licensee shall use and operate said apparatus only in accordance with the following terms:

1. On a frequency of 1310 kHz.
2. With nominal power of   -   watts nighttime and 5 kilo watts day time,
   with antenna input power of - watts - directional
   antenna nighttime. . . . . [-   current   -   amperes
                                                                                                                      resistance -   ohms
   and antenna input power of 5 kilo watts non directional   [antenna current   8.90 amperes
   antenna daytime. . . . . antenna resistance 63.0 ohms
3. Hours of operation: Daytime as follows:
   Jan. 7:30am to 5:30pm; Feb. 7.15am to 6:00pm;
   Mar. 6:30am to 6:30pm; Apr. 5:45am to 7:00pm;
   May 5:15am to 7:30pm;   June 5:00am to 7:45pm;
   July 5:15am to 7:45pm; Aug. 5:30am to 7:15pm;
   Sep. 6:00am to 6:30pm; Oct. 6:30am to 5:45pm;
   Nov. 7:00am to 5:15pm; Dec. 7:30am to 5:00pm;
   Eastern Standard Time (non-advanced)

4. With the station located at: White Sulphur Springs, West Virginia
5. With the main studio located at:
   73 East Main Street
   White Sulphur Springs, West Virginia

Transmitter may be operated by remote control from 73 East Main Street, White Sulphur Springs, West Virginia

6. The apparatus herein authorized to be used and operated is located at: North Latitude: 37° 48' 34.5"
   Rural area 0.75 mi. North of White Sulphur Springs, West Virginia   West Longitude: 80° 17'   59"

7. Transmitter(s): BAUER, FB-5V

(or other transmitter currently listed in the Commission's "Radio Equipment List, Part B, Aural Broadcast Equipment" for the power herein authorized).**

8. Obstruction marking specifications in accordance with the following paragraphs of FCC Form 715: 1, 3, 11, and 21
9. Conditions:
   **ANTENNA: 190' (193' overall height) uniform cross section, guyed, series excited vertical radiator.
   Ground system consists of 120 equally spaced, buried copper radials 106 to 190 feet in length plus 120 interspaced radials 50 to 106 feet in length.
   The Commission reserves the right during said license period of terminating this license or making effective any changes or modification of this license which may be necessary to comply with any decision of the Commission rendered as a result of any hearing held under the rules of the Commission prior to the commencement of the license period or any decision rendered as a result of any such hearing which has been designated but not held, prior to the commencement of this license period.
   This license is issued on the licensee's representation that the statements contained in licensee's application are true and that the undertakings therein contained so far as they are consistent herewith, will be carried out in good faith. The license shall, during the term of this license, render such broadcasting service as will serve public interest, convenience, or necessity to the full extent of the privileges herein conferred.
   This license shall not vest in the licensee any right to operate nor any right in the use of the frequency designated in the license beyond the term hereof, nor in any other manner than authorized herein. Neither the license nor the right granted hereunder shall be assigned or otherwise transferred in violation of the Communications Act of 1934. This license is subject to the right of use or control by the Government of the United States conferred by Section 606 of the Communications Act of 1934.

[1]This license consists of this page and pages ____

Dated: NOVEMBER 4, 1971

FEDERAL
COMMUNICATIONS
COMMISSION

Appendix C
Elements of a
directional AM station

Elements of a Directional AM Station

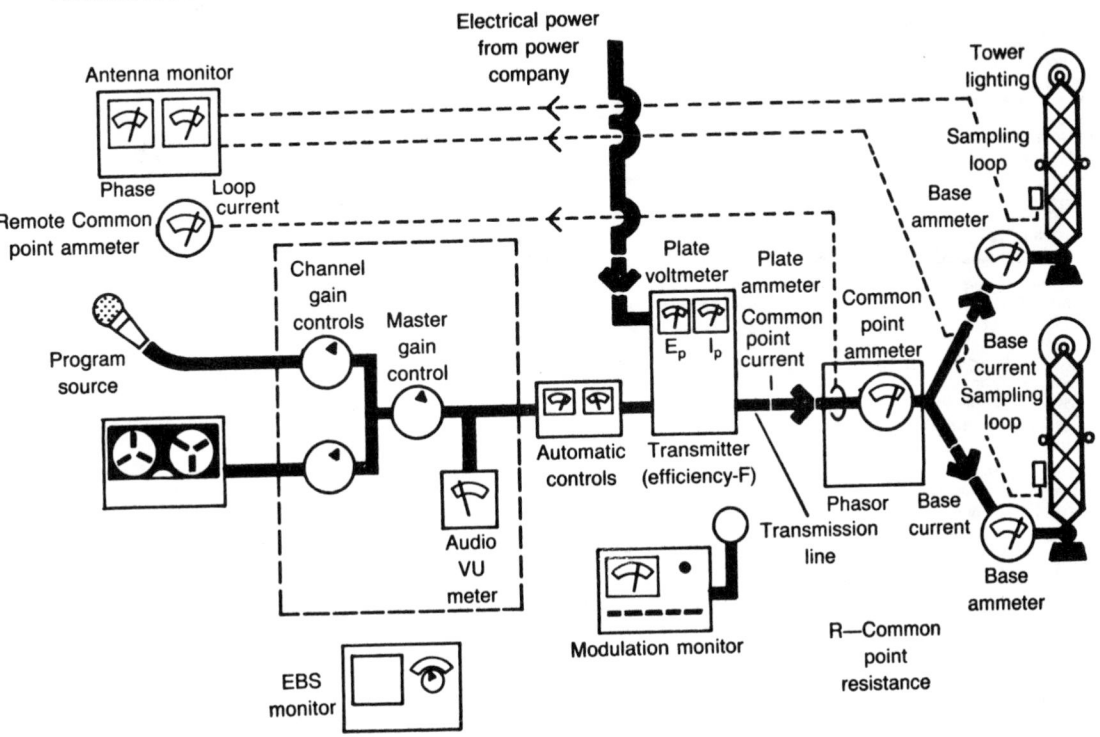

Appendix D
Directional AM station license

FCC Form 352

UNITED STATES OF AMERICA
FEDERAL COMMUNICATIONS COMMISSION

File No.: BR-989

STANDARD BROADCAST STATION LICENSE
MAIN AND AUXILIARY TRANSMITTERS

Call Sign: K X X O

Subject to the provisions of the Communications Act of 1934, subsequent Acts, and Treaties, and Commission Rules made thereunder, and further subject to conditions set forth in this license, [1]the LICENSEE
SAN ANTONIO BROADCASTING, INC.
is hereby authorized to use and operate the radio transmitting apparatus hereinafter described for the purpose of broadcasting for the term ending 3 a.m. Local Time JUNE 1, 1977
The licensee shall use and operate said apparatus only in accordance with the following terms:

1. On a frequency of 1300 kHz.
2. With nominal power of 1 kilo watts nighttime and 5 kilo watts daytime,

with antenna input power of 1.08 kilowatts - directional	Common Point	current	3.93	amperes
antenna nighttime..............................	Common Point	resistance	70	ohms
and antenna input power of 5.4 kilo watts directional	Common Point	current	8.79	amperes
antenna daytime...............................	Common Point	resistance	70	ohms

3. Hours of operation: Unlimited Time.

Average hours of sunrise and sunset.

Jan. 7:30 am to 5:30 pm; Feb. 7.15 am to 6:00 pm;
Mar. 6:30 am to 6:30 pm; Apr. 6:00 am to 7:00 pm;
May 5:15 am to 7:30 pm; June 5:00 am to 7:45 pm;
July 5:15 am to 7:45 pm; Aug. 5:45 am to 7:15 pm;
Sep. 6:00 am to 6:30 pm; Oct. 6:30 am to 5:45 pm;
Nov. 7:00 am to 5:15 pm; Dec. 7:30 am to 5:15 pm;
Central Standard Time (Non-Advanced).

Transmitters may be operated by remote control from 2805 East Skelly Drive, Tulsa, Oklahoma.

4. With the station located at: Tulsa, Oklahoma
5. With the main studio located at:
2805 East Skelly Drive
Tulsa, Oklahoma
6. The apparatus herein authorized to be used and operated is located at: North Latitude: 36° 02' 19"
8601 South Harvard   West Longitude: 95° 56' 07"
Tulsa, Oklahoma

7. Transmitter(s): COLLINS, 820E-1 (Main)

   WESTERN ELECTRIC, 405-B2 (Auxiliary)

(or other transmitter currently listed in the Commission's "Radio Equipment List, Part B, Aural Broadcast Equipment" for the power herein authorized).

8. Obstruction marking specifications in accordance with the following paragraphs of FCC Form 715:**
9. Conditions: (See Page 1A.)
**TOWERS 1, 2, & 4: Paragraphs 1, 3, 12 & 21. Beacons and all obstruction lights
shall be flashed, with flashing of towers synchronized so that
at any instant two towers are lighted and one tower is not.

Tower 3: Paragraph 1.

The Commission reserves the right during said license period of terminating this license or making effective any changes or modification of this license which may be necessary to comply with any decision of the Commission rendered as a result of any hearing held under the rules of the Commission prior to the commencement of the license period or any decision rendered as a result of any such hearing which has been designated but not held, prior to the commencement of this license period.

This license is issued on the licensee's representation that the statements contained in licensee's application are true and that the undertakings therein contained so far as they are consistent herewith, will be carried out in good faith. The license shall, during the term of this license, render such broadcasting service as will serve public interest, convenience, or necessity to the full extent of the privileges herein conferred.

This license shall not vest in the licensee any right to operate nor any right in the use of the frequency designated in the license beyond the term hereof, nor in any other manner than authorized herein. Neither the license nor the right granted hereunder shall be assigned or otherwise transferred in violation of the Communications Act of 1934. This license is subject to the right of use or control by the Government of the United States conferred by Section 606 of the Communications Act of 1934.

[1]This license consists of this page and pages 1a, 2, 3, & 4.

Dated: DECEMBER 17, 1975

FEDERAL
COMMUNICATIONS
COMMISSION

Date: 12-17-75
DA- 2

File No. BR-989   Call Sign: K X X O

1. DESCRIPTION OF DIRECTIONAL ANTENNA SYSTEM

No. and Type of Elements:	Four, triangular cross-section, guyed, series-excited vertical towers. Two towers used daytime, three used nighttime. A communications type antenna is side-mounted near the top of the W (No. 4) tower.
Height above Insulators:	284' (135°)
Overall Height:	288' (Towers 1 and 3); 290' (Tower 4); 289' (Tower 2)
Spacing Orientation:	West to West Center Tower, 273.5' (130°) - Day; West to East Center Tower, 547' (260°) East Center to East Tower, 547' (260°) - Night Line of towers bears 72° true.
Non-Directional Antenna:	None used.

Ground System consists of 240 - 43' buried copper wire radials equally spaced about each tower; 120 - 43' to 284' buried copper wire radials alternately spaced. All radials bonded together by a copper wire at a radius of 43' from each tower. Copper strap between transmitter ground and bond straps at each tower.

2. THEORETICAL SPECIFICATIONS

		E(No.1)	EC(No.2)	WC(No.3)	W(No.4)
Phasing:	Night	0°	−9.44°	-	0°
	Day	-	-	0°	−52°
Field Ratio:	Night	0.85	1.36	-	0.65
	Day	-	-	1.0	0.80

3. CREATING SPECIFICATIONS

		E(No.1)	EC(No.2)	WC(No.3)	W(No.4)
Phase Indication*	Night	17°	0°	-	8°
	Day	-	-	0°	57°
Antenna Base Current Ratio:	Night	0.608	1.0	-	0.473
	Day	-	-	1.0	0.818
Antenna Monitor Sample Current Ratio	Night	56	100	-	40
	Day	-	-	100	80

*As indicated by Potomac Instruments Antenna Monitor. PM-112

Appendix E
Elements of an FM station

Elements of an FM Station

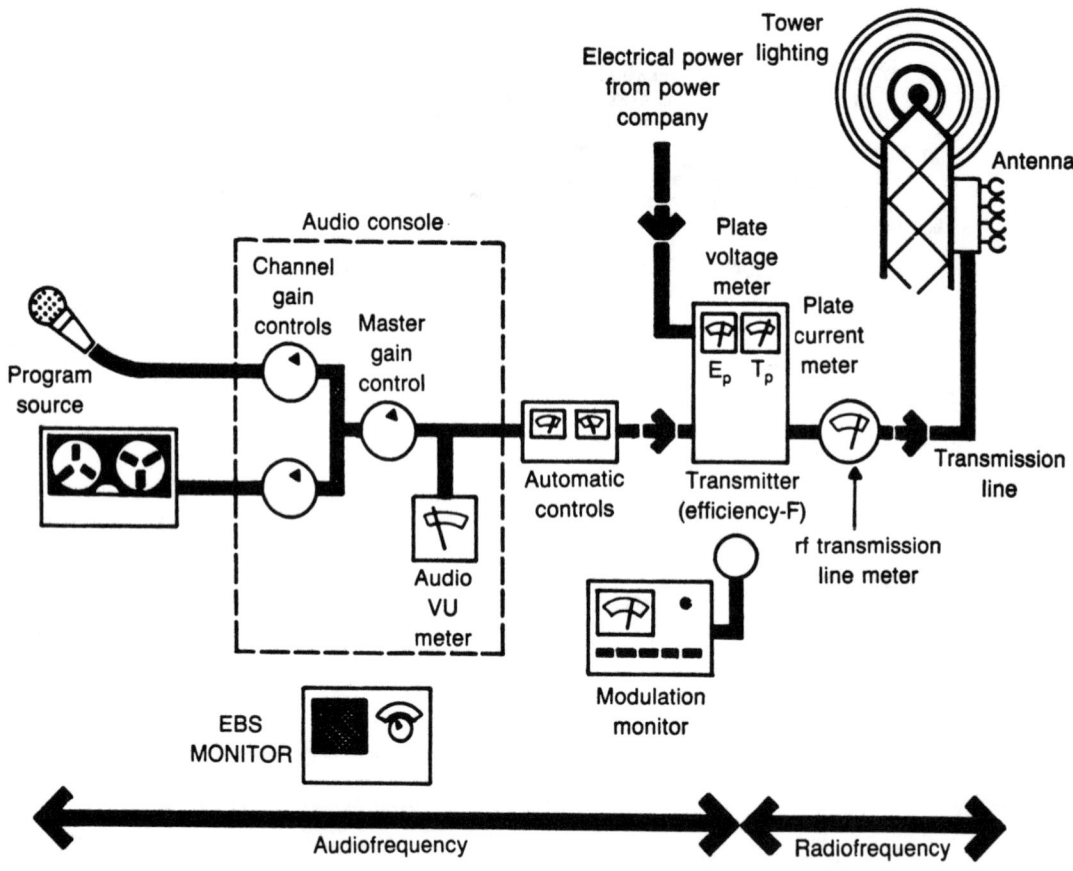

Appendix F
FM station license

FCC Form 352-A

United States of America
FEDERAL COMMUNICATIONS COMMISSION

File No.: BRH-2019

Call Sign: W F Y N-FM

FM BROADCAST STATION LICENSE

Subject to the provisions of the Communications Act of 1934, as amended, treaties, and Commission Rules, and further subject to conditions set forth in this license, [1]the LICENSEE

FLORIDA KEYS BROADCASTING CORPORATION

is hereby authorized to use and operate the radio transmitting apparatus hereinafter described for the purpose of broadcasting for the term ending 3 a.m. Local Time FEBRUARY 1, 1979

The licensee shall use and operate said apparatus only in accordance with the following terms:

1. Frequency (MHz)........: 92.5
2. Transmitter output power..: 10 kilowatts
3. Effective radiated power..: 25 kilowatts (Horiz.) & 23.5 kilowatts (Vert.)
4. Antenna height above
   average terrain (feet)...: 135′ (Horiz.) & 130′ (Vert.)
5. Hours of operation.......: Unlimited
6. Station location.........: Key West, Florida
7. Main studio location......:
   Fifth Avenue Stock Island
   Key West, Florida
8. Remote Control point.....:

9. Antenna & supporting structure: North Latitude 24° 34′ 01″
West Longitude 81° 44′ 54″

ANTENNA: COLLINS, 37M-5/300-C-5, Five-sections (Horiz. & Vert.), FM antenna side-mounted near the top of the north tower of WKIZ(AM) directional array. Overall height above ground 155 feet.

10. Transmitter location......:

Fifth Avenue Stock Island
Key West, Florida

11. Transmitter(s) ....: COLLINS, 830-F-1A

12. Obstruction marking specifications in accordance with the following paragraphs of FCC Form 715: 1, 3, 11 & 21.
13. Conditions:

The Commission reserves the right during said license period of terminating this license or making effective any changes or modification of this license which may be necessary to comply with any decision of the Commission rendered as a result of any hearing held under the rules of the Commission prior to the commencement of the license period or any decision rendered as a result of any such hearing which has been designated but not held, prior to the commencement of this license period.

This license is issued on the licensee's representation that the statements contained in licensee's application are true and that the undertakings therein contained so far as they are consistent herewith, will be carried out in good faith. The license shall, during the term of this license, render such broadcasting service as will serve public interest, convenience, or necessity to the full extent of the privileges herein conferred.

This license shall not vest in the licensee any right to operate nor any right in the use of the frequency designated in the license beyond the term hereof, nor in any other manner than authorized herein. Neither the license nor the right granted hereunder shall be assigned or otherwise transferred in violation of the Communications Act of 1934. This license is subject to the right of use or control by the Government of the United States conferred by Section 606 of the Communications Act of 1934.

[1]This license consists of this page and pages —

Dated: January 28, 1976

Federal
Communications
Commission

Appendix G
FCC emission designations

In 1987 the FCC introduced new emission designations into its rules and regulations. Each type of emission is now designated by three symbols of which the first and third symbols are letters and the second or middle symbol is a number. The meanings of these symbols are:

First Symbol (letter)   Type of Modulation.
Second Symbol (number)   Nature of Signal.
Third Symbol (letter)   Type of Information.

FIRST SYMBOL

A   Amplitude Modulation. Double Sideband. Full Carrier.
C   Vestigial Sideband.
F   Frequency Modulation.
G   Phase Modulation.
H   Single Sideband. Full Carrier.
J   Single Sideband. Suppressed Carrier.
K   Pulse Amplitude Modulation.
L   Pulse Width (Duration) Modulation.
M   Pulse Position Modulation.
N   Unmodulated Carrier.
P   Unmodulated Pulse Sequence.
R   Single Sideband. Reduced Carrier.

SECOND SYMBOL

0   Absence of Any Modulation.
1   Telegraphy On-Off Keying without the use of a Modulating AF Tone.
2   Telegraphy by On-Off Keying of a Modulating AF Tone, or by the On-Off Keying of the Modulated Emission.
3   Analog Voice Communication.

THIRD SYMBOL

A   Telegraphy (aural reception).
B   Telegraphy (reception by automatic machine).
C   Facsimile.
D   Telemetry, Data Transmission.
E   Telephony.
F   Television (video signal).
N   No Information.

EXAMPLES OF EMISSIONS

A1A   Telegraphy by On-Off Keying. Previously Designated as A1.
A3C   AM Facsimile. Previously Designated as A4.
A3E   Amplitude Modulated, Double Sideband, Telephony. Previously Designated as A3.
C3F   Vestigial Sideband Transmission for Television's Video Signal. Previously Designated as A5C.
F1B   Frequency Shift Keying (FSK). Previously Designated as F1.
F3C   FM Facsimile. Previously Designated as F4.
F3E   FM Telephony. Previously Designated as F3.

G3E   Phase Modulated (PM) Telephony. Previously Designated as F3.
H3E   Single Sideband, Full Carrier. Previously Designated as A3H.
J3E   Single Sideband, Suppressed Carrier (SSSC). Previously Designated as A3J.
NON   Unmodulated Carrier. No Information. Previously Designated as A0.
R3E   Single Sideband, Reduced Carrier. Previously Designated as A3A.

Note that sometimes a designation is preceded by a number that represents the allowed bandwidth in kilohertz.

Appendix H
FCC tolerance and standards

Carrier frequency

Standard AM broadcast stations. . . . . ±20 Hz
Commercial FM broadcast stations. . . . . ±2 kHz
Television broadcast stations—aural and visual transmitters. . . . . ±1 kHz
Non-commercial educational FM broadcast stations:

    (1) Licensed for power of more than 10 watts. . . . . ±2 kHz
    (2) Licensed for power of 10 watts or less. . . . . ±3 kHz

Studio transmitter link (STL). 0.005%
International broadcast stations. 0.0015%
Public Safety Radio Services
Frequency range

All mobile stations

MHz	All fixed and base stations	Over 3 W	3 W or less
	Percent	Percent	Percent
Below 25.	0.01	0.01	0.02
25 to 50.	.002	.002	.005
50 to 450[1].	.0005	.0005	.005
450 to 470[2,3].	.00025	.0005	.0005
470 to 512.	.00025	.0005	.0005
806 to 820.	.00015	.00025	.00025
351 to 888.	.00015	.00025	.00025
250 to 1,427[2]			
1,427 to 1,435[4].	.03	.03	.03
Above 1,435[2]			

. [1]Stations authorized for operation on or before Dec. 1, 1961, in the frequency band 73.0–74.6 MHz may operate with a frequency tolerance of 0.005 percent.

[2]Radiolocation equipment using pulse modulation shall meet the following frequency tolerance: the frequency at which maximum emission occurs shall be within the authorized frequency band and shall not be closer than $1.5/T$ MHz to the upper and lower limits of the authorized frequency band where T is the pulse duration in microseconds. For other radiolocation equipment, tolerances will be specified in the station authorization.

[3]Operational fixed stations controlling mobile relay stations, through the use of the associated mobile frequency, may operate with a frequency tolerance of 0.0005 percent.

⁽⁴⁾For fixed stations with power above 200 watts, the frequency tolerance is 0.01 percent if the necessary bandwidth of the emission does not exceed 3 kHz. For fixed station transmitters with a power of 200 watts or less and using time division multiplex, the frequency tolerance can be increased to 0.05 percent.

Power
Transmitters of standard AM and FM commercial broadcast stations............10% below and
5% above

Aural and visual transmitters of TV broadcast stations........................20% below and
10% above

Current
All currents..5%

Modulation—Standard AM Broadcast
Minimum modulation on *average modulation peaks*..................................85%
Maximum modulation on *positive modulation peaks*................................125%
Maximum modulation on *negative modulation peaks*...............................100%
Maximum carrier shift allowed...5%

Temperature for Master-Oscillator Crystals
X-cut and Y-cut crystals.. ±0.1° C
Low temperature-coefficient crystals......................................±1.0° C

Final rf stage: Plate Voltage and Plate Current Meters
Accuracy at full-scale reading..2%
Maximum permissible full-scale reading: 5 times minimum normal reading

Meters: Recording Antenna Current
Accuracy at full-scale reading..2%
Maximum permissible full-scale reading for the scale of current-squared meters................
....3 times minimum normal reading
Portion of scale used for accuracy with current-squared meters..............Upper two-thirds

FCC STANDARDS

Standard AM Broadcast
Band..535-1605 kHz
Channel width...10 kHz

FM Commercial Broadcast
Band..88-108 MHz
Channel width..200 kHz
Transmitted AF range (main channel)....................................50 to 15000 Hz
100% modulation (deviation ratio = 5)..................................±75 kHz swing
Time constant for pre-emphasis and de-emphasis.........................75 microseconds

Television Broadcast
Bands
  Channels 2 through 4...54 to 72 MHz

Channels 5 and 6 . . . . .76 to 88 MHz
Channels 7 through 13 . . . . .174 to 216 MHz
Channels 14 through 83 . . . . .470 to 890 MHz
Channel width . . . . .6 MHz
Field frequency . . . . .60 Hz
Frame frequency . . . . .30 Hz
Lines per frame 525
Horizontal scanning frequency . . . . .15750 Hz
Aspect ratio . . . . .4 to 3
Visual bandwidth . . . . .4.9 MHz
Frequency separation between aural carrier frequency and channel upper limit . . . . . .0.25 MHz
Frequency separation between visual carrier frequency (below) and aural carrier frequency. 4.5 MHz
Frequency separation between visual carrier frequency (below) and chrominance sub-carrier frequency
. . . . 3.579545 MHz ±10 Hz
Reference white level . . . . .12.5% of peak carrier level (±2.5%)
Reference black level . . . . .70% of peak carrier level
Blanking level in a monochrome TV signal . . . . .75% of peak
carrier level (±2.5%)
100% modulation for the aural FM transmission . . . . ±25 kHz swing
Transmitted AF range (main channel) . . . . .50 Hz to 15 kHz
Deviation ratio 1.667

Public Safety Radio Services
Maximum audio frequency . . . . .3 kHz
A1A emission—maximum bandwidth . . . . .0.25 kHz
A3E emission—maximum bandwidth . . . . .8 kHz
Minimum modulation on average modulation peaks 70%
Maximum modulation on negative modulation peaks 100%
F3E emission

Frequency band (MHz)	Authorized bandwidth (kHz)	Frequency deviation (kHz)
25 to 50	20	5
50 to 150	*20	*5
150 to 450	20	5
450 to 470	20	5
470 to 512	20	5
806 to 821	20	5
831 to 866	20	5

In each frequency band the deviation ratio is 1.667
*Stations authorized for operation on or before Dec. 1, 1961, in the frequency band 73.0-74.6 MHz may continue to operate with a bandwidth of 40 kHz and a deviation of 15 kHz.

Harmonic Attenuation
The mean power of emissions shall be attenuated below the mean power output of the transmitter in accordance with the following schedule:

(1) On any frequency removed from the assigned frequency by more than 50% up to and including 100% of the authorized bandwidth: at least 25 decibels.

(2) On any frequency removed from the assigned frequency by more than 250% of the authorized bandwidth: at least 35 decibels.

(3) On any frequency removed from the assigned frequency by more than 250 percent of the authorized bandwidth: at least 43 plus 10 log (mean output power in watts) decibels or 80 decibels, whichever is the lesser attenuation.

Appendix I
Multiple – choice question answers

Chapter 1   Resonance and filters

1.E	2.E	3.B	4.D	5.C	6.C	7.A	8.D	9.A	10.C
11.D	12.C	13.C	14.E	15.E	16.A	17.D	18.C	19.B	20.A
21.E	22.D	23.E	24.D	25.D	26.A	27.D	28.B	29.E	30.E

Chapter 2   Solid state devices

1.D	2.B	3.E	4.E	5.D	6.C	7.E	8.B	9.D	10.E
11.A	12.D	13.C	14.C	15.E	16.E	17.E	18.E	19.D	20.D
21.E	22.B	23.B	24.D	25.D	26.E	27.E	28.C	29.C	30.E

Chapter 3   Tubes

1.A	2.D	3.B	4.E	5.B	6.D	7.E	8.B	9.B	10.B
11.D	12.B	13.B	14.D	15.B	16.E	17.E	18.C	19.E	20.D
21.D	22.C	23.E	24.E	25.B	26.D	27.D	28.A	29.A	30.D

Chapter 4   Basic amplifiers

1.B	2.D	3.B	4.E	5.E	6.A	7.E	8.E	9.A	10.E
11.C	12.B	13.C	14.E	15.E	16.C	17.D	18.D	19.E	20.B
21.C	22.C	23.A	24.A	25.C	26.B	27.A	28.D	29.A	30.B

Chapter 5   rf amplifiers

1.B	2.A	3.E	4.D	5.D	6.B	7.C	8.D	9.E	10.E
11.D	12.A	13.B	14.E	15.D	16.B	17.B	18.E	19.A	20.E
21.E	22.D	23.E	24.E	25.E	26.B	27.E	28.E	29.D	30.C

Chapter 6   Oscillators

1.A	2.B	3.D	4.C	5.A	6.D	7.E	8.D	9.B	10.E
11.A	12.E	13.E	14.E	15.E	16.D	17.C	18.B	19.C	20.C
21.A	22.E	23.E	24.C	25.E	26.D	27.C	28.A	29.B	30.A

Chapter 7   Digital technology

1.D	2.C	3.B	4.E	5.C	6.A	7.C	8.B	9.B	10.C
11.C	12.A	13.B	14.B	15.C	16.C	17.C	18.C	19.E	20.E
21.E	22.C	23.C	24.E	25.A	26.D	27.D	28.D	29.B	30.E

Chapter 8   Amplitude modulation

1.D	2.B	3.D	4.A	5.E	6.C	7.D	8.B	9.E	10.E
11.B	12.D	13.E	14.A	15.B	16.A	17.C	18.B	19.E	20.C
21.B	22.D	23.B	24.D	25.D	26.A	27.B	28.D	29.A	30.C

Chapter 9   Frequency modulation

1.B	2.C	3.D	4.B	5.B	6.A	7.E	8.E	9.E	10.B
11.C	12.E	13.D	14.A	15.B	16.D	17.B	18.B	19.E	20.B
21.E	22.D	23.E	24.D	25.B	26.C	27.B	28.E	29.C	30.A

Chapter 10   Transmitters

1.D	2.B	3.E	4.C	5.E	6.D	7.A	8.B	9.E	10.B
11.D	12.C	13.C	14.E	15.A	16.C	17.B	18.E	19.C	20.B
21.C	22.E	23.C	24.D	25.E	26.D	27.E	28.E	29.D	30.D

Chapter 11   Receivers

1.C	2.D	3.A	4.E	5.A	6.B	7.A	8.D	9.B	10.E
11.E	12.A	13.B	14.C	15.D	16.B	17.D	18.B	19.E	20.D
21.D	22.D	23.B	24.D	25.B	26.E	27.C	28.A	29.D	30.D

Chapter 12   Television

1.D	2.B	3.A	4.B	5.D	6.B	7.B	8.C	9.B	10.E
11.E	12.E	13.C	14.D	15.B	16.C	17.C	18.A	19.E	20.B
21.E	22.A	23.C	24.D	25.C	26.D	27.C	28.D	29.D	30.B

Chapter 13   Transmission lines and antennas

1.D	2.B	3.A	4.A	5.B	6.D	7.C	8.A	9.D	10.A
11.A	12.D	13.C	14.A	15.E	16.C	17.D	18.A	19.D	20.E
21.C	22.E	23.B	24.A	25.A	26.D	27.E	28.B	29.D	30.D

Chapter 14   Microwave techniques

1.A	2.B	3.A	4.E	5.E	6.C	7.A	8.C	9.E	10.E
11.C	12.E	13.E	14.A	15.C	16.E	17.D	18.E	19.B	20.D
21.B	22.D	23.B	24.E	25.C	26.B	27.E	28.A	29.D	30.D

Appendix J
Basic and advanced problem answers

Chapter 1   Resonance and filters
Basic problems
1. 503 kHz   **2.** V_R = 15 V; $V_C = V_L$ = 9.49 V   **3.** L = 100 μH; C = 0.001 μF

4. 61.6 dB   **5.** 7120 Hz   **6.** 15.9 kHz   **7.** 7.482 kHz

8. 17.7 kHz   **9.** Surface acoustic wave   **10.** R_P = 130 Ω; X_P = 26 Ω

Advanced problems
1. 40 μ H   **2.** Original Bandwidth = 159 kHz; New Bandwidth = 39.8 kHz

3. 159 kHz   **4.** L = 10 μH; C = 100 pF; R = 10 kΩ **5.** 16 dB

6. $mL/2$ = 5.53 mH; mC = 0.0553 μF; $(1 - m^2)L/4m$ = 6.28 mH

7. f_1 = 8400 Hz; f_2 = 9400 Hz   **8.** Series: 5.14 MHz;

9. Programming and hardware   **10.** Parallel: 5.14 MHz + 3.2 kHz

Chapter 2   Solid-state devices
Basic problems
1. The most distinctive feature of a semiconductor crystal is its conductivity, which lies between that of an insulator—glass, and a conductor—copper. (Also read Atomic Structure).

2. A donor atom has one extra electron that is not locked into the crystalline structure, hence it promotes electrical conduction.

3. $\alpha = I_C/I_E = I_C/(I_C + I_B) = \beta\, I_B/(\beta I_B + I_B)$; $= \beta\,/\,(1 + \beta\,)$

4. In the depletion-mode FET the channel is created in the fabrication process. In direct contrast, the channel does not exist for the enhancement-mode FET until a voltage of the proper polarity is applied to the gate.

5. A digital circuit has two possible stable states and must be in one or the other. Analog circuits, such as a linear amplifier, have a continuum of states as, for example, the range of output voltage levels corresponding to the input signal level.

6. Programmable, high current, high voltage, comparators, voltage follower, Norton, etc.

7. It is a voltage reference.   **8.** See chapter 2.

9. Cost and matching parameters.   **10.** Fabricated on single crystal.

Advanced problems

1. Raising the temperature effectively increases the supply of free electrons.

2. In a good conductor the valence band and the conduction band overlap. The two bands are separated by a forbidden band in an insulator.

3. Acceptor atoms leave unoccupied points, called holes, in the crystal. It is the movement of holes which accounts for the major part of the current in this doped semiconductor.

4. Read chapter 2 for a brief description of the role.

5. The application of an external bias of one polarity increases the potential barrier, which restricts conduction and vice versa.

6. The capture of electrons in the base region of an npn bipolar transistor, for example, tends to create a net negative electrical charge that opposes the flow of additional electrons from the emitter. Therefore, the flow of base current is necessary to control the buildup of charge and the amount of emitter current.

7. CMOS has greatly reduced power dissipation. Also read MOS logic.

8. The several ways are outlined in chapter 2. You should be aware that the practical op amp closely approximates the functions of the ideal op amp in real electronic circuits.

9. A gate array is a predesigned, preprocessed matrix of transistors that are subsequently interconnected uniquely for each application.

10. Briefly, gate arrays permit the design to be implemented with standard ICs. Whenever quantity production justifies the additional development expense, the same design is transferred to the application of custom ICs.

Chapter 3   Tubes

Basic problems

1. $R_P = 667\ \Omega;\ r_p = 500\ \Omega$ **2.** $r_p = 8\ k\Omega$ **3.** $g_m = 3000\ \mu S$

4. $\mu = 16.7$ **5.** $r_p = 5\ k\Omega$ **6.** $G_v = 16$

7. Peak value = 45.25 V
dc level = 130 V

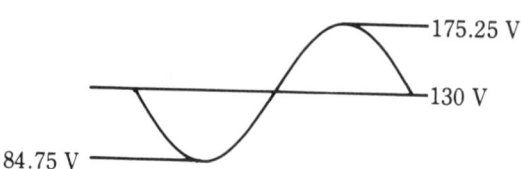

175.25 V

130 V

84.75 V

8. Screen resistor value = 80 kΩ

9. G_v = 94.5   **10.** Input dc voltage = 350 V

Advanced problems

1. dc resistance = 31.25 kΩ; r_p = 16.7 kΩ   **2.** g_m = 2500 μS; μ = 41.7

3. 310 V; −15 V   **4.** 10 kΩ; 1 W   **5.** G_v = 15.2; 64.5 V

6. Plate potential = +151 V; Screen potential = +115 V; G_v = 82.5

7. rms input signal = 0.536 V   **8.** μ = 12   **9.** 1250 μS; 200 μS

10. + 260 V

Chapter 4   Basic amplifiers

Basic problems

1. 250   **2.** 1.98   **3.** 0.996   **4.** class B   **5.** 100 kHz   **6.** −3 V

7. +5 V   **8.** 78 μF   **9.** Grid current must flow.   **10.** 15.3

Advanced problems

1. 150

2. A change in the operating point, for whatever reason, can cause the transistor to be cutoff or driven into saturation, with undesirable distortion of the output.

3. 50   **4.** 9.9   **5.** 99 Ω   **6.** −6 V; −9 V

7. +10.8 V; +7.2 V   **8.** R_I = 10 kΩ; R_F = 50 kΩ

9. Triangular output

10. Capacitor C_C is a coupling capacitor that has negligible reactance at midfrequencies and higher frequencies. Its purpose is to isolate the dc level of the plate circuit in one amplifier stage from the different dc level in the grid circuit of the following stage.

Chapter 5   rf amplifiers

Basic problems

1. The rf amplifier bandwidth must be sufficient to accommodate the sidebands of the modulated carrier.

2. See the formula for the resonant frequency of an LC circuit.

3. Capacitance between the higher signal level collector and the lower signal level base might be involved in amplifier instability and sustained oscillations. The effects of such capacitance can be cancelled by the selective feedback of output signals of the proper phase to neutralize the regenerative feedback.

4. The output of the amplifier must have harmonics.

5. 52.1%

6. Because of the lower impedance levels and subsequent higher current levels, there are additional power losses and lowered efficiency.

7. By placing one, or more, additional resonant circuits in the amplifier, the collector voltage can be modified to more closely resemble that of the class-D or class-E amplifier, thereby contributing to the efficiency.

8. 81%

9. A linear amplifier requires that the output be directly proportional to the input.

10. Causes negative feedback.

Advanced problems

1. 333 Ω

2. See the question in the text.

3. See the question in the text.

4. See the question in the text.

5. Complementary voltage switching establishes a square wave drive so that the output transistors are either cutoff or saturated for nearly the entire rf period.

6. The "flywheel effect" refers to the energy transfer between the capacitive element and the inductor element in the tank circuit that occurs at the resonant frequency. The sinusoidal voltage and current waveforms suffer little distortion for the high values of Q in practical loaded amplifier tank circuits.

7. 80.4%

8. See the description of the linear power amplifier.

9. $R_K = 750$ Ω; $C_K = 42.4$ μF

10. $R_G = 100$ kΩ

Chapter 6   Oscillators

Basic problems

1. 100   **2.** Required % positive feedback = 2%   **3.** 75 kHz; 237 kHz

4. 159 kHz; 4000   **5.** 3.0003 MHz   **6.** 541 Hz

7. 103 pF gives 2.2 MHz; 238 pF gives 1.45 MHz   **8.** 2550 Hz

9. 482 Hz   **10.** 903 kHz

Advanced problems

1. 25 pF   **2.** 591 kHz   **3.** 31.4 kHz   **4.** 32.0032 MHz

5. 4.3 to 10.8 kHz   **6.** 4.4 to 44.2 kHz   **7.** 12.8 pF

8. 258 kHz; 4 μH

9. 42 pF   **10.** 29 pF to 228 pF

Chapter 7   Digital technology

Basic problems

1. XOR   **2.** See Fig. 7-4   **3.** See Fig. 7-5A

4. The half-adder has two binary inputs and provides the binary sum together with the carry as the output. The full adder differs only in having three binary inputs, one of which can be the carry from a preceding addition stage for multibit binary numbers.

5.
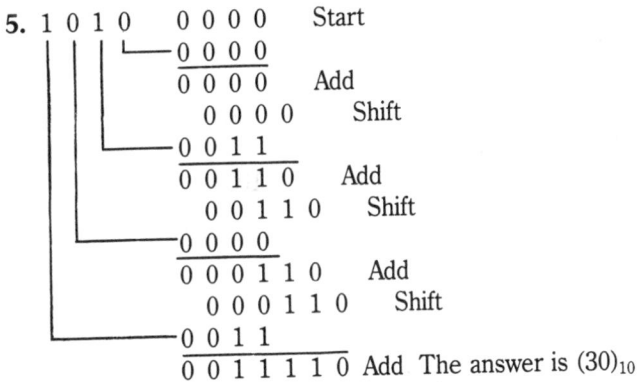

6. The word "static" is used in two different senses. The static RAM remains in one binary state indefinitely, so long as the dc power remains uninterrupted. The dynamic RAM stores a static electrical charge that will leak away unless an automatic refresh method is used to restore it periodically.

7. The proper choice is the erasable PROM for which the stored program is unchanged during normal operation. However, the program can be changed or modified, but not during the normal operation.

8. Various hardware elements are required depending on the application. Most of these are related to the means of getting data into the microcomputer and getting the results out again to achieve the desired objectives. Additionally, the software is directly related to the specific application.

9. The address bus has 16 bits so that $2^{16} = 65,536$ memory locations that can be addressed. The data bus has 8 bits, which is the word length for most personal computers at this time. More powerful microcomputers in this era have as many as 32 bits for the data word.

10. In the absence of a jump instruction, the instructions in the stored program are executed in sequence. The use of jump instructions, of which there are a number having different rules for execution, greatly increases the flexibility and power of digital computers as we know them today.

Advanced problems

1.

Decimal	Binary	Gray	Decimal	Binary	Gray
00	00000	00000	16	10000	11000
01	00001	00001	17	10001	11001
02	00010	00011	18	10010	11011
03	00011	00010	19	10011	11010
04	00100	00110	20	10100	11110
05	00101	00111	21	10101	11111
06	00110	00101	22	10110	11101
07	00111	00100	23	10111	11100
08	01000	01100	24	11000	10100
09	01001	01101	25	11001	10101
10	01010	01111	26	11010	10111
11	01011	01110	27	11011	10110
12	01100	01010	28	11100	10010
13	01101	01011	29	11101	10011
14	01110	01001	30	11110	10001
15	01111	01000	31	11111	10000

2. The Gray code is superior because there is only one bit position transition at a time as the quantity is incremented by one.

3.

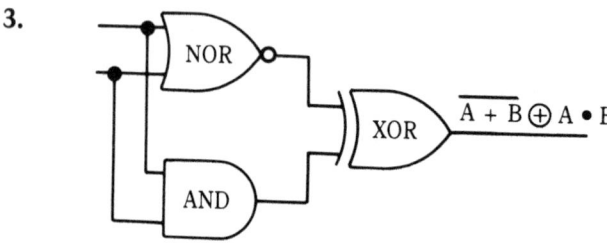

4. The count cannot proceed at a rate faster than the time required for a transition to ripple through all the flip-flops in the counter.

5. See Fig. 7-14.

6. In a stable state a conducting path never exists from the power supply through the MOS elements.

7. Mask ROM.

8. The microprocessor is the key element in virtually every operation of the micro-computer. The other parts support the microprocessor functions.

9. One, two, or three locations are required depending on the specific instruction.

10. The number of times is indefinite. It continues counting in much the same way as an automobile odometer recycles through 00000, but never stops.

Chapter 8   Amplitude modulation

Basic problems

1. An electronic communication system can be divided into the information source, transmitter, channel, receiver, and information user.

2. See Fig. 8-2.

3. The standard broadcast frequency band extends from 535 kHz to 1605 kHz with 107 assigned carrier frequencies at 530 kHz, 550 kHz . . . 1600 kHz, adjacent carrier frequencies being separated by 10 kHz.

4. Broadcast equipment will usually include some form of audio limiter implemented by a method for gain reduction automatically activated by peak sound levels. For a more complete treatment of this subject, read Sound processing and control.

5. 66.7%   **6.** 900 W   **7.** 24.9 kW   **8.** 66.7%   **9.** 1200 Hz

10. The spectrum of the recovered modulating waveform will be shifted 5 Hz. A listener gifted with absolute pitch might sense that the piano requires tuning. Most listeners will probably be unaware of the defect.

Advanced problems

1. 387 μV   **2.** 50 %   **3.** 70.7%   **4.** 1098 W   **5.** 16.7%

6. 83.3%   **7.** (a) 1385 W   (b) 80 W, 125 W, 180 W   **8.** 10 mV

9. Since the object of envelope detection is to recover the audio signal waveform, the frequency components above the audio range will be attenuated. In particular, components that are even harmonics of 455 kHz, the standard intermediate frequency, will effectively disappear in the output.

10. The phase-locked loop can be made to produce a signal at the exact carrier frequency and is essentially coherent (in phase) with that carrier component.

Chapter 9   Frequency modulation

Basic problems

1. 37.5 kHz   **2.** 12

3. Noise, and impulse noise in particular, cause the envelope to deviate from the desired amplitude-modulated form. After demodulation, the corrupting noise appears, together with the modulating signal, in the output.

4. Impulse noise has more influence on the amplitude (envelope) than on the phase. It is more apparent in its nefarious effects on AM than on FM.

5. The noise triangle illustrates the greater significance of higher-noise frequency components in the modulating frequency band.

6. Indirect FM requires that the modulating signal be processed so that each frequency component amplitude reappears inversely proportional to its frequency.

7. A phase deviation of 1 radian is independent of the frequency at which it occurs. However, when viewed as FM, the maximum frequency deviation for the 1 radian phase deviation is exactly 1 kHz, corresponding to the question.

phase deviation = frequency deviation/modulating frequency

8. 22.1 pF   **9.** 0.075 μF   **10.** 75 mH

Advanced problems

1. 18.75 kHz

2. 25 kHz for 200 kHz station spacing using Carsons rule

3. 10 kHz   **4.** 90 kHz

5. With indirect FM the phase modulation method creates FM after signal pre-processing. The direct FM method forces the frequency deviation to follow the modulating signal.

6. 79   **7.** 100 μV   **8.** 75 mV   **9.** 2122 Hz

10. A. 0.4125 V; B. 0.3375 V

Chapter 10   Transmitters

Basic problems

1. 11.995 MHz; 12.000 MHz; 12.005 MHz

2. Any odd harmonic frequency within reason is possible such as the 3rd, 5th, etc. harmonics. The small conduction angle simply ensures strong harmonic content.

3. 0.01 Hz

4. Improper pulse shaping can result in keying harmonics that exceed the FCC limits for spurious emissions.

5. A filter is required after each multiplication and translation to eliminate the undesired products of the process.

6. δ = 18.75 kHz

7. The low power, combined with line-of-sight propagation, will limit the operating range to a few miles between mobile units in adverse terrain with hills, buildings, etc.

8. 806 kW   **9.** 12.48 kW

10. See the discussion of directional AM stations in Broadcast station operations.

Advanced problems

1. 2.166 MHz

2. The maximum count is 18 for frequency division by 19. This is a modulo 19 counter.

3. If the voltage applied to the VCO is constant, the frequency of the VCO must also be constant and exactly matched to the reference so that the phase error is also fixed at a constant value. Normally, there will be some fluctuations and drift as the phase-locked loop automatically adjusts to changes of the reference signal frequency or aging of its own components.

4. Multiplication factor = 88/0.100 = 880
Frequency deviation multiplication = 880 × 14.468 = 12,731 kHz,
this is far greater than the 75 kHz allowed

5. The crystal labeled TXCO in Fig. 10-16.

6. m = 3.75

7. Salient features: Subdivision of geographical area into cells, each having a low-power transmitter. Reuse of frequencies. Automatic switching as the mobile unit moves from one cell to another.

8. 112 kW   **9.** 38.7 A

10. Tower	#1	#2	#3	#4
Ratio	0.52	1.26	1.40	1.0

Chapter 11   Receivers

Basic problems

1. 995-2055 kHz   **2.** Doubled   **3.** Changed by a factor of $\sqrt{2}$

4. 122.5 MHz   **5.** 79.7 MHz   **6.** 4.47   **7.** 3.59   **8.** 13 dB   **9.** 2.1

10. This chapter does not give an explicit answer. The purpose of the attenuator could be to protect the receiver front end in the presence of a nearby, powerful transmitter. Normally, the attenuation would be set to 0 dB and reset to another value when protection is needed.

Advanced problems

1. The ratio of the greater local oscillator frequency to the lesser local oscillator frequency is 13.5. Compare this with the ratio for the conventional arrangement in which the local oscillator frequency is always tuned to be above that of the received signal as analyzed in Basic problem 1 giving a ratio of 2.07. The conclusion to be drawn is that the tuning range for the present alternate arrangement cannot be achieved with practical tuning elements.

2. 8.09×10^{-18} W   **3.** 2.01×10^{-8} V

4. 4 dB   **5.** 2.34   **6.** 32.8 dB   **7.** $Q = 52$ **8.** 1.2

9. 1.12 = 1.00 dB   **10.** A. 77.055 MHz;   B. 47.6 MHz

Chapter 12 Television

Basic problems

1. A frame consists of 525 lines. Its rate is 30 Hz. Each frame is subdivided into two fields, the field rate being 60 Hz, and each field having 262½ lines, alternately even and odd-numbered.

2. The audio carrier is situated 4.5 MHz above the video carrier. Frequency modulation is used with a maximum frequency deviation of 25 kHz.

3. From Table 12-1, the assigned frequency is 210 MHz. See Fig. 12-11.
$f_v = 210 + 1.25 = 211.25$ MHz; $f_a = 211.25 + 4.5 = 215.75$ MHz

4. This frequency arises from the product
(525 lines/frame) (30 frames/second) = 15,750 lines/second

5. Efficiency = $(55+11) \times 100 / 135 = 49\%$

6. Without further information, a complete answer is not possible. You can, however, state that 15.75 kHz and its harmonics will be present.

7. *Y* is the luminance signal, while R, G, and B are outputs of the respective color cameras.

8. Chrominance modulation is double-sideband-suppressed-carrier in which two quadrature subcarriers are employed for *I* and *Q*. The *I*-signal has unsymmetrical sidebands.

9. Read the description in Color television entitled *Color CRT*.

10. The transmitters remain separate and the respective modulated carriers are combined in a diplexer to feed a common antenna.

Advanced problems

1. The base of a horizontal blanking pulse cuts off the electron beam during retrace. Positioned on the blanking pulse is the synchronizing pulse whose maximum width is about 5 μs as compared with the blanking pulse whose width is about twice as much.
  The leading edge of the synchronizing pulse follows that of the blanking pulse by about 1μs, forming what is referred to as the front porch. Similarly, a wider back porch follows the synchronizing pulse. (See Fig. 12-3.) An eight cycle color burst of 3,579,545 Hz is placed on the back porch for NTSC color transmission.

2. This group of pulses is quite complex, as shown in Fig. 12-4. The overall duration of the group is 833 μs to 1333 μs, the exact width depends on the number of lines that are to be blanked by the transmitting station.

3. A simplified drawing of the vidicon appears in Fig. 12-5. It has an electron gun, electrodes for accelerating and decelerating the electron beam, and coils surrounding the tube for focusing and deflection.
  The electron target of the vidicon is a photoconductive mosaic layer separated from a transparent conductive film by a semiconductor photoresistive layer. When the target is dark, the electrons are stored on the inner target surface, making the target potential

roughly the same as that of the cathode. A point of light, corresponding to a picture element, will cause electrons to leak from the target, so the corresponding point on the target surface will become more positive. These electrons will be replaced by the electron beam on its next scan, causing a pulse whose amplitude is a function of the picture element spot light intensity.

4. Vestigial sideband modulation is employed for the video information. The upper sidebands extend 4 MHz above the picture carrier, whereas the lower bands are partially suppressed. (See Fig. 12-11.) Therefore, the highest video frequency component is about 4 MHz.

5. Examining the formula that is implemented in the matrix for luminance, $Y = 0.299R + 0.587G + 0.114B$ tells us that all three, R, G, and B must be zero for $Y = 0$, the condition for black.

6. Relative to the lower edge of the assigned spectrum:
Picture carrier frequency = 1.25 MHz
Color burst frequency = 3.579545 MHz
Sound carrier frequency = 5.75 MHz

7. Color burst frequency = $455 \times f_H/2 = 455 \times 15{,}734.3 = 3{,}579{,}545$ Hz

8. $Y = 0.587G + 0.114B$

$$\begin{aligned} I &= -0.27 \, (B - (0.587G + 0.114B)) \\ &= -0.27B + 0.158G - 0.031B \\ &= -.301B + 0.158G \end{aligned}$$

$$\begin{aligned} Q &= 0.41 \, (B - (0.587G + 0.114B)) \\ &= 0.41B - 0.241G - 0.047B \\ &= 0.363B - 0.241G \end{aligned}$$

9. A charge-coupled device analog shift register of 910 sections driven by a clock of 4×3.579545 MHz is used. This automatically creates a time delay of $1/f_H$.

10. After demodulation.

Chapter 13
Transmission lines and antennas

Basic problems
1. 864 Ω; 311 W   **2.** 300 Ω; 75 W   **3.** 2.55 m

4. 0.5 m; 9 λ/8; 100 Ω   **5.** 2 m; 9λ/8; 50 Ω   **6.** 3; 0.5

7. 12.5 W; 37.5 W   **8.** 8.2 ft.   **9.** 26 ft.   **10.** 83.7 Ω

Advanced problems
1. 200 Ω; 0.625 A; 78 W   **2.** 81.4 Ω; 63.8 V; 0.78 A

3. 461 Ω; 1.04×10^{-3} dB/m; 6.74 rad/m; 0.93 m; 280 Mm/s

4. 0.592 $\underline{|-62.91°}$; 3.9   **5.** 31.5 W; 58.5 W; 3 m; 1.5 m

6. 275 μV/m; 458 μV/m   **7.** 155 μV/m; 12.9 miles

8. 184 kW   **9.** 47.8 kW   **10.** 5.4 kW; 1.8

Chapter 14   Microwave Techniques

Basic problems

1. 4.84 cm   **2.** 9.61 cm; 3.12 GHz; C band

3. 7.14 cm; 48.7°   **4.** 9.09 cm; 39.06°; 11.7 cm

5. 232.9 Mm/s; 386.3 Mm/s; 0.537 rad/m

6. 1.38 cm; 40.3°; 1.81 cm; 494 Ω   **7.** 314 Ω

8. 9.02 cm; 3.33 GHz   **9.** 4.17 cm; 7.19 GHz

10. 1.25 MW; 1406 W

Advanced problems

1. TE_{10} mode (wide dimension); 14.428 cm; 2.08 GHz
TE_{20} mode (wide dimension); 7.214 cm; 4.16 GHz
TE_{10} mode (narrow dimension); 6.808 cm; 4.41 GHz

2. 5.1 dB; 0.286; 1.22 W; 13.78 W

3. 3.26 cm; 45.5°; 4.65 cm; 210 Mm/s; 428 Mm/s; 1.35 rad/m; 4.572 cm; 6.56 GHz; 1.4

4. 538 Ω; 375 Ω   **5.** 0.283; 1.79

6. 6.25 cm; 41.1°; 8.29 cm; 4.145 cm

7. 2.37 cm; 12.66 GHz

8. 12.77 cm; 3.19 cm from the sealed end.

9. 5 cm; 3.06 cm; 52.3°; 3.35 cm

10. 0.363; 0.602; 4.025

Index